FLUID MECHANICS

FLUID MECHANICS

Seventh Edition

Victor L. Streeter

Professor Emeritus of Hydraulics
University of Michigan

E. Benjamin Wylie

Professor of Civil Engineering
University of Michigan

McGraw-Hill Book Company

New York   St. Louis   San Francisco   Auckland   Bogotá   Düsseldorf
Johannesburg   London   Madrid   Mexico   Montreal   New Delhi
Panama   Paris   São Paulo   Singapore   Sydney   Tokyo   Toronto

FLUID MECHANICS

4567890   DODO   8321

This book was set in Times Roman.
The editors were Julienne V. Brown and Madelaine Eichberg;
the cover was designed by Rafael Hernandez;
the production supervisor was Dominick Petrellese.
New drawings were done by J & R Services, Inc.
R. R. Donnelley & Sons Company was printer and binder.

Library of Congress Cataloging in Publication Data

Streeter, Victor Lyle, date
    Fluid mechanics.

    Includes bibliographies and index.
    1. Fluid mechanics. I. Wylie, E. Benjamin,
joint author.
TA357.S8 1979   532   78-6264
ISBN 0-07-062232-9

CONTENTS

A continuing effort has been made to improve the readability of the text; it includes rewrites of some sections and reorganization of some of the chapters. The most significant amount of new material is the discussion of molecular and turbulent diffusion and dispersion in Chapter 5. In an effort to reduce the length (and weight) of the text, three-dimensional ideal-flow cases have been omitted from Chapter 7, and Chapter 12 now has reduced coverage of waterhammer and unsteady open-channel flow.

The trend to metric usage has continued, and about three-fourths of the examples and problems are now in SI notation. Objective-type problems have been introduced after the sections in which the material is discussed in an effort to make them more useful.

Emphasis continues to be placed on use of the digital computer for the longer and more tedious problems of Part 2. Use of the programmable hand calculator has been suggested for many of the longer, or iterative type, problems throughout the text, including simple steady-state networks and waterhammer problems. One example of this is in Chapter 5, where Swamee and Jain's approach to direct solution of single pipes for flow and diameter is treated, including iterative solutions for diameter and flow with minor losses. Programs are not given, but the equations are developed in convenient form for programming.

Victor L. Streeter
E. Benjamin Wylie

FLUID MECHANICS

PART
ONE

FUNDAMENTALS OF
FLUID MECHANICS

In the first three chapters of Part 1, the properties of fluids, fluid statics, and the underlying framework of concepts, definitions, and basic equations for fluid dynamics are discussed. Dimensionless parameters are next introduced, including dimensional analysis and dynamic similitude. Chapter 5 deals with real fluids and the introduction of experimental data into fluid-flow calculations. Compressible flow of both real and frictionless fluids is then treated. The final chapter on fundamentals deals with two-dimensional ideal-fluid flow. The theory has been illustrated with elementary applications throughout Part 1.

ONE
FLUID PROPERTIES

The engineering science of fluid mechanics has been developed through an understanding of fluid properties, the application of the basic laws of mechanics and thermodynamics, and orderly experimentation. The properties of density and viscosity play principal roles in open- and closed-channel flow and in flow around immersed objects. Surface-tension effects are important in the formation of droplets, in the flow of small jets, and in situations where liquid-gas-solid or liquid-liquid-solid interfaces occur, as well as in the formation of capillary waves. The property of vapor pressure, which accounts for changes of phase from liquid to gas, becomes important when reduced pressures are encountered.

In this chapter a fluid is defined and consistent systems of force, mass, length, and time units are discussed before the discussion of properties and definition of terms is taken up.

1.1 DEFINITION OF A FLUID

A fluid is a substance that deforms continuously when subjected to a shear stress, no matter how small that shear stress may be. A shear force is the force component tangent to a surface, and this force divided by the area of the surface is the average shear stress over the area. Shear stress at a point is the limiting value of shear force to area as the area is reduced to the point.

In Fig. 1.1 a substance is placed between two closely spaced parallel plates so large that conditions at their edges may be neglected. The lower plate is fixed, and a force F is applied to the upper plate, which exerts a shear stress F/A on any

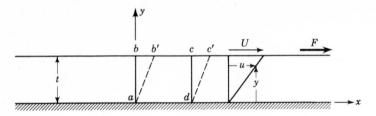

Figure 1.1 Deformation resulting from application of constant shear force.

substance between the plates. A is the area of the upper plate. When the force F causes the upper plate to move with a steady (nonzero) velocity, no matter how small the magnitude of F, one may conclude that the substance between the two plates is a fluid.

The fluid in immediate contact with a solid boundary has the same velocity as the boundary; i.e., there is no slip at the boundary.† This is an experimental fact which has been verified in countless tests with various kinds of fluids and boundary materials. The fluid in the area $abcd$ flows to the new position $ab'c'd$, each fluid particle moving parallel to the plate and the velocity u varying uniformly from zero at the stationary plate to U at the upper plate. Experiments show that, other quantities being held constant, F is directly proportional to A and to U and is inversely proportional to thickness t. In equation form

$$F = \mu \frac{AU}{t}$$

in which μ is the proportionality factor and includes the effect of the particular fluid. If $\tau = F/A$ for the shear stress,

$$\tau = \mu \frac{U}{t}$$

The ratio U/t is the angular velocity of line ab, or it is the *rate of angular deformation* of the fluid, i.e., the rate of decrease of angle bad. The angular velocity may also be written du/dy, as both U/t and du/dy express the velocity change divided by the distance over which the change occurs. However, du/dy is more general, as it holds for situations in which the angular velocity and shear stress change with y. The velocity gradient du/dy may also be visualized as the rate at which one layer moves relative to an adjacent layer. In differential form,

$$\tau = \mu \frac{du}{dy} \tag{1.1.1}$$

---

† S. Goldstein, "Modern Developments in Fluid Dynamics," vol. II, pp. 676–680, Oxford University Press, London, 1938.

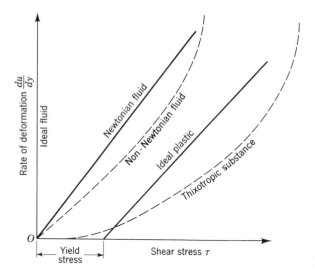

Figure 1.2 Rheological diagram.

is the relation between shear stress and rate of angular deformation for one-dimensional flow of a fluid. The proportionality factor μ is called the *viscosity* of the fluid, and Eq. (1.1.1) is *Newton's law of viscosity*.

Materials other than fluids cannot satisfy the definition of a fluid. A plastic substance will deform a certain *amount* proportional to the force, but not continuously when the stress applied is below its yield shear stress. A complete vacuum between the plates would cause deformation at an ever-increasing rate. If sand were placed between the two plates, Coulomb friction would require a finite force to cause a continuous motion. Hence, plastics and solids are excluded from the classification of fluids.

Fluids may be classified as Newtonian or non-Newtonian. In Newtonian fluid there is a linear relation between the magnitude of applied shear stress and the resulting rate of deformation [μ constant in Eq. (1.1.1)], as shown in Fig. 1.2. In non-Newtonian fluid there is a nonlinear relation between the magnitude of applied shear stress and the rate of angular deformation. An *ideal plastic* has a definite yield stress and a constant linear relation of τ to du/dy. A *thixotropic* substance, such as printer's ink, has a viscosity that is dependent upon the immediately prior angular deformation of the substance and has a tendency to take a set when at rest. Gases and thin liquids tend to be Newtonian fluids, while thick, long-chained hydrocarbons may be non-Newtonian.

For purposes of analysis, the assumption is frequently made that a fluid is nonviscous. With zero viscosity the shear stress is always zero, regardless of the motion of the fluid. If the fluid is also considered to be incompressible, it is then called an *ideal* fluid and plots as the ordinate in Fig. 1.2.

EXERCISES

1.1.1 A fluid is a substance that (a) always expands until it fills any container; (b) is practically incompressible; (c) cannot be subjected to shear forces; (d) cannot remain at rest under action of any shear force; (e) has the same shear stress at a point regardless of its motion.

1.1.2 Newton's law of viscosity relates (a) pressure, velocity, and viscosity; (b) shear stress and rate of angular deformation in a fluid; (c) shear stress, temperature, viscosity, and velocity; (d) pressure, viscosity, and rate of angular deformation; (e) yield shear stress, rate of angular deformation, and viscosity.

1.2 FORCE, MASS, LENGTH, AND TIME UNITS

Consistent units of force, mass, length, and time greatly simplify problem solutions in mechanics; also, by the use of consistent units, derivations may be carried out without reference to any particular consistent system. A system of mechanics units is said to be consistent when unit force causes unit mass to undergo unit acceleration. The International System (SI) has been adopted in many countries and is expected to be adopted in the United States within a few years. This system has the newton (N) as unit of force, the kilogram (kg) as unit of mass, the meter (m) as unit of length, and the second (s) as unit of time. With the kilogram, meter, and second as defined units, the newton is derived to exactly satisfy Newton's second law of motion

$$1 \text{ N} \equiv 1 \text{ kg } \frac{1 \text{ m}}{\text{s}^2} \tag{1.2.1}$$

In the United States the consistent set of units at present is the pound (lb) force, the slug mass, the foot (ft) length, and the second (s) time. The slug is the derived unit; it is taken at such size that one pound force accelerates the unit of mass at one foot per second per second, or

$$1 \text{ lb} \equiv 1 \text{ slug } \frac{1 \text{ ft}}{\text{s}^2} \tag{1.2.2}$$

Some professional engineering groups in the United States use the inconsistent set of units: pound (lb) force, pound (lb_m) mass, foot (ft) length, and second (s) time. With inconsistent units a proportionality constant is required in Newton's second law, usually written as

$$F = \frac{m}{g_0} a \tag{1.2.3}$$

By substitution of the set of units into the situation of one pound force acting on one pound mass at standard gravity in a vacuum, we know the mass is accelerated 32.174 ft/s², or

$$1 \text{ lb} = \frac{1 \text{ lb}_m}{g_0} 32.174 \ \frac{\text{ft}}{\text{s}^2}$$

Table 1.1 Values of g_0 for common systems of units

System	Mass	Length	Time	Force	g_0
SI	kg	m	s	N	1 kg·m/N·s^2
U.S. customary	slug	ft	s	lb	1 slug·ft/lb·s^2
U.S. inconsistent	lb_m	ft	s	lb	$32.174 \text{ lb}_m\text{·ft/lb·s}^2$
Metric, cgs	g	cm	s	dyne	1 g·cm/dyne·s^2
Metric, mks	kg	m	s	kg_f	$9.806 \text{ kg·m/kg}_f\text{·s}^2$

from which g_0 may be determined:

$$g_0 = 32.174 \text{ lb}_m\text{·ft/lb·s}^2 \tag{1.2.4}$$

g_0 has this fixed value for this set of units, whether applied under standard conditions or on the moon.

The mass M of a body does not change with location, but the weight W of a body is determined by the product of the mass and the local acceleration of gravity g:

$$W = Mg \tag{1.2.5}$$

For example, where $g = 32.174 \text{ ft/s}^2$, a body weighing 10 lb has a mass $M = 10/32.174$ slug. At a location where $g = 31.5 \text{ ft/s}^2$, the weight of the body is

$$W = \frac{10 \text{ lb}}{32.174 \text{ ft/s}^2} 31.5 \text{ ft/s}^2 = 9.791 \text{ lb}$$

Standard gravity in SI is 9.806 m/s^2. On the inside front cover, many conversions for various units are given. Since they are presented in the form of dimensionless ratios equal to 1, they can be used on one side of an equation, as a multiplier or as a divisor, to convert units.

The pound-foot-slug-second system will be referred to as the U.S. customary system in this text.

In Table 1.1 the units of g_0 are shown for several common systems.

Abbreviations of SI units are written in lowercase (small) letters for terms like hours (h), meters (m), and seconds (s). When a unit is named after a person, the abbreviation (but not the spelled form) is capitalized; examples are watt (W), pascal (Pa), and newton (N). Multiples and submultiples in powers of 10^3 are

Table 1.2 Selected prefixes for powers of 10 in SI units

Multiple	SI prefix	Abbreviation	Multiple	SI prefix	Abbreviation
10^9	giga	G	10^{-3}	milli	m
10^6	mega	M	10^{-6}	micro	μ
10^3	kilo	k	10^{-9}	nano	n
10^{-2}	centi	c	10^{-12}	pico	p

indicated by prefixes, which also are abbreviated. Common prefixes are shown in Table 1.2. Note that prefixes may not be doubled up: the correct form for 10^{-9} is the prefix n-, as in nanometers; combinations of millimicro-, formerly acceptable, are no longer to be used.

EXERCISES

1.2.1 An object has a mass of 2 kg and weighs 19 N on a spring balance. The value of gravity at this location, in meters per second per second, is (a) 0.105; (b) 2; (c) 9.5; (d) 19; (e) none of these answers.

1.2.2 At a location where $g = 30.00$ ft/s², 2.0 slugs is equivalent to how many pounds mass? (a) 60.0; (b) 62.4; (c) 64.35; (d) not equivalent units; (e) none of these answers.

1.2.3 The weight, in newtons, of 3 kg mass on a planet where $g = 10$ m/s² is (a) 0.30; (b) 3.33; (c) 29.42; (d) 30; (e) none of these answers.

1.2.4 A pressure intensity of 10^9 Pa may be written (a) gPa; (b) GPa; (c) kMPa; (d) μPa; (e) none of these answers.

1.3 VISCOSITY

Of all the fluid properties, viscosity requires the greatest consideration in the study of fluid flow. The nature and characteristics of viscosity are discussed in this section, as well as dimensions and conversion factors for both absolute and kinematic viscosity. Viscosity is that property of a fluid by virtue of which it offers resistance to shear. Newton's law of viscosity [Eq. (1.1.1)] states that for a given rate of angular deformation of fluid the shear stress is directly proportional to the viscosity. Molasses and tar are examples of highly viscous liquids; water and air have very small viscosities.

The viscosity of a gas increases with temperature, but the viscosity of a liquid decreases with temperature. The variation in temperature trends can be explained by examining the causes of viscosity. The resistance of a fluid to shear depends upon its cohesion and upon its rate of transfer of molecular momentum. A liquid, with molecules much more closely spaced than a gas, has cohesive forces much larger than a gas. Cohesion appears to be the predominant cause of viscosity in a liquid; and since cohesion decreases with temperature, the viscosity does likewise. A gas, on the other hand, has very small cohesive forces. Most of its resistance to shear stress is the result of the transfer of molecular momentum.

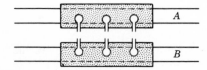

Figure 1.3 Model illustrating transfer of momentum.

As a rough model of the way in which momentum transfer gives rise to an apparent shear stress, consider two idealized railroad cars loaded with sponges and on parallel tracks, as in Fig. 1.3. Assume each car has a water tank and pump so arranged that the water is directed by nozzles at right angles to the track. First, consider A stationary and B moving to the right, with the water from its nozzles striking A and being absorbed by the sponges. Car A will be set in motion owing to the component of the momentum of the jets which is parallel to the tracks, giving rise to an apparent shear stress between A and B. Now if A is pumping water back into B at the same rate, its action tends to slow down B and equal and opposite apparent shear forces result. When both A and B are stationary or have the same velocity, the pumping does not exert an apparent shear stress on either car.

Within fluid there is always a transfer of molecules back and forth across any fictitious surface drawn in it. When one layer moves relative to an adjacent layer, the molecular transfer of momentum brings momentum from one side to the other so that an apparent shear stress is set up that resists the relative motion and tends to equalize the velocities of adjacent layers in a manner analogous to that of Fig. 1.3. The measure of the motion of one layer relative to an adjacent layer is du/dy.

Molecular activity gives rise to an apparent shear stress in gases which is more important than the cohesive forces, and since molecular activity increases with temperature, the viscosity of a gas also increases with temperature.

For ordinary pressures viscosity is independent of pressure and depends upon temperature only. For very great pressures, gases and most liquids have shown erratic variations of viscosity with pressure.

A fluid at rest or in motion so that no layer moves relative to an adjacent layer will not have apparent shear forces set up, regardless of the viscosity, because du/dy is zero throughout the fluid. Hence, in the study of *fluid statics*, no shear forces can be considered because they do not occur in a static fluid, and the only stresses remaining are normal stresses, or pressures. This greatly simplifies the study of fluid statics, since any free body of fluid can have only gravity forces and normal surface forces acting on it.

The dimensions of viscosity are determined from Newton's law of viscosity [Eq. (1.1.1)]. Solving for the viscosity μ

$$\mu = \frac{\tau}{du/dy}$$

and inserting dimensions F, L, T for force, length, and time,

$$\tau : FL^{-2} \qquad u : LT^{-1} \qquad y : L$$

shows that μ has the dimensions $FL^{-2}T$. With the force dimension expressed in terms of mass by use of Newton's second law of motion, $F = MLT^{-2}$, the dimensions of viscosity may be expressed as $ML^{-1}T^{-1}$.

The SI unit of viscosity, the newton-second per square meter $(N \cdot s/m^2)$ or the kilogram per meter per second $(kg/m \cdot s)$, has no name. The U.S. customary unit of viscosity (which has no name) is 1 lb·s/ft^2 or 1 slug/ft·s (these are identical). A common unit of viscosity is the cgs unit, called the poise (P); it is 1 dyne·s/cm^2 or 1 g/cm·s. The SI unit is 10 times larger than the poise unit.†

Kinematic Viscosity

The viscosity μ is frequently referred to as the *absolute* viscosity or the *dynamic* viscosity to avoid confusing it with the *kinematic* viscosity v, which is the ratio of viscosity to mass density:

$$v = \frac{\mu}{\rho} \qquad (1.3.1)$$

The kinematic viscosity occurs in many applications, e.g., in the dimensionless Reynolds number for motion of a body through a fluid, Vl/v, in which V is the body velocity and l is a representative linear measure of the body size. The dimensions of v are L^2T^{-1}. The SI unit of kinematic viscosity is 1 m^2/s, and the U.S. customary unit is 1 ft^2/s. The cgs unit, called the stoke (St), is 1 cm^2/s.

In SI units, to convert from v to μ, it is necessary to multiply v by ρ, the mass density in kilograms per cubic meter. In U.S. customary units μ is obtained from v by multiplying by the mass density in slugs per cubic foot. To change from the stoke to the poise, one multiplies by mass density in grams per cubic centimeter, which is numerically equal to specific gravity.

Example 1.1 A liquid has a viscosity of 0.005 kg/m·s and a density of 850 kg/m^3. Calculate (a) the kinematic viscosity in SI, (b) the kinematic viscosity in U.S. customary units, and (c) the viscosity in U.S. customary units.

(a) $$v = \frac{\mu}{\rho} = \frac{0.005 \text{ kg/m·s}}{850 \text{ kg/m}^3} = 5.882 \ \mu\text{m}^2/\text{s}$$

(b) $$v = 5.882 \times 10^{-6} \text{ m}^2/\text{s} \times \left(\frac{1 \text{ ft}}{0.3048 \text{ m}}\right)^2 = 6.331 \times 10^{-5} \text{ ft}^2/\text{s}$$

(c) $$\mu = 0.005 \text{ kg/m·s} \times \frac{1 \text{ slug/ft·s}}{47.9 \text{ kg/m·s}} = 0.0001044 \text{ slug/ft·s}$$

Viscosity is practically independent of pressure and depends upon temperature only. The kinematic viscosity of liquids, and of gases at a given pressure, is

† The conversion from the U.S. customary unit of viscosity to the SI unit is

$$\frac{1 \text{ slug}}{\text{ft·s}} \cdot \frac{14.594 \text{ kg}}{\text{slug}} \cdot \frac{1 \text{ ft}}{0.3048 \text{ m}} = 47.9 \text{ kg/m·s}$$

or

$$\frac{1 \text{ U.S. customary unit viscosity}}{47.9 \text{ units SI viscosity}} = 1$$

substantially a function of temperature. Charts for the determination of absolute viscosity and kinematic viscosity are given in Appendix C, Figs. C.1 and C.2, respectively.

EXERCISES

1.3.1 Viscosity has the dimensions (a) $FL^{-2}T$; (b) $FL^{-1}T^{-1}$; (c) FLT^{-2}; (d) FL^2T; (e) FLT^2.

1.3.2 Select the *incorrect* completion. Apparent shear forces (a) can never occur when the fluid is at rest; (b) may occur owing to cohesion when the liquid is at rest; (c) depend upon molecular interchange of momentum; (d) depend upon cohesive forces; (e) can never occur in a frictionless fluid, regardless of its motion.

1.3.3 Correct units for dynamic viscosity are (a) $m \cdot s/kg$; (b) $N \cdot m/s^2$; (c) $kg \cdot s/N$; (d) $kg/m \cdot s$; (e) $N \cdot s/m$.

1.3.4 Viscosity, expressed in poises, is converted to the U.S. customary unit of viscosity by multiplication by (a) $\frac{1}{479}$; (b) 479; (c) ρ; (d) $1/\rho$; (e) none of these answers.

1.3.5 The dimensions for kinematic viscosity are (a) $FL^{-2}T$; (b) $ML^{-1}T^{-1}$; (c) L^2T^2; (d) L^2T^{-1}; (e) L^2T^{-2}.

1.3.6 The viscosity of kerosene at 20°C, from Fig. C.1, in newton-seconds per square meter, is (a) 4×10^{-5}; (b) 4×10^{-4}; (c) 1.93×10^{-3}; (d) 1.93×10^{-2}; (e) 1.8×10^{-2}.

1.3.7 The kinematic viscosity of dry air at 25°F and 29.4 psia, in square feet per second, is (a) 6.89×10^{-5}; (b) 1.4×10^{-4}; (c) 6.89×10^{-4}; (d) 1.4×10^{-3}; (e) none of these answers.

1.3.8 For $\mu = 0.06$ kg/m·s, sp gr $= 0.60$, v is, in stokes, (a) 2.78; (b) 1.0; (c) 0.60; (d) 0.36; (e) none of these answers.

1.3.9 For $\mu = 2.0 \times 10^{-4}$ slug/ft·s, the value of μ, in pound-seconds per square foot, is (a) 1.03×10^{-4}; (b) 2.0×10^{-4}; (c) 6.21×10^{-4}; (d) 6.44×10^{-3}; (e) none of these answers.

1.3.10 For $v = 3 \times 10^{-8}$ m²/s and $\rho = 800$ kg/m³, μ in SI equals (a) 3.75×10^{-11}; (b) 2.4×10^{-5}; (c) 2.4×10^5; (d) 2.4×10^{12}; (e) none of these answers.

1.4 CONTINUUM

In dealing with fluid-flow relations on a mathematical or analytical basis, it is necessary to consider that the actual molecular structure is replaced by a hypothetical continuous medium, called the *continuum*. For example, velocity at a point in space is indefinite in a molecular medium, as it would be zero at all times except when a molecule occupied this exact point, and then it would be the velocity of the molecule and not the mean mass velocity of the particles in the neighborhood. This dilemma is avoided if one considers velocity at a point to be the average or mass velocity of all molecules surrounding the point, say, within a small sphere with radius large compared with the *mean distance between molecules*. With n molecules per cubic centimeter, the mean distance between molecules is of the order $n^{-1/3}$ cm. Molecular theory, however, must be used to calculate fluid properties (e.g., viscosity) which are associated with molecular motions, but continuum equations can be employed with the results of molecular calculations.

In rarefied gases, such as the atmosphere at 50 mi above sea level, the ratio of the mean free path† of the gas to a characteristic length for a body or conduit is used to distinguish the type of flow. The flow regime is called *gas dynamics* for very small values of the ratio; the next regime is called *slip flow*; and for large values of the ratio it is *free molecule flow*. In this text only the gas-dynamics regime is studied.

The quantities density, specific volume, pressure, velocity, and acceleration are assumed to vary continuously throughout a fluid (or be constant).

EXERCISE

1.4.1 Under which two of the following flow regimes would the assumption of a continuum be reasonable? (1) free molecule flow, (2) slip flow, (3) gas dynamics, (4) complete vacuum, (5) liquid flow. (*a*) 1, 2; (*b*) 1, 4; (*c*) 2, 3; (*d*) 3, 5; (*e*) 1, 5.

1.5 DENSITY, SPECIFIC VOLUME, SPECIFIC WEIGHT, SPECIFIC GRAVITY, PRESSURE

The *density* ρ of a fluid is defined as its mass per unit volume. To define density at a point, the mass Δm of fluid in a small volume $\Delta \mho$ surrounding the point is divided by $\Delta \mho$ and the limit is taken as $\Delta \mho$ becomes a value ϵ^3 in which ϵ is still large compared with the mean distance between molecules,

$$\rho = \lim_{\Delta \mho \to \epsilon^3} \frac{\Delta m}{\Delta \mho} \tag{1.5.1}$$

For water at standard pressure (760 mm Hg) and 4°C (39.2°F), $\rho = 1.94$ slugs/ft³, or 1000 kg/m³.

The *specific volume* v_s is the reciprocal of the density ρ; that is, it is the volume occupied by unit mass of fluid. Hence

$$v_s = \frac{1}{\rho} \tag{1.5.2}$$

The *specific weight* γ of a substance is its weight per unit volume. It changes with location,

$$\gamma = \rho g \tag{1.5.3}$$

depending upon gravity. It is a convenient property when dealing with fluid statics or with liquids with a free surface.

The *specific gravity* S of a substance is the ratio of its weight to the weight of an equal volume of water at standard conditions. It may also be expressed as a ratio of its density or specific weight to that of water.

---

† The mean free path is the average distance a molecule travels between collisions.

The normal force pushing against a plane area divided by the area is the average *pressure*. The pressure at a point is the ratio of normal force to area as the area approaches a small value enclosing the point. If a fluid exerts a pressure against the walls of a container, the container will exert a reaction on the fluid which will be compressive. Liquids can sustain very high compressive pressures, but unless they are extremely pure, they are very weak in tension. It is for this reason that the absolute pressures used in this book are never negative, since this would imply that fluid is sustaining a tensile stress. Pressure p has the units force per area, which may be newtons per square meter, called pascals (Pa), or pounds per square foot (psf), or pounds per square inch (psi). Pressure may also be expressed in terms of an equivalent height h of a fluid column, $p = \gamma h$, as shown in Sec. 2.3.

1.6 PERFECT GAS

In this treatment, thermodynamic relations and compressible-fluid-flow cases have been limited generally to perfect gases. The perfect gas is defined in this section, and its various interrelations with specific heats are treated in Sec. 6.1.

The perfect gas, as used herein, is defined as a substance that satisfies the *perfect-gas law*

$$pv_s = RT \qquad (1.6.1)$$

and that has constant specific heats. p is the absolute pressure; v_s is the specific volume; R is the gas constant; and T is the absolute temperature. The perfect gas must be carefully distinguished from the ideal fluid. An ideal fluid is frictionless and incompressible. The perfect gas has viscosity and can therefore develop shear stresses, and it is compressible according to Eq. (1.6.1).

Equation (1.6.1) is the equation of state for a perfect gas. It may be written

$$p = \rho RT \qquad (1.6.2)$$

The units of R can be determined from the equation when the other units are known. For p in pascals, ρ in kilograms per cubic meter, and T in degrees kelvin (K) (°C + 273)†

$$R = \frac{N}{m^2} \frac{m^3}{kg \cdot K} = \frac{m \cdot N}{kg \cdot K} \qquad \text{or } m \cdot N/kg \cdot K$$

For U.S. customary units, °R = °F + 459.6

$$R = \frac{lb}{ft^2} \frac{ft^3}{slug \cdot °R} = \frac{ft \cdot lb}{slug \cdot °R} \qquad \text{or } ft \cdot lb/slug \cdot °R$$

† In 1967 the name *degree Kelvin* (°K) was changed to *kelvin* (K).

For ρ in pounds mass per cubic foot

$$R = \frac{lb}{ft^2} \frac{ft^3}{lb_m \cdot {}^\circ R} = \frac{ft \cdot lb}{lb_m \cdot {}^\circ R} \quad \text{or } ft \cdot lb/lb_m \cdot {}^\circ R$$

The magnitude of R in slugs is 32.174 times greater than in pounds mass. Values of R for several common gases are given in Table C.3 of Appendix C.

Real gases below critical pressure and above the critical temperature tend to obey the perfect-gas law. As the pressure increases, the discrepancy increases and becomes serious near the critical point. The perfect-gas law encompasses both Charles' law and Boyle's law. Charles' law states that for constant pressure the volume of a given mass of gas varies as its absolute temperature. Boyle's law (isothermal law) states that for constant temperature the density varies directly as the absolute pressure. The volume $\mho$ of m mass units of gas is mv_s; hence

$$p\mho = mRT \tag{1.6.3}$$

Certain simplifications result from writing the perfect-gas law on a mole basis. A kilogram mole of gas is the number of kilograms mass of gas equal to the molecular weight; e.g., a kilogram mole of oxygen O_2 is 32 kg. With $\bar{v}_s$ being the volume per mole, the perfect-gas law becomes

$$p\bar{v}_s = MRT \tag{1.6.4}$$

if M is the molecular weight. In general, if n is the number of moles of the gas in volume $\mho$,

$$p\mho = nMRT \tag{1.6.5}$$

since $nM = m$. Now, from Avogadro's law, equal volumes of gases at the same absolute temperature and pressure have the same number of molecules; hence their masses are proportional to the molecular weights. From Eq. (1.6.5) it is seen that MR must be constant, since $p\mho/nT$ is the same for any perfect gas. The product MR, called the universal gas constant, has a value depending only upon the units employed. It is

$$MR = 8312 \ m \cdot N/kg \cdot mole \cdot K \tag{1.6.6}$$

The gas constant R can then be determined from

$$R = \frac{8312}{M} \ m \cdot N/kg \cdot K \tag{1.6.7}$$

In U.S. customary units

$$R = \frac{49{,}709}{M} \ ft \cdot lb/slug \cdot {}^\circ R \tag{1.6.8}$$

In pound mass units

$$R = \frac{1545}{M} \ ft \cdot lb/lb_m \cdot {}^\circ R \tag{1.6.9}$$

so that knowledge of molecular weight leads to the value of R. In Table C.3 of Appendix C molecular weights of some common gases are listed.

The *specific heat* c_v of a gas is the number of units of heat added per unit mass to raise the temperature of the gas one degree when the volume is held constant. The *specific heat* c_p is the number of heat units added per unit mass to raise the temperature one degree when the pressure is held constant. The *specific heat ratio* k is c_p/c_v. The *intrinsic energy u* (dependent upon p, ρ, and T) is the energy per unit mass due to molecular spacing and forces. The *enthalpy h* is an important property of a gas given by $h = u + p/\rho$.

c_v and c_p have the units kilocalorie per kilogram per kelvin (kcal/kg·K) or Btu per pound mass per degree Rankine (Btu/lb$_m$·°R). One kilocalorie of heat added raises the temperature of one kilogram of water one degree Celsius at standard conditions. One Btu of heat added raises the temperature of one pound mass of water one degree Fahrenheit. Because of these definitions of kilocalorie and Btu, the numerical values of c_v and c_p are the same in both systems of units (see Appendix C, Table C.3). R is related to c_v and c_p by

$$c_p = c_v + R$$

in which all quantities must be in either mechanical or thermal units. If the slug unit is used, c_p, c_v, and R are 32.174 times greater than with the pound mass unit.

Additional relations and definitions used in perfect-gas flow are introduced in Chaps. 3 and 6.

Example 1.2 A gas with molecular weight of 44 is at a pressure of 0.9 MPa and a temperature of 20°C. Determine its density.
From Eq. (1.6.7),

$$R = \frac{8312}{44} = 188.91 \text{ m·N/kg·K}$$

Then, from Eq. (1.6.2),

$$\rho = \frac{p}{RT} = \frac{0.9 \times 10^6 \text{ N/m}^2}{(188.91 \text{ m·N/kg·K})(273 + 20 \text{ K})} = 16.26 \text{ kg/m}^3$$

EXERCISES

1.6.1 A perfect gas (*a*) has zero viscosity; (*b*) has constant viscosity; (*c*) is incompressible; (*d*) satisfies $p\rho = RT$; (*e*) fits none of these statements.

1.6.2 The molecular weight of a gas is 28. The value of R in foot-pounds per slug per degree Rankine is (*a*) 53.3; (*b*) 55.2; (*c*) 1545; (*d*) 1775; (*e*) none of these answers.

1.6.3 The density of air at 10°C and 1 MPa abs in SI is (*a*) 1.231; (*b*) 12.31; (*c*) 65.0; (*d*) 118.4; (*e*) none of these answers.

1.6.4 How many pounds mass of carbon monoxide gas at 20°F and 30 psia is contained in a volume of 4.0 ft^3? (*a*) 0.00453; (*b*) 0.0203; (*c*) 0.652; (*d*) 2.175; (*e*) none of these answers.

1.6.5 A container holds 1 kg air at 30°C and 9 MPa abs. If 1.5 kg air is added and the final temperature is 110°C, the final absolute pressure is (*a*) 7.26 MPa; (*b*) 25.3 MPa; (*c*) 73.4 MPa; (*d*) indeterminable; (*e*) none of these answers.

1.7 BULK MODULUS OF ELASTICITY

In the preceding section the compressibility of a perfect gas is described by the perfect-gas law. For most purposes a liquid may be considered as incompressible, but for situations involving either sudden or great changes in pressure, its compressibility becomes important. Liquid (and gas) compressibility also becomes important when temperature changes are involved, e.g., free convection. The compressibility of a liquid is expressed by its *bulk modulus of elasticity*. If the pressure of a unit volume of liquid is increased by dp, it will cause a volume decrease $-d\mathcal{V}$; the ratio $-dp/d\mathcal{V}$ is the bulk modulus of elasticity K. For any volume $\mathcal{V}$ of liquid,

$$K = -\frac{dp}{d\mathcal{V}/\mathcal{V}} \qquad (1.7.1)$$

Since $d\mathcal{V}/\mathcal{V}$ is dimensionless, K is expressed in units of p. For water at 20°C (Table C.1, Appendix C) $K = 2.2$ GPa, or, from Table C.2, $K = 311,000$ lb/in² for water at 60°F.

To gain some idea about the compressibility of water, consider the application of 0.1 MPa (about one atmosphere) to a cubic meter of water.

$$-d\mathcal{V} = \frac{\mathcal{V}\,dp}{K} = \frac{(1.0\ \text{m}^3)(0.1\ \text{MPa})}{2.2\ \text{GPa}} = \frac{1}{22,000}\ \text{m}^3$$

or about 45.5 cm³. As a liquid is compressed, its resistance to further compression increases. At 3000 atmospheres the value of K for water has doubled.

Example 1.3 A liquid compressed in a cylinder has a volume of 1 liter (l) (1000 cm³) at 1 MN/m² and a volume of 995 cm³ at 2 MN/m². What is its bulk modulus of elasticity?

$$K = -\frac{\Delta p}{\Delta \mathcal{V}/\mathcal{V}} = -\frac{2 - 1\ \text{MN/m}^2}{(995 - 1000)/1000} = 200\ \text{MPa}$$

EXERCISES

1.7.1 The bulk modulus of elasticity K for a gas at constant temperature T_0 is given by (a) p/ρ; (b) RT_0; (c) pp; (d) ρRT_0; (e) none of these answers.

1.7.2 The bulk modulus of elasticity (a) is independent of temperature; (b) increases with the pressure; (c) has the dimensions of $1/p$; (d) is larger when the fluid is more compressible; (e) is independent of pressure and viscosity.

1.7.3 For 70 atm increase in pressure the density of water has increased, in percent, by about (a) $\frac{1}{300}$; (b) $\frac{1}{30}$; (c) $\frac{1}{3}$; (d) $\frac{1}{2}$; (e) none of these answers.

1.7.4 A pressure of 150 psi applied to 10 ft³ liquid causes a volume reduction of 0.02 ft³. The bulk modulus of elasticity in pounds per square inch is (a) -750; (b) 750; (c) 7500; (d) 75,000; (e) none of these answers.

1.8 VAPOR PRESSURE

Liquids evaporate because of molecules escaping from the liquid surface. The vapor molecules exert a partial pressure in the space, known as *vapor pressure*. If the space above the liquid is confined, after a sufficient time the number of vapor molecules striking the liquid surface and condensing is just equal to the number escaping in any interval of time, and equilibrium exists. Since this phenomenon depends upon molecular activity, which is a function of temperature, the vapor pressure of a given fluid depends upon temperature and increases with it. When the pressure above a liquid equals the vapor pressure of the liquid, boiling occurs. Boiling of water, for example, may occur at room temperature if the pressure is reduced sufficiently. At 20°C water has a vapor pressure of 2.447 kPa and mercury has a vapor pressure of 0.173 Pa.

In many situations involving the flow of liquids it is possible that very low pressures are produced at certain locations in the system. Under such circumstances the pressures may be equal to or less than the vapor pressure. When this occurs, the liquid flashes into vapor. This is the phenomenon of cavitation. A rapidly expanding vapor pocket, or cavity, forms, which is usually swept away from its point of origin and enters regions of the flow where the pressure is greater than vapor pressure. The cavity collapses. This growth and decay of the vapor bubbles affects the operating performance of hydraulic pumps and turbines and can result in erosion of the metal parts in the region of cavitation.

EXERCISE

1.8.1 The vapor pressure of water at 30°C, in pascals, is (*a*) 0.44; (*b*) 7.18; (*c*) 223; (*d*) 4315; (*e*) none of these answers.

1.9 SURFACE TENSION

Capillarity

At the interface between a liquid and a gas, or two immiscible liquids, a film or special layer seems to form on the liquid, apparently owing to attraction of liquid molecules below the surface. It is a simple experiment to place a small needle on a quiet water surface and observe that it is supported there by the film.

The formation of this film may be visualized on the basis of *surface energy* or work per unit area required to bring the molecules to the surface. The surface tension is then the stretching force required to form the film, obtained by dividing the surface-energy term by unit length of the film in equilibrium. The surface tension of water varies from about 0.074 N/m at 20°C to 0.059 N/m at 100°C. Surface tensions, along with other properties, are given for a few common liquids in Table 1.3.

Table 1.3 Approximate properties of common liquids at 20°C and standard atmospheric pressure

Liquid	Specific gravity, S	Bulk modulus of elasticity, K GPa	Vapor pressure, p_v kPa	Surface tension,* σ N/m
Alcohol, ethyl	0.79	1.21	5.86	0.0223
Benzene	0.88	1.03	10.0	0.0289
Carbon tetrachloride	1.59	1.10	13.1	0.0267
Kerosene	0.81	. . . .	. . . . . .	0.023–0.032
Mercury	13.57	26.20	0.00017	0.51
Oil:				
Crude	0.85–0.93	. . . .	. . . . . .	0.023–0.038
Lubricating	0.85–0.88	. . . .	. . . . . .	0.035–0.038
Water	1.00	2.07	2.45	0.074

* In contact with air.

The action of surface tension is to increase the pressure within a droplet of liquid or within a small liquid jet. For a small spherical droplet of radius r the internal pressure p necessary to balance the tensile force due to the surface tension σ is calculated in terms of the forces which act on a hemispherical free body (see Sec. 2.6),

$$p\pi r^2 = 2\pi r\sigma \qquad \text{or} \qquad p = \frac{2\sigma}{r}$$

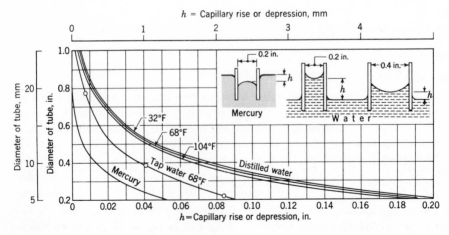

Figure 1.4 Capillarity in circular glass tubes. (*By permission from R. L. Daugherty, "Hydraulics," McGraw-Hill Book Company, New York, 1937.*)

For the cylindrical liquid jet of radius r, the pipe-tension equation applies:

$$p = \frac{\sigma}{r}$$

Both equations show that the pressure becomes large for a very small radius of droplet or cylinder.

Capillary attraction is caused by surface tension and by the relative value of adhesion between liquid and solid to cohesion of the liquid. A liquid that *wets* the solid has a greater adhesion than cohesion. The action of surface tension in this case is to cause the liquid to rise within a small vertical tube that is partially immersed in it. For liquids that do not wet the solid, surface tension tends to depress the meniscus in a small vertical tube. When the contact angle between liquid and solid is known, the capillary rise can be computed for an assumed shape of the meniscus. Figure 1.4 shows the capillary rise for water and mercury in circular glass tubes in air.

PROBLEMS

1.1 Classify the substance that has the following rates of deformation and corresponding shear stresses:

du/dy, rad/s	0	1	3	5
τ, lb/ft²	15	20	30	40

1.2 Classify the following substances (maintained at constant temperature):

(a) du/dy, rad/s	0	3	4	6	5	4
τ, lb/ft²	2	4	6	8	6	4

(b) du/dy, rad/s	0	0.5	1.1	1.8
τ, N/m²	0	2	4	6

(c) du/dy, rad/s	0	0.3	0.6	0.9	1.2
τ, N/m²	0	2	4	6	8

1.3 A Newtonian liquid flows down an inclined plane in a thin sheet of thickness t (Fig. 1.5). The upper surface is in contact with air, which offers almost no resistance to the flow. Using Newton's law of viscosity, decide what the value of du/dy, y measured normal to the inclined plane, must be at the upper surface. Would a linear variation of u with y be expected?

1.4 What kinds of rheological materials are paint and grease?

1.5 A Newtonian fluid is in the clearance between a shaft and a concentric sleeve. When a force of 500 N is applied to the sleeve parallel to the shaft, the sleeve attains a speed of 1 m/s. If a 1500-N force is applied, what speed will the sleeve attain? The temperature of the sleeve remains constant.

1.6 Determine the weight in pounds of 3 slugs mass at a place where $g = 31.7$ ft/s².

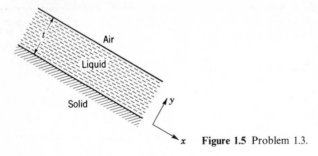

x   **Figure 1.5** Problem 1.3.

1.7 When standard scale weights and a balance are used, a body is found to be equivalent in pull of gravity to two of the 1-lb scale weights at a location where $g = 31.5$ ft/s². What would the body weigh on a correctly calibrated spring balance (for sea level) at this location?

1.8 Determine the value of proportionality constant g_0 needed for the following set of units: kip (1000 lb), slug, foot, second.

1.9 On another planet, where standard gravity is 3 m/s², what would be the value of the proportionality constant g_0 in terms of the kilogram force, gram, millimeter, and second?

1.10 A correctly calibrated spring scale records the weight of a 2-kg body as 17.0 N at a location away from the earth. What is the value of g at this location?

1.11 Does the weight of a 20-N bag of flour at sea level denote a force or the mass of the flour? What is the mass of the flour in kilograms? What are the mass and weight of the flour at a location where the gravitational acceleration is one-seventh that of the earth's standard?

1.12 Convert 10.4 SI units of kinematic viscosity to U.S. customary units of dynamic viscosity if $S = 0.85$.

1.13 A shear stress of 4 dynes/cm² causes a Newtonian fluid to have an angular deformation of 1 rad/s. What is its viscosity in centipoises?

1.14 A plate, 0.5 mm distant from a fixed plate, moves at 0.25 m/s and requires a force per unit area of 2 Pa (N/m²) to maintain this speed. Determine the fluid viscosity of the substance between the plates in SI units.

1.15 Determine the viscosity of fluid between shaft and sleeve in Fig. 1.6.

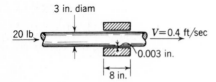

Figure 1.6 Problem 1.15.

1.16 A flywheel weighing 600 N has a radius of gyration of 300 mm. When it is rotating 600 rpm, its speed reduces 1 rpm/s owing to fluid viscosity between sleeve and shaft. The sleeve length is 50 mm; shaft diameter is 20 mm; and radial clearance is 0.05 mm. Determine the fluid viscosity.

1.17 A 1-in-diameter steel cylinder 12 in long falls, because of its own weight, at a uniform rate of 0.5 ft/s inside a tube of slightly larger diameter. A castor-oil film of constant thickness is between the cylinder and the tube. Determine the clearance between the tube and the cylinder. The temperature is 100°F. Specific gravity of steel = 7.85.

1.18 A piston of diameter 50.00 mm moves within a cylinder of 50.10 mm. Determine the percent decrease in force necessary to move the piston when the lubricant warms up from 0 to 120°C. Use crude-oil viscosity from Fig. C.1, Appendix C.

1.19 How much greater is the viscosity of water at 0°C than at 100°C? How much greater is its kinematic viscosity for the same temperature range?

1.20 A fluid has a viscosity of 6 cP and a density of 50 lb_m/ft^3. Determine its kinematic viscosity in U.S. customary units and in stokes.

1.21 A fluid has a specific gravity of 0.83 and a kinematic viscosity of 4 St. What is its viscosity in U.S. customary units and in SI units?

1.22 A body weighing 120 lb with a flat surface area of 2 ft^2 slides down a lubricated inclined plane making a 30° angle with the horizontal. For viscosity of 1 P and body speed of 3 ft/s, determine the lubricant film thickness.

1.23 What is the viscosity of gasoline at 25°C in poises?

1.24 Determine the kinematic viscosity of benzene at 80°F in stokes.

1.25 Calculate the value of the gas constant R in SI units, starting with $R = 1545/M$ ft·lb/lb_m·°R.

1.26 What is the specific volume in cubic feet per pound mass and cubic feet per slug of a substance of specific gravity 0.75?

1.27 What is the relation between specific volume and specific weight?

1.28 The density of a substance is 2.94 g/cm^3. In SI, what is its (a) specific gravity, (b) specific volume, and (c) specific weight?

1.29 A force, expressed by $F = 4i + 3j + 9k$, acts upon a square area, 2 by 2 in, in the xy plane. Resolve this force into a normal-force and a shear-force component. What are the pressure and the shear stress? Repeat the calculations for $F = -4i + 3j - 9k$.

1.30 A gas at 20°C and 0.2 MPa abs has a volume of 40 l and a gas constant $R = 210$ m·N/kg·K. Determine the density and mass of the gas.

1.31 What is the specific weight of air at 60 psia and 90°F?

1.32 What is the density of water vapor at 0.3 MPa abs and 15°C in SI units?

1.33 A gas with molecular weight 28 has a volume of 4.0 ft^3 and a pressure and temperature of 2000 psfa (lb/ft^2 abs) and 600°R, respectively. What are its specific volume and specific weight?

1.34 One kilogram of hydrogen is confined in a volume of 150 l at −40°C. What is the pressure?

1.35 Express the bulk modulus of elasticity in terms of density change rather than volume change.

1.36 For constant bulk modulus of elasticity, how does the density of a liquid vary with the pressure?

1.37 What is the bulk modulus of a liquid that has a density increase of 0.02 percent for a pressure increase of 1000 lb/ft^2? For a pressure increase of 60 kPa?

1.38 For $K = 2.2$ GPa for bulk modulus of elasticity of water what pressure is required to reduce its volume by 0.5 percent?

1.39 A steel container expands in volume 1 percent when the pressure within it is increased by 10,000 psi. At standard pressure, 14.7 psia, it holds 1000 lb_m water, $\rho = 62.4$ lb_m/ft^3. For $K = 300,000$ psi, when it is filled, how many pounds mass water need be added to increase the pressure to 10,000 psi?

1.40 What is the isothermal bulk modulus for air at 0.4 MPa abs?

1.41 At what pressure can cavitation be expected at the inlet of a pump that is handling water at 20°C?

1.42 What is the pressure within a droplet of water of 0.002-in diameter at 68°F if the pressure outside the droplet is standard atmospheric pressure of 14.7 psi?

1.43 A small circular jet of mercury 0.1 mm in diameter issues from an opening. What is the pressure difference between the inside and outside of the jet when at 20°C?

1.44 Determine the capillary rise for distilled water at 104°F in a circular $\frac{1}{4}$-in-diameter glass tube.

1.45 What diameter of glass tube is required if the capillary effects on the water within are not to exceed 0.5 mm?

1.46 Using the data given in Fig. 1.4, estimate the capillary rise of tap water between two parallel glass plates 0.20 in apart.

1.47 A method of determining the surface tension of a liquid is to find the force needed to pull a platinum wire ring from the surface (Fig. 1.7). Estimate the force necessary to remove a 20-mm-diameter ring from the surface of water at 20°C. Why is platinum used as the material for the ring?

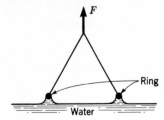

Figure 1.7 Problem 1.47.

TWO
FLUID STATICS

The science of fluid statics will be treated in two parts: the study of pressure and its variation throughout a fluid and the study of pressure forces on finite surfaces. Special cases of fluids moving as solids are included in the treatment of statics because of the similarity of forces involved. Since there is no motion of a fluid layer relative to an adjacent layer, there are no shear stresses in the fluid. Hence, all free bodies in fluid statics have only normal pressure forces acting on their surfaces.

2.1 PRESSURE AT A POINT

The average pressure is calculated by dividing the normal force pushing against a plane area by the area. The pressure at a point is the limit of the ratio of normal force to area as the area approaches zero size at the point. At a point a fluid at rest has the same pressure in all directions. This means that an element δA of very small area, free to rotate about its center when submerged in a fluid at rest, will have a force of constant magnitude acting on either side of it, regardless of its orientation.

To demonstrate this, a small wedge-shaped free body of unit width is taken at the point (x, y) in a fluid at rest (Fig. 2.1). Since there can be no shear forces, the

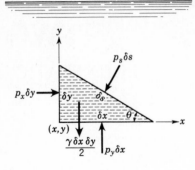

Figure 2.1 Free-body diagram of wedge-shaped particle.

only forces are the normal surface forces and gravity. So, the equations of motion in the x and y directions are, respectively,

$$\Sigma F_x = p_x \, \delta y - p_s \, \delta s \sin \theta = \frac{\delta x \, \delta y}{2} \rho a_x = 0$$

$$\Sigma F_y = p_y \, \delta x - p_s \, \delta s \cos \theta - \gamma \, \frac{\delta x \, \delta y}{2} = \frac{\delta x \, \delta y}{2} \rho a_y = 0$$

in which p_x, p_y, p_s are the average pressures on the three faces, γ is the specific weight of the fluid, ρ is its density, and a_x, a_y are the accelerations. When the limit is taken as the free body is reduced to zero size by allowing the inclined face to approach (x, y) while maintaining the same angle θ, and using the geometric relations

$$\delta s \sin \theta = \delta y \qquad \delta s \cos \theta = \delta x$$

the equations simplify to

$$p_x \, \delta y - p_s \, \delta y = 0 \qquad p_y \, \delta x - p_s \, \delta x - \gamma \, \frac{\delta x \, \delta y}{2} = 0$$

The last term of the second equation is an infinitestimal of higher order of smallness and may be neglected. When divided by δy and δx, respectively, the equations can be combined:

$$p_s = p_x = p_y \tag{2.1.1}$$

Since θ is any arbitrary angle, this equation proves that the pressure is the same in all directions at a point in a static fluid. Although the proof was carried out for a two-dimensional case, it may be demonstrated for the three-dimensional case with the equilibrium equations for a small tetrahedron of fluid with three faces in the coordinate planes and the fourth face inclined arbitrarily.

If the fluid is in motion so that one layer moves relative to an adjacent layer, shear stresses occur and the normal stresses are, in general, no longer the same in all directions at a point. The pressure is then defined as the average of any three mutually perpendicular normal compressive stresses at a point,

$$p = \frac{p_x + p_y + p_z}{3}$$

In a fictitious fluid of zero viscosity, i.e., a frictionless fluid, no shear stresses can occur for any motion of the fluid, and so at a point the pressure is the same in all directions.

EXERCISE

2.1.1 The normal stress is the same in all directions at a point in a fluid (*a*) *only* when the fluid is frictionless; (*b*) *only* when the fluid is frictionless and incompressible; (*c*) *only* when the fluid has zero viscosity and is at rest; (*d*) when there is no motion of one fluid layer relative to an adjacent layer; (*e*) regardless of the motion of one fluid layer relative to an adjacent layer.

2.2 BASIC EQUATION OF FLUID STATICS

Pressure Variation in a Static Fluid

The forces acting on an element of fluid at rest, Fig. 2.2, consist of surface forces and body forces. With gravity the only body force acting, and by taking the *y* axis vertically upward, it is $-\gamma\,\delta x\,\delta y\,\delta z$ in the *y* direction. With pressure *p* at its center

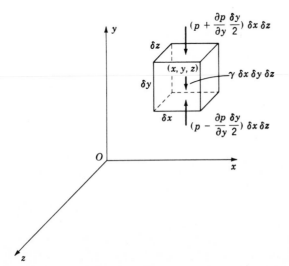

Figure 2.2 Rectangular parallelepiped element of fluid at rest.

(x, y, z), the approximate force exerted on the side normal to the y axis closest to the origin is approximately

$$\left(p - \frac{\partial p}{\partial y}\frac{\delta y}{2}\right)\delta x\ \delta z$$

and the force exerted on the opposite side is

$$\left(p + \frac{\partial p}{\partial y}\frac{\delta y}{2}\right)\delta x\ \delta z$$

where $\delta y/2$ is the distance from center to a face normal to y. Summing the forces acting on the element in the y direction gives

$$\delta F_y = -\frac{\partial p}{\partial y}\delta x\ \delta y\ \delta z - \gamma\ \delta x\ \delta y\ \delta z$$

For the x and z directions, since no body forces act,

$$\delta F_x = -\frac{\partial p}{\partial x}\delta x\ \delta y\ \delta z \qquad \delta F_z = -\frac{\partial p}{\partial z}\delta x\ \delta y\ \delta z$$

The elemental force vector $\delta\mathbf{F}$ is given by

$$\delta\mathbf{F} = \mathbf{i}\ \delta F_x + \mathbf{j}\ \delta F_y + \mathbf{k}\ \delta F_z = -\left(\mathbf{i}\ \frac{\partial p}{\partial x} + \mathbf{j}\ \frac{\partial p}{\partial y} + \mathbf{k}\ \frac{\partial p}{\partial z}\right)\delta x\ \delta y\ \delta z - \mathbf{j}\gamma\ \delta x\ \delta y\ \delta z$$

If the element is reduced to zero size, after dividing through by $\delta x\ \delta y\ \delta z = \delta\mho$, the expression becomes exact.

$$\frac{\delta\mathbf{F}}{\delta\mho} = -\left(\mathbf{i}\ \frac{\partial}{\partial x} + \mathbf{j}\ \frac{\partial}{\partial y} + \mathbf{k}\ \frac{\partial}{\partial z}\right)p - \mathbf{j}\gamma \qquad \lim \delta\mho \rightarrow 0 \qquad (2.2.1)$$

This is the resultant force per unit volume at a point, which must be equated to zero for a fluid at rest. The quantity in parentheses is the *gradient*, called ∇ (del), Sec. 7.2,

$$\nabla = \mathbf{i}\ \frac{\partial}{\partial x} + \mathbf{j}\ \frac{\partial}{\partial y} + \mathbf{k}\ \frac{\partial}{\partial z} \qquad (2.2.2)$$

and the negative gradient of p, $-\nabla p$, is the vector field $\mathbf{f}$ of the surface pressure force per unit volume,

$$\mathbf{f} = -\nabla p \qquad (2.2.3)$$

The fluid static law of variation of pressure is then

$$\mathbf{f} - \mathbf{j}\gamma = 0 \qquad (2.2.4)$$

For an inviscid fluid in motion, or a fluid so moving that the shear stress is everywhere zero, Newton's second law takes the form

$$\mathbf{f} - \mathbf{j}\gamma = \rho\mathbf{a} \qquad (2.2.5)$$

where **a** is the acceleration of the fluid element. $\mathbf{f} - \mathbf{j}\gamma$ is the resultant fluid force when gravity is the only body force acting. Equation (2.2.5) is used to study relative equilibrium in Sec. 2.9 and in the derivation of Euler's equations in Chaps. 3 and 7.

In component form, Eq. (2.2.4) becomes

$$\frac{\partial p}{\partial x} = 0 \qquad \frac{\partial p}{\partial y} = -\gamma \qquad \frac{\partial p}{\partial z} = 0 \tag{2.2.6}$$

The partials, for variation in horizontal directions, are one form of Pascal's law; they state that two points at the same elevation in the same continuous mass of fluid at rest have the same pressure.

Since p is a function of y only,

$$dp = -\gamma \, dy \tag{2.2.7}$$

This simple differential equation relates the change of pressure to specific weight and change of elevation and holds for both compressible and incompressible fluids.

For fluids that may be considered homogeneous and incompressible, γ is constant, and Eq. (2.2.7), when integrated, becomes

$$p = -\gamma y + c$$

in which c is the constant of integration. The hydrostatic law of variation of pressure is frequently written in the form

$$p = \gamma h \tag{2.2.8}$$

in which h is measured vertically downward ($h = -y$) from a free-liquid surface and p is the increase in pressure from that at the free surface. Equation (2.2.8) may be derived by taking as fluid free body a vertical column of liquid of finite height h with its upper surface in the free surface. This is left as an exercise for the student.

Example 2.1 An oceanographer is to design a sea lab 5 m high to withstand submersion to 100 m, measured from sea level to the top of the sea lab. Find the pressure variation on a side of the container and the pressure on the top if the specific gravity of salt water is 1.020.

$$\gamma = 1.020 \times 9806 \text{ N/m}^3 = 10 \text{ kN/m}^3$$

At the top, $h = 100$ m, and

$$p = \gamma h = 1 \text{ MN/m}^2 = 1 \text{ MPa}$$

If y is measured from the top of the sea lab downward, the pressure variation is

$$p = 10(y + 100) \text{ kN/m}^2$$

Pressure Variation in a Compressible Fluid

When the fluid is a perfect gas at rest at constant temperature, from Eq. (1.6.2)

$$\frac{p}{\rho} = \frac{p_0}{\rho_0} \tag{2.2.9}$$

When the value of γ in Eq. (2.2.7) is replaced by ρg and ρ is eliminated between Eqs. (2.2.7) and (2.2.9),

$$dy = \frac{-p_0}{g\rho_0} \frac{dp}{p} \tag{2.2.10}$$

It must be remembered that, if ρ is in pounds mass per cubic foot, then $\gamma = g\rho/g_0$ with $g_0 = 32.174$ $lb_m \cdot ft/lb \cdot s^2$. If $p = p_0$ when $\rho = \rho_0$, integration between limits

$$\int_{y_0}^{y} dy = -\frac{p_0}{g\rho_0} \int_{p_0}^{p} \frac{dp}{p}$$

yields

$$y - y_0 = -\frac{p_0}{g\rho_0} \ln \frac{p}{p_0} \tag{2.2.11}$$

in which ln is the natural logarithm. Then

$$p = p_0 \exp\left(-\frac{y - y_0}{p_0/g\rho_0}\right) \tag{2.2.12}$$

which is the equation for variation of pressure with elevation in an isothermal gas.

The atmosphere frequently is assumed to have a constant temperature gradient expressed by

$$T = T_0 + \beta y \tag{2.2.13}$$

For the standard atmosphere, $\beta = -0.00357°F/ft$ $(-0.00651°C/m)$ up to the stratosphere. The density may be expressed in terms of pressure and elevation from the perfect-gas law:

$$\rho = \frac{p}{RT} = \frac{p}{R(T_0 + \beta y)} \tag{2.2.14}$$

Substitution into $dp = -\rho g \, dy$ [Eq. (2.2.7)] permits the variables to be separated and p to be found in terms of y by integration.

Example 2.2 Assuming isothermal conditions to prevail in the atmosphere, compute the pressure and density at 2000 m elevation if $p = 10^5$ Pa abs, $\rho = 1.24$ kg/m³ at sea level. From Eq. (2.2.12)

$$p = 10^5 \text{ N/m}^2 \exp\left\{-\frac{2000 \text{ m}}{(10^5 \text{ N/m}^2)/[(9.806 \text{ m/s}^2)(1.24 \text{ kg/m}^3)]}\right\}$$

$$= 78.4 \text{ kPa abs}$$

Then, from Eq. (2.2.9)

$$\rho = \frac{\rho_0}{p_0} p = (1.24 \text{ kg/m}^3) \frac{78,400}{100,000} = 0.972 \text{ kg/m}^3$$

When compressibility of a liquid in static equilibrium is taken into account, Eqs. (2.2.7) and (1.7.1) are utilized.

EXERCISES

2.2.1 The pressure in the air space above an oil (sp gr 0.75) surface in a tank is 2 psi. The pressure 5.0 ft below the surface of the oil, in feet of water, is (a) 7.0; (b) 8.37; (c) 9.62; (d) 11.16; (e) none of these answers.

2.2.2 The pressure, in millimeters of mercury gage, equivalent to 80 mm H_2O plus 60 mm manometer fluid, sp gr 2.94, is (a) 10.3; (b) 18.8; (c) 20.4; (d) 30.6; (e) none of these answers.

2.2.3 The differential equation for pressure variation in a static fluid may be written (y measured vertically upward) (a) $dp = -\gamma\, dy$; (b) $d\rho = -\gamma\, dy$; (c) $dy = -\rho\, dp$; (d) $dp = -\rho\, dy$; (e) $dp = -y\, d\rho$.

2.2.4 In an isothermal atmosphere, the pressure (a) remains constant; (b) decreases linearly with elevation; (c) increases exponentially with elevation; (d) varies in the same way as the density; (e) remains constant, as does the density.

2.3 UNITS AND SCALES OF PRESSURE MEASUREMENT

Pressure may be expressed with reference to any arbitrary datum. The usual data are *absolute zero* and *local atmospheric pressure*. When a pressure is expressed as a difference between its value and a complete vacuum, it is called an *absolute pressure*. When it is expressed as a difference between its value and the local atmospheric pressure, it is called a *gage pressure*.

The *bourdon gage* (Fig. 2.3) is typical of the devices used for measuring gage pressures. The pressure element is a hollow, curved, flat metallic tube closed at one end; the other end is connected to the pressure to be measured. When the internal pressure is increased, the tube tends to straighten, pulling on a linkage to which is attached a pointer and causing the pointer to move. The dial reads zero when the inside and outside of the tube are at the same pressure, regardless of its particular value. The dial may be graduated to any convenient units, common ones being pascals, pounds per square inch, pounds per square foot, inches of mercury, feet of water, centimeters of mercury, and millimeters of mercury. Owing to its inherent

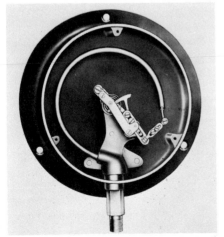

Figure 2.3 Bourdon gage. (*Crosby Steam Gage and Valve Co.*)

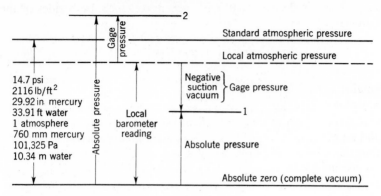

Figure 2.4 Units and scales for pressure measurement.

construction, the gage measures pressure relative to the pressure of the medium surrounding the tube, which is the local atmosphere.

Figure 2.4 illustrates the data and the relations of the common units of pressure measurement. Standard atmospheric pressure is the mean pressure at sea level, 29.92 in Hg. A pressure expressed in terms of *the length of a column* of liquid is equivalent to the force per unit area at the base of the column. The relation for variation of pressure with altitude in a liquid $p = \gamma h$ [Eq. (2.2.8)] shows the relation between head h, in length of a fluid column of specific weight γ, and the pressure p. In consistent units, p is in pounds per square foot, γ in pounds per cubic foot, and h in feet or p in pascals, γ in newtons per cubic meter, and h in meters. With the specific weight of any liquid expressed as its specific gravity S times the specific weight of water, Eq. (2.2.8) becomes

$$p = \gamma_w Sh \qquad (2.3.1)$$

For water γ_w may be taken as 62.4 lb/ft^3 or 9806 N/m^3.

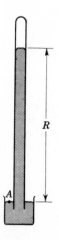

Figure 2.5 Mercury barometer.

When the pressure is desired in pounds per square inch, both sides of the equation are divided by 144:

$$p_{psi} = \frac{62.4}{144} Sh = 0.433Sh \tag{2.3.2}$$

in which h remains in feet.†

Local atmospheric pressure is measured by a mercury barometer (Fig. 2.5) or by an *aneroid* barometer, which measures the difference in pressure between the atmosphere and an evacuated box or tube in a manner analogous to the bourdon gage except that the tube is evacuated and sealed.

A mercury barometer consists of a glass tube closed at one end, filled with mercury, and inverted so that the open end is submerged in mercury. It has a scale so arranged that the height of column R (Fig. 2.5) can be determined. The space above the mercury contains mercury vapor. If the pressure of the mercury vapor h_v is given in millimeters of mercury and R is measured in the same units, the pressure at A may be expressed as

$$h_v + R = h_A \qquad \text{mm Hg}$$

Although h_v is a function of temperature, it is very small at usual atmospheric temperatures. The barometric pressure varies with location, i.e., elevation, and with weather conditions.

In Fig. 2.4 a pressure may be located vertically on the chart, which indicates its relation to absolute zero and to local atmospheric pressure. If the point is below the local-atmospheric-pressure line and is referred to gage datum, it is called *negative, suction,* or *vacuum.* For example, the pressure 460 mm Hg abs, as at 1, with barometer reading 720 mm, may be expressed as -260 mm Hg, 11 in Hg suction, or 11 in Hg vacuum. It should be noted that

$$p_{abs} = p_{bar} + p_{gage}$$

To avoid any confusion, the convention is adopted throughout this text that a *pressure is gage unless specifically marked absolute,* with the exception of the *atmosphere,* which is an absolute pressure unit.

† In Eq. (2.3.2) the standard atmospheric pressure may be expressed in pounds per square inch,

$$p_{psi} = \frac{62.4}{144} (13.6)\left(\frac{29.92}{12}\right) = 14.7$$

when $S = 13.6$ for mercury. When 14.7 is multiplied by 144, the standard atmosphere becomes 2116 lb/ft². Then 2116 divided by 62.4 yields 33.91 ft H₂O. Any of these designations is for the standard atmosphere, and may be called *one atmosphere,* if it is always understood that it is a standard atmosphere and is measured from absolute zero. These various designations of a standard atmosphere (Fig. 2.4) are equivalent and provide a convenient means of converting from one set of units to another. For example, to express 100 ft H₂O in pounds per square inch,

$$\frac{100}{33.91} \times 14.7 = 43.3 \text{ psi}$$

since 100/33.91 is the number of standard atmospheres and each standard atmosphere is 14.7 psi.

Example 2.3 The rate of temperature change in the atmosphere with change in elevation is called its *lapse rate*. The motion of a parcel of air depends on the density of the parcel relative to the density of the surrounding (ambient) air. However, as the parcel ascends through the atmosphere, the air pressure decreases, the parcel expands, and its temperature decreases at a rate known as the *dry adiabatic lapse rate*. A firm wants to burn a large quantity of refuse. It is estimated that the temperature of the smoke plume at 30 ft above the ground will be 20°F greater than that of the ambient air. For the following conditions determine what will happen to the smoke.
(a) At standard atmospheric lapse rate $\beta = -0.00357$°F per foot and $t_0 = 70$°.
(b) At an inverted lapse rate $\beta = 0.002$°F per foot.
By combining Eqs. (2.2.7) and (2.2.14),

$$\int_{p_0}^{p} \frac{dp}{p} = -\frac{g}{R} \int_{0}^{y} \frac{dy}{T_0 + \beta y} \quad \text{or} \quad \frac{p}{p_0} = \left(1 + \frac{\beta y}{T_0}\right)^{-g/R\beta}$$

The relation between pressure and temperature for a mass of gas expanding without heat transfer (isentropic relation, Sec. 6.1) is

$$\frac{T}{T_1} = \left(\frac{p}{p_0}\right)^{(k-1)/k}$$

in which T_1 is the initial smoke absolute temperature and p_0 the initial absolute pressure; k is the specific heat ratio, 1.4 for air and other diatomic gases.
    Eliminating p/p_0 in the last two equations gives

$$T = T_1\left(1 + \frac{\beta y}{T_0}\right)^{-[(k-1)/k](g/R\beta)}$$

Since the gas will rise until its temperature is equal to the ambient temperature,

$$T = T_0 + \beta y$$

the last two equations may be solved for y. Let

$$a = \frac{-1}{(k-1)g/kR\beta + 1}$$

Then
$$y = \frac{T_0}{\beta}\left[\left(\frac{T_0}{T_1}\right)^a - 1\right]$$

For $\beta = -0.00357$°F per foot, $R = 53.3g$ ft·lb/slug·°R, $a = 1.994$, and $y = 10{,}570$ ft. For the atmospheric temperature inversion $\beta = 0.002$°F per foot, $a = -0.2717$, and $y = 2680$ ft.

EXERCISES

2.3.1 Select the correct statement: (a) Local atmospheric pressure is always below standard atmospheric pressure. (b) Local atmospheric pressure depends upon elevation of locality only. (c) Standard atmospheric pressure is the mean local atmospheric pressure at sea level. (d) A barometer reads the difference between local and standard atmospheric pressure. (e) Standard atmospheric pressure is 34 in Hg abs.

2.3.2 Select the three pressures that are equivalent: (a) 10.0 psi, 23.1 ft H_2O, 4.91 in Hg; (b) 10.0 psi, 4.33 ft H_2O, 20.3 in Hg; (c) 10.0 psi, 20.3 ft H_2O, 23.1 in Hg; (d) 4.33 psi, 10.0 ft H_2O, 20.3 in Hg; (e) 4.33 psi, 10.0 ft H_2O, 8.83 in Hg.

2.3.3 When the barometer reads 730 mm Hg, 10 kPa suction is the same as (a) -10.2 m H_2O; (b) 0.075 m Hg; (c) 8.91 m H_2O abs; (d) 107 kPa abs; (e) none of these answers.

2.3.4 With the barometer reading 29 in Hg, 7.0 psia is equivalent to (a) 0.476 atm; (b) 0.493 atm; (c) 7.9 psi suction; (d) 7.7 psi; (e) 13.8 in Hg abs.

2.4 MANOMETERS

Manometers are devices that employ liquid columns for determining differences in pressure. The most elementary manometer, usually called a *piezometer*, is illustrated in Fig. 2.6a; it measures the pressure in a liquid when it is above zero gage. A glass tube is mounted vertically so that it is connected to the space within the container. Liquid rises in the tube until equilibrium is reached. The pressure is then given by the vertical distance h from the meniscus (liquid surface) to the point where the pressure is to be measured, expressed in units of length of the liquid in the container. It is obvious that the piezometer would not work for negative gage pressures, because air would flow into the container through the tube. It is also impractical for measuring large pressures at A, since the vertical tube would need to be very long. If the specific gravity of the liquid is S, the pressure at A is hS units of length of water.

For measurement of small negative or positive gage pressures in a liquid the tube may take the form shown in Fig. 2.6b. With this arrangement the meniscus may come to rest below A, as shown. Since the pressure at the meniscus is zero gage and since pressure *decreases* with elevation,

$$h_A = -hS \qquad \text{units of length } H_2O$$

For greater negative or positive gage pressures a second liquid of greater specific gravity is employed (Fig. 2.6c). It must be immiscible in the first fluid, which may now be a gas. If the specific gravity of the fluid at A is S_1 (based on water) and the specific gravity of the manometer liquid is S_2, the equation for pressure at A may be written thus, starting at either A or the upper meniscus and proceeding through the manometer,

$$h_A + h_2 S_1 - h_1 S_2 = 0$$

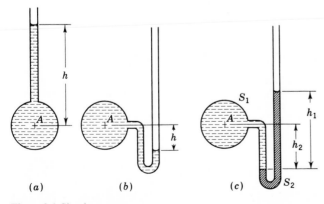

(a)   (b)   (c)

Figure 2.6 Simple manometers.

in which h_A is the unknown pressure, expressed in length units of water, and h_1, h_2 are in length units. If A contains a gas, S_1 is generally so small that $h_2 S_1$ may be neglected.

A general procedure should be followed in working all manometer problems:

1. Start at one end (or any meniscus if the circuit is continuous) and write the pressure there in an appropriate unit (say pascals) or in an appropriate symbol if it is unknown.
2. Add to this the change in pressure, in the same unit, from one meniscus to the next (plus if the next meniscus is lower, minus if higher). (For pascals this is the product of the difference in elevation in meters and the specific weight of the fluid in newtons per cubic meter.)
3. Continue until the other end of the gage (or the starting meniscus) is reached and equate the expression to the pressure at that point, known or unknown.

The expression will contain one unknown for a simple manometer or will give a difference in pressures for the differential manometer. In equation form,

$$p_0 - (y_1 - y_0)\gamma_0 - (y_2 - y_1)\gamma_1 - (y_3 - y_2)\gamma_2$$
$$- (y_4 - y_3)\gamma_3 - \cdots - (y_n - y_{n-1})\gamma_{n-1} = p_n$$

in which y_0, y_1, ..., y_n are elevations of each meniscus in length units and γ_0, γ_1, γ_2, ..., γ_{n-1} are specific weights of the fluid columns. The above expression yields the answer in force per unit area and may be converted to other units by use of the conversions in Fig. 2.4.

A differential manometer (Fig. 2.7) determines the difference in pressures at two points A and B when the actual pressure at any point in the system cannot be determined. Application of the procedure outlined above to Fig. 2.7a produces

$$p_A - h_1 \gamma_1 - h_2 \gamma_2 + h_3 \gamma_3 = p_B \quad \text{or} \quad p_A - p_B = h_1 \gamma_1 + h_2 \gamma_2 - h_3 \gamma_3$$

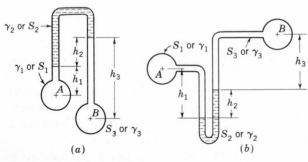

(a)   (b)

Figure 2.7 Differential manometers.

Similarly, for Fig. 2.7b,

$$p_A + h_1 \gamma_1 - h_2 \gamma_2 - h_3 \gamma_3 = p_B \qquad \text{or} \qquad p_A - p_B = -h_1 \gamma_1 + h_2 \gamma_2 + h_3 \gamma_3$$

No formulas for particular manometers should be memorized. It is much more satisfactory to work them out from the general procedure for each case as needed.

If the pressures at A and B are expressed in length of the water column, the above results can be written, for Fig. 2.7a,

$$h_A - h_B = h_1 S_1 + h_2 S_2 - h_3 S_3 \qquad \text{units of length } H_2O$$

Similarly, for Fig. 2.7b,

$$h_A - h_B = -h_1 S_1 + h_2 S_2 + h_3 S_3$$

in which S_1, S_2, and S_3 are the applicable specific gravities of the liquids in the system.

Example 2.4 In Fig. 2.7a the liquids at A and B are water and the manometer liquid is oil. $S = 0.80$; $h_1 = 300$ mm; $h_2 = 200$ mm; and $h_3 = 600$ mm. (a) Determine $p_A - p_B$, in pascals. (b) If $p_B = 50$ kPa and the barometer reading is 730 mm Hg, find the pressure at A, in meters of water absolute.

(a)
$$h_A \text{ (m } H_2O) - h_1 S_{H_2O} - h_2 S_{\text{oil}} + h_3 S_{H_2O} = h_B \text{ (m } H_2O)$$

$$h_A - 0.3(1) - 0.2(0.8) + 0.6(1) = h_B$$

$$h_A - h_B = -0.14 \text{ m } H_2O$$

$$p_A - p_B = \gamma(h_A - h_B) = (9806 \text{ N/m}^3)(-0.14 \text{ m}) = -1373 \text{ Pa}$$

(b)
$$h_B = \frac{p_B}{\gamma} = \frac{5 \times 10^4 \text{ N/m}^2}{9806 \text{ N/m}^3} = 5.099 \text{ m } H_2O$$

$$h_B \text{ (m } H_2O \text{ abs)} = h_B \text{ (m } H_2O \text{ gage)} + (0.73 \text{ m})(13.6)$$

$$= 5.099 + 9.928 = 15.027 \text{ m } H_2O \text{ abs}$$

From (a)

$$h_{A_{\text{abs}}} = h_{B_{\text{abs}}} - 0.14 = 15.027 - 0.14 \text{ m} = 14.89 \text{ m } H_2O \text{ abs}$$

Micromanometers

Several types of manometers are on the market for determining very small differences in pressure or determining large pressure differences precisely. One type very accurately measures the differences in elevation of two menisci of a manometer. By means of small telescopes with horizontal cross hairs mounted along the tubes on a rack which is raised and lowered by a pinion and slow-motion screw so that the cross hairs can be set accurately, the difference in elevation of menisci (the gage difference) can be read with verniers.

With two gage liquids, immiscible in each other and in the fluid to be measured, a large gage difference R (Fig. 2.8) can be produced for a small pressure

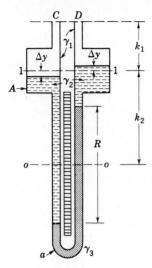

Figure 2.8 Micromanometer using two gage liquids.

difference. The heavier gage liquid fills the lower U tube up to 0-0; then the lighter gage liquid is added to both sides, filling the larger reservoirs up to 1-1. The gas or liquid in the system fills the space above 1-1. When the pressure at C is slightly greater than at D, the menisci move as indicated in Fig. 2.8. The volume of liquid displaced in each reservoir equals the displacement in the U tube; thus

$$\Delta y A = \frac{R}{2} a$$

in which A and a are the cross-sectional areas of reservoir and U tube, respectively. The manometer equation may be written, starting at C, in force per unit area,

$$p_C + (k_1 + \Delta y)\gamma_1 + \left(k_2 - \Delta y + \frac{R}{2}\right)\gamma_2 - R\gamma_3$$

$$- \left(k_2 - \frac{R}{2} + \Delta y\right)\gamma_2 - (k_1 - \Delta y)\gamma_1 = p_D$$

in which γ_1, γ_2, and γ_3 are the specific weights as indicated in Fig. 2.8. Simplifying and substituting for Δy gives

$$p_C - p_D = R\left[\gamma_3 - \gamma_2\left(1 - \frac{a}{A}\right) - \gamma_1\frac{a}{A}\right] \tag{2.4.1}$$

The quantity in brackets is a constant for specified gage and fluids; hence, the pressure difference is directly proportional to R.

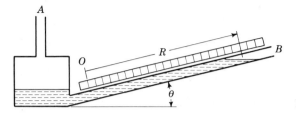

Figure 2.9 Inclined manometer.

Example 2.5 In the micromanometer of Fig. 2.8 the pressure difference is wanted, in pascals, when air is in the system, $S_2 = 1.0$, $S_3 = 1.10$, $a/A = 0.01$, $R = 5$ mm, $t = 20°C$, and the barometer reads 760 mm Hg.

$$\rho_{air} = \frac{p}{RT} = \frac{(0.76 \text{ m})(13.6 \times 9806 \text{ N/m}^3)}{(287 \text{ N}\cdot\text{m/kg}\cdot\text{K})(273 + 20 \text{ K})} = 1.205 \text{ kg/m}^3$$

$$\gamma_1 \frac{a}{A} = (1.205 \text{ kg/m}^3)(9.806 \text{ m/s}^2)(0.01) = 0.118 \text{ N/m}^3$$

$$\gamma_3 - \gamma_2\left(1 - \frac{a}{A}\right) = (9806 \text{ N/m}^3)(1.10 - 0.99) = 1079 \text{ N/m}^3$$

The term $\gamma_1(a/A)$ may be neglected. Substituting into Eq. (2.4.1) gives

$$p_C - p_D = (0.005 \text{ m})(1079 \text{ N/m}^3) = 5.39 \text{ Pa}$$

The inclined manometer (Fig. 2.9) is frequently used for measuring small differences in gas pressures. It is adjusted to read zero, by moving the inclined scale, when A and B are open. Since the inclined tube requires a greater displacement of the meniscus for given pressure difference than a vertical tube, it affords greater accuracy in reading the scale.

Surface tension causes a capillary rise in small tubes. If a U tube is used with a meniscus in each leg, the surface-tension effects cancel. The capillary rise is negligible in tubes with a diameter of 0.5 in or greater.

EXERCISES

2.4.1 In Fig. 2.6b the liquid is oil, sp gr 0.80. When $h = 2$ ft, the pressure at A may be expressed as (a) -1.6 ft H_2O abs; (b) 1.6 ft H_2O; (c) 1.6 ft H_2O suction; (d) 2.5 ft H_2O vacuum; (e) none of these answers.

2.4.2 In Fig. 2.6c air is contained in the pipe, water is the manometer liquid, and $h_1 = 500$ mm, $h_2 = 200$ mm. The pressure at A is (a) 10.14 m H_2O abs; (b) 0.2 m H_2O vacuum; (c) 0.2 m H_2O; (d) 4901 Pa; (e) none of these answers.

2.4.3 In Fig. 2.7a, $h_1 = 2.0$ ft, $h_2 = 1.0$ ft, $h_3 = 4.0$ ft, $S_1 = 0.80$, $S_2 = 0.65$, $S_3 = 1.0$. Then $h_B - h_A$ in feet of water is (a) -3.05; (b) -1.75; (c) 3.05; (d) 6.25; (e) none of these answers.

2.4.4 In Fig. 2.7b, $h_1 = 1.5$ ft, $h_2 = 1.0$ ft, $h_3 = 2.0$ ft, $S_1 = 1.0$, $S_2 = 3.0$, $S_3 = 1.0$. Then $p_A - p_B$ in pounds per square inch is (a) -1.08; (b) 1.52; (c) 8.08; (d) 218; (e) none of these answers.

2.4.5 A mercury-water manometer has a gage difference of 500 mm (difference in elevation of menisci). The difference in pressure, measured in meters of water, is (a) 0.5; (b) 6.3; (c) 6.8; (d) 7.3; (e) none of these answers.

2.4.6 In the inclined manometer of Fig. 2.9 the reservoir is so large that its surface may be assumed to remain at a fixed elevation. $\theta = 30°$. Used as a simple manometer for measuring air pressure, it contains water, and $R = 1.2$ ft. The pressure at A, in inches of water, is (a) 7.2; (b) 7.2 vacuum; (c) 12.5; (d) 14.4; (e) none of these answers.

2.5 FORCES ON PLANE AREAS

In the preceding sections variations of pressure throughout a fluid have been considered. The distributed forces resulting from the action of fluid on a finite area may be conveniently replaced by a resultant force, insofar as external reactions to the force system are concerned. In this section the magnitude of resultant force and its line of action (pressure center) are determined by integration, by formula, and by use of the concept of the pressure prism.

Horizontal Surfaces

A plane surface in a horizontal position in a fluid at rest is subjected to a constant pressure. The magnitude of the force acting on one side of the surface is

$$\int p \, dA = p \int dA = pA$$

The elemental forces $p \, dA$ acting on A are all parallel and in the same sense; therefore, a scalar summation of all such elements yields the magnitude of the resultant force. Its direction is normal to the surface and *toward* the surface if p is *positive*. To find the line of action of the resultant, i.e., the point in the area where the moment of the distributed force about any axis through the point is zero, arbitrary xy axes may be selected, as in Fig. 2.10. Then, since the moment of the resultant must equal the moment of the distributed force system about any axis, say the y axis,

$$pAx' = \int_A xp \, dA$$

in which x' is the distance from the y axis to the resultant. Since p is constant,

$$x' = \frac{1}{A} \int_A x \, dA = \bar{x}$$

in which $\bar{x}$ is the distance to the centroid of the area (see Appendix A). Hence, for a horizontal area subjected to static fluid pressure, the resultant passes through the centroid of the area.

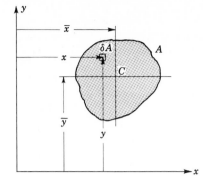

Figure 2.10 Notation for determining the line of action of a force.

Inclined Surfaces

In Fig. 2.11 a plane surface is indicated by its trace $A'B'$. It is inclined $\theta°$ from the horizontal. The intersection of the plane of the area and the free surface is taken as the x axis. The y axis is taken in the plane of the area, with origin O, as shown, in the free surface. The xy plane portrays the arbitrary inclined area. The magnitude, direction, and line of action of the resultant force *due to the liquid*, acting on one side of the area, are sought.

For an element with area δA as a strip with thickness δy with long edges horizontal, the magnitude of force δF acting on it is

$$\delta F = p \, \delta A = \gamma h \, \delta A = \gamma y \sin \theta \, \delta A \qquad (2.5.1)$$

Since all such elemental forces are parallel, the integral over the area yields the magnitude of force F, acting on one side of the area,

$$F = \int p \, dA = \gamma \sin \theta \int y \, dA = \gamma \sin \theta \bar{y} A = \gamma \bar{h} A = p_G A \qquad (2.5.2)$$

with the relations from Fig. 2.11, $\bar{y} \sin \theta = \bar{h}$ and $p_G = \gamma \bar{h}$, the pressure at the centroid of the area. In words, the magnitude of force exerted on one side of a plane area submerged in a liquid is the product of the area and the pressure at its centroid. In this form, it should be noted, the presence of a free surface is unnecessary. Any means for determining the pressure at the centroid may be used. The sense of the force is to push against the area if p_G is positive. As all force elements are normal to the surface, the line of action of the resultant also is normal to the surface. Any surface may be rotated about any axis through its centroid without changing the magnitude of the resultant if the total area remains submerged in the static liquid.

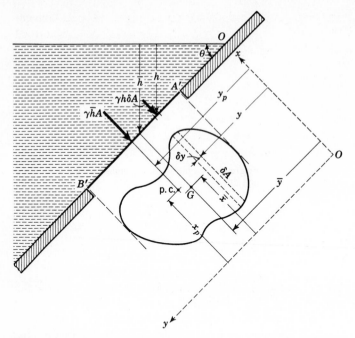

Figure 2.11 Notation for force of liquid on one side of a plane inclined area.

Center of Pressure

The line of action of the resultant force has its piercing point in the surface at a point called the *pressure center*, with coordinates (x_p, y_p) (Fig. 2.11). Unlike that for the horizontal surface, the center of pressure of an inclined surface is not at the centroid. To find the pressure center, the moments of the resultant $x_p F$, $y_p F$ are equated to the moment of the distributed forces about the y axis and x axis, respectively; thus

$$x_p F = \int_A xp \, dA \tag{2.5.3}$$

$$y_p F = \int_A yp \, dA \tag{2.5.4}$$

The area element in Eq. (2.5.3) should be $\delta x \, \delta y$, and not the strip shown in Fig. 2.11.

Solving for the coordinates of the pressure center results in

$$x_p = \frac{1}{F} \int_A xp \, dA \tag{2.5.5}$$

$$y_p = \frac{1}{F} \int_A yp \, dA \tag{2.5.6}$$

In many applications Eqs. (2.5.5) and (2.5.6) may be evaluated most conveniently through graphical integration; for simple areas they may be transformed into general formulas as follows (see Appendix A):

$$x_p = \frac{1}{\gamma \bar{y} A \sin \theta} \int_A x\gamma y \sin \theta \, dA = \frac{1}{\bar{y} A} \int_A xy \, dA = \frac{I_{xy}}{\bar{y} A} \qquad (2.5.7)$$

In Eqs. (A.10), of Appendix A, and (2.5.7),

$$x_p = \frac{\bar{I}_{xy}}{\bar{y} A} + \bar{x} \qquad (2.5.8)$$

When either of the centroidal axes, $x = \bar{x}$ or $y = \bar{y}$, is an axis of symmetry for the surface, $\bar{I}_{xy}$ vanishes and the pressure center lies on $x = \bar{x}$. Since $\bar{I}_{xy}$ may be either positive or negative, the pressure center may lie on either side of the line $x = \bar{x}$. To determine y_p by formula, with Eqs. (2.5.2) and (2.5.6),

$$y_p = \frac{1}{\gamma \bar{y} A \sin \theta} \int_A y\gamma y \sin \theta \, dA = \frac{1}{\bar{y} A} \int_A y^2 \, dA = \frac{I_x}{\bar{y} A} \qquad (2.5.9)$$

In the parallel-axis theorem for moments of inertia

$$I_x = I_G + \bar{y}^2 A$$

in which I_G is the second moment of the area about its horizontal centroidal axis. If I_x is eliminated from Eq. (2.5.9),

$$y_p = \frac{I_G}{\bar{y} A} + \bar{y} \qquad (2.5.10)$$

or

$$y_p - \bar{y} = \frac{I_G}{\bar{y} A} \qquad (2.5.11)$$

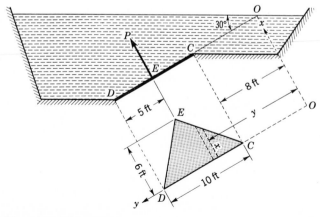

Figure 2.12 Triangular gate.

I_G is always positive; hence, $y_p - \bar{y}$ is always positive and the pressure center is always below the centroid of the surface. It should be emphasized that $\bar{y}$ and $y_p - \bar{y}$ are distances in the plane of the surface.

Example 2.6 The triangular gate CDE (Fig. 2.12) is hinged along CD and is opened by a normal force P applied at E. It holds oil, sp gr 0.80, above it and is open to the atmosphere on its lower side. Neglecting the weight of the gate, find (a) the magnitude of force exerted on the gate by integration and by Eq. (2.5.2); (b) the location of pressure center; (c) the force P needed to open the gate.

(a) By integration with reference to Fig. 2.12,

$$F = \int_A p \, dA = \gamma \sin \theta \int yx \, dy = \gamma \sin \theta \int_8^{13} xy \, dy + \gamma \sin \theta \int_{13}^{18} xy \, dy$$

When $y = 8$, $x = 0$, and when $y = 13$, $x = 6$, with x varying linearly with y; thus

$$x = ay + b \qquad 0 = 8a + b \qquad 6 = 13a + b$$

in which the coordinates have been substituted to find x in terms of y. Solving for a and b gives

$$a = \tfrac{6}{5} \qquad b = -\tfrac{48}{5} \qquad x = \tfrac{6}{5}(y - 8)$$

Similarly, $y = 13$, $x = 6$; $y = 18$, $x = 0$; and $x = \tfrac{6}{5}(18 - y)$. Hence,

$$F = \gamma \sin \theta \, \frac{6}{5} \left[\int_8^{13} (y - 8)y \, dy + \int_{13}^{18} (18 - y)y \, dy \right]$$

Integrating and substituting for $\gamma \sin \theta$ leads to

$$F = 62.4 \times 0.8 \times 0.50 \times \frac{6}{5} \left[\left(\frac{y^3}{3} - 4y^2 \right)_8^{13} + \left(9y^2 - \frac{y^3}{3} \right)_{13}^{18} \right] = 9734.4 \text{ lb}$$

By Eq. (2.5.2),

$$F = p_G A = \gamma \bar{y} \sin \theta \, A = 62.4 \times 0.80 \times 13 \times 0.50 \times 30 = 9734.4 \text{ lb}$$

(b) With the axes as shown, $\bar{x} = 2.0$, $\bar{y} = 13$. In Eq. (2.5.8),

$$x_p = \frac{I_{xy}}{\bar{y}A} + \bar{x}$$

I_{xy} is zero owing to symmetry about the centroidal axis parallel to the x axis; hence, $\bar{x} = x_p = 2.0$ ft. In Eq. (2.5.11),

$$y_p - \bar{y} = \frac{I_G}{\bar{y}A} = 2 \times \frac{1 \times 6 \times 5^3}{12 \times 13 \times 30} = 0.32 \text{ ft}$$

i.e., the pressure center is 0.32 ft below the centroid, measured in the plane of the area.

(c) When moments about CD are taken and the action of the oil is replaced by the resultant,

$$P \times 6 = 9734.4 \times 2 \qquad P = 3244.8 \text{ lb}$$

The Pressure Prism

Another approach to the problem of determining the resultant force and line of action of the force on a plane surface is given by the concept of a pressure prism. It is a prismatic volume with its base the given surface area and with altitude at any point of the base given by $p = \gamma h$. h is the vertical distance to the free surface, Fig. 2.13. (An imaginary free surface may be used to define h if no real free surface exists.) In the figure, γh may be laid off to any convenient scale such that its trace is OM. The force acting on an elemental area δA is

$$\delta F = \gamma h \, \delta A = \delta \mathcal{V} \tag{2.5.12}$$

which is an element of volume of the pressure prism. After integrating, $F = \mathcal{V}$, the volume of the pressure prism equals the magnitude of the resultant force acting on one side of the surface.

From Eqs. (2.5.5) and (2.5.6),

$$x_p = \frac{1}{\mathcal{V}} \int_{\mathcal{V}} x \, d\mathcal{V} \qquad y_p = \frac{1}{\mathcal{V}} \int_{\mathcal{V}} y \, d\mathcal{V} \tag{2.5.13}$$

which show that x_p, y_p are distances to the *centroid* of the pressure prism [Appendix A, Eq. (A.5)]. Hence, the line of action of the resultant passes through the centroid of the pressure prism. For some simple areas the pressure prism is more convenient than either integration or formula. For example, a rectangular area with one edge in the free surface has a wedge-shaped prism. Its centroid is one-third the altitude from the base; hence, the pressure center is one-third the altitude from its lower edge.

Example 2.7 A structure is so arranged along a channel that it will spill the water out if a certain height y (Fig. 2.14a) is reached. The gate is made of steel plate weighing 2500 N/m². Determine the height y.

Using pressure prism concepts, for unit width normal to the page, the force on the horizontal leaf (Fig. 2.14b) is given by the volume of a pressure prism of base 1.2 m² and constant altitude

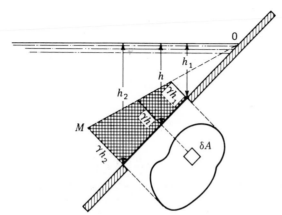

Figure 2.13 Pressure prism.

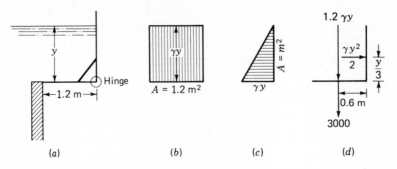

Figure 2.14 Flashboard arrangement on side of channel.

γy N/m², which yields $F_y = 1.2\,\gamma y$ N acting through the center of the base. The pressure prism for the vertical face (Fig. 2.14c) is a wedge of base y m² and altitude varying from 0 to γy N/m². The average altitude is $\gamma y/2$, so $F_x = \gamma y^2/2$ N. The centroid of the wedge prism is $y/3$ from the hinge. The weight of gate floor exerts a force of 3000 N at its center. Figure 2.14d shows all forces and moment arms. For equilibrium, i.e., the value of y for tipping, moments about the hinge must be zero.

$$M = 3000 \text{ N} \times 0.6 \text{ m} + 1.2\gamma y \text{ N} \times 0.6 \text{ m} - \frac{\gamma y^2}{2} \text{ N} \times \frac{y}{3} \text{ m} = 0$$

or
$$M = y^3 - 4.32y - 1.1014 = 0$$

This equation has only one positive root, easily seen to be between $y = 2$ and $y = 3$. By use of the Newton-Raphson method (Appendix B)

$$y = y - \frac{M(y)}{M'(y)} = y - \frac{y^3 - 4.32y - 1.1014}{3y^2 - 4.32}$$

which is an iterative procedure. A trial value of y is assumed, say $y = 2.5$. Substitution into the right-hand side yields an improved value of y. By repeating this procedure three times, the root is found to be $y = 2.196$ m. A cubic equation is easily solved by use of a programmable calculator. Preparing the program requires no more effort than solving the equation once by calculator.

Effects of Atmospheric Pressure on Forces on Plane Areas

In the discussion of pressure forces the pressure datum was not mentioned. The pressures were computed by $p = \gamma h$, in which h is the vertical distance below the free surface. Therefore, the datum taken was gage pressure zero, or the local atmospheric pressure. When the opposite side of the surface is open to the atmosphere, a force is exerted on it by the atmosphere equal to the product of the atmospheric pressure p_0 and the area, or $p_0 A$, based on absolute zero as datum. On the liquid side the force is

$$\int (p_0 + \gamma h)\, dA = p_0 A + \gamma \int h\, dA$$

The effect $p_0 A$ of the atmosphere acts equally on both sides and in no way contributes to the resultant force or its location.

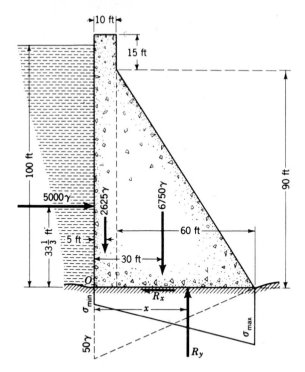

Figure 2.15 Concrete gravity dam.

So long as the same pressure datum is selected for all sides of a free body, the resultant and moment can be determined by constructing a free surface at pressure zero on this datum and using the above methods.

Example 2.8 An application of pressure forces on plane areas is given in the design of a gravity dam. The maximum and minimum compressive stresses in the base of the dam are computed from the forces which act on the dam. Figure 2.15 shows a cross section through a concrete dam where the specific weight of concrete has been taken as 2.5γ and γ is the specific weight of water. A 1-ft section of dam is considered as a free body; the forces are due to the concrete, the water, the foundation pressure, and the hydrostatic uplift. Determining amount of hydrostatic uplift is beyond the scope of this treatment, but it will be assumed to be one-half the hydrostatic head at the upstream edge, decreasing linearly to zero at the downstream edge of the dam. Enough friction or shear stress must be developed at the base of the dam to balance the thrust due to the water; that is, $R_x = 5000\gamma$. The resultant upward force on the base equals the weight of the dam less the hydrostatic uplift, $R_y = 6750\gamma + 2625\gamma - 1750\gamma = 7625\gamma$ lb. The position of R_y is such that the free body is in equilibrium. For moments around O,

$$\Sigma M_0 = 0 = R_y x - 5000\gamma(33.33) - 2625\gamma(5) - 6750\gamma(30) + 1750\gamma(23.33)$$

and

$$x = 44.8 \text{ ft}$$

It is customary to assume that the foundation pressure varies linearly over the base of the dam, i.e., that the pressure prism is a trapezoid with a volume equal to R_y; thus

$$\frac{\sigma_{max} + \sigma_{min}}{2} 70 = 7625\gamma$$

in which σ_{max}, σ_{min} are the maximum and minimum compressive stresses in pounds per square foot. The centroid of the pressure prism is at the point where $x = 44.8$ ft. By taking moments about O to express the position of the centroid in terms of σ_{max} and σ_{min},

$$44.8 = \frac{(\sigma_{min})70 \times \frac{70}{2} + (\sigma_{max} - \sigma_{min})\frac{70}{2} \times \frac{2}{3}(70)}{(\sigma_{max} + \sigma_{min})\frac{70}{2}}$$

Simplifying gives

$$\sigma_{max} = 11.75\sigma_{min}$$

Then

$$\sigma_{max} = 210\gamma = 12,500 \text{ lb/ft}^2 \qquad \sigma_{min} = 17.1\gamma = 1067 \text{ lb/ft}^2$$

When the resultant falls within the middle third of the base of the dam, σ_{min} will always be a compressive stress. Owing to the poor tensile properties of concrete, good design requires the resultant to fall within the middle third of the base.

EXERCISES

2.5.1 The magnitude of force on one side of a circular surface of unit area, with centroid 10 ft below a free-water surface, is (a) less than 10γ; (b) dependent upon orientation of the area; (c) greater than 10γ; (d) the product of γ and the vertical distance from free surface to pressure center; (e) none of the above.

2.5.2 A rectangular surface 3 by 4 ft has the lower 3-ft edge horizontal and 6 ft below a free-oil surface, sp gr 0.80. The surface is inclined 30° with the horizontal. The force on one side of the surface is (a) 38.4γ; (b) 48γ; (c) 51.2γ; (d) 60γ; (e) none of these answers.

2.5.3 The pressure center of the surface of Exercise 2.5.2 is vertically below the liquid surface (a) 10.133 ft; (b) 5.133 ft; (c) 5.067 ft; (d) 5.00 ft; (e) none of these answers.

2.5.4 The pressure center is (a) at the centroid of the submerged area; (b) the centroid of the pressure prism; (c) independent of the orientation of the area; (d) a point on the line of action of the resultant force; (e) always above the centroid of the area.

2.5.5 What is the force exerted on the vertical annular area enclosed by concentric circles of radii 1.0 and 2.0 m? The center is 3.0 m below a free-water surface. $\gamma =$ sp wt. (a) $3\pi\gamma$; (b) $9\pi\gamma$; (c) $10.25\pi\gamma$; (d) $12\pi\gamma$; (e) none of these answers.

2.5.6 The pressure center for the annular area of Exercise 2.5.5 is below the centroid of the area (a) 0 m; (b) 0.42 m; (c) 0.44 m; (d) 0.47 m; (e) none of these answers.

2.5.7 A vertical triangular area has one side in a free surface, with vertex downward. Its altitude is h. The pressure center is below the free surface (a) $h/4$; (b) $h/3$; (c) $h/2$; (d) $2h/3$; (e) $3h/4$.

2.5.8 A vertical gate 4 by 4 m holds water with free surface at its top. The moment about the bottom of the gate is (a) 42.7γ; (b) 57γ; (c) 64γ; (d) 85.3γ; (e) none of these answers.

2.6 FORCE COMPONENTS ON CURVED SURFACES

When the elemental forces $p \, \delta A$ vary in direction, as in the case of a curved surface, they must be added as vector quantities; i.e., their components in three mutually perpendicular directions are added as scalars, and then the three components are added vectorially. With two horizontal components at right angles and with the vertical component—all easily computed for a curved surface—the resultant can be determined. The lines of action of the components also are readily determined.

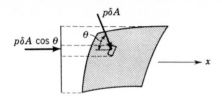

Figure 2.16 Horizontal component of force on a curved surface.

Horizontal Component of Force on a Curved Surface

The horizontal component of pressure force on a curved surface is equal to the pressure force exerted on a projection of the curved surface. The vertical plane of projection is normal to the direction of the component. The surface of Fig. 2.16 represents any three-dimensional surface, and δA an element of its area, its normal making the angle θ with the negative x direction. Then

$$\delta F_x = p\, \delta A \cos \theta$$

is the x component of force exerted on one side of δA. Summing up the x components of force over the surface gives

$$F_x = \int_A p \cos \theta\, dA \tag{2.6.1}$$

$\cos \theta\, \delta A$ is the projection of δA onto a plane perpendicular to x. The element of force on the projected area is $p \cos \theta\, \delta A$, which is also in the x direction. Projecting each element on a plane perpendicular to x is equivalent to projecting the curved surface as a whole onto the vertical plane. Hence, the force acting on this projection of the curved surface is the horizontal component of force exerted on the curved surface in the direction normal to the plane of projection. To find the horizontal component at right angles to the x direction, the curved surface is projected onto a vertical plane parallel to x and the force on the projection is determined.

When the horizontal component of pressure force on a closed body is to be found, the projection of the curved surface on a vertical plane is always zero, since on opposite sides of the body the area-element projections have opposite signs as indicated in Fig. 2.17. Let a small cylinder of cross section δA with axis parallel to

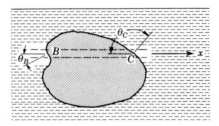

Figure 2.17 Projections of area elements on opposite sides of a body.

x intersect the closed body at B and C. If the element of area of the body cut by the prism at B is δA_B and at C is δA_C, then

$$\delta A_B \cos \theta_B = -\delta A_C \cos \theta_C = \delta A$$

as $\cos \theta_C$ is negative. Hence, with the pressure the same at each end of the cylinder,

$$p \,\delta A_B \cos \theta_B + p \,\delta A_C \cos \theta_C = 0$$

and similarly for all other area elements.

To find the line of action of a horizontal component of force on a curved surface, the resultant of the parallel force system composed of the force components from each area element is required. This is exactly the resultant of the force on the projected area, since the two force systems have an identical distribution of elemental horizontal force components. Hence, the pressure center is located on the projected area by the methods of Sec. 2.5.

Example 2.9 The equation of an ellipsoid of revolution submerged in water is $x^2/4 + y^2/4 + z^2/9 = 1$. The center of the body is located 2 m below the free surface. Find the horizontal force components acting on the curved surface that is located in the first octant. Consider the xz plane to be horizontal and y to be positive upward.

The projection of the surface on the yz plane has an area of $(\pi/4) \times 2 \times 3$ m^2. Its centroid is located $2 - (4/3\pi) \times 2$ m below the free surface. Hence,

$$F_x = -\left(\frac{\pi}{4} \times 6\right)\left(2 - \frac{8}{3\pi}\right)\gamma = (-5.425 \text{ m}^3)(9806 \text{ N/m}^3) = -53.2 \text{ kN}$$

Similarly,

$$F_z = -\left(\frac{\pi}{4} \times 4\right)\left(2 - \frac{8}{3\pi}\right)\gamma = (-3.617 \text{ m}^3)(9806 \text{ N/m}^3) = -35.4 \text{ kN}$$

Vertical Component of Force on a Curved Surface

The vertical component of pressure force on a curved surface is equal to the weight of liquid vertically above the curved surface and extending up to the free surface. The

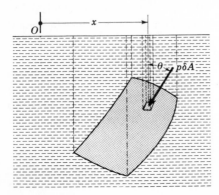

Figure 2.18 Vertical component of force on a curved surface.

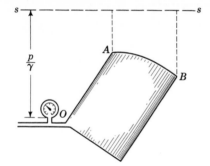

Figure 2.19 Liquid with equivalent free surface.

vertical component of force on a curved surface can be determined by summing up the vertical components of pressure force on elemental areas δA of the surface. In Fig. 2.18 an area element is shown with the force $p\, \delta A$ acting normal to it. Let θ be the angle the normal to the area element makes with the vertical. Then the vertical component of force acting on the area element is $p \cos \theta\, \delta A$, and the vertical component of force on the curved surface is given by

$$F_v = \int_A p \cos \theta\, dA \tag{2.6.2}$$

When p is replaced by its equivalent γh, in which h is the distance from the area element to the free surface, and it is noted that $\cos \theta\, \delta A$ is the projection of δA on a horizontal plane, Eq. (2.6.2) becomes

$$F_v = \gamma \int_A h \cos \theta\, dA = \gamma \int_\mho d\mho \tag{2.6.3}$$

in which $\delta\mho$ is the volume of the prism of height h and base $\cos \theta\, \delta A$, or the volume of liquid vertically above the area element. Integrating gives

$$F_v = \gamma\mho \tag{2.6.4}$$

When the liquid is below the curved surface (Fig. 2.19) and the pressure magnitude is known at some point, for example, O, an *imaginary* or equivalent free surface s-s can be constructed p/γ above O, so that the product of specific weight and vertical distance to any point in the tank is the pressure at the point. The weight of the imaginary volume of liquid vertically above the curved surface is then the vertical component of pressure force on the curved surface. In constructing an imaginary free surface, the imaginary liquid must be of the same specific weight as the liquid in contact with the curved surface; otherwise, the pressure distribution over the surface will not be correctly represented. With an imaginary liquid above a surface, the pressure at a point on the curved surface is equal on both sides, but the elemental force components in the vertical direction are opposite in sign. Hence, the direction of the vertical force component is reversed when an imaginary fluid is above the surface. In some cases a confined liquid may be above the curved surface, and an imaginary liquid must be added (or subtracted) to determine the free surface.

The line of action of the vertical component is determined by equating moments of the elemental vertical components about a convenient axis with the moment of the resultant force. With the axis at O (Fig. 2.18),

$$F_v \bar{x} = \gamma \int_\mathcal{V} x \, d\mathcal{V}$$

in which $\bar{x}$ is the distance from O to the line of action. Then, since $F_v = \gamma \mathcal{V}$,

$$\bar{x} = \frac{1}{\mathcal{V}} \int_\mathcal{V} x \, d\mathcal{V}$$

the distance to the centroid of the volume. Therefore, the line of action of the vertical force passes through the centroid of the volume, real or imaginary, that extends above the curved surface up to the real or imaginary free surface.

Example 2.10 A cylindrical barrier (Fig. 2.20) holds water as shown. The contact between cylinder and wall is smooth. Considering a 1-m length of cylinder, determine (*a*) its weight and (*b*) the force exerted against the wall.

(*a*) For equilibrium the weight of the cylinder must equal the vertical component of force exerted on it by the water. (The imaginary free surface for *CD* is at elevation *A*.) The vertical force on *BCD* is

$$F_{v_{BCD}} = \left(\frac{\pi r^2}{2} + 2r^2 \right) \gamma = (2\pi + 8)\gamma$$

The vertical force on *AB* is

$$F_{v_{AB}} = -\left(r^2 - \frac{\pi r^2}{4} \right) \gamma = -(4 - \pi)\gamma$$

Hence, the weight per meter of length is

$$F_{v_{BCD}} + F_{v_{AB}} = (3\pi + 4)\gamma = 0.132 \text{ MN}$$

(*b*) The force exerted against the wall is the horizontal force on *ABC* minus the horizontal force on *CD*. The horizontal components of force on *BC* and *CD* cancel; the projection of *BCD* on a vertical plane is zero. Hence,

$$F_H = F_{H_{AB}} = 2\gamma = 19.6 \text{ kN}$$

since the projected area is 2 m^2 and the pressure at the centroid of the projected area is 9806 Pa.

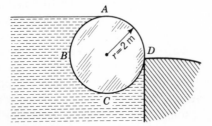

Figure 2.20 Semifloating body.

To find external reactions due to pressure forces, the action of the fluid may be replaced by the two horizontal components and one vertical component acting along their lines of action.

Tensile Stress in a Pipe and Spherical Shell

A circular pipe under the action of an internal pressure is in tension around its periphery. Assuming that no longitudinal stress occurs, the walls are in tension, as shown in Fig. 2.21. A section of pipe of unit length is considered, i.e., the ring between two planes normal to the axis and unit length apart. If one-half of this ring is taken as a free body, the tensions per unit length at top and bottom are respectively T_1 and T_2, as shown in the figure. The horizontal component of force acts through the pressure center of the projected area and is $2pr$, in which p is the pressure at the centerline and r is the internal pipe radius.

For high pressures the pressure center may be taken at the pipe center; then $T_1 = T_2$, and

$$T = pr \tag{2.6.5}$$

in which T is the tensile force per unit length. For wall thickness e, the *tensile stress* in the pipe wall is

$$\sigma = \frac{T}{e} = \frac{pr}{e} \tag{2.6.6}$$

For larger variations in pressure between top and bottom of pipe, the location of pressure center y is computed. Two equations are needed,

$$T_1 + T_2 = 2pr \qquad 2rT_1 - 2pry = 0$$

The second equation is the moment equation about the lower end of the free body, neglecting the vertical component of force. Solving gives

$$T_1 = py \qquad T_2 = p(2r - y)$$

Example 2.11 A 4.0-in-ID steel pipe has a $\frac{1}{4}$-in wall thickness. For an allowable tensile stress of 10,000 lb/in², what is the maximum pressure?
From Eq. (2.6.6)

$$p = \frac{\sigma e}{r} = \frac{(10{,}000 \text{ lb/in}^2)(0.25 \text{ in})}{2.0 \text{ in}} = 1250 \text{ lb/in}^2$$

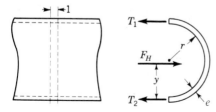

Figure 2.21 Tensile stress in pipe.

If a thin spherical shell is subjected to an internal pressure, neglecting the weight of the fluid within the sphere, the stress in its walls can be found by considering the forces on a free body consisting of a hemisphere cut from the sphere by a vertical plane. The fluid component of force normal to the plane acting on the inside of the hemisphere is $p\pi r^2$, with r the radius. The stress σ times the cut wall area $2\pi re$, with e the thickness, must balance the fluid force; hence,

$$\sigma = \frac{pr}{2e}$$

EXERCISES

2.6.1 The horizontal component of force on a curved surface is equal to the (a) weight of liquid vertically above the curved surface; (b) weight of liquid retained by the curved surface; (c) product of pressure at its centroid and area; (d) force on a projection of the curved surface onto a vertical plane; (e) scalar sum of all elemental horizontal components.

2.6.2 A pipe 5 m in diameter is to carry water at 1.4 MPa. For an allowable tensile stress of 55 MPa, the thickness of pipe wall, in millimeters, is (a) 32; (b) 42; (c) 64; (d) 80; (e) none of these answers.

2.6.3 The vertical component of pressure force on a submerged curved surface is equal to (a) its horizontal component; (b) the force on a vertical projection of the curved surface; (c) the product of pressure at centroid and surface area; (d) the weight of liquid vertically above the curved surface; (e) none of these answers.

2.6.4 The vertical component of force on the upper half of a horizontal right-circular cylinder, 3 ft in diameter and 10 ft long, filled with water and with a pressure of 0.433 psi at the axis, is (a) -458 lb; (b) -333 lb; (c) 124.8 lb; (d) 1872 lb; (e) none of these answers.

2.6.5 A cylindrical wooden barrel is held together by hoops at top and bottom. When the barrel is filled with liquid, the ratio of tension in the top hoop to tension in the bottom hoop, due to the liquid, is (a) $\frac{1}{2}$; (b) 1; (c) 2; (d) 3; (e) none of these answers.

2.6.6 A 50-mm-ID pipe with 5-mm wall thickness carries water at 0.89 MPa. The tensile stress in the pipe wall, in megapascals, is (a) 4.9; (b) 9.8; (c) 19.6; (d) 39.2; (e) none of these answers.

2.7 BUOYANT FORCE

The resultant force exerted on a body *by a static fluid* in which it is submerged or floating is called the *buoyant force*. The buoyant force always acts vertically upward. There can be no horizontal component of the resultant because the projection of the submerged body or submerged portion of the floating body on a vertical plane is always zero.

The buoyant force on a submerged body is the difference between the vertical component of pressure force on its underside and the vertical component of pressure force on its upper side. In Fig. 2.22 the upward force on the bottom is equal to the weight of liquid, real or imaginary, which is vertically above the surface ABC, indicated by the weight of liquid within $ABCEFA$. The downward force on the upper surface equals the weight of liquid $ADCEFA$. The difference

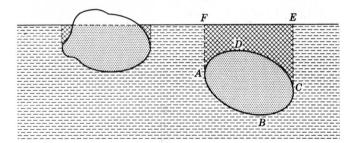

Figure 2.22 Buoyant force on floating and submerged bodies.

between the two forces is a force, vertically upward, due to the weight of fluid *ABCD* that is *displaced* by the solid. In equation form

$$F_B = \mathfrak{V}\gamma \qquad (2.7.1)$$

in which F_B is the buoyant force, $\mathfrak{V}$ is the volume of fluid displaced, and γ is the specific weight of fluid. The same formula holds for floating bodies when $\mathfrak{V}$ is taken as the volume of liquid displaced. This is evident from inspection of the floating body in Fig. 2.22.

In Fig. 2.23 the vertical force exerted on an element of the body in the form of a vertical prism of cross section δA is

$$\delta F_B = (p_2 - p_1)\,\delta A = \gamma h\,\delta A = \gamma\,\delta\mathfrak{V}$$

in which $\delta\mathfrak{V}$ is the volume of the prism. Integrating over the complete body gives

$$F_B = \gamma \int_{\mathfrak{V}} d\mathfrak{V} = \gamma\mathfrak{V}$$

when γ is considered constant throughout the volume.

To find the line of action of the buoyant force, moments are taken about a convenient axis O and are equated to the moment of the resultant; thus,

$$\gamma \int_{\mathfrak{V}} x\,d\mathfrak{V} = \gamma\mathfrak{V}\bar{x} \qquad \text{or} \qquad \bar{x} = \frac{1}{\mathfrak{V}} \int_{\mathfrak{V}} x\,d\mathfrak{V}$$

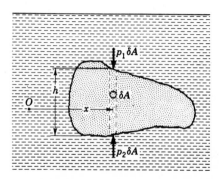

Figure 2.23 Vertical force components on element of body.

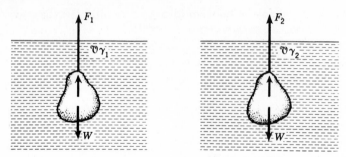

Figure 2.24 Free-body diagrams for body suspended in a fluid.

in which $\bar{x}$ is the distance from the axis to the line of action. This equation yields the distance to the centroid of the volume; hence, *the buoyant force acts through the centroid of the displaced volume of fluid*. This holds for both submerged and floating bodies. The centroid of the displaced volume of fluid is called the *center of buoyancy*.

In solving a statics problem involving submerged or floating objects, the object is generally taken as a free body, and a free-body diagram is drawn. The action of the fluid is replaced by the buoyant force. The weight of the object must be shown (acting through its center of gravity) as well as all other contact forces.

Weighing an odd-shaped object suspended in two different fluids yields sufficient data to determine its weight, volume, specific weight, and specific gravity. Figure 2.24 shows two free-body diagrams for the same object suspended and weighed in two fluids. F_1, F_2 are the weights submerged; γ_1, γ_2 are the specific weights of the fluids. W and $\mathcal{V}$, the weight and volume of the object, are to be found.

The equations of equilibrium are written

$$F_1 + \mathcal{V}\gamma_1 = W \qquad F_2 + \mathcal{V}\gamma_2 = W$$

and solved,

$$\mathcal{V} = \frac{F_1 - F_2}{\gamma_2 - \gamma_1} \qquad W = \frac{F_1\gamma_2 - F_2\gamma_1}{\gamma_2 - \gamma_1}$$

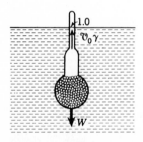
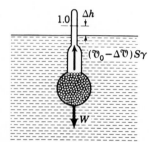

Figure 2.25 Hydrometer in water and in liquid of specific gravity S.

A *hydrometer* uses the principle of buoyant force to determine specific gravities of liquids. Figure 2.25 shows a hydrometer in two liquids. It has a stem of prismatic cross section a. Considering the liquid on the left to be distilled water, $S = 1.00$, the hydrometer floats in equilibrium when

$$\mathcal{V}_0 \gamma = W \tag{2.7.2}$$

in which $\mathcal{V}_0$ is the volume submerged, γ is the specific weight of water, and W is the weight of hydrometer. The position of the liquid surface is marked 1.00 on the stem to indicate unit specific gravity S. When the hydrometer is floated in another liquid, the equation of equilibrium becomes

$$(\mathcal{V}_0 - \Delta\mathcal{V})S\gamma = W \tag{2.7.3}$$

in which $\Delta\mathcal{V} = a\,\Delta h$. Solving for Δh with Eqs. (2.7.2) and (2.7.3) gives

$$\Delta h = \frac{\mathcal{V}_0}{a}\frac{S-1}{S} \tag{2.7.4}$$

from which the stem can be marked off to read specific gravities.

Example 2.12 A piece of ore weighing 1.5 N in air is found to weigh 1.1 N when submerged in water. What is its volume, in cubic centimeters, and what is its specific gravity? The buoyant force due to air may be neglected. From Fig. 2.24

$$1.5\ \text{N} = 1.1\ \text{N} + (9806\ \text{N/m}^3)\mathcal{V}$$

$$\mathcal{V} = 0.0000408\ \text{m}^3 = 40.8\ \text{cm}^3$$

$$S = \frac{W}{\gamma\mathcal{V}} = \frac{1.5\ \text{N}}{(9806\ \text{N/m}^3)(0.0000408\ \text{m}^3)} = 3.75$$

EXERCISES

2.7.1 A slab of wood 4 by 4 by 1 ft, sp gr 0.50, floats in water with a 400-lb load on it. The volume of slab submerged, in cubic feet, is (*a*) 1.6; (*b*) 6.4; (*c*) 8.0; (*d*) 14.4; (*e*) none of these answers.

2.7.2 The line of action of the buoyant force acts through the (*a*) center of gravity of any submerged body; (*b*) centroid of the volume of any floating body; (*c*) centroid of the displaced volume of fluid; (*d*) centroid of the volume of fluid vertically above the body; (*e*) centroid of the horizontal projection of the body.

2.7.3 Buoyant force is (*a*) the resultant force on a body due to the fluid surrounding it; (*b*) the resultant force acting on a floating body; (*c*) the force necessary to maintain equilibrium of a submerged body; (*d*) a nonvertical force for nonsymmetrical bodies; (*e*) equal to the volume of liquid displaced.

2.8 STABILITY OF FLOATING AND SUBMERGED BODIES

A body floating in a static liquid has vertical stability. A small upward displacement decreases the volume of liquid displaced, resulting in an unbalanced downward force which tends to return the body to its original position. Similarly, a

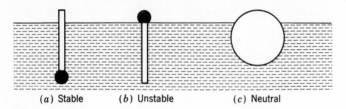

(a) Stable   (b) Unstable   (c) Neutral

Figure 2.26 Examples of stable, unstable, and neutral equilibrium.

small downward displacement results in a greater buoyant force, which causes an unbalanced upward force.

A body has linear stability when a small linear displacement in any direction sets up restoring forces tending to return it to its original position. It has rotational stability when a restoring couple is set up by any small angular displacement.

Methods for determining rotational stability are developed in the following discussion. A body may float in stable, unstable, or neutral equilibrium. When a body is in unstable equilibrium, any small angular displacement sets up a couple that tends to increase the angular displacement. With the body in neutral equilibrium, any small angular displacement sets up no couple whatever. Figure 2.26 illustrates the three cases of equilibrium: (a) a light piece of wood with a metal weight at its bottom is stable; (b) when the metal weight is at the top, the body is in equilibrium but any slight angular displacement causes it to assume the position in a; (c) a homogeneous sphere or right-circular cylinder is in equilibrium for any angular rotation; i.e., no couple results from an angular displacement.

A completely submerged object is rotationally stable only when its center of gravity is below the center of buoyancy, as in Fig. 2.27a. When the object is rotated counterclockwise, as in Fig. 2.27b, the buoyant force and weight produce a couple in the clockwise direction.

Normally, when a body is too heavy to float, it submerges and goes down until it rests on the bottom. Although the specific weight of a liquid increases slightly with depth, the higher pressure tends to cause the liquid to compress the

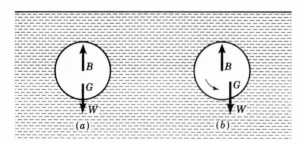

Figure 2.27 Rotationally stable submerged body.

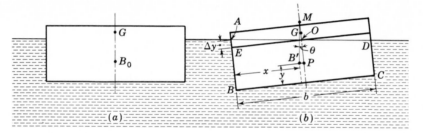

Figure 2.28 Stability of a prismatic body.

body or to penetrate into pores of solid substances and thus decrease the buoyancy of the body. A ship, for example, is sure to go to the bottom once it is completely submerged, owing to compression of air trapped in its various parts.

Determination of Rotational Stability of Floating Objects

Any floating object with center of gravity below its center of buoyancy (centroid of displaced volume) floats in stable equilibrium, as in Fig. 2.26a. Certain floating objects, however, are in stable equilibrium when their center of gravity is above the center of buoyancy. The stability of prismatic bodies is first considered, followed by an analysis of general floating bodies for small angles of tip.

 Figure 2.28a is a cross section of a body with all other parallel cross sections identical. The center of buoyancy is always at the centroid of the displaced volume, which is at the centroid of the cross-sectional area below liquid surface in this case. Hence, when the body is tipped, as in Fig. 2.28b, the center of buoyancy is at the centroid B' of the trapezoid $ABCD$, the buoyant force acts upward through B', and the weight acts downward through G, the center of gravity of the body. When the vertical through B' intersects the original centerline above G, as at M, a restoring couple is produced and the body is in stable equilibrium. The intersection of the buoyant force and the centerline is called the *metacenter*, designated M. When M is above G, the body is stable; when below G, it is unstable; and when at G, it is in neutral equilibrium. The distance $\overline{MG}$ is called the *metacentric height* and is a direct measure of the stability of the body. The restoring couple is

$$W\overline{MG} \sin \theta$$

in which θ is the angular displacement and W the weight of the body.

 Example 2.13 In Fig. 2.28 a scow 20 ft wide and 60 ft long has a gross weight of 225 short tons (2000 lb). Its center of gravity is 1.0 ft above the water surface. Find the metacentric height and restoring couple when $\Delta y = 1.0$ ft.
 The depth of submergence h in the water is

$$h = \frac{225 \times 2000}{20 \times 60 \times 62.4} = 6.0 \text{ ft}$$

The centroid in the tipped position is located with moments about AB and BC,

$$x = \frac{5 \times 20 \times 10 + 2 \times 20 \times \frac{1}{2} \times \frac{20}{3}}{6 \times 20} = 9.46 \text{ ft}$$

$$y = \frac{5 \times 20 \times \frac{5}{2} + 2 \times 20 \times \frac{1}{2} \times 5\frac{2}{3}}{6 \times 20} = 3.03 \text{ ft}$$

By similar triangles AEO and $B'PM$,

$$\frac{\Delta y}{b/2} = \frac{\overline{B'P}}{\overline{MP}}$$

$\Delta y = 1$, $b/2 = 10$, $\overline{B'P} = 10 - 9.46 = 0.54$ ft; then

$$\overline{MP} = \frac{0.54 \times 10}{1} = 5.40 \text{ ft}$$

G is 7.0 ft from the bottom; hence

$$\overline{GP} = 7.00 - 3.03 = 3.97 \text{ ft}$$

and

$$\overline{MG} = \overline{MP} - \overline{GP} = 5.40 - 3.97 = 1.43 \text{ ft}$$

The scow is stable, since $\overline{MG}$ is positive; the righting moment is

$$W\overline{MG} \sin \theta = 225 \times 2000 \times 1.43 \times \frac{1}{\sqrt{101}} = 64{,}000 \text{ lb} \cdot \text{ft}$$

Nonprismatic Cross Sections

For a floating object of variable cross section, such as a ship (Fig. 2.29a), a convenient formula can be developed for determination of metacentric height for very small angles of rotation θ. The horizontal shift in center of buoyancy r (Fig. 2.29b) is determined by the change in buoyant forces due to the wedge being submerged, which causes an upward force on the left, and by the other wedge decreasing the buoyant force by an equal amount ΔF_B on the right. The force system, consisting of the original buoyant force at B and the couple $\Delta F_B \times s$ due to the wedges, must have as resultant the equal buoyant force at B'. With moments about B to determine the shift r,

$$\Delta F_B \times s = Wr \tag{2.8.1}$$

The amount of the couple can be determined with moments about O, the centerline of the body at the liquid surface. For an element of area δA on the horizontal section through the body at the liquid surface, an element of volume of the wedge is $x\theta \, \delta A$. The buoyant force due to this element is $\gamma x\theta \, \delta A$, and its moment about O is $\gamma \theta x^2 \, \delta A$, in which θ is the small angle of tip in radians. By integrating over the complete original horizontal area at the liquid surface, the couple is determined to be

$$\Delta F_B \times s = \gamma \theta \int_A x^2 \, dA = \gamma \theta I \tag{2.8.2}$$

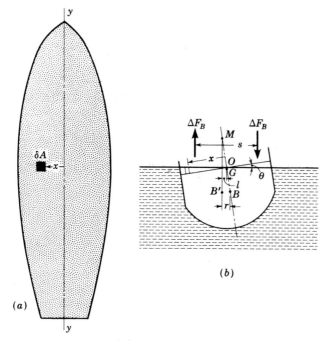

Figure 2.29 Stability relations in a body of variable cross section.

in which I is the moment of inertia of the area about the axis y-y (Fig. 2.29a). Substitution into Eq. (2.8.1) produces

$$\gamma\theta I = Wr = \mathbb{U}\gamma r$$

in which $\mathbb{U}$ is the total volume of liquid displaced.

Since θ is very small,

$$\overline{MB}\sin\theta = \overline{MB}\theta = r \qquad \text{or} \qquad \overline{MB} = \frac{r}{\theta} = \frac{I}{\mathbb{U}}$$

The metacentric height is then

$$\overline{MG} = \overline{MB} \mp \overline{GB}$$

or

$$\overline{MG} = \frac{I}{\mathbb{U}} \mp \overline{GB} \tag{2.8.3}$$

The minus sign is used if G is above B, the plus sign if G is below B.

Example 2.14 A barge displacing 1 Mkg has the horizontal cross section at the waterline shown in Fig. 2.30. Its center of buoyancy is 2.0 m below the water surface, and its center of gravity is

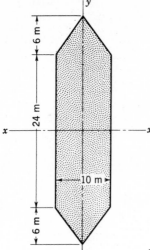

Figure 2.30 Horizontal cross section of a ship at the waterline.

0.5 m below the water surface. Determine its metacentric height for rolling (about y-y axis) and for pitching (about x-x axis).

$$\overline{GB} = 2 - 0.5 = 1.5 \text{ m}$$

$$\mho = \frac{\text{mass displacement}}{\text{density}} = \frac{1 \text{ Mkg}}{1000 \text{ kg/m}^3} = 1000 \text{ m}^3$$

$$I_{yy} = \tfrac{1}{12}(24 \text{ m})(10 \text{ m})^3 + 4(\tfrac{1}{12})(6 \text{ m})(5 \text{ m})^3 = 2250 \text{ m}^4$$

$$I_{xx} = \tfrac{1}{12}(10 \text{ m})(24 \text{ m})^3 + 2(\tfrac{1}{36})(10 \text{ m})(6 \text{ m})^3 + (60 \text{ m}^2)(14 \text{ m})^2 = 23{,}400 \text{ m}^4$$

For rolling

$$\overline{MG} = \frac{I}{\mho} - \overline{GB} = \frac{2250}{1000} - 1.5 = 0.75 \text{ m}$$

For pitching

$$\overline{MG} = \frac{I}{\mho} - GB = \frac{23{,}400}{1000} - 1.5 = 21.9 \text{ m}$$

EXERCISES

2.8.1 A body floats in stable equilibrium (a) when its metacentric height is zero; (b) *only* when its center of gravity is below its center of buoyancy; (c) when $\overline{GB} - I/\mho$ is positive and G is above B; (d) when $I/\mho$ is positive; (e) when the metacenter is above the center of gravity.

2.8.2 A closed cubical metal box 3 ft on an edge is made of uniform sheet and weighs 1200 lb. Its metacentric height when placed in oil, sp gr 0.90, with sides vertical, is (a) 0 ft; (b) −0.08 ft; (c) 0.62 ft; (d) 0.78 ft; (e) none of these answers.

2.9 RELATIVE EQUILIBRIUM

In fluid statics the variation of pressure is simple to compute, thanks to the absence of shear stresses. For fluid motion such that no layer moves relative to an adjacent layer, the shear stress is also zero throughout the fluid. A fluid with a translation at uniform velocity still follows the laws of static variation of pressure. When a fluid is being accelerated so that no layer moves relative to an adjacent one, i.e., when the fluid moves as if it were a solid, no shear stresses occur and variation in pressure can be determined by writing the equation of motion for an appropriate free body. Two cases are of interest, a uniform linear acceleration and a uniform rotation about a vertical axis. When moving thus, the fluid is said to be in *relative equilibrium*.

Although relative equilibrium is not a fluid statics phenomenon, it is discussed here because of the similarity of the relations.

Uniform Linear Acceleration

A liquid in an open vessel is given a uniform linear acceleration **a** as in Fig. 2.31. After some time the liquid adjusts to the acceleration so that it moves as a solid; i.e., the distance between any two fluid particles remains fixed, and hence no shear stresses occur.

By selecting a cartesian coordinate system with y vertical and x such that the acceleration vector **a** is in the xy plane (Fig. 2.31a), the z axis is normal to **a** and there is no acceleration component in that direction. Equation (2.2.5) applies to this situation,

$$\mathbf{f} - \mathbf{j}\gamma = -\nabla p - \mathbf{j}\gamma = \rho\mathbf{a} \qquad (2.2.5)$$

The pressure gradient ∇p is then the vector sum of $-\rho\mathbf{a}$ and $-\mathbf{j}\gamma$ as shown in Fig. 2.31b. Since ∇p is in the direction of maximum change in p (the gradient), at right angles to ∇p there is no change in p. Surfaces of constant pressure, including the free surface, must therefore be normal to ∇p. To obtain a convenient algebraic

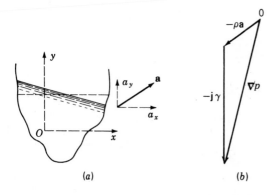

(a)   (b)

Figure 2.31 Acceleration with free surface.

expression for variation of p with x, y, and z, that is, $p = p(x, y, z)$, Eq. (2.2.5) is written in component form:

$$\nabla p = \mathbf{i} \frac{\partial p}{\partial x} + \mathbf{j} \frac{\partial p}{\partial y} + \mathbf{k} \frac{\partial p}{\partial z} = -\mathbf{j}\gamma - \frac{\gamma}{g}(\mathbf{i}a_x + \mathbf{j}a_y)$$

or

$$\frac{\partial p}{\partial x} = -\frac{\gamma}{g}a_x \qquad \frac{\partial p}{\partial y} = -\gamma\left(1 + \frac{a_y}{g}\right) \qquad \frac{\partial p}{\partial z} = 0$$

Since p is a function of position (x, y, z), its total differential is

$$dp = \frac{\partial p}{\partial x} dx + \frac{\partial p}{\partial y} dy + \frac{\partial p}{\partial z} dz$$

Substituting for the partial differentials gives

$$dp = -\gamma\frac{a_x}{g} dx - \gamma\left(1 + \frac{a_y}{g}\right)dy \tag{2.9.1}$$

which can be integrated for an incompressible fluid,

$$p = -\gamma\frac{a_x}{g}x - \gamma\left(1 + \frac{a_y}{g}\right)y + c$$

To evaluate the constant of integration c, let $x = 0$, $y = 0$, $p = p_0$; then $c = p_0$ and

$$p = p_0 - \gamma\frac{a_x}{g}x - \gamma\left(1 + \frac{a_y}{g}\right)y \tag{2.9.2}$$

When the accelerated incompressible fluid has a free surface, its equation is given by setting $p = 0$ in Eq. (2.9.2). Solving Eq. (2.9.2) for y gives

$$y = -\frac{a_x}{a_y + g}x + \frac{p_0 - p}{\gamma(1 + a_y/g)} \tag{2.9.3}$$

The lines of constant pressure, $p = $ const, have the slope

$$-\frac{a_x}{a_y + g}$$

and are parallel to the free surface. The y intercept of the free surface is

$$\frac{p_0}{\gamma(1 + a_y/g)}$$

Example 2.15 The tank in Fig. 2.32 is filled with oil, sp gr 0.8, and accelerated as shown. There is a small opening in the tank at A. Determine the pressure at B and C; and the acceleration a_x required to make the pressure at B zero.

By selecting point A as origin and by applying Eq. (2.9.2) for $a_y = 0$,

$$p = -\gamma\frac{a_x}{g}x - \gamma y = -\frac{0.8(9806 \text{ N/m}^3)(4.903 \text{ m/s}^2)}{9.806 \text{ m/s}^2}x - 0.8(9806 \text{ N/m}^3)y$$

or

$$p = -3922.4x - 7844.8y \text{ Pa}$$

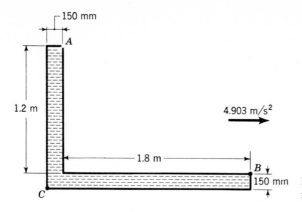

Figure 2.32 Tank completely filled with liquid.

At B, $x = 1.8$ m, $y = -1.2$ m, and $p = 2.35$ kPa. At C, $x = -0.15$ m, $y = -1.35$ m, and $p = 11.18$ kPa. For zero pressure at B, from Eq. (2.9.2) with origin at A,

$$0.0 = 0.0 - \frac{0.8(9806 \text{ N/m}^3)}{9.806 \text{ m/s}^2} 1.8a_x - 0.8(9806 \text{ N/m}^3)(-1.2)$$

or

$$a_x = 6.537 \text{ m/s}^2$$

Example 2.16 A closed box with horizontal base 6 by 6 units and a height of 2 units is half-filled with liquid (Fig. 2.33). It is given a constant linear acceleration $a_x = g/2$, $a_y = -g/4$. Develop an equation for variation of pressure along its base.
The free surface has the slope

$$\frac{-a_x}{a_y + g} = \frac{-g/2}{-g/4 + g} = -\frac{2}{3}$$

hence, the free surface is located as shown in the figure. When the origin is taken at 0, Eq. (2.9.2) becomes

$$p = p_0 - \frac{\gamma}{2}x - \gamma\left(1 - \frac{1}{4}\right)y = p_0 - \frac{\gamma}{2}\left(x + \frac{3}{2}y\right)$$

$p = 0$ for $y = 0$, $x = 4.5$; so $p_0 = 2.25\gamma$. Then, for $y = 0$, along the bottom,

$$p = 2.25\gamma - 0.5\gamma x \qquad 0 \le x \le 4.5$$

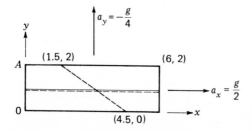

Figure 2.33 Uniform linear acceleration of container.

Uniform Rotation about a Vertical Axis

Rotation of a fluid, moving as a solid, about an axis is called *forced-vortex motion*. Every particle of fluid has the same angular velocity. This motion is to be distinguished from *free-vortex motion*, in which each particle moves in a circular path with a speed varying inversely as the distance from the center. Free-vortex motion is discussed in Chaps. 7 and 9. A liquid in a container, when rotated about a vertical axis at constant angular velocity, moves like a solid after some time interval. No shear stresses exist in the liquid, and the only acceleration that occurs is directed radially inward toward the axis of rotation. By selecting a coordinate system (Fig. 2.34*a*) with the unit vector **i** in the *r* direction and **j** in the vertical upward direction with *y* the axis of rotation, Eq. (2.2.5) may be applied to determine pressure variation throughout the fluid:

$$\nabla p = -\mathbf{j}\gamma - \rho\mathbf{a} \qquad (2.2.5)$$

For constant angular velocity ω, any particle of fluid P has an acceleration $\omega^2 r$ directed radially inward, as $\mathbf{a} = -\mathbf{i}\omega^2 r$. Vector addition of $-\mathbf{j}\gamma$ and $-\rho\mathbf{a}$ (Fig. 2.34*b*) yields ∇p, the pressure gradient. The pressure does not vary normal to this line at a point; hence, if P is taken at the surface, the free surface is normal to ∇p. Expanding Eq. (2.2.5)

$$\mathbf{i}\,\frac{\partial p}{\partial r} + \mathbf{j}\,\frac{\partial p}{\partial y} + \mathbf{k}\,\frac{\partial p}{\partial z} = -\mathbf{j}\gamma + \mathbf{i}\rho\omega^2 r$$

k is the unit vector along the *z* axis (or tangential direction). Then

$$\frac{\partial p}{\partial r} = \frac{\gamma}{g}\omega^2 r \qquad \frac{\partial p}{\partial y} = -\gamma \qquad \frac{\partial p}{\partial z} = 0$$

Since *p* is a function of *y* and *r* only, the total differential *dp* is

$$dp = \frac{\partial p}{\partial y}\,dy + \frac{\partial p}{\partial r}\,dr$$

Substituting for $\partial p/\partial y$ and $\partial p/\partial r$ results in

$$dp = -\gamma\,dy + \frac{\gamma}{g}\omega^2 r\,dr \qquad (2.9.4)$$

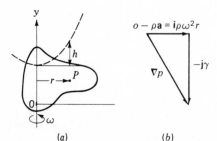

(a)   (b)   **Figure 2.34** Rotation of a fluid about a vertical axis.

For a liquid ($\gamma \approx$ const) integration yields

$$p = \frac{\gamma}{g}\omega^2\frac{r^2}{2} - \gamma y + c$$

in which c is the constant of integration. If the value of pressure at the origin ($r = 0$, $y = 0$) is p_0, then $c = p_0$ and

$$p = p_0 + \gamma\frac{\omega^2 r^2}{2g} - \gamma y \tag{2.9.5}$$

When the particular horizontal plane ($y = 0$) for which $p_0 = 0$ is selected and Eq. (2.9.5) is divided by γ,

$$h = \frac{p}{\gamma} = \frac{\omega^2 r^2}{2g} \tag{2.9.6}$$

which shows that the head, or vertical depth, varies as the square of the radius. The surfaces of equal pressure are paraboloids of revolution.

When a free surface occurs in a container that is being rotated, the fluid volume underneath the paraboloid of revolution is the original fluid volume. The shape of the paraboloid depends only upon the angular velocity ω.

For a circular cylinder rotating about its axis (Fig. 2.35) the rise of liquid from its vertex to the wall of the cylinder is, from Eq. (2.9.6), $\omega^2 r_0^2/2g$. Since a paraboloid of revolution has a volume equal to one-half its circumscribing cylinder, the volume of the liquid above the horizontal plane through the vertex is

$$\pi r_0^2 \times \frac{1}{2}\frac{\omega^2 r_0^2}{2g}$$

When the liquid is at rest, this liquid is also above the plane through the vertex to a uniform depth of

$$\frac{1}{2}\frac{\omega^2 r_0^2}{2g}$$

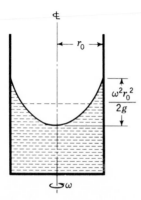

Figure 2.35 Rotation of circular cylinder about its axis.

Hence, the liquid rises along the walls the same amount as the center drops, thereby permitting the vertex to be located when ω, r_0, and depth before rotation are given.

Example 2.17 A liquid, sp gr 1.2, is rotated at 200 rpm about a vertical axis. At one point A in the fluid 1 m from the axis, the pressure is 70 kPa. What is the pressure at a point B, which is 2 m higher than A and 1.5 m from the axis?

When Eq. (2.9.5) is written for the two points,

$$p_A = p_0 + \gamma \frac{\omega^2 r_A^2}{2g} - \gamma y \qquad p_B = p_0 + \gamma \frac{\omega^2 r_B^2}{2g} - \gamma(y + 2)$$

Then $\omega = 200 \times 2\pi/60 = 20.95$ rad/s, $\gamma = 1.2 \times 9806 = 11{,}767$ N/m³, $r_A = 1$ m, and $r_B = 1.5$ m. When the second equation is subtracted from the first and the values are substituted,

$$70{,}000 - p_B = (2 \text{ m})(11{,}767 \text{ N/m}^3) + \frac{11{,}767 \text{ N/m}^3}{2(9.806 \text{ m/s}^2)} (20.95/\text{s})^2 \times [1 \text{ m}^2 - (1.5 \text{ m})^2]$$

Hence

$$p_B = 375.6 \text{ kPa}$$

If a closed container with no free surface or with a partially exposed free surface is rotated uniformly about some vertical axis, an *imaginary* free surface can be constructed; it consists of a paraboloid of revolution of shape given by Eq. (2.9.6). The vertical distance from any point in the fluid to this free surface is the pressure head at the point.

Example 2.18 A straight tube 4 ft long, closed at the bottom and filled with water, is inclined 30° with the vertical and rotated about a vertical axis through its midpoint 8.02 rad/s. Draw the paraboloid of zero pressure, and determine the pressure at the bottom and midpoint of the tube.

In Fig. 2.36, the zero-pressure paraboloid passes through point A. If the origin is taken at the vertex, that is, $p_0 = 0$, Eq. (2.9.6) becomes

$$h = \frac{\omega^2 r^2}{2g} = \frac{8.02^2}{64.4} (2 \sin 30°)^2 = 1.0 \text{ ft}$$

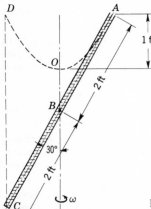

Figure 2.36 Rotation of inclined tube of liquid about a vertical axis.

which locates the vertex at O, 1.0 ft below A. The pressure at the bottom of the tube is $\gamma \times \overline{CD}$, or

$$4 \cos 30° \times 62.4 = 216 \text{ lb/ft}^2$$

At the midpoint, $\overline{OB} = 0.732$ ft and

$$p_B = 0.732 \times 62.4 = 45.6 \text{ lb/ft}^2$$

Fluid Pressure Forces in Relative Equilibrium

The magnitude of the force acting on a plane area in contact with a liquid accelerating as a rigid body can be obtained by integration over the surface

$$F = \int p \, dA$$

The nature of the acceleration and orientation of the surface governs the particular variation of p over the surface. When the pressure varies linearly over the plane surface (linear acceleration), the magnitude of force is given by the product of pressure at the centroid and area, since the volume of the pressure prism is given by $p_G A$. For nonlinear distributions the magnitude and line of action can be found by integration.

EXERCISES

2.9.1 A closed cubical box, 1 m on each edge, is half filled with water, the other half being filled with oil, sp gr 0.75. When it is accelerated vertically upward 4.903 m/s², the pressure difference between bottom and top, in kilopascals, is (a) 4.9; (b) 11; (c) 12.9; (d) 14.7; (e) none of these answers.

2.9.2 When the box of Exercise 2.9.1 is accelerated uniformly in a horizontal direction parallel to one side, 16.1 ft/s², the slope of the interface is (a) 0; (b) $-\frac{1}{4}$; (c) $-\frac{1}{2}$; (d) -1; (e) none of these answers.

2.9.3 When the minimum pressure in the box of Exercise 2.9.2 is zero gage, the maximum pressure in meters of water is (a) 0.94; (b) 1.125; (c) 1.31; (d) 1.5; (e) none of these answers.

2.9.4 Liquid in a cylinder 10 m long is accelerated horizontally $20g$ m/s² along the axis of the cylinder. The difference in pressure intensities at the ends of the cylinder, in pascals, if $\gamma = $ sp wt of liquid is (a) 20γ; (b) 200γ; (c) $20g\gamma$; (d) $200\gamma/g$; (e) none of these answers.

2.9.5 When a liquid rotates at constant angular velocity about a vertical axis as a rigid body, the pressure (a) decreases as the square of the radial distance; (b) increases linearly as the radial distance; (c) decreases as the square of increase in elevation along any vertical line; (d) varies inversely as the elevation along any vertical line; (e) varies as the square of the radial distance.

2.9.6 When a liquid rotates about a vertical axis as a rigid body so that points on the axis have the same pressure as points 2 ft higher and 2 ft from the axis, the angular velocity in radians per second is (a) 8.02; (b) 11.34; (c) 64.4; (d) not determinable from data given; (e) none of these answers.

2.9.7 A right-circular cylinder, open at the top, is filled with liquid, sp gr 1.2, and rotated about its vertical axis at such speed that half the liquid spills out. The pressure at the center of the bottom is (a) zero; (b) one-fourth its value when cylinder was full; (c) indeterminable for reason of insufficient data; (d) greater than a similar case with water as liquid; (e) none of these answers.

2.9.8 A forced vortex (a) turns in an opposite direction to a free vortex; (b) always occurs in conjunction with a free vortex; (c) has the velocity decreasing with the radius; (d) occurs when fluid rotates as a solid; (e) has the velocity decreasing inversely with the radius.

PROBLEMS

2.1 Prove that the pressure is the same in all directions at a point in a static fluid for the three-dimensional case.

2.2 The container of Fig. 2.37 holds water and air as shown. What is the pressure at A, B, C, and D in pounds per square foot and in pascals?

2.3 The tube in Fig. 2.38 is filled with oil. Determine the pressure at A and B in meters of water.

2.4 Calculate the pressure at A, B, C, and D of Fig. 2.39 in pascals.

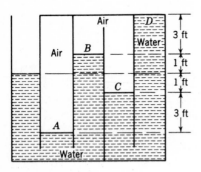

Figure 2.37 Problem 2.2.

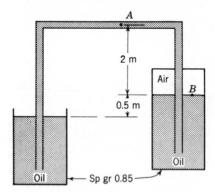

Figure 2.38 Problem 2.3.

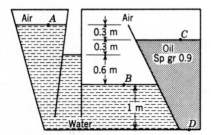

Figure 2.39 Problem 2.4.

2.5 Derive the equations that give the pressure and density at any elevation in a static gas when conditions are known at one elevation and the temperature gradient β is known.

2.6 By a limiting process as $\beta \to 0$, derive the isothermal case from the results of Prob. 2.5.

2.7 By use of the results of Prob. 2.5, determine the pressure and density at 3000 m elevation when $p = 100$ kPa abs, $t = 20°C$, at elevation 300 m for air and $\beta = -0.005°C/m$.

2.8 For isothermal air at 0°C, determine the pressure and density at 3000 m when the pressure is 0.1 MPa abs at sea level.

2.9 In isothermal air at 80°F what is the vertical distance for reduction of density by 10 percent?

2.10 Express a pressure of 8 psi in (a) inches of mercury, (b) feet of water, (c) feet of acetylene tetrabromide, sp gr 2.94, (d) pascals.

2.11 A bourdon gage reads 2 psi suction, and the barometer is 29.5 in Hg. Express the pressure in six other customary ways.

2.12 Express 3 atm in meters of water gage, barometer reading 750 mm.

2.13 Bourdon gage A inside a pressure tank (Fig. 2.40) reads 12 psi. Another bourdon gage B outside the pressure tank and connected with it reads 20 psi, and an aneroid barometer reads 30 in Hg. What is the absolute pressure measured by A in inches of mercury?

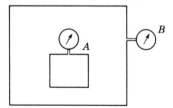

Figure 2.40 Problem 2.13.

2.14 Determine the heights of columns of water; kerosene, sp gr 0.83; and acetylene tetrabromide, sp gr 2.94, equivalent to 200 mm Hg.

2.15 In Fig. 2.6a, for a reading $h = 20$ in, determine the pressure at A in pounds per square inch. The liquid has a specific gravity of 1.90.

2.16 Determine the reading h in Fig. 2.6b for $p_A = 30$ kPa suction if the liquid is kerosene, sp gr 0.83.

2.17 In Fig. 2.6b, for $h = 6$ in and barometer reading 29 in, with water the liquid, find p_A in feet of water absolute.

2.18 In Fig. 2.6c $S_1 = 0.86$, $S_2 = 1.0$, $h_2 = 90$ mm, $h_1 = 150$ mm. Find p_A in millimeters of mercury gage. If the barometer reading is 720 mm, what is p_A in meters of water absolute?

2.19 Gas is contained in vessel A of Fig. 2.6c. With water the manometer fluid and $h_1 = 75$ mm, determine the pressure at A in inches of mercury.

2.20 In Fig. 2.7a $S_1 = 1.0$, $S_2 = 0.95$, $S_3 = 1.0$, $h_1 = h_2 = 280$ mm, and $h_3 = 1$ m. Compute $p_A - p_B$ in millimeters of water.

2.21 In Prob. 2.20 find the gage difference h_2 for $p_A - p_B = -350$ mm H_2O.

2.22 In Fig. 2.7b $S_1 = S_3 = 0.83$, $S_2 = 13.6$, $h_1 = 150$ mm, $h_2 = 70$ mm, and $h_3 = 120$ mm. (a) Find p_A if $p_B = 10$ psi. (b) For $p_A = 20$ psia and a barometer reading of 720 mm, find p_B in meters of water gage.

2.23 Find the gage difference h_2 in Prob. 2.22 for $p_A = p_B$.

2.24 In Fig. 2.41, A contains water, and the manometer fluid has a specific gravity of 2.94. When the left meniscus is at zero on the scale, $p_A = 90$ mm H_2O. Find the reading of the right meniscus for $p_A = 8$ kPa with no adjustment of the U tube or scale.

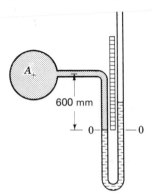

Figure 2.41 Problem 2.24.

2.25 The Empire State Building is 1250 ft high. What is the pressure difference in pounds per square inch of a water column of the same height?

2.26 What is the pressure at a point 10 m below the free surface in a fluid that has a variable density in kilograms per cubic meter given by $\rho = 450 + ah$, in which $a = 12$ kg/m^4 and h is the distance in meters measured from the free surface?

2.27 A vertical gas pipe in a building contains gas, $\rho = 0.002$ slug/ft^3 and $p = 3.0$ in H$_2$O gage in the basement. At the top of the building 800 ft higher, determine the gas pressure in inches water gage for two cases: (a) gas assumed incompressible and (b) gas assumed isothermal. Barometric pressure 34 ft H$_2$O; $t = 70°$F.

2.28 In Fig. 2.8 determine R, the gage difference, for a difference in gas pressure of 9 mm H$_2$O. $\gamma_2 = 9.8$ kN/m^3; $\gamma_3 = 10.5$ kN/m^3; $a/A = 0.01$.

2.29 The inclined manometer of Fig. 2.9 reads zero when A and B are at the same pressure. The diameter of reservoir is 2.0 in, and that of the inclined tube $\frac{1}{4}$ in. For $\theta = 30°$, gage fluid sp gr 0.832, find $p_A - p_B$ in pounds per square inch as a function of gage reading R in feet.

2.30 Determine the weight W that can be sustained by the force acting on the piston of Fig. 2.42.

2.31 Neglecting the weight of the container (Fig. 2.43), find (a) the force tending to lift the circular top CD and (b) the compressive load on the pipe wall at A-A.

2.32 Find the force of oil on the top surface CD of Fig. 2.43 if the liquid level in the open pipe is reduced by 1 m.

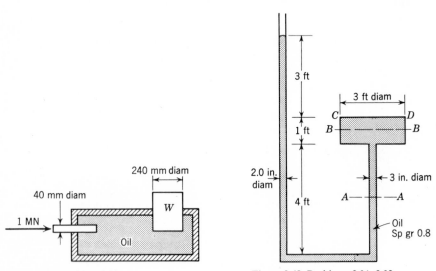

Figure 2.42 Problem 2.30.   **Figure 2.43** Problems 2.31, 2.32.

2.33 The container shown in Fig. 2.44 has a circular cross section. Determine the upward force on the surface of the cone frustum $ABCD$. What is the downward force on the plane EF? Is this force equal to the weight of the fluid? Explain.

2.34 The cylindrical container of Fig. 2.45 weighs 400 N when empty. It is filled with water and supported on the piston. (a) What force is exerted on the upper end of the cylinder? (b) If an additional 600-N weight is placed on the cylinder, how much will the water force against the top of the cylinder be increased?

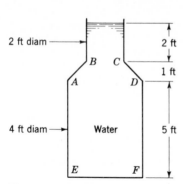

2 ft diam —

B   C

A   D

4 ft diam —   Water

E   F

2 ft

1 ft

5 ft

Figure 2.44 Problem 2.33.

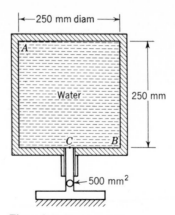

←— 250 mm diam —→

A

Water   250 mm

C   B

←— 500 mm²

Figure 2.45 Problem 2.34.

2.35 A barrel 2 ft in diameter filled with water has a vertical pipe of 0.50 in diameter attached to the top. Neglecting compressibility, how many pounds of water must be added to the pipe to exert a force of 1000 lb on the top of the barrel?

2.36 A vertical right-angled triangular surface has a vertex in the free surface of a liquid (Fig. 2.46). Find the force on one side (*a*) by integration and (*b*) by formula.

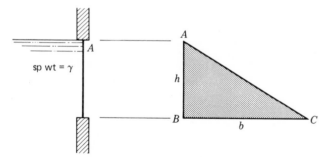

sp wt = γ

A

A

h

B   C

b

Figure 2.46 Problems 2.36, 2.38, 2.49, 2.50.

2.37 Determine the magnitude of the force acting on one side of the vertical triangle ABC of Fig. 2.47 (*a*) by integration and (*b*) by formula.

2.38 Find the moment about AB of the force acting on one side of the vertical surface ABC of Fig. 2.46. $\gamma = 9000$ N/m³.

2.39 Find the moment about AB of the force acting on one side of the vertical surface ABC of Fig. 2.47.

2.40 Locate a horizontal line below AB of Fig. 2.47 such that the magnitude of pressure force on the vertical surface ABC is equal above and below the line.

2.41 Determine the force acting on one side of the vertical surface $OABCO$ of Fig. 2.48. $\gamma = 9$ kN/m³.

2.42 Calculate the force exerted by water on one side of the vertical annular area shown in Fig. 2.49.

2.43 Determine the moment at A required to hold the gate as shown in Fig. 2.50.

2.44 If there is water on the other side of the gate (Fig. 2.50) up to A, determine the resultant force due to water on both sides of the gate, including its line of action.

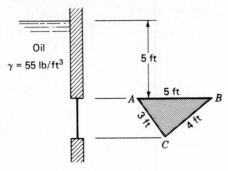

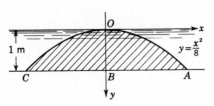

Figure 2.47 Problems 2.37, 2.39, 2.40, 2.47, 2.48.

Figure 2.48 Problems 2.41, 2.56, 2.83.

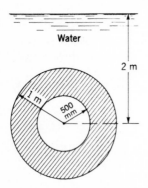

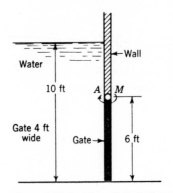

Figure 2.49 Problems 2.42, 2.51.

Figure 2.50 Problems 2.43, 2.44, 2.52.

2.45 The shaft of the gate in Fig. 2.51 will fail at a moment of 145 kN·m. Determine the maximum value of liquid depth h.

2.46 The dam of Fig. 2.52 has a strut AB every 6 m. Determine the compressive force in the strut, neglecting the weight of the dam.

2.47 Locate the distance of the pressure center below the liquid surface in the triangular area ABC of Fig. 2.47 by integration and by formula.

2.48 By integration locate the pressure center horizontally in the triangular area ABC of Fig. 2.47.

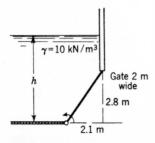

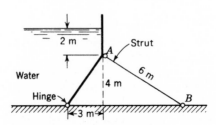

Figure 2.51 Problems 2.45, 2.55.

Figure 2.52 Problem 2.46.

2.49 By using the pressure prism, determine the resultant force and location for the triangle of Fig. 2.46.

2.50 By integration, determine the pressure center for Fig. 2.46.

2.51 Locate the pressure center for the annular area of Fig. 2.49.

2.52 Locate the pressure center for the gate of Fig. 2.50.

2.53 A vertical square area 6 by 6 ft is submerged in water with upper edge 3 ft below the surface. Locate a horizontal line on the surface of the square such that (a) the force on the upper portion equals the force on the lower portion and (b) the moment of force about the line due to the upper portion equals the moment due to the lower portion.

2.54 An equilateral triangle with one edge in a water surface extends downward at a 45° angle. Locate the pressure center in terms of the length of a side b.

2.55 In Fig. 2.51 develop the expression for y_p in terms of h.

2.56 Locate the pressure center of the vertical area $OABCO$ of Fig. 2.48.

2.57 Locate the pressure center for the vertical area of Fig. 2.53.

2.58 Demonstrate the fact that the magnitude of the resultant force on a totally submerged plane area is unchanged if the area is rotated about an axis through its centroid.

2.59 The gate of Fig. 2.54 weighs 300 lb/ft normal to the paper. Its center of gravity is 1.5 ft from the left face and 2.0 ft above the lower face. It is hinged at O. Determine the water-surface position for the gate just to start to come up. (Water surface is below the hinge.)

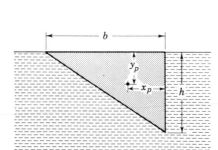

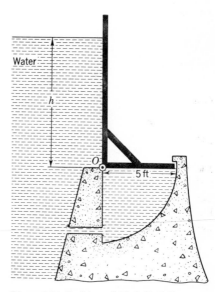

Figure 2.53 Problem 2.57.   **Figure 2.54** Problems 2.59, 2.60, 2.61.

2.60 Find h of Prob. 2.59 for the gate just to come up to the vertical position shown.

2.61 Determine the value of h and the force against the stop when this force is a maximum for the gate of Prob. 2.59.

2.62 Determine y of Fig. 2.55 so that the flashboards will tumble when water reaches their top.

2.63 Determine the hinge location y of the rectangular gate of Fig. 2.56 so that it will open when the liquid surface is as shown.

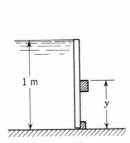

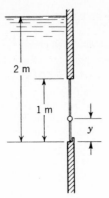

Figure 2.55 Problem 2.62.   **Figure 2.56** Problem 2.63.

2.64 By use of the pressure prism, show that the pressure center approaches the centroid of an area as its depth of submergence is increased.

2.65 (a) Find the magnitude and line of action of force on each side of the gate of Fig. 2.57. (b) Find the resultant force due to the liquid on both sides of the gate. (c) Determine F to open the gate if it is uniform and weighs 5000 lb.

2.66 For linear stress variation over the base of the dam of Fig. 2.58, (a) locate where the resultant crosses the base and (b) compute the maximum and minimum compressive stresses at the base. Neglect hydrostatic uplift.

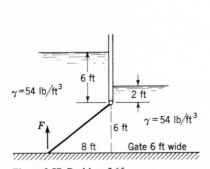

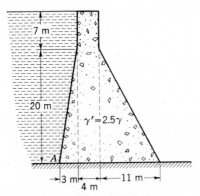

Figure 2.57 Problem 2.65.   **Figure 2.58** Problems 2.66, 2.67.

2.67 Work Prob. 2.66 with the addition that the hydrostatic uplift varies linearly from 20 m at A to zero at the toe of the dam.

2.68 Find the moment M at O (Fig. 2.59) to hold the gate closed.

2.69 The gate shown in Fig. 2.60 is in equilibrium. Compute W, the weight of counterweight per meter of width, neglecting the weight of the gate. Is the gate in stable equilibrium?

2.70 How high (h) will the water on the right have to rise to open the gate shown in Fig. 2.61? The gate is 5 ft wide, and it is constructed of material with specific gravity $S = 2.5$. Use the pressure prism method.

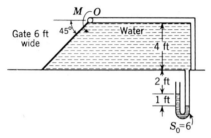

Figure 2.59 Problem 2.68.

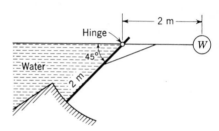

Figure 2.60 Problem 2.69.

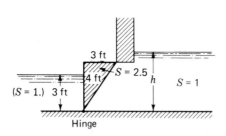

Figure 2.61 Problem 2.70.

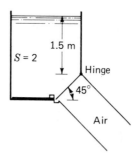

Figure 2.62 Problem 2.71.

2.71 Compute the air pressure required to keep the 700-mm-diameter gate of Fig. 2.62 closed. The gate is a circular plate that weighs 1800 N.

2.72 A 15-ft-diameter pressure pipe carries liquid at 200 psi. What pipe-wall thickness is required for maximum stress of 10,000 psi?

2.73 To obtain the same flow area, which pipe system requires the least steel, a single pipe or four pipes having half the diameter? The maximum allowable pipe-wall stress is the same in each case.

2.74 A thin-walled hollow sphere 3 m in diameter holds gas at 1.5 MPa. For allowable stress of 60 MPa determine the minimum wall thickness.

2.75 A cylindrical container 7 ft high and 4 ft in diameter provides for pipe tension with two hoops a foot from each end. When it is filled with water, what is the tension in each hoop due to the water?

2.76 A 20-mm-diameter steel ball covers a 10-mm-diameter hole in a pressure chamber where the pressure is 30 MPa. What force is required to lift the ball from the opening?

2.77 If the horizontal component of force on a curved surface did *not* equal the force on a projection of the surface onto a vertical plane, what conclusions could you draw regarding the propulsion of a boat (Fig. 2.63)?

Figure 2.63 Problem 2.77.

2.78 (*a*) Determine the horizontal component of force acting on the radial gate (Fig. 2.64) and its line of action. (*b*) Determine the vertical component of force and its line of action. (*c*) What force *F* is required to open the gate, neglecting its weight? (*d*) What is the moment about an axis normal to the paper and through point *O*?

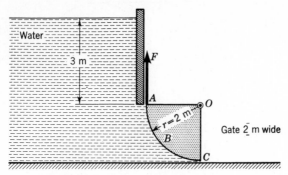

Figure 2.64 Problem 2.78.

2.79 Calculate the force F required to hold the gate of Fig. 2.65 in a closed position, $R = 2$ ft.

2.80 Calculate the force F required to open or hold closed the gate of Fig. 2.65 when $R = 1.5$ ft.

2.81 What is R of Fig. 2.65 for no force F required to hold the gate closed or to open it?

2.82 Find the vertical component of force on the curved gate of Fig. 2.66, including its line of action.

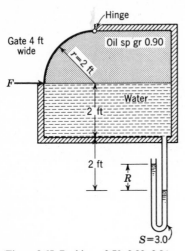

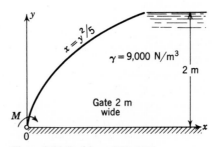

Figure 2.65 Problems 2.79, 2.80, 2.81.   **Figure 2.66** Problems 2.82, 2.86.

2.83 What is the force on the surface whose trace is OA of Fig. 2.48? The length normal to the paper is 3 m, $\gamma = 9$ kN/m^3.

2.84 A right-circular cylinder is illustrated in Fig. 2.67. The pressure, in pounds per square foot, due to flow around the cylinder varies over the segment ABC as $p = 2\rho(1 - 4 \sin^2 \theta) + 10$. Calculate the force on ABC.

2.85 If the pressure variation on the cylinder in Fig. 2.67 is $p = 2\rho \times [1 - 4(1 + \sin \theta)^2] + 10$, determine the force on the cylinder.

2.86 Determine the moment M to hold the gate of Fig. 2.66, neglecting its weight.

2.87 Find the resultant force, including its line of action, acting on the outer surface of the first quadrant of a spherical shell of radius 600 mm with center at the origin. Its center is 1.2 m below the water surface.

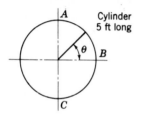

Figure 2.67 Problems 2.84, 2.85.

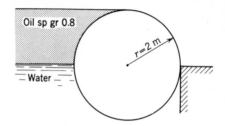

Figure 2.68 Problem 2.89.

2.88 The volume of the ellipsoid given by $x^2/a^2 + y^2/b^2 + z^2/c^2 = 1$ is $4\pi abc/3$, and the area of the ellipse $x^2/a^2 + z^2/c^2 = 1$ is πac. Determine the vertical force on the surface given in Example 2.9.

2.89 A log holds the water as shown in Fig. 2.68. Determine (a) the force per meter pushing it against the dam, (b) the weight of the cylinder per meter of length, and (c) its specific gravity.

2.90 The cylinder of Fig. 2.69 is filled with liquid as shown. Find (a) the horizontal component of force on AB per unit of length, including its line of action, and (b) the vertical component of force on AB per unit of length, including its line of action.

2.91 The cylinder gate of Fig. 2.70 is made up from a circular cylinder and a plate hinged at the dam. The gate position is controlled by pumping water into or out of the cylinder. The center of gravity of the empty gate is on the line of symmetry 4 ft from the hinge. It is in equilibrium when empty in the position shown. How many cubic feet of water must be added per foot of cylinder to hold the gate in its position when the water surface is raised 3 ft?

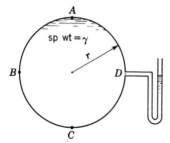

Figure 2.69 Problem 2.90.

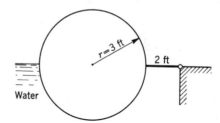

Figure 2.70 Problem 2.91.

2.92 A hydrometer weighs 0.035 N and has a stem 5 mm in diameter. Compute the distance between specific gravity markings 1.0 and 1.1.

2.93 Design a hydrometer to read specific gravities in the range from 0.80 to 1.10 when the scale is to be 75 mm long.

2.94 A sphere 250 mm in diameter, sp gr 1.4, is submerged in a liquid having a density varying with the depth y below the surface given by $\rho = 1000 + 0.03y$ kg/m^3. Determine the equilibrium position of the sphere in the liquid.

2.95 Repeat the calculations for Prob. 2.94 for a horizontal circular cylinder with a specific gravity of 1.4 and a diameter of 250 mm.

2.96 A cube, 2 ft on an edge, has its lower half of sp gr 1.4 and upper half of sp gr 0.6. It is submerged into a two-layered fluid, the lower sp gr 1.2 and the upper sp gr 0.9. Determine the height of the top of the cube above the interface.

2.97 Determine the density, specific volume, and volume of an object that weighs 3 N in water and 4 N in oil, sp gr 0.83.

2.98 Two cubes of the same size, 1 m³, one of sp gr 0.80, the other of sp gr 1.1, are connected by a short wire and placed in water. What portion of the lighter cube is above the water surface, and what is the tension in the wire?

2.99 In Fig. 2.71 the hollow triangular prism is in equilibrium as shown when $z = 1$ ft and $y = 0$. Find the weight of prism per foot of length and z in terms of y for equilibrium. Both liquids are water. Determine the value of y for $z = 1.5$ ft.

2.100 How many pounds of concrete, $\gamma = 25$ kN/m³, must be attached to a beam having a volume of 0.1 m³ and sp gr 0.65 to cause both to sink in water?

2.101 The gate of Fig. 2.72 weighs 150 lb/ft normal to the page. It is in equilibrium as shown. Neglecting the weight of the arm and brace supporting the counterweight, (a) find W and (b) determine whether the gate is in stable equilibrium. The weight is made of concrete, sp gr 2.50.

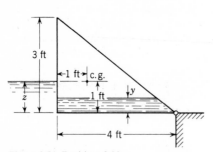

Figure 2.71 Problem 2.99.   **Figure 2.72** Problem 2.101.

2.102 A wooden cylinder 600 mm in diameter, sp gr 0.50, has a concrete cylinder 600 mm long of the same diameter, sp gr 2.50, attached to one end. Determine the length of wooden cylinder for the system to float in stable equilibrium with axis vertical.

2.103 What are the proportions r_0/h of a right-circular cylinder of specific gravity S so that it will float in water with end faces horizontal in stable equilibrium?

2.104 Will a beam 4 m long with square cross section, sp gr 0.75, float in stable equilibrium in water with two sides horizontal?

2.105 Determine the metacentric height of the torus shown in Fig. 2.73.

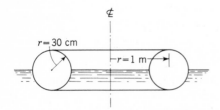

Figure 2.73 Problem 2.105.

2.106 The plane gate (Fig. 2.74) weighs 2000 N/m normal to the paper, and its center of gravity is 2 m from the hinge at O. (a) Find h as a function of θ for equilibrium of the gate. (b) Is the gate in stable equilibrium for any values of θ?

2.107 A spherical balloon 15 m in diameter is open at the bottom and filled with hydrogen. For barometer reading of 28 in Hg and 20°C, what is the total weight of the balloon and the load to hold it stationary?

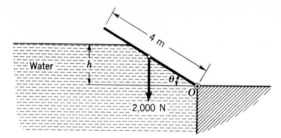

Figure 2.74 Problem 2.106.

2.108 A tank of liquid $S = 0.86$ is accelerated uniformly in a horizontal direction so that the pressure decreases within the liquid 20 kPa/m in the direction of motion. Determine the acceleration.

2.109 The free surface of a liquid makes an angle of 20° with the horizontal when accelerated uniformly in a horizontal direction. What is the acceleration?

2.110 In Fig. 2.75, $a_x = 12.88$ ft/s², $a_y = 0$. Find the imaginary free liquid surface and the pressure at B, C, D, and E.

2.111 In Fig. 2.75, $a_x = 0$, $a_y = -8.05$ ft/s². Find the pressure at B, C, D, and E.

2.112 In Fig. 2.75, $a_x = 8.05$ ft/s², $a_y = 16.1$ ft/s². Find the imaginary free surface and the pressure at B, C, D, and E.

2.113 In Fig. 2.76, $a_x = 9.806$ m/s², $a_y = 0$. Find the pressure at A, B, and C.

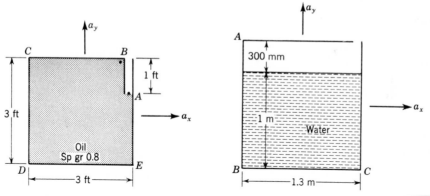

Figure 2.75 Problems 2.110, 2.111, 2.112, 2.116.   **Figure 2.76** Problems 2.113, 2.114.

2.114 In Fig. 2.76, $a_x = 4.903$ m/s², $a_y = 9.806$ m/s². Find the pressure at A, B, and C.

2.115 A circular cross-sectional tank of 6-ft depth and 4-ft diameter is filled with liquid and accelerated uniformly in a horizontal direction. If one-third of the liquid spills out, determine the acceleration.

2.116 Determine a_x and a_y in Fig. 2.75 for pressure at A, B, and C to be the same.

2.117 The tube of Fig. 2.77 is filled with liquid, sp gr 2.40. When it is accelerated to the right 8.05 ft/s², draw the imaginary free surface and determine the pressure at A. For $p_A = 8$ psi vacuum determine a_x.

2.118 A cubical box 1 m on an edge, open at the top and half filled with water, is placed on an inclined plane making a 30° angle with the horizontal. The box alone weighs 500 N and has a coefficient of friction with the plane of 0.30. Determine the acceleration of the box and the angle the free-water surface makes with the horizontal.

2.119 Show that the pressure is the same in all directions at a point in a liquid moving as a solid.

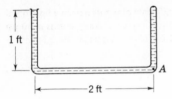

Figure 2.77 Problems 2.117, 2.123, 2.124, 2.134.

2.120 A closed box contains two immiscible liquids. Prove that, when it is accelerated uniformly in the x direction, the interface and zero-pressure surface are parallel.

2.121 Verify the statement made in Sec. 2.9 on uniform rotation about a vertical axis that, when a fluid rotates in the manner of a solid body, no shear stresses exist in the fluid.

2.122 A vessel containing liquid, sp gr 1.2, is rotated about a vertical axis. The pressure at one point 0.6 m radially from the axis is the same as at another point 1.2 m from the axis and with elevation 0.6 m higher. Calculate the rotational speed.

2.123 The U tube of Fig. 2.77 is rotated about a vertical axis 6 in to the right of A at such a speed that the pressure at A is zero gage. What is the rotational speed?

2.124 Locate the vertical axis of rotation and the speed of rotation of the U tube of Fig. 2.77 so that the pressure of liquid at the midpoint of the U tube and at A are both zero.

2.125 An incompressible fluid of density ρ moving as a solid rotates at speed ω about an axis inclined at $\theta°$ with the vertical. Knowing the pressure at one point in the fluid, how do you find the pressure at any other point?

2.126 A right-circular cylinder of radius r_0 and height h_0 with axis vertical is open at the top and filled with liquid. At what speed must it rotate so that half the area of the bottom is exposed?

2.127 A liquid rotating about a *horizontal* axis as a solid has a pressure of 10 psi at the axis. Determine the pressure variation along a vertical line through the axis for density ρ and speed ω.

2.128 Determine the equation for the surfaces of constant pressure for the situation described in Prob. 2.127.

2.129 Prove by integration that a paraboloid of revolution has a volume equal to half its circumscribing cylinder.

2.130 A tank containing two immiscible liquids is rotated about a vertical axis. Prove that the interface has the same shape as the zero-pressure surface.

2.131 A hollow sphere of radius r_0 is filled with liquid and rotated about its vertical axis at speed ω. Locate the circular line of maximum pressure.

2.132 A gas following the law $p\rho^{-n} = \text{const}$ is rotated about a vertical axis as a solid. Derive an expression for pressure in a radial direction for speed ω, pressure p_0, and density ρ_0 at a point on the axis.

2.133 A vessel containing water is rotated about a vertical axis with an angular velocity of 50 rad/s. At the same time the container has a downward acceleration of 16.1 ft/s². What is the equation for a surface of constant pressure?

2.134 The U tube of Fig. 2.77 is rotated about a vertical axis through A at such a speed that the water in the tube begins to vaporize at the closed end above A, which is at 70°F. What is the angular velocity? What would happen if the angular velocity were increased?

2.135 A cubical box 1.3 m on an edge is open at the top and filled with water. When it is accelerated upward 2.45 m/s², find the magnitude of water force on one side of the box.

2.136 A cube 1 m on an edge is filled with liquid, sp gr 0.65, and is accelerated downward 2.45 m/s². Find the resultant force on one side of the cube due to liquid pressure.

2.137 A cylinder 2 ft in diameter and 6 ft long is accelerated uniformly along its axis in a horizontal direction 16.1 ft/s². It is filled with liquid, $\gamma = 50$ lb/ft³, and it has a pressure along its axis of 10 psi before acceleration starts. Find the horizontal net force exerted against the liquid in the cylinder.

2.138 A closed cube, 300 mm on an edge, has a small opening at the center of its top. When it is filled with water and rotated uniformly about a vertical axis through its center at ω rad/s, find the force on a side due to the water in terms of ω.

THREE
FLUID-FLOW CONCEPTS AND
BASIC EQUATIONS

The statics of fluids, treated in the preceding chapter, is almost an exact science, specific weight (or density) being the only quantity that must be determined experimentally. On the other hand, the nature of *flow* of a real fluid is very complex. Since the basic laws describing the complete motion of a fluid are not easily formulated and handled mathematically, recourse to experimentation is required. By an analysis based on mechanics, thermodynamics, and orderly experimentation, large hydraulic structures and efficient fluid machines have been produced.

This chapter introduces the concepts needed for analysis of fluid motion. The basic equations that enable us to predict fluid behavior are stated or derived; they are equations of motion, continuity, and momentum and the first and second laws of thermodynamics as applied to steady flow of a perfect gas. In this chapter the control-volume approach is utilized in the derivation of the continuity, energy, and momentum equations. Viscous effects, the experimental determination of losses, and the dimensionless presentation of loss data are presented in Chap. 5 after dimensional analysis has been introduced in Chap. 4. In general, one-dimensional-flow theory is developed in this chapter, with applications limited to incompressible cases where viscous effects do not predominate. Chapter 6 deals with compressible flow, and Chap. 7 with two-dimensional flow.

3.1 FLOW CHARACTERISTICS; DEFINITIONS

Flow may be classified in many ways such as turbulent, laminar; real, ideal; reversible, irreversible; steady, unsteady; uniform, nonuniform; rotational, irrotational. In this and the following section various types of flow are distinguished.

Turbulent flow situations are most prevalent in engineering practice. In turbulent flow the fluid particles (small molar masses) move in very irregular paths, causing an exchange of momentum from one portion of the fluid to another in a manner somewhat similar to the molecular momentum transfer described in Sec. 1.3 but on a much larger scale. The fluid particles can range in size from very small (say a few thousand molecules) to very large (thousands of cubic feet in a large swirl in a river or in an atmospheric gust). In a situation in which the flow could be either turbulent or nonturbulent (laminar), the turbulence sets up greater shear stresses throughout the fluid and causes more irreversibilities or losses. Also, in turbulent flow, the losses vary about as the 1.7 to 2 power of the velocity; in laminar flow, they vary as the first power of the velocity.

In *laminar* flow, fluid particles move along smooth paths in laminas, or layers, with one layer gliding smoothly over an adjacent layer. Laminar flow is governed by Newton's law of viscosity [Eq. (1.1.1) or extensions of it to three-dimensional flow], which relates shear stress to rate of angular deformation. In laminar flow, the action of viscosity damps out turbulent tendencies (see Sec. 5.3 for criteria for laminar flow). Laminar flow is not stable in situations involving combinations of low viscosity, high velocity, or large flow passages and breaks down into turbulent flow. An equation similar in form to Newton's law of viscosity may be written for turbulent flow:

$$\tau = \eta \, \frac{du}{dy} \tag{3.1.1}$$

The factor η, however, is not a fluid property alone; it depends upon the fluid motion and the density. It is called the *eddy viscosity*.

In many practical flow situations, both viscosity and turbulence contribute to the shear stress:

$$\tau = (\mu + \eta) \, \frac{du}{dy} \tag{3.1.2}$$

Experimentation is required to determine this type of flow.

An *ideal fluid* is frictionless and incompressible and should not be confused with a perfect gas (Sec. 1.6). The assumption of an ideal fluid is helpful in analyzing flow situations involving large expanses of fluids, as in the motion of an airplane or a submarine. A frictionless fluid is nonviscous, and its flow processes are reversible.

The layer of fluid in the immediate neighborhood of an actual flow boundary that has had its velocity relative to the boundary affected by viscous shear is called the *boundary layer*. Boundary layers may be laminar or turbulent, depending generally upon their length, the viscosity, the velocity of the flow near them, and the boundary roughness.

Adiabatic flow is that flow of a fluid in which no heat is transferred to or from the fluid. *Reversible adiabatic* (frictionless adiabatic) *flow* is called *isentropic flow*.†

---

† An isentropic process, however, can occur in irreversible flow with the proper amount of heat transfer (isentropic = constant entropy).

To proceed in an orderly manner into the analysis of fluid flow requires a clear understanding of the terminology involved. Several of the more important technical terms are defined and illustrated in this section.

Steady flow occurs when conditions at any point in the fluid do not change with the time. For example, if the velocity at a certain point is 3 m/s in the $+x$ direction in steady flow, it remains exactly that amount and in that direction indefinitely. This can be expressed as $\partial\mathbf{v}/\partial t = 0$, in which space $(x, y, z$ coordinates of the point) is held constant. Likewise, in steady flow there is no change in density ρ, pressure p, or temperature T with time at any point; thus

$$\frac{\partial \rho}{\partial t} = 0 \qquad \frac{\partial p}{\partial t} = 0 \qquad \frac{\partial T}{\partial t} = 0$$

In turbulent flow, owing to the erratic motion of the fluid particles, there are always small fluctuations occurring at any point. The definition for steady flow must be generalized somewhat to provide for these fluctuations. To illustrate this, a plot of velocity against time, at some point in turbulent flow, is given in Fig. 3.1. When the temporal mean velocity

$$v_t = \frac{1}{t}\int_0^t v \, dt$$

indicated in the figure by the horizontal line, does not change with the time, the flow is said to be steady. The same generalization applies to density, pressure, temperature, etc., when they are substituted for v in the above formula.

The flow is *unsteady* when conditions at any point change with the time, $\partial\mathbf{v}/\partial t \neq 0$. Water being pumped through a fixed system at a constant rate is an example of steady flow. Water being pumped through a fixed system at an increasing rate is an example of unsteady flow.

Uniform flow occurs when, at every point, the velocity vector is identically the same (in magnitude and direction) for any given instant. In equation form, $\partial\mathbf{v}/\partial s = 0$, in which time is held constant and δs is a displacement in any direction. The equation states that there is no change in the velocity vector in any direction

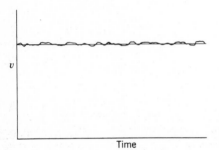

v

Time

Figure 3.1 Velocity at a point in steady turbulent flow.

throughout the fluid at any one instant. It says nothing about the change in velocity at a point with time.

In flow of a real fluid in an open or closed conduit, the definition of uniform flow may also be extended in most cases even though the velocity vector at the boundary is always zero. When all parallel cross sections through the conduit are identical (i.e., when the conduit is prismatic) and the average velocity at each cross section is the same at any given instant, the flow is said to be *uniform*.

Flow such that the velocity vector varies from place to place at any instant $(\partial v/\partial s \neq 0)$ is *nonuniform flow*. A liquid being pumped through a long straight pipe has uniform flow. A liquid flowing through a reducing section or through a curved pipe has nonuniform flow.

Examples of steady and unsteady flow and of uniform and nonuniform flow are liquid flow through a long pipe at a constant rate is *steady uniform flow*; liquid flow through a long pipe at a decreasing rate is *unsteady uniform flow*; flow through an expanding tube at a constant rate is *steady nonuniform flow*; and flow through an expanding tube at an increasing rate is *unsteady nonuniform flow*.

Rotation of a fluid particle about a given axis, say the z axis, is defined as the average angular velocity of two infinitesimal line elements in the particle that are at right angles to each other and to the given axis. If the fluid particles within a region have rotation about any axis, the flow is called *rotational flow*, or *vortex flow*. If the fluid within a region has no rotation, the flow is called *irrotational flow*. It is shown in texts on hydrodynamics that if a fluid is at rest and is frictionless, any later motion of this fluid will be irrotational.

One-dimensional flow neglects variations or changes in velocity, pressure, etc., transverse to the main flow direction. Conditions at a cross section are expressed in terms of average values of velocity, density, and other properties. Flow through a pipe, for example, may usually be characterized as one dimensional. Many practical problems can be handled by this method of analysis, which is much simpler than two- and three-dimensional methods of analysis. In *two-dimensional flow* all particles are assumed to flow in parallel planes along identical paths in each of these planes; hence, there are no changes in flow normal to these planes. The flow net, developed in Chap. 7, is the most useful method for analysis of two-dimensional-flow situations. *Three-dimensional flow* is the most general flow in which the velocity components u, v, w in mutually perpendicular directions are functions of space coordinates and time x, y, z, and t. Methods of analysis are generally complex mathematically, and only simple geometrical flow boundaries can be handled.

A *streamline* is a continuous line drawn through the fluid so that it has the direction of the velocity vector at every point. There can be no flow across a streamline. Since a particle moves in the direction of the streamline at any instant, its displacement δs, having components δx, δy, δz, has the direction of the velocity vector $\mathbf{q}$ with components u, v, w in the x, y, z directions, respectively. Then

$$\frac{\delta x}{u} = \frac{\delta y}{v} = \frac{\delta z}{w}$$

states that the corresponding components are proportional and hence that δs and $\mathbf{q}$ have the same direction. Expressing the displacements in differential form

$$\frac{dx}{u} = \frac{dy}{v} = \frac{dz}{w} \tag{3.1.3}$$

produces the differential equations of a streamline. Equations (3.1.3) are two independent equations. Any continuous line that satisfies them is a streamline.

In steady flow, since there is no change in direction of the velocity vector at any point, the streamline has a fixed inclination at every point and is, therefore, *fixed in space*. A particle always moves tangent to the streamline; hence, in steady flow the *path of a particle* is a streamline. In unsteady flow, since the direction of the velocity vector at any point may change with time, a streamline may shift in space from instant to instant. A particle then follows one streamline one instant, another one the next instant, and so on, so that the path of the particle may have no resemblance to any given instantaneous streamline.

A dye or smoke is frequently injected into a fluid in order to trace its subsequent motion. The resulting dye or smoke trails are called *streak lines*. In steady flow a streak line is a streamline and the path of a particle.

Streamlines in two-dimensional flow can be obtained by inserting fine, bright particles (aluminum dust) into the fluid, brilliantly lighting one plane, and taking a photograph of the streaks made in a short time interval. Tracing on the picture continuous lines that have the direction of the streaks at every point portrays the streamlines for either steady or unsteady flow.

In illustration of an incompressible two-dimensional flow, as in Fig. 3.2, the streamlines are drawn so that, per unit time, the volume flowing between adjacent streamlines is the same if unit depth is considered normal to the plane of the figure. Hence, when the streamlines are closer together, the velocity must be greater, and vice versa. If v is the average velocity between two adjacent streamlines at some position where they are h apart, the flow rate Δq is

$$\Delta q = vh \tag{3.1.4}$$

At any other position on the chart where the distance between streamlines is h_1, the average velocity is $v_1 = \Delta q/h_1$. By increasing the number of streamlines drawn, i.e., by decreasing Δq, in the limiting case the velocity at a point is obtained.

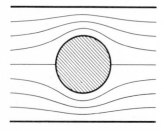

Figure 3.2 Streamlines for steady flow around a cylinder between parallel walls.

A *stream tube* is the tube made by all the streamlines passing through a small, closed curve. In steady flow it is fixed in space and can have no flow through its walls because the velocity vector has no component normal to the tube surface.

Example 3.1 In two-dimensional, incompressible steady flow around an airfoil the streamlines are drawn so that they are 10 mm apart at a great distance from the airfoil, where the velocity is 40 m/s. What is the velocity near the airfoil, where the streamlines are 7.5 mm apart?

$$(40 \text{ m/s})(0.01 \text{ m})(1 \text{ m}) = 0.40 \text{ m}^3/\text{s} = v0.0075 \text{ m}^2$$

and
$$v = \frac{0.40 \text{ m}^3/\text{s}}{0.0075 \text{ m}^2} = 53.3 \text{ m/s}$$

EXERCISES

3.1.1 One-dimensional flow is (a) steady uniform flow; (b) uniform flow; (c) flow which neglects changes in a transverse direction; (d) restricted to flow in a straight line; (e) none of these answers.

3.1.2 Isentropic flow is (a) irreversible adiabatic flow; (b) perfect-gas flow; (c) ideal-fluid flow; (d) reversible adiabatic flow; (e) frictionless reversible flow.

3.1.3 In turbulent flow (a) the fluid particles move in an orderly manner; (b) cohesion is more effective than momentum transfer in causing shear stress; (c) momentum transfer is on a molecular scale only; (d) one lamina of fluid glides smoothly over another; (e) the shear stresses are generally larger than in a similar laminar flow.

3.1.4 The ratio $\eta = \tau/(du/dy)$ for turbulent flow is (a) a physical property of the fluid; (b) dependent upon the flow and the density; (c) the viscosity divided by the density; (d) a function of temperature and pressure of fluid; (e) independent of the nature of the flow.

3.1.5 Turbulent flow generally occurs for cases involving (a) very viscous fluids; (b) very narrow passages or capillary tubes; (c) very slow motions; (d) combinations of (a), (b), and (c); (e) none of these answers.

3.1.6 In laminar flow (a) experimentation is required for the simplest flow cases; (b) Newton's law of viscosity applies; (c) the fluid particles move in irregular and haphazard paths; (d) the viscosity is unimportant; (e) the ratio $\tau/(du/dy)$ depends upon the flow.

3.1.7 An ideal fluid is (a) very viscous; (b) one which obeys Newton's law of viscosity; (c) a useful assumption in problems in conduit flow; (d) frictionless and incompressible; (e) none of these answers.

3.1.8 Which of the following must be fulfilled by the flow of any fluid, real or ideal?
1. Newton's law of viscosity.
2. Newton's second law of motion.
3. The continuity equation.
4. $\tau = (\mu + \eta) \, du/dy$.
5. Velocity at boundary must be zero relative to boundary.
6. Fluid cannot penetrate a boundary.
(a) 1, 2, 3; (b) 1, 3, 6; (c) 2, 3, 5; (d) 2, 3, 6; (e) 2, 4, 5.

3.1.9 Steady flow occurs when (a) conditions do not change with time at any point; (b) conditions are the same at adjacent points at any instant; (c) conditions change steadily with the time; (d) $\partial v/\partial t$ is constant; (e) $\partial v/\partial s$ is constant.

3.1.10 Uniform flow occurs (a) whenever the flow is steady; (b) when $\partial \mathbf{v}/\partial t$ is everywhere zero; (c) only when the velocity vector at any point remains constant; (d) when $\partial \mathbf{v}/\partial s = 0$; (e) when the discharge through a curved pipe of constant cross-sectional area is constant.

3.1.11 Select the correct practical example of steady nonuniform flow: (*a*) motion of water around a ship in a lake; (*b*) motion of a river around bridge piers; (*c*) steadily increasing flow through a pipe; (*d*) steadily decreasing flow through a reducing section; (*e*) constant discharge through a long, straight pipe.

3.1.12 A streamline (*a*) is the line connecting the midpoints of flow cross sections; (*b*) is defined for uniform flow only; (*c*) is drawn normal to the velocity vector at every point; (*d*) is always the path of a particle; (*e*) is fixed in space in steady flow.

3.1.13 In two-dimensional flow around a cylinder the streamlines are 2 in apart at a great distance from the cylinder, where the velocity is 100 ft/s. At one point near the cylinder the streamlines are 1.5 in apart. The average velocity there is (*a*) 75 ft/s; (*b*) 133 ft/s; (*c*) 150 ft/s; (*d*) 200 ft/s; (*e*) 300 ft/s.

3.1.14 The head loss in turbulent flow in a pipe (*a*) varies directly as the velocity; (*b*) varies inversely as the square of the velocity; (*c*) varies inversely as the square of the diameter; (*d*) depends upon the orientation of the pipe; (*e*) varies approximately as the square of the velocity.

3.2 THE CONCEPTS OF SYSTEM AND CONTROL VOLUME

The free-body diagram was used in Chap. 2 as a convenient way to show forces exerted on some arbitrary fixed mass. This is a special case of a *system*. A system refers to a definite mass of material and distinguishes it from all other matter, called its *surroundings*. The boundaries of a system form a closed surface. This surface may vary with time, so that it contains the same mass during changes in its condition. For example, a kilogram of gas may be confined in a cylinder and be compressed by motion of a piston; the system boundary coinciding with the end of the piston then moves with the piston. The system may contain an infinitesimal mass or a large finite mass of fluids and solids at the will of the investigator.

The law of conservation of mass states that the mass within a system remains constant with time (disregarding relativity effects). In equation form

$$\frac{dm}{dt} = 0 \tag{3.2.1}$$

where m is the total mass.

Newton's second law of motion is usually expressed for a system as

$$\Sigma \mathbf{F} = \frac{d}{dt}(m\mathbf{v}) \tag{3.2.2}$$

in which it must be remembered that m is the constant mass of the system. $\Sigma \mathbf{F}$ refers to the resultant of all external forces acting on the system, including body forces such as gravity, and $\mathbf{v}$ is the velocity of the center of mass of the system.

A *control volume* refers to a region in space and is useful in the analysis of situations where flow occurs into and out of the space. The boundary of a control volume is its *control surface*. The size and shape of the control volume are entirely arbitrary, but frequently they are made to coincide with solid boundaries in parts; in other parts they are drawn normal to the flow directions as a matter of simplification. By superposition of a uniform velocity on a system and its sur-roundings a convenient situation for application of the control volume may sometimes be found, e.g., determination of sound-wave velocity in a medium. The control-volume concept is used in the derivation of continuity, momentum, and energy equations, as well as in the solution of many types of problems. The control volume is also referred to as an *open system*.

Regardless of the nature of the flow, all flow situations are subject to the following relations, which may be expressed in analytic form:

1. Newton's laws of motion, which must hold for every particle at every instant
2. The continuity relation, i.e., the law of conservation of mass
3. The first and second laws of thermodynamics
4. Boundary conditions; analytical statements that a real fluid has zero velocity relative to a boundary at a boundary or that frictionless fluids cannot penetrate a boundary

Other relations and equations, such as an equation of state or Newton's law of viscosity, may enter.

In the derivation that follows the control-volume concept is related to the system in terms of a general property of the system. It is then applied specifically to obtain continuity, energy, and linear-momentum relations.

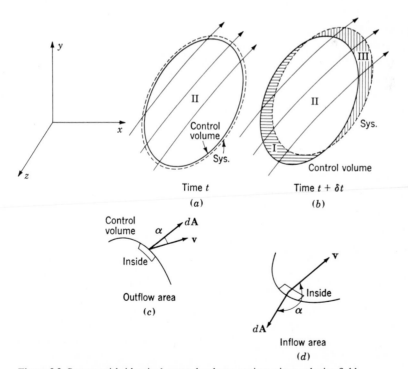

Figure 3.3 System with identical control volume at time t in a velocity field.

To formulate the relation between equations applied to a system and those applied to a control volume, consider some general flow situation, Fig. 3.3, in which the velocity of a fluid is given relative to an xyz coordinate system. At time t consider a certain mass of fluid that is contained within a system, having the dotted-line boundaries indicated. Also consider a control volume, fixed relative to the xyz axes, that exactly coincides with the system at time t. At time $t + \delta t$ the system has moved somewhat, since each mass particle moves at the velocity associated with its location.

Let N be the total amount of some property (mass, energy, momentum) within the system at time t, and let η be the amount of this property, per unit mass, throughout the fluid. The time rate of increase of N for the system is now formulated in terms of the control volume.

At $t + \delta t$, Fig. 3.3b, the system comprises volumes II and III, while at time t it occupies volume II, Fig. 3.3a. The increase in property N in the system in time δt is given by

$$N_{\text{sys}_{t+\delta t}} - N_{\text{sys}_t} = \left(\int_{\text{II}} \eta\rho\, d\mathcal{V} + \int_{\text{III}} \eta\rho\, d\mathcal{V} \right)_{t+\delta t} - \left(\int_{\text{II}} \eta\rho\, d\mathcal{V} \right)_t$$

in which $d\mathcal{V}$ is the element of volume. Rearrangement, after adding and subtracting

$$\left(\int_{\text{I}} \eta\rho\, d\mathcal{V} \right)_{t+\delta t}$$

to the right, then dividing through by δt leads to

$$\frac{N_{\text{sys}_{t+\delta t}} - N_{\text{sys}_t}}{\delta t} = \frac{(\int_{\text{II}} \eta\rho\, d\mathcal{V} + \int_{\text{I}} \eta\rho\, d\mathcal{V})_{t+\delta t} - (\int_{\text{II}} \eta\rho\, d\mathcal{V})_t}{\delta t}$$

$$+ \frac{(\int_{\text{III}} \eta\rho\, d\mathcal{V})_{t+\delta t}}{\delta t} - \frac{(\int_{\text{I}} \eta\rho\, d\mathcal{V})_{t+\delta t}}{\delta t} \tag{3.2.3}$$

The term on the left is the average time rate of increase of N within the system during time δt. In the limit as δt approaches zero, it becomes dN/dt. If the limit is taken as δt approaches zero for the first term on the right-hand side of the equation, the first two integrals are the amount of N in the control volume at $t + \delta t$ and the third integral is the amount of N in the control volume at time t. The limit is

$$\frac{\partial}{\partial t} \int_{cv} \eta\rho\, d\mathcal{V}$$

the partial being needed as the volume is held constant (the control volume) as $\delta t \to 0$.

The next term, which is the time rate of flow of N out of the control volume, in the limit, may be written

$$\lim_{\delta t \to 0} \frac{(\int_{\text{III}} \eta\rho\, d\mathcal{V})_{t+\delta t}}{\delta t} = \int_{\text{outflow area}} \eta\rho\mathbf{v} \cdot d\mathbf{A} = \int \eta\rho v \cos \alpha\, dA \tag{3.2.4}$$

in which $d\mathbf{A}$, Fig. 3.3c, is the vector representing an area element of the outflow area. It has a direction normal to the surface-area element of the control volume, positive outward; α is the angle between the velocity vector and the elemental area vector.

Similarly, the last term of Eq. (3.2.3), which is the rate of flow of N into the control volume, is, in the limit,

$$\lim_{\delta t \to 0} \frac{(\int_I \eta\rho \, d\mho)_{t+\delta t}}{\delta t} = -\int_{\text{inflow area}} \eta\rho\mathbf{v} \cdot d\mathbf{A} = -\int \eta\rho v \cos \alpha \, dA \qquad (3.2.5)$$

The minus sign is needed as $\mathbf{v} \cdot d\mathbf{A}$ (or $\cos \alpha$) is negative for inflow, Fig. 3.3d. The last two terms of Eq. (3.2.3), given by Eqs. (3.2.4) and (3.2.5), may be combined into the single term which is an integral over the complete control-volume surface (cs)

$$\lim_{\delta t \to 0} \left(\frac{(\int_{III} \eta\rho \, d\mho)_{t+\delta t}}{\delta t} - \frac{(\int_I \eta\rho \, d\mho)_{t+\delta t}}{\delta t} \right) = \int_{cs} \eta\rho\mathbf{v} \cdot d\mathbf{A} = \int_{cs} \eta\rho v \cos \alpha \, dA$$

Where there is no inflow or outflow, $\mathbf{v} \cdot d\mathbf{A} = 0$; hence, the equation can be evaluated over the whole control surface.†

Collecting the reorganized terms of Eq. (3.2.3) gives

$$\frac{dN}{dt} = \frac{\partial}{\partial t} \int_{cv} \eta\rho \, d\mho + \int_{cs} \eta\rho\mathbf{v} \cdot d\mathbf{A} \qquad (3.2.6)$$

In words, Eq. (3.2.6) states that the time rate of increase of N within a system is just equal to the time rate of increase of the property N within the control volume (fixed relative to xyz) plus the net rate of efflux of N across the control-volume boundary.

Equation (3.2.6) is used throughout this chapter in converting laws and principles from the system form to the control-volume form. The system form, which in effect follows the motion of the particles, is referred to as the Lagrangian method of analysis; the control-volume approach is called the Eulerian method of analysis, as it observes flow from a reference system fixed relative to the control volume.

Since the xyz frame of reference may be given an arbitrary constant velocity without affecting the dynamics of the system and its surroundings, Eq. (3.2.6) is valid if the control volume, fixed in size and shape, has a uniform velocity of translation.

EXERCISES

3.2.1 An *open system* implies (*a*) the presence of a free surface; (*b*) that a specified mass is considered; (*c*) the use of a control volume; (*d*) no interchange between system and surroundings; (*e*) none of these answers.

---

† This derivation was developed by Professor William Mirsky of the Department of Mechanical Engineering, The University of Michigan.

3.2.2 A *control volume* refers to (a) a fixed region in space; (b) a specified mass; (c) an isolated system; (d) a reversible process only; (e) a closed system.

3.3 APPLICATION OF THE CONTROL VOLUME TO CONTINUITY, ENERGY, AND MOMENTUM

In this section the general relation of system and control volume to a property, developed in Sec. 3.2, is applied first to continuity, then to energy, and finally to linear momentum. In the following sections the uses of equations are brought out and illustrated.

Continuity

The continuity equations are developed from the general principle of conservation of mass, Eq. (3.2.1), which states that the mass within a system remains constant with time; i.e.,

$$\frac{dm}{dt} = 0$$

In Eq. (3.2.6) let N be the mass of the system m. Then η is the mass per unit mass, or $\eta = 1$

$$0 = \frac{\partial}{\partial t} \int_{cv} \rho \, d\mathcal{V} + \int_{cs} \rho \mathbf{v} \cdot d\mathbf{A} \qquad (3.3.1)$$

In words, the continuity equation for a control volume states that the time rate of increase of mass within a control volume is just equal to the net rate of mass inflow to the control volume. This equation is examined further in Sec. 3.4.

Energy Equation

The first law of thermodynamics for a system states that the heat Q_H added to a system minus the work W done by the system depends only upon the initial and final states of the system. The difference in states of the system, being independent of the path from initial to final state, must be a property of the system. It is called the internal energy E. The first law in equation form is

$$Q_H - W = E_2 - E_1 \qquad (3.3.2)$$

The internal energy per unit mass is called e; hence, applying Eq. (3.2.6), $N = E$ and $\eta = \rho e/\rho$,

$$\frac{dE}{dt} = \frac{\partial}{\partial t} \int_{cv} \rho e \, d\mathcal{V} + \int_{cs} \rho e \mathbf{v} \cdot d\mathbf{A} \qquad (3.3.3)$$

or by use of Eq. (3.3.2)

$$\frac{\delta Q_H}{\delta t} - \frac{\delta W}{\delta t} = \frac{dE}{dt} = \frac{\partial}{\partial t} \int_{cv} \rho e \, d\mathcal{V} + \int_{cs} \rho e \mathbf{v} \cdot d\mathbf{A} \tag{3.3.4}$$

The work done by the system on its surroundings may be broken into two parts: the work W_{pr} done by pressure forces on the moving boundaries and the work W_S done by shear forces such as the torque exerted on a rotating shaft. The work done by pressure forces in time δt is

$$\delta W_{pr} = \delta t \int p\mathbf{v} \cdot d\mathbf{A} \tag{3.3.5}$$

By use of the definitions of the work terms, Eq. (3.3.4) becomes

$$\frac{\delta Q_H}{\delta t} - \frac{\delta W_S}{\delta t} = \frac{\partial}{\partial t} \int_{cv} \rho e \, d\mathcal{V} + \int_{cs} \left(\frac{p}{\rho} + e \right) \rho \mathbf{v} \cdot d\mathbf{A} \tag{3.3.6}$$

In the absence of nuclear, electrical, magnetic, and surface-tension effects, the internal energy e of a pure substance is the sum of potential, kinetic, and "intrinsic" energies. The intrinsic energy u per unit mass is due to molecular spacing and forces (dependent upon p, ρ, or T):

$$e = gz + \frac{v^2}{2} + u \tag{3.3.7}$$

Linear-Momentum Equation

Newton's second law for a system, Eq. (3.2.2), is used as the basis for finding the linear-momentum equation for a control volume by use of Eq. (3.2.6). Let N be the linear momentum $m\mathbf{v}$ of the system, and let η be the linear momentum per unit mass $\rho \mathbf{v}/\rho$. Then by use of Eqs. (3.2.2) and (3.2.6)

$$\Sigma \mathbf{F} = \frac{d(m\mathbf{v})}{dt} = \frac{\partial}{\partial t} \int_{cv} \rho \mathbf{v} \, d\mathcal{V} + \int_{cs} \rho \mathbf{v} \mathbf{v} \cdot d\mathbf{A} \tag{3.3.8}$$

In words, the resultant force acting on a control volume is equal to the time rate of increase of linear momentum within the control volume plus the net efflux of linear momentum from the control volume.

Equations (3.3.1), (3.3.6), and (3.3.8) provide the relations for analysis of many of the problems of fluid mechanics. In effect, they provide a bridge from the solid-dynamics relations of the system to the convenient control-volume relations of fluid flow.

The basic control-volume equations are examined and applied next.

EXERCISE

3.3.1 The first law of thermodynamics, for steady flow, (a) accounts for all energy entering and leaving a control volume; (b) is an energy balance for a specified mass of fluid; (c) is an expression of the conservation of linear momentum; (d) is primarily concerned with heat transfer; (e) is restricted in its application to perfect gases.

3.4 CONTINUITY EQUATION

The use of Eq. (3.3.1) is developed in this section. First, consider steady flow through a portion of the stream tube of Fig. 3.4. The control volume comprises the walls of the stream tube between sections 1 and 2, plus the end areas of sections 1 and 2. Because the flow is steady, the first term of Eq. (3.3.1) is zero; hence

$$\int_{cs} \rho \mathbf{v} \cdot d\mathbf{A} = 0 \tag{3.4.1}$$

which states that the net mass outflow from the control volume must be zero. At section 1 the net mass outflow is $\rho_1 \mathbf{v}_1 \cdot d\mathbf{A}_1 = -\rho_1 v_1 \, dA_1$, and at section 2 it is $\rho_2 \mathbf{v}_2 \cdot d\mathbf{A}_2 = \rho_2 v_2 \, dA_2$. Since there is no flow through the wall of the stream tube,

$$\rho_1 v_1 \, dA_1 = \rho_2 v_2 \, dA_2 \tag{3.4.2}$$

is the continuity equation applied to two sections along a stream tube in steady flow.

For a collection of stream tubes, as in Fig. 3.5, if ρ_1 is the average density at section 1 and ρ_2 the average density at section 2,

$$\dot{m} = \rho_1 V_1 A_1 = \rho_2 V_2 A_2 \tag{3.4.3}$$

in which V_1, V_2 represent average velocities over the cross sections and $\dot{m}$ is the rate of mass flow. The average velocity over a cross section is given by

$$V = \frac{1}{A} \int v \, dA$$

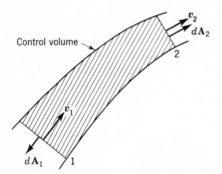

Control volume

Figure 3.4 Steady flow through a stream tube.

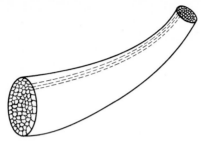

Figure 3.5 Collection of stream tubes between fixed boundaries.

If the *discharge Q* (also called *volumetric flow rate*, or *flow*) is defined as

$$Q = AV \qquad (3.4.4)$$

the continuity equation may take the form

$$\dot{m} = \rho_1 Q_1 = \rho_2 Q_2 \qquad (3.4.5)$$

For incompressible, steady flow

$$Q = A_1 V_1 = A_2 V_2 \qquad (3.4.6)$$

is a useful form of the equation.

For constant-density flow, steady or unsteady, Eq. (3.3.1) becomes

$$\int_{cs} \mathbf{v} \cdot d\mathbf{A} = 0 \qquad (3.4.7)$$

which states that the net volume efflux is zero (this implies that the control volume is filled with liquid at all times).

Example 3.2 At section 1 of a pipe system carrying water (Fig. 3.6) the velocity is 3.0 ft/s and the diameter is 2.0 ft. At section 2 the diameter is 3.0 ft. Find the discharge and the velocity at section 2.
From Eq. (3.4.6)

$$Q = V_1 A_1 = 3.0\pi = 9.42 \text{ ft}^3/\text{s}$$

and

$$V_2 = \frac{Q}{A_2} = \frac{9.42}{2.25\pi} = 1.33 \text{ ft/s}$$

For two- and three-dimensional-flow studies, differential expressions of the continuity equation must be used. For three-dimensional cartesian coordinates,

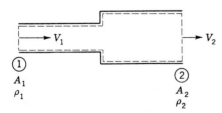

Figure 3.6 Control volume for flow through series pipes.

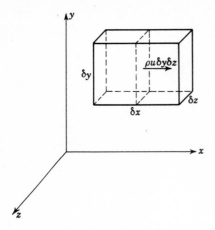

Figure 3.7 Control volume for derivation of three-dimensional continuity equation in cartesian coordinates.

Eq. (3.3.1) is applied to the control-volume element $\delta x\, \delta y\, \delta z$ of Fig. 3.7 with center at (x, y, z), where the velocity components in the x, y, z directions are u, v, w, respectively, and ρ is the density. Consider first the flux through the pair of faces normal to the x direction. On the right-hand face the flux outward is

$$\left[\rho u + \frac{\partial}{\partial x} (\rho u) \frac{\delta x}{2} \right] \delta y\, \delta z$$

since both ρ and u are assumed to vary continuously throughout the fluid. In the expression, $\rho u\, \delta y\, \delta z$ is the mass flux through the center face normal to the x axis. The second term is the rate of increase of mass flux, with respect to x, multiplied by the distance $\delta x/2$ to the right-hand face. Similarly, on the left-hand face the flux into the volume is

$$\left[\rho u - \frac{\partial}{\partial x} (\rho u) \frac{\delta x}{2} \right] \delta y\, \delta z$$

since the step is $-\delta x/2$. The net flux out through these two faces is

$$\frac{\partial}{\partial x} (\rho u)\, \delta x\, \delta y\, \delta z$$

The other two directions yield similar expressions; hence, the net mass outflow is

$$\left[\frac{\partial}{\partial x} (\rho u) + \frac{\partial}{\partial y} (\rho v) + \frac{\partial}{\partial z} (\rho w) \right] \delta x\, \delta y\, \delta z$$

which takes the place of the right-hand part of Eq. (3.3.1). The left-hand part of Eq. (3.3.1) becomes, for an element,

$$-\frac{\partial \rho}{\partial t} \delta x\, \delta y\, \delta z$$

When these two expressions are used in Eq. (3.3.1), after dividing through by the volume element and taking the limit as $\delta x\,\delta y\,\delta z$ approaches zero, the continuity equation at a point becomes

$$\frac{\partial}{\partial x}(\rho u)+\frac{\partial}{\partial y}(\rho v)+\frac{\partial}{\partial z}(\rho w)=-\frac{\partial\rho}{\partial t} \tag{3.4.8}$$

which must hold for every point in the flow, steady or unsteady, compressible or incompressible.† For incompressible flow, however, it simplifies to

$$\frac{\partial u}{\partial x}+\frac{\partial v}{\partial y}+\frac{\partial w}{\partial z}=0 \tag{3.4.9}$$

Equations (3.4.8) and (3.4.9) may be compactly written in vector notation. By using fixed unit vectors in x, y, z directions, $\mathbf{i}$, $\mathbf{j}$, $\mathbf{k}$, respectively, the operator ∇ is defined as (Sec. 2.2)

$$\nabla=\mathbf{i}\,\frac{\partial}{\partial x}+\mathbf{j}\,\frac{\partial}{\partial y}+\mathbf{k}\,\frac{\partial}{\partial z} \tag{3.4.10}$$

and the velocity vector $\mathbf{q}$ is given by

$$\mathbf{q}=\mathbf{i}u+\mathbf{j}v+\mathbf{k}w \tag{3.4.11}$$

Then

$$\nabla\cdot(\rho\mathbf{q})=\left(\mathbf{i}\,\frac{\partial}{\partial x}+\mathbf{j}\,\frac{\partial}{\partial y}+\mathbf{k}\,\frac{\partial}{\partial z}\right)\cdot(\mathbf{i}\rho u+\mathbf{j}\rho v+\mathbf{k}\rho w)$$

$$=\frac{\partial}{\partial x}(\rho u)+\frac{\partial}{\partial y}(\rho v)+\frac{\partial}{\partial z}(\rho w)$$

because $\mathbf{i}\cdot\mathbf{i}=1$, $\mathbf{i}\cdot\mathbf{j}=0$, etc. Equation (3.4.8) becomes

$$\nabla\cdot\rho\mathbf{q}=-\frac{\partial\rho}{\partial t} \tag{3.4.12}$$

and Eq. (3.4.9) becomes

$$\nabla\cdot\mathbf{q}=0 \tag{3.4.13}$$

The dot product $\nabla\cdot\mathbf{q}$ is called the *divergence* of the velocity vector $\mathbf{q}$. In words, it is the net volume efflux per unit volume at a point and must be zero for incompressible flow.

In two-dimensional flow, generally assumed to be in planes parallel to the xy plane, $w=0$, and there is no change with respect to z, so $\partial/\partial z=0$, which reduces the three-dimensional equations given for continuity.

---

† Equation (3.4.8) can be derived from Eq. (3.3.1) by application of Gauss' theorem. See L. Page, "Introduction to Theoretical Physics," 2d ed., pp. 32–36, Van Nostrand, Princeton, N.J., 1935.

Example 3.3 The velocity distribution for a two-dimensional incompressible flow is given by

$$u = -\frac{x}{x^2 + y^2} \qquad v = -\frac{y}{x^2 + y^2}$$

Show that it satisfies continuity.

In two dimensions the continuity equation is, from Eq. (3.4.9),

$$\frac{\partial u}{\partial x} + \frac{\partial v}{\partial y} = 0$$

Then

$$\frac{\partial u}{\partial x} = -\frac{1}{x^2 + y^2} + \frac{2x^2}{(x^2 + y^2)^2} \qquad \frac{\partial v}{\partial y} = -\frac{1}{x^2 + y^2} + \frac{2y^2}{(x^2 + y^2)^2}$$

and their sum does equal zero, satisfying continuity.

EXERCISES

3.4.1 The continuity equation may take the form (a) $Q = \rho A v$; (b) $\rho_1 A_1 = \rho_2 A_2$; (c) $p_1 A_1 v_1 = p_2 A_2 v_2$; (d) $\nabla \cdot \mathbf{p} = 0$; (e) $A_1 v_1 = A_2 v_2$.

3.4.2 The continuity equation (a) requires that Newton's second law of motion be satisfied at every point in the fluid; (b) expresses the relation between energy and work; (c) states that the velocity at a boundary must be zero relative to the boundary for a real fluid; (d) relates the momentum per unit volume for two points on a streamline; (e) relates mass rate of flow along a stream tube.

3.4.3 Water has an average velocity of 10 ft/s through a 24-in pipe. The discharge through the pipe, in cubic feet per second, is (a) 7.85; (b) 31.42; (c) 40; (d) 125.68; (e) none of these answers.

3.4.4 The continuity equation in ideal-fluid flow (a) states that the net rate of inflow into any small volume must be zero; (b) states that the energy is constant along a streamline; (c) states that the energy is constant everywhere in the fluid; (d) applies to irrotational flow only; (e) implies the existence of a velocity potential.

3.5 EULER'S EQUATION OF MOTION ALONG A STREAMLINE

In addition to the continuity equation, other general controlling equations are Euler's equation, Bernoulli's equation, the momentum equations, and the energy equation based on the first and second laws of thermodynamics. In this section Euler's equation is derived in differential form. In Sec. 3.6 it is integrated to obtain Bernoulli's equation. The first law of thermodynamics is then developed for steady flow, and some of the interrelations of the equations are explored, including an introduction to the second law of thermodynamics. In Chap. 7 Euler's equation is derived for general three-dimensional flow. Here it is restricted to flow along a streamline. Two derivations of Euler's equation of motion are presented; the first one is developed by use of the control volume for a small cylindrical element of fluid with axis along a streamline. This approach to a differential equation usually requires both the linear-momentum and the continuity equations to be utilized.

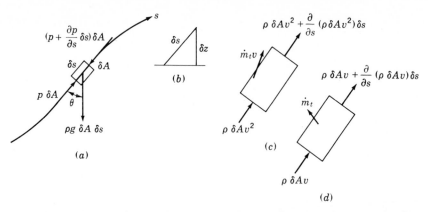

Figure 3.8 Application of continuity and momentum to flow through a control volume in the s direction.

The second approach uses Eq. (2.2.5), which is Newton's second law of motion in the form force equals mass times acceleration.

In Fig. 3.8 a prismatic control volume of very small size, with cross-sectional area δA and length δs, is selected. Fluid velocity is along the streamline s. By assuming that the viscosity is zero, or that the flow is frictionless, the only forces acting on the control volume in the x direction are the end forces and the weight. The momentum equation [Eq. (3.3.8)] is applied to the control volume for the s component

$$\Sigma F_s = \frac{\partial}{\partial t}(\rho v)\,\delta s\,\delta A + \sum_{cs} \rho v \mathbf{v} \cdot d\mathbf{A} \tag{3.5.1}$$

δs and δA are not functions of time. The forces acting are

$$\Sigma F_s = p\,\delta A - \left(p\,\delta A + \frac{\partial p}{\partial s}\,\delta s\,\delta A\right) - \rho g\,\delta s\,\delta A\cos\theta$$

$$= -\frac{\partial p}{\partial s}\,\delta s\,\delta A - \rho g\,\frac{\partial z}{\partial s}\,\delta s\,\delta A \tag{3.5.2}$$

since, as s increases, the vertical coordinate increases in such a manner that $\cos\theta = \partial z/\partial s$.

The net efflux of s momentum must consider flow through the cylindrical surface $\dot m_t$, as well as flow through the end faces (Fig. 3.8c).

$$\sum_{cs} \rho v \mathbf{v} \cdot d\mathbf{A} = \dot m_t v - \rho\,\delta A\,v^2 + \left[\rho\,\delta A\,v^2 + \frac{\partial}{\partial s}(\rho\,\delta A\,v^2)\,\delta s\right] \tag{3.5.3}$$

To determine the value of $\dot{m}_t$, the continuity equation (3.3.1) is applied to the control volume (Fig. 3.8d).

$$0 = \frac{\partial \rho}{\partial t} \delta A \, \delta s + \dot{m}_t + \frac{\partial}{\partial s}(\rho v) \, \delta A \, \delta s \tag{3.5.4}$$

Now, by eliminating $\dot{m}_t$ in Eqs. (3.5.3) and (3.5.4) and simplifying,

$$\sum_{cs} \rho v v \cdot d\mathbf{A} = \left(\rho v \frac{\partial v}{\partial s} - v \frac{\partial \rho}{\partial t} \right) \delta A \, \delta s \tag{3.5.5}$$

Next, by substituting Eqs. (3.5.2) and Eq. (3.5.5) in Eq. (3.5.1)

$$\left(\frac{\partial p}{\partial s} + \rho g \frac{\partial z}{\partial s} + \rho v \frac{\partial v}{\partial s} + \rho \frac{\partial v}{\partial t} \right) \delta A \, \delta s = 0$$

After dividing through by $\rho \, \delta A \, \delta s$ and by taking the limit as δA and δs approach zero, the equation reduces to

$$\frac{1}{\rho} \frac{\partial p}{\partial s} + g \frac{\partial z}{\partial s} + v \frac{\partial v}{\partial s} + \frac{\partial v}{\partial t} = 0 \tag{3.5.6}$$

Two assumptions have been made: (1) that the flow is along a streamline and (2) that the flow is frictionless. If the flow is also steady, Eq. (3.5.6) reduces to

$$\frac{1}{\rho} \frac{\partial p}{\partial s} + g \frac{\partial z}{\partial s} + v \frac{\partial v}{\partial s} = 0 \tag{3.5.7}$$

Now s is the only independent variable, and total differentials may replace the partials,

$$\frac{dp}{\rho} + g \, dz + v \, dv = 0 \tag{3.5.8}$$

Alternative derivation of Euler's equation along a streamline At a point in the fluid construct an element δs of the streamline, with δz taken in the vertical upward direction. The component of Eq. (2.2.5) in the s direction is

$$\frac{\partial p}{\partial s} = -\gamma \frac{\partial z}{\partial s} - \rho a_s \tag{3.5.9}$$

Since the acceleration component a_s of the particle is a function of distance s along the streamline and time,

$$a_s = \frac{dv}{dt} = \frac{\partial v}{\partial s} \frac{ds}{dt} + \frac{\partial v}{\partial t} = v \frac{\partial v}{\partial s} + \frac{\partial v}{\partial t}$$

as ds/dt is the time rate of displacement of the particle, which is its velocity v. After rearranging Eq. (3.5.9) with substitution of a_s, Eq. (3.5.6) is obtained. A frictionless fluid was assumed in deriving Eq. (2.2.5), and the component along s, the streamline, was taken in Eq. (3.5.9); hence, the same assumptions were made in obtaining Eq. (3.5.6).

Equation (3.5.8) is one form of Euler's equation, which required the three assumptions (1) frictionless, (2) along a streamline, and (3) steady flow. It can be integrated if ρ is a function of p or is constant. When ρ is constant, the Bernoulli equation is obtained. It is developed and applied in the next section.

EXERCISE

3.5.1 The assumptions about flow required in deriving the equation $gz + v^2/2 + \int dp/\rho = \text{const}$ are that it is (a) steady, frictionless, incompressible, along a streamline; (b) uniform, frictionless, along a streamline, ρ a function of p; (c) steady, uniform, incompressible, along a streamline; (d) steady, frictionless, ρ a function of p, along a streamline; (e) none of these answers.

3.6 THE BERNOULLI EQUATION

Integration of Eq. (3.5.8) for constant density yields the Bernoulli equation

$$gz + \frac{v^2}{2} + \frac{p}{\rho} = \text{const} \tag{3.6.1}$$

The constant of integration (called the *Bernoulli constant*) in general varies from one streamline to another but remains constant along a *streamline* in *steady*, *frictionless*, *incompressible* flow. These four assumptions are needed and must be kept in mind when applying this equation. Each term has the dimensions $(L/T)^2$ or the units meter-newtons per kilogram:

$$\frac{\text{m} \cdot \text{N}}{\text{kg}} = \frac{\text{m} \cdot \text{kg} \cdot \text{m/s}^2}{\text{kg}} = \frac{\text{m}^2}{\text{s}^2}$$

because $1 \text{ N} = 1 \text{ kg} \cdot \text{m/s}^2$. Therefore, Eq. (3.6.1) is energy per unit mass. When it is divided by g,

$$z + \frac{v^2}{2g} + \frac{p}{\gamma} = \text{const} \tag{3.6.2}$$

it can be interpreted as energy per unit weight, meter-newtons per newton (or foot-pounds per pound). This form is particularly convenient for dealing with liquid problems with a free surface. Multiplying Eq. (3.6.1) by ρ gives

$$\gamma z + \frac{\rho v^2}{2} + p = \text{const} \tag{3.6.3}$$

which is convenient for gas flow, since elevation changes are frequently unimportant and γz may be dropped out. In this form each term is meter-newtons per cubic meter, foot-pounds per cubic foot, or energy per unit volume.

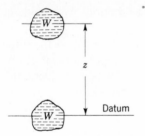

Figure 3.9 Potential energy.

Each of the terms of Bernoulli's equation may be interpreted as a form of energy. In Eq. (3.6.1) the first term is potential energy per unit mass. With reference to Fig. 3.9, the work needed to lift W newtons a distance z meters is Wz. The mass of W newtons is W/g kg; hence, the potential energy, in meter-newtons per kilogram, is

$$\frac{Wz}{W/g} = gz$$

The next term, $v^2/2$, is interpreted as follows. Kinetic energy of a particle of mass is $\delta m\, v^2/2$. To place this on a unit mass basis, divide by δm; thus $v^2/2$ is meter-newtons per kilogram kinetic energy.

The last term, p/ρ, is the *flow work* or *flow energy* per unit mass. Flow work is net work done by the fluid element on its surroundings while it is flowing. For example, in Fig. 3.10, imagine a turbine consisting of a vaned unit that rotates as fluid passes through it, exerting a torque on its shaft. For a small rotation the pressure drop across a vane times the exposed area of vane is a force on the rotor. When multiplied by the distance from center of force to axis of the rotor, a torque is obtained. Elemental work done is $p\,\delta A\,ds$ by $\rho\,\delta A\,ds$ units of mass of flowing fluid; hence, the work per unit mass is p/ρ. The three energy terms in Eq. (3.6.1) are referred to as *available energy*.

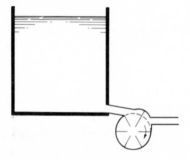

Figure 3.10 Work done by sustained pressure.

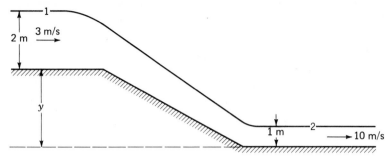

Figure 3.11 Open-channel flow.

By applying Eq. (3.6.2) to two points on a streamline,

$$z_1 + \frac{p_1}{\gamma} + \frac{v_1^2}{2g} = z_2 + \frac{p_2}{\gamma} + \frac{v_2^2}{2g} \tag{3.6.4}$$

or

$$z_1 - z_2 + \frac{p_1 - p_2}{\gamma} + \frac{v_1^2 - v_2^2}{2g} = 0$$

This equation shows that it is the difference in potential energy, flow energy, and kinetic energy that actually has significance in the equation. Thus $z_1 - z_2$ is independent of the particular elevation datum, as it is difference in elevation of the two points. Similarly, $p_1/\gamma - p_2/\gamma$ is the difference in pressure heads expressed in units of length of the fluid flowing and is not altered by the particular pressure datum selected. Since the velocity terms are not linear, their datum is fixed.

Example 3.4 Water is flowing in an open channel (Fig. 3.11) at a depth of 2 m and a velocity of 3 m/s. It then flows down a contracting chute into another channel where the depth is 1 m and the velocity is 10 m/s. Assuming frictionless flow, determine the difference in elevation of the channel floors.

The velocities are assumed to be uniform over the cross sections, and the pressures hydrostatic. The points 1 and 2 may be selected on the free surface, as shown, or they could be selected at other depths. If the difference in elevation of floors is y, Bernoulli's equation is

$$\frac{V_1^2}{2g} + \frac{p_1}{\gamma} + z_1 = \frac{V_2^2}{2g} + \frac{p_2}{\gamma} + z_2$$

Then $z_1 = y + 2$, $z_2 = 1$, $V_1 = 3$ m/s, $V_2 = 10$ m/s, and $p_1 = p_2 = 0$,

$$\frac{3^2}{2 \times 9.806} + 0 + y + 2 = \frac{10^2}{2 \times 9.806} + 0 + 1$$

and $y = 3.64$ m.

Modification of Assumptions Underlying Bernoulli's Equation

Under special conditions each of the four assumptions underlying Bernoulli's equation may be waived.

1. When all streamlines originate from a reservoir, where the energy content is everywhere the same, the constant of integration does not change from one streamline to another and points 1 and 2 for application of Bernoulli's equation may be selected arbitrarily, i.e., not necessarily on the same streamline.
2. In the flow of a gas, as in a ventilation system, where the change in pressure is only a small fraction (a few percent) of the absolute pressure, the gas may be considered incompressible. Equation (3.6.4) may be applied, with an average specific weight γ.
3. For unsteady flow with gradually changing conditions, e.g., emptying a reservoir, Bernoulli's equation may be applied without appreciable error.
4. Bernoulli's equation is of use in analyzing real-fluid cases by first neglecting viscous shear to obtain theoretical results. The resulting equation may then be modified by a coefficient, determined by experiment, which corrects the theoretical equation so that it conforms to the actual physical case. In general, losses are handled by use of the energy equation developed in Secs. 3.8 and 3.9.

Example 3.5 (*a*) Determine the velocity of efflux from the nozzle in the wall of the reservoir of Fig. 3.12. (*b*) Find the discharge through the nozzle.
(*a*) The jet issues as a cylinder with atmospheric pressure around its periphery. The pressure along its centerline is at atmospheric pressure for all practical purposes. Bernoulli's equation is applied between a point on the water surface and a point downstream from the nozzle,

$$\frac{V_1^2}{2g} + \frac{p_1}{\gamma} + z_1 = \frac{V_2^2}{2g} + \frac{p_2}{\gamma} + z_2$$

With the pressure datum as local atmospheric pressure, $p_1 = p_2 = 0$; with the elevation datum through point 2, $z_2 = 0$, $z_1 = H$. The velocity on the surface of the reservoir is zero (practically); hence,

$$0 + 0 + H = \frac{V_2^2}{2g} + 0 + 0$$

and
$$V_2 = \sqrt{2gH} = \sqrt{2 \times 9.806 \times 4} = 8.86 \text{ m/s}$$

which states that the velocity of efflux is equal to the velocity of free fall from the surface of the reservoir. This is known as *Torricelli's theorem*.
(*b*) The discharge Q is the product of velocity of efflux and area of stream,

$$Q = A_2 V_2 = \pi(0.05 \text{ m})^2(8.86 \text{ m/s}) = 0.07 \text{ m}^3/\text{s} = 70 \text{ l/s}$$

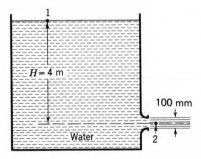

Figure 3.12 Flow through nozzle from reservoir.

Figure 3.13 Venturi meter.

Example 3.6 A venturi meter, consisting of a converging portion followed by a throat portion of constant diameter and then a gradually diverging portion, is used to determine rate of flow in a pipe (Fig. 3.13). The diameter at section 1 is 6.0 in, and at section 2 it is 4.0 in. Find the discharge through the pipe when $p_1 - p_2 = 3$ psi and oil, sp gr 0.90, is flowing.
From the continuity equation, Eq. (3.4.6),

$$Q = A_1 V_1 = A_2 V_2 = \frac{\pi}{16} V_1 = \frac{\pi}{36} V_2$$

in which Q is the discharge (volume per unit time flowing). By applying Eq. (3.6.4) for $z_1 = z_2$,

$$p_1 - p_2 = 3 \times 144 = 432 \text{ lb/ft}^2 \qquad \gamma = 0.90 \times 62.4 = 56.16 \text{ lb/ft}^3$$

$$\frac{p_1 - p_2}{\gamma} = \frac{V_2^2}{2g} - \frac{V_1^2}{2g} \qquad \text{or} \qquad \frac{432}{56.16} = \frac{Q^2}{\pi^2} \frac{1}{2g} (36^2 - 16^2)$$

Solving for discharge gives $Q = 2.20$ cfs.

EXERCISES

3.6.1 The equation $z + p/\gamma + v^2/2g = C$ has the units of (a) $m \cdot N/s$; (b) N; (c) $m \cdot N/kg$; (d) $m \cdot N/m^3$; (e) $m \cdot N/N$.
3.6.2 The work that a liquid is capable of doing by virtue of its sustained pressure is, in foot-pounds per pound, (a) z; (b) p; (c) p/γ; (d) $v^2/2g$; (e) $\sqrt{2gh}$.
3.6.3 The velocity head is (a) $v^2/2g$; (b) z; (c) v; (d) $\sqrt{2gh}$; (e) none of these answers.

3.7 REVERSIBILITY, IRREVERSIBILITY, AND LOSSES

A process may be defined as the path of the succession of states through which the system passes, such as the changes in velocity, elevation, pressure, density, temperature, etc. The expansion of air in a cylinder as the piston moves out and heat is transferred through the walls is an example of a process. Normally, the process causes some change in the surroundings, e.g., displacing it or transferring heat to or from its boundaries. When a process can be made to take place in such a manner that it can be *reversed*, i.e., made to return to its original state without a final change in either the system or its surroundings, it is said to be *reversible*. In any actual flow of a real fluid or change in a mechanical system, the effects of viscous friction, Coulomb friction, unrestrained expansion, hysteresis, etc., prohibit the process from being reversible. It is, however, an ideal to be strived for in design processes, and their efficiency is usually defined in terms of their nearness to reversibility.

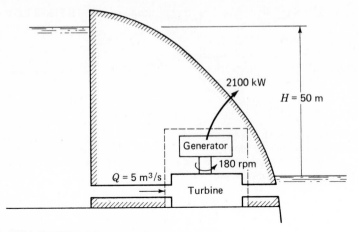

Figure 3.14 Irreversibility in hydroelectric plant.

When a certain process has a sole effect upon its surroundings that is equivalent to the raising of a weight, it is said to have done *work* on its surroundings. Any actual process is *irreversible*. The difference between the amount of work a substance can do by changing from one state to another state along a path reversibly and the actual work it produces for the same path is the *irreversibility* of the process. It may be defined in terms of work per unit mass or weight or work per unit time. Under certain conditions the irreversibility of a process is referred to as its *lost work*,† i.e., the loss of ability to do work because of friction and other causes. In the Bernoulli equation (3.6.4), in which all losses are neglected, all terms are *available-energy terms*, or *mechanical-energy terms*, in that they are directly able to do work by virtue of potential energy, kinetic energy, or sustained pressure. In this book, when losses are referred to, they mean irreversibility, or lost work, or the transformation of available energy into thermal energy.

Example 3.7 A hydroelectric plant (Fig. 3.14) has a difference in elevation from head water to tail water of $H = 50$ m and a flow $Q = 5$ m³/s of water through the turbine. The turbine shaft rotates at 180 rpm, and the torque in the shaft is measured to be $T = 1.16 \times 10^5$ N·m. Output of the generator is 2100 kW. Determine (*a*) the reversible power for the system; (*b*) the irreversibility, or losses, in the system; (*c*) the losses and the efficiency in the turbine and in the generator.
(*a*) The potential energy of the water is 50 m·N/N. Hence, for perfect conversion the reversible power is

$$\gamma QH = (9806 \text{ N/m}^3)(5 \text{ m}^3/\text{s})(50 \text{ m·N/N}) = 2{,}451{,}500 \text{ N·m/s} = 2451.5 \text{ kW}$$

(*b*) The irreversibility, or lost power, in the system is the difference between the power into and out of the system, or

$$2451.5 - 2100 = 351.5 \text{ kW}$$

---

† Reference to a text on thermodynamics is advised for a full discussion of these concepts.

(c) The rate of work by the turbine is the product of the shaft torque and the rotational speed:

$$T\omega = 1.16 \times 10^5 \text{ N} \cdot \text{m} \frac{180(2\pi)}{60} \text{ s}^{-1} = 2186.5 \text{ kW}$$

The irreversibility through the turbine is then $2451.5 - 2186.5 = 265.0$ kW, or, when expressed as lost work per unit weight of fluid flowing,

$$(265.0 \text{ kW}) \frac{1000 \text{ N} \cdot \text{m/s}}{1 \text{ kW}} \frac{1}{9806 \text{ N/m}^3} \frac{1}{5 \text{ m}^3/\text{s}} = 5.4 \text{ m} \cdot \text{N/N}$$

The generator power loss is $2186.5 - 2100 = 86.5$ kW, or

$$\frac{86.5(1000)}{9806(5)} = 1.76 \text{ m} \cdot \text{N/N}$$

Efficiency of the turbine η_t is

$$\eta_t = 100 \frac{50 \text{ m} \cdot \text{N/N} - 5.4 \text{ m} \cdot \text{N/N}}{50 \text{ m} \cdot \text{N/N}} = 89.19\%$$

and efficiency of the generator η_g is

$$\eta_g = 100 \frac{50 - 5.4 - 1.76}{50 - 5.4} = 96.05\%$$

3.8 THE STEADY-STATE ENERGY EQUATION

When Eq. (3.3.6) is applied to steady flow through a control volume similar to Fig. 3.15, the volume integral drops out and it becomes

$$\frac{\delta Q_H}{\delta t} + \left(\frac{p_1}{\rho_1} + g z_1 + \frac{v_1^2}{2} + u_1 \right) \rho_1 v_1 A_1 = \frac{\delta W_s}{\delta t} + \left(\frac{p_2}{\rho_2} + g z_2 + \frac{v_2^2}{2} + u_2 \right) \rho_2 v_2 A_2$$

Since the flow is steady in this equation, it is convenient to divide through by the mass per second flowing through the system $\rho_1 A_1 v_1 = \rho_2 A_2 v_2$, getting

$$q_H + \frac{p_1}{\rho_1} + g z_1 + \frac{v_1^2}{2} + u_1 = w_s + \frac{p_2}{\rho_2} + g z_2 + \frac{v_2^2}{2} + u_2 \qquad (3.8.1)$$

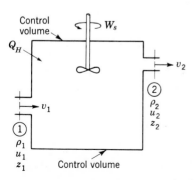

Figure 3.15 Control volume with flow across control surface normal to surface.

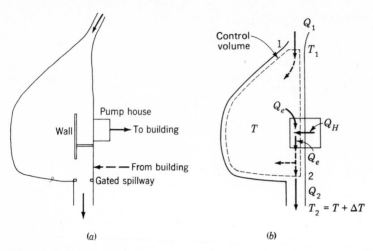

Figure 3.16 Cooling-water system.

q_H is the heat added per unit mass of fluid flowing, and w_s is the shaft work per unit mass of fluid flowing. This is the *energy* equation for steady flow through a control volume.

> **Example 3.8** The cooling-water plant for a large building is located on a small lake fed by a stream, as shown in Fig. 3.16a. The design low stream flow is 5 cfs, and at this condition the only outflow from the lake is 5 cfs via a gated structure near the discharge channel for the cooling-water system. The temperature of the incoming stream is 80°F. The flow rate of the cooling system is 10 cfs, and the building's heat exchanger raises the cooling-water temperature by 10°F. What is the temperature of the cooling water recirculated through the lake, neglecting heat losses to the atmosphere and lake bottom, if these conditions exist for a prolonged period?
>
> The control volume is shown in Fig. 3.16b with the variables volumetric flow rate Q and temperature T. There is no change in pressure, density, velocity, or elevation from section 1 to 2. Equation (3.3.6) applied to the control volume is
>
> $$\frac{\delta Q_H}{\delta t} + u_1 \rho Q_1 = u_2 \rho Q_2$$
>
> in which $\delta Q_H/\delta t$ is the time rate of heat addition by the heat exchanger. The intrinsic energy per unit mass at constant pressure and density is a function of temperature only; it is $u_2 - u_1 = c(T_2 - T_1)$, in which c is the specific heat or heat capacity of water. Hence, the energy equation applied to the control volume is
>
> $$\frac{\delta Q_H}{\delta t} = c(T_2 - T_1)\rho Q_1$$
>
> Similarly, the heat added in the heat exchanger is given by
>
> $$\frac{\delta Q_H}{\delta t} = c\,\Delta T\,\rho Q_e$$

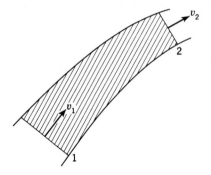

Figure 3.17 Steady-stream tube as control volume.

in which $\Delta T = 10$ is the temperature rise and $Q_e = 10$ cfs is the volumetric flow rate through the heat exchanger. Thus

$$c \, \Delta T \, \rho Q_e = c(T_2 - T_1)\rho Q_1$$

or
$$T_2 = T_1 + \frac{\Delta T \, Q_e}{Q_1} = 80 + \frac{10(10)}{5} = 100°F$$

Since $T_2 = T + \Delta T$, the lake temperature T is 90°F.

The energy equation (3.8.1) in differential form, for flow through a stream tube (Fig. 3.17) with no shaft work, is

$$d\frac{p}{\rho} + g \, dz + v \, dv + du - dq_H = 0 \qquad (3.8.2)$$

Rearranging gives

$$\frac{dp}{\rho} + g \, dz + v \, dv + du + p \, d\frac{1}{\rho} - dq_H = 0 \qquad (3.8.3)$$

For frictionless flow the sum of the first three terms equals zero from the Euler equation (3.5.8); the last three terms are one form of the first law of thermodynamics for a system,

$$dq_H = p \, d\frac{1}{\rho} + du \qquad (3.8.4)$$

Now, for reversible flow, *entropy s* per unit mass is defined by

$$ds = \left(\frac{dq_H}{T}\right)_{rev} \qquad (3.8.5)$$

in which T is the absolute temperature. Entropy is shown to be a fluid property in texts on thermodynamics. In this equation it may have the units Btu per slug per degree Rankine or foot-pounds per slug per degree Rankine, as heat may be expressed in foot-pounds (1 Btu = 778 ft·lb). In SI units s is in kilocalories per kilogram per kelvin or joules per kilogram per kelvin (1 kcal = 4187 J). Since

Eq. (3.8.4) is for a frictionless fluid (reversible), dq_H can be eliminated from Eqs. (3.8.4) and (3.8.5),

$$T \, ds = du + p \, d \frac{1}{\rho} \qquad (3.8.6)$$

which is a very important thermodynamic relation. Although it was derived for a reversible process, since all terms are thermodynamic properties, it must also hold for irreversible-flow cases as well. By use of Eq. (3.8.6), together with the *Clausius inequality*† and various combinations of Euler's equation and the first law, a clearer understanding of entropy and losses is gained. Equation (3.8.6) is one form of the second law of thermodynamics.

EXERCISES

3.8.1 A *reversible* process requires that (a) there be no heat transfer; (b) Newton's law of viscosity be satisfied; (c) temperature of system and surroundings be equal; (d) there be no viscous or Coulomb friction in the system; (e) heat transfer occurs from surroundings to system only.

3.8.2 Entropy, for reversible flow, is defined by the expression (a) $ds = du + p \, d(1/\rho)$; (b) $ds = T \, dq_H$; (c) $s = u + pv_s$; (d) $ds = dq_H/T$; (e) none of these answers.

3.9 INTERRELATIONS BETWEEN EULER'S EQUATION AND THE THERMODYNAMIC RELATIONS

The first law in differential form, from Eq. (3.8.3), with shaft work included, is

$$dw_s + \frac{dp}{\rho} + v \, dv + g \, dz + du + p \, d \frac{1}{\rho} - dq_H = 0 \qquad (3.9.1)$$

Substituting for $du + p \, d(1/\rho)$ in Eq. (3.8.6) gives

$$dw_s + \frac{dp}{\rho} + v \, dv + g \, dz + T \, ds - dq_H = 0 \qquad (3.9.2)$$

The Clausius inequality states that

$$ds \geq \frac{dq_H}{T}$$

or
$$T \, ds \geq dq_H \qquad (3.9.3)$$

Thus $T \, ds - dq_H \geq 0$. The equals sign applies to a reversible process. If the quantity called losses or irreversibilities is identified as

$$d \text{ (losses)} \equiv T \, ds - dq_H \qquad (3.9.4)$$

---

† See any text on thermodynamics.

it is seen that d (losses) is positive in irreversible flow, is zero in reversible flow, and can never be negative. Substituting Eq. (3.9.4) into Eq. (3.9.2) yields

$$dw_s + \frac{dp}{\rho} + v\,dv + g\,dz + d\,(\text{losses}) = 0 \qquad (3.9.5)$$

This is a most important form of the energy equation. In general, the losses must be determined by experimentation. It implies that some of the available energy is converted into intrinsic energy during an irreversible process. This equation, in the absence of the shaft work, differs from Euler's equation by the loss term only. In integrated form,

$$\frac{v_1^2}{2} + gz_1 = \int_1^2 \frac{dp}{\rho} + \frac{v_2^2}{2} + gz_2 + w_s + \text{losses}_{1-2} \qquad (3.9.6)$$

If work is done on the fluid in the control volume, as with a pump, then w_s is negative. Section 1 is upstream, and section 2 is downstream.

EXERCISE

3.9.1 The equation d (losses) $= T\,ds$ is restricted to (a) isentropic flow; (b) reversible flow; (c) adiabatic flow; (d) perfect-gas flow; (e) none of these answers.

3.10 APPLICATION OF THE ENERGY EQUATION TO STEADY FLUID-FLOW SITUATIONS

For an incompressible fluid Eq. (3.9.6) may be simplified to

$$\frac{p_1}{\gamma} + \frac{v_1^2}{2g} + z_1 = \frac{p_2}{\gamma} + \frac{v_2^2}{2g} + z_2 + \text{losses}_{1-2} \qquad (3.10.1)$$

in which each term now is energy in foot-pounds per pound or meter-newtons per newton, including the loss term. The work term has been omitted but may be inserted if needed.

Kinetic-Energy Correction Factor

In dealing with flow situations in open- or closed-channel flow, the so-called *one-dimensional* form of analysis is frequently used. The whole flow is considered to be one large stream tube with average velocity V at each cross section. The kinetic energy per unit weight given by $V^2/2g$, however, is not the average of $v^2/2g$ taken over the cross section. It is necessary to compute a correction factor α for $V^2/2g$, so that $\alpha V^2/2g$ is the average kinetic energy per unit weight passing the

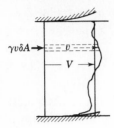

Figure 3.18 Velocity distribution and average velocity.

section. Referring to Fig. 3.18, the kinetic energy passing the cross section per unit time is

$$\gamma \int_A \frac{v^2}{2g} v \, dA$$

in which $\gamma v \, \delta A$ is the weight per unit time passing δA and $v^2/2g$ is the kinetic energy per unit weight. By equating this to the kinetic energy per unit time passing the section, in terms of $\alpha V^2/2g$,

$$\alpha \frac{V^2}{2g} \gamma V A = \gamma \int_A \frac{v^3}{2g} \, dA$$

By solving for α, the *kinetic-energy correction factor*,

$$\alpha = \frac{1}{A} \int_A \left(\frac{v}{V}\right)^3 dA \tag{3.10.2}$$

The energy equation becomes

$$z_1 + \frac{p_1}{\gamma} + \alpha_1 \frac{V_1^2}{2g} = z_2 + \frac{p_2}{\gamma} + \alpha_2 \frac{V_2^2}{2g} + \text{losses}_{1-2} \tag{3.10.3}$$

For laminar flow in a pipe, $\alpha = 2$, as shown in Sec. 5.2. For turbulent flow[†] in a pipe, α varies from about 1.01 to 1.10 and is usually neglected except for precise work.

Example 3.9 The velocity distribution in turbulent flow in a pipe is given approximately by Prandtl's one-seventh-power law,

$$\frac{v}{v_{max}} = \left(\frac{y}{r_0}\right)^{1/7}$$

with y the distance from the pipe wall and r_0 the pipe radius. Find the kinetic-energy correction factor.

The average velocity V is expressed by

$$\pi r_0^2 V = 2\pi \int_0^{r_0} rv \, dr$$

† V. L. Streeter, The Kinetic Energy and Momentum Correction Factors for Pipes and Open Channels of Great Width, *Civ. Eng. N.Y.*, vol. 12, no. 4, pp. 212–213, 1942.

in which $r = r_0 - y$. By substituting for r and v,

$$\pi r_0^2 V = 2\pi v_{max} \int_0^{r_0} (r_0 - y)\left(\frac{y}{r_0}\right)^{1/7} dy = \pi r_0^2 v_{max} \frac{98}{120}$$

or

$$V = \frac{98}{120} v_{max} \qquad \frac{v}{V} = \frac{120}{98}\left(\frac{y}{r_0}\right)^{1/7}$$

By substituting into Eq. (3.10.2)

$$\alpha = \frac{1}{\pi r_0^2} \int_0^{r_0} 2\pi r \left(\frac{120}{98}\right)^3 \left(\frac{y}{r_0}\right)^{3/7} dr$$

$$= 2\left(\frac{120}{98}\right)^3 \frac{1}{r_0^2} \int_0^{r_0} (r_0 - y)\left(\frac{y}{r_0}\right)^{3/7} dy = 1.06$$

All the terms in the energy equation (3.10.1) except the term losses are *available energy*. For real fluids flowing through a system, the available energy decreases in the downstream direction; it is available to do work, as in passing through a water turbine. A plot showing the available energy along a stream tube portrays the *energy grade line* (see Sec. 10.2). A plot of the two terms $z + p/\gamma$ along a stream tube portrays the *piezometric head*, or *hydraulic grade line*. The energy grade line always slopes downward in real-fluid flow, except at a pump or other source of energy. Reductions in energy grade line are also referred to as *head losses*.

Example 3.10 The siphon of Fig. 3.19 is filled with water and is discharging at 150 l/s. Find the losses from point 1 to point 3 in terms of velocity head $V^2/2g$. Find the pressure at point 2 if two-thirds of the losses occur between points 1 and 2.

The energy equation (3.10.1) is first applied to the control volume consisting of all the water in the system upstream from point 3, with elevation datum at point 3 and gage pressure zero for pressure datum:

$$\frac{V_1^2}{2g} + \frac{p_1}{\gamma} + z_1 = \frac{V_3^2}{2g} + \frac{p_3}{\gamma} + z_3 + losses_{1-3}$$

or

$$0 + 0 + 1.5 = \frac{V_3^2}{2g} + 0 + 0 + K\frac{V_3^2}{2g}$$

in which the losses from 1 to 3 are expressed as $KV_3^2/2g$. From the discharge

$$V_3 = \frac{Q}{A} = \frac{150 \text{ l/s}}{\pi(0.1^2)} \frac{1 \text{ m}^3/\text{s}}{1000 \text{ l/s}} = 4.77 \text{ m/s}$$

and $V_3^2/2g = 1.16$ m. Hence, $K = 0.29$, and the losses are $0.29 \, V_3^2/2g = 0.34$ m·N/N.

The energy equation applied to the control volume between points 1 and 2, with losses $\frac{2}{3}KV_3^2/2g = 0.23$ m, is

$$0 + 0 + 0 = 1.16 + \frac{p_2}{\gamma} + 2 + 0.23$$

The pressure head at 2 is -3.39 m H_2O, or $p_2 = -33.2$ kPa.

Example 3.11 The device shown in Fig. 3.20 is used to determine the velocity of liquid at point 1. It is a tube with its lower end directed upstream and its other leg vertical and open to the

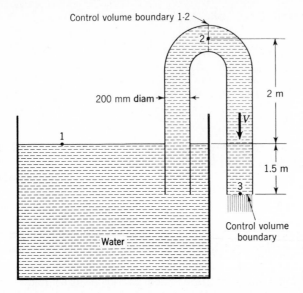

Control volume boundary 1-2

200 mm diam

2 m

1.5 m

Water

Control volume boundary

Figure 3.19 Siphon.

atmosphere. The impact of liquid against the opening 2 forces liquid to rise in the vertical leg to the height Δz above the free surface. Determine the velocity at 1.

Point 2 is a stagnation point, where the velocity of the flow is reduced to zero. This creates an impact pressure, called the dynamic pressure, which forces the fluid into the vertical leg. By writing the energy equation between points 1 and 2, neglecting losses, which are very small,

$$\frac{V_1^2}{2g} + \frac{p_1}{\gamma} + 0 = 0 + \frac{p_2}{\gamma} + 0$$

p_1/γ is given by the height of fluid above point 1 and equals k ft of fluid flowing. p_2/γ is given by the manometer as $k + \Delta z$, neglecting capillary rise. After substituting these values into the equation,

$$\frac{V_1^2}{2g} = \Delta z \qquad \text{and} \qquad V_1 = \sqrt{2g\,\Delta z}$$

This is the pitot tube in a simple form.

Examples of compressible flow are given in Chap. 6.

EXERCISES

3.10.1 The kinetic-energy correction factor (a) applies to the continuity equation; (b) has the units of velocity head;

(c) is expressed by $\dfrac{1}{A} \int_A \dfrac{v}{V} \, dA$; (d) is expressed by $\dfrac{1}{A} \int_A \left(\dfrac{v}{V}\right)^2 \, dA$;

(e) is expressed by $\dfrac{1}{A} \int_A \left(\dfrac{v}{V}\right)^3 \, dA$.

3.10.2 The kinetic-energy correction factor for the velocity distribution given by Fig. 1.1 is (a) 0; (b) 1; (c) $\frac{4}{3}$; (d) 2; (e) none of these answers.

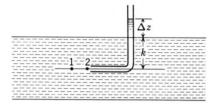

Figure 3.20 Pitot tube.

3.10.3 A glass tube with a 90° bend is open at both ends. It is inserted into a flowing stream of oil, sp gr 0.90, so that one opening is directed upstream and the other is directed upward. Oil inside the tube is 50 mm higher than the surface of flowing oil. The velocity measured by the tube is, in meters per second, (a) 0.89; (b) 0.99; (c) 1.10; (d) 1.40; (e) none of these answers.

3.10.4 In Fig. 8.3a the gage difference R' for $v_1 = 5$ ft/s, $S = 0.08$, $S_0 = 1.2$ is, in feet, (a) 0.39; (b) 0.62; (c) 0.78; (d) 1.17; (e) none of these answers.

3.10.5 The theoretical velocity of oil, sp gr 0.75, flowing from an orifice in a reservoir under a head of 4 m is, in meters per second, (a) 6.7; (b) 8.86; (c) 11.8; (d) not determinable from data given; (e) none of these answers.

3.10.6 In which of the following cases is it possible for flow to occur from low pressure to high pressure? (a) Flow through a converging section; (b) adiabatic flow in a horizontal pipe; (c) flow of a liquid upward in a vertical pipe; (d) flow of air downward in a pipe; (e) impossible in a constant-cross-section conduit.

3.10.7 If all losses are neglected, the pressure at the summit of a siphon (a) is a minimum for the siphon; (b) depends upon height of summit above upstream reservoir only; (c) is independent of the length of the downstream leg; (d) is independent of the discharge through the siphon; (e) is independent of the liquid density.

3.11 APPLICATIONS OF THE LINEAR-MOMENTUM EQUATION

Newton's second law, the equation of motion, was developed into the linear-momentum equation in Sec. 3.3,

$$\Sigma \mathbf{F} = \frac{\partial}{\partial t} \int_{cv} \rho \mathbf{v} \, d\mathcal{V} + \int_{cs} \rho \mathbf{v} \mathbf{v} \cdot d\mathbf{A} \qquad (3.11.1)$$

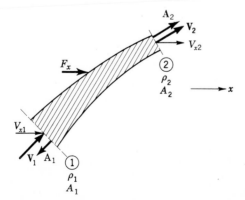

Figure 3.21 Control volume with uniform inflow and outflow normal to control surface.

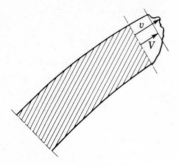

Figure 3.22 Nonuniform flow through a control surface.

This vector relation may be applied for any component, say the x direction, reducing to

$$\Sigma F_x = \frac{\partial}{\partial t} \int_{cv} \rho v_x \, d\mathcal{U} + \int_{cs} \rho v_x \mathbf{v} \cdot d\mathbf{A} \qquad (3.11.2)$$

In selecting the arbitrary control volume, it is generally advantageous to take the surface normal to the velocity wherever it cuts across the flow. In addition, if the velocity is constant over the surface, the surface integral can be dispensed with. In Fig. 3.21, with the control surface as shown and with steady flow, the resultant force F_x acting on the control volume is given by Eq. (3.11.2) as

$$F_x = \rho_2 A_2 V_2 V_{x2} - \rho_1 A_1 V_1 V_{x1}$$

or
$$F_x = \rho Q(V_{x2} - V_{x1})$$

as the mass per second entering and leaving is $\rho Q = \rho_1 Q_1 = \rho_2 Q_2$.

When the velocity varies over a plane cross section of the control surface, by introduction of a momentum correction factor β, the average velocity may be utilized, Fig. 3.22,

$$\int_A \rho v^2 \, dA = \beta \rho V^2 A \qquad (3.11.3)$$

in which β is dimensionless. Solving for β yields

$$\beta = \frac{1}{A} \int_A \left(\frac{v}{V}\right)^2 dA \qquad (3.11.4)$$

which is analogous to α, the kinetic-energy correction factor, Eq. (3.10.2). For laminar flow in a straight round tube, β may be shown to equal $\frac{4}{3}$ in Sec. 5.2. It equals 1 for uniform flow and cannot have a value less than 1.

In applying Eq. (3.11.1) or a component equation such as Eq. (3.11.2), care should be taken to define the control volume and the forces acting on it clearly. Also, the sign of the inflow or outflow term must be carefully evaluated. The first example is an unsteady one using Eq. (3.11.2) and the general continuity equation (3.3.1).

Example 3.12 The horizontal pipe of Fig. 3.23 is filled with water for the distance x. A jet of constant velocity V_1 impinges against the filled portion. Fluid frictional force on the pipe wall is

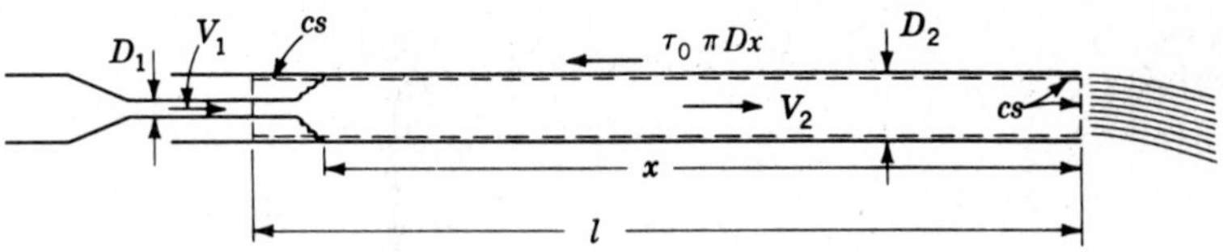

Figure 3.23 Jet impact on pipe flowing full over partial length.

given by $\tau_0 \pi Dx$, with $\tau_0 = \rho f V_2^2 / 8$ [see Eq. (5.10.2)]. Determine the equations to analyze this flow condition when initial conditions are known; that is, $t = 0$, $x = x_0$, $V_2 = V_{2_0}$. Specifically, for $V_1 = 20$ m/s, $D_1 = 60$ mm, $V_{2_0} = 500$ mm/s, $D_2 = 250$ mm, $x_0 = 100$ m, $\rho = 1000$ kg/m^3, and $f = 0.02$, find the rate of change of V_2 and x with time.

The continuity and momentum equations are used to analyze this unsteady-flow problem. Take as control volume the inside surface of the pipe, with the two end sections l ft apart, as shown. The continuity equation

$$0 = \frac{\partial}{\partial t} \int_{cv} \rho \, d\mathcal{V} + \int_{cs} \rho \mathbf{v} \cdot d\mathbf{A}$$

becomes, using $A_1 = \pi D_1^2 / 4$, $A_2 = \pi D_2^2 / 4$,

$$\frac{\partial}{\partial t} [\rho A_2 x + \rho A_1 (l - x)] + \rho (V_2 A_2 - V_1 A_1) = 0$$

After simplifying,

$$\frac{\partial x}{\partial t} (A_2 - A_1) + V_2 A_2 - V_1 A_1 = 0$$

The momentum equation for the horizontal direction x,

$$\Sigma F_x = \frac{\partial}{\partial t} \int_{cv} \rho v_x \, d\mathcal{V} + \int_{cs} \rho v_x \mathbf{v} \cdot d\mathbf{A}$$

becomes

$$-\rho \frac{f V_2^2 \pi D_2 x}{8} = \frac{\partial}{\partial t} [\rho A_2 x V_2 + \rho A_1 (l - x) V_1] + \rho A_2 V_2^2 - \rho A_1 V_1^2$$

which simplifies to

$$\frac{f V_2^2 \pi D_2 x}{8} + A_2 \frac{\partial}{\partial t} (x V_2) - A_1 V_1 \frac{\partial x}{\partial t} + A_2 V_2^2 - A_1 V_1^2 = 0$$

As t is the only independent variable, the partials may be replaced by totals. The continuity equation is

$$\frac{dx}{dt} = -\frac{V_2 A_2 - V_1 A_1}{A_2 - A_1}$$

By expanding the momentum equation, and substituting for dx/dt, it becomes

$$\frac{dV_2}{dt} = \frac{1}{x A_2} \left[A_1 V_1^2 - A_2 V_2^2 - \frac{f V_2^2 \pi D_2 x}{8} + \frac{(A_2 V_2 - A_1 V_1)^2}{A_2 - A_1} \right]$$

These two equations, being nonlinear, can be solved simultaneously by numerical methods, such as Runge-Kutta methods described in Appendix B, when initial conditions are known. The rate of change of x and V_2 can be found directly from the equations for the specific problem

$$\frac{dx}{dt} = 0.692 \text{ m/s} \qquad \frac{dV_2}{dt} = 0.0496 \text{ m/s}^2$$

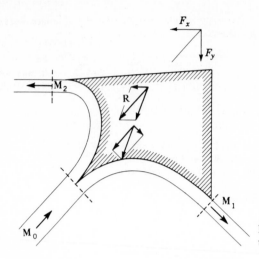

Figure 3.24 Solution of linear-momentum problem by addition of vectors.

Example 3.13 In Fig. 3.24 a fluid jet impinges on a body as shown; the momentum per second of each of the jets is given by $\mathbf{M}$ and is the vector located at the center of the jets. By vector addition find the resultant force needed to hold the body at rest.

The vector form of the linear-momentum equation (3.11.1) is to be applied to a control volume comprising the fluid bounded by the body and the three dotted cross sections. As the problem is steady, Eq. (3.11.1) reduces to

$$\Sigma\mathbf{F} = \int \rho\mathbf{v}\mathbf{v}\cdot d\mathbf{A} = \sum_{\text{out}} \mathbf{M}_i$$

By taking $\mathbf{M}_1$ and $\mathbf{M}_0$ first, the vector $\mathbf{M}_1 - \mathbf{M}_0$ is the net momentum efflux for these two vectors, shown graphically on their lines of action. The resultant of these two vectors is then added to momentum efflux $\mathbf{M}_2$, along its line of action, to obtain $\mathbf{R}$. $\mathbf{R}$ is the momentum efflux over the

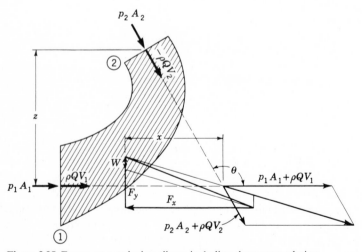

Figure 3.25 Forces on a reducing elbow, including the vector solution.

control surface and is just equal to the force that must be exerted on the control surface. The same force must then be exerted on the body to resist the control-volume force on it.

Example 3.14 The reducing bend of Fig. 3.25 is in a vertical plane. Water is flowing, $D_1 = 6$ ft, $D_2 = 4$ ft, $Q = 300$ cfs, $W = 18,000$ lb, $z = 10$ ft, $\theta = 120°$, $p_1 = 40$ psi, $x = 6$ ft, and losses through the bend are $0.5\ V_2^2/2g$ ft·lb/lb. Determine F_x, F_y, and the line of action of the resultant force. $\beta_1 = \beta_2 = 1$.

The inside surface of the reducing bend comprises the control-volume surface for the portion of the surface with no flow across it. The normal sections 1 and 2 complete the control surface.

$$V_1 = \frac{Q}{A_1} = \frac{300}{\pi(6^2)/4} = 10.61 \text{ ft/s} \qquad V_2 = \frac{Q}{A_2} = \frac{300}{\pi(4^2)/4} = 23.87 \text{ ft/s}$$

By application of the energy equation, Eq. (3.10.1),

$$\frac{p_1}{\gamma} + \frac{V_1^2}{2g} + z_1 = \frac{p_2}{\gamma} + \frac{V_2^2}{2g} + z_2 + \text{losses}_{1-2}$$

$$\frac{40 \times 144}{62.4} + \frac{10.61^2}{64.4} + 0 = \frac{p_2}{62.4} + \frac{23.87^2}{64.4} + 10 + 0.5 \times \frac{23.87^2}{64.4}$$

from which $p_2 = 4420$ lb/ft² $= 30.7$ psi.

To determine F_x, Eq. (3.11.2) yields

$$p_1 A_1 - p_2 A_2 \cos\theta - F_x = \rho Q(V_2 \cos\theta - V_1)$$

$$40 \times 144\pi(3^2) - 4420\pi(2^2)\cos 120° - F_x = 1.935 \times 300(23.87 \cos 120° - 10.61)$$

Since $\cos 120° = -0.5$,

$$162,900 + 27,750 - F_x = 580.5(-11.94 - 10.61)$$

$$F_x = 203,740 \text{ lb}$$

For the y direction

$$\Sigma F_y = \rho Q(V_{y2} - V_{y1})$$

$$F_y - W - p_2 A_2 \sin\theta = \rho Q V_2 \sin\theta$$

$$F_y - 18,000 - 4420\pi(2^2)\sin 120° = 1.935 \times 300 \times 23.87 \sin 120°$$

$$F_y = 78,100 \text{ lb}$$

To find the line of action of the resultant force, using the momentum flux vectors (Fig. 3.25), $\rho Q V_1 = 6160$ lb, $\rho Q V_2 = 13,860$ lb, $p_1 A_1 = 162,900$ lb, $p_2 A_2 = 55,560$ lb. Combining these vectors and the weight W in Fig. 3.25 yields the final force, 218,000 lb, which must be opposed by F_x and F_y.

As demonstrated in Example 3.14, a change in direction of a pipeline causes forces to be exerted on the line unless the bend or elbow is anchored in place. These forces are due to both static pressure in the line and dynamic reactions in the turning fluid stream. Expansion joints are placed in large pipelines to avoid stress in the pipe in an axial direction, whether caused by fluid or by temperature change. These expansion joints permit relatively free movement of the line in an axial direction, and hence the static and dynamic forces must be provided for at the bends.

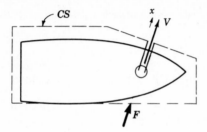

Figure 3.26 Nozzle mounted on boat.

Example 3.15 A jet of water 80 mm in diameter with a velocity of 40 m/s is discharged in a horizontal direction from a nozzle mounted on a boat. What force is required to hold the boat stationary?

When the control volume is selected as shown in Fig. 3.26, the net efflux of momentum is [Eq. (3.11.2)]

$$F_x = \rho Q(V_{x2} - V_{x1}) = \rho Q V = 1000 \text{ kg/m}^3 \times \frac{\pi}{4}(0.08 \text{ m})^2(40 \text{ m/s})^2 = 8.04 \text{ kN}$$

The force exerted against the boat is 8.04 kN in the x direction.

Example 3.16 Find the force exerted by the nozzle on the pipe of Fig. 3.27a. Neglect losses. The fluid is oil, sp gr 0.85, and $p_1 = 100$ psi.

To determine the discharge, Bernoulli's equation is written for the stream from section 1 to the downstream end of the nozzle, where the pressure is zero.

$$z_1 + \frac{V_1^2}{2g} + \frac{(100 \text{ lb/in}^2)(144 \text{ in}^2/\text{ft}^2)}{0.85(62.4 \text{ lb/ft}^3)} = z_2 + \frac{V_2^2}{2g} + 0$$

Since $z_1 = z_2$ and $V_2 = (D_1/D_2)^2 V_1 = 9V_1$, after substituting,

$$\frac{V_1^2}{2g}(1 - 81) + \frac{(100 \text{ lb/in}^2)(144 \text{ in}^2/\text{ft}^2)}{0.85(62.4 \text{ lb/ft}^3)} = 0$$

and

$$V_1 = 14.78 \text{ ft/s} \qquad V_2 = 133 \text{ ft/s} \qquad Q = 14.78 \frac{\pi}{4}\left(\frac{1}{4}\right)^2 = 0.725 \text{ ft}^3/\text{s}$$

Let P_x (Fig. 3.27b) be the force exerted on the liquid control volume by the nozzle; then, with Eq. (3.11.2),

$$(100 \text{ lb/in}^2)\frac{\pi}{4}(3 \text{ in})^2 - P_x = (1.935 \text{ slugs/ft}^3)(0.85)(0.725 \text{ ft}^3/\text{s})(133 \text{ ft/s} - 14.78 \text{ ft/s})$$

or $P_x = 565$ lb. The oil exerts a force on the nozzle of 565 lb to the right, and a tension force of 565 lb is exerted by the nozzle on the pipe.

In many situations an unsteady-flow problem can be converted to a steady-flow problem by superposing a constant velocity upon the system and its surroundings, i.e., by changing the reference velocity. The dynamics of a system and its surroundings are unchanged by the superposition of a constant velocity; hence, pressures and forces are unchanged. In the next flow situation studied, advantage is taken of this principle.

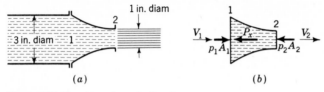

Figure 3.27 Nozzle at the end of a pipe.

The Momentum Theory for Propellers

The action of a propeller is to change the momentum of the fluid within which it is submerged and thus to develop a thrust that is used for propulsion. Propellers cannot be designed according to the momentum theory, although some of the relations governing them are made evident by its application. A propeller, with its slipstream and velocity distributions at two sections a fixed distance from it, is shown in Fig. 3.28. The propeller may be either (1) stationary in a flow as indicated or (2) moving to the left with a velocity V_1 through a stationary fluid, since the relative picture is the same. The fluid is assumed to be frictionless and incompressible.

The flow is undisturbed at section 1 upstream from the propeller and is accelerated as it approaches the propeller, owing to the reduced pressure on its upstream side. In passing through the propeller, the fluid has its pressure increased, which further accelerates the flow and reduces the cross section at 4. The velocity V does not change across the propeller, from 2 to 3. The pressure at 1 and 4 is that of the undisturbed fluid, which is also the pressure along the slipstream boundary.

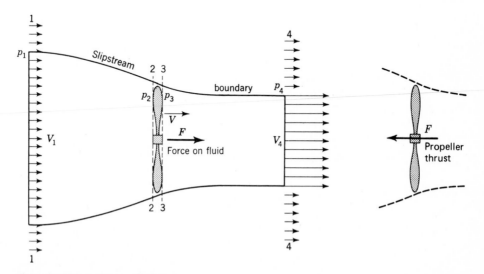

Figure 3.28 Propeller in a fluid stream.

When the momentum equation (3.11.2) is applied to the control volume within sections 1 and 4 and the slipstream boundary of Fig. 3.28, the force F exerted by the propeller is the only external force acting in the axial direction, since the pressure is everywhere the same on the control surface. Therefore,

$$F = \rho Q(V_4 - V_1) = (p_3 - p_2)A \tag{3.11.5}$$

in which A is the area swept over by the propeller blades. The propeller thrust must be equal and opposite to the force on the fluid. After substituting $Q = AV$ and simplifying,

$$\rho V(V_4 - V_1) = p_3 - p_2 \tag{3.11.6}$$

When Bernoulli's equation is written for the stream between sections 1 and 2 and between sections 3 and 4.

$$p_1 + \tfrac{1}{2}\rho V_1^2 = p_2 + \tfrac{1}{2}\rho V^2 \qquad p_3 + \tfrac{1}{2}\rho V^2 = p_4 + \tfrac{1}{2}\rho V_4^2$$

since $z_1 = z_2 = z_3 = z_4$. In solving for $p_3 - p_2$, with $p_1 = p_4$,

$$p_3 - p_2 = \tfrac{1}{2}\rho(V_4^2 - V_1^2) \tag{3.11.7}$$

Eliminating $p_3 - p_2$ in Eqs. (3.11.6) and (3.11.7) gives

$$V = \frac{V_1 + V_4}{2} \tag{3.11.8}$$

which shows that the velocity through the propeller area is the average of the velocities upstream and downstream from it.

The useful work per unit time done by a propeller moving through still fluid (power transferred) is the product of propeller thrust and velocity, i.e.,

$$\text{Power} = FV_1 = \rho Q(V_4 - V_1)V_1 \tag{3.11.9}$$

The power input is that required to increase the velocity of fluid from V_1 to V_4. Since Q is the volumetric flow rate,

$$\text{Power input} = \frac{\rho Q(V_4^2 - V_1^2)}{2} \tag{3.11.10}$$

Power input may also be expressed as the useful work (power output) plus the kinetic energy per unit time remaining in the slipstream (power loss)

$$\text{Power input} = \rho Q(V_4 - V_1)V_1 + \frac{\rho Q(V_4 - V_1)^2}{2} \tag{3.11.11}$$

The theoretical mechanical efficiency e_t is given by the ratio of Eqs. (3.11.9) and (3.11.10) or (3.11.11)

$$e_t = \frac{\text{output}}{\text{output} + \text{loss}} = \frac{2V_1}{V_4 + V_1} = \frac{V_1}{V} \tag{3.11.12}$$

If $\Delta V = V_4 - V_1$ is the increase in slipstream velocity, substituting into Eq. (3.11.12) produces

$$e_t = \frac{1}{1 + \Delta V/2V_1} \tag{3.11.13}$$

which shows that maximum efficiency is obtained with a propeller that increases the velocity of slipstream as little as possible, or for which $\Delta V/V_1$ is a minimum.

Owing to compressibility effects, the efficiency of an airplane propeller drops rapidly with speeds above 400 mph. Airplane propellers under optimum conditions have actual efficiencies close to the theoretical efficiencies, in the neighborhood of 85 percent. Ship propeller efficiencies are less, around 60 percent, owing to restrictions in diameter.

The windmill can be analyzed by application of the momentum relations. The jet has its speed reduced, and the diameter of slipstream is increased.

Example 3.17 An airplane traveling 400 km/h through still air, $\gamma = 12$ N/m³, discharges 1000 m³/s through its two 2.25-m-diameter propellers. Determine (a) the theoretical efficiency, (b) the thrust, (c) the pressure difference across the propellers, and (d) the theoretical power required.

(a)
$$V_1 = \frac{400 \text{ km}}{1 \text{ h}} \frac{1000 \text{ m}}{1 \text{ km}} \frac{1 \text{ h}}{3600 \text{ s}} = 111.11 \text{ m/s}$$

$$V = \frac{500 \text{ m}^3/\text{s}}{(\pi/4)(2.25^2)} = 126 \text{ m/s}$$

From Eq. (3.11.11)

$$e_t = \frac{V_1}{V} = \frac{111.11}{126} = 88.2\%$$

(b) From Eq. (3.11.8)

$$V_4 = 2V - V_1 = 2 \times 126 - 111.11 = 140.9 \text{ m/s}$$

The thrust from the propellers is, from Eq. (3.11.5),

$$F = \frac{12 \text{ N/m}^3}{9.806 \text{ m/s}^2} (1000 \text{ m}^3/\text{s})(140.9 \text{ m/s} - 111.11 \text{ m/s}) = 36.5 \text{ kN}$$

(c) The pressure difference, from Eq. (3.11.6), is

$$p_3 - p_2 = \frac{12 \text{ N/m}^3}{9.806 \text{ m/s}^2} (126 \text{ m/s})(140.9 \text{ m/s} - 111.11 \text{ m/s}) = 4.6 \text{ kPa}$$

(d) The theoretical power is

$$\frac{FV_1}{e_t} = (36,500 \text{ N}) \frac{111.11 \text{ m/s}}{0.882} \frac{1 \text{ kW}}{1000 \text{ N} \cdot \text{m/s}} = 4.6 \text{ MW}$$

Jet Propulsion

The propeller is one form of jet propulsion in that it creates a jet and by so doing has a thrust exerted upon it that is the propelling force. In jet engines, air (initially

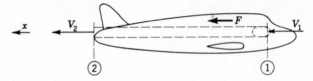

Figure 3.29 Walls of flow passages through jet engines taken as inpenetrable part of control surface for plane when viewed as a steady-state problem.

at rest) is taken into the engine and burned with a small amount of fuel; the gases are then ejected with a much higher velocity than in a propeller slipstream. The jet diameter is necessarily smaller than the propeller slipstream. If the mass of fuel burned is neglected, the propelling force F [Eq. (3.11.5)] is

$$F = \rho Q(V_2 - V_1) = \rho Q V_{abs} \tag{3.11.14}$$

in which $V_{abs} = \Delta V$ (Fig. 3.29) is the absolute velocity of fluid in the jet and ρQ is the mass per unit time being discharged. The theoretical mechanical efficiency is the same expression as that for efficiency of the propeller, Eq. (3.11.13). It is obvious that, other things being equal, V_{abs}/V_1 should be as small as possible. For a given speed V_1, the resistance force F is determined by the body and fluid in which it moves; hence, for V_{abs} in Eq. (3.11.13) to be very small, ρQ must be very large.

An example is the type of propulsion system to be used on a boat. If the boat requires a force of 400 lb to move it through water at 15 mph, first a method of jet propulsion can be considered in which water is taken in at the bow and discharged out the stern by a 100 percent efficient pumping system. To analyze the propulsion system, the problem is converted to steady state by superposition of the boat speed $-V_1$ on boat and surroundings (Fig. 3.30).

If a 6-in-diameter jet pipe is used, $V_2 = 16Q/\pi$. By use of Eq. (3.11.2), for $V_1 = 15$ mi/h $= 22$ ft/s,

$$400 = 1.935Q\left(\frac{16Q}{\pi} - 22\right)$$

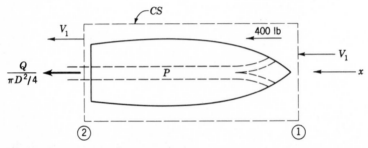

Figure 3.30 Steady-state flow around a boat.

Hence $Q = 8.89$ ft^3/s, $V_{abs} = 23.2$, and the efficiency is

$$e_t = \frac{1}{1 + V_{abs}/2V_1} = \frac{1}{1 + 23.2/44} = 65.5\%$$

The horsepower required is

$$\frac{FV_1}{550e_t} = \frac{400 \times 22}{550 \times 0.655} = 24.4$$

With an 8-in-diameter jet pipe, $V_2 = 9Q/\pi$, and

$$400 = 1.935Q\left(\frac{9Q}{\pi} - 22\right)$$

so that $Q = 13.14$ ft^3/s, $V_2 = 15.72$, $e_t = 73.7$ percent, and the horsepower required is 21.7.

With additional enlarging of the jet pipe and the pumping of more water with less velocity head, the efficiency can be further increased. The type of pump best suited for large flows at small head is the axial-flow propeller pump. Increasing the size of pump and jet pipe would increase weight greatly and take up useful space in the boat; the logical limit is to drop the propeller down below or behind the boat and thus eliminate the jet pipe, which is the usual propeller for boats. Jet propulsion of a boat by a jet pipe is practical, however, in very shallow water where a propeller would be damaged by striking bottom or other obstructions.

To take the weight of fuel into account in jet propulsion of aircraft, let $\dot{m}_{air}$ be the mass of air per unit time and r the ratio of mass of fuel burned to mass of air. Then (Fig. 3.29), the propulsive force F is

$$F = \dot{m}_{air}(V_2 - V_1) + r\dot{m}_{air}V_2$$

The second term on the right is the mass of fuel per unit time multiplied by its change in velocity. Rearranging gives

$$F = \dot{m}_{air}[V_2(1 + r) - V_1] \tag{3.11.15}$$

Defining the mechanical efficiency again as the useful work divided by the sum of useful work and kinetic energy remaining gives

$$e_t = \frac{FV_1}{FV_1 + \dot{m}_{air}(1 + r)(V_2 - V_1)^2/2}$$

and by Eq. (3.11.15)

$$e_t = \frac{1}{1 + (1 + r)[(V_2/V_1) - 1]^2/2[(1 + r)(V_2/V_1) - 1]} \tag{3.11.16}$$

The efficiency becomes unity for $V_1 = V_2$, as the combustion products are then brought to rest and no kinetic energy remains in the jet.

Example 3.18 An airplane consumes 1 kg fuel for each 20 kg air and discharges hot gases from the tailpipe at $V_2 = 1800$ m/s. Determine the mechanical efficiency for airplane speeds of 300 and 150 m/s.

For 300 m/s, $V_2/V_1 = 1800/300 = 6$, $r = 0.05$. From Eq. (3.11.16),

$$e_t = \frac{1}{1 + (1 + 0.05)(6 - 1)^2/2[6(1 + 0.05) - 1]} = 0.287$$

For 150 m/s, $V_2/V_1 = 1800/150 = 12$, and

$$e_t = \frac{1}{1 + (1 + 0.05)(12 - 1)^2/2[12(1 + 0.05) - 1]} = 0.154$$

Propulsion through air or water in each case is caused by reaction to the formation of a jet behind the body. The various means include the propeller, turbojet, turboprop, ram jet, and rocket motor, which are briefly described in the following paragraphs.

The momentum relations for a propeller determine that its theoretical efficiency increases as the speed of the aircraft increases and the absolute velocity of the slipstream decreases. As the speed of the blade tips approaches the speed of sound, however, compressibility effects greatly increase the drag on the blades and thus decrease the overall efficiency of the propulsion system.

A *turbojet* is an engine consisting of a compressor, a combustion chamber, a turbine, and a jet pipe. Air is scooped through the front of the engine and is compressed, and fuel is added and burned with a great excess of air. The air and combustion gases then pass through a gas turbine that drives the compressor. Only a portion of the energy of the hot gases is removed by the turbine, since the only means of propulsion is the issuance of the hot gas through the jet pipe. The overall efficiency of a jet engine increases with speed of the aircraft. Although there is very little information available on propeller systems near the speed of sound, it appears that the overall efficiencies of the turbojet and propeller systems are about the same at the speed of sound.

The *turboprop* is a system combining thrust from a propeller with thrust from the ejection of hot gases. The gas turbine must drive both compressor and propeller. The proportion of thrust between the propeller and the jet may be selected arbitrarily by the designer.

The *ram jet* is a high-speed engine that has neither compressor nor turbine. The ram pressure of the air forces air into the front of the engine, where some of the kinetic energy is converted into pressure energy by enlarging the flow cross section. It then enters a combustion chamber, where fuel is burned, and the air and gases of combustion are ejected through a jet pipe. It is a supersonic device requiring very high speed for compression of the air. An intermittent ram jet was used by the Germans in the V-1 *buzz bomb*. Air is admitted through spring-closed flap valves in the nose. Fuel is ignited to build up pressure that closes the flap valves and ejects the hot gases as a jet. The ram pressure then opens the valves in the nose to repeat the cycle. The cyclic rate is around 40 s^{-1}. Such a device must be launched at high speed to initiate operation of the ram jet.

Rocket Mechanics

The rocket motor carries with it an oxidizing agent to mix with its fuel so that it develops a thrust that is independent of the medium through which it travels. In contrast, a gas turbine can eject a mass many times the mass of fuel it carries because it takes in air to mix with the fuel.

To determine the acceleration of a rocket during flight, Fig. 3.31, it is convenient to take the control volume as the outer surface of the rocket, with a plane area normal to the jet across the nozzle exit. The control volume has a velocity equal to the velocity of the rocket at the instant the analysis is made. Let R be the air resistance, m_R the mass of the rocket body, m_f the mass of fuel, $\dot{m}$ the rate at which fuel is being burned, and v_r the exit-gas velocity relative to the rocket. V_1 is the actual velocity of the rocket (and of the frame of reference), and V is the velocity of the rocket relative to the frame of reference. V is zero, but $dV/dt = dV_1/dt$ is the rocket acceleration. The basic linear-momentum equation for the y direction (vertical motion)

$$\Sigma F_y = \frac{\partial}{\partial t} \int_{cv} \rho v_y \, d\mathcal{V} + \int_{cs} \rho v_y \mathbf{v} \cdot d\mathbf{A} \qquad (3.11.17)$$

becomes

$$- R - (m_R + m_f)g = \frac{\partial}{\partial t}[(m_R + m_f)V] - \dot{m}v_r \qquad (3.11.18)$$

Since V is a function of t only, the equation can be written as a total differential equation

$$\frac{dV}{dt} = \frac{dV_1}{dt} = \frac{\dot{m}v_r - g(m_f + m_R) - R}{m_R + m_f} \qquad (3.11.19)$$

The mass of propellant reduces with time; for constant burning rate $\dot{m}$, $m_f = m_{f_0} - \dot{m}t$, with m_{f_0} the initial mass of fuel and oxidizer. Gravity is a function

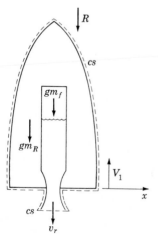

Figure 3.31 Control surface for analysis of rocket acceleration. Frame of reference has the velocity V_1 of the rocket.

of y, and the air resistance depends on the Reynolds number and Mach number (Chap. 4), as well as on the shape and size of the rocket.

By considering the mass of rocket and fuel together [Eq. (3.11.19)], the thrust $\dot{m}v_r$ minus the weight and the air resistance just equals the combined mass times its acceleration.

The theoretical efficiency of a rocket motor (based on available energy) is shown to increase with rocket speed. E represents the available energy in the propellant per unit mass. When the propellant is ignited, its available energy is converted into kinetic energy; $E = v_r^2/2$, in which v_r is the jet velocity relative to the rocket, or $v_r = \sqrt{2E}$. For rocket speed V_1 referred to axes fixed in the earth, the useful power is $\dot{m}v_r V_1$. The kinetic energy being used up per unit time is due to mass loss $\dot{m}V_1^2/2$ of the unburned propellant and to the burning $\dot{m}E$, or

$$\text{Available energy input per unit time} = \dot{m}\left(E + \frac{V_1^2}{2}\right) \qquad (3.11.20)$$

The mechanical efficiency e is

$$e = \frac{\dot{m}V_1\sqrt{2E}}{\dot{m}(E + V_1^2/2)} = \frac{2v_r/V_1}{1 + (v_r/V_1)^2} \qquad (3.11.21)$$

When $v_r/V_1 = 1$, the maximum efficiency $e = 1$ is obtained. In this case the absolute velocity of ejected gas is zero.

When the thrust on a vertical rocket is greater than the total weight plus resistance, the rocket accelerates. Its mass is continuously reduced. To lift a rocket off its pad, its thrust $\dot{m}v_r$ must exceed its total weight.

Example 3.19 (*a*) Determine the burning time for a rocket that initially weighs 4.903 MN, of which 70 percent is propellant. It consumes fuel at a constant rate, and its initial thrust is 10 percent greater than its weight. $v_r = 3300$ m/s. (*b*) Considering g constant at 9.8 m/s² and the flight to be vertical without air resistance, find the speed of the rocket at burnout time, its height above ground, and the maximum height it will attain.
(*a*) From the thrust relation

$$\dot{m}v_r = 1.1W_0 = 1.1(4.903 \text{ MN}) = 5.393 \text{ MN} = \dot{m}3300$$

and $\dot{m} = 1634.3$ kg/s. The available mass of propellant is 350,000 kg; hence, the burning time is

$$\frac{350{,}000 \text{ kg}}{1634.3 \text{ kg/s}} = 214.2 \text{ s}$$

(*b*) From Eq. (3.11.19)

$$\frac{dV_1}{dt} = \frac{(1634.3 \text{ kg/s})(3300 \text{ m/s}) - (9.8 \text{ m/s}^2)[350{,}000 \text{ kg} - (1634.3 \text{ kg/s})t + 150{,}000 \text{ kg}]}{150{,}000 \text{ kg} + 350{,}000 \text{ kg} - (1634.3 \text{ kg/s})t}$$

Simplifying gives

$$\frac{dV_1}{dt} = \frac{299.95 + 9.8t}{305.94 - t} = -9.8 + \frac{3298.16}{305.94 - t}$$

$$V_1 = -9.8t - 3298.16 \ln(305.94 - t) + \text{const}$$

When $t = 0$, $V_1 = 0$; hence,

$$V_1 = -9.8t - 3298.16 \ln\left(1 - \frac{t}{305.94}\right)$$

When $t = 214.2$, $V_1 = 1873.24$ m/s. The height at $t = 214.2$ s is

$$y = \int_0^{214.2} V \, dt = -9.8\frac{t^2}{2}\bigg]_0^{214.2} - 3298.16 \int_0^{214.2} \ln\left(1 - \frac{t}{305.94}\right) dt$$

$$= 117.22 \text{ km}$$

The rocket will glide $V_1^2/2g$ ft higher after burnout, or

$$117{,}220 \text{ m} + \frac{1873.24^2}{2 \times 9.8} \text{ m} = 296.25 \text{ km}$$

Fixed and Moving Vanes

The theory of turbomachines is based on the relations between jets and vanes. The mechanics of transfer of work and energy from fluid jets to moving vanes is studied as an application of the momentum principles. When a free jet impinges onto a smooth vane that is curved, as in Fig. 3.32, the jet is deflected, its momentum is changed, and a force is exerted on the vane. The jet is assumed to flow onto the vane in a tangential direction, without shock; and furthermore the frictional resistance between jet and vane is neglected. The velocity is assumed to be uniform throughout the jet upstream and downstream from the vane. Since the jet is open to the air, it has the same pressure at each end of the vane. When the small change in elevation between ends, if any, is neglected, application of Bernoulli's equation shows that the magnitude of the velocity is unchanged for *fixed* vanes.

Example 3.20 Find the force exerted on a fixed vane when a jet discharging 60 l/s water at 50 m/s is deflected through 135°.

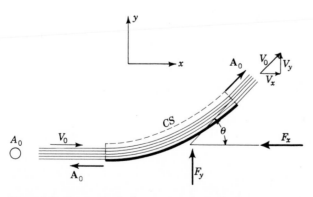

Figure 3.32 Free jet impinging on smooth, fixed vane.

By referring to Fig. 3.32 and by applying Eq. (3.11.2) in the x and y directions, it is found that

$$-F_x = \rho V_0 \cos \theta \, V_0 A_0 + \rho V_0(-V_0 A_0) \qquad F_y = \rho V_0 \sin \theta \, V_0 A_0$$

Hence,

$$F_x = -(1000 \text{ kg/m}^3)(0.06 \text{ m}^3/\text{s})(50 \cos 135° - 50 \text{ m/s}) = 5.121 \text{ kN}$$

$$F_y = (1000 \text{ kg/m}^3)(0.06 \text{ m}^3/\text{s})(50 \sin 135°) = 2.121 \text{ kN}$$

The force components on the fixed vane are then equal and opposite to F_x and F_y.

Example 3.21 Fluid issues from a long slot and strikes against a smooth inclined flat plate (Fig. 3.33). Determine the division of flow and the force exerted on the plate, neglecting losses due to impact.

As there are no changes in elevation or pressure before and after impact, the magnitude of the velocity leaving is the same as the initial speed of the jet. The division of flow Q_1, Q_2 can be computed by applying the momentum equation in the s direction, parallel to the plate. No force is exerted on the fluid by the plate in this direction; hence, the final momentum component must equal the initial momentum component in the s direction. The steady-state momentum equation for the s direction, from Eq. (3.11.2), yields

$$\Sigma F_s = \int_{cs} \rho v_s \mathbf{V} \cdot d\mathbf{A} = 0 = \rho V_0 \, V_0 \, A_1 + \rho V_0 \cos \theta(-V_0 A_0) + \rho(-V_0)V_0 \, A_2$$

By substituting $Q_1 = V_0 A_1$, $Q_2 = V_0 A_2$, and $Q_0 = V_0 A_0$, it reduces to

$$Q_1 - Q_2 = Q_0 \cos \theta$$

and with the continuity equation,

$$Q_1 + Q_2 = Q_0$$

The two equations can be solved for Q_1 and Q_2:

$$Q_1 = \frac{Q_0}{2} (1 + \cos \theta) \qquad Q_2 = \frac{Q_0}{2} (1 - \cos \theta)$$

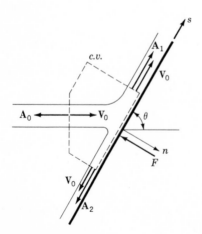

Figure 3.33 Two-dimensional jet impinging on an inclined fixed plane surface.

The force F exerted on the plate must be normal to it. For the momentum equation normal to the plate, Fig. 3.33,

$$\Sigma F_n = \int_{cs} \rho v_n \mathbf{V} \cdot d\mathbf{A} = -F = \rho V_0 \sin \theta (-V_0 A_0)$$

$$F = \rho Q_0 V_0 \sin \theta$$

Moving Vanes

Turbomachinery utilizes the forces resulting from the motion over moving vanes. No work can be done on or by a fluid that flows over a fixed vane. When vanes can be displaced, work can be done either on the vane or on the fluid. In Fig. 3.34a a moving vane is shown with fluid flowing onto it tangentially. Forces exerted on the fluid by the vane are indicated by F_x and F_y. To analyze the flow, the problem is reduced to steady state by superposition of vane velocity u to the left (Fig. 3.34b) on both vane and fluid. The control volume then encloses the fluid in contact with the vane, with its control surface normal to the flow at sections 1 and 2. Figure 3.34c shows the *polar vector diagram* for flow through the vane. The absolute-velocity vectors originate at the origin O, and the relative-velocity vector $\mathbf{V}_0 - \mathbf{u}$ is turned through the angle θ of the vane as shown. $\mathbf{V}_2$ is the final absolute velocity leaving the vane. The relative velocity $v_r = V_0 - u$ is unchanged in magnitude as it traverses the vane. The mass per unit time is given by $\rho A_0 v_r$ and is not the mass rate being discharged from the nozzle. If a *series of vanes* is employed, as on the periphery of a wheel, so arranged that one or another of the jets intercept all flow from the nozzle and the velocity is substantially u, then the mass per second is the

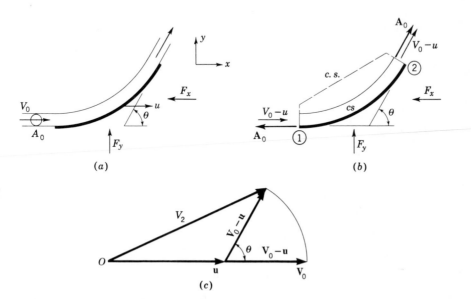

Figure 3.34 (a) Moving vane. (b) Vane flow viewed as steady-state problem by superposition of velocity u to the left. (c) Polar vector diagram.

total mass per second being discharged. Application of Eq. (3.11.2) to the control volume of Fig. 3.34b yields

$$\Sigma F_x = \int_{cs} \rho v_x \mathbf{V} \cdot d\mathbf{A} = -F_x = \rho(V_0 - u)\cos\theta[(V_0 - u)A_0]$$

$$+ \rho(V_0 - u)[-(V_0 - u)A_0]$$

or $\qquad F_x = \rho(V_0 - u)^2 A_0 (1 - \cos\theta)$

$$\Sigma F_y = \int_{cs} \rho v_y \mathbf{V} \cdot d\mathbf{A} = F_y = \rho(V_0 - u)\sin\theta[(V_0 - u)A_0]$$

or $\qquad F_y = \rho(V_0 - u)^2 A_0 \sin\theta$

These relations are for the single vane. For a series of vanes they become

$$F_x = \rho Q_0(V_0 - u)(1 - \cos\theta) \qquad F_y = \rho Q_0(V_0 - u)\sin\theta$$

Example 3.22 Determine for a single moving vane of Fig. 3.35a the force components due to the water jet and the rate of work done on the vane.

Figure 3.35b is the steady-state reduction with a control volume shown. The polar vector diagram is shown in Fig. 3.35c. By applying Eq. (3.11.2) in the x and y directions to the control volume of Fig. 3.35b.

$$-F_x = (1000 \text{ kg/m}^3)(60 \text{ m/s})(\cos 170°)(60 \text{ m/s})(0.001 \text{ m}^2)$$

$$+ (1000 \text{ kg/m}^3)(60 \text{ m/s})(-60 \text{ m/s})(0.001 \text{ m}^2)$$

$$F_x = 7.145 \text{ kN}$$

$$F_y = (1000 \text{ kg/m}^3)(60 \text{ m/s})(\sin 170°)(60 \text{ m/s})(0.001 \text{ m}^2) = 625 \text{ N}$$

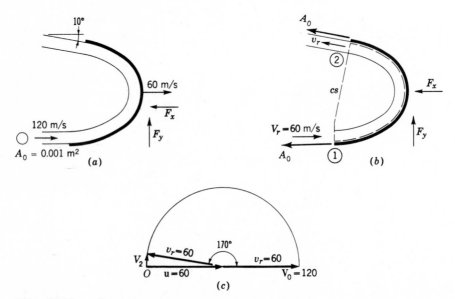

Figure 3.35 Jet acting on a moving vane.

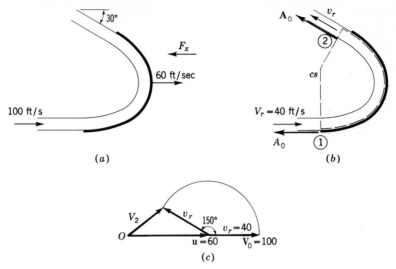

Figure 3.36 Flow through moving vanes.

The power exerted on the vane is

$$uF_x = (60 \text{ m/s})(7.145 \text{ kN}) = 428.7 \text{ kW}$$

Example 3.23 Determine the horsepower that can be obtained from a series of vanes (Fig. 3.36a), curved through 150°, moving 60 ft/s away from a 3.0-cfs water jet having a cross section of 0.03 ft². Draw the polar vector diagram and calculate the energy remaining in the jet.
The jet velocity is $V_0 = 3/0.03 = 100$ ft/s. The steady-state vane control volume is shown in Fig. 3.36b and the polar vector diagram in Fig. 3.36c. The force on the series of vanes in the x direction is

$$F_x = (1.94 \text{ slugs/ft}^3)(3 \text{ ft}^3/\text{s})(40 \text{ ft/s})(1 - \cos 150°) = 434 \text{ lb}$$

The horsepower is

$$\frac{(434 \text{ lb})(60 \text{ ft/s})}{550 \text{ ft} \cdot \text{lb/s/hp}} = 47.4 \text{ hp}$$

The components of absolute velocity leaving the vane are, from Fig. 3.36c,

$$V_{2x} = 60 - 40 \cos 30° = 25.4 \text{ ft/s} \qquad V_{2y} = 40 \sin 30° = 20 \text{ ft/s}$$

and the exit-velocity head is

$$\frac{V_2^2}{2g} = \frac{25.4^2 + 20^2}{64.4} = 16.2 \text{ ft} \cdot \text{lb/lb}$$

The kinetic energy remaining in the jet, in foot-pounds per second, is

$$Q\gamma \frac{V_2^2}{2g} = (3 \text{ ft}^3/\text{s})(62.4 \text{ lb/ft}^3)(16.2 \text{ ft}) = 3030 \text{ ft} \cdot \text{lb/s}$$

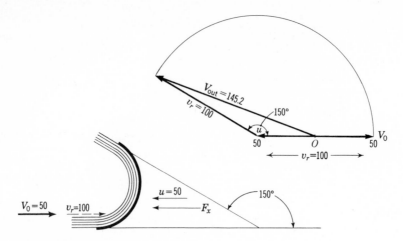

Figure 3.37 Vector diagram for vane doing work on a jet.

The initial kinetic energy available was

$$(3 \text{ ft}^3/\text{s})(62.4 \text{ lb/ft}^3)\frac{100^2}{64.4} \text{ ft} = 29{,}070 \text{ ft} \cdot \text{lb/s}$$

which is the sum of the work done and the energy remaining per second.

When a vane or series of vanes moves toward a jet, work is done by the vane system on the fluid, thereby increasing the energy of the fluid. Figure 3.37 illustrates this situation; the polar vector diagram shows the exit velocity to be greater than the entering velocity.

In turbulent flow, losses generally must be determined from experimental tests on the system or a geometrically similar model of the system. In the following two cases, application of the continuity, energy, and momentum equations permits the losses to be evaluated analytically.

Losses Due to Sudden Expansion in a Pipe

The losses due to sudden enlargement in a pipeline may be calculated with both the energy and momentum equations. For steady, incompressible, turbulent flow through the control volume between sections 1 and 2 of the sudden expansion of Fig. 3.38a, b, the small shear force exerted on the walls between the two sections may be neglected. By assuming uniform velocity over the flow cross sections, which is approached in turbulent flow, application of Eq. (3.11.2) produces

$$p_1 A_2 - p_2 A_2 = \rho V_2 (V_2 A_2) + \rho V_1 (-V_1 A_1)$$

At section 1 the radial acceleration of fluid particles in the eddy along the surface is small, and so generally a hydrostatic pressure variation occurs across

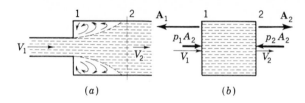

Figure 3.38 Sudden expansion in a pipe.

the section. The energy equation (3.10.1) applied to sections 1 and 2, with the loss term h_l, is (for $\alpha = 1$)

$$\frac{V_1^2}{2g} + \frac{p_1}{\gamma} = \frac{V_2^2}{2g} + \frac{p_2}{\gamma} + h_l$$

Solving for $(p_1 - p_2)/\gamma$ in each equation and equating the results give

$$\frac{V_2^2 - V_2 V_1}{g} = \frac{V_2^2 - V_1^2}{2g} + h_l$$

As $V_1 A_1 = V_2 A_2$,

$$h_l = \frac{(V_1 - V_2)^2}{2g} = \frac{V_1^2}{2g}\left(1 - \frac{A_1}{A_2}\right)^2 \tag{3.11.22}$$

which indicates that the losses in turbulent flow are proportional to the square of the velocity.

Hydraulic Jump

The hydraulic jump is the second application of the basic equations to determine losses due to a turbulent flow situation. Under proper conditions a rapidly flowing stream of liquid in an open channel suddenly changes to a slowly flowing stream with a larger cross-sectional area and a sudden rise in elevation of liquid surface. This phenomenon, known as the *hydraulic jump*, is an example of steady nonuniform flow. In effect, the rapidly flowing liquid jet expands (Fig. 3.39) and converts kinetic energy into potential energy and losses or irreversibilities. A roller develops on the inclined surface of the expanding liquid jet and draws air into the liquid. The surface of the jump is very rough and turbulent, the losses being greater as the jump height is greater. For small heights, the form of the jump changes to a standing wave (Fig. 3.40). The jump is discussed further in Sec. 11.4.

The relations between the variables for the hydraulic jump in a horizontal rectangular channel are easily obtained by use of the continuity, momentum, and

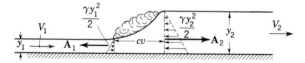

Figure 3.39 Hydraulic jump in a rectangular channel.

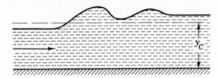

Figure 3.40 Standing wave.

energy equations. For convenience, the width of channel is taken as unity. The continuity equation (Fig. 3.39) is $(A_1 = y_1,\ A_2 = y_2)$

$$V_1 y_1 = V_2 y_2$$

The momentum equation is

$$\frac{\gamma y_1^2}{2} - \frac{\gamma y_2^2}{2} = \rho V_2(y_2 V_2) + \rho V_1(-y_1 V_1)$$

and the energy equation (for points on the liquid surface) is

$$\frac{V_1^2}{2g} + y_1 = \frac{V_2^2}{2g} + y_2 + h_j$$

in which h_j represents losses due to the jump. Eliminating V_2 in the first two equations leads to

$$y_2 = -\frac{y_1}{2} + \sqrt{\left(\frac{y_1}{2}\right)^2 + \frac{2V_1^2 y_1}{g}} \tag{3.11.23}$$

in which the plus sign has been taken before the radical (a negative y_2 has no physical significance). The depths y_1 and y_2 are referred to as *conjugate* depths. Solving the energy equation for h_j and eliminating V_1 and V_2 give

$$h_j = \frac{(y_2 - y_1)^3}{4 y_1 y_2} \tag{3.11.24}$$

The hydraulic jump, which is a very effective device for creating irreversibilities, is commonly used at the ends of chutes or the bottoms of spillways to destroy much of the kinetic energy in the flow. It is also an effective mixing chamber, because of the violent agitation that takes place in the roller. Experimental measurements of hydraulic jumps show that the equations yield the correct value of y_2 to within 1 percent.

Example 3.24 If 12 m³/s of water per meter of width flows down a spillway onto a horizontal floor and the velocity is 20 m/s, determine the downstream depth required to cause a hydraulic jump and the losses in power by the jump per meter of width

$$y_1 = \frac{12 \text{ m}^2/\text{s}}{20 \text{ m/s}} = 0.6 \text{ m}$$

Substituting into Eq. (3.11.23) gives

$$y_2 = -0.3 + \sqrt{0.3^2 + \frac{2 \times 20^2 \times 0.6}{9.806}} = 7 \text{ m}$$

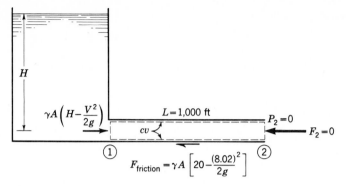

Figure 3.41 Acceleration of liquid in a pipe.

With Eq. (3.11.24),

$$\text{Losses} = \frac{(7 - 0.6)^3}{4 \times 0.6 \times 7} = 15.6 \text{ m} \cdot \text{N/N}$$

$$\text{Power/m} = \gamma Q(\text{losses}) = (9806 \text{ N/m}^3)(12 \text{ m}^3/\text{s})(15.6 \text{ m}) = 1836 \text{ kW}$$

Example 3.25 Find the head H in the reservoir of Fig. 3.41 needed to accelerate the flow of oil, $S = 0.85$, at the rate of 0.5 ft/s^2 when the flow is 8.02 ft/s. At 8.02 ft/s the steady-state head on the pipe is 20 ft. Neglect entrance loss.

The oil may be considered to be incompressible and to be moving uniformly in the pipeline. By applying Eq. (3.11.2), the last term is zero, as the net efflux is zero,

$$\gamma A\left(H - \frac{V^2}{2g}\right) - \gamma A\left(20 - \frac{8.02^2}{2g}\right) = \frac{\partial}{\partial t}\left(\frac{\gamma}{g} ALV\right)$$

or

$$H = 20 + \frac{1000}{32.2} \times 0.5 = 35.52 \text{ ft}$$

EXERCISES

3.11.1 The equation $\Sigma F_x = \rho Q(V_{x_{out}} - V_{x_{in}})$ requires which two of the following assumptions for its derivation?
1. Velocity constant over the end cross sections
2. Steady flow
3. Uniform flow
4. Compressible fluid
5. Frictionless fluid

(*a*) 1, 2; (*b*) 1, 5; (*c*) 1, 3; (*d*) 3, 5; (*e*) 2, 4.

3.11.2 The momentum correction factor is expressed by

(*a*) $\dfrac{1}{A}\displaystyle\int_A \dfrac{v}{V}\, dA$ (*b*) $\dfrac{1}{A}\displaystyle\int_A \left(\dfrac{v}{V}\right)^2 dA$ (*c*) $\dfrac{1}{A}\displaystyle\int_A \left(\dfrac{v}{V}\right)^3 dA$ (*d*) $\dfrac{1}{A}\displaystyle\int_A \left(\dfrac{v}{V}\right)^4 dA$

(*e*) none of these answers

3.11.3 The momentum correction factor for the velocity distribution given by Fig. 1.1 is (a) 0; (b) 1; (c) $\frac{4}{3}$; (d) 2; (e) none of these answers.

3.11.4 The velocity is zero over one-third of a cross section and is uniform over the remaining two-thirds of the area. The momentum correction factor is (a) 1; (b) $\frac{4}{3}$; (c) $\frac{3}{2}$; (d) $\frac{9}{4}$; (e) none of these answers.

3.11.5 The magnitude of the resultant force necessary to hold a 200-mm-diameter 90° elbow under no-flow conditions when the pressure is 0.98 MPa is, in kilonewtons, (a) 61.5; (b) 43.5; (c) 30.8; (d) 0; (e) none of these answers.

3.11.6 A 12-in-diameter 90° elbow carries water with average velocity of 15 ft/s and pressure of −5 psi. The force component in the direction of the approach velocity necessary to hold the elbow in place is, in pounds, (a) −342; (b) 223; (c) 565; (d) 907; (e) none of these answers.

3.11.7 A 50-mm-diameter 180° bend carries a liquid, $\rho = 1000$ kg/m³ at 6 m/s, at a pressure of zero gage. The force tending to push the bend off the pipe is, in newtons, (a) 0; (b) 70.5; (c) 141; (d) 515; (e) none of these answers.

3.11.8 The thickness of wall for a large high-pressure pipeline is determined by consideration of (a) axial tensile stresses in the pipe; (b) forces exerted by dynamic action at bends; (c) forces exerted by static and dynamic action at bends; (d) circumferential pipe-wall tension; (e) temperature stresses.

3.11.9 Select from the following list the correct assumptions for analyzing flow of a jet that is deflected by a fixed or moving vane:
1. The momentum of the jet is unchanged.
2. The absolute speed does not change along the vane.
3. The fluid flows onto the vane without shock.
4. The flow from the nozzle is steady.
5. Friction between jet and vane is neglected.
6. The jet leaves without velocity.
7. The velocity is uniform over the cross section of the jet before and after contacting the vane.
(a) 1, 3, 4, 5; (b) 2, 3, 5, 6; (c) 3, 4, 5, 6; (d) 3, 4, 5, 7; (e) 3, 5, 6, 7.

3.11.10 When a steady jet impinges on a fixed inclined plane surface (a) the momentum in the direction of the approach velocity is unchanged; (b) no force is exerted on the jet by the vane; (c) the flow is divided into parts directly proportional to the angle of inclination of the surface; (d) the speed is reduced for that portion of the jet turned through more than 90° and increased for the other portion; (e) the momentum component parallel to the surface is unchanged.

3.11.11 A force of 250 N is exerted upon a moving blade in the direction of its motion, $u = 20$ m/s. The power obtained in kilowatts is (a) 0.5; (b) 30; (c) 50; (d) 100; (e) none of these answers.

3.11.12 A series of moving vanes, $u = 50$ ft/s, $\theta = 90°$, intercepts a jet, $Q = 1$ ft³/s, $\rho = 1.5$ slugs/ft³, $V_0 = 100$ ft/s. The work done on the vanes, in foot-pounds per second, is (a) 1875; (b) 2500; (c) 3750; (d) 7500; (e) none of these answers.

3.11.13 The kilowatts available in a water jet of cross-sectional area 0.004 m² and velocity 20 m/s is (a) 0.495; (b) 16.0; (c) 17.2; (d) 32; (e) none of these answers.

3.11.14 A ship moves through water at 30 ft/s. The velocity of water in the slipstream behind the boat is 20 ft/s, and the propeller diameter is 3.0 ft. The theoretical efficiency of the propeller is, in percent, (a) 0; (b) 60; (c) 75; (d) 86; (e) none of these answers.

3.11.15 The losses due to a sudden expansion are expressed by

(a) $\dfrac{V_1^2 - V_2^2}{2g}$ (b) $\dfrac{V_1 - V_2}{2g}$ (c) $\dfrac{V_2^2 - V_1^2}{g}$ (d) $\dfrac{(V_1 - V_2)^2}{g}$ (e) $\dfrac{(V_1 - V_2)^2}{2g}$

3.11.16 The depth conjugate to $y = 3$ m and $V = 8$ m/s is (a) 4.55 m; (b) 4.9 m; (c) 7.04 m; (d) 9.16 m; (e) none of these answers.

3.11.17 The depth conjugate to $y = 10$ ft and $V = 1$ ft/s is (a) 0.06 ft; (b) 1.46 ft; (c) 5.06 ft; (d) 10.06 ft; (e) none of these answers.

3.12 THE MOMENT-OF-MOMENTUM EQUATION

The general unsteady linear-momentum equation applied to a control volume, Eq. (3.11.1), is

$$\mathbf{F} = \frac{\partial}{\partial t} \int_{cv} \rho \mathbf{v} \, d\mathcal{V} + \int_{cs} \rho \mathbf{v} \mathbf{v} \cdot d\mathbf{A} \tag{3.12.1}$$

The moment of a force $\mathbf{F}$ about a point O (Fig. 3.42) is given by

$$\mathbf{r} \times \mathbf{F}$$

which is the cross, or vector, product of $\mathbf{F}$ and the position vector $\mathbf{r}$ of a point on the line of action of the vector from O. The cross product of two vectors is a vector at right angles to the plane defined by the first two vectors and with magnitude

$$Fr \sin \theta$$

which is the product of F and the shortest distance from O to the line of action of $\mathbf{F}$. The sense of the final vector follows the right-hand rule. In Fig. 3.42 the force tends to cause a counterclockwise rotation around O. If this were a right-hand screw thread turning in this direction, it would tend to come up, and so the vector is likewise directed up out of the paper. If one curls the fingers of the right hand in the direction the force tends to cause rotation, the thumb yields the direction, or sense, of the vector.

By taking $\mathbf{r} \times \mathbf{F}$, using Eq. (3.12.1),

$$\mathbf{r} \times \mathbf{F} = \frac{\partial}{\partial t} \int_{cv} \rho \mathbf{r} \times \mathbf{v} \, d\mathcal{V} + \int_{cs} (\rho \mathbf{r} \times \mathbf{v})(\mathbf{v} \cdot d\mathbf{A}) \tag{3.12.2}$$

The left-hand side of the equation is the torque exerted by any forces on the control volume, and terms on the right-hand side represent the rate of change of *moment of momentum* within the control volume plus the net efflux of moment of momentum from the control volume. This is the general moment-of-momentum equation for a control volume. It has great value in the analysis of certain flow problems, e.g., in turbomachinery, where torques are more significant than forces.

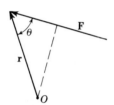

Figure 3.42 Notation for moment of a vector.

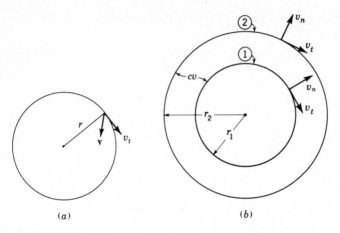

Figure 3.43 Two-dimensional flow in a centrifugal pump impeller.

When Eq. (3.12.2) is applied to a case of flow in the xy plane, with r the shortest distance to the tangential component of the velocity v_t, as in Fig. 3.43a, and v_n is the normal component of velocity,

$$F_t r = T_z = \int_{cs} \rho r v_t v_n \, dA + \frac{\partial}{\partial t} \int_{cv} \rho r v_t \, d\mathbb{U} \qquad (3.12.3)$$

in which T_z is the torque. A useful form of Eq. (3.12.3) applied to an annular control volume, in steady flow (Fig. 3.43b), is

$$T_z = \int_{A_2} \rho_2 r_2 v_{t_2} v_{n_2} \, dA_2 - \int_{A_1} \rho_1 r_1 v_{t_1} v_{n_1} \, dA_1 \qquad (3.12.4)$$

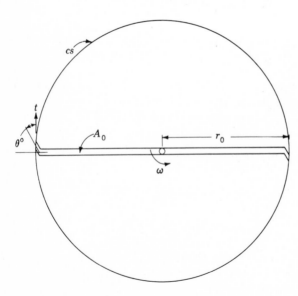

Figure 3.44 Plan view of sprinkler and control surface.

For complete circular symmetry, where r, ρ, v_t, and v_n are constant over the inlet and outlet control surfaces, it takes the simple form

$$T_z = \rho Q[(rv_t)_2 - (rv_t)_1] \tag{3.12.5}$$

since $\int \rho v_n \, dA = \rho Q$, the same at inlet or outlet.

Example 3.26 The sprinkler shown in Fig. 3.44 discharges water upward and outward from the horizontal plane so that it makes an angle of $\theta°$ with the t axis when the sprinkler arm is at rest. It has a constant cross-sectional flow area of A_0 and discharges q cfs starting with $\omega = 0$ and $t = 0$. The resisting torque due to bearings and seals is the constant T_0, and the moment of inertia of the rotating empty sprinkler head is I_s. Determine the equation for ω as a function of time.

Equation (3.12.2) may be applied. The control volume is the cylindrical area enclosing the rotating sprinkler head. The inflow is along the axis, so that it has no moment of momentum; hence, the torque $-T_0$ due to friction is equal to the time rate of change of moment of momentum of sprinkler head and fluid within the sprinkler head plus the net efflux of moment of momentum from the control volume. Let $V_r = q/2A_0$

$$-T_0 = 2\frac{d}{dt}\int_0^{r_0} A_0 \rho \omega r^2 \, dr + I_s \frac{d\omega}{dt} - \frac{2\rho q r_0}{2}(V_r \cos\theta - \omega r_0)$$

The total derivative may be used. Simplifying gives

$$\frac{d\omega}{dt}\left(I_s + \frac{2}{3}\rho A_0 r_0^3\right) = \rho q r_0(V_r \cos\theta - \omega r_0) - T_0$$

For rotation to start, $\rho q r_0 V_r \cos\theta$ must be greater than T_0. The equation is easily integrated to find ω as a function of t. The final value of ω is obtained by setting $d\omega/dt = 0$ in the equation.

Example 3.27 A turbine discharging 10 m^3/s is to be so designed that a torque of 10,000 N·m is to be exerted on an impeller turning at 200 rpm that takes all the moment of momentum out of the fluid. At the outer periphery of the impeller, $r = 1$ m. What must the tangential component of velocity be at this location?

Equation (3.12.5) is

$$T = \rho Q(rv_t)_{in}$$

in this case, since the outflow has $v_t = 0$. By solving for $v_{t_{in}}$,

$$v_{t_{in}} = \frac{T}{\rho Q r} = \frac{10,000 \text{ N·m}}{(1000 \text{ kg/m}^3)(10 \text{ m}^3/\text{s})(1 \text{ m})} = 1.000 \text{ m/s}$$

Example 3.28 The sprinkler of Fig. 3.45 discharges 0.01 cfs through each nozzle. Neglecting friction, find its speed of rotation. The area of each nozzle opening is 0.001 ft^2.

The fluid entering the sprinkler has no moment of momentum, and no torque is exerted on the system externally; hence the moment of momentum of fluid leaving must be zero. Let ω be the speed of rotation; then the moment of momentum leaving is

$$\rho Q_1 r_1 v_{t1} + \rho Q_2 r_2 v_{t2}$$

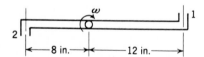

Figure 3.45 Rotating jet system.

in which v_{t1} and v_{t2} are absolute velocities. Then

$$v_{t1} = v_{r1} - \omega r_1 = \frac{Q_1}{0.001} - \omega r_1 = 10 - \omega$$

and

$$v_{t2} = v_{r2} - \omega r_2 = 10 - \tfrac{2}{3}\omega$$

For moment of momentum to be zero,

$$\rho Q(r_1 v_{t1} + r_2 v_{t2}) = 0 \qquad \text{or} \qquad 10 - \omega + \tfrac{2}{3}(10 - \tfrac{2}{3}\omega) = 0$$

and $\omega = 11.54$ rad/s, or 110.2 rpm.

PROBLEMS

3.1 In a flow of liquid through a pipeline the losses are 3 kW for average velocity of 2 m/s and 6 kW for 3 m/s. What is the nature of the flow?

3.2 When tripling the flow in a line causes the losses to increase by 7.64 times, how do the losses vary with velocity and what is the nature of the flow?

3.3 In two-dimensional flow around a circular cylinder (Fig. 3.2), the discharge between streamlines is 0.01 cfs per foot of width. At a great distance the streamlines are 0.20 in apart, and at a point near the cylinder they are 0.12 in apart. Calculate the magnitude of the velocity at these two points.

3.4 A three-dimensional velocity distribution is given by $u = -x, v = 2y, w = 5 - z$. Find the equation of the streamline through (2,1,1).

3.5 A two-dimensional flow can be described by $u = -y/b^2, v = x/a^2$. Verify that this is the flow of an incompressible fluid and that the ellipse $x^2/a^2 + y^2/b^2 = 1$ is a streamline.

3.6 A pipeline carries oil, sp gr 0.86, at $V = 2$ m/s through 200-mm-ID pipe. At another section the diameter is 60 mm. Find the velocity at this section and the mass rate of flow in kilograms per second.

3.7 Hydrogen is flowing in a 2.0-in-diameter pipe at the mass rate of 0.03 lb_m/s. At section 1 the pressure is 40 psia and $t = 80°$F. What is the average velocity?

3.8 A nozzle with a base diameter of 80 mm and with 30-mm-diameter tip discharges 10 l/s. Derive an expression for the fluid velocity along the axis of the nozzle. Measure the distance x along the axis from the plane of the larger diameter.

3.9 Does the velocity distribution of Prob. 3.4 for incompressible flow satisfy the continuity equation?

3.10 Does the velocity distribution

$$\mathbf{q} = \mathbf{i}(5x) + \mathbf{j}(5y) + \mathbf{k}(-10z)$$

satisfy the law of mass conservation for incompressible flow?

3.11 Consider a cube with 1-m edges parallel to the coordinate axes located in the first quadrant with one corner at the origin. By using the velocity distribution of Prob. 3.10, find the flow through each face and show that no mass is being accumulated within the cube if the fluid is of constant density.

3.12 Find the flow (per foot in the z direction) through each side of the square with corners at (0, 0), (0, 1), (1, 1), (1, 0) due to

$$\mathbf{q} = \mathbf{i}(16y - 12x) + \mathbf{j}(12y - 9x)$$

and show that continuity is satisfied.

3.13 Show that the velocity

$$\mathbf{q} = \mathbf{i}\,\frac{4x}{x^2 + y^2} + \mathbf{j}\,\frac{4y}{x^2 + y^2}$$

satisfies continuity at every point except the origin.

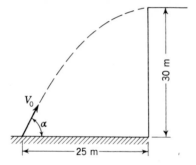

Figure 3.46 Problem 3.17.

3.14 Problem 3.13 is a velocity distribution that is everywhere radial from the origin with magnitude $v_r = 4/r$. Show that the flow through each circle concentric with the origin (per unit length in the z direction) is the same.

3.15 Perform the operation $\mathbf{V} \cdot \mathbf{q}$ on the velocity vectors of Probs. 3.10, 3.12, and 3.13.

3.16 By introducing the following relations between cartesian coordinates and plane polar coordinates, obtain a form of the the continuity equation in plane polar coordinates:

$$x^2 + y^2 = r^2 \qquad \frac{y}{x} = \tan \theta \qquad \frac{\partial}{\partial x} = \frac{\partial}{\partial r}\frac{\partial r}{\partial x} + \frac{\partial}{\partial \theta}\frac{\partial \theta}{\partial x}$$

$$u = v_r \cos \theta - v_\theta \sin \theta \qquad v = v_r \sin \theta + v_\theta \cos \theta$$

Does the velocity given in Prob. 3.14 satisfy the equation that has been derived?

3.17 What angle α of jet is required to reach the roof of the building of Fig. 3.46 with minimum jet velocity V_0 at the nozzle? What is the value of V_0?

3.18 A standpipe 20 ft in diameter and 40 ft high is filled with water. How much potential energy is in this water if the elevation datum is taken 10 ft below the base of the standpipe?

3.19 How much work could be obtained from the water of Prob. 3.18 if run through a 100 percent efficient turbine that discharged into a reservoir with elevation 30 ft below the base of the standpipe?

3.20 What is the kinetic-energy flux, in meter-newtons per second, of 0.01 m³/s of oil, sp gr 0.80, discharging through a 40-mm-diameter nozzle?

3.21 By neglecting air resistance, determine the height a vertical jet of water will rise with velocity 50 ft/s.

3.22 If the water jet of Prob. 3.21 is directed upward 45° with the horizontal and air resistance is neglected, how high will it rise and what is the velocity at its high point?

3.23 Show that the work a liquid can do by virtue of its pressure is $\int p \, d\mathcal{V}$, in which $\mathcal{V}$ is the volume of liquid displaced.

3.24 The velocity distribution between two parallel plates separated by a distance a is

$$u = -10\frac{y}{a} + 20\frac{y}{a}\left(1 - \frac{y}{a}\right)$$

in which u is the velocity component parallel to the plate and y is measured from, and normal to, the lower plate. Determine the volume rate of flow and the average velocity. What is the time rate of flow of kinetic energy between the plates? In what direction is the kinetic energy flowing?

3.25 What is the efflux of kinetic energy out of the cube given by Prob. 3.11 for the velocity prescribed in Prob. 3.10?

3.26 Water is flowing in a channel, as shown in Fig. 3.47. Neglecting all losses, determine the two possible depths of flow y_1 and y_2.

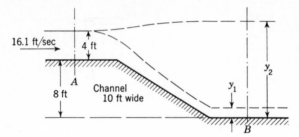

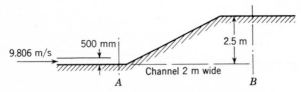

Figure 3.47 Problems 3.26, 3.28, 3.43.

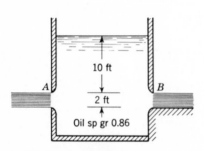

Figure 3.48 Problems 3.27, 3.44, 3.45.

3.27 High-velocity water flows up an inclined plane as shown in Fig. 3.48. Neglecting all losses, calculate the two possible depths of flow at section B.

3.28 Neglecting all losses, in Fig. 3.47 the channel narrows in the drop to 6 ft wide at section B. For uniform flow across section B, determine the two possible depths of flow.

3.29 Some steam locomotives had scoops installed that took water from a tank between the tracks and lifted it into the water reservoir in the tender. To lift the water 12 ft with a scoop, neglecting all losses, what speed is required? *Note:* Consider the locomotive stationary and the water moving toward it to reduce to a steady-flow situation.

3.30 In Fig. 3.49 oil discharges from a "two-dimensional" slot as indicated at A into the air. At B oil discharges from under a gate onto a floor. Neglecting all losses, determine the discharges at A and at B per foot of width. Why do they differ?

3.31 Neglecting losses, determine the discharge in Fig. 3.50.

3.32 Neglecting losses and surface-tension effects, derive an equation for the water surface r of the jet of Fig. 3.51 in terms of y/H.

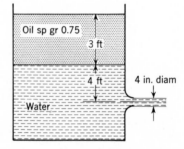

Figure 3.49 Problem 3.30.

Figure 3.50 Problem 3.31.

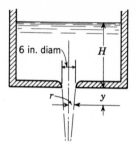

Figure 3.51 Problem 3.32.

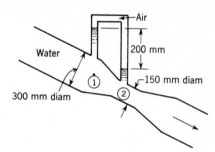

Figure 3.52 Problems 3.33, 3.62.

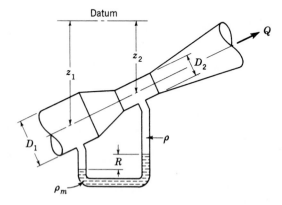

Figure 3.53 Problem 3.34.

3.33 Neglecting losses, find the discharge through the venturi meter of Fig. 3.52.

3.34 For the venturi meter and manometer installation shown in Fig. 3.53 derive an expression relating the volume rate of flow with the manometer reading.

3.35 In Fig. 3.54 determine V for $R = 15$ in.

3.36 Neglecting losses, calculate H in terms of R for Fig. 3.55.

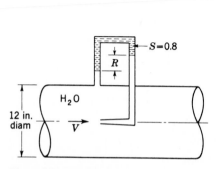

Figure 3.54 Problem 3.35.

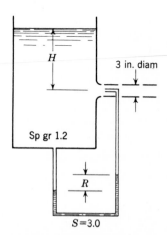

Figure 3.55 Problems 3.36, 3.64.

3.37 A pipeline leads from one water reservoir to another which has its water surface 10 m lower. For a discharge of 0.6 m³/s determine the losses in meter-newtons per kilogram and in kilowatts.

3.38 A pump which is located 10 ft above the surface of a lake expels a jet of water vertically upward a distance of 50 ft. If 0.5 cfs is being pumped by a 5-hp electric motor running at rated capacity, what is the efficiency of the motor-pump combination? What is the irreversibility of the pump system when comparing the zenith of the jet and the lake surface? What is the irreversibility after the water falls to the lake surface?

3.39 A blower delivers 2 m³/s air, $\rho = 1.3$ kg/m³, at an increase in pressure of 150 mm water. It is 72 percent efficient. Determine the irreversibility of the blower in meter-newtons per kilogram and in kilowatts, and determine the torque in the shaft if the blower turns at 1800 rpm.

3.40 An 18-ft-diameter pressure pipe has a velocity of 10 ft/s. After passing through a reducing bend, the flow is in a 16-ft-diameter pipe. If the losses vary as the square of the velocity, how much greater are they through the 16-ft pipe than through the 18-ft pipe per 1000 ft of pipe?

3.41 The velocity distribution in laminar flow in a pipe is given by

$$v = V_{max}[1 - (r/r_0)^2]$$

Determine the average velocity and the kinetic-energy correction factor.

3.42 For highly turbulent flow the velocity distribution in a pipe is given by

$$\frac{v}{v_{max}} = \left(\frac{y}{r_0}\right)^{1/9}$$

with y the wall distance and r_0 the pipe radius. Determine the kinetic-energy correction factor for this flow.

3.43 If the losses from section A to section B of Fig. 3.47 are 1.9 ft·lb/lb, determine the two possible depths at section B.

3.44 In the situation shown in Fig. 3.48 each kilogram of water increases in temperature 0.0006°C because of losses incurred in flowing between A and B. Determine the lower depth of flow at section B.

3.45 In Fig. 3.48 the channel changes in width from 2 m at section A to 3 m at section B. For losses of 0.3 m·N/N between sections A and B, find the two possible depths at section B.

3.46 At point A in a pipeline carrying water the diameter is 1 m, the pressure 98 kPa, and the velocity 1 m/s. At point B, 2 m higher than A, the diameter is 0.5 m, and the pressure 20 kPa. Determine the direction of flow.

3.47 For losses of 0.1 m·N/N, find the velocity at A in Fig. 3.56. Barometer reading is 750 mm Hg.

3.48 The losses in Fig. 3.57 for $H = 25$ ft are $3V^2/2g$ ft·lb/lb. What is the discharge?

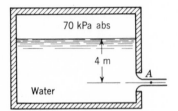

Figure 3.56 Problem 3.47.

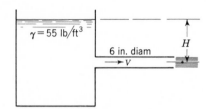

Figure 3.57 Problems 3.48, 3.49, 3.50.

3.49 For flow of 750 gpm in Fig. 3.57, determine H for losses of $10V^2/2g$ ft·lb/lb.

3.50 For 1500-gpm flow and $H = 32$ ft in Fig. 3.57, calculate the losses through the system in velocity heads, $KV^2/2g$.

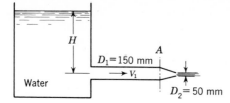

Figure 3.58 Problems 3.51, 3.52.

3.51 In Fig. 3.58 the losses up to section A are $5V_1^2/2g$ and the nozzle losses are $0.05V_2^2/2g$. Determine the discharge and the pressure at A. $H = 8$ m.

3.52 For pressure at A of 25 kPa in Fig. 3.58 with the losses in Prob. 3.51, determine the discharge and the head H.

3.53 The pumping system shown in Fig. 3.59 must have pressure of 5 psi in the discharge line when cavitation is incipient at the pump inlet. Calculate the length of pipe from the reservoir to the pump for this operating condition if the loss in this pipe can be expressed as $(V_1^2/2g)(0.03L/D)$. What horsepower is being supplied to the fluid by the pump? What percent of this power is being used to overcome losses? Barometer reads 30 in Hg.

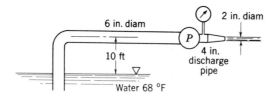

Figure 3.59 Problem 3.53.

3.54 In the siphon of Fig. 3.60, $h_1 = 1$ m, $h_2 = 3$ m, $D_1 = 3$ m, $D_2 = 5$ m, and the losses to section 2 are $2.6V_2^2/2g$, with 10 percent of the losses occurring before section 1. Find the discharge and the pressure at section 1.

3.55 Find the pressure at A of Prob. 3.54 if it is a stagnation point (velocity zero).

3.56 The siphon of Fig. 3.19 has a nozzle 150 mm long attached at section 3, reducing the diameter to 150 mm. For no losses, compute the discharge and the pressure at sections 2 and 3.

3.57 With exit velocity V_E in Prob. 3.56 and losses from 1 to 2 of $1.7V_2^2/2g$, from 2 to 3 of $0.9V_2^2/2g$, and through the nozzle $0.06V_E^2/2g$, calculate the discharge and the pressure at sections 2 and 3.

3.58 Determine the shaft horsepower for an 80 percent efficient pump to discharge 30 l/s through the system of Fig. 3.61. The system losses, exclusive of pump losses, are $12V^2/2g$, and $H = 16$ m.

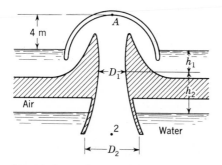

Figure 3.60 Problems 3.54, 3.55.

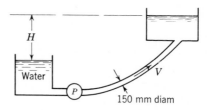

Figure 3.61 Problems 3.58, 3.59.

3.59 The fluid horsepower $(Q\gamma H_p/550)$ produced by the pump of Fig. 3.61 is 10. For $H = 60$ ft and system losses of $8\ V^2/2g$, determine the discharge and the pump head H_p. Draw the hydraulic and energy grade lines.

3.60 If the overall efficiency of the system and turbine in Fig. 3.62 is 80 percent, what horsepower is produced for $H = 200$ ft and $Q = 1000$ cfs?

3.61 Losses through the system of Fig. 3.62 are $4V^2/2g$, exclusive of the turbine. The turbine is 90 percent efficient and runs at 240 rpm. To produce 1000 hp for $H = 300$ ft, determine the discharge and torque in the turbine shaft. Draw the energy grade line and the hydraulic grade line.

3.62 With losses of $0.2V_1^2/2g$ between sections 1 and 2 of Fig. 3.52, calculate the flow in gallons per minute.

3.63 In Fig. 3.63 $H = 6$ m and $h = 5.75$ m. Calculate the discharge and the losses in meter-newtons per newton and in watts.

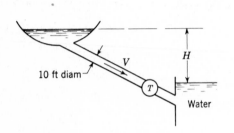

Figure 3.62 Problems 3.60, 3.61.   **Figure 3.63** Problem 3.63.

3.64 For losses of $0.1H$ through the nozzle of Fig. 3.55, what is the gage difference R in terms of H?

3.65 A liquid flows through a long pipeline with losses of 5 ft·lb/lb per 100 ft of pipe. What is the slope of the hydraulic and energy grade lines?

3.66 In Fig. 3.64, 100 l/s water flows from section 1 to section 2 with losses of $0.4(V_1 - V_2)^2/2g$; $p_1 = 80$ kPa. Compute p_2 and plot the energy and hydraulic grade lines through the diffuser.

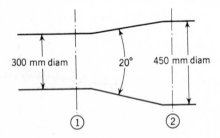

Figure 3.64 Problem 3.66.

3.67 In an isothermal, reversible flow at 200°F, 3 Btu/s heat is added to 14 slugs/s flowing through a control volume. Calculate the entropy increase, in foot-pounds per slug per degree Rankine.

3.68 In isothermal flow of a real fluid through a pipe system the losses are 20 m·N/kg per 100 m, and 0.02 kcal/s per 100 m heat transfer from the fluid is required to hold the temperature at 10°C. What is the entropy change Δs in meter-newtons per kilogram per kelvin of pipe system for 4 kg/s flowing?

3.69 Determine the momentum correction factor for the velocity distribution of Prob. 3.41.

3.70 Calculate the average velocity and momentum correction factor for the velocity distribution in a pipe,

$$\frac{v}{v_{max}} = \left(\frac{y}{r_0}\right)^{1/n}$$

with y the wall distance and r_0 the pipe radius.

3.71 By introducing $V + v'$ for v into Eq. (3.11.4) show that $\beta \geq 1$. The term v' is the variation of v from the average velocity V and can be positive or negative.

3.72 Determine the time rate of x momentum passing out of the cube of Prob. 3.11. (*Hint:* Consider all six faces of the cube.)

3.73 Calculate the y-momentum efflux from the figure described in Prob. 3.12 for the velocity given there.

3.74 If gravity acts in the negative z direction, determine the z component of the force acting on the fluid within the cube described in Prob. 3.11 for the velocity specified there.

3.75 Find the y component of the force acting on the control volume given in Prob. 3.12 for the velocity given there. Consider gravity to be acting in the negative y direction.

3.76 What force components F_x, F_y are required to hold the black box of Fig. 3.65 stationary? All pressures are zero gage.

3.77 What force F (Fig. 3.66) is required to hold the plate for oil flow, sp gr 0.83, for $V_0 = 25$ m/s?

3.78 How much is the apparent weight of the tank full of water (Fig. 3.67) increased by the steady jet flow into the tank?

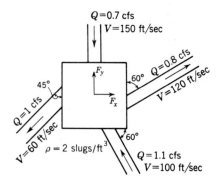

Figure 3.65 Problem 3.76.

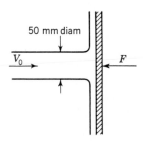

Figure 3.66 Problem 3.77.

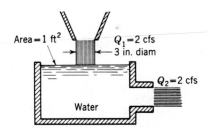

Figure 3.67 Problems 3.78, 3.139.

3.79 Does a nozzle on a fire hose place the hose in tension or in compression?

3.80 When a jet from a nozzle is used to aid in maneuvering a fireboat, can more force be obtained by directing the jet against a solid surface such as a wharf than by allowing it to discharge into air?

3.81 Work Example 3.16 with the flow direction reversed, and compare results.

3.82 In the reducing bend of Fig. 3.25, $D_1 = 4$ m, $D_2 = 3$ m, $\theta = 135°$, $Q = 50$ m³/s, $W = 392.2$ kN, $z = 2$ m, $p_2 = 1.4$ MPa, $x = 2.2$ m, and losses may be neglected. Find the force components and the line of action of the force which must be resisted by an anchor block.

3.83 20 ft³/s of water flows through an 18-in-diameter pipeline that contains a horizontal 90° bend. The pressure at the entrance to the bend is 20 psi. Determine the force components, parallel and normal to the approach velocity, required to hold the bend in place. Neglect losses.

3.84 Oil, sp gr 0.83, flows through a 90° expanding pipe bend from 400- to 600-mm-diameter pipe. The pressure at the bend entrance is 130 kPa, and losses are to be neglected. For 0.6 m³/s, determine the force components (parallel and normal to the approach velocity) necessary to support the bend.

3.85 Work Prob. 3.84 with elbow losses of $0.6V_1^2/2g$, and with V_1 the approach velocity, and compare results.

3.86 A 4-in-diameter steam line carries saturated steam at 1400 ft/s velocity. Water is entrained by the steam at the rate of 0.3 lb/s. What force is required to hold a 90° bend in place owing to the entrained water?

3.87 Neglecting losses, determine the x and y components of force needed to hold the Y (Fig. 3.68) in place. The plane of the Y is horizontal.

3.88 Determine the net force on the sluice gate shown in Fig. 3.69. Neglect losses. By noting that the pressure at A and B is atmospheric, sketch the pressure distribution on the surface AB. Is it a hydrostatic distribution? How is it related to the force just calculated?

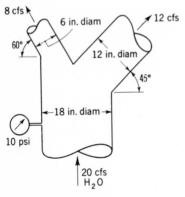

Figure 3.68 Problem 3.87.

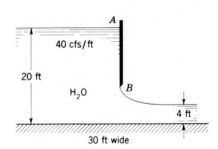

Figure 3.69 Problem 3.88.

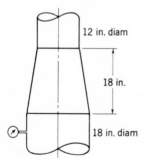

Figure 3.70 Problem 3.89.

3.89 The vertical reducing section shown in Fig. 3.70 contains oil, sp gr 0.86, flowing upward at the rate of 0.5 m³/s. The pressure at the larger section is 200 kPa. Neglecting losses but including gravity, determine the force on the contraction.

3.90 Apply the momentum and energy equations to a windmill as if it were a propeller, noting that the slipstream is slowed down and expands as it passes through the blades. Show that the velocity through the plane of the blades is the average of the velocities in the slipstream at the downstream and upstream sections. By defining the theoretical efficiency (neglecting all losses) as the power output divided by the power available in an undisturbed jet having the area at the plane of the blades, determine the maximum theoretical efficiency of a windmill.

3.91 An airplane with propeller diameter of 8.0 ft travels through still air ($\rho = 0.0022$ slug/ft³) at 200 mph. The speed of air through the plane of the propeller is 280 mph relative to the airplane. Calculate (*a*) the thrust on the plane, (*b*) the kinetic energy per second remaining in the slipstream, (*c*) the theoretical horsepower required to drive the propeller, (*d*) the propeller efficiency, and (*e*) the pressure difference across the blades.

3.92 A boat traveling at 40 km/h has a 500-mm-diameter propeller that discharges 4.5 m³/s through its blades. Determine the thrust on the boat, the theoretical efficiency of the propulsion system, and the power input to the propeller.

3.93 A ship propeller has a theoretical efficiency of 60 percent. If it is 3.5 ft in diameter and the ship travels 20 mph, what is the thrust developed and what is the theoretical horsepower required?

3.94 A jet-propelled airplane traveling 1000 km/h takes in 40 kg/s air and discharges it at 550 m/s relative to the airplane. Neglecting the weight of fuel, find the thrust produced.

3.95 A jet-propelled airplane travels 700 mph. It takes in 165 lb_m/s of air, burns 3 lb_m/s of fuel, and develops 8000 lb of thrust. What is the velocity of the exhaust gas?

3.96 What is the theoretical mechanical efficiency of the jet engine of Prob. 3.95?

3.97 A boat requires an 18-kN thrust to keep it moving at 25 km/h. How many cubic meters of water per second must be taken in and ejected through a 500-mm pipe to maintain this motion? What is the overall efficiency if the pumping system is 60 percent efficient?

3.98 In Prob. 3.97 what would be the required discharge if water were taken from a tank inside the boat and ejected from the stern through a 430-mm pipe?

3.99 Determine the size of jet pipe and the theoretical power necessary to produce a thrust of 10 kN on a boat moving 12 m/s when the propulsive efficiency is 68 percent.

3.100 In Fig. 3.71, a jet, $\rho = 2$ slugs/ft³, is deflected by a vane through 180°. Assume that the cart is frictionless and free to move in a horizontal direction. The cart weighs 200 lb. Determine the velocity and the distance traveled by the cart 10 s after the jet is directed against the vane. $A_0 = 0.02$ ft²; $V_0 = 100$ ft/s.

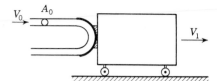

Figure 3.71 Problem 3.100.

3.101 A rocket burns 260 lb_m/s fuel, ejecting hot gases at 6560 ft/s relative to the rocket. How much thrust is produced at 1500 and 3000 mph?

3.102 What is the mechanical efficiency of a rocket moving at 1200 m/s that ejects gas at 1800 m/s relative to the rocket?

3.103 Can a rocket travel faster than the velocity of ejected gas? What is the mechanical efficiency when it travels 4200 m/s and the gas is ejected at 2800 m/s relative to the rocket? Is a positive thrust developed?

3.104 In Example 3.19 what is the thrust just before the completion of burning?

3.105 Neglecting air resistance, what velocity will a vertically directed V-2 rocket attain in 68 s if it starts from rest, initially has a mass of 13,000 kg, burns 124 kg/s fuel, and ejects gas at $v_r = 1950$ m/s? Consider $g = 9.8$ m/s².

3.106 What altitude has the rocket of Prob. 3.105 reached after 68 s?

3.107 If the fuel supply is exhausted after 68 s (burnout), what is the maximum height of the rocket of Prob. 3.105?

3.108 What is the thrust of the rocket of Prob. 3.105 after 68 s?

3.109 Draw the polar vector diagram for a vane, angle θ, doing work on a jet. Label all vectors.

3.110 Determine the resultant force exerted on the vane of Fig. 3.32. $A_0 = 0.1$ ft²; $V_0 = 100$ ft/s; $\theta = 60°$, $\gamma = 60$ lb/ft³. How can the line of action be determined?

3.111 In Fig. 3.33, 45 percent of the flow is deflected in one direction. What is the plate angle θ?

3.112 A flat plate is moving with velocity u into a jet, as shown in Fig. 3.72. Derive the expression for power required to move the plate.

3.113 At what speed u should the cart of Fig. 3.72 move away from the jet in order to produce maximum power from the jet?

3.114 Calculate the force components F_x, F_y needed to hold the stationary vane of Fig. 3.73. $Q_0 = 80$ l/s; $\rho = 1000$ kg/m³; $V_0 = 120$ m/s.

3.115 If the vane of Fig. 3.73 moves in the x direction at $u = 40$ ft/s, for $Q_0 = 2$ ft³/s, $\rho = 1.935$ slugs/ft³, $V_0 = 120$ ft/s, what are the force components F_x, F_y?

3.116 For the flow divider of Fig. 3.74 find the force components for the following conditions: $Q_0 = 10$ l/s, $Q_1 = 3$ l/s, $\theta_0 = 45°$, $\theta_1 = 30°$, $\theta_2 = 120°$, $V_0 = 10$ m/s, $\rho = 830$ kg/m³.

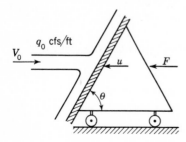

Figure 3.72 Problems 3.112, 3.113.

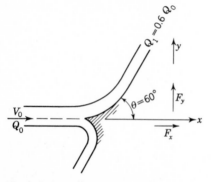

Figure 3.73 Problems 3.114, 3.115.

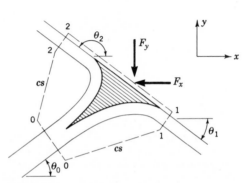

Figure 3.74 Problems 3.116, 3.117.

3.117 Solve the preceding problem by graphical vector addition.

3.118 At what speed u should the vane of Fig. 3.34 travel for maximum power from the jet? What should be the angle θ for maximum power?

3.119 Draw the polar vector diagram for the moving vane of Fig. 3.34 for $V_0 = 30$ m/s, $u = 20$ m/s, and $\theta = 160°$.

3.120 Draw the polar vector diagram for the moving vane of Fig. 3.34 for $V_0 = 40$ m/s, $u = -20$ m/s, and $\theta = 150°$.

3.121 What horsepower can be developed from (a) a single vane and (b) a series of vanes (Fig. 3.34) when $A_0 = 10$ in^2, $V_0 = 240$ ft/s, $u = 80$ ft/s, and $\theta = 173°$, for water flowing?

3.122 Determine the blade angles θ_1 and θ_2 of Fig. 3.75 so that the flow enters the vane tangent to its leading edge and leaves with no x component of absolute velocity.

3.123 Determine the vane angle required to deflect the absolute velocity of a jet 130° (Fig. 3.76).

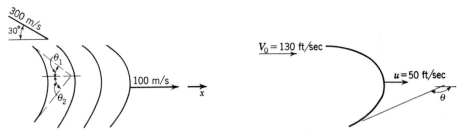

Figure 3.75 Problem 3.122. Figure 3.76 Problem 3.123.

3.124 In Prob. 3.29 for pickup of 30 l/s water at locomotive speed of 60 km/h, what force is exerted parallel to the tracks?

3.125 Figure 3.77 shows an orifice called a Borda mouthpiece. The tube is long enough for the fluid velocity near the bottom of the tank to be nearly zero. Calculate the ratio of the jet area to the tube area.

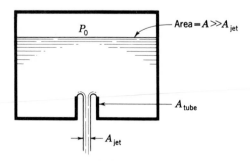

Figure 3.77 Problem 3.125.

3.126 Determine the irreversibility in foot-pounds per pound mass for 5 ft^3/s flow of liquid, $\rho = 1.6$ slugs/ft^3, through a sudden expansion from a 12- to 24-in-diameter pipe. $g = 30$ ft/s^2.

3.127 Air flows through a 650-mm-diameter duct at $p = 70$ kPa, $t = 10°C$, $V = 60$ m/s. The duct suddenly expands to 800 mm diameter. Considering the gas as incompressible, calculate the losses, in meter-newtons per newton of air, and the pressure difference, in centimeters of water.

3.128 What are the losses when 4 m^3/s water discharges from a submerged 1.5-m-diameter pipe into a reservoir?

3.129 Show that, in the limiting case, as $y_1 = y_2$ in Eq. (3.11.23), the relation $V = \sqrt{gy}$ is obtained.

3.130 A jump occurs in a 6-m-wide channel carrying 15 m³/s water at a depth of 300 mm. Determine y_2, V_2, and the losses in meter-newtons per newton, in kilowatts, and in kilocalories per kilogram.

3.131 Derive an expression for a hydraulic jump in a channel having an equilateral triangle as its cross section (symmetric with the vertical).

3.132 Derive Eq. (3.11.24).

3.133 Assuming no losses through the gate of Fig. 3.78 and neglecting $V_0^2/2g$, for $y_0 = 20$ ft and $y_1 = 2$ ft, find y_2 and losses through the jump. What is the basis for neglecting $V_0^2/2g$?

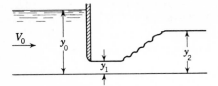

Figure 3.78 Problem 3.133.

3.134 Under the same assumptions as in Prob. 3.133, for $y_1 = 400$ mm and $y_2 = 2$ m, determine y_0.

3.135 Under the same assumptions as in Prob. 3.133, $y_0 = 20$ ft and $y_2 = 8$ ft. Find the discharge per foot.

3.136 For losses down the spillway of Fig. 3.79 of 2 m·N/N and discharge per meter of 10 m³/s, determine the floor elevation for the jump to occur.

3.137 Water is flowing through the pipe of Fig. 3.80 with velocity $V = 8.02$ ft/s and losses of 10 ft·lb/lb up to section 1. When the obstruction at the end of the pipe is removed, calculate the acceleration of water in the pipe.

3.138 Water fills the piping system of Fig. 3.81. At one instant $p_1 = 10$ psi, $p_2 = 0$, $V_1 = 10$ ft/s, and the flow rate is increasing by 3000 gpm/min. Find the force F_x required to hold the piping system stationary.

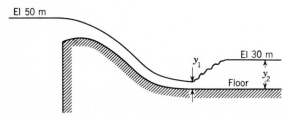

Figure 3.79 Problem 3.136.

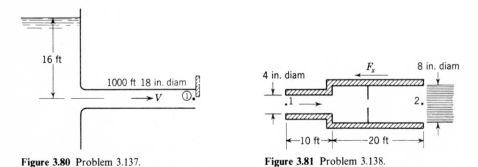

Figure 3.80 Problem 3.137.

Figure 3.81 Problem 3.138.

3.139 If in Fig. 3.67 Q_2 is 1.0 cfs, what is the vertical force to support the tank? Assume that overflow has not occurred. The tank weighs 20 lb, and water depth is 1 ft.

3.140 In Fig. 3.43b, $r_1 = 120$ mm, $r_2 = 160$ mm, $v_{t1} = 0$, and $v_{t2} = 3$ m/s for a centrifugal pump impeller discharging 0.2 m³/s of water. What torque must be exerted on the impeller?

3.141 In a centrifugal pump 400 gpm water leaves an 8-in-diameter impeller with a tangential velocity component of 30 ft/s. It enters the impeller in a radial direction. For pump speed of 1200 rpm and neglecting all losses, determine the torque in the pump shaft, the horsepower input, and the energy added to the flow in foot-pounds per pound.

3.142 A water turbine at 240 rpm discharges 40 m³/s. To produce 42 MW, what must be the tangential component of velocity at the entrance to the impeller at $r = 1.6$ m? All whirl is taken from the water when it leaves the turbine. Neglect all losses. What head is required for the turbine?

3.143 The symmetrical sprinkler of Fig. 3.82 has a total discharge of 14 gpm and is frictionless. Determine its rpm if the nozzle tips are $\frac{1}{4}$ in diameter.

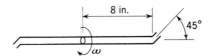

Figure 3.82 Problems 3.143, 3.144, 3.145, 3.146.

3.144 What torque would be required to hold the sprinkler of Prob. 3.143 stationary? Total flow 2 l/s water.

3.145 If there is a torque resistance of 0.50 lb·ft in the shaft of Prob. 3.143, what is its speed of rotation?

3.146 For torque resistance of $0.01\omega^2$ in the shaft, determine the speed of rotation of the sprinkler of Prob. 3.143.

FOUR

DIMENSIONAL ANALYSIS AND DYNAMIC SIMILITUDE

Dimensionless parameters significantly deepen our understanding of fluid-flow phenomena in a way which is analogous to the case of a hydraulic jack, where the ratio of piston diameters determines the mechanical advantage, a dimensionless number which is independent of the overall size of the jack. They permit limited experimental results to be applied to situations involving different physical dimensions and often different fluid properties. The concepts of dimensional analysis introduced in this chapter plus an understanding of the mechanics of the type of flow under study make possible this generalization of experimental data. The consequence of such generalization is manifold, since one is now able to describe the phenomenon in its entirety and is not restricted to discussing the specialized experiment that was performed. Thus, it is possible to conduct fewer, although highly selective, experiments to uncover the hidden facets of the problem and thereby achieve important savings in time and money. The results of an investigation can also be presented to other engineers and scientists in a more compact and meaningful way to facilitate their use. Equally important is the fact that, through such incisive and uncluttered presentations of information, researchers are able to discover new features and missing areas of knowledge of the problem at hand. This directed advancement of our understanding of a phenomenon would be impaired if the tools of dimensional analysis were not available. In the following chapter, dealing primarily with viscous effects, one parameter is highly significant, viz., the Reynolds number. In Chap. 6, dealing with compressible flow, the Mach number is the most important dimensionless parameter. In Chap. 11, dealing with open channels, the Froude number has the greatest significance.

Many of the dimensionless parameters may be viewed as a ratio of a pair of fluid forces, the relative magnitude indicating the relative importance of one of the forces with respect to the other. If some forces in a particular flow situation are very much larger than a few others, it is often possible to neglect the effect of the smaller forces and treat the phenomenon as though it were completely determined by the major forces. This means that simpler, although not necessarily easy, mathematical and experimental procedures can be used to solve the problem. For situations with several forces of the same magnitude, such as inertial, viscous, and gravitational forces, special techniques are required. After a discussion of dimensions, dimensional analysis, and dimensionless parameters, dynamic similitude and model studies are presented.

4.1 DIMENSIONAL HOMOGENEITY AND DIMENSIONLESS RATIOS

Solving practical design problems in fluid mechanics usually requires both theoretical developments and experimental results. By grouping significant quantities into dimensionless parameters, it is possible to reduce the number of variables appearing and to make this compact result (equations or data plots) applicable to all similar situations.

If one were to write the equation of motion $\Sigma F = ma$ for a fluid particle, including all types of force terms that could act, such as gravity, pressure, viscous, elastic, and surface-tension forces, an equation of the sum of these forces equated to ma, the inertial force, would result. As with all physical equations, each term must have the same dimensions, in this case, force. The division of each term of the equation by any one of the terms would make the equation dimensionless. For example, dividing through by the inertial force term would yield a sum of dimensionless parameters equated to unity. The relative size of any one parameter, compared with unity, would indicate its importance. If one were to divide the force equation through by a different term, say the viscous force term, another set of dimensionless parameters would result. Without experience in the flow case it is difficult to determine which parameters will be most useful.

An example of the use of dimensional analysis and its advantages is given by considering the hydraulic jump, treated in Sec. 3.11. The momentum equation for this case

$$\frac{\gamma y_1^2}{2} - \frac{\gamma y_2^2}{2} = \frac{V_1 y_1 \gamma}{g} (V_2 - V_1) \tag{4.1.1}$$

can be rewritten as

$$\frac{\gamma}{2} y_1^2 \left[1 - \left(\frac{y_2}{y_1} \right)^2 \right] = V_1^2 \frac{\gamma}{g} y_1 \left(1 - \frac{y_2}{y_1} \right) \frac{y_1}{y_2}$$

Clearly, the right-hand side represents the inertial forces and the left-hand side represents the pressure forces due to gravity. These two forces are of equal magnitude, since one determines the other in this equation. Furthermore, the term $\gamma y_1^2 /2$ has the dimensions of force per unit width, and it multiplies a dimensionless number which is specified by the geometry of the hydraulic jump.

If one divides this equation by the geometric term $1 - y_2/y_1$ and a number representative of the gravity forces, one has

$$\frac{V_1^2}{gy_1} = \frac{1}{2}\frac{y_2}{y_1}\left(1 + \frac{y_2}{y_1}\right) \tag{4.1.2}$$

It is now clear that the left-hand side is the ratio of the inertia and gravity forces, even though the explicit representation of the forces has been obscured through the cancellation of terms that are common in both the numerator and denominator. This ratio is equivalent to a dimensionless parameter, actually the square of the *Froude number*, which will be discussed in further detail later in this chapter. It is also interesting to note that this ratio of forces is known once the ratio y_2/y_1 is given, regardless of what the values y_2 and y_1 are. From this observation one can obtain an appreciation of the increased scope that Eq. (4.1.2) affords over Eq. (4.1.1) even though one is only a rearrangement of the other.

In writing the momentum equation which led to Eq. (4.1.2) only inertia and gravity forces were included in the original problem statement. But other forces, such as surface tension and viscosity, are present. These were neglected as being small in comparison with gravity and inertia forces; however, only experience with the phenomenon, or with phenomena similar to it, would justify such an initial simplification. For example, if viscosity had been included because one was not sure of the magnitude of its effect, the momentum equation would become

$$\frac{\gamma y_1^2}{2} - \frac{\gamma y_2^2}{2} - F_{\text{viscous}} = V_1 \frac{y_1 \gamma}{g}(V_2 - V_1)$$

with the result that

$$\frac{V_1^2}{gy_1} + \frac{F_{\text{viscous}} y_2}{\gamma y_1^2(y_1 - y_2)} = \frac{1}{2}\frac{y_2}{y_1}\left(1 + \frac{y_2}{y_1}\right)$$

This statement is more complete than that given by Eq. (4.1.2). However, experiments would show that the second term on the left-hand side is usually a small fraction of the first term and could be neglected in making initial tests on a hydraulic jump.

In the last equation one can consider the ratio y_2/y_1 to be a dependent variable which is determined for each of the various values of the force ratios, V_1^2/gy_1 and $F_{\text{viscous}}/\gamma y_1^2$, which are the independent variables. From the previous discussion it appears that the latter variable plays only a minor role in determining the values of y_2/y_1. Nevertheless, if one observed that the ratios of the forces, V_1^2/gy_1 and $F_{\text{viscous}}/\gamma y_1^2$, had the same values in two different tests, one would expect, on the basis of the last equation, that the values of y_2/y_1 would be the same in the two situations. If the ratio of V_1^2/gy_1 was the same in the two tests but the ratio $F_{\text{viscous}}/\gamma y_1$, which has only a minor influence for this case, was not, one would conclude that the values of y_2/y_1 for the two cases would be almost the same.

This is the key to much of what follows. For if one can create in a model experiment the same geometric and force ratios that occur on the full-scale unit, then the dimensionless solution for the model is valid for the prototype also. Often, as will be seen, it is not possible to have all the ratios equal in the model and prototype. Then one attempts to plan the experimentation in such a way that the dominant force ratios are as nearly equal as possible. The results obtained with such incomplete modeling are often sufficient to describe the phenomenon in the detail that is desired.

Writing a force equation for a complex situation may not be feasible, and another process, *dimensional analysis*, is then used if one knows the pertinent quantities that enter into the problem.

In a given situation several of the forces may be of little significance, leaving perhaps two or three forces of the same order of magnitude. With three forces of the same order of magnitude, two dimensionless parameters are obtained; one set of experimental data on a geometrically similar model provides the relations between parameters holding for all other similar flow cases.

EXERCISE

4.1.1 Select a common dimensionless parameter in fluid mechanics from the following: (*a*) angular velocity; (*b*) kinematic viscosity; (*c*) specific gravity; (*d*) specific weight; (*e*) none of these answers.

4.2 DIMENSIONS AND UNITS

The dimensions of mechanics are force, mass, length, and time; they are related by Newton's second law of motion,

$$\mathbf{F} = m\mathbf{a} \tag{4.2.1}$$

Force and mass units are discussed in Sec. 1.2. For all physical systems, it would probably be necessary to introduce two more dimensions, one dealing with electromagnetics and the other with thermal effects. For the compressible work in this text, it is unnecessary to include a thermal unit, because the equations of state link pressure, density, and temperature.

Newton's second law of motion in dimensional form is

$$F = MLT^{-2} \tag{4.2.2}$$

which shows that only three of the dimensions are independent. F is the force dimension, M the mass dimension, L the length dimension, and T the time dimension. One common system employed in dimensional analysis is the MLT system.

Table 4.1 Dimensions of physical quantities used in fluid mechanics

Quantity	Symbol	Dimensions (M, L, T)
Length	l	L
Time	t	T
Mass	m	M
Force	F	MLT^{-2}
Velocity	V	LT^{-1}
Acceleration	a	LT^{-2}
Area	A	L^2
Discharge	Q	$L^3 T^{-1}$
Pressure	Δp	$ML^{-1}T^{-2}$
Gravity	g	LT^{-2}
Density	ρ	ML^{-3}
Specific weight	γ	$ML^{-2}T^{-2}$
Dynamic viscosity	μ	$ML^{-1}T^{-1}$
Kinematic viscosity	ν	$L^2 T^{-1}$
Surface tension	σ	MT^{-2}
Bulk modulus of elasticity	K	$ML^{-1}T^{-2}$

Table 4.1 lists some of the quantities used in fluid flow, together with their symbols and dimensions.

EXERCISE

4.2.1 A dimensionless combination of Δp, ρ, l, Q is

(a) $\sqrt{\dfrac{\Delta p}{\rho}\dfrac{Q}{l^2}}$ (b) $\dfrac{\rho Q}{\Delta p l^2}$ (c) $\dfrac{\rho l}{\Delta p Q^2}$ (d) $\dfrac{\Delta p l Q}{\rho}$ (e) $\sqrt{\dfrac{\rho}{\Delta p}\dfrac{Q}{l^2}}$

4.3 THE Π THEOREM

The Buckingham† Π theorem proves that, in a physical problem including n quantities in which there are m dimensions, the quantities can be arranged into $n - m$ independent dimensionless parameters. Let $A_1, A_2, A_3, \ldots, A_n$ be the quantities involved, such as pressure, viscosity, velocity, etc. All the quantities are known to be essential to the solution, and hence some functional relation must exist

$$F(A_1, A_2, A_3, \ldots, A_n) = 0 \qquad (4.3.1)$$

† E. Buckingham, Model Experiments and the Form of Empirical Equations, *Trans. ASME*, vol. 37, pp. 263–296, 1915.

If $\Pi_1, \Pi_2, \ldots$, represent dimensionless groupings of the quantities $A_1, A_2, A_3, \ldots$, then with m dimensions involved, an equation of the form

$$f(\Pi_1, \Pi_2, \Pi_3, \ldots, \Pi_{n-m}) = 0 \qquad (4.3.2)$$

exists.

Proof of the Π theorem may be found in Buckingham's paper, as well as in Sedov's book listed in the references at the end of this chapter. The method of determining the Π parameters is to select m of the A quantities, with different dimensions, that contain among them the m dimensions, and to use them as repeating variables† together with one of the other A quantities for each Π. For example, let A_1, A_2, A_3 contain M, L, and T, not necessarily in each one, but collectively. Then the first Π parameter is made up as

$$\Pi_1 = A_1^{x_1} A_2^{y_1} A_3^{z_1} A_4 \qquad (4.3.3)$$

the second one as

$$\Pi_2 = A_1^{x_2} A_2^{y_2} A_3^{z_2} A_5$$

and so on, until

$$\Pi_{n-m} = A_1^{x_{n-m}} A_2^{y_{n-m}} A_3^{z_{n-m}} A_n$$

In these equations the exponents are to be determined so that each Π is dimensionless. The dimensions of the A quantities are substituted, and the exponents of M, L, and T are set equal to zero respectively. These produce three equations in three unknowns for each Π parameter, so that the x, y, z exponents can be determined, and hence the Π parameter.

If only two dimensions are involved, then two of the A quantities are selected as repeating variables, and two equations in the two unknown exponents are obtained for each Π term.

In many cases the grouping of A terms is such that the dimensionless arrangement is evident by inspection. The simplest case is that when two quantities have the same dimensions, e.g., length, the ratio of these two terms is the Π parameter.

The procedure is best illustrated by several examples.

Example 4.1 The discharge through a horizontal capillary tube is thought to depend upon the pressure drop per unit length, the diameter, and the viscosity. Find the form of the equation. The quantities are listed with their dimensions:

Quantity	Symbol	Dimensions
Discharge	Q	$L^3 T^{-1}$
Pressure drop/length	$\Delta p/l$	$M L^{-2} T^{-2}$
Diameter	D	L
Viscosity	μ	$M L^{-1} T^{-1}$

† It is essential that no one of the m selected quantities used as repeating variables be derivable from the other repeating variables.

Then

$$F\left(Q, \frac{\Delta p}{l}, D, \mu\right) = 0$$

Three dimensions are used, and with four quantities there will be one Π parameter:

$$\Pi = Q^{x_1}\left(\frac{\Delta p}{l}\right)^{y_1} D^{z_1}\mu$$

Substituting in the dimensions gives

$$\Pi = (L^3 T^{-1})^{x_1}(ML^{-2}T^{-2})^{y_1}L^{z_1}ML^{-1}T^{-1} = M^0 L^0 T^0$$

The exponents of each dimension must be the same on both sides of the equation. With L first,

$$3x_1 - 2y_1 + z_1 - 1 = 0$$

and similarly for M and T

$$y_1 + 1 = 0$$

$$-x_1 - 2y_1 - 1 = 0$$

from which $x_1 = 1$, $y_1 = -1$, $z_1 = -4$, and

$$\Pi = \frac{Q\mu}{D^4 \, \Delta p/l}$$

After solving for Q,

$$Q = C \, \frac{\Delta p}{l} \frac{D^4}{\mu}$$

from which dimensional analysis yields no information about the numerical value of the dimensionless constant C; experiment (or analysis) shows that it is $\pi/128$ [Eq. (5.2.10a)].

When dimensional analysis is used, the variables in a problem must be known. In the last example if kinematic viscosity had been used in place of dynamic viscosity, an incorrect formula would have resulted.

Example 4.2 A V-notch weir is a vertical plate with a notch of angle ϕ cut into the top of it and placed across an open channel. The liquid in the channel is backed up and forced to flow through the notch. The discharge Q is some function of the elevation H of upstream liquid surface above the bottom of the notch. In addition, the discharge depends upon gravity and upon the velocity of approach V_0 to the weir. Determine the form of discharge equation.
A functional relation

$$F(Q, H, g, V_0, \phi) = 0$$

is to be grouped into dimensionless parameters. ϕ is dimensionless; hence, it is one Π parameter. Only two dimensions are used, L and T. If g and H are the repeating variables,

$$\Pi_1 = H^{x_1}g^{y_1}Q = L^{x_1}(LT^{-2})^{y_1}L^3 T^{-1}$$

$$\Pi_2 = H^{x_2}g^{y_2}V_0 = L^{x_2}(LT^{-2})^{y_2}LT^{-1}$$

Then

$$x_1 + y_1 + 3 = 0 \qquad x_2 + y_2 + 1 = 0$$

$$-2y_1 - 1 = 0 \qquad -2y_2 - 1 = 0$$

from which $x_1 = -\frac{5}{2}$, $y_1 = -\frac{1}{2}$, $x_2 = -\frac{1}{2}$, $y_2 = -\frac{1}{2}$, and

$$\Pi_1 = \frac{Q}{\sqrt{g}\,H^{5/2}} \qquad \Pi_2 = \frac{V_0}{\sqrt{gH}} \qquad \Pi_3 = \phi$$

or

$$f\left(\frac{Q}{\sqrt{g}\,H^{5/2}}, \frac{V_0}{\sqrt{gH}}, \phi\right) = 0$$

This may be written

$$\frac{Q}{\sqrt{g}\,H^{5/2}} = f_1\left(\frac{V_0}{\sqrt{gH}}, \phi\right)$$

in which both f, f_1 are unknown functions. After solving for Q,

$$Q = \sqrt{g}\,H^{5/2}f_1\left(\frac{V_0}{\sqrt{gH}}, \phi\right)$$

Either experiment or analysis is required to yield additional information about the function f_1. If H and V_0 were selected as repeating variables in place of g and H,

$$\Pi_1 = H^{x_1}V_0^{y_1}Q = L^{x_1}(LT^{-1})^{y_1}L^3T^{-1}$$

$$\Pi_2 = H^{x_2}V_0^{y_2}g = L^{x_2}(LT^{-1})^{y_2}LT^{-2}$$

Then

$$x_1 + y_1 + 3 = 0 \qquad x_2 + y_2 + 1 = 0$$

$$-y_1 - 1 = 0 \qquad -y_2 - 2 = 0$$

from which $x_1 = -2$, $y_1 = -1$, $x_2 = 1$, $y_2 = -2$, and

$$\Pi_1 = \frac{Q}{H^2V_0} \qquad \Pi_2 = \frac{gH}{V_0^2} \qquad \Pi_3 = \phi$$

or

$$f\left(\frac{Q}{H^2V_0}, \frac{gH}{V_0^2}, \phi\right) = 0$$

Since any of the Π parameters may be inverted or raised to any power without affecting their dimensionless status,

$$Q = V_0 H^2 f_2\left(\frac{V_0}{\sqrt{gH}}, \phi\right)$$

The unknown function f_2 has the same parameters as f_1, but it could not be the same function. The last form is not very useful, in general, because frequently V_0 may be neglected with V-notch weirs. This shows that a term of minor importance should not be selected as a repeating variable.

Another method of determining alternative sets of Π parameters would be the arbitrary recombination of the first set. If four independent Π parameters $\Pi_1, \Pi_2, \Pi_3, \Pi_4$ are known, the term

$$\Pi_a = \Pi_1^{a_1}\Pi_2^{a_2}\Pi_3^{a_3}\Pi_4^{a_4}$$

with the exponents chosen at will, would yield a new parameter. Then Π_a, Π_2, Π_3, Π_4 would constitute a new set. This procedure may be continued to find all possible sets.

Example 4.3 The losses $\Delta h/l$ per unit length of pipe in turbulent flow through a smooth pipe depend upon velocity V, diameter D, gravity g, dynamic viscosity μ, and density ρ. With dimensional analysis, determine the general form of the equation

$$F\left(\frac{\Delta h}{l}, V, D, \rho, \mu, g\right) = 0$$

Clearly, $\Delta h/l$ is a Π parameter. If V, D, and ρ are repeating variables,

$$\Pi_1 = V^{x_1} D^{y_1} \rho^{z_1} \mu = (LT^{-1})^{x_1} L^{y_1} (ML^{-3})^{z_1} ML^{-1}T^{-1}$$

$$x_1 + y_1 - 3z_1 - 1 = 0$$

$$-x_1 \qquad\qquad -1 = 0$$

$$z_1 + 1 = 0$$

from which $x_1 = -1$, $y_1 = -1$, $z_1 = -1$.

$$\Pi_2 = V^{x_2} D^{y_2} \rho^{z_2} g = (LT^{-1})^{x_2} L^{y_2} (ML^{-3})^{z_2} LT^{-2}$$

$$x_2 + y_2 - 3z_2 + 1 = 0$$

$$-x_2 \qquad\qquad -2 = 0$$

$$z_2 \qquad = 0$$

from which $x_2 = -2$, $y_2 = 1$, $z_2 = 0$.

$$\Pi_1 = \frac{\mu}{VD\rho} \qquad \Pi_2 = \frac{gD}{V^2} \qquad \Pi_3 = \frac{\Delta h}{l}$$

or

$$f\left(\frac{VD\rho}{\mu}, \frac{V^2}{gD}, \frac{\Delta h}{l}\right) = 0$$

since the Π quantities may be inverted if desired. The first parameter, $VD\rho/\mu$, is the *Reynolds number* **R**, one of the most important of the dimensionless parameters in fluid mechanics. The size of the Reynolds number determines the nature of the flow. It is discussed in Sec. 5.3. Solving for $\Delta h/l$ gives

$$\frac{\Delta h}{l} = f_1\left(\mathbf{R}, \frac{V^2}{gD}\right)$$

The usual formula is

$$\frac{\Delta h}{l} = f(\mathbf{R}) \frac{1}{D} \frac{V^2}{2g}$$

Example 4.4 A fluid-flow situation depends upon the velocity V, the density ρ, several linear dimensions l, l_1, l_2, pressure drop Δp, gravity g, viscosity μ, surface tension σ, and bulk modulus of elasticity K. Apply dimensional analysis to these variables to find a set of Π parameters.

$$F(V, \rho, l, l_1, l_2, \Delta p, g, \mu, \sigma, K) = 0$$

As three dimensions are involved, three repeating variables are selected. For complex situations, V, ρ, and l are generally helpful. There are seven Π parameters:

$$\Pi_1 = V^{x_1}\rho^{y_1}l^{z_1}\,\Delta p \qquad \Pi_2 = V^{x_2}\rho^{y_2}l^{z_2}g$$

$$\Pi_3 = V^{x_3}\rho^{y_3}l^{z_3}\mu \qquad \Pi_4 = V^{x_4}\rho^{y_4}l^{z_4}\sigma$$

$$\Pi_5 = V^{x_5}\rho^{y_5}l^{z_5}K \qquad \Pi_6 = \frac{l}{l_1}$$

$$\Pi_7 = \frac{l}{l_2}$$

By expanding the Π quantities into dimensions,

$$\Pi_1 = (LT^{-1})^{x_1}(ML^{-3})^{y_1}L^{z_1}ML^{-1}T^{-2}$$

$$x_1 - 3y_1 + z_1 - 1 = 0$$

$$-x_1 \qquad\qquad -2 = 0$$

$$y_1 \qquad +1 = 0$$

from which $x_1 = -2$, $y_1 = -1$, $z_1 = 0$.

$$\Pi_2 = (LT^{-1})^{x_2}(ML^{-3})^{y_2}L^{z_2}LT^{-2}$$

$$x_2 - 3y_2 + z_2 + 1 = 0$$

$$-x_2 \qquad\qquad -2 = 0$$

$$y_2 \qquad\qquad = 0$$

from which $x_2 = -2$, $y_2 = 0$, $z_2 = 1$.

$$\Pi_3 = (LT^{-1})^{x_3}(ML^{-3})^{y_3}L^{z_3}ML^{-1}T^{-1}$$

$$x_3 - 3y_3 + z_3 - 1 = 0$$

$$-x_3 \qquad\qquad -1 = 0$$

$$y_3 \qquad +1 = 0$$

from which $x_3 = -1$, $y_3 = -1$, $z_3 = -1$.

$$\Pi_4 = (LT^{-1})^{x_4}(ML^{-3})^{y_4}L^{z_4}MT^{-2}$$

$$x_4 - 3y_4 + z_4 \qquad = 0$$

$$-x_4 \qquad\qquad -2 = 0$$

$$y_4 \qquad +1 = 0$$

from which $x_4 = -2$, $y_4 = -1$, $z_4 = -1$.

$$\Pi_5 = (LT^{-1})^{x_5}(ML^{-3})^{y_5}L^{z_5}ML^{-1}T^{-2}$$

$$x_5 - 3y_5 + z_5 - 1 = 0$$

$$-x_5 \qquad\qquad -2 = 0$$

$$y_5 \qquad +1 = 0$$

from which $x_5 = -2$, $y_5 = -1$, $z_5 = 0$.

$$\Pi_1 = \frac{\Delta p}{\rho V^2} \qquad \Pi_2 = \frac{gl}{V^2} \qquad \Pi_3 = \frac{\mu}{Vl\rho} \qquad \Pi_4 = \frac{\sigma}{V^2\rho l}$$

$$\Pi_5 = \frac{K}{\rho V^2} \qquad \Pi_6 = \frac{l}{l_1} \qquad \Pi_7 = \frac{l}{l_2}$$

and

$$f\left(\frac{\Delta p}{\rho V^2}, \frac{gl}{V^2}, \frac{\mu}{Vl\rho}, \frac{\sigma}{V^2\rho l}, \frac{K}{\rho V^2}, \frac{l}{l_1}, \frac{l}{l_2}\right) = 0$$

It is convenient to invert some of the parameters and to take some square roots,

$$f_1\left(\frac{\Delta p}{\rho V^2}, \frac{V}{\sqrt{gl}}, \frac{Vl\rho}{\mu}, \frac{V^2l\rho}{\sigma}, \frac{V}{\sqrt{K/\rho}}, \frac{l}{l_1}, \frac{l}{l_2}\right) = 0$$

The first parameter, usually written $\Delta p/(\rho V^2/2)$, is the *pressure coefficient;* the second parameter is the *Froude* number $\mathbf{F}$; the third is the *Reynolds* number $\mathbf{R}$; the fourth is the *Weber* number $\mathbf{W}$; and the fifth is the *Mach* number $\mathbf{M}$. Hence,

$$f_1\left(\frac{\Delta p}{\rho V^2}, \mathbf{F}, \mathbf{R}, \mathbf{W}, \mathbf{M}, \frac{l}{l_1}, \frac{l}{l_2}\right) = 0$$

After solving for pressure drop,

$$\Delta p = \rho V^2 f_2\left(\mathbf{F}, \mathbf{R}, \mathbf{W}, \mathbf{M}, \frac{l}{l_1}, \frac{l}{l_2}\right)$$

in which f_1, f_2 must be determined from analysis or experiment. By selecting other repeating variables, a different set of Π parameters could be obtained.

Figure 5.32 is a representation of a functional relation of the type just given as it applies to the flow in pipes. Here the parameters $\mathbf{F}$, $\mathbf{W}$, and $\mathbf{M}$ are neglected as being unimportant; l is the pipe diameter D, l_1 is the length of the pipe L, and l_2 is a dimension which is representative of the effective height of the surface roughness of the pipe and is given by ϵ. Thus

$$\frac{\Delta p}{\rho V^2} = f_3\left(\mathbf{R}, \frac{L}{D}, \frac{\epsilon}{D}\right)$$

The fact that the pressure drop in the pipeline varies linearly with the length (i.e., doubling the length of pipe doubles the loss in pressure) appears reasonable, so that one has

$$\frac{\Delta p}{\rho V^2} = \frac{L}{D} f_4\left(\mathbf{R}, \frac{\epsilon}{D}\right) \qquad \text{or} \qquad \frac{\Delta p}{\rho V^2 (L/D)} = f_4\left(\mathbf{R}, \frac{\epsilon}{D}\right)$$

The term on the left-hand side is commonly given the notation $f/2$, as in Fig. 5.32. The curves shown in this figure have f and $\mathbf{R}$ as ordinate and abscissa, respectively, with ϵ/D a parameter which assumes a given value for each curve. The nature of these curves was determined through experiment. Such experiments show that, when the parameter **R** is below the value of 2000, all the curves for the various values of ϵ/D coalesce into one. Hence f is independent of ϵ/D, and the result is

$$f = f_5(\mathbf{R})$$

This relation will be predicted in Chap. 5 on the basis of theoretical considerations, but it remained for an experimental verification of these predictions to indicate the power of the theoretical methods.

Example 4.5 The thrust due to any one of a family of geometrically similar airplane propellers is to be determined experimentally from a wind tunnel test on a model. By means of dimensional analysis find suitable parameters for plotting test results.

The thrust F_T depends upon speed of rotation ω, speed of advance V_0, diameter D, air viscosity μ, density ρ, and speed of sound c. The function

$$F(F_T, V_0, D, \omega, \mu, \rho, c) = 0$$

is to be arranged into four dimensionless parameters, since there are seven quantities and three dimensions. Starting first by selecting ρ, ω, and D as repeating variables,

$$\Pi_1 = \rho^{x_1}\omega^{y_1}D^{z_1}F_T = (ML^{-3})^{x_1}(T^{-1})^{y_1}L^{z_1}MLT^{-2}$$

$$\Pi_2 = \rho^{x_2}\omega^{y_2}D^{z_2}V_0 = (ML^{-3})^{x_2}(T^{-1})^{y_2}L^{z_2}LT^{-1}$$

$$\Pi_3 = \rho^{x_3}\omega^{y_3}D^{z_3}\mu = (ML^{-3})^{x_3}(T^{-1})^{y_3}L^{z_3}ML^{-1}T^{-1}$$

$$\Pi_4 = \rho^{x_4}\omega^{y_4}D^{z_4}c = (ML^{-3})^{x_4}(T^{-1})^{y_4}L^{z_4}LT^{-1}$$

By writing the simultaneous equations in x_1, y_1, z_1, etc., as before and solving them,

$$\Pi_1 = \frac{F_T}{\rho\omega^2 D^2} \qquad \Pi_2 = \frac{V_0}{\omega D} \qquad \Pi_3 = \frac{\mu}{\rho\omega D^2} \qquad \Pi_4 = \frac{c}{\omega D}$$

Solving for the thrust parameter leads to

$$\frac{F_T}{\rho\omega^2 D^4} = f_1\left(\frac{V_0}{\omega D}, \frac{\rho\omega D^2}{\mu}, \frac{c}{\omega D}\right)$$

Since the parameters may be recombined to obtain other forms, the second term is replaced by the product of the first and second terms, $VD\rho/\mu$, and the third term is replaced by the first term divided by the third term, V_0/c; thus

$$\frac{F_T}{\rho\omega^2 D^4} = f_2\left(\frac{V_0}{\omega D}, \frac{V_0 D\rho}{\mu}, \frac{V_0}{c}\right)$$

Of the dimensionless parameters, the first is probably of the most importance, since it relates speed of advance to speed of rotation. The second parameter is a Reynolds number and accounts for viscous effects. The last parameter, speed of advance divided by speed of sound, is a Mach number, which would be important for speeds near or higher than the speed of sound. Reynolds effects are usually small, so that a plot of $F_T/\rho\omega^2 D^4$ against $V_0/\omega D$ should be most informative.

The steps in a dimensional analysis may be summarized as follows:

1. Select the pertinent variables. This requires some knowledge of the process.
2. Write the functional relations, e.g.,

$$F(V, D, \rho, \mu, c, H) = 0$$

3. Select the repeating variables. (Do not make the dependent quantity a repeating variable.) These variables should contain all the m dimensions of the problem. Often one variable is chosen because it specifies the scale, another the kinematic conditions; and in the cases of major interest in this chapter one variable which is related to the forces or mass of the system, for example, D, V, ρ, is chosen.

4. Write the Π parameters in terms of unknown exponents, e.g.,

$$\Pi_1 = V^{x_1}D^{y_1}\rho^{z_1}\mu = (LT^{-1})^{x_1}L^{y_1}(ML^{-3})^{z_1}ML^{-1}T^{-1}$$

5. For each of the Π expressions write the equations of the exponents, so that the sum of the exponents of each dimension will be zero.

6. Solve the equations simultaneously.

7. Substitute back into the Π expressions of step 4 the exponents to obtain the dimensionless Π parameters.

8. Establish the functional relation

$$f_1(\Pi_1, \Pi_2, \Pi_3, \ldots, \Pi_{n-m}) = 0$$

or solve for one of the Π's explicitly:

$$\Pi_2 = f(\Pi_1, \Pi_3, \ldots, \Pi_{n-m})$$

9. Recombine, if desired, to alter the forms of the Π parameters, keeping the same number of independent parameters.

Alternate Formulation of Π Parameters

A rapid method for obtaining Π parameters, developed by Hunsaker and Rightmire (referenced at end of chapter), uses the repeating variables as primary quantities and solves for M, L, and T in terms of them. In Example 4.3 the repeating variables are V, D, and ρ; therefore,

$$V = LT^{-1} \qquad D = L \qquad \rho = ML^{-3}$$

$$L = D \qquad T = DV^{-1} \qquad M = \rho D^3 \tag{4.3.4}$$

Now, by use of Eqs. (4.3.4),

$$\mu = ML^{-1}T^{-1} = \rho D^3 D^{-1}D^{-1}V = \rho DV$$

hence the Π parameter is

$$\Pi_1 = \frac{\mu}{\rho DV}$$

Equations (4.3.4) may be used directly to find the other Π parameters. For Π_2

$$g = LT^{-2} = DD^{-2}V^2 = V^2D^{-1}$$

and

$$\Pi_2 = \frac{g}{V^2D^{-1}} = \frac{gD}{V^2}$$

This method does not require the repeated solution of three equations in three unknowns for each Π parameter determination.

EXERCISES

4.3.1 An *incorrect* arbitrary recombination of the Π parameters

$$F\left(\frac{V_0}{\omega D}, \frac{\rho \omega D^2}{\mu}, \frac{c}{\omega D}\right) = 0$$

is

(a) $F\left(\dfrac{c}{V_0}, \dfrac{\rho c D}{\mu}, \dfrac{c}{\omega D}\right) = 0$ (b) $F\left(\dfrac{V_0}{\omega D}, \dfrac{\rho c D^2}{\mu}, \dfrac{c}{\omega D}\right) = 0$

(c) $F\left(\dfrac{V_0}{\omega D}, \dfrac{V_0 c \rho}{\omega \mu}, \dfrac{\rho c D}{\mu}\right) = 0$ (d) $F\left(\dfrac{V_0 \mu}{\omega^2 D^3 \rho}, \dfrac{V_0 \rho D}{\mu}, \dfrac{c}{\omega D}\right) = 0$

(e) none of these answers

4.3.2 The repeating variables in a dimensional analysis should (a) include the dependent variable; (b) have two variables with the same dimensions if possible; (c) exclude one of the dimensions from each variable if possible; (d) include the variables not considered very important factors; (e) satisfy none of these answers.

4.3.3 Select the quantity in the following that is *not* a dimensionless parameter: (a) pressure coefficient; (b) Froude number; (c) Darcy-Weisbach friction factor; (d) kinematic viscosity; (e) Weber number.

4.3.4 How many Π parameters are needed to express the function $F(a, V, t, v, L) = 0$? (a) 5; (b) 4; (c) 3; (d) 2; (e) 1.

4.3.5 Which of the following could be a Π parameter of the function $F(Q, H, g, V_0, \phi) = 0$ when Q and g are taken as repeating variables? (a) Q^2/gH^4; (b) V_0^2/g^2Q; (c) $Q/g\phi^2$; (d) $Q/\sqrt{gH}$; (e) none of these answers.

4.4 DISCUSSION OF DIMENSIONLESS PARAMETERS

The five dimensionless parameters—pressure coefficient, Reynolds number, Froude number, Weber number, and Mach number—are of importance in correlating experimental data. They are discussed in this section, with particular emphasis placed on the relation of pressure coefficient to the other parameters.

Pressure Coefficient

The pressure coefficient $\Delta p/(\rho V^2/2)$ is the ratio of pressure to dynamic pressure. When multiplied by area, it is the ratio of pressure force to inertial force, as $(\rho V^2/2)A$ would be the force needed to reduce the velocity to zero. It may also be written as $\Delta h/(V^2/2g)$ by division by γ. For pipe flow the Darcy-Weisbach equation relates losses h_l to length of pipe L, diameter D, and velocity V by a dimensionless friction factor† f

$$h_l = f \frac{L}{D} \frac{V^2}{2g} \quad \text{or} \quad \frac{fL}{D} = \frac{h_l}{V^2/2g} = f_2\left(\mathbf{R}, \mathbf{F}, \mathbf{W}, \mathbf{M}, \frac{l}{l_1}, \frac{l}{l_2}\right)$$

as fL/D is shown to be equal to the pressure coefficient (see Example 4.4). In pipe flow, gravity has no influence on losses; therefore, $\mathbf{F}$ may be dropped out. Similarly, surface tension has no effect, and $\mathbf{W}$ drops out. For steady liquid flow, compressibility is not important, and $\mathbf{M}$ is dropped. l may refer to D; l_1 to roughness height projection ϵ in the pipe wall; and l_2 to their spacing ϵ'; hence,

$$\frac{fL}{D} = f_2\left(\mathbf{R}, \frac{\epsilon}{D}, \frac{\epsilon'}{D}\right) \tag{4.4.1}$$

Pipe flow problems are discussed in Chaps. 5, 6, and 10. If compressibility is important,

$$\frac{fL}{D} = f_2\left(\mathbf{R}, \mathbf{M}, \frac{\epsilon}{D}, \frac{\epsilon'}{D}\right) \tag{4.4.2}$$

Compressible flow problems are studied in Chap. 6. With orifice flow, studied in Chap. 8, $V = C_v \sqrt{2gH}$,

$$\frac{H}{V^2/2g} = \frac{1}{C_v^2} = f_2\left(\mathbf{R}, \mathbf{W}, \mathbf{M}, \frac{l}{l_1}, \frac{l}{l_2}\right) \tag{4.4.3}$$

in which l may refer to orifice diameter and l_1 and l_2 to upstream dimensions. Viscosity and surface tension are unimportant for large orifices and low-viscosity fluids. Mach number effects may be very important for gas flow with large pressure drops, i.e., Mach numbers approaching unity.

In steady, uniform open-channel flow, discussed in Chap. 5, the Chézy formula relates average velocity V, slope of channel S, and hydraulic radius of cross section R (area of section divided by wetted perimeter) by

$$V = C\sqrt{RS} = C\sqrt{R\frac{\Delta h}{L}} \tag{4.4.4}$$

C is a coefficient depending upon size, shape, and roughness of channel. Then

$$\frac{\Delta h}{V^2/2g} = \frac{2gL}{R}\frac{1}{C^2} = f_2\left(\mathbf{F}, \mathbf{R}, \frac{l}{l_1}, \frac{l}{l_2}\right) \tag{4.4.5}$$

since surface tension and compressible effects are usually unimportant.

---

† There are several friction factors in general use. This is the Darcy-Weisbach friction factor, which is four times the size of the *Fanning friction factor*, also called f.

The drag F on a body is expressed by $F = C_D A\rho V^2/2$, in which A is a typical area of the body, usually the projection of the body onto a plane normal to the flow. Then F/A is equivalent to Δp, and

$$\frac{F}{A\rho V^2/2} = C_D = f_2\left(\mathbf{R}, \mathbf{F}, \mathbf{M}, \frac{l}{l_1}, \frac{l}{l_2}\right) \tag{4.4.6}$$

The term $\mathbf{R}$ is related to *skin friction* drag due to viscous shear as well as to *form*, or *profile*, drag resulting from *separation* of the flow streamlines from the body; $\mathbf{F}$ is related to wave drag if there is a free surface; for large Mach numbers C_D may vary more markedly with $\mathbf{M}$ than with the other parameters; the length ratios may refer to shape or roughness of the surface.

The Reynolds Number

The Reynolds number $VD\rho/\mu$ is the ratio of inertial forces to viscous forces. A *critical* Reynolds number distinguishes among flow regimes, such as laminar or turbulent flow in pipes, in the boundary layer, or around immersed objects. The particular value depends upon the situation. In compressible flow, the Mach number is generally more significant than the Reynolds number.

The Froude Number

The Froude number $V/\sqrt{gl}$, when squared and then multiplied and divided by ρA, is a ratio of dynamic (or inertial) force to weight. With free liquid-surface flow the nature of the flow (rapid† or tranquil) depends upon whether the Froude number is greater or less than unity. It is useful in calculations of hydraulic jump, in design of hydraulic structures, and in ship design.

The Weber Number

The Weber number $V^2 l\rho/\sigma$ is the ratio of inertial forces to surface-tension forces (evident when numerator and denominator are multiplied by l). It is important at gas-liquid or liquid-liquid interfaces and also where these interfaces are in contact with a boundary. Surface tension causes small (capillary) waves and droplet formation and has an effect on discharge of orifices and weirs at very small heads. The effect of surface tension on wave propagation is shown in Fig. 4.1. To the left of the curve's minimum the wave speed is controlled by surface tension (the waves are called ripples), and to the right of the curve's minimum gravity effects are dominant.

---

† Open-channel flow at depth y is *rapid* when the flow velocity is greater than the speed $\sqrt{gy}$ of an elementary wave in quiet liquid. *Tranquil* flow occurs when the flow velocity is less than $\sqrt{gy}$.

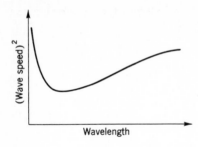

Figure 4.1 Wave speed vs. wavelength for surface waves.

The Mach Number

The speed of sound in a liquid is written $\sqrt{K/\rho}$ if K is the bulk modulus of elasticity (Secs. 1.7 and 6.2) or $c = \sqrt{kRT}$ (k is the specific heat ratio and T the absolute temperature for a perfect gas). V/c or $V/\sqrt{K/\rho}$ is the Mach number. It is a measure of the ratio of inertial forces to elastic forces. By squaring V/c and multiplying by $\rho A/2$ in numerator and denominator, the numerator is the dynamic force and the denominator is the dynamic force at sonic flow. It may also be shown to be a measure of the ratio of kinetic energy of the flow to internal energy of the fluid. It is the most important correlating parameter when velocities are near or above local sonic velocities.

EXERCISES

4.4.1 Which of the following has the form of a Reynolds number?

(a) $\dfrac{ul}{v}$ (b) $\dfrac{VD\mu}{\rho}$ (c) $\dfrac{uv}{l}$ (d) $\dfrac{V}{gD}$ (e) $\dfrac{\Delta p}{\rho V^2}$

4.4.2 The Reynolds number may be defined as the ratio of (a) viscous forces to inertial forces; (b) viscous forces to gravity forces; (c) gravity forces to inertial forces; (d) elastic forces to pressure forces; (e) none of these answers.

4.4.3 The pressure coefficient may take the form

(a) $\dfrac{\Delta p}{\gamma H}$ (b) $\dfrac{\Delta p}{\rho V^2/2}$ (c) $\dfrac{\Delta p}{l\mu V}$ (d) $\Delta p \dfrac{\rho}{\mu^2 l^4}$ (e) none of these answers

4.4.4 The pressure coefficient is a ratio of pressure forces to (a) viscous forces; (b) inertial forces; (c) gravity forces; (d) surface-tension forces; (e) elastic-energy forces.

4.4.5 Select the situation in which inertial forces would be *unimportant*: (a) flow over a spillway crest; (b) flow through an open-channel transition; (c) waves breaking against a sea wall; (d) flow through a long capillary tube; (e) flow through a half-opened valve.

4.4.6 Which two forces are most important in laminar flow between closely spaced parallel plates? (a) inertial, viscous; (b) pressure, inertial; (c) gravity, pressure; (d) viscous, pressure; (e) none of these answers.

4.4.7 If the capillary rise Δh of a liquid in a circular tube of diameter D depends upon surface tension σ and specific weight γ, the formula for capillary rise could take the form

(a) $\Delta h = \sqrt{\dfrac{\sigma}{\gamma}} F\left(\dfrac{\sigma}{\gamma D^2}\right)$ (b) $\Delta h = c\left(\dfrac{\sigma}{\gamma D^2}\right)^n$ (c) $\Delta h = cD\left(\dfrac{\sigma}{\gamma}\right)^n$

(d) $\Delta h = \sqrt{\dfrac{\gamma}{\sigma}} F\left(\dfrac{\gamma D^2}{\sigma}\right)$ (e) none of these answers

4.5 SIMILITUDE; MODEL STUDIES

Model studies of proposed hydraulic structures and machines are frequently undertaken as an aid to the designer. They permit visual observation of the flow and make it possible to obtain certain numerical data, e.g., calibrations of weirs and gates, depths of flow, velocity distributions, forces on gates, efficiencies and capacities of pumps and turbines, pressure distributions, and losses.

If accurate quantitative data are to be obtained from a model study, there must be dynamic similitude between model and prototype. This similitude requires (1) that there be exact geometric similitude and (2) that the ratio of dynamic pressures at corresponding points be a constant. The second requirement may also be expressed as a kinematic similitude, i.e., the streamlines must be geometrically similar.

Geometric similitude extends to the actual surface roughness of model and prototype. If the model is one-tenth the size of the prototype in every linear dimension, then the height of roughness projections must be in the same ratio. For dynamic pressures to be in the same ratio at corresponding points in model and prototype, the ratios of the various types of forces must be the same at corresponding points. Hence, for strict dynamic similitude, the Mach, Reynolds, Froude, and Weber numbers must be the same in both model and prototype.

Strict fulfillment of these requirements is generally impossible to achieve, except with a 1 : 1 scale ratio. Fortunately, in many situations only two of the forces are of the same magnitude. Discussion of a few cases will make this clear.

Wind- and Water-Tunnel Tests

This equipment is used to examine the streamlines and the forces that are induced as the fluid flows past a fully submerged body. The type of test that is being conducted and the availability of the equipment determine which kind of tunnel will be used. Because the kinematic viscosity of water is about one-tenth that of air, a water tunnel can be used for model studies at relatively high Reynolds numbers. The drag effect of various parachutes was studied in a water tunnel! At very high air velocities the effects of compressibility, and consequently Mach number, must be taken into consideration, and indeed may be the chief reason for undertaking an investigation. Figure 4.2 shows a model of an aircraft carrier being

Figure 4.2 Wind tunnel tests on an aircraft carrier superstructure. Model is inverted and suspended from ceiling. (*Photograph taken in Aeronautical and Astronautical Laboratories of The University of Michigan for the Dyna-sciences Corp.*)

tested in a low-speed tunnel to study the flow pattern around the ship's superstructure. The model has been inverted and suspended from the ceiling so that the wool tufts can be used to give an indication of the flow direction. Behind the model there is an apparatus for sensing the air speed and direction at various locations along an aircraft's glide path.

Pipe Flow

In steady flow in a pipe, viscous and inertial forces are the only ones of consequence; hence, when geometric similitude is observed, the same Reynolds number in model and prototype provides dynamic similitude. The various corresponding pressure coefficients are the same. For testing with fluids having the same kinematic viscosity in model and prototype, the product, VD, must be the same. Frequently this requires very high velocities in small models.

Open Hydraulic Structures

Structures such as spillways, stilling pools, channel transitions, and weirs generally have forces due to gravity (from changes in elevation of liquid surfaces) and inertial forces that are greater than viscous and turbulent shear forces. In these cases geometric similitude and the same value of Froude's number in model and prototype produce a good approximation to dynamic similitude; thus

$$\frac{V_m^2}{g_m l_m} = \frac{V_p^2}{g_p l_p}$$

Since gravity is the same, the velocity ratio varies as the square root of the scale ratio $\lambda = l_p/l_m$,

$$V_p = V_m \sqrt{\lambda}$$

The corresponding times for events to take place (as time for passage of a particle through a transition) are related; thus

$$t_m = \frac{l_m}{V_m} \qquad t_p = \frac{l_p}{V_p} \qquad \text{and} \qquad t_p = t_m \frac{l_p}{l_m} \frac{V_m}{V_p} = t_m \sqrt{\lambda}$$

Figure 4.3 Model test on a harbor to determine the effect of a breakwater. (*Department of Civil Engineering, The University of Michigan.*)

The discharge ratio Q_p/Q_m is

$$\frac{Q_p}{Q_m} = \frac{l_p^3/t_p}{l_m^3/t_m} = \lambda^{5/2}$$

Force ratios, e.g., on gates, F_p/F_m, are

$$\frac{F_p}{F_m} = \frac{\gamma h_p l_p^2}{\gamma h_m l_m^2} = \lambda^3$$

where h is the head. In a similar fashion other pertinent ratios can be derived so that model results can be interpreted as prototype performance.

Figure 4.3 shows a model test conducted to determine the effect of a breakwater on the wave formation in a harbor.

Ship's Resistance

The resistance to motion of a ship through water is composed of pressure drag, skin friction, and wave resistance. Model studies are complicated by the three types of forces that are important, inertia, viscosity, and gravity. Skin friction studies should be based on equal Reynolds numbers in model and prototype, but wave resistance depends upon the Froude number. To satisfy both requirements, model and prototype must be the same size.

The difficulty is surmounted by using a small model and measuring the total drag on it when towed. The skin friction is then computed for the model and subtracted from the total drag. The remainder is stepped up to prototype size by Froude's law, and the prototype skin friction is computed and added to yield total resistance due to the water. Figure 4.4 shows the dramatic change in the wave profile which resulted from a redesigned bow. From such tests it is possible to predict through Froude's law the wave formation and drag that would occur on the prototype.

Hydraulic Machinery

The moving parts in a hydraulic machine require an extra parameter to ensure that the streamline patterns are similar in model and prototype. This parameter must relate the throughflow (discharge) to the speed of moving parts. For geometrically similar machines, if the vector diagrams of velocity entering or leaving the moving parts are similar, the units are *homologous*; i.e., for practical purposes dynamic similitude exists. The Froude number is unimportant, but the Reynolds number effects (called *scale effects* because it is impossible to maintain the same Reynolds number in homologous units) may cause a discrepancy of 2 or 3 percent in efficiency between model and prototype. The Mach number is also of importance in axial-flow compressors and gas turbines.

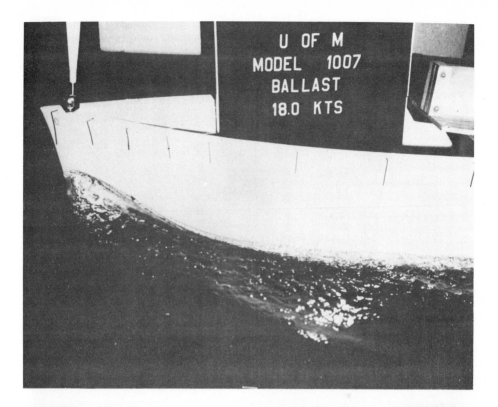

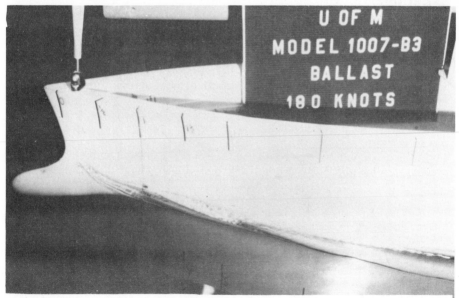

Figure 4.4 Model tests showing the influence of a bulbous bow on bow wave formation. (*Department of Naval Architecture and Marine Engineering, The University of Michigan.*)

Example 4.6 The valve coefficients $K = \Delta p/(\rho V^2/2)$ for a 600-mm-diameter valve are to be determined from tests on a geometrically similar 300-mm-diameter valve using atmospheric air at 80°F. The ranges of tests should be for flow of water at 70°F at 1 to 2.5 m/s. What ranges of airflows are needed?

The Reynolds number range for the prototype valve is

$$\left(\frac{VD}{v}\right)_{min} = \frac{(1 \text{ m/s})(0.6 \text{ m})}{(1.059 \times 10^{-5} \text{ ft}^2/\text{s})(0.3048 \text{ m/ft})^2} = 610,000$$

$$\left(\frac{VD}{v}\right)_{max} = 610,000 \times 2.5 = 1,525,000$$

For testing with air at 80°F

$$v = (1.8 \times 10^{-4} \text{ ft}^2/\text{s})(0.3048 \text{ m/ft})^2 = 1.672 \times 10^{-5} \text{ m}^2/\text{s}$$

Then the ranges of air velocities are

$$\frac{(V_{min})(0.3 \text{ m})}{1.672 \times 10^{-5} \text{ m}^2/\text{s}} = 610,000 \qquad V_{min} = 30.6 \text{ m/s}$$

$$\frac{(V_{max})(0.3 \text{ m})}{1.672 \times 10^{-5} \text{ m}^2/\text{s}} = 1,525,000 \qquad V_{max} = 85 \text{ m/s}$$

$$Q_{min} = \frac{\pi}{4} (0.3 \text{ m})^2 (30.6 \text{ m/s}) = 2.16 \text{ m}^3/\text{s}$$

$$Q_{max} = \frac{\pi}{4} (0.3 \text{ m})^2 (85 \text{ m/s}) = 6.0 \text{ m}^3/\text{s}$$

EXERCISES

4.5.1 What velocity of oil, $\rho = 1.6$ slugs/ft^3, $\mu = 0.20$ P, must occur in a 1-in-diameter pipe to be dynamically similar to 10 ft/s water velocity at 68°F in a $\frac{1}{4}$-in-diameter tube? (a) 0.60 ft/s; (b) 9.6 ft/s; (c) 4.0 ft/s; (d) 60 ft/s; (e) none of these answers.

4.5.2 The velocity at a point on a model dam crest was measured to be 1 m/s. The corresponding prototype velocity for $\lambda = 25$ is, in meters per second, (a) 25; (b) 5; (c) 0.2; (d) 0.04; (e) none of these answers.

4.5.3 The height of a hydraulic jump in a stilling pool was found to be 4.0 in in a model, $\lambda = 36$. The prototype jump height is (a) 12 ft; (b) 2 ft; (c) not determinable from data given; (d) less than 4 in; (e) none of these answers.

4.5.4 A ship's model, scale 1 : 100, had a wave resistance of 10 N at its design speed. The corresponding prototype wave resistance is, in kilonewtons, (a) 10; (b) 100; (c) 1000; (d) 10,000; (e) none of these answers.

4.5.5 A 1 : 5 scale model of a projectile has a drag coefficient of 3.5 at $M = 2.0$. How many times greater would the prototype resistance be when fired at the same Mach number in air of the same temperature and half the density? (a) 0.312; (b) 3.12; (c) 12.5; (d) 25; (e) none of these answers.

PROBLEMS

4.1 Show that Eqs. (3.8.6), (3.6.3), and (3.11.13) are dimensionally homogeneous.

4.2 Arrange the following groups into dimensionless parameters: (a) Δp, ρ, V; (b) ρ, g, V, F; (c) μ, F, Δp, t.

4.3 By inspection, arrange the following groups into dimensionless parameters: (a) a, l, t; (b) v, l, t; (c) A, Q, ω; (d) K, σ, A.

4.4 Derive the unit of mass consistent with the units inches, minutes, tons.

4.5 In terms of M, L, T, determine the dimensions of radians, angular velocity, power, work, torque, and moment of momentum.

4.6 Find the dimensions of the quantities in Prob. 4.5 in the FLT system.

4.7 Work Example 4.2 using Q and H as repeating variables.

4.8 Using the variables $Q, D, \Delta H/l, \rho, \mu, g$ as pertinent to smooth-pipe flow, arrange them into dimensionless parameters with Q, ρ, μ as repeating variables.

4.9 If the shear stress τ is known to depend upon viscosity and rate of angular deformation du/dy in one-dimensional laminar flow, determine the form of Newton's law of viscosity by dimensional reasoning.

4.10 The variation Δp of pressure in static liquids is known to depend upon specific weight γ and elevation difference Δz. By dimensional reasoning determine the form of the hydrostatic law of variation of pressure.

4.11 When viscous and surface-tension effects are neglected, the velocity V of efflux of liquid from a reservoir is thought to depend upon the pressure drop Δp of the liquid and its density ρ. Determine the form of expression for V.

4.12 The buoyant force F_B on a body is thought to depend upon its volume submerged $\mho$ and the gravitational body force acting on the fluid. Determine the form of the buoyant-force equation.

4.13 In a fluid rotated as a solid about a vertical axis with angular velocity ω, the pressure rise p in a radial direction depends upon speed ω, radius r, and fluid density ρ. Obtain the form of equation for p.

4.14 In Example 4.3, work out two other sets of dimensionless parameters by recombination of the dimensionless parameters given.

4.15 Find the dimensionless parameters of Example 4.4 using $\Delta p, \rho$, and l as repeating variables.

4.16 The Mach number **M** for flow of a perfect gas in a pipe depends upon the specific-heat ratio k (dimensionless), the pressure p, the density ρ, and the velocity V. Obtain by dimensional reasoning the form of the Mach number expression.

4.17 Work out the scaling ratio for torque T on a disk of radius r that rotates in fluid of viscosity μ with angular velocity ω and clearance y between disk and fixed plate.

4.18 The velocity at a point in a model of a spillway for a dam is 1 m/s. For a ratio of prototype to model of 10 : 1, what is the velocity at the corresponding point in the prototype under similar conditions?

4.19 The power input to a pump depends upon the discharge Q, the pressure rise Δp, the fluid density ρ, size D, and efficiency e. Find the expression for power by the use of dimensional analysis.

4.20 The torque delivered by a water turbine depends upon discharge Q, head H, specific weight γ, angular velocity ω, and efficiency e. Determine the form of equation for torque.

4.21 A model of a venturi meter has linear dimensions one-fifth those of the prototype. The prototype operates on water at 20°C, and the model on water at 95°C. For a throat diameter of 600 mm and a velocity at the throat of 6 m/s in the prototype, what discharge is needed through the model for similitude?

4.22 The drag F on a high-velocity projectile depends upon speed V of projectile, density of fluid ρ, acoustic velocity c, diameter of projectile D, and viscosity μ. Develop an expression for the drag.

4.23 The wave drag on a model of a ship is 16 N at a speed of 3 m/s. For a prototype fifteen times as long, what will the corresponding speed and wave drag be if the liquid is the same in each case?

4.24 Determine the specific gravity of spherical particles, $D = \frac{1}{200}$ in, which drop through air at 33°F at a speed U of 0.3 ft/s. The drag force on a small sphere in laminar motion is given by $3\pi\mu DU$.

4.25 A small liquid sphere of radius r_0 and density ρ_0 settles at velocity U in a second liquid of density ρ and viscosity μ. The tests are conducted inside vertical tubes of radius r. By dimensional analysis determine a set of dimensionless parameters to be used in determining the influence of the tube wall on the settling velocity.

4.26 The losses in a Y in a 1.2-m-diameter pipe system carrying gas ($\rho = 40$ kg/m^3, $\mu = 0.002$ P, $V = 25$ m/s) are to be determined by testing a model with water at 20°C. The laboratory has a water capacity of 75 l/s. What model scale should be used, and how are the results converted into prototype losses?

4.27 Ripples have a velocity of propagation that is dependent upon the surface tension and density of the fluid as well as the wavelength. By dimensional analysis justify the shape of Fig. 4.1 for small wavelengths.

4.28 In very deep water the velocity of propagation of waves depends upon the wavelength, but in shallow water it is independent of this dimension. Upon what variables does the speed of advance depend for shallow-water waves? Is Fig. 4.1 in agreement with this problem?

4.29 If a vertical circular conduit which is not flowing full is rotated at high speed, the fluid will attach itself uniformly to the inside wall as it flows downward (see Sec. 2.9). Under these conditions the radial acceleration of the fluid yields a radial force field which is similar to gravitational attraction and a hydraulic jump can occur on the inside of the tube, whereby the fluid thickness suddenly changes. Determine a set of dimensionless parameters for studying this rotating hydraulic jump.

4.30 A nearly spherical fluid drop oscillates as it falls. Surface tension plays a dominant role. Determine a meaningful dimensionless parameter for this natural frequency.

4.31 The lift and drag coefficients for a wing are shown in Fig. 5.23. If the wing has a chord of 10 ft, determine the lift and drag per foot of length when the wing is operating at zero angle of attack at a Reynolds number, based on the chord length, of 4.5×10^7 in air at 50°F. What force would be on a 1 : 20 scale model if the tests were conducted in water at 70°F? What would be the speed of the water? Comment on the desirability of conducting the model tests in water.

4.32 A 1 : 5 scale model of a water pumping station piping system is to be tested to determine overall head losses. Air at 25°C, 1 atm is available. For a prototype velocity of 500 mm/s in a 4-m-diameter section with water at 15°C, determine the air velocity and quantity needed and how losses determined from the model are converted into prototype losses.

4.33 Full-scale wind tunnel tests of the lift and drag on hydrofoils for a boat are to be made. The boat will travel at 35 mph through water at 60°F. What velocity of air ($p = 30$ psia, $t = 90$°F) is required to determine the lift and drag? *Note:* The lift coefficient C_L is dimensionless. Lift = $C_L A \rho V^2 / 2$.

4.34 The resistance to ascent of a balloon is to be determined by studying the ascent of a 1 : 50 scale model in water. How would such a model study be conducted and the results converted to prototype behavior?

4.35 The moment exerted on a submarine by its rudder is to be studied with a 1 : 20 scale model in a water tunnel. If the torque measured on the model is 5 N·m for a tunnel velocity of 15 m/s, what are the corresponding torque and speed for the prototype?

4.36 For two hydraulic machines to be homologous they must (a) be geometrically similar, (b) have the same discharge coefficient when viewed as an orifice, $Q_1/(A_1\sqrt{2gH_1}) = Q_2/(A_2\sqrt{2gH_2})$, and (c) have the same ratio of peripheral speed to fluid velocity, $\omega D/(Q/A)$. Show that the scaling ratios may be expressed as $Q/ND^3 = $ const and $H/(ND)^2 = $ const. N is the rotational speed.

4.37 By use of the scaling ratios of Prob. 4.36, determine the head and discharge of a 1 : 4 model of a centrifugal pump that produces 600 l/s at 30 m head when turning 240 rpm. The model operates at 1200 rpm.

REFERENCES

Bridgman, P. W.: "Dimensional Analysis," Yale University Press, New Haven, Conn., 1931, Paperback Y-82, 1963.

Holt, M.: Dimensional Analysis, sec. 15 in V. L. Streeter (ed.), "Handbook of Fluid Dynamics," McGraw-Hill, New York, 1961.

Hunsaker, J. C., and B. G. Rightmire: "Engineering Applications of Fluid Mechanics," pp. 110, 111, McGraw-Hill, New York, 1947.

Hydraulic Models, *ASCE Man. Eng. Pract.* 25, 1942.

Ipsen, D. C.: "Units, Dimensions, and Dimensionless Numbers," McGraw-Hill, New York, 1960.

Klien, S. J.: "Similitude and Approximation Theory," McGraw-Hill, New York, 1965.

Langhaar, H. L.: "Dimensional Analysis and Theory of Models," Wiley, New York, 1951.

Sedov, L. I.: "Similarity and Dimensional Methods in Mechanics," English trans. ed. by M. Holt, Academic, New York, 1959.

FIVE

VISCOUS EFFECTS: FLUID RESISTANCE

In Chap. 3 the basic equations used in the analysis of fluid-flow situations were discussed. The fluid was considered frictionless, or in some cases losses were assumed or computed without probing into their underlying causes. This chapter deals with real fluids, i.e., with situations in which irreversibilities are important. Viscosity is the fluid property that causes shear stresses in a moving fluid; it is also one means by which irreversibilities or losses are developed. Without viscosity in a fluid there is no fluid resistance. Simple cases of steady, laminar, incompressible flow are first developed in this chapter, since in these cases the losses can be computed. The concept of the Reynolds number, introduced in Chap. 4, is then further developed. Turbulent flow shear relations are introduced by use of the Prandtl mixing-length theory and are applied to turbulent velocity distributions. This is followed by boundary-layer concepts and by drag on immersed bodies. Resistance to steady, uniform, incompressible, turbulent flow is then examined for open and closed conduits, with a section devoted to open channels and to pipe flow. The chapter closes with a section on lubrication mechanics.

The equations of motion for a real fluid can be developed from consideration of the forces acting on a small element of the fluid, including the shear stresses generated by fluid motion and viscosity. The derivation of these equations, called the *Navier-Stokes equations*, is beyond the scope of this treatment. They are listed, however, for the sake of completeness, and many of the developments of this chapter could be made directly from them. First, Newton's law of viscosity, Eq. (1.1.1), for one-dimensional laminar flow can be generalized to three-dimensional flow (Stokes' law of viscosity)

$$\tau_{xy} = \mu\left(\frac{\partial u}{\partial y} + \frac{\partial v}{\partial x}\right) \qquad \tau_{yz} = \mu\left(\frac{\partial v}{\partial z} + \frac{\partial w}{\partial y}\right) \qquad \tau_{zx} = \mu\left(\frac{\partial w}{\partial x} + \frac{\partial u}{\partial z}\right)$$

The first subscript of the shear stress is the normal direction to the face over which the stress component is acting. The second subscript is the direction of the stress component.

In Chap. 3, in developing the Euler and energy equations, z was taken as the vertical coordinate, so that z was a measure of potential energy per unit weight. In dealing with problems in Chaps. 5 and 7 it is convenient to allow the x, y, z system of right-angular coordinates to take on any arbitrary orientation. Since gravity, the only body force considered, always acts vertically downward, h is taken as a coordinate which is positive vertically upward; then $\partial h/\partial x$ is the cosine of the angle between the x axis and the h axis, and similarly for the y and z axes. By limiting the Navier-Stokes equations to incompressible fluids, they become

$$-\frac{1}{\rho}\frac{\partial}{\partial x}(p + \gamma h) + \nu \nabla^2 u = \frac{du}{dt}$$

$$-\frac{1}{\rho}\frac{\partial}{\partial y}(p + \gamma h) + \nu \nabla^2 v = \frac{dv}{dt}$$

$$-\frac{1}{\rho}\frac{\partial}{\partial z}(p + \gamma h) + \nu \nabla^2 w = \frac{dw}{dt}$$

in which ν is the kinematic viscosity, assumed to be constant, d/dt is differentiation with respect to the motion

$$\frac{d}{dt} = u\frac{\partial}{\partial x} + v\frac{\partial}{\partial y} + w\frac{\partial}{\partial z} + \frac{\partial}{\partial t}$$

as explained in Sec. 7.2, and the operator ∇^2 is

$$\nabla^2 = \frac{\partial^2}{\partial x^2} + \frac{\partial^2}{\partial y^2} + \frac{\partial^2}{\partial z^2}$$

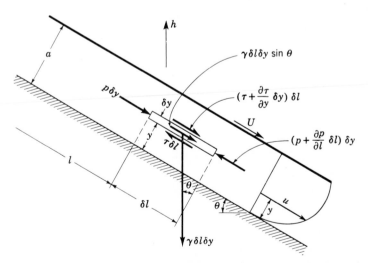

Figure 5.1 Flow between inclined parallel plates with the upper plate in motion.

For a nonviscous fluid, the Navier-Stokes equations reduce to the Euler equations of motion in three dimensions, given by Eqs. (7.2.3), (7.2.4), and (7.2.5).

For one-dimensional flow of a real fluid in the l direction (Fig. 5.1) with h vertically upward and y normal to l ($v = 0$, $w = 0$, $\partial u/\partial l = 0$), the Navier-Stokes equations reduce to

$$-\frac{1}{\rho}\frac{\partial}{\partial l}(p + \gamma h) + \frac{\mu}{\rho}\frac{\partial^2 u}{\partial y^2} = \frac{\partial u}{\partial t} \qquad \frac{\partial}{\partial y}(p + \gamma h) = 0$$

$$\frac{\partial}{\partial z}(p + \gamma h) = 0$$

For steady flow

$$\frac{\partial}{\partial l}(p + \gamma h) = \mu \frac{\partial^2 u}{\partial y^2}$$

and $p + \gamma h$ is a function of l only. Since u is a function of y only, $\tau = \mu\, du/dy$ for one-dimensional flow, and

$$\frac{d\tau}{dy} = \frac{d}{dl}(p + \gamma h)$$

5.1 LAMINAR, INCOMPRESSIBLE, STEADY FLOW BETWEEN PARALLEL PLATES

The general case of steady flow between parallel inclined plates is first developed for laminar flow, with the upper plate having a constant velocity U (Fig. 5.1). Flow between fixed plates is a special case obtained by setting $U = 0$. In Fig. 5.1 the upper plate moves parallel to the flow direction, and there is a pressure variation in the l direction. The flow is analyzed by taking a thin lamina of unit width as a free body. In steady flow the lamina moves at constant velocity u. The equation of motion yields

$$p\,\delta y - \left(p\,\delta y + \frac{dp}{dl}\,\delta l\,\delta y\right) - \tau\,\delta l + \left(\tau\,\delta l + \frac{d\tau}{dy}\,\delta y\,\delta l\right) + \gamma\,\delta l\,\delta y \sin\theta = 0$$

Dividing through by the volume of the element, using $\sin\theta = -\partial\tilde{h}/\partial l$, and simplifying yields

$$\frac{\partial\tau}{\partial y} = \frac{\partial}{\partial l}(p + \gamma h)$$

Since u is a function of y only, $\partial\tau/\partial y = d\tau/dy$; and since $p + \gamma h$ does not change value in the y direction (no acceleration), $p + \gamma h$ is a function of l only. Hence, $\partial(p + \gamma h)/\partial l = d(p + \gamma h)/dl$, and

$$\frac{d\tau}{dy} = \mu\frac{d^2 u}{dy^2} = \frac{d}{dl}(p + \gamma h) \qquad (5.1.1)$$

This equation was also determined from the Navier-Stokes equations in the preceding paragraph.

Integrating Eq. (5.1.1) with respect to y yields

$$\mu \frac{du}{dy} = y \frac{d}{dl}(p + \gamma h) + A$$

Integrating again with respect to y leads to

$$u = \frac{1}{2\mu} \frac{d}{dl}(p + \gamma h)y^2 + \frac{A}{\mu}y + B$$

in which A and B are constants of integration. To evaluate them, take $y = 0, u = 0$ and $y = a, u = U$ and obtain

$$B = 0 \qquad U = \frac{1}{2\mu} \frac{d}{dl}(p + \gamma h)a^2 + \frac{Aa}{\mu} + B$$

Eliminating A and B results in

$$u = \frac{Uy}{a} - \frac{1}{2\mu} \frac{d}{dl}(p + \gamma h)(ay - y^2) \tag{5.1.2}$$

For horizontal plates, $h = C$; for no gradient due to pressure or elevation, i.e., hydrostatic pressure distribution, $p + \gamma h = C$ and the velocity has a straight-line distribution. For fixed plates, $U = 0$, and the velocity distribution is parabolic.

The discharge past a fixed cross section is obtained by integration of Eq. (5.1.2) with respect to y:

$$Q = \int_0^a u \, dy = \frac{Ua}{2} - \frac{1}{12\mu} \frac{d}{dl}(p + \gamma h)a^3 \tag{5.1.3}$$

In general, the maximum velocity is not at the midplane.

Example 5.1 In Fig. 5.2 one plate moves relative to the other as shown. $\mu = 0.80$ P; $\rho = 850$ kg/m³. Determine the velocity distribution, the discharge, and the shear stress exerted on the upper plate.

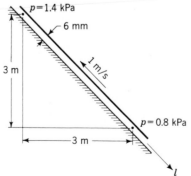

$p = 1.4$ kPa

6 mm

3 m

1 m/s

$p = 0.8$ kPa

3 m

l

Figure 5.2 Flow between inclined flat plates.

At the upper plate

$$p + \gamma h = 1400 \text{ Pa} + (850 \text{ kg/m}^3)(9.806 \text{ m/s}^2)(3 \text{ m}) = 26{,}405 \text{ Pa}$$

and at the lower point

$$p + \gamma h = 800 \text{ Pa}$$

to the same datum. Hence,

$$\frac{d}{dl}(p + \gamma h) = \frac{800 \text{ Pa} - 26{,}405 \text{ Pa}}{3\sqrt{2} \text{ m}} = -6035 \text{ N/m}^3$$

From the figure, $a = 0.006$ m, $U = -1$ m/s; and from Eq. (5.1.2)

$$u = \frac{(-1 \text{ m/s})(y \text{ m})}{0.006 \text{ m}} + \frac{6035 \text{ N/m}^3}{2(0.08 \text{ N·s/m}^2)}(0.006y - y^2 \text{ m}^2)$$

$$= 59.646y - 37{,}718y^2 \text{ m/s}$$

The maximum velocity occurs where $du/dy = 0$, or $y = 0.00079$ m, and it is $u_{max} = 0.0236$ m/s. The discharge per meter of width is

$$Q = \int_0^{0.006} u \, dy = 29.823y^2 - 12{,}573y^3 \Big]_0^{0.006} = -0.00164 \text{ m}^2/\text{s}$$

which is upward. To find the shear stress on the upper plate,

$$\frac{du}{dy}\bigg|_{y=0.006} = 59.646 - 75{,}436y \bigg|_{y=0.006} = -392.97 \text{ s}^{-1}$$

and

$$\tau = \mu \frac{du}{dy} = 0.08(-392.97) = -31.44 \text{ Pa}$$

This is the fluid shear at the upper plate; hence, the shear force on the plate is 31.44 Pa resisting the motion of the plate.

Losses in Laminar Flow

Expressions for irreversibilities are developed for one-dimensional, incompressible, steady, laminar flow. For steady flow in a tube, between parallel plates, or in a film flow at constant depth, the kinetic energy does not change and the reduction in $p + \gamma h$ represents the work done on the fluid per unit volume. The work done is converted into irreversibilities by the action of viscous shear. The losses in length L are $Q \, \Delta(p + \gamma h)$ per unit time.

If u is a function of y, the transverse direction, and the change in $p + \gamma h$ is a function of distance x in the direction of flow, total derivatives may be used throughout the development. First, from Eq. (5.1.1)

$$\frac{d(p + \gamma h)}{dx} = \frac{d\tau}{dy} \tag{5.1.4}$$

With reference to Fig. 5.3, a particle of fluid of rectangular shape of unit width has its center at (x, y), where the shear is τ, the pressure p, the velocity u, and the elevation h. It moves in the x direction. In unit time it has work done on it by the

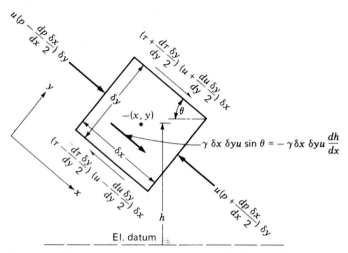

Figure 5.3 Work done and loss of potential energy for a fluid particle in one-dimensional flow.

surface boundaries as shown, and it gives up potential energy $\gamma\,\delta y\,\delta y\,u\,\sin\theta$. As there is no change in kinetic energy of the particle, the net work done and the loss of potential energy represent the losses per unit time due to irreversibilities. Collecting the terms from Fig. 5.3, dividing through by the volume $\delta x\,\delta y$, and taking the limit as $\delta x\,\delta y$ goes to zero yields

$$-u\frac{dp}{dx} - \gamma u\frac{dh}{dx} + \tau\frac{du}{dy} + u\frac{d\tau}{dy} = \frac{\text{net power input}}{\text{unit volume}} \qquad (5.1.5)$$

By combining with Eq. (5.1.4)

$$\frac{\text{Net power input}}{\text{Unit volume}} = \tau\frac{du}{dy} = \mu\left(\frac{du}{dy}\right)^2 = \frac{\tau^2}{\mu} \qquad (5.1.6)$$

Integrating this expression over a length L between two parallel plates, with Eq. (5.1.2) for $U = 0$, gives

$$\text{Net power input} = \int_0^a \mu\left(\frac{du}{dy}\right)^2 L\,dy$$

$$= \mu L \int_0^a \left[\frac{1}{2\mu}\frac{d(p + \gamma h)}{dx}(2y - a)\right]^2 dy$$

$$= \left[\frac{d(p + \gamma h)}{dx}\right]^2 \frac{a^3 L}{12\mu}$$

Substituting for Q from Eq. (5.1.3) for $U = 0$ yields

$$\text{Losses} = \text{net power input} = -Q\frac{d(p + \gamma h)}{dx}L = Q\,\Delta(p + \gamma h)$$

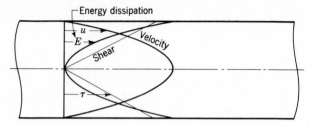

Figure 5.4 Distribution of velocity, shear, and losses per unit volume for a round tube.

in which $\Delta(p + \gamma h)$ is the drop in $p + \gamma h$ in the length L. The expression for power input per unit volume [Eq. (5.1.6)] is also applicable to laminar flow in a tube. The irreversibilities are greatest when du/dy is greatest. The distribution of shear stress, velocity, and losses per unit volume is shown in Fig. 5.4 for a round tube.

Example 5.2 A conveyor-belt device, illustrated in Fig. 5.5, is mounted on a ship and used to pick up undesirable surface contaminants, e.g., oil, from the surface of the sea. Assume the oil film to be thick enough for the supply to be unlimited with respect to the operation of the device. Assume the belt to operate at a steady velocity U and to be long enough for a uniform flow depth to exist. Determine the rate at which oil can be carried up the belt per unit width, in terms of θ, U, and the oil properties μ and γ.

A thin lamina of unit width that moves at velocity u is shown in Fig. 5.5. With the free surface as shown on the belt, and for steady flow at constant depth, the end-pressure effects on the lamina cancel. The equation of motion applied to the element yields

$$-\left(\tau + \frac{d\tau}{dy}\,\delta y\right)\delta l + \tau\,\delta l - \gamma\,\delta y\,\delta l \sin\theta = 0 \qquad \text{or} \qquad \frac{d\tau}{dy} = -\gamma \sin\theta$$

When the shear stress at the surface is recognized as zero, integration yields

$$\tau = \gamma \sin\theta(a - y)$$

This equation can be combined with Newton's law of viscosity, $\tau = -\mu\,du/dy$, to give

$$\int_{U}^{u} du = -\frac{\gamma \sin\theta}{\mu}\int_{0}^{y}(a - y)\,dy$$

or

$$u = U - \frac{\gamma \sin\theta}{\mu}\left(ay - \frac{y^2}{2}\right)$$

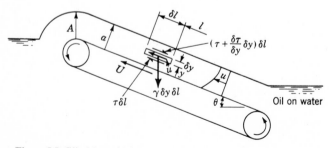

Figure 5.5 Oil pickup device.

The flow rate per unit width up the belt can be determined by integration:

$$q = \int_0^a u \, dy = Ua - \frac{\gamma \sin \theta}{\mu} \frac{a^3}{3}$$

This expression shows the flow rate to vary with a. However, a is still a dependent variable that is not uniquely defined by the above equations. The actual depth of flow on the belt is controlled by the end conditions.

The depth for maximum flow rate can be obtained by setting the derivative dq/da to zero and solving for the particular a

$$a = \bar{a} = \left(\frac{U\mu}{\gamma \sin \theta}\right)^{1/2}$$

To attach some physical significance to this particular depth, the influence of alternative crest depths may be considered. If the crest depth A, Fig. 5.5, is such that $\bar{a}$ occurs on the belt, then the maximum flow for that belt velocity and slope will be achieved. If A is physically controlled at a depth greater than $\bar{a}$, more flow will temporarily be supplied by the belt than can get away at the crest. That will cause the belt depth to increase, and the flow to decrease correspondingly, until either an equilibrium condition is realized or A is lowered. Alternatively, if $A < \bar{a}$, flow off the belt will be less than the maximum flow up the belt at depth $\bar{a}$ and the crest depth will increase to $\bar{a}$. At all times it is assumed that an unlimited supply is available at the bottom. By this reasoning it is seen that $\bar{a}$ is the only physical flow depth that can exist on the belt if the crest depth is free to seek its own level. A similar reasoning at the base leads to the same conclusion.

The discharge, as a function of fluid properties and U and θ, is given by

$$q = U\left(\frac{U\mu}{\gamma \sin \theta}\right)^{1/2} - \frac{\gamma \sin \theta}{3\mu}\left(\frac{U\mu}{\gamma \sin \theta}\right)^{3/2}$$

or

$$q = \left(\frac{\mu}{\gamma \sin \theta}\right)^{1/2} \frac{2}{3} U^{3/2}$$

EXERCISES

5.1.1 The shear stress in a fluid flowing between two fixed parallel plates (a) is constant over the cross section; (b) is zero at the plates and increases linearly to the midpoint; (c) varies parabolically across the section; (d) is zero at the midplane and varies linearly with distance from the midplane; (e) is none of these answers.

5.1.2 The velocity distribution for flow between two fixed parallel plates (a) is constant over the cross section; (b) is zero at the plates and increases linearly to the midplane; (c) varies parabolically across the section; (d) varies as the three-halves power of the distance from the midpoint; (e) is none of these answers.

5.1.3 The discharge between two parallel plates, distance a apart, when one has the velocity U and the shear stress is zero at the fixed plate, is (a) $Ua/3$; (b) $Ua/2$; (c) $2Ua/3$; (d) Ua; (e) none of these answers.

5.1.4 Fluid is in laminar motion between two parallel plates, with one plate in motion, and is under the action of a pressure gradient so that the discharge through any fixed cross section is zero. The minimum velocity occurs at a point which is distant from the fixed plate (a) $a/6$; (b) $a/3$; (c) $a/2$; (d) $2a/3$; (e) none of these answers.

5.1.5 In Exercise 5.1.4 the value of the minimum velocity is (a) $-3U/4$; (b) $-2U/3$; (c) $-U/2$; (d) $-U/3$; (e) $-U/6$.

5.1.6 The relation between pressure and shear stress in one-dimensional laminar flow in the x direction is given by (a) $dp/dx = \mu \, d\tau/dy$; (b) $dp/dy = d\tau/dx$; (c) $dp/dy = \mu \, d\tau/dx$; (d) $dp/dx = d\tau/dy$; (e) none of these answers.

5.1.7 The expression for power input per unit volume to a fluid in one-dimensional laminar motion in the x direction is (a) $\tau\, du/dy$; (b) τ/μ^2; (c) $\mu\, du/dy$; (d) $\tau(du/dy)^2$; (e) none of these answers.

5.1.8 When liquid is in laminar motion at constant depth and flowing down an inclined plate (y measured normal to surface), (a) the shear is zero throughout the liquid; (b) $d\tau/dy = 0$ at the plate; (c) $\tau = 0$ at the surface of the liquid; (d) the velocity is constant throughout the liquid; (e) there are no losses.

5.1.9 A 4-in-diameter shaft rotates at 240 rpm in a bearing with a radial clearance of 0.006 in. The shear stress in an oil film, $\mu = 0.1$ P, is, in pounds per square foot, (a) 0.15; (b) 1.75; (c) 3.50; (d) 16.70; (e) none of these answers.

5.2 LAMINAR FLOW THROUGH CIRCULAR TUBES AND CIRCULAR ANNULI

For steady, incompressible, laminar flow through a circular tube or an annulus, a cylindrical infinitesimal sleeve (Fig. 5.6) is taken as a free body. The equation of motion is applied in the l direction, with acceleration equal to zero. From the figure,

$$2\pi r\, \delta rp - \left(2\pi r\, \delta rp + 2\pi r\, \delta r\, \frac{dp}{dl}\, \delta l\right) + 2\pi r\, \delta l\tau$$

$$- \left[2\pi r\, \delta l\tau + \frac{d}{dr}(2\pi r\, \delta l\tau)\, \delta r\right] + \gamma 2\pi r\, \delta r\, \delta l \sin\theta = 0$$

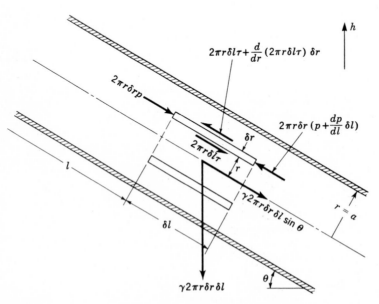

Figure 5.6 Free-body diagram of cylindrical sleeve element for laminar flow in an inclined circular tube.

Replacing $\sin \theta$ by $-dh/dl$ and dividing by the volume of the free body, $2\pi r \, \delta r \, \delta l$, gives

$$\frac{d}{dl}(p + \gamma h) + \frac{1}{r}\frac{d}{dr}(\tau r) = 0 \tag{5.2.1}$$

Since $d(p + \gamma h)/dl$ is not a function of r, the equation may be multiplied by $r \, \delta r$ and integrated with respect to r, yielding

$$\frac{r^2}{2}\frac{d}{dl}(p + \gamma h) + \tau r = A \tag{5.2.2}$$

in which A is the constant of integration. For a circular tube this equation must be satisfied when $r = 0$; hence, $A = 0$ for this case. Substituting

$$\tau = -\mu \, \frac{du}{dr}$$

note that the minus sign is required to obtain the sign of the τ term in Fig. 5.6. (u is considered to decrease with r; hence, du/dr is negative.)

$$du = \frac{1}{2\mu}\frac{d}{dl}(p + \gamma h)r \, dr - \frac{A}{\mu}\frac{dr}{r}$$

Another integration gives

$$u = \frac{r^2}{4\mu}\frac{d}{dl}(p + \gamma h) - \frac{A}{\mu}\ln r + B \tag{5.2.3}$$

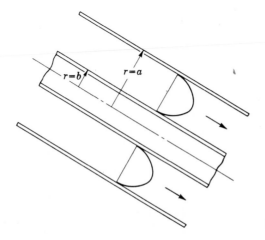

Figure 5.7 Flow through an annulus.

For the annular case, to evaluate A and B, $u = 0$ when $r = b$, the inner tube radius, and $u = 0$ when $r = a$ (Fig. 5.7). When A and B are eliminated,

$$u = -\frac{1}{4\mu}\frac{d}{dl}(p + \gamma h)\left(a^2 - r^2 + \frac{a^2 - b^2}{\ln b/a}\ln\frac{a}{r}\right) \tag{5.2.4}$$

and for discharge through an annulus (Fig. 5.7),

$$Q = \int_b^a 2\pi r u\, dr = -\frac{\pi}{8\mu}\frac{d}{dl}(p + \gamma h)\left[a^4 - b^4 - \frac{(a^2 - b^2)^2}{\ln a/b}\right] \tag{5.2.5}$$

Circular Tube; Hagen-Poiseuille Equation

For the circular tube, $A = 0$ in Eq. (5.2.3) and $u = 0$ for $r = a$,

$$u = -\frac{a^2 - r^2}{4\mu}\frac{d}{dl}(p + \gamma h) \tag{5.2.6}$$

The maximum velocity u_{max} is given for $r = 0$ as

$$u_{max} = -\frac{a^2}{4\mu}\frac{d}{dl}(p + \gamma h) \tag{5.2.7}$$

Since the velocity distribution is a paraboloid of revolution (Fig. 5.4), its volume is one-half that of its circumscribing cylinder; therefore, the average velocity is one-half of the maximum velocity,

$$V = -\frac{a^2}{8\mu}\frac{d}{dl}(p + \gamma h) \tag{5.2.8}$$

The discharge Q is equal to $V\pi a^2$,

$$Q = -\frac{\pi a^4}{8\mu}\frac{d}{dl}(p + \gamma h) \tag{5.2.9}$$

The discharge can also be obtained by integration of the velocity u over the area, i.e.,

$$Q = \int_0^a 2\pi r u\, dr$$

For a horizontal tube, $h = $ const; writing the pressure drop Δp in the length L gives

$$\frac{\Delta p}{L} = -\frac{dp}{dl}$$

and substituting diameter D leads to

$$Q = \frac{\Delta p\pi D^4}{128\mu L} \quad \text{(horizontal only)} \tag{5.2.10a}$$

In terms of average velocity,

$$V = \frac{\Delta p D^2}{32\mu L} \qquad \text{(horizontal only)} \qquad (5.2.10b)$$

Equation (5.2.10a) can then be solved for pressure drop, which represents losses per unit volume,

$$\Delta p = \frac{128\mu L Q}{\pi D^4} \qquad \text{(horizontal only)} \qquad (5.2.11)$$

The losses are seen to vary directly as the viscosity, the length, and the discharge and to vary inversely as the fourth power of the diameter. It should be noted that tube roughness does not enter into the equations. Equation (5.2.10a) is known as the *Hagen-Poiseuille equation;* it was determined experimentally by Hagen in 1839 and independently by Poiseuille in 1840. The analytical derivation was made by Wiedemann in 1856.

The results as given by Eqs. (5.2.1) to (5.2.10) are not valid near the entrance of a pipe. If the flow enters the pipe from a reservoir through a well-rounded entrance, the velocity at first is almost uniform over the cross section. The action of wall shear stress (as the velocity must be zero at the wall) is to slow down the fluid near the wall. As a consequence of continuity, the velocity must then increase in the central region. The transition length L' for the characteristic parabolic velocity distribution to develop is a function of the Reynolds number. Langhaar† developed the theoretical formula

$$\frac{L'}{D} = 0.058\mathbf{R}$$

which agrees well with observation.

Example 5.3 Determine the direction of flow through the tube shown in Fig. 5.8, in which $\gamma = 8000$ N/m^3 and $\mu = 0.04$ kg/m·s. Find the quantity flowing in liters per second and calculate the Reynolds number for the flow.
At section 1

$$p + \gamma h = 200{,}000 \text{ N/m}^2 + (8000 \text{ N/m}^3)(5 \text{ m}) = 240 \text{ kPa}$$

and at section 2

$$p + \gamma h = 300 \text{ kPa}$$

if the elevation datum is taken through section 2. The flow is from 2 to 1, since the energy is greater at 2 (kinetic energy must be the same at each section) than at 1. To determine the quantity flowing, the expression is written

$$\frac{d}{dl}(p + \gamma h) = \frac{300{,}000 - 240{,}000}{10 \text{ m}} \text{ N/m}^2 = 6000 \text{ N/m}^3$$

† H. L. Langhaar, Steady Flow in the Transition Length of a Straight Tube, *J. Appl. Mech.*, vol. 9, pp. 55–58, 1942.

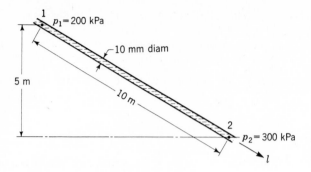

Figure 5.8 Flow through an inclined tube.

with l positive from 1 to 2. Substituting into Eq. (5.2.9) gives

$$Q = -\frac{\pi(0.005 \text{ m})^4}{8(0.04 \text{ N} \cdot \text{s/m}^2)} 6000 \text{ N/m}^3 = -0.0000368 \text{ m}^3/\text{s} = -0.0368 \text{ l/s}$$

The average velocity is

$$V = \frac{0.0000368 \text{ m}^3/\text{s}}{\pi(0.005 \text{ m})^2} = 0.4686 \text{ m/s}$$

and the Reynolds number (Sec. 4.4) is

$$\mathbf{R} = \frac{VD\rho}{\mu} = \frac{(0.4686 \text{ m/s})(0.01 \text{ m})}{(0.04 \text{ N} \cdot \text{s/m}^2)}\left(\frac{8000 \text{ N/m}^3}{9.806 \text{ m/s}^2}\right) = 95.6$$

If the Reynolds number had been above 2000, the Hagen-Poiseuille equation would no longer apply, as discussed in Sec. 5.3.

The kinetic-energy correction factor α [Eq. (3.10.2)] can be determined for laminar flow in a tube by use of Eqs. (5.2.6) and (5.2.7),

$$\frac{u}{V} = 2\frac{u}{u_{\text{max}}} = 2\left[1 - \left(\frac{r}{a}\right)^2\right] \tag{5.2.12}$$

Substituting into the expression for α gives

$$\alpha = \frac{1}{A}\int\left(\frac{u}{V}\right)^3 dA = \frac{1}{\pi a^2}\int_0^a \left\{2\left[1 - \left(\frac{r}{a}\right)^2\right]\right\}^3 2\pi r \, dr = 2 \tag{5.2.13}$$

There is twice as much energy in the flow as in uniform flow at the same average velocity. The momentum correction factor is obtained by replacing the exponent 3 with the exponent 2, yielding $\beta = \frac{4}{3}$.

EXERCISES

5.2.1 The shear stress in a fluid flowing in a round pipe (a) is constant over the cross section; (b) is zero at the wall and increases linearly to the center; (c) varies parabolically across the section; (d) is zero at the center and varies linearly with the radius; (e) is none of these answers.

5.2.2 When the pressure drop in a 24-in-diameter pipeline is 10 psi in 100 ft, the wall shear stress in pounds per square foot is (a) 0; (b) 7.2; (c) 14.4; (d) 720; (e) none of these answers.

5.2.3 In laminar flow through a round tube the discharge varies (a) linearly as the viscosity; (b) as the square of the radius; (c) inversely as the pressure drop; (d) inversely as the viscosity; (e) as the cube of the diameter.

5.2.4 When a tube is inclined, the term $-dp/dl$ is replaced by (a) $-dh/dl$; (b) $-\gamma \, dh/dl$; (c) $-d(p + h)/dl$; (d) $-d(p + \rho h)/dl$; (e) $-d(p + \gamma h)/dl$.

5.3 THE REYNOLDS NUMBER

Laminar flow is defined as flow in which the fluid moves in layers, or laminas, one layer gliding smoothly over an adjacent layer with only a molecular interchange of momentum. Any tendencies toward instability and turbulence are damped out by viscous shear forces that resist relative motion of adjacent fluid layers. Turbulent flow, however, has very erratic motion of fluid particles, with a violent transverse interchange of momentum. The nature of the flow, i.e., whether laminar or turbulent, and its relative position along a scale indicating the relative importance of turbulent to laminar tendencies are indicated by the *Reynolds number*. The concept of the Reynolds number and its interpretation are discussed in this section. In Sec. 3.5 an equation of motion was developed with the assumption that the fluid is frictionless, i.e., that the viscosity is zero. More general equations that include viscosity have been developed by including shear stresses. These equations (see introduction to this chapter) are complicated, nonlinear, partial differential equations for which no general solution has been obtained. In the last century Osborne Reynolds† studied them to try to determine when two different flow situations would be similar.

Two flow cases are said to be *dynamically similar* when

1. They are geometrically similar, i.e., corresponding linear dimensions have a constant ratio.
2. The corresponding streamlines are geometrically similar, or pressures at corresponding points have a constant ratio.

In considering two geometrically similar flow situations, Reynolds deduced that they would be dynamically similar if the general differential equations describing their flow were identical. By changing the units of mass, length, and time in one set of equations and determining the condition that must be satisfied to make them identical to the original equations, Reynolds found that the dimensionless group $ul\rho/\mu$ must be the same for both cases. Of these, u is a characteristic

---

† O. Reynolds, An Experimental Investigation of the Circumstances Which Determine Whether the Motion of Water Shall Be Direct or Sinuous, and of the Laws of Resistance in Parallel Channels, *Trans. R. Soc. Lond.*, vol. 174, 1883.

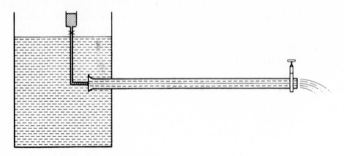

Figure 5.9 Reynolds' apparatus.

velocity, l a characteristic length, ρ the mass density, and μ the viscosity. This group, or parameter, is now called the Reynolds number **R**,

$$\mathbf{R} = \frac{ul\rho}{\mu} \tag{5.3.1}$$

To determine the significance of the dimensionless group, Reynolds conducted his experiments on flow of water through glass tubes, illustrated in Fig. 5.9. A glass tube was mounted horizontally with one end in a tank and a valve on the opposite end. A smooth bellmouth entrance was attached to the upstream end, with a dye jet so arranged that a fine stream of dye could be ejected at any point in front of the bellmouth. Reynolds took the average velocity V as characteristic velocity and the diameter of tube D as characteristic length, so that $\mathbf{R} = VD\rho/\mu$.

For small flows the dye stream moved as a straight line through the tube, showing that the flow was laminar. As the flow rate increased, the Reynolds number increased, since D, ρ, μ were constant and V was directly proportional to the rate of flow. With increasing discharge a condition was reached at which the dye stream wavered and then suddenly broke up and was diffused throughout the tube. The flow had changed to turbulent flow with its violent interchange of momentum that had completely disrupted the orderly movement of laminar flow. By careful manipulation Reynolds was able to obtain a value $\mathbf{R} = 12{,}000$ before turbulence set in. A later investigator, using Reynolds' original equipment, obtained a value of 40,000 by allowing the water to stand in the tank for several days before the experiment and by taking precautions to avoid vibration of the water or equipment. These numbers, referred to as the *Reynolds upper critical numbers*, have no practical significance in that the ordinary pipe installation has irregularities that cause turbulent flow at a much smaller value of the Reynolds number.

Starting with turbulent flow in the glass tube, Reynolds found that it always becomes laminar when the velocity is reduced to make **R** less than 2000. This is the *Reynolds lower critical number* for pipe flow and is of practical importance. With the usual piping installation, the flow will change from laminar to turbulent

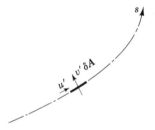

Figure 5.10 Notation for shear stress due to turbulent flow.

in the range of Reynolds numbers from 2000 to 4000. For the purpose of this treatment it is assumed that the change occurs at $\mathbf{R} = 2000$. In laminar flow the losses are directly proportional to the average velocity, while in turbulent flow the losses are proportional to the velocity to a power varying from 1.7 to 2.0.

There are many Reynolds numbers in use today in addition to the one for straight round tubes. For example, the motion of a sphere through a fluid may be characterized by $UD\rho/\mu$, in which U is the velocity of sphere, D is the diameter of sphere, and ρ and μ are the fluid density and viscosity.

The Reynolds number may be viewed as a ratio of shear stress τ_t due to turbulence to shear stress τ_v due to viscosity. By applying the momentum equation to the flow through an element of area δA (Fig. 5.10), the apparent shear stress due to turbulence can be determined. If v' is the velocity normal to δA and u' is the difference in velocity, or the velocity fluctuation, on the two sides of the area, then, with Eq. (3.11.1), the shear force δF acting is computed to be

$$\delta F = \rho v' \,\delta A \, u'$$

in which $\rho v' \, \delta A$ is the mass per second having its momentum changed and u' is the final velocity minus the initial velocity in the s direction. By dividing through by δA, the shear stress τ_t due to turbulent fluctuations is obtained,

$$\tau_t = \rho u'v' \tag{5.3.2}$$

The shear stress due to viscosity may be written

$$\tau_v = \frac{\mu u'}{l} \tag{5.3.3}$$

in which u' is interpreted as the change in velocity in the distance l, measured normal to the velocity. Then the ratio

$$\frac{\tau_t}{\tau_v} = \frac{v'l\rho}{\mu}$$

has the form of a Reynolds number.

The *nature* of a given flow of an incompressible fluid is characterized by its Reynolds number. For large values of $\mathbf{R}$ one or all of the terms in the numerator are large compared with the denominator. This implies a large expanse of fluid, high velocity, great density, extremely small viscosity, or combinations of these extremes. The numerator terms are related to *inertial forces*, or to forces set up by acceleration or deceleration of the fluid. The denominator term is the cause of viscous shear forces. Thus the Reynolds number parameter may also be considered as a ratio of inertial to viscous forces. A large **R** indicates a highly turbulent flow with losses proportional to the square of the velocity. The turbulence may be *fine scale*, composed of a great many small eddies that rapidly convert mechanical energy into irreversibilities through viscous action; or it may be *large scale*, like the huge vortices and swirls in a river or gusts in the atmosphere. The large eddies generate smaller eddies, which in turn create fine-scale turbulence. Turbulent flow may be thought of as a smooth, possibly uniform flow, with a secondary flow superposed on it. A fine-scale turbulent flow has small fluctuations in velocity that occur with high frequency. The root-mean-square value of the fluctuations and the frequency of change of sign of the fluctuations are quantitative measures of turbulence. In general, the intensity of turbulence increases as the Reynolds number increases.

For intermediate values of **R** both viscous and inertial effects are important, and changes in viscosity change the velocity distribution and the resistance to flow.

For the same **R**, two geometrically similar closed-conduit systems (one, say, twice the size of the other) will have the same ratio of losses to velocity head. The Reynolds number provides a means of using experimental results with one fluid to predict results in a similar case with another fluid.

In addition to the applications of laminar flow shown in this and the preceding section, the results may apply to greatly different situations, because the equations describing the cases are analogous. As an example, the two-dimensional laminar flow between closely spaced plates is called Hele-Shaw flow.† If some of the space is filled between the plates, by use of dye in the fluid the streamlines for flow around the obstructions are made visible. These streamlines in laminar flow are the same as the streamlines for similar flow of a frictionless (irrotational) fluid around the same obstructions. Likewise, the two-dimensional frictionless flow cases (Chap. 7) are analogous and similar to two-dimensional percolation through porous media.

EXERCISES

5.3.1 The upper critical Reynolds number is (*a*) important from a design viewpoint; (*b*) the number at which turbulent flow changes to laminar flow; (*c*) about 2000; (*d*) not more than 2000; (*e*) of no practical importance in pipe flow problems.

5.3.2 The Reynolds number for pipe flow is given by (*a*) VD/ν; (*b*) $VD\mu/\rho$; (*c*) $VD\rho/\nu$; (*d*) VD/μ; (*e*) none of these answers.

5.3.3 The lower critical Reynolds number has the value (*a*) 200; (*b*) 1200; (*c*) 12,000; (*d*) 40,000; (*e*) none of these answers.

---

† H. J. S. Hele-Shaw, Investigation of the Nature of the Surface Resistance of Water and of Streamline Motion under Certain Experimental Conditions, *Trans. Inst. Nav. Archit.*, vol. 40, 1898.

5.3.4 The Reynolds number for a 30-mm-diameter sphere moving 3 m/s through oil, sp gr 0.90, $\mu = 0.10$ kg/m·s, is (a) 404; (b) 808; (c) 900; (d) 8080; (e) none of these answers.

5.3.5 The Reynolds number for 10 cfs discharge of water at 68°F through a 12-in-diameter pipe is (a) 2460; (b) 980,000; (c) 1,178,000; (d) 14,120,000; (e) none of these answers.

5.4 PRANDTL MIXING LENGTH; VELOCITY DISTRIBUTION IN TURBULENT FLOW

Pressure drop and velocity distribution for several cases of laminar flow were worked out in the preceding sections. In this section the mixing-length theory of turbulence is developed, including its application to several flow situations. The apparent shear stress in turbulent flow is expressed by [Eq. (3.1.2)]

$$\tau = (\mu + \eta) \frac{du}{dy} \tag{5.4.1}$$

including direct viscous effects. Prandtl† has developed a most useful theory of turbulence called the *mixing-length theory*. In Sec. 5.3 the shear stress, τ, due to turbulence, was shown to be

$$\tau_t = \rho u'v' \tag{5.3.2}$$

in which u', v' are the velocity fluctuations at a point. In Prandtl's‡ theory, expressions for u' and v' are obtained in terms of a mixing-length distance l and the velocity gradient du/dy, in which u is the temporal mean velocity at a point and y is the distance normal to u, usually measured from the boundary. In a gas, one molecule, before striking another, travels an average distance known as the *mean free path* of the gas. Using this as an analogy (Fig. 5.11a), Prandtl assumed that a particle of fluid is displaced a distance l before its momentum is changed by the new environment. The fluctuation u' is then related to l by

$$u' \sim l \, \frac{du}{dy}$$

which means that the amount of the change in velocity depends upon the changes in temporal mean velocity at two points distant l apart in the y direction. From the continuity equation, he reasoned that there must be a correlation between u' and v' (Fig. 5.11b), so that v' is proportional to u',

$$v' \sim u' \sim l \, \frac{du}{dy}$$

---

† For an account of the development of turbulence theory the reader is referred to L. Prandtl, "Essentials of Fluid Dynamics," pp. 105–145, Hafner, New York, 1952.

‡ L. Prandtl, Bericht über Untersuchungen zur ausgebildeten Turbulenz, *Z. Angew. Math. Mech.*, vol. 5, no. 2, p. 136, 1925.

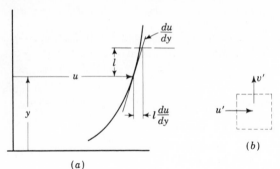

(b)

(a)

Figure 5.11 Notation for mixing-length theory.

By substituting for u' and v' in Eq. (5.3.2) and by letting l absorb the proportionality factor, the defining equation for mixing length is obtained:

$$\tau = \rho l^2 \left(\frac{du}{dy}\right)^2 \tag{5.4.2}$$

τ always acts in the sense that causes the velocity distribution to become more uniform. When Eq. (5.4.2) is compared with Eq. (3.1.1), it is found that

$$\eta = \rho l^2 \frac{du}{dy} \tag{5.4.3}$$

But η is not a fluid property like dynamic viscosity; instead η depends upon the density, the velocity gradient, and the mixing length l. In turbulent flow there is a violent interchange of globules of fluid except at a boundary, or very near to it, where this interchange is reduced to zero; hence, l must approach zero at a fluid boundary. The particular relation of l to wall distance y is not given by Prandtl's derivation. Von Kármán† suggested, after considering similitude relations in a turbulent fluid, that

$$l = \kappa \frac{du/dy}{d^2u/dy^2} \tag{5.4.4}$$

in which κ is a universal constant in turbulent flow, regardless of the boundary configuration or value of the Reynolds number.

In turbulent flows, η, sometimes referred to as the *eddy viscosity*, is generally much larger than μ. It may be considered as a coefficient of momentum transfer, expressing the transfer of momentum from points where the concentration is high to points where it is lower. It is convenient to utilize a *kinematic eddy viscosity* $\epsilon = \eta/\rho$ which is a property of the flow alone and is analogous to kinematic viscosity.

† T. von Kármán, Turbulence and Skin Friction, *J. Aeronaut. Sci.*, vol. 1, no. 1, p. 1, 1934.

Velocity Distributions

The mixing-length concept is used to discuss turbulent velocity distributions for the flat plate and the pipe. For turbulent flow over a smooth plane surface (such as the wind blowing over smooth ground) the shear stress in the fluid is constant, say τ_0. Equation (5.4.1) is applicable, but η approaches zero at the surface and μ becomes unimportant away from the surface. If η is negligible for the film thickness $y = \delta$, in which μ predominates, Eq. (5.4.1) becomes

$$\frac{\tau_0}{\rho} = \frac{\mu}{\rho}\frac{u}{y} = \nu\frac{u}{y} \qquad y \le \delta \tag{5.4.5}$$

The term $\sqrt{\tau_0/\rho}$ has the dimensions of a velocity and is called the shear-stress velocity u_*. Hence,

$$\frac{u}{u_*} = \frac{u_* y}{\nu} \qquad y \le \delta \tag{5.4.6}$$

shows a linear relation between u and y in the laminar film. For $y > \delta$, μ is neglected, and Eq. (5.4.1) produces

$$\tau_0 = \rho l^2 \left(\frac{du}{dy}\right)^2 \tag{5.4.7}$$

Since l has the dimensions of a length and from dimensional consideration would be proportional to y (the only significant linear dimension), assume $l = \kappa y$. Substituting into Eq. (5.4.7) and rearranging gives

$$\frac{du}{u_*} = \frac{1}{\kappa}\frac{dy}{y} \tag{5.4.8}$$

and integration leads to

$$\frac{u}{u_*} = \frac{1}{\kappa}\ln y + \text{const} \tag{5.4.9}$$

It is to be noted that this value of u substituted in Eq. (5.4.4) also determines l proportional to y (d^2u/dy^2 is negative, since the velocity gradient decreases as y increases). Equation (5.4.9) agrees well with experiment and, in fact, is also useful when τ is a function of y, because most of the velocity change occurs near the wall, where τ is substantially constant. It is quite satisfactory to apply the equation to turbulent flow in pipes.

Example 5.4 By integration of Eq. (5.4.9) find the relation between the average velocity V and the maximum velocity u_m in turbulent flow in a pipe.
When $y = r_0$, $u = u_m$, so that

$$\frac{u}{u_*} = \frac{u_m}{u_*} + \frac{1}{\kappa}\ln\frac{y}{r_0}$$

The discharge $V \pi r_0^2$ is obtained by integrating the velocity over the area,

$$V \pi r_0^2 = 2\pi \int_0^{r_0 - \delta} ur \, dr = 2\pi \int_\delta^{r_0} \left(u_m + \frac{u_*}{\kappa} \ln \frac{y}{r_0} \right)(r_0 - y) \, dy$$

The integration cannot be carried out to $y = 0$, since the equation holds in the turbulent zone only. The volume per second flowing in the laminar zone is so small that it may be neglected. Then

$$V = 2 \int_{\delta/r_0}^1 \left(u_m + \frac{u_*}{\kappa} \ln \frac{y}{r_0} \right)\left(1 - \frac{y}{r_0} \right) d\frac{y}{r_0}$$

in which the variable of integration is y/r_0. Integrating gives

$$V = 2\left\{ u_m \left[\frac{y}{r_0} - \frac{1}{2}\left(\frac{y}{r_0}\right)^2 \right] + \frac{u_*}{\kappa} \left[\frac{y}{r_0} \ln \frac{y}{r_0} - \frac{y}{r_0} - \frac{1}{2}\left(\frac{y}{r_0}\right)^2 \ln \frac{y}{r_0} + \frac{1}{4}\left(\frac{y}{r_0}\right)^2 \right] \right\}\Big|_{\delta/r_0}^1$$

Since δ/r_0 is very small, such terms as δ/r_0 and $\delta/r_0 \ln (\delta/r_0)$ become negligible $(\lim_{x \to 0} x \ln x = 0)$; thus

$$V = u_m - \frac{3}{2}\frac{u_*}{\kappa} \qquad \text{or} \qquad \frac{u_m - V}{u_*} = \frac{3}{2\kappa}$$

In evaluating the constant in Eq. (5.4.9) following the methods of Bakhmeteff,[†] $u = u_w$, the *wall velocity*, when $y = \delta$. According to Eq. (5.4.6),

$$\frac{u_w}{u_*} = \frac{u_* \delta}{\nu} = N \tag{5.4.10}$$

from which it is reasoned that $u_* \delta/\nu$ should have a critical value N at which flow changes from laminar to turbulent, since it is a Reynolds number in form. Substituting $u = u_w$ when $y = \delta$ into Eq. (5.4.9) and using Eq. (5.4.10) yields

$$\frac{u_w}{u_*} = N = \frac{1}{\kappa} \ln \delta + \text{const} = \frac{1}{\kappa} \ln \frac{N\nu}{u_*} + \text{const}$$

Eliminating the constant gives

$$\frac{u}{u_*} = \frac{1}{\kappa} \ln \frac{yu_*}{\nu} + N - \frac{1}{\kappa} \ln N$$

or

$$\frac{u}{u_*} = \frac{1}{\kappa} \ln \frac{yu_*}{\nu} + A \tag{5.4.11}$$

in which $A = N - (1/\kappa) \ln N$ has been found experimentally by plotting u/u_* against $\ln yu_*/\nu$. For flat plates $\kappa = 0.417$, $A = 5.84$, but for smooth-wall pipes Nikuradse's[‡] experiments yield $\kappa = 0.40$ and $A = 5.5$.

† B. A. Bakhmeteff, "The Mechanics of Turbulent Flow," Princeton University Press, Princeton, 1941.

‡ J. Nikuradse, Gesetzmassigkeiten der turbulenten Strömung in glatten Rohren, *Ver. Dtsch. Ing. Forschungsh.*, vol. 356, 1932.

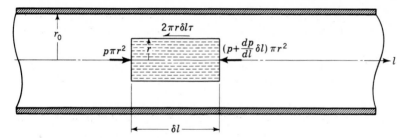

Figure 5.12 Free-body diagram for steady flow through a round tube.

Prandtl has developed a convenient exponential velocity-distribution formula for turbulent pipe flow,

$$\frac{u}{u_m} = \left(\frac{y}{r_0}\right)^n \tag{5.4.12}$$

in which n varies with the Reynolds number. This empirical equation is valid only at some distance from the wall. For $\mathbf{R}$ less than 100,000, $n = \frac{1}{7}$, and for greater values of $\mathbf{R}$, n decreases. The velocity-distribution equations, Eqs. (5.4.11) and (5.4.12), both have the fault of a nonzero value of du/dy at the center of the pipe.

Example 5.5 Find an approximate expression for mixing-length distribution in turbulent flow in a pipe from Prandtl's one-seventh-power law.

Writing a force balance for steady flow in a round tube (Fig. 5.12) gives

$$\tau = -\frac{dp}{dl}\frac{r}{2}$$

At the wall

$$\tau_0 = -\frac{dp}{dl}\frac{r_0}{2}$$

hence,

$$\tau = \tau_0 \frac{r}{r_0} = \tau_0 \left(1 - \frac{y}{r_0}\right) = \rho l^2 \left(\frac{du}{dy}\right)^2$$

Solving for l gives

$$l = \frac{u_* \sqrt{1 - y/r_0}}{du/dy}$$

From Eq. (5.4.12)

$$\frac{u}{u_m} = \left(\frac{y}{r_0}\right)^{1/7}$$

the approximate velocity gradient is obtained,

$$\frac{du}{dy} = \frac{u_m}{r_0}\frac{1}{7}\left(\frac{y}{r_0}\right)^{-6/7} \quad \text{and} \quad \frac{l}{r_0} = \frac{u_*}{u_m}7\left(\frac{y}{r_0}\right)^{6/7}\sqrt{1 - \frac{y}{r_0}}$$

The dimensionless *velocity deficiency*, $(u_m - u)/u_*$, is a function of y/r_0 only for large Reynolds numbers (Example 5.4) whether the pipe surface is smooth or rough. From Eq. (5.4.9), evaluating the constant for $u = u_m$ when $y = r_0$ gives

$$\frac{u_m - u}{u_*} = \frac{1}{\kappa} \ln \frac{r_0}{y} \qquad (5.4.13)$$

For rough pipes, the velocity may be assumed to be u_w at the wall distance $y_w = m\epsilon'$, in which ϵ' is a typical height of the roughness projections and m is a form coefficient depending upon the nature of the roughness. Substituting into Eq. (5.4.13) and eliminating u_m/u_* between the two equations leads to

$$\frac{u}{u_*} = \frac{1}{\kappa} \ln \frac{y}{\epsilon'} + \frac{u_w}{u_*} - \frac{1}{\kappa} \ln m \qquad (5.4.14)$$

in which the last two terms on the right-hand side are constant for a given type of roughness,

$$\frac{u}{u_*} = \frac{1}{\kappa} \ln \frac{y}{\epsilon'} + B \qquad (5.4.15)$$

In Nikuradse's experiments with sand-roughened pipes, constant-size sand particles (those passing a given screen and being retained on a slightly finer screen) were glued to the inside pipe walls. If ϵ' represents the diameter of sand grains, experiment shows that $\kappa = 0.40$, $B = 8.48$.

Spreading of a fluid jet. A free jet of fluid issuing into a large space containing the same fluid otherwise at rest is acted upon by frictional forces between the jet and the surrounding fluid. The jet velocity reduces and additional fluid is set in motion in the axial direction. The pressure is substantially constant throughout the jet and surroundings, so that the momentum in the axial direction remains constant. The turbulent mixing length within the jet can be taken as proportional to its breadth b (Fig. 5.13) $l = \alpha b$. Experiments show that $\alpha = \frac{1}{8}$. A conclusion from the constancy of momentum within the jet is that the maximum velocity (at

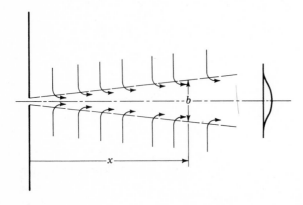

Figure 5.13 Fluid jet issuing into same-fluid medium.

the centerline) varies inversely as the axial distance x along the jet. Both theory†
and experiment show that the breadth varies linearly with axial distance, $b = x/8$.
Turbulent shear forces reduce the jet velocity within the central cone, and equal
turbulent shear forces act to increase velocity in the outer portions of the jet.

EXERCISES

5.4.1 The Prandtl mixing length is (a) independent of radial distance from pipe axis; (b) independent of
the shear stress; (c) zero at the pipe wall; (d) a universal constant; (e) useful for computing laminar
flow problems.

5.4.2 The average velocity divided by the maximum velocity, as given by the one-seventh-power law, is
(a) $\frac{49}{120}$; (b) $\frac{1}{2}$; (c) $\frac{6}{7}$; (d) $\frac{98}{120}$; (e) none of these answers.

5.5 TRANSPORT PHENOMENA

The atmosphere, rivers, lakes and oceans have always been used as waste recep-
tacles. They were quite capable of absorbing wastes until population concentrations, together with industrial and agricultural growth, overloaded them and
caused a rapid deterioration of the environment. It is appropriate in fluid
mechanics to look into the mechanisms by which these materials (including heat
and momentum) are carried, diffused, and dispersed. This class of problems comes
under the heading of *transport phenomena*.

Diffusion

It is convenient to think of the contaminant as marked fluid particles, or markers.
There is a movement of markers in the fluid when a nonuniform distribution of
markers, or a concentration gradient, exists. *Diffusion* is the process by which the
marker concentration is changed when the fluid is at rest or, if there is flow, when
the velocity distribution over any cross section is uniform. The diffusive process
may be the result of molecular activity, called *molecular diffusion*, or the action of
turbulence, called *turbulent diffusion*. Eddy velocities during turbulent flows cause
turbulent diffusion to generally be much greater than molecular diffusion.

Energy, momentum, and mass (matter) may be transported by diffusion.
Examples of molecular diffusion include the following. (1) Energy in the form of
heat is transported through a thin layer of fluid between fixed parallel plates of
different temperatures. (2) Momentum transport occurs in fluid between two
parallel plates when one plate moves relative to the other. The shear stress at the
moving plate, by molecular action, causes adjacent fluid layers to be set into
motion as a consequence of Newton's law of viscosity. (3) The placement of a

† H. Schlichting, "Boundary Layer Theory," pp. 681–685, McGraw-Hill Book Company, New
York, 1968.

permanganate crystal in a container of quiescent water results in molecular mass transport as the dye particles slowly spread throughout the fluid.

The following are examples of turbulent diffusion of energy, momentum, and mass. (1) Turbulent energy diffusion occurs during convection of heat from a paved surface during a hot summer day. (2) In Sec. 5.4, turbulent momentum transport, along with the mixing-length concept, was used in developing the logarithmic velocity distribution. (3) Automobile exhaust emissions, such as carbon monoxide, spreading into the atmosphere represent turbulent diffusion of mass.

Molecular Diffusion

The transport of substances or properties in a quiescent fluid, or relative to a fluid having a uniform velocity distribution, requires that uneven marker concentrations exist. Fick's law

$$P = -D_m \frac{\partial C}{\partial x} \tag{5.5.1}$$

states that the transfer rate P of the substance per unit area normal to the x direction varies directly as the coefficient of molecular diffusion D_m and the negative gradient of the concentration of the substance. The dimensions of D_m, P, and C are $L^2 T^{-1}$, $ML^{-2}T^{-1}$, and ML^{-3}, respectively, if the markers are identified in units of mass. The markers can be identified equally well as a certain number of particles, or by weight, etc.

A long pipe containing gas at rest may be examined as an example of molecular diffusion by using Fick's law. If at some cross section within the pipe, at time zero, a large concentration C of molecules are marked so that their movement may be traced, a continuity equation, stating that the net number transferred out of a unit volume is just equal to the reduction in number remaining, is

$$\frac{\partial P}{\partial x} = -\frac{\partial C}{\partial t} \tag{5.5.2}$$

By taking the partial of Eq. (5.5.1) with respect to x and substituting into Eq. (5.5.2),

$$\frac{\partial C}{\partial t} = D_m \frac{\partial^2 C}{\partial x^2} \tag{5.5.3}$$

This equation may be integrated to obtain

$$C = \frac{B}{\sqrt{t}} e^{-x^2/4D_m t} \tag{5.5.4}$$

in which B is the constant of integration. That the integration is correct is easily seen by differentiation of Eq. (5.5.4) and substitution into Eq. (5.5.3). The integration constant may be evaluated by consideration of a specific example. If $x \neq 0$, $C = 0$ for $t = 0$. Let M be the quantity of markers placed at $x = 0$ in the stationary fluid at time $t = 0$. Since only conservative materials are being considered, the amount of material M must be present in the system at any stage during the diffusion process. That is,

$$M = \int_{-\infty}^{\infty} CA \, dx \qquad (5.5.5)$$

in which A is the cross-sectional area of the conduit. The integration of Eq. (5.5.5), with Eq. (5.5.4) substituted, makes use of the probability integral

$$\int_{-\infty}^{\infty} e^{-y^2} \, dy = \sqrt{\pi}$$

The constant B is determined to be $M/(A\sqrt{4\pi D_m})$, which, when substituted into Eq. (5.5.4), yields

$$C = \frac{M}{A\sqrt{4\pi D_m t}} e^{-x^2/4D_m t} \qquad (5.5.6)$$

If, for example, 10^6 marked molecules are released at $x = 0$ at $t = 0$ in a long tube of 1-m^2 cross-sectional area in which $D_m = 10^{-5}$ m^2/s, then, by substitution into Eq. (5.5.6),

$$C = \frac{8.92(10)^7}{\sqrt{t}} e^{-25,000x^2/t}$$

By substitution, the concentration of markers one meter from the origin after one hour is 1.43×10^3 m^{-3}. At one hour the concentration at the origin is 1.49×10^6 m^{-3}. The time for the concentration at the origin to fall to 1.43×10^3 m^{-3} is 123 yr. In this example, by using differential equations, it was tacitly assumed that we were dealing with continuous quantities. By a molecular physics approach,[†] substantially the same result is obtained.

Turbulent Diffusion

The empirical equation (5.5.1) may be applied by analogy to turbulent transport by replacing D_m by an appropriate turbulent coefficient of diffusion D_t. It should be pointed out that D_t is not a material property, as is D_m, but depends upon the character of the turbulence. If one could have turbulent flow in a tube, with a uniform velocity distribution (i.e., as **R** becomes very large), assuming Fick's law applies and by use of the continuity equation, Eq. (5.5.3) now becomes

$$\frac{\partial C}{\partial t} + U \frac{\partial C}{\partial x} = D_t \frac{\partial^2 C}{\partial x^2} \qquad (5.5.7)$$

† A. J. Raudkivi and R. A. Callander, "Advanced Fluid Mechanics," Edward Arnold Ltd., London, 1975.

in which U is the uniform velocity. The terms on the left represent differentiation with respect to the motion U. If there were no turbulent diffusion, $D_t = 0$, then the equation would state that the markers move along the tube with unchanged concentration.

The equation

$$C = \frac{M}{A\sqrt{4\pi D_t t}} \exp \left[-\frac{(x - Ut)^2}{4D_t t} \right] \tag{5.5.8}$$

may be shown to satisfy Eq. (5.5.7) by substitution and to satisfy Eq. (5.5.5) when the probability integral is used.

The transfer of heat by turbulent diffusion is given by

$$H = -c_p \eta \frac{\partial T}{\partial y} = -c_p \rho l^2 \frac{\partial u}{\partial y} \frac{\partial T}{\partial y} \tag{5.5.9}$$

in which $c_p \eta$ is the eddy conductivity, c_p is the specific heat at constant pressure, T is the temperature, and H is the heat transfer per unit area per unit time.

An interesting case of steady-state particle transport is that of silt being carried by a river. All particles in suspension tend to settle to the bottom of the river at fall velocities dependent upon the particle size, shape, and weight. The action of turbulent diffusion causes a flow of particles upward that is due to the larger concentration near the bottom. If the two tendencies are equated, a steady-state relation is established showing how the concentration of silt varies vertically. It is necessary, however, to know the concentration at one level as a boundary condition. Fick's law for this case is

$$P = -\epsilon_c \frac{\partial C}{\partial y} \tag{5.5.10}$$

with P the rate of transfer (number of particles per unit area per unit time), C the concentration in number of particles per unit volume at level y, and ϵ_c assumed proportional to ϵ the kinematic eddy viscosity. For particles of nonuniform size having different settling velocities, Eq. (5.5.10) would be applied separately to groups of particles having about the same settling velocities and the resulting concentrations would be added together. For rivers ϵ_c would vary with distance y above the bottom.

Example 5.6 A tank of liquid containing fine solid particles of uniform size is agitated by shaking an internal wire lattice, so that the kinematic eddy viscosity may be considered constant. If the fall velocity of the particles in still liquid is v_f and the concentration of particles is C_0 at $y = y_0$ (y measured from the bottom), find the distribution of solid particles vertically throughout the liquid.

By using Eq. (5.5.10) to determine the rate per second carried upward by turbulence per unit of area at the level y, the amount per second falling across this surface by settling is equated to it for steady conditions. Those particles in the height $v_f \times 1$ s above the unit area will fall out in a second; that is, Cv_f particles cross the level downward per second per unit area. From

Eq. (5.5.10) $-\epsilon_c\, dC/dy$ particles are carried upward owing to the turbulence and the higher concentration below; hence,

$$P = Cv_f = -\epsilon_c \frac{dC}{dy} \quad \text{or} \quad \frac{dC}{C} = -\frac{v_f}{\epsilon_c}\, dy$$

Integrating gives

$$\ln C = -\frac{v_f}{\epsilon_c} y + \text{const}$$

For $C = C_0$, $y = y_0$,

$$C = C_0 \exp\left[-\frac{v_f}{\epsilon_c}(y - y_0)\right]$$

Dispersion

There is a spreading of markers in the longitudinal direction in a river or a pipe when the velocity distribution over the cross section is nonuniform. Inasmuch as fluid elements in the same cross section travel at different speeds, they will separate and thereby tend to spread markers greatly in the longitudinal direction.[†] This will cause a large transverse gradient of the concentration C of the marker, and turbulent and molecular diffusion will act to make the concentration more uniform over the cross section. This longitudinal *dispersion* may be defined as the spreading of marked fluid particles by the combined action of a nonuniform velocity distribution and diffusion.[‡] In the presence of the nonuniform velocity distribution much greater longitudinal spreading of the markers occurs than under the action of diffusion alone. A single injection of salt into a pipeline with flowing water is an example of dispersion along with molecular diffusion if the flow is laminar and dispersion together with both turbulent and molecular diffusion if the flow is turbulent.

Equation (5.5.7) may be used for analysis of longitudinal dispersion if the coefficient is appropriately increased to include the dispersive action of the velocity distribution. G. I. Taylor[§] has suggested a dispersion coefficient

$$D_d = 10.1a \sqrt{\frac{\tau_0}{\rho}} \tag{5.5.11}$$

for turbulent pipe flow, in which a is the pipe radius and τ_0 is the wall shear stress. For open channels the equation becomes

$$D_d = kd \sqrt{\frac{\tau_0}{\rho}} \tag{5.5.12}$$

[†] E. R. Holley, Unified View of Diffusion and Dispersion, *ASCE, Jour. of Hyd. Div.*, vol. 95, no. HY2, pp. 621–631, March 1969.
[‡] H. B. Fischer, Longitudinal Dispersion and Turbulent Mixing in Open Channel Flow, *Annual Review of Fluid Mechanics*, vol. 5, pp. 59–77, 1973.
[§] Ibid.

where d is the depth of channel. Experimental values[†] of k for a smooth laboratory channel vary from $k = 6$ to $k = 35$. For roughened laboratory channels the variation is from $k = 150$ to $k = 392$. For an irrigation canal k has been measured as 8.6. For large rivers two values, 74 and 7500, are known.

Consider the case of flow of water through a 500-mm-diameter pipe at 0.5 m/s with an average turbulent dispersion coefficient $D_d = 0.4$ m²/s. At $t = 0$, 10 N of salt is added at the origin. At any time $t > 0$, the maximum concentration [from Eq. (5.5.8)] is located at $x - Ut = 0$; it is given by

$$C_{max} = \frac{M}{A\sqrt{4\pi D_d t}}$$

At $t = 60$ s, substitution yields $x = 30$ m, $C_{max} = 2.933$ N/m³. For the same instant the concentration at $x = 29$ m is 2.90 N/m³; and at $x = 20$ m or 40 m, $C = 1.03$ N/m³.

5.6 BOUNDARY-LAYER CONCEPTS

In 1904 Prandtl[‡] developed the concept of the boundary layer. It provides an important link between ideal-fluid flow and real-fluid flow. *For fluids having relatively small viscosity, the effect of internal friction in a fluid is appreciable only in a narrow region surrounding the fluid boundaries.* From this hypothesis, the flow outside the narrow region near the solid boundaries may be considered as ideal flow or potential flow. Relations within the boundary-layer region can be computed from the general equations for viscous fluids, but use of the momentum equation permits the developing of approximate equations for boundary-layer growth and drag. In this section the boundary layer is described and the momentum equation is applied to it. Two-dimensional flow along a flat plate is studied by means of the momentum relations for both the laminar and the turbulent boundary layer. The phenomenon of separation of the boundary layer and formation of the wake is described.

Description of the Boundary Layer

When motion is started in a fluid having very small viscosity, the flow is essentially irrotational (Sec. 3.1) in the first instants. Since the fluid at the boundaries has zero velocity relative to the boundaries, there is a steep velocity gradient from the boundary into the flow. This velocity gradient in a real fluid sets up near the

---

† Ibid.

‡ L. Prandtl, Über Flussigkeitsbewegung bei sehr kleiner Reibung, *Verh. III Int. Math.-Kongr.*, *Heidelb*, 1904.

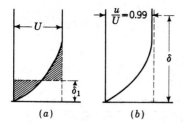

Figure 5.14 Definitions of boundary-layer thickness.

boundary shear forces that reduce the flow relative to the boundary. That fluid layer which has had its velocity affected by the boundary shear is called the *boundary layer*. The velocity in the boundary layer approaches the velocity in the main flow asymptotically. The boundary layer is very thin at the upstream end of a streamlined body at rest in an otherwise uniform flow. As this layer moves along the body, the continual action of shear stress tends to slow down additional fluid particles, causing the thickness of the boundary layer to increase with distance from the upstream point. The fluid in the layer is also subjected to a pressure gradient, determined from the potential flow, that increases the momentum of the layer if the pressure decreases downstream and decreases its momentum if the pressure increases downstream (*adverse* pressure gradient). The flow outside the boundary layer may also bring momentum into the layer.

For smooth upstream boundaries the boundary layer starts out as a *laminar boundary layer* in which the fluid particles move in smooth layers. As the laminar boundary layer increases in thickness, it becomes unstable and finally transforms into a *turbulent boundary layer* in which the fluid particles move in haphazard paths, although their velocity has been reduced by the action of viscosity at the boundary. When the boundary layer has become turbulent, there is still a very thin layer next to the boundary that has laminar motion. It is called the *laminar sublayer*.

Various definitions of boundary-layer thickness δ have been suggested. The most basic definition refers to the displacement of the main flow due to slowing down of fluid particles in the boundary zone. This thickness δ_1, called the *displacement thickness*, is expressed by

$$U\delta_1 = \int_0^\delta (U - u)\, dy \qquad (5.6.1)$$

in which δ is that value of y at which $u = U$ in the undisturbed flow. In Fig. 5.14a, the line $y = \delta_1$ is so drawn that the shaded areas are equal. This distance is, in itself, not the distance that is strongly affected by the boundary but is the amount the main flow must be shifted away from the boundary. In fact, that region is frequently taken as $3\delta_1$. Another definition, expressed by Fig. 5.14b, is the distance to the point where $u/U = 0.99$.

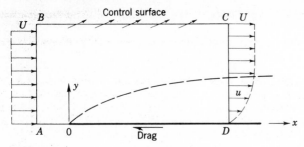

Figure 5.15 Control volume applied to fluid flowing over one side of a flat plate.

Momentum Equation Applied to the Boundary Layer

By following von Kármán's method† the principle of momentum can be applied directly to the boundary layer in steady flow along a flat plate. In Fig. 5.15 a control volume is taken enclosing the fluid above the plate, as shown, extending the distance x along the plate. In the y direction it extends a distance h so great that the velocity is undisturbed in the x direction, although some flow occurs along the upper surface, leaving the control volume.

The momentum equation for the x direction is

$$\Sigma F_x = \frac{\partial}{\partial t} \int_{cv} \rho u \, d\mathcal{V} + \int_{cs} \rho u \mathbf{v} \cdot d\mathbf{A}$$

It will be applied to the case of incompressible steady flow. The only force acting is due to drag or shear at the plate, since the pressure is constant around the periphery of the control volume. For unit widths of plate normal to the paper,

$$-\text{Drag} = \rho \int_0^h u^2 \, dy - \rho U^2 h + U\rho \int_0^h (U - u) \, dy$$

The first term on the right-hand side of the equation is the efflux of x momentum from CD, and the second term is the x-momentum influx through AB. The integral in the third term is the net volume influx through AB and CD, which, by continuity, must just equal the volume efflux through BC. It is multiplied by $U\rho$ to yield x-momentum efflux through BC. Combining the integrals gives

$$\text{Drag} = \rho \int_0^h u(U - u) \, dy \tag{5.6.2}$$

The drag $D(x)$ on the plate is in the reverse direction, so that

$$D(x) = \rho \int_0^h u(U - u) \, dy \tag{5.6.3}$$

† T. von Kármán, On Laminar and Turbulent Friction, Z. Angew. Math. Mech., vol. 1, pp. 235–236, 1921.

The drag on the plate may also be expressed as an integral of the shear stress along the plate,

$$D(x) = \int_0^x \tau_0 \, dx \qquad (5.6.4)$$

Equating the last two expressions and then differentiating with respect to x leads to

$$\tau_0 = \rho \, \frac{\partial}{\partial x} \int_0^h u(U - u) \, dy \qquad (5.6.5)$$

which is the momentum equation for two-dimensional flow along a flat plate.

Calculations of boundary-layer growth, in general, are complex and require advanced mathematical treatment. The parallel-flow cases, laminar or turbulent, along a flat plate may be worked out approximately by use of momentum methods that do not give any detail regarding the velocity distribution—in fact, a velocity distribution must be assumed. The results can be shown to agree closely with the more exact approach obtained from general viscous flow differential equations.

For an assumed distribution which satisfies the boundary conditions $u = 0$, $y = 0$ and $u = U$, $y = \delta$, the boundary-layer thickness as well as the shear at the boundary can be determined. The velocity distribution is assumed to have the same form at each value of x,

$$\frac{u}{U} = F\left(\frac{y}{\delta}\right) = F(\eta) \qquad \eta = \frac{y}{\delta}$$

when δ is unknown. For the laminar boundary layer Prandtl assumed that

$$\frac{u}{U} = F = \frac{3}{2}\eta - \frac{\eta^3}{2} \qquad 0 \le y \le \delta \qquad \text{and} \qquad F = 1 \qquad \delta \le y$$

which satisfy the boundary conditions. Equation (5.6.5) may be rewritten

$$\tau_0 = \rho U^2 \, \frac{\partial \delta}{\partial x} \int_0^1 \left(1 - \frac{u}{U}\right) \frac{u}{U} \, d\eta$$

and $\qquad \tau_0 = \rho U^2 \, \dfrac{\partial \delta}{\partial x} \displaystyle\int_0^1 \left(1 - \frac{3}{2}\eta + \frac{\eta^3}{2}\right)\left(\frac{3}{2}\eta - \frac{\eta^3}{2}\right) d\eta = 0.139 \rho U^2 \, \dfrac{\partial \delta}{\partial x}$

At the boundary

$$\tau_0 = \mu \left.\frac{\partial u}{\partial y}\right|_{y=0} = \mu \frac{U}{\delta} \left.\frac{\partial F}{\partial \eta}\right|_{\eta=0} = \mu \frac{U}{\delta} \left.\frac{\partial}{\partial \eta}\left(\frac{3}{2}\eta - \frac{\eta^3}{2}\right)\right|_{\eta=0} = \frac{3}{2}\mu\frac{U}{\delta} \qquad (5.6.6)$$

Equating the two expressions for τ_0 yields

$$\frac{3}{2}\mu\frac{U}{\delta} = 0.139 \rho U^2 \, \frac{\partial \delta}{\partial x}$$

and rearranging gives

$$\delta \, d\delta = 10.78 \frac{\mu \, dx}{\rho U}$$

since δ is a function of x only in this equation. Integrating gives

$$\frac{\delta^2}{2} = 10.78 \frac{v}{U} x + \text{const}$$

If $\delta = 0$, for $x = 0$, the constant of integration is zero. Solving for δ/x leads to

$$\frac{\delta}{x} = 4.65 \sqrt{\frac{v}{Ux}} = \frac{4.65}{\sqrt{\mathbf{R}_x}} \tag{5.6.7}$$

in which $\mathbf{R}_x = Ux/v$ is a Reynolds number based on the distance x from the leading edge of the plate. This equation for boundary-layer thickness in laminar flow shows that δ increases as the square root of the distance from the leading edge.

Substituting the value of δ into Eq. (5.6.6) yields

$$\tau_0 = 0.322 \sqrt{\frac{\mu \rho U^3}{x}} \tag{5.6.8}$$

The shear stress varies inversely as the square root of x and directly as the three-halves power of the velocity. The drag on one side of the plate, of unit width, is

$$\text{Drag} = \int_0^l \tau_0 \, dx = 0.644 \sqrt{\mu \rho U^3 l} \tag{5.6.9}$$

Selecting other velocity distributions does not radically alter these results. The exact solution, worked out by Blasius from the general equations of viscous motion, yields the coefficients 0.332 and 0.664 for Eqs. (5.6.8) and (5.6.9), respectively.

The drag can be expressed in terms of a drag coefficient C_D times the stagnation pressure $\rho U^2/2$ and the area of plate l (per unit breadth),

$$\text{Drag} = C_D \frac{\rho U^2}{2} l$$

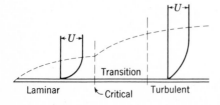

Laminar   Critical   Turbulent

Figure 5.16 Boundary-layer growth. (The vertical scale is greatly enlarged.)

in which, for the laminar boundary layer,

$$C_D = \frac{1.328}{\sqrt{\mathbf{R}_l}} \qquad (5.6.10)$$

and $\mathbf{R}_l = Ul/v$.

When the Reynolds number for the plate reaches a value between 500,000 and 1,000,000, the boundary layer becomes turbulent. Figure 5.16 indicates the growth and transition from laminar to turbulent boundary layer. The critical Reynolds number depends upon the initial turbulence of the fluid stream, the upstream edge of the plate, and the plate roughness.

Turbulent Boundary Layer

The momentum equation can be used to determine turbulent boundary-layer growth and shear stress along a smooth plate in a manner analogous to the treatment of the laminar boundary layer. The universal velocity-distribution law for smooth pipes, Eq. (5.4.11), provides the best basis, but the calculations are involved. A simpler approach is to use Prandtl's *one-seventh-power* law. It is $u/u_{max} = (y/r_0)^{1/7}$, in which y is measured from the wall of the pipe and r_0 is the pipe radius. Applying it to flat plates produces

$$F = \frac{u}{U} = \left(\frac{y}{\delta}\right)^{1/7} = \eta^{1/7}$$

and

$$\tau_0 = 0.0228 \rho U^2 \left(\frac{v}{U\delta}\right)^{1/4} \qquad (5.6.11)$$

in which the latter expression is the shear stress at the wall of a smooth plate with a turbulent boundary layer.† The method used to calculate the laminar boundary layer gives

$$\tau_0 = \rho U^2 \frac{d\delta}{dx} \int_0^1 (1 - \eta^{1/7})\eta^{1/7}\, d\eta = \frac{7}{72}\rho U^2 \frac{d\delta}{dx} \qquad (5.6.12)$$

By equating the expressions for shear stress, the differential equation for boundary-layer thickness δ is obtained:

$$\delta^{1/4}\, d\delta = 0.234\left(\frac{v}{U}\right)^{1/4} dx$$

After integrating, and then by assuming that the boundary layer is turbulent over the whole length of the plate so that the initial conditions $x = 0, \delta = 0$ can be used,

$$\delta^{5/4} = 0.292\left(\frac{v}{U}\right)^{1/4} x$$

---

† Equation (5.6.11) is obtained from the following pipe equations: $\tau_0 = \rho f V^2/8$, $f = 0.316/\mathbf{R}^{1/4}$ (Blasius eq.), $\mathbf{R} = V2r_0 \rho/\mu$, and $V = u_m/1.235$. To transfer to the flat plate $r_0 \sim \delta$, $u_m \sim U$.

Solving for δ gives

$$\delta = 0.37\left(\frac{\nu}{U}\right)^{1/5}x^{4/5} = \frac{0.37x}{(Ux/\nu)^{1/5}} = \frac{0.37x}{\mathbf{R}_x^{1/5}} \tag{5.6.13}$$

The thickness increases more rapidly in the turbulent boundary layer. In it the thickness increases as $x^{4/5}$, but in the laminar boundary layer δ varies as $x^{1/2}$.

To determine the drag on a smooth, flat plate, δ is eliminated in Eqs. (5.6.11) and (5.6.13), and

$$\tau_0 = 0.029\rho U^2\left(\frac{\nu}{Ux}\right)^{1/5} \tag{5.6.14}$$

The drag for unit width on one side of the plate is

$$\text{Drag} = \int_0^l \tau_0\,dx = 0.036\rho U^2 l\left(\frac{\nu}{Ul}\right)^{1/5} = \frac{0.036\rho U^2 l}{\mathbf{R}_l^{1/5}} \tag{5.6.15}$$

In terms of the drag coefficient,

$$C_D = 0.072\mathbf{R}_l^{-1/5} \tag{5.6.16}$$

in which $\mathbf{R}_l$ is the Reynolds number based on the length of plate.

The above equations are valid only for the range in which the Blasius resistance equation holds. For larger Reynolds numbers in smooth-pipe flow, the exponent in the velocity-distribution law is reduced. For $\mathbf{R} = 400{,}000$, $n = \frac{1}{8}$, and for $\mathbf{R} = 4{,}000{,}000$, $n = \frac{1}{10}$. The drag law, Eq. (5.6.15), is valid for a range

$$5 \times 10^5 < \mathbf{R}_l < 10^7$$

Experiment shows that the drag is slightly higher than is predicted by Eq. (5.6.16),

$$C_D = 0.074\mathbf{R}_l^{-1/5} \tag{5.6.17}$$

The boundary layer is actually laminar along the upstream part of the plate. Prandtl[†] has subtracted the drag from the equation for the upstream end of the plate up to the critical Reynolds number and then added the drag as given by the laminar equation for this portion of the plate, producing the equation

$$C_D = 0.074\mathbf{R}_l^{-1/5} - \frac{1700}{\mathbf{R}_l} \qquad 5 \times 10^5 < \mathbf{R}_l < 10^7 \tag{5.6.18}$$

In Fig. 5.17 a log-log plot of C_D vs. $\mathbf{R}_l$ shows the trend of the drag coefficients.

Use of the logarithmic velocity distribution for pipes produces

$$C_D = \frac{0.455}{(\log \mathbf{R}_l)^{2.58}} \qquad 10^6 < \mathbf{R}_l < 10^9 \tag{5.6.19}$$

in which the constant term has been selected for best agreement with experimental results.

† L. Prandtl, Über den Reibungswiderstand strömender Luft, *Result. Aerodyn. Test Inst., Goett.* III Lieferung, 1927.

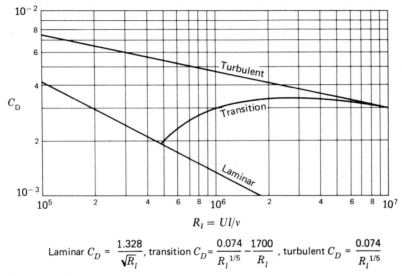

$$\text{Laminar } C_D = \frac{1.328}{\sqrt{R_l}}, \text{ transition } C_D = \frac{0.074}{R_l^{1/5}} - \frac{1700}{R_l}, \text{ turbulent } C_D = \frac{0.074}{R_l^{1/5}}$$

Figure 5.17 The drag law for smooth plates.

Example 5.7 A smooth, flat plate 3 m wide and 30 m long is towed through still water at 20°C with a speed of 6 m/s. Determine the drag on one side of the plate and the drag on the first 3 m of the plate.

For the whole plate

$$\mathbf{R} = \frac{Vl}{\nu} = \frac{(6 \text{ m/s})(30 \text{ m})}{1.007 \times 10^{-6} \text{ m}^2/\text{s}} = 1.787 \times 10^8$$

From Eq. (5.6.19)

$$C_D = \frac{0.455}{[\log (1.787 \times 10^8)]^{2.58}} = 0.00196$$

The drag on one side is

$$\text{Drag} = C_D b l \rho \frac{U^2}{2} = 0.00196(3 \text{ m})(30 \text{ m})(998.2 \text{ kg/m}^3)\frac{(6 \text{ m/s})^2}{2} = 3169 \text{ N}$$

in which b is the plate width. ν and ρ came from Appendix C, Table C.1. If the critical Reynolds number occurs at 5×10^5, the length l_0 to the transition is

$$\frac{(6 \text{ m/s})(l_0 \text{ m})}{1.007 \times 10^{-6} \text{ m}^2/\text{s}} = 5 \times 10^5 \qquad l_0 = 0.084 \text{ m}$$

For the first 3 m of the plate $\mathbf{R}_l = 1.787 \times 10^7$, and, using Eq. (5.6.19) again,

$$\text{Drag} = C_D b l \rho \frac{U^2}{2} = \frac{0.455(3 \text{ m})}{[\log (1.787 \times 10^7)]^{2.58}}(3 \text{ m})$$

$$\times (998.2 \text{ kg/m}^3)\frac{(6 \text{ m/s})^2}{2} = 443 \text{ N}$$

Calculation of the turbulent boundary layer over rough plates proceeds in similar fashion, starting with the rough-pipe tests using sand roughnesses. At the upstream end of the flat plate, the flow may be laminar; then, in the turbulent boundary layer, where the boundary layer is still thin and the ratio of roughness height to boundary-layer thickness ϵ/δ is significant, the region of fully developed roughness occurs and the drag is proportional to the square of the velocity. For long plates, this region is followed by a transition region where ϵ/δ becomes increasingly smaller, and eventually the plate becomes hydraulically smooth; i.e., the loss would not be reduced by reducing the roughness. Prandtl and Schlichting† have carried through these calculations, which are too complicated for reproduction here.

Separation; Wake

Along a flat plate the boundary layer continues to grow in the downstream direction, regardless of the length of the plate, when the pressure gradient remains zero. With the pressure decreasing in the downstream direction, as in a conical reducing section, the boundary layer tends to be reduced in thickness.

For *adverse* pressure gradients, i.e., with pressure increasing in the downstream direction, the boundary layer thickens rapidly. The adverse gradient and the boundary shear decrease the momentum in the boundary layer; and if both act over a sufficient distance, they cause the boundary layer to come to rest. This phenomenon is called *separation*. Figure 5.18a illustrates this case. The boundary streamline must leave the boundary at the separation point, and downstream from this point the adverse pressure gradient causes backflow near the wall. This region downstream from the streamline that separates from the boundary is known as the *wake*. The effect of separation is to decrease the net amount of flow work that can be done by a fluid element on the surrounding fluid at the expense of its kinetic energy, with the net result that pressure recovery is incomplete and flow losses (drag) increase. Figure 5.18b and c illustrate actual flow cases, the first with a very small adverse pressure gradient, which causes thickening of the boundary layer, and the second with a large diffuser angle, which causes separation and backflow near the boundaries.

Streamlined bodies (Fig. 5.19) are so designed that the separation point occurs as far downstream along the body as possible. If separation can be avoided, the boundary layer remains thin and the pressure is almost recovered downstream along the body. The only loss or drag is due to shear stress in the boundary layer, called *skin friction*. In the wake, the pressure is not recovered, and a *pressure drag* results. Reduction of wake size reduces the pressure drag on a body. In general, the drag is caused by both skin friction and pressure drag.

---

† L. Prandtl and H. Schlichting, Das Widerstandsgesetz rauher Platten, *Werft, Reederei, Hafen*, p. 1, 1934; see also *NACA Tech. Mem.* 1218, pt. II.

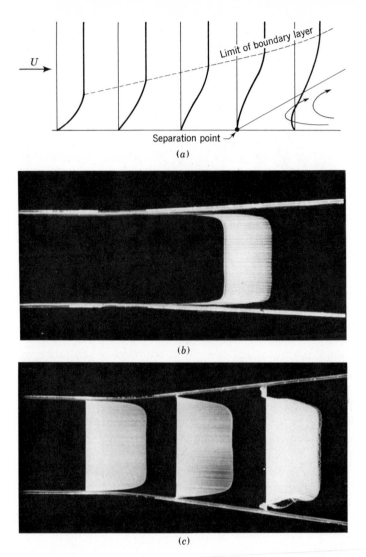

(a)

(b)

(c)

Figure 5.18 (a) Effect of adverse pressure gradient on boundary layer. Separation. (b) Boundary-layer growth in a small-angle diffuser. (c) Boundary-layer separation in a large-angle diffuser. [*Parts* (b) *and* (c) *from the film "Fundamentals of Boundary Layers," by the National Committee for Fluid Mechanics Films and the Education Development Center.*]

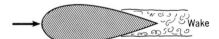

Figure 5.19 Streamlined body.

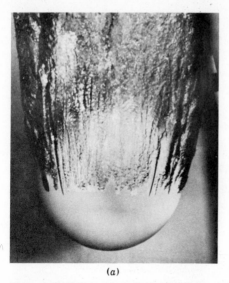
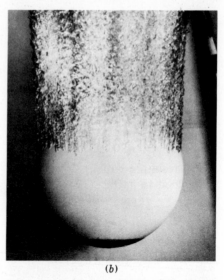

(a)   (b)

Figure 5.20 Shift in separation point due to induced turbulence: (a) 8.5-in bowling ball, smooth surface, 25 ft/s entry velocity into water; (b) same except for 4-in-diameter patch of sand on nose. (*Official U.S. Navy photograph made at Navy Ordnance Test Station, Pasadena Annex.*)

Flow around a sphere is an excellent example of the effect of separation on drag. For very small Reynolds numbers, $VD/v < 1$, the flow is everywhere nonturbulent, and the drag is referred to as *deformation drag*. Stokes' law (Sec. 5.7) gives the drag force for this case. For large Reynolds numbers, the flow may be considered potential flow except in the boundary layer and the wake. The boundary

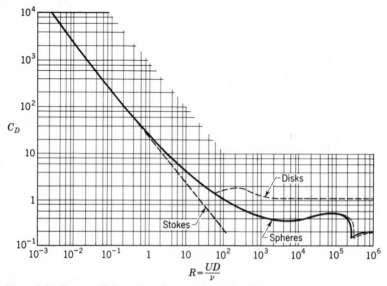

Figure 5.21 Drag coefficients for spheres and circular disks.

layer forms at the forward stagnation point and is generally laminar. In the laminar boundary layer, an adverse pressure gradient causes separation more readily than in a turbulent boundary layer because of the small amount of momentum brought into the laminar layer. If separation occurs in the laminar boundary layer, the location is farther upstream on the sphere than it is when the boundary layer becomes turbulent first and then separation occurs.

In Fig. 5.20 this is graphically portrayed by the photographs of the two spheres dropped into water at 25 ft/s. In *a*, separation occurs in the laminar boundary layer that forms along the smooth surface and causes a very large wake with a resulting large pressure drag. In *b*, the nose of the sphere, roughened by sand glued to it, induced an early transition to turbulent boundary layer before separation occurred. The high-momentum transfer in the turbulent boundary layer delayed the separation so that the wake is substantially reduced, resulting in a total drag on the sphere less than half that occurring in *a*.

A plot of drag coefficient against Reynolds number (Fig. 5.21) for smooth spheres shows that the shift to turbulent boundary layer (before separation) occurs by itself at a sufficiently high Reynolds number, as evidenced by the sudden drop in drag coefficient. The exact Reynolds number for the sudden shift depends upon the smoothness of the sphere and the turbulence in the fluid stream. In fact, the sphere is frequently used as a turbulence meter by determining the Reynolds number at which the drag coefficient is 0.30, a point located in the center of the sudden drop (Fig. 5.21). By use of the hot-wire anemometer, Dryden[†] has correlated the turbulence level of the fluid stream to the Reynolds number for the sphere at $C_D = 0.30$. The greater the turbulence of the fluid stream, the smaller the Reynolds number for shift in separation point.

EXERCISES

5.6.1 The displacement thickness of the boundary layer is (*a*) the distance from the boundary affected by boundary shear; (*b*) one-half the actual thickness of the boundary layer; (*c*) the distance to the point where $u/U = 0.99$; (*d*) the distance the main flow is shifted; (*e*) none of these answers.

5.6.2 The shear stress at the boundary of a flat plate is (*a*) $\partial p/\partial x$; (*b*) $\mu \left. \partial u/\partial y \right|_{y=0}$; (*c*) $\rho \left. \partial u/\partial y \right|_{y=0}$; (*d*) $\mu \left. \partial u/\partial y \right|_{y=\delta}$; (*e*) none of these answers.

5.6.3 Which of the following velocity distributions u/U satisfies the boundary conditions for flow along a flat plate? $\eta = y/\delta$. (*a*) e^{η}; (*b*) $\cos \pi\eta/2$; (*c*) $\eta - \eta^2$; (*d*) $2\eta - \eta^3$; (*e*) none of these answers.

5.6.4 The drag coefficient for a flat plate is $(D = \text{drag})$ (*a*) $2D/\rho U^2 l$; (*b*) $\rho U l/D$; (*c*) $\rho U l/2D$; (*d*) $\rho U^2 l/2D$; (*e*) none of these answers.

5.6.5 The laminar-boundary-layer thickness varies as (*a*) $1/x^{1/2}$; (*b*) $x^{1/7}$; (*c*) $x^{1/2}$; (*d*) $x^{6/7}$; (*e*) none of these answers.

5.6.6 The turbulent-boundary-layer thickness varies as (*a*) $1/x^{1/5}$; (*b*) $x^{1/5}$; (*c*) $x^{1/2}$; (*d*) $x^{4/5}$; (*e*) none of these answers.

5.6.7 In flow along a rough plate, the order of flow type from upstream to downstream is (*a*) laminar, fully developed wall roughness, transition region, hydraulically smooth; (*b*) laminar, transition region,

† H. Dryden, Reduction of Turbulence in Wind Tunnels, *NACA Tech. Rep.* 392, 1931.

hydraulically smooth, fully developed wall roughness; (c) laminar, hydraulically smooth, transition region, fully developed wall roughness; (d) laminar, hydraulically smooth, fully developed wall roughness, transition region; (e) laminar, fully developed wall roughness, hydraulically smooth, transition region.

5.6.8 Separation is caused by (a) reduction of pressure to vapor pressure; (b) reduction of pressure gradient to zero; (c) an adverse pressure gradient; (d) the boundary-layer thickness reducing to zero; (e) none of these answers.

5.6.9 Separation occurs when (a) the cross section of a channel is reduced; (b) the boundary layer comes to rest; (c) the velocity of sound is reached; (d) the pressure reaches a minimum; (e) a valve is closed.

5.6.10 The wake (a) is a region of high pressure; (b) is the principal cause of skin friction; (c) always occurs when deformation drag predominates; (d) always occurs after a separation point; (e) is none of these answers.

5.6.11 A sudden change in position of the separation point in flow around a sphere occurs at a Reynolds number of about (a) 1; (b) 300; (c) 30,000; (d) 3,000,000; (e) none of these answers.

5.7 DRAG ON IMMERSED BODIES

The principles of potential flow around bodies are developed in Chap. 7, and principles of the boundary layer, separation, and wake in Sec. 5.6. In this section drag is defined, some experimental drag coefficients are listed, the effect of compressibility on drag is discussed, and Stokes' law is presented. Lift is defined, and the lift and drag coefficients for an airfoil are given.

Drag is defined as the force component, parallel to the relative approach velocity, exerted on the body by the moving fluid. The drag-coefficient curves for spheres and circular disks are shown in Fig. 5.21. In Fig. 5.22 the drag coefficient for an infinitely long circular cylinder (two-dimensional case) is plotted against

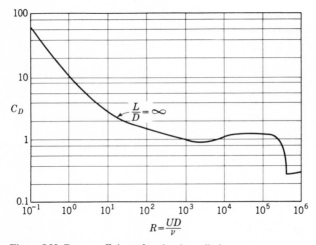

Figure 5.22 Drag coefficients for circular cylinders.

Table 5.1 Typical drag coefficients for various cylinders in two-dimensional flow†

Body shape			C_D	Reynolds number
Circular cylinder	→	○	1.2	10^4 to 1.5×10^5
Elliptical cylinder	→	⬭	0.6	4×10^4
		2:1	0.46	10^5
	→	⬭	0.32	2.5×10^4 to 10^5
	→	4:1	0.29	2.5×10^4
		8:1	0.20	2×10^5
Square cylinder	→	□	2.0	3.5×10^4
	→	◇	1.6	10^4 to 10^5
Triangular cylinders	→	120° ◁	2.0	10^4
	→	◁120°	1.72	10^4
	→	90° ◁	2.15	10^4
	→	◁90°	1.60	10^4
	→	60° ▷	2.20	10^4
	→	◁60°	1.39	10^4
	→	30° ▷	1.8	10^5
	→	◁30°	1.0	10^5
Semitubular	→	)	2.3	4×10^4
	→	(	1.12	4×10^4

† Data from W. F. Lindsey, *NACA Tech. Rep.* 619, 1938.

the Reynolds number. Like the sphere, this case also has the sudden shift in separation point. In each case, the drag coefficient C_D is defined by

$$\text{Drag} = C_D A \frac{\rho U^2}{2}$$

in which A is the projected area of the body on a plane normal to the flow.

In Table 5.1 typical drag coefficients are shown for several cylinders. In general, the values given are for the range of Reynolds numbers in which the coefficient changes little with the Reynolds number.

A typical lift and drag curve for an airfoil section is shown in Fig. 5.23. *Lift* is the fluid force component on a body at right angles to the relative approach velocity. The lift coefficient C_L is defined by

$$\text{Lift} = C_L A \frac{\rho U^2}{2}$$

in which A refers to the chord length times the wing length for lift and drag for airfoil sections.

Effect of Compressibility on Drag

To determine drag in high-speed gas flow the effects of compressibility, as expressed by the *Mach* number, are more important than the Reynolds number. The

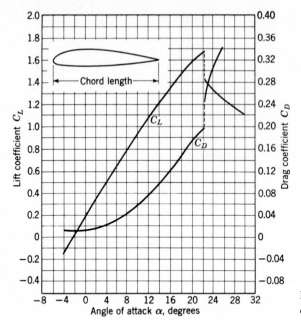

Figure 5.23 Typical lift and drag coefficients for an airfoil.

Mach number **M** is defined as the ratio of fluid velocity to velocity of sound in the fluid medium. When flow is at the critical velocity c, it has exactly the speed of the sound wave, so that small pressure waves cannot travel upstream. For this condition **M** = 1. When **M** is greater than unity, the flow is supersonic; and when **M** is less than unity, it is subsonic.

Any small disturbance is propagated with the speed of sound (Sec. 6.2). For example, a disturbance in still air travels outward as a spherical pressure wave. When the source of the disturbance moves with a velocity less than c, as in Fig. 5.24a, the wave travels ahead of the disturbing body and gives the fluid a

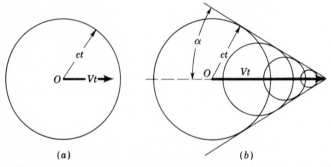

Figure 5.24 Wave propagation produced by a particle moving at (a) subsonic velocity and (b) supersonic velocity.

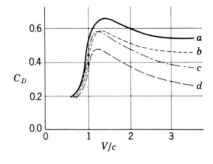

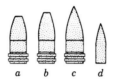

Figure 5.25 Drag coefficients for projectiles as a function of the Mach number. (*From L. Prandtl, "Abriss der Strömungslehre," Friedrich Vieweg und Sohn, Brunswick, Germany, 1935.*)

chance to adjust itself to the oncoming body. By the time the particle has moved a distance Vt, the disturbance wave has moved out as far as $r = ct$ from the point O. As the disturbing body moves along, new spherical waves are sent out, but in all subsonic cases they are contained within the initial spherical wave shown. In supersonic motion of a particle (Fig. 5.24b) the body moves faster than the spherical wave emitted from it, yielding a cone-shaped wavefront with vertex at the body, as shown. The half angle of cone α is called the *Mach angle*,

$$\alpha = \sin^{-1} \frac{ct}{Vt} = \sin^{-1} \frac{c}{V}$$

The conical pressure front extends out behind the body, and it is called a *Mach wave* (Sec. 6.4). There is a sudden small change in velocity and pressure across a Mach wave.

The drag on bodies varies greatly with the Mach number and becomes relatively independent of the Reynolds number when compressibility effects become important. In Fig. 5.25 the drag coefficients for four projectiles are plotted against the Mach number.

For low Mach numbers, a body should be rounded in front, with a blunt nose and a long-tapering afterbody for minimum drag. For high Mach numbers (0.7 and over), the drag rises very rapidly owing to formation of the vortices behind the projectile and to formation of the shock waves; the body should have a tapered nose or thin forward edge. As the Mach numbers increase, the curves tend to drop and to approach a constant value asymptotically. This appears to be due to the fact that the reduction of pressure behind the projectile is limited to absolute zero, and hence its contribution to the total drag tends to become constant. The pointed projectile creates a narrower shock front that tends to reduce the limiting value of the drag coefficient.

Stokes' Law

The flow of a viscous incompressible fluid around a sphere has been solved by Stokes† for values of the Reynolds number UD/ν below 1. The derivation is beyond the scope of this treatment; the results, however, are of value in such problems as the settling of dust particles. Stokes found the drag (force exerted on the sphere by flow of fluid around it) to be

$$\text{Drag} = 6\pi a\mu U$$

in which a is the radius of the sphere and U the velocity of the sphere relative to the fluid at a great distance. To find the terminal velocity for a sphere dropping through a fluid that is otherwise at rest, the buoyant force plus the drag force must just equal its weight, or

$$\tfrac{4}{3}\pi a^3 \gamma + 6\pi a\mu U = \tfrac{4}{3}\pi a^3 \gamma_s$$

in which γ is the specific weight of liquid and γ_s is the specific weight of the sphere. By solving for U, the terminal velocity is found to be

$$U = \frac{2}{9}\frac{a^2}{\mu}(\gamma_s - \gamma) \tag{5.7.1}$$

The straight-line portion of Fig. 5.21 represents Stokes' law.

The drag relations on particles, as given by Stokes' law and by the experimental results of Fig. 5.21, are useful in the design of settling basins for separating small solid particles from fluids. Applications include separating coolants from metal chips in machining operations, desilting river flow, and sanitary engineering applications to treatment of raw water and sewage.

Example 5.8 A jet aircraft discharges solid particles of matter 10 μm in diameter, $S = 2.5$, at the base of the stratosphere at 11,000 m. Assume the viscosity μ of air, in poises, to be expressed by the relation

$$\mu = 1.78 \times 10^{-4} - 3.06 \times 10^{-9}y$$

where y in meters is measured from sea level. Estimate the time for these particles to reach sea level. Neglect air currents and wind effects.

Writing $U = -dy/dt$ in Eq. (5.7.1) and recognizing the unit weight of air to be much smaller than the unit weight of the solid particles, one has

$$-\frac{dy}{dt} = \frac{2}{9}\frac{a^2\gamma_s}{\mu}$$

$$\int_0^T dt = -\int_{11,000}^0 \tfrac{9}{2}(1.78 \times 10^{-4} - 3.06 \times 10^{-9}y \text{ P})\frac{0.1 \text{ N·s/m}^2}{1 \text{ P}}$$

$$\times \frac{1}{(5 \times 10^{-6} \text{ m})^2}\frac{1}{2.5 \times 9806 \text{ N/m}^3}\,dy \text{ m}$$

$$T = \frac{1 \text{ d}}{86,400 \text{ s}}\left[1.78y - \frac{3.06 \times 10^{-5}y^2}{2}\right]_0^{11,000} \times 73.45 \text{ s}$$

$$= 15.07 \text{ d}$$

where d is the abbreviation for day.

---

† G. Stokes, *Trans. Camb. Phil. Soc.*, vol. 8, 1845; vol. 9, 1851.

EXERCISES

5.7.1 In a fluid stream of low viscosity (*a*) the effect of viscosity does not appreciably increase the drag on a body; (*b*) the potential theory yields the drag force on a body; (*c*) the effect of viscosity is limited to a narrow region surrounding a body; (*d*) the deformation drag on a body always predominates; (*e*) the potential theory contributes nothing of value regarding flow around bodies.

5.7.2 The lift on a body immersed in a fluid stream is (*a*) due to buoyant force; (*b*) always opposite in direction to gravity; (*c*) the resultant fluid force on the body; (*d*) the dynamic fluid force component exerted on the body normal to the approach velocity; (*e*) the dynamic fluid force component exerted on the body parallel to the approach velocity.

5.7.3 Pressure drag results from (*a*) skin friction; (*b*) deformation drag; (*c*) breakdown of potential flow near the forward stagnation point; (*d*) occurrence of a wake; (*e*) none of these answers.

5.7.4 A body with a rounded nose and long, tapering tail is usually best suited for (*a*) laminar flow; (*b*) turbulent subsonic flow; (*c*) supersonic flow; (*d*) flow at speed of sound; (*e*) none of these answers.

5.7.5 The effect of compressibility on the drag force is to (*a*) increase it greatly near the speed of sound; (*b*) decrease it near the speed of sound; (*c*) cause it asymptotically to approach a constant value for large Mach numbers; (*d*) cause it to increase more rapidly than the square of the speed at high Mach numbers; (*e*) reduce it throughout the whole flow range.

5.7.6 The terminal velocity of a small sphere settling in a viscous fluid varies as the (*a*) first power of its diameter; (*b*) inverse of the fluid viscosity; (*c*) inverse square of the diameter; (*d*) inverse of the diameter; (*e*) square of the difference in specific weights of solid and fluid.

5.8 RESISTANCE TO TURBULENT FLOW IN OPEN AND CLOSED CONDUITS

In steady, uniform turbulent incompressible flow in conduits of constant cross section the wall shear stress varies about proportional to the square of the average velocity

$$\tau_0 = \lambda \frac{\rho}{2} V^2 \tag{5.8.1}$$

in which λ is a dimensionless coefficient. In open channels and noncircular closed conduits the shear stress is not constant over the surface. In these cases, τ_0 is used as the average wall shear stress. Secondary flows† occurring in noncircular conduits act to equalize the wall shear stress.

In Fig. 5.26 a steady uniform flow is indicated in either an open or a closed conduit. For an open channel p_1 and p_2 are equal and flow occurs as a result of reduction in potential energy $z_1 - z_2$ m·N/N. For closed-conduit flow, energy for flow could be supplied by the potential energy drop, as well as by a drop in pressure $p_1 - p_2$. With flow vertically downward in a pipe, p_2 could increase in the flow direction but potential energy drop $z_1 - z_2$ would have to be greater than $(p_2 - p_1)/\gamma$ to supply energy to overcome the wall shear stress.

---

† Secondary flows, not wholly understood, are transverse components that cause the main central flow to spread out into corners or near walls.

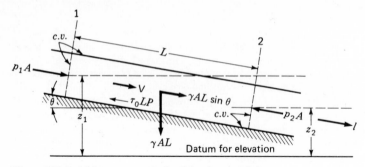

Figure 5.26 Axial forces on control volume in a conduit.

We may write the energy equation (3.10.1) to relate losses to available energy reduction

$$\frac{p_1}{\gamma} + \frac{V_1^2}{2g} + z_1 = \frac{p_2}{\gamma} + \frac{V_2^2}{2g} + z_2 + \text{losses}_{1-2}$$

Since the velocity head $V^2/2g$ is the same,

$$\text{Losses}_{1-2} = \frac{p_1 - p_2}{\gamma} + z_1 - z_2 \qquad (5.8.2)$$

Owing to the uniform assumption, the linear-momentum equation (3.11.2) applied in the l direction yields

$$\Sigma F_l = 0 = (p_1 - p_2)A + \gamma AL \sin\theta - \tau_0 LP$$

in which P is the *wetted perimeter* of the conduit, i.e., the portion of the perimeter where the wall is in contact with the fluid (free-liquid surface excluded). Since $L \sin\theta = z_1 - z_2$,

$$\frac{p_1 - p_2}{\gamma} + z_1 - z_2 = \frac{\tau_0 LP}{\gamma A} \qquad (5.8.3)$$

From Eqs. (5.8.2) and (5.8.3), using Eq. (5.8.1)

$$\text{Losses}_{1-2} = \frac{\tau_0 LP}{\gamma A} = \lambda \frac{\rho}{2} V^2 \frac{LP}{\gamma A} = \lambda \frac{L}{R} \frac{V^2}{2g} \qquad (5.8.4)$$

in which $R = A/P$ has been substituted. R, called the *hydraulic radius* of the conduit, is most useful in dealing with open channels. For a pipe $R = D/4$.

The loss term in Eq. (5.8.4) is in units of meter-newtons per newton or foot-pounds per pound. It is given the name h_f, *head loss due to friction*. By defining S as the losses per unit weight per unit length of channel,

$$S = \frac{h_f}{L} = \frac{\lambda}{R} \frac{V^2}{2g} \qquad (5.8.5)$$

After solving for V,

$$V = \sqrt{\frac{2g}{\lambda}} \sqrt{RS} = C\sqrt{RS} \qquad (5.8.6)$$

The *coefficient* λ, or coefficient C, must be found by experiment. This is the Chézy formula, in which originally the Chézy coefficient C was thought to be a constant for any size conduit or wall-surface condition. Various formulas for C are now generally used.

For pipes, when $\lambda = f/4$ and $R = D/4$, the Darcy-Weisbach equation is obtained:

$$h_f = f \frac{L}{D} \frac{V^2}{2g} \qquad (5.8.7)$$

in which D is the pipe inside diameter. This equation may be applied to open channels in the form

$$V = \sqrt{\frac{8g}{f}} \sqrt{RS} \qquad (5.8.8)$$

with values of f determined from pipe experiments.

EXERCISES

5.8.1 The hydraulic radius is given by (a) wetted perimeter divided by area; (b) area divided by square of wetted perimeter; (c) square root of area; (d) area divided by wetted perimeter; (e) none of these answers.

5.8.2 The hydraulic radius of a 60-mm-wide by 120-mm-deep open channel is, in millimeters, (a) 20; (b) 24; (c) 40; (d) 60; (e) none of these answers.

5.9 STEADY UNIFORM FLOW IN OPEN CHANNELS

For incompressible, steady flow at constant depth in a prismatic open channel, the *Manning* formula is widely used. It can be obtained from the Chézy formula [Eq. (5.8.6)] by setting

$$C = \frac{C_m}{n} R^{1/6} \qquad (5.9.1)$$

so that

$$V = \frac{C_m}{n} R^{2/3} S^{1/2} \qquad (5.9.2)$$

which is the Manning formula.

The value of C_m is 1.49 and 1.0 for U.S. customary and SI units, respectively; V is the average velocity at a cross section; R is the hydraulic radius (Sec. 5.8); and S is the losses per unit weight per unit length of channel or the slope of the bottom

Table 5.2 Average values of the Manning roughness factor for various boundary materials†

Boundary material	Manning n
Planed wood	0.012
Unplaned wood	0.013
Finished concrete	0.012
Unfinished concrete	0.014
Cast iron	0.015
Brick	0.016
Riveted steel	0.018
Corrugated metal	0.022
Rubble	0.025
Earth	0.025
Earth, with stones or weeds	0.035
Gravel	0.029

† Work by the U.S. Bureau of Reclamation and other government agencies indicates that the Manning roughness factor should be increased (say, 10 to 15 percent) for hydraulic radii greater than about 10 ft. The loss in capacity of large channels is due to the roughening of the surfaces with age, marine and plant growths, deposits, and the addition of bridge piers as the highway system is expanded.

of the channel. It is also the slope of the water surface, which is parallel to the channel bottom. The coefficient n was thought to be an absolute roughness coefficient, i.e., dependent upon surface roughness only, but it actually depends upon the size and shape of channel cross section in some unknown manner. Values of the coefficient n, determined by many tests on actual canals, are given in Table 5.2. Equation (5.9.2) must have consistent English or SI units as indicated for use with the values in Table 5.2.†

When Eq. (5.9.2) is multiplied by the cross-sectional area A, the Manning formula takes the form

$$Q = \frac{C_m}{n} A R^{2/3} S^{1/2} \qquad (5.9.3)$$

When the cross-sectional area is known, any one of the other quantities can be obtained from Eq. (5.9.3) by direct solution.

Example 5.9 Determine the discharge for a trapezoidal channel (Fig. 5.27) with a bottom width $b = 8$ ft and side slopes 1 on 1. The depth is 6 ft, and the slope of the bottom is 0.0009. The channel has a finished concrete lining.

† To convert the empirical equation in U.S. customary units to SI units, n is taken to be dimensionless; then the constant has dimensions and $(1.49 \text{ ft}^{1/3}/\text{s})(0.3048 \text{ m/ft})^{1/3} = 1.0 \text{ m}^{1/3}/\text{s}$.

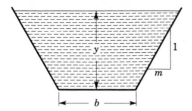

Figure 5.27 Notation for trapezoidal cross section.

From Table 5.2, $n = 0.012$. The area is

$$A = 8 \times 6 + 6 \times 6 = 84 \text{ ft}^2$$

and the wetted perimeter is

$$P = 8 + 2 \times 6\sqrt{2} = 24.96$$

By substituting into Eq. (5.9.3),

$$Q = \frac{1.49}{0.012} 84 \left(\frac{84}{24.96} \right)^{2/3} (0.0009^{1/2}) = 703 \text{ cfs}$$

Trial solutions are required in some instances when the cross-sectional area is unknown. Expressions for both the hydraulic radius and the area contain the depth in a form that cannot be solved explicitly.

Example 5.10 What depth is required for 4 m³/s flow in a rectangular planed-wood channel 2 m wide with a bottom slope of 0.002?
If the depth is y, $A = 2y$, $P = 2 + 2y$, and $n = 0.012$. By substituting in Eq. (5.9.3),

$$4 \text{ m}^3/\text{s} = \frac{1.00}{0.012} 2y \left(\frac{2y}{2 + 2y} \right)^{2/3} 0.002^{1/2}$$

Simplifying gives $\qquad f(y) = y \left(\frac{y}{1 + y} \right)^{2/3} = 0.536$

Assume $y = 1$ m; then $f(y) = 0.63$. Assume $y = 0.89$ m, then $f(y) = 0.538$. The correct depth is about 0.89 m.

Example 5.11 Riprap problem. A developer has been required by environmental regulatory authorities to line an open channel to prevent erosion. The channel is trapezoidal in cross section and has a slope of 0.0009. The bottom width is 10 ft and side slopes are 2 : 1 (horizontal to vertical). If he uses roughly spherical rubble ($\gamma_s = 135$ lb/ft³) for the lining, what is the minimum D_{50} of the rubble that can be used? The design flow is 1000 cfs. Assume the shear that rubble can withstand is described by

$$\tau = 0.040(\gamma_s - \gamma)D_{50} \qquad \text{lb/ft}^2$$

in which γ_s is the unit weight of rock and D_{50} is the average rock diameter in feet.
A Manning n of 0.03 is appropriate for the rubble. To find the size of channel, from Eq. (5.9.3)

$$1000 = \frac{1.49}{0.03} \frac{[y(10 + 2y)]^{5/3}}{(10 + 2\sqrt{5}\,y)^{2/3}} 0.03$$

By trial solution the depth is $y = 8.62$ ft, and the hydraulic radius $R = 4.84$ ft. From Eqs. (5.8.4) and (5.8.5)

$$\tau_0 = \gamma RS = 62.4 \times 4.84 \times 0.0009 = 0.272 \text{ lb/ft}^2$$

To find the D_{50} size for incipient movement $\tau = \tau_0$, and

$$0.040(135 - 62.4)D_{50} = 0.272$$

Hence $D_{50} = 0.0936$ ft.

More general cases of open-channel flow are considered in Chap. 11.

EXERCISES

5.9.1 The losses in open-channel flow generally vary as the (*a*) first power of the roughness; (*b*) inverse of the roughness; (*c*) square of the velocity; (*d*) inverse square of the hydraulic radius; (*e*) velocity.

5.9.2 The most simple form of open-channel-flow computation is (*a*) steady uniform; (*b*) steady nonuniform; (*c*) unsteady uniform; (*d*) unsteady nonuniform; (*e*) gradually varied.

5.9.3 In an open channel of great width the hydraulic radius equals (*a*) $y/3$; (*b*) $y/2$; (*c*) $2y/3$; (*d*) y; (*e*) none of these answers.

5.9.4 The Manning roughness coefficient for finished concrete is (*a*) 0.002; (*b*) 0.020; (*c*) 0.20; (*d*) dependent upon hydraulic radius; (*e*) none of these answers.

5.10 STEADY INCOMPRESSIBLE FLOW THROUGH SIMPLE PIPE SYSTEMS

Colebrook Formula

A force balance for steady flow (no acceleration) in a horizontal pipe (Fig. 5.28) yields

$$\Delta p \pi r_0^2 = \tau_0 2\pi r_0 \, \Delta L$$

This simplifies to
$$\tau_0 = \frac{\Delta p}{\Delta L} \frac{r_0}{2} \qquad (5.10.1)$$

which holds for laminar or turbulent flow. The Darcy-Weisbach equation (5.8.7) may be written

$$\Delta p = \gamma h_f = f \frac{\Delta L}{2r_0} \rho \frac{V^2}{2}$$

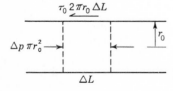

Figure 5.28 Equilibrium conditions for steady flow in a pipe.

Eliminating Δp in the two equations and simplifying gives

$$\sqrt{\frac{\tau_0}{\rho}} = \sqrt{\frac{f}{8}} V \qquad (5.10.2)$$

which relates wall shear stress, friction factor, and average velocity. The average velocity V may be obtained from Eq. (5.4.11) by integrating over the cross section. Substituting for V in Eq. (5.10.2) and simplifying produces the equation for friction factor in smooth-pipe flow,

$$\frac{1}{\sqrt{f}} = A_s + B_s \ln (\mathbf{R}\sqrt{f}) \qquad (5.10.3)$$

With the Nikuradse† data for smooth pipes the equation becomes

$$\frac{1}{\sqrt{f}} = 0.86 \ln (\mathbf{R}\sqrt{f}) - 0.8 \qquad (5.10.4)$$

For rough pipes in the complete turbulence zone,

$$\frac{1}{\sqrt{f}} = F_2\left(m, \frac{\epsilon'}{D}\right) + B_r \ln \frac{\epsilon}{D} \qquad (5.10.5)$$

in which F_2 is, in general, a constant for a given form and spacing of the roughness elements. For the Nikuradse sand-grain roughness (Fig. 5.31) Eq. (5.10.5) becomes

$$\frac{1}{\sqrt{f}} = 1.14 - 0.86 \ln \frac{\epsilon}{D} \qquad (5.10.6)$$

The roughness height ϵ for sand-roughened pipes may be used as a measure of the roughness of commercial pipes. If the value of f is known for a commercial pipe in the fully developed wall turbulence zone, i.e., large Reynolds numbers and the loss proportional to the square of the velocity, the value of ϵ can be computed by Eq. (5.10.6). In the transition region, where f depends upon both ϵ/D and $\mathbf{R}$, sand-roughened pipes produce different results from commercial pipes. This is made evident by a graph based on Eqs. (5.10.4) and (5.10.6) with both sand-roughened and commercial-pipe-test results shown. Rearranging Eq. (5.10.6) gives

$$\frac{1}{\sqrt{f}} + 0.86 \ln \frac{\epsilon}{D} = 1.14$$

and adding $0.86 \ln (\epsilon/D)$ to each side of Eq. (5.10.4) leads to

$$\frac{1}{\sqrt{f}} + 0.86 \ln \frac{\epsilon}{D} = 0.86 \ln \left(\mathbf{R}\sqrt{f}\,\frac{\epsilon}{D}\right) - 0.8$$

---

† J. Nikuradse, Gesetzmässigkeiten der turbulenten Strömung in glatten Rohren, *Ver. Dtsch. Ing. Forschungsh.*, vol. 356, 1932.

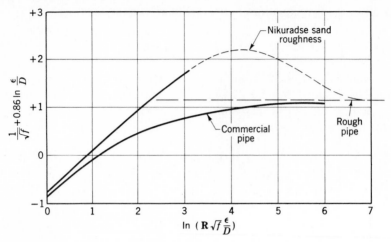

Figure 5.29 Colebrook transition function.

By selecting $1/\sqrt{f} + 0.86 \ln (\epsilon/D)$ as ordinate and $\ln (\mathbf{R}\sqrt{f}\,\epsilon/D)$ as abscissa (Fig. 5.29), smooth-pipe-test results plot as a straight line with slope $+0.86$ and rough-pipe-test results in the complete turbulence zone plot as the horizontal line. Nikuradse sand-roughness-test results plot along the dashed line in the transition region, and commercial-pipe-test results plot along the lower curved line.

The explanation of the difference in shape of the artificial roughness curve of Nikuradse and the commercial roughness curve is that the laminar sublayer, or laminar film, covers all the artificial roughness or allows it to protrude uniformly as the film thickness decreases. With commercial roughness, which varies greatly in uniformity, small portions extend beyond the film first, as the film decreases in thickness with increasing Reynolds number. An empirical transition function for commercial pipes for the region between smooth pipes and the complete turbulence zone has been developed by Colebrook,†

$$\frac{1}{\sqrt{f}} = -0.86 \ln \left(\frac{\epsilon/D}{3.7} + \frac{2.51}{\mathbf{R}\sqrt{f}} \right) \qquad (5.10.7)$$

which is the basis for the Moody diagram (Fig. 5.32).

Pipe Flow

In steady incompressible flow in a pipe the irreversibilities are expressed in terms of a head loss, or drop in *hydraulic grade line* (Sec. 10.2). The hydraulic grade line is p/γ above the center of the pipe, and if z is the elevation of the center of the pipe,

---

† C. F. Colebrook, Turbulent Flow in Pipes, with Particular Reference to the Transition Region between the Smooth and Rough Pipe Laws, *J. Inst. Civ. Eng. Lond.*, vol. 11, pp. 133–156, 1938–1939.

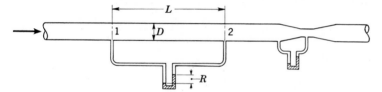

Figure 5.30 Experimental arrangement for determining head loss in a pipe.

then $z + p/\gamma$ is the elevation of a point on the hydraulic grade line. The locus of values of $z + p/\gamma$ along the pipeline gives the hydraulic grade line. Losses, or irreversibilities, cause this line to drop in the direction of flow. The Darcy-Weisbach equation

$$h_f = f \, \frac{L}{D} \frac{V^2}{2g} \tag{5.8.7}$$

is generally adopted for pipe flow calculations. h_f is the head loss, or drop in hydraulic grade line, in the pipe length L, having an inside diameter D and an average velocity V. h_f has the dimension length and is expressed in terms of foot-pounds per pound or meter-newtons per newton. The friction factor f is a dimensionless factor that is required to make the equation produce the correct value for losses. All quantities in Eq. (5.8.7) except f can be measured experimentally. A typical setup is shown in Fig. 5.30. By measuring the discharge and inside diameter, the average velocity can be computed. The head loss h_f is measured by a differential manometer attached to piezometer openings at sections 1 and 2, distance L apart.

Experimentation shows the following to be true in turbulent flow:

1. The head loss varies directly as the length of the pipe.
2. The head loss varies almost as the square of the velocity.
3. The head loss varies almost inversely as the diameter.
4. The head loss depends upon the surface roughness of the interior pipe wall.
5. The head loss depends upon the fluid properties of density and viscosity.
6. The head loss is independent of the pressure.

The friction factor f must be so selected that Eq. (5.8.7) correctly yields the head loss; hence, f cannot be a constant but must depend upon velocity V, diameter D, density ρ, viscosity μ, and certain characteristics of the wall roughness signified by ϵ, ϵ', and m, where ϵ is a measure of the *size* of the roughness projections and has the dimensions of a length, ϵ' is a measure of the *arrangement* or *spacing* of the roughness elements and also has the dimensions of a length, and m is a form factor, depending upon the *shape* of the individual roughness elements and is dimensionless. The term f, instead of being a simple constant, turns out to depend upon seven quantities:

$$f = f(V, D, \rho, \mu, \epsilon, \epsilon', m) \tag{5.10.8}$$

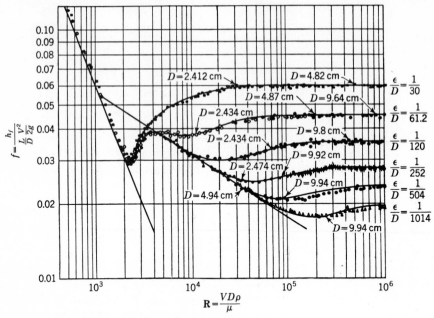

Figure 5.31 Nikuradse's sand-roughened-pipe tests.

Since f is a dimensionless factor, it must depend upon the grouping of these quantities into dimensionless parameters. For *smooth* pipe $\epsilon = \epsilon' = m = 0$, leaving f dependent upon the first four quantities. They can be arranged in only one way to make them dimensionless, namely, $VD\rho/\mu$, which is the Reynolds number. For *rough* pipes the terms ϵ, ϵ' may be made dimensionless by dividing by D. Therefore, in general,

$$f = f\left(\frac{VD\rho}{\mu}, \frac{\epsilon}{D}, \frac{\epsilon'}{D}, m\right) \tag{5.10.9}$$

The proof of this relation is left to experimentation. For smooth pipes a plot of all experimental results shows the functional relation, subject to a scattering of ± 5 percent. The plot of friction factor against the Reynolds number on a log-log chart is called a *Stanton diagram*. Blasius[†] was the first to correlate the smooth-pipe experiments in turbulent flow. He presented the results by an empirical formula that is valid up to about $\mathbf{R} = 100,000$. The Blasius formula is

$$f = \frac{0.316}{\mathbf{R}^{1/4}} \tag{5.10.10}$$

In rough pipes the term ϵ/D is called the *relative roughness*. Nikuradse[‡] proved the validity of the relative-roughness concept by his tests on sand-

---

[†] H. Blasius, Das Ähnlichkeitsgesetz bei Reibungsvorgängen in Flüssigkeiten, *Ver. Dtsch. Ing. Forschungsh.*, vol. 131, 1913.

[‡] J. Nikuradse, Strömungsgesetze in rauhen Rohren, *Ver. Dtsch. Ing. Forschungsh.*, vol. 361, 1933.

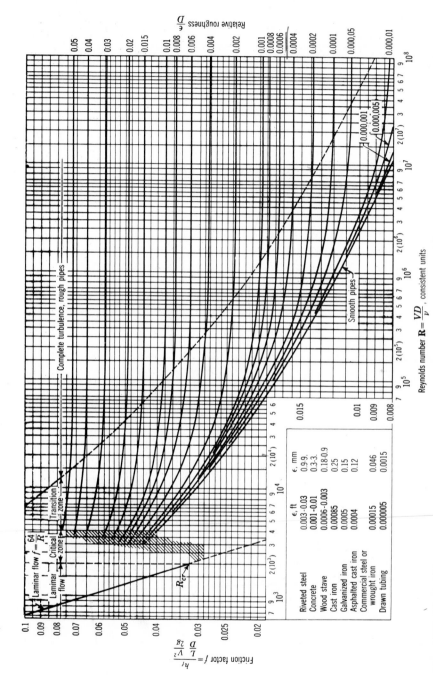

Figure 5.32 Moody diagram.

	ϵ, ft	ϵ, mm
Riveted steel	0.003–0.03	0.9–9.9
Concrete	0.001–0.01	0.3–3.3
Wood stave	0.0006–0.003	0.18–0.9
Cast iron	0.00085	0.25
Galvanized iron	0.0005	0.15
Asphalted cast iron	0.0004	0.12
Commercial steel or wrought iron	0.00015	0.046
Drawn tubing	0.000005	0.0015

Friction factor $f = \dfrac{h_f}{\dfrac{L}{D}\dfrac{V^2}{2g}}$

Relative roughness $\dfrac{\epsilon}{D}$

Reynolds number $\mathbf{R} = \dfrac{VD}{\nu}$, consistent units

Laminar flow $f = \dfrac{64}{\mathbf{R}}$

Laminar flow

Critical zone

Transition zone

Complete turbulence, rough pipes

Smooth pipes

$\mathbf{R}_{cr}$

roughened pipes. He used three sizes of pipes and glued sand grains (ϵ = diameter of the sand grains) of practically constant size to the interior walls so that he had the same values of ϵ/D for different pipes. These experiments (Fig. 5.31) show that for one value of ϵ/D the f, **R** curve is smoothly connected regardless of the actual pipe diameter. These tests did not permit variation of ϵ'/D or m but proved the validity of the equation

$$f = f\left(\mathbf{R}, \frac{\epsilon}{D}\right)$$

for one type of roughness.

Because of the extreme complexity of naturally rough surfaces, most of the advances in understanding the basic relations have been developed around experiments on artificially roughened pipes. Moody† has constructed one of the most convenient charts for determining friction factors in clean, commercial pipes. This chart, presented in Fig. 5.32, is the basis for pipe flow calculations in this chapter. The chart is a Stanton diagram that expresses f as a function of relative roughness and the Reynolds number. The values of absolute roughness of the commercial pipes are determined by experiment in which f and **R** are found and substituted into the Colebrook formula, Eq. (5.10.7), which closely represents natural pipe trends. These are listed in the table in the lower left-hand corner of Fig. 5.32. The Colebrook formula provides the shape of the ϵ/D = const curves in the transition region.

The straight line marked "laminar flow" in Fig. 5.32 is the Hagen-Poiseuille equation. Equation (5.2.10b),

$$V = \frac{\Delta p r_0^2}{8\mu L}$$

may be transformed into Eq. (5.8.7) with $\Delta p = \gamma h_f$ and by solving for h_f,

$$h_f = \frac{V 8 \mu L}{\gamma r_0^2} = \frac{64\mu}{\rho D} \frac{L}{D} \frac{V}{2g} = \frac{64}{\rho D V/\mu} \frac{L}{D} \frac{V^2}{2g}$$

or

$$h_f = f \frac{L}{D} \frac{V^2}{2g} = \frac{64}{\mathbf{R}} \frac{L}{D} \frac{V^2}{2g} \qquad (5.10.11)$$

from which

$$f = \frac{64}{\mathbf{R}} \qquad (5.10.12)$$

This equation, which plots as a straight line with slope -1 on a log-log chart, may be used for the solution of laminar flow problems in pipes. It applies to all roughnesses, as the head loss in laminar flow is independent of wall roughness.

---

† L. F. Moody, Friction Factors for Pipe Flow, *Trans. ASME*, November 1944.

The Reynolds critical number is about 2000, and the *critical zone*, where the flow may be either laminar or turbulent, is about 2000 to 4000.

It should be noted that the relative-roughness curves $\epsilon/D = 0.001$ and smaller approach the smooth-pipe curve for decreasing Reynolds numbers. This can be explained by the presence of a laminar film at the wall of the pipe that decreases in thickness as the Reynolds number increases. For certain ranges of Reynolds numbers in the transition zone, the film completely covers small roughness projections, and the pipe has a friction factor the same as that of a smooth pipe. For larger Reynolds numbers, projections protrude through laminar film, and each projection causes extra turbulence that increases the head loss. For the zone marked "complete turbulence, rough pipes," the film thickness is negligible compared with the height of roughness projections, and each projection contributes fully to the turbulence. Viscosity does not affect the head loss in this zone, as evidenced by the fact that the friction factor does not change with the Reynolds number. In this zone the loss follows the V^2 law; i.e., it varies directly as the square of the velocity.

Simple Pipe Problems

By "simple pipe problems" we refer to pipes or pipelines in which pipe friction is the only loss. The pipe may be laid at any angle with the horizontal. Six variables enter into the problems (the fluid is treated as incompressible): $Q, L, D, h_f, v, \epsilon$. In general, L, v, and ϵ, the length, kinematic viscosity of fluid, and absolute roughness, are given or may be determined. The simple pipe problems may then be treated as three types:

Given	To find
I. Q, L, D, v, ϵ	h_f
II. h_f, L, D, v, ϵ	Q
III. h_f, Q, L, v, ϵ	D

In each of these cases the Darcy-Weisbach equation, the continuity equation, and the Moody diagram are used to find the unknown quantity. In place of the Moody diagram, the following explicit formula† for f may be utilized with the restrictions placed on it.

$$f = \frac{1.325}{[\ln{(\epsilon/3.7D + 5.74/\mathbf{R}^{0.9})}]^2} \qquad \begin{array}{c} 10^{-6} \leq \dfrac{\epsilon}{D} \leq 10^{-2} \\[2mm] 5000 \leq \mathbf{R} \leq 10^8 \end{array} \qquad (5.10.13)$$

This equation yields an f within 1 percent of the Colebrook equation (5.10.7), and may be conveniently used with an electronic calculator.

---

† P. K. Swamee and A. K. Jain, Explicit Equations for Pipe-Flow Problems, *J. Hydr. Div., Proc. ASCE*, pp. 657–664, May 1976.

Solution for h_f For the type I solution, with Q, ϵ, and D known, $\mathbf{R} = VD/\nu = 4Q/\pi D\nu$ and f may be looked up in Fig. 5.32 or calculated from Eq. (5.10.13). Substitution into Eq. (5.8.7) yields h_f, the energy loss due to flow through the pipe per unit weight of fluid.

Example 5.12 Determine the head (energy) loss for flow of 140 l/s of oil, $\nu = 0.00001$ m^2/s, through 400 m of 200-mm-diameter cast-iron pipe.

$$\mathbf{R} = \frac{4Q}{\pi D\nu} = \frac{4(0.140 \text{ m}^3/\text{s})}{\pi(0.2 \text{ m})(0.00001 \text{ m}^2/\text{s})} = 89{,}127$$

The relative roughness is $\epsilon/D = 0.25$ mm/200 mm $= 0.00125$. From Fig. 5.32, by interpolation, $f = 0.023$. By solution of Eq. (5.10.13), $f = 0.0234$; hence,

$$h_f = f\frac{L}{D}\frac{V^2}{2g} = 0.023\frac{400 \text{ m}}{0.2 \text{ m}}\left[\frac{0.14}{(\pi/4)(0.2 \text{ m})^2}\right]^2\frac{1}{2(9.806 \text{ m/s}^2)}$$

$$= 46.58 \text{ m} \cdot \text{N/N}$$

Solution for discharge Q In the second case, V and f are unknowns, and the Darcy-Weisbach equation and Moody diagram must be used simultaneously to find their values. Since ϵ/D is known, a value of f may be assumed by inspection of the Moody diagram. Substitution of this trial f into the Darcy-Weisbach equation produces a trial value of V, from which a trial Reynolds number is computed. With the Reynolds number an improved value of f is found from the Moody diagram. When f has been found correct to two significant figures, the corresponding V is the value sought and Q is determined by multiplying by the area.

Example 5.13 Water at 15°C flows through a 300-mm-diameter riveted steel pipe, $\epsilon = 3$ mm, with a head loss of 6 m in 300 m. Determine the flow.

The relative roughness is $\epsilon/D = 0.003/0.3 = 0.01$, and from Fig. 5.32 a trial f is taken as 0.04. By substituting into Eq. (5.8.7).

$$6 \text{ m} = 0.04\frac{300 \text{ m}}{0.3 \text{ m}}\frac{(V \text{ m/s})^2}{2(9.806 \text{ m/s}^2)}$$

from which $V = 1.715$ m/s. From Appendix C, $\nu = 1.13 \times 10^{-6}$ m^2/s, and so

$$\mathbf{R} = \frac{VD}{\nu} = \frac{(1.715 \text{ m/s})(0.30 \text{ m})}{1.13 \times 10^{-6} \text{ m}^2/\text{s}} = 455{,}000$$

From the Moody diagram $f = 0.038$, and

$$Q = AV = \pi(0.15 \text{ m})^2\sqrt{\frac{(6 \text{ m} \times 0.3 \text{ m})(2)(9.806 \text{ m/s}^2)}{0.038(300 \text{ m})}} = 0.1245 \text{ m}^3/\text{s}$$

An explicit solution for discharge Q may be obtained from the Colebrook equation (5.10.7) and the Darcy-Weisbach equation (5.8.7). From Eq. (5.8.7)

$$h_f = f\frac{L}{D}\frac{Q^2}{2g[(\pi/4)D^2]^2} \qquad (5.10.14)$$

Solving for $1/\sqrt{f}$

$$\frac{1}{\sqrt{f}} = \frac{\sqrt{8}\,Q}{\pi\sqrt{gh_f\,D^5/L}}$$

By substitution of $1/\sqrt{f}$ into Eq. (5.10.7) and simplifying

$$Q = -0.955D^2\sqrt{gDh_f/L}\,\ln\left(\frac{\epsilon}{3.7D} + \frac{1.775v}{D\sqrt{gDh_f/L}}\right) \qquad (5.10.15)$$

This equation, first derived by Swamee and Jain,† is as accurate as the Colebrook equation and holds for the same range of values of ϵ/D and R. Substitution of the variables from Example 5.13, $D = 0.3$ m, $g = 9.806$ m/s^2, $h_f/L = 0.02$, $\epsilon/D = 0.01$, and $v = 1.13 \times 10^{-6}$ m^2/s, yields $Q = 0.1231$ m^3/s.

Solution for diameter D In the third case, with D unknown, there are three unknowns in Eq. (5.8.7), f, V, D; two in the continuity equation, V, D; and three in the Reynolds number equation, V, D, R. The relative roughness also is unknown. Using the continuity equation to eliminate the velocity in Eq. (5.8.7) and in the expression for R simplifies the problem. Equation (5.8.7) becomes

$$h_f = f\,\frac{L}{D}\,\frac{Q^2}{2g(D^2\pi/4)^2}$$

or

$$D^5 = \frac{8LQ^2}{h_f g\pi^2}f = C_1 f \qquad (5.10.16)$$

in which C_1 is the known quantity $8LQ^2/h_f g\pi^2$. As $VD^2 = 4Q/\pi$ from continuity,

$$R = \frac{VD}{v} = \frac{4Q}{\pi v}\frac{1}{D} = \frac{C_2}{D} \qquad (5.10.17)$$

in which C_2 is the known quantity $4Q/\pi v$. The solution is now effected by the following procedure:

1. Assume a value of f.
2. Solve Eq. (5.10.16) for D.
3. Solve Eq. (5.10.17) for R.
4. Find the relative roughness ϵ/D.
5. With R and ϵ/D, look up a new f from Fig. 5.32.
6. Use the new f, and repeat the procedure.
7. When the value of f does not change in the first two significant figures, all equations are satisfied and the problem is solved.

   Normally only one or two trials are required. Since standard pipe sizes are usually selected, the next larger size of pipe than that given by the computation is taken.

† Ibid.

Example 5.14 Determine the size of clean wrought-iron pipe required to convey 4000 gpm oil, $v = 0.0001$ ft^2/s, 10,000 ft with a head loss of 75 ft·lb/lb.
The discharge is

$$Q = \frac{4000}{448.8} = 8.93 \text{ cfs}$$

From Eq. (5.10.16)

$$D^5 = \frac{8 \times 10,000 \times 8.93^2}{75 \times 32.2 \times \pi^2} f = 267.0 f$$

and from Eq. (5.10.17)

$$R = \frac{4 \times 8.93}{\pi 0.0001} \frac{1}{D} = \frac{113,800}{D}$$

and from Fig. 5.32, $\epsilon = 0.00015$ ft.
If $f = 0.02$, $D = 1.398$ ft, $R = 81,400$, $\epsilon/D = 0.00011$, and from Fig. 5.32, $f = 0.019$. In repeating the procedure, $D = 1.382$, $R = 82,300$, $f = 0.019$. Therefore, $D = 1.382 \times 12 = 16.6$ in.

Following Swamee and Jain,[†] an empirical equation to determine diameter directly by using dimensionless relations and an approach similar to development of the Colebrook equation yields

$$D = 0.66 \left[\epsilon^{1.25} \left(\frac{LQ^2}{gh_f} \right)^{4.75} + vQ^{9.4} \left(\frac{L}{gh_f} \right)^{5.2} \right]^{0.04} \tag{5.10.18}$$

Solution of Example 5.14 by use of Eq. (5.10.18) for $Q = 8.93$ cfs, $\epsilon = 0.00015$ ft, $L = 10,000$ ft, $h_f = 75$ ft·lb/lb, $v = 0.0001$ ft^2/s, and $g = 32.2$ ft/s^2 yields $D = 1.404$ ft.
Equation (5.10.18) is valid for the following ranges:

$$3 \times 10^3 \leq R \leq 3 \times 10^8$$

$$10^{-6} \leq \frac{\epsilon}{D} \leq 2 \times 10^{-2}$$

and will yield a D within 2 percent of the value obtained by the method using the Colebrook equation.

In each of the cases considered, the loss has been expressed in units of energy per unit weight. For horizontal pipes, this loss shows up as a gradual reduction in pressure along the line. For nonhorizontal cases, the energy equation (3.10.1) is applied to the two end sections of the pipe, and the loss term is included; thus

$$\frac{V_1^2}{2g} + \frac{p_1}{\gamma} + z_1 = \frac{V_2^2}{2g} + \frac{p_2}{\gamma} + z_2 + h_f \tag{3.10.1}$$

in which the kinetic-energy correction factors have been taken as unity. The upstream section is given the subscript 1 and the downstream section the subscript

† Ibid.

2. The total head at section 1 is equal to the sum of the total head at section 2 and all the head losses between the two sections.

Example 5.15 In the preceding example, for $D = 16.6$ in, if the specific gravity is 0.85, $p_1 = 40$ psi, $z_1 = 200$ ft, and $z_2 = 50$ ft, determine the pressure at section 2. In Eq. (3.10.1) $V_1 = V_2$; hence,

$$\frac{40 \text{ psi}}{0.85(0.433 \text{ psi/ft})} + 200 \text{ ft} = \frac{p_2 \text{ psi}}{0.85(0.433 \text{ psi/ft})} + 50 \text{ ft} + 75 \text{ ft}$$

and

$$p_2 = 67.6 \text{ psi}$$

Minor Losses

Losses which occur in pipelines because of bends, elbows, joints, valves, etc., are called *minor losses*. This is a misnomer, because in many situations they are more important than the losses due to pipe friction considered so far in this section, but the name is conventional. In almost all cases the minor loss is determined by experiment. However, one important exception is the head loss due to a *sudden expansion* in a pipeline (Sec. 3.11).

Equation (3.11.22) may also be written

$$h_e = K \frac{V_1^2}{2g} = \left[1 - \left(\frac{D_1}{D_2} \right)^2 \right]^2 \frac{V_1^2}{2g} \qquad (5.10.19)$$

in which

$$K = \left[1 - \left(\frac{D_1}{D_2} \right)^2 \right]^2 \qquad (5.10.20)$$

From Eq. (5.10.19) it is obvious that the head loss varies as the square of the velocity. This is substantially true for all minor losses in turbulent flow. A convenient method of expressing the minor losses in flow is by means of the coefficient K, usually determined by experiment.

If the sudden expansion is from a pipe to a reservoir, $D_1/D_2 = 0$ and the loss becomes $V_1^2/2g$; that is, the complete kinetic energy in the flow is converted into thermal energy.

The head loss h_c due to a *sudden contraction* in the pipe cross section, illustrated in Fig. 5.33, is subject to the same analysis as the sudden expansion, provided that the amount of contraction of the jet is known. The process of converting pressure head into velocity head is very efficient; hence, the head loss

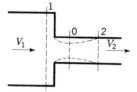

Figure 5.33 Sudden contraction in a pipeline.

from section 1 to the *vena contracta*† is small compared with the loss from section 0 to section 2, where velocity head is being reconverted into pressure head. By applying Eq. (3.11.22) to this expansion, the head loss is computed to be

$$h_c = \frac{(V_0 - V_2)^2}{2g}$$

With the continuity equation $V_0 C_c A_2 = V_2 A_2$, in which C_c is the contraction coefficient, i.e., the area of jet at section 0 divided by the area of section 2, the head loss is computed to be

$$h_c = \left(\frac{1}{C_c} - 1\right)^2 \frac{V_2^2}{2g} \qquad (5.10.21)$$

The contraction coefficient C_c for water, determined by Weisbach,‡ is presented in the tabulation.

A_2/A_1	0.1	0.2	0.3	0.4	0.5	0.6	0.7	0.8	0.9	1.0
C_c	0.624	0.632	0.643	0.659	0.681	0.712	0.755	0.813	0.892	1.00

The head loss at the entrance to a pipeline from a reservoir is usually taken as $0.5V^2/2g$ if the opening is square-edged. For well-rounded entrances, the loss is between $0.01V^2/2g$ and $0.05V^2/2g$ and may usually be neglected. For re-entrant openings, as with the pipe extending into the reservoir beyond the wall, the loss is taken as $1.0V^2/2g$ for thin pipe walls, Fig. 5.34.

The head loss due to gradual expansions (including pipe friction over the length of the expansion) has been investigated experimentally by Gibson,§ whose results are given in Fig. 5.35.

A summary of representative head loss coefficients K for typical fittings, published by the Crane Company,¶ is given in Table 5.3.

---

† The *vena contracta* is the section of greatest contraction of the jet.
‡ Julius Weisbach, "Die Experimental-Hydraulik," p. 133, Englehardt, Freiburg, 1855.
§ A. H. Gibson, The Conversion of Kinetic to Pressure Energy in the Flow of Water through Passages Having Divergent Boundaries, *Engineering*, vol. 93, p. 205, 1912.
¶ Crane Company, Flow of Fluids, *Tech. Pap.* 409, May 1942.

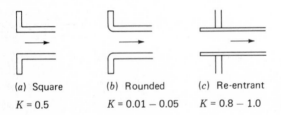

(a) Square   (b) Rounded   (c) Re-entrant

$K = 0.5$ $K = 0.01 - 0.05$ $K = 0.8 - 1.0$

Figure 5.34 Head loss coefficient K, in number of velocity heads, $V^2/2g$, for a pipe entrance.

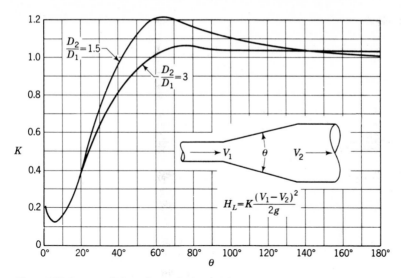

Figure 5.35 Loss coefficients for conical expansions.

Minor losses may be expressed in terms of the equivalent length L_e of pipe that has the same head loss in meter-newtons per newton (foot-pounds per pound) for the same discharge; thus

$$f \frac{L_e}{D} \frac{V^2}{2g} = K \frac{V^2}{2g}$$

in which K may refer to one minor head loss or to the sum of several losses. Solving for L_e gives

$$L_e = \frac{KD}{f} \qquad (5.10.22)$$

Table 5.3 Head-loss coefficients K for various fittings

Fitting	K
Globe valve (fully open)	10.0
Angle valve (fully open)	5.0
Swing check valve (fully open)	2.5
Gate valve (fully open)	0.19
Close return bend	2.2
Standard tee	1.8
Standard elbow	0.9
Medium sweep elbow	0.75
Long sweep elbow	0.60

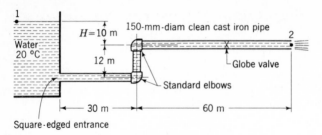

Figure 5.36 Pipeline with minor losses.

For example, if the minor losses in a 12-in pipeline add to $K = 20$, and if $f = 0.020$ for the line, then to the actual length of line may be added $20(1/0.020) = 1000$ ft, and this additional or equivalent length causes the same resistance to flow as the minor losses.

Example 5.16 Find the discharge through the pipeline in Fig. 5.36 for $H = 10$ m, and determine the head loss H for $Q = 60$ l/s.

The energy equation applied between points 1 and 2, including all the losses, may be written

$$H_1 + 0 + 0 = \frac{V_2^2}{2g} + 0 + 0 + \frac{1}{2}\frac{V_2^2}{2g} + f\frac{102 \text{ m}}{0.15 \text{ m}}\frac{V_2^2}{2g} + 2 \times 0.9\frac{V_2^2}{2g} + 10\frac{V_2^2}{2g}$$

in which the entrance loss coefficient is $\frac{1}{2}$, each elbow 0.9, and the globe valve 10. Then

$$H_1 = \frac{V_2^2}{2g}(13.3 + 680f)$$

When the head is given, this problem is solved as the second type of simple pipe problem. If $f = 0.022$,

$$10 = \frac{V_2^2}{2g}(13.3 + 680 \times 0.022)$$

and $V_2 = 2.63$ m/s. From Appendix C, $v = 1.01 \times 10^{-6}$ m^2/s, $\epsilon/D = 0.0017$, $\mathbf{R} = (2.63 \text{ m/s}) \times (0.15 \text{ m})/(1.01 \times 10^{-6}$ m^2/s$) = 391{,}000$. From Fig. 5.32, $f = 0.023$. Repeating the procedure gives $V_2 = 2.60$ m/s, $\mathbf{R} = 380{,}000$, and $f = 0.023$. The discharge is

$$Q = V_2 A_2 = (2.60 \text{ m/s})(\pi/4)(0.15 \text{ m})^2 = 45.9 \text{ l/s}$$

For the second part, with Q known, the solution is straightforward:

$$V_2 = \frac{Q}{A} = \frac{0.06 \text{ m}^3/\text{s}}{(\pi/4)(0.15 \text{ m})^2} = 3.40 \text{ m/s} \qquad \mathbf{R} = 505{,}000 \qquad f = 0.023$$

and

$$H_1 = \frac{(3.4 \text{ m/s})^2}{2(9.806 \text{ m/s}^2)}(13.3 + 680 \times 0.023) = 17.06 \text{ m}$$

With equivalent lengths [Eq. (5.10.22)] the value of f is approximated, say $f = 0.022$. The sum of minor losses is $K = 13.3$, in which the kinetic energy at 2 is considered a minor loss,

$$L_e = \frac{13.3 \times 0.15}{0.022} = 90.7 \text{ m}$$

Hence, the total length of pipe is $90.7 + 102 = 192.7$ m. The first part of the problem is solved by this method,

$$10 \text{ m} = f \; \frac{L + L_e}{D} \frac{V_2^2}{2g} = f \; \frac{192.7 \text{ m}}{0.15 \text{ m}} \frac{(V_2 \text{ m/s})^2}{2g \text{ m/s}^2}$$

If $f = 0.022$, $V_2 = 2.63$ m/s, $\mathbf{R} = 391{,}000$, and $f = 0.023$, then $V_2 = 2.58$ m/s and $Q = 45.6$ l/s. Normally, it is not necessary to use the new f to improve L_e.

Minor losses may be neglected when they comprise only 5 percent or less of the head losses due to pipe friction. The friction factor, at best, is subject to about 5 percent error, and it is meaningless to select values to more than three significant figures. In general, minor losses may be neglected when, on the average, there is a length of 1000 diameters between each minor loss.

Compressible flow in pipes is treated in Chap. 6. Complex pipe flow situations are treated in Chap. 10.

Iterative solution of minor loss problems by programmable calculator The solution for head loss is straightforward, since with D, Q, v, ϵ, L, and K known, $\mathbf{R}$, ϵ/D, A, and f may be calculated, and

$$h_f = \frac{f}{D}\left(L + \frac{KD}{f}\right)\frac{Q^2}{2gA^2} \tag{5.10.23}$$

Solution for discharge By replacing L in Eq. (5.10.15) by $L + KD/f$, an equation for Q in terms of an unknown f is obtained. Let

$$Y = \sqrt{\frac{gDh_f}{L + KD/f}} = \sqrt{\frac{R_3}{1 + R_4/f}} \tag{5.10.24}$$

with $R_3 = gDh_f/L$ and $R_4 = KD/L$. Then Eq. (5.10.15) becomes

$$Q = -0.955D^2 Y \left[\ln\left(\frac{\epsilon}{3.7D} + \frac{1.775v}{Dy}\right)\right] = R_2 Y \ln\left(R_1 + \frac{R_0}{y}\right) \tag{5.10.25}$$

with $R_0 = 1.775v/D$, $R_1 = \epsilon/(3.7D)$, and $R_2 = -0.955D^2$. The Reynolds number is given by

$$\mathbf{R} = \frac{4Q}{\pi Dv} = R_5 Q \tag{5.10.26}$$

with $R_5 = 4/\pi Dv$. The friction factor Eq. (5.10.13) becomes

$$f = \frac{R_7}{\left[\ln\left(R_1 + \frac{R_6}{\mathbf{R}^{0.9}}\right)\right]^2} \tag{5.10.27}$$

with $R_7 = 1.325$ and $R_6 = 5.74$. An assumed value of f, say $f = 0.022$, is keyed into the calculator, with R_0 to R_7 the stored constants, and the equations (5.10.24),

(5.10.25), (5.10.26), and (5.10.27) are solved in sequence. This procedure is continued until f and Q do not change (to four significant figures), and f and Q have been determined. Example 5.16 yields $f = 0.0231$, $Q = 45.6$ l/s after three iterations. The program converges satisfactorily when KD/f is much larger than L.

Solution for diameter In Eq. (5.10.18) L may be replaced by $L + KD/f$, yielding the following sequence of equations

$$\mathbf{R} = \frac{R_5}{D} \tag{5.10.28}$$

$$f = \frac{R_7}{[\ln \ (R_3/D + R_2/\mathbf{R}^{0.9})]^2} \tag{5.10.29}$$

$$x = R_6 + R_4 D/f \tag{5.10.30}$$

$$D = R_0(x^{4.75} + R_1 x^{5.2})^{0.04} \tag{5.10.31}$$

where $R_0 = 0.66(\epsilon^{1.25}Q^{9.5})^{0.04}$
$\quad R_1 = v/(\epsilon^{1.25}Q^{0.1})$
$\quad R_2 = 5.74$
$\quad R_3 = \epsilon/3.7$
$\quad R_4 = K/gh_f$
$\quad R_5 = 4Q/\pi v$
$\quad R_6 = L/gh_f$
$\quad R_7 = 1.325$

The program solves Eqs. (5.10.28) to (5.10.31) in sequence after a trial D has been keyed in.

Example 5.17 300 l/s water at 10°C is to be conveyed 500 m through commercial steel pipe with a total head drop of 6 m. Minor losses are $12V^2/2g$. Determine the required diameter.
With $v = 1.308 \times 10^{-6}$ m^2/s and $\epsilon = 0.000046$ m, the constants in Eqs. (5.10.28) to (5.10.31) become $R_0 = 0.25351$, $R_1 = 0.38945$, $R_3 = 1.2432 \times 10^{-5}$, $R_4 = 0.20396$, $R_5 = 292,027$, and $R_6 = 8.4982$. Assume $D = 1$ m and then solve the equations in order (or run the program), yielding $D = 438$ mm, $f = 0.0141$ after four iterations. The hand-held calculators have sufficient capacity for solution of the four equations.

EXERCISES

5.10.1 In turbulent flow a rough pipe has the same friction factor as a smooth pipe (a) in the zone of complete turbulence, rough pipes; (b) when the friction factor is independent of the Reynolds number; (c) when the roughness projections are much smaller than the thickness of the laminar film; (d) everywhere in the transition zone; (e) when the friction factor is constant.

5.10.2 The friction factor in turbulent flow in smooth pipes depends upon the following: (a) V, D, ρ, L, μ; (b) Q, L, μ, ρ; (c) V, D, ρ, p, μ; (d) V, D, μ, ρ; (e) p, L, D, Q, V.

5.10.3 In a given rough pipe, the losses depend upon (a) f, V; (b) μ, ρ; (c) $\mathbf{R}$; (d) Q only; (e) none of these answers.

5.10.4 In the complete-turbulence zone, rough pipes, (a) rough and smooth pipes have the same friction factor; (b) the laminar film covers the roughness projections; (c) the friction factor depends upon Reynolds number only; (d) the head loss varies as the square of the velocity; (e) the friction factor is independent of the relative roughness.

5.10.5 The friction factor for flow of water at 60°F through a 2-ft-diameter cast-iron pipe with a velocity of 5 ft/s is (a) 0.013; (b) 0.017; (c) 0.019; (d) 0.021; (e) none of these answers.

5.10.6 The procedure to follow in solving for losses when $Q, L, D, v,$ and ϵ are given is to (a) assume an f, look up $\mathbf{R}$ on Moody diagram, etc.; (b) assume an h_f, solve for f, check against $\mathbf{R}$ on Moody diagram; (c) assume an f, solve for h_f, compute $\mathbf{R}$, etc.; (d) compute $\mathbf{R}$, look up f for ϵ/D, solve for h_f; (e) assume an $\mathbf{R}$, compute V, look up f, solve for h_f.

5.10.7 The procedure to follow in solving for discharge when $h_f, L, D, v,$ and ϵ are given is to (a) assume an f, compute $V, \mathbf{R}, \epsilon/D$, look up f, and repeat if necessary; (b) assume an $\mathbf{R}$, compute f, check ϵ/D, etc.; (c) assume a V, compute $\mathbf{R}$, look up f, compute V again, etc.; (d) solve Darcy-Weisbach for V, compute Q; (e) assume a Q, compute $V, \mathbf{R}$, look up f, etc.

5.10.8 The procedure to follow in solving for pipe diameter when $h_f, Q, L, v,$ and ϵ are given is to (a) assume a D, compute $V, \mathbf{R}, \epsilon/D$, look up f, and repeat; (b) compute V from continuity, assume an f, solve for D; (c) eliminate V in $\mathbf{R}$ and Darcy-Weisbach, using continuity, assume an f, solve for $D, \mathbf{R}$, look up f, and repeat; (d) assume an $\mathbf{R}$ and an ϵ/D, look up f, solve Darcy-Weisbach for V^2/D, and solve simultaneously with continuity for V and D, compute new $\mathbf{R}$, etc.; (e) assume a V, solve for $D, \mathbf{R}, \epsilon/D$, look up f, and repeat.

5.10.9 The losses due to a sudden contraction are given by

(a) $\left(\dfrac{1}{C_c^2} - 1\right)\dfrac{V_2^2}{2g}$ (b) $(1 - C_c^2)\dfrac{V_2^2}{2g}$ (c) $\left(\dfrac{1}{C_c} - 1\right)^2\dfrac{V_2^2}{2g}$

(d) $(C_c - 1)^2\dfrac{V_2^2}{2g}$ (e) none of these answers

5.10.10 The losses at the exit of a submerged pipe in a reservoir are (a) negligible; (b) $0.05(V^2/2g)$; (c) $0.5(V^2/2g)$; (d) $V^2/2g$; (e) none of these answers.

5.10.11 Minor losses usually may be neglected when (a) there is 100 ft of pipe between special fittings; (b) their loss is 5 percent or less of the friction loss; (c) there are 500 diameters of pipe between minor losses; (d) there are no globe valves in the line; (e) rough pipe is used.

5.10.12 The length of pipe ($f = 0.025$) in diameters, equivalent to a globe valve, is (a) 40; (b) 200; (c) 300; (d) 400; (e) not determinable; insufficient data.

5.11 LUBRICATION MECHANICS

The effect of viscosity on flow and its effects on head losses have been examined in the preceding sections of this chapter. A laminar flow case of great practical importance is the hydrodynamic theory of lubrication. Simple aspects of this theory are developed in this section.

Large forces are developed in small clearances when the surfaces are slightly inclined and one is in motion so that fluid is " wedged " into the decreasing space. The slipper bearing, which operates on this principle, is illustrated in Fig. 5.37a. The journal bearing (Fig. 5.37b) develops its force by the same action, except that the surfaces are curved.

The laminar flow equations may be used to develop the theory of lubrication. The assumption is made that there is no flow out of the ends of the bearing normal to the plane of Fig. 5.37a. From Eq. (5.1.4), which relates pressure drop and shear

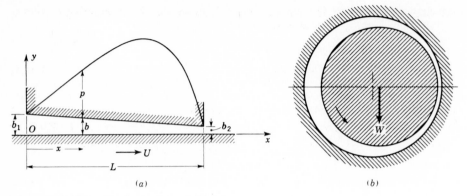

Figure 5.37 Sliding bearing and journal bearing.

stress, the equation for the force P that the bearing will support is worked out, and the drag on the bearing is computed.

Substituting Newton's law of viscosity into Eq. (5.1.4) produces

$$\frac{dp}{dx} = \mu \, \frac{d^2u}{dy^2} \qquad (5.11.1)$$

Since the inclination of the upper portion of the bearing (Fig. 5.37a) is very slight, it may be assumed that the velocity distribution is the same as if the plates were parallel and that p is independent of y. Integrating Eq. (5.11.1) twice with respect to y, with dp/dx constant, produces

$$\frac{dp}{dx} \int dy = \mu \int \frac{d^2u}{dy^2} \, dy + A \qquad \text{or} \qquad \frac{dp}{dx} y = \mu \, \frac{du}{dy} + A$$

and the second time

$$\frac{dp}{dx} \int y \, dy = \mu \int \frac{du}{dy} \, dy + A \int dy + B \qquad \text{or} \qquad \frac{dp}{dx} \frac{y^2}{2} = \mu u + Ay + B$$

The constants of integration A, B are determined from the conditions $u = 0$, $y = b$; $u = U$, $y = 0$. Substituting in turn produces

$$\frac{dp}{dx} \frac{b^2}{2} = Ab + B \qquad \mu U + B = 0$$

Eliminating A and B and solving for u results in

$$u = \frac{y}{2\mu} \frac{dp}{dx} (y - b) + U\left(1 - \frac{y}{b}\right) \qquad (5.11.2)$$

The discharge Q must be the same at each cross section. By integrating over a typical section, again with dp/dx constant,

$$Q = \int_0^b u \, dy = \frac{Ub}{2} - \frac{b^3}{12\mu} \frac{dp}{dx} \qquad (5.11.3)$$

Now, since Q cannot vary with x, b may be expressed in terms of x, $b = b_1 - \alpha x$, in which $\alpha = (b_1 - b_2)/L$, and the equation is integrated with respect to x to determine the pressure distribution. Solving Eq. (5.11.3) for dp/dx produces

$$\frac{dp}{dx} = \frac{6\mu U}{(b_1 - \alpha x)^2} - \frac{12\mu Q}{(b_1 - \alpha x)^3} \qquad (5.11.4)$$

Integrating gives

$$\int \frac{dp}{dx} dx = 6\mu U \int \frac{dx}{(b_1 - \alpha x)^2} - 12\mu Q \int \frac{dx}{(b_1 - \alpha x)^3} + C$$

or $\qquad p = \dfrac{6\mu U}{\alpha(b_1 - \alpha x)} - \dfrac{6\mu Q}{\alpha(b_1 - \alpha x)^2} + C$

In this equation Q and C are unknowns. Since the pressure must be the same, say zero, at the ends of the bearing, namely, $p = 0$, $x = 0$; $p = 0$, $x = L$, the constants can be determined,

$$Q = \frac{Ub_1 b_2}{b_1 + b_2} \qquad C = -\frac{6\mu U}{\alpha(b_1 + b_2)}$$

With these values inserted, the equation for pressure distribution becomes

$$p = \frac{6\mu Ux(b - b_2)}{b^2(b_1 + b_2)} \qquad (5.11.5)$$

This equation shows that p is positive between $x = 0$ and $x = L$ if $b > b_2$. It is plotted in Fig. 5.37a to show the distribution of pressure throughout the bearing. With this one-dimensional method of analysis the very slight change in pressure along a vertical line $x = \text{const}$ is neglected.

The total force P that the bearing will sustain, per unit width, is

$$P = \int_0^L p \, dx = \frac{6\mu U}{b_1 + b_2} \int_0^L \frac{x(b - b_2) \, dx}{b^2}$$

After substituting the value of b in terms of x and performing the integration,

$$P = \frac{6\mu U L^2}{(b_1 - b_2)^2} \left(\ln \frac{b_1}{b_2} - 2 \frac{b_1 - b_2}{b_1 + b_2} \right) \qquad (5.11.6)$$

The drag force D required to move the lower surface at speed U is expressed by

$$D = \int_0^L \tau \bigg|_{y=0} dx = - \int_0^L \mu \frac{du}{dy} \bigg|_{y=0} dx$$

By evaluating du/dy from Eq. (5.11.2), for $y = 0$,

$$\frac{du}{dy} \bigg|_{y=0} = - \frac{b}{2\mu} \frac{dp}{dx} - \frac{U}{b}$$

This value in the integral, along with the value of dp/dx from Eq. (5.11.4), gives

$$D = \frac{2\mu UL}{b_1 - b_2} \left(2 \ln \frac{b_1}{b_2} - 3 \frac{b_1 - b_2}{b_1 + b_2} \right) \tag{5.11.7}$$

The maximum load P is computed with Eq. (5.11.6) when $b_1 = 2.2b_2$. With this ratio,

$$P = 0.16 \frac{\mu U L^2}{b_2^2} \qquad D = 0.75 \frac{\mu U L}{b_2} \tag{5.11.8}$$

The ratio of load to drag for optimum load is

$$\frac{P}{D} = 0.21 \frac{L}{b_2} \tag{5.11.9}$$

which can be very large, since b_2 can be very small.

Example 5.18 A vertical turbine shaft carries a load of 80,000 lb on a thrust bearing consisting of 16 flat rocker plates, 3 by 9 in, arranged with their long dimensions radial from the shaft and their centers on a circle of radius 1.5 ft. The shaft turns at 120 rpm; $\mu \doteq 0.002$ lb·s/ft^2. If the plates take the angle for maximum load, neglecting effects of curvature of path and radial lubricant flow, find (a) the clearance between rocker plate and fixed plate; (b) the torque loss due to the bearing.
(a) Since the motion is considered straight line.

$$U = 1.5(\tfrac{120}{60})(2\pi) = 18.85 \text{ ft/s} \qquad L = 0.25 \text{ ft}$$

The load is 5000 lb for each plate, which is $5000/0.75 = 6667$ lb for unit width. By solving for the clearance b_2, from Eq. (5.11.8),

$$b_2 = \sqrt{\frac{0.16\mu U L^2}{P}} = 0.4 \times 0.25 \sqrt{\frac{0.002 \times 18.85}{6667}} = 2.38 \times 10^{-4} = 0.0029 \text{ in}$$

(b) The drag due to one rocker plate is, per foot of width,

$$D = 0.75 \frac{\mu U L}{b_2} = \frac{0.75 \times 0.002 \times 18.85 \times 0.25}{2.38 \times 10^{-4}} = 29.6 \text{ lb}$$

For a 9-in plate, $D = 29.6 \times 0.75 = 22.2$ lb. The torque loss due to the 16 rocker plates is

$$16 \times 22.2 \times 1.5 = 533 \text{ ft·lb}$$

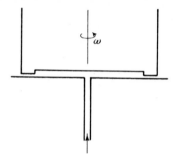

Figure 5.38 Hydrostatic lubrication by high-pressure pumping of oil.

Another form of lubrication, called *hydrostatic lubrication*,† has many important applications. It involves the continuous high-pressure pumping of oil under a step bearing, as illustrated in Fig. 5.38. The load may be lifted by the lubrication before rotation starts, which greatly reduces starting friction.

EXERCISE

5.11.1 In the theory of lubrication the assumption is made that (*a*) the velocity distribution is the same at all cross sections; (*b*) the velocity distribution at any section is the same as if the plates were parallel; (*c*) the pressure variation along the bearing is the same as if the plates were parallel; (*d*) the shear stress varies linearly between the two surfaces; (*e*) the velocity varies linearly between the two surfaces.

PROBLEMS

5.1 Determine the formulas for shear stress on each plate and for the velocity distribution for flow in Fig. 5.1 when an adverse pressure gradient exists such that $Q = 0$.

5.2 In Fig. 5.1, with U positive as shown, find the expression for $d(p + \gamma h)/dl$ such that the shear is zero at the fixed plate. What is the discharge for this case?

5.3 In Fig. 5.39*a*, $U = 0.7$ m/s. Find the rate at which oil is carried into the pressure chamber by the piston and the shear force and total force F acting on the piston.

5.4 Determine the force on the piston of Fig. 5.39*a* due to shear and the leakage from the pressure chamber for $U = 0$.

† For further information on hydrostatic lubrication see D. D. Fuller, Lubrication Mechanics, in V. L. Streeter (ed.), "Handbook of Fluid Dynamics," pp. 22–21 to 22–30, McGraw-Hill, New York, 1961.

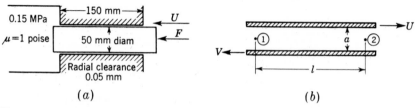

(*a*)   (*b*)

Figure 5.39 Problems 5.3 to 5.7.

5.5 Find F and U in Fig. 5.39a such that no oil is lost through the clearance from the pressure chamber.

5.6 Derive an expression for the flow past a fixed cross section of Fig. 5.39b for laminar flow between the two moving plates.

5.7 In Fig. 5.39b, for $p_1 = p_2 = 0.1$ MPa, $U = 2V = 2$ m/s, $a = 1.5$ mm, $\mu = 0.5$ P, find the shear stress at each plate.

5.8 Compute the kinetic-energy and momentum correction factors for laminar flow between fixed parallel plates.

5.9 Determine the formula for angle θ for fixed parallel plates so that laminar flow at constant pressure takes place.

5.10 With a free body, as in Fig. 5.40, for uniform flow of a thin lamina of liquid down an inclined plane, show that the velocity distribution is

$$u = \frac{\gamma}{2\mu}(b^2 - s^2)\sin\theta$$

and that the discharge per unit width is

$$Q = \frac{\gamma}{3\mu}b^3 \sin\theta$$

5.11 Derive the velocity distribution of Prob. 5.10 by inserting into the appropriate equation prior to Eq. (5.1.2) the condition that the shear at the free surface must be zero.

5.12 In Fig. 5.41, $p_1 = 6$ psi, $p_2 = 8$ psi, $l = 4$ ft, $a = 0.006$ ft, $\theta = 30°$, $U = 4$ ft/s, $\gamma = 50$ lb/ft^3, and $\mu = 0.8$ P. Determine the tangential force per square foot exerted on the upper plate and its direction.

5.13 For $\theta = 90°$ in Fig. 5.41, what speed U is required for no discharge? $S = 0.87$, $a = 3$ mm, $p_1 = p_2$, and $\mu = 0.2$ kg/m·s.

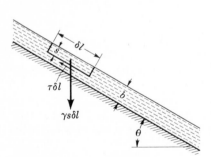

Figure 5.40 Problems 5.10, 5.11.

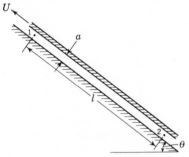

Figure 5.41 Problems 5.12, 5.13.

5.14 The belt conveyor (Fig. 5.42) delivers fluid to a reservoir of such depth that the velocity on the free-liquid surface on the belt is zero. By considering only the work done by the belt on the fluid in shear, find how efficient this device is in transferring energy to the fluid.

5.15 What is the velocity distribution of the fluid on the belt and the volume rate of fluid being transported in Prob. 5.14?

5.16 What is the time rate of momentum and kinetic energy passing through a cross section that is normal to the flow if in Eq. (5.1.3) $Q = 0$?

5.17 A film of fluid 0.005 ft thick flows down a fixed vertical surface with a surface velocity of 2 ft/s. Determine the fluid viscosity. $\gamma = 55$ lb/ft^3.

5.18 Determine the momentum correction factor for laminar flow in a round tube.

5.19 Water at standard conditions is in laminar flow in a tube at pressure p_1 and diameter d_1. This

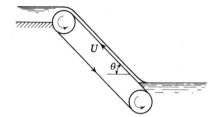

Figure 5.42 Problems 5.14, 5.15.

tube expands to a diameter of $2d_1$ and pressure p_2, and the flow is again described by Eq. (5.2.6) some distance downstream of the expansion. Determine the force on the tube which results from the expansion.

5.20 At what distance r from the center of a tube of radius r_0 does the average velocity occur in laminar flow?

5.21 Determine the maximum wall shear stress for laminar flow in a tube of diameter D with fluid properties μ and ρ given.

5.22 Show that laminar flow between parallel plates may be used in place of flow through an annulus for 2 percent accuracy if the clearance is no more than 4 percent of the inner radius.

5.23 What are the losses per kilogram per meter of tubing for flow of mercury at 35°C through a 0.6-mm-diameter tube at $\mathbf{R} = 1600$?

5.24 Determine the shear stress at the wall of a $\frac{1}{16}$-in-diameter tube when water at 80°F flows through it with a velocity of 1 ft/s.

5.25 Determine the pressure drop per meter of 3-mm-ID horizontal tubing for flow of liquid, $\mu = 60$ cP, sp gr $= 0.83$, at $\mathbf{R} = 200$.

5.26 Glycerin at 100°F flows through a horizontal $\frac{3}{8}$-in-diameter pipe with a pressure drop of 5 psi/ft. Find the discharge and the Reynolds number.

5.27 Calculate the diameter of vertical pipe needed for flow of liquid at $\mathbf{R} = 1600$ when the pressure remains constant, $v = 1.5 \times 10^{-6}\text{m}^2/\text{s}$.

5.28 Calculate the discharge of the system in Fig. 5.43, neglecting all losses except through the pipe.

5.29 In Fig. 5.44, $H = 12$ m, $L = 20$ m, $\theta = 30°$, $D = 8$ mm, $\gamma = 10$ kN/m³, and $\mu = 0.08$ kg/m·s. Find the head loss per unit length of pipe and the discharge, in liters per minute.

5.30 In Fig. 5.44 and Prob. 5.29, find H if the velocity is 0.1 m/s.

5.31 Oil, sp gr 0.85, $\mu = 0.06$ N·s/m², flows through an annulus $a = 15$ mm, $b = 7$ mm. When the shear stress at the outer wall is 12 Pa, calculate (a) the pressure drop per meter for a horizontal system, (b) the discharge, in liters per hour, and (c) the axial force exerted on the inner tube per meter of length.

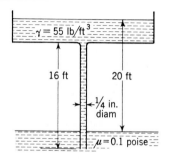

Figure 5.43 Problems 5.28, 5.118.

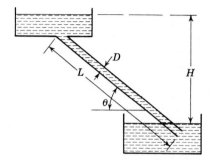

Figure 5.44 Problems 5.29, 5.30, 5.108.

5.32 What is the Reynolds number of flow of 0.3 m³/s oil, sp gr 0.86, $\mu = 0.025$ N·s/m², through a 450-mm-diameter pipe?

5.33 Show that the power input for laminar flow in a round tube is $Q \, \Delta p$ by integration of Eq. (5.1.6).

5.34 By use of the one-seventh-power law of velocity distribution, $u/u_{max} = (y/r_0)^{1/7}$, determine the mixing-length distribution l/r_0 in terms of y/r_0 from Eq. (5.4.4).

5.35 A pollutant is released in quiescent fluid in a long uniform tank at $x = 0$ at $t = 0$. The tank cross-sectional area is 25 m², $D_m = 10^{-9}$ m²/s, and $M = 1000$ kg. What is the concentration at $x = 0$ at $t = 4$ days and at $t = 30$ days? What is the concentration at $x = 1$ m at $t = 365$ days?

5.36 At an instant, 20 kg of salt is released into a stream. What will the distribution of the salt be 30 min later? The stream's velocity is 0.4 m/s; its cross-sectional area is 10 m²; and the dispersion coefficient is 40 m²/s.

5.37 If a contaminant is fed into a flowing stream at a steady rate, then the concentration of contaminant downstream is given by the rate of contaminant input divided by the volume of fluid flow per unit time. For 0.3 l/s contaminant inflow to a river discharging 30 m³/s, determine the contamination, in parts per million.

5.38 A fluid is agitated so that the kinematic eddy viscosity increases linearly from zero ($y = 0$) at the bottom of the tank to 0.2 m²/s at $y = 600$ mm. For uniform particles with fall velocities of 300 mm/s in still fluid, find the concentration at $y = 350$ mm if it is 10 per liter at $y = 600$ mm.

5.39 Plot a curve of $\epsilon/u_* r_0$ as a function of y/r_0 using Eq. (5.4.9) for velocity distribution in a pipe.

5.40 Find the value of y/r_0 in a pipe where the velocity equals the average velocity for turbulent flow.

5.41 Plot the velocity profiles for Prandtl's exponential velocity formula for values of n of $\frac{1}{7}$, $\frac{1}{8}$, and $\frac{1}{9}$.

5.42 Estimate the skin-friction drag on an airship 100 m long, average diameter 20 m, with velocity of 130 km/h traveling through air at 90 kPa abs and 25°C.

5.43 The velocity distribution in a boundary layer is given by $u/U = 3(y/\delta) - 2(y/\delta)^2$. Show that the displacement thickness of the boundary layer is $\delta_1 = \delta/6$.

5.44 Using the velocity distribution $u/U = \sin(\pi y/2\delta)$, determine the equation for growth of the laminar boundary layer and for shear stress along a smooth flat plate in two-dimensional flow.

5.45 Compare the drag coefficients that are obtained with the velocity distributions given in Probs. 5.43 and 5.44.

5.46 Work out the equations for growth of the turbulent boundary layer, based on the exponential law $u/U = (y/\delta)^{1/9}$ and $f = 0.185/R^{1/5}$. $(\tau_0 = \rho f V^2/8.)$

5.47 Air at 20°C, 100 kPa abs flows along a smooth plate with a velocity of 150 km/h. How long does the plate have to be to obtain a boundary-layer thickness of 8 mm?

5.48 The walls of a wind tunnel are sometimes made divergent to offset the effect of the boundary layer in reducing the portion of the cross section in which the flow is of constant speed. At what angle must plane walls be set so that the displacement thickness does not encroach upon the tunnel's constant-speed cross section at distances greater than 0.8 ft from the leading edge of the wall? Use the data of Prob. 5.47.

5.49 What is the terminal velocity of a 2-in-diameter metal ball, sp gr 3.5, dropped in oil, sp gr 0.80, $\mu = 1$ P? What would be the terminal velocity for the same-size ball but with a 7.0 sp gr? How do these results agree with the experiments attributed to Galileo at the Leaning Tower of Pisa?

5.50 At what speed must a 120-mm sphere travel through water at 10°C to have a drag of 5 N?

5.51 A spherical balloon that contains helium ascends through air at 14 psia, 40°F. Balloon and payload weigh 300 lb. What diameter permits ascension at 10 ft/s? $C_D = 0.21$. If the balloon is tethered to the ground in a 10-mph wind, what is the angle of inclination of the retaining cable?

5.52 How many 30-m-diameter parachutes $(C_D = 1.2)$ should be used to drop a bulldozer weighing 45 kN at a terminal speed of 10 m/s through air at 100 kPa abs at 20°C?

5.53 An object weighing 400 lb is attached to a circular disk and dropped from a plane. What diameter should the disk be to have the object strike the ground at 72 ft/s? The disk is so attached that it is normal to direction of motion. $p = 14.7$ psia; $t = 70°F$.

5.54 A circular disk 3 m in diameter is held normal to a 100 km/h airstream ($\rho = 1.1$ kg/m^3). What force is required to hold it at rest?

5.55 Discuss the origin of the drag on a disk when its plane is parallel to the flow and when it is normal to it.

5.56 A semitubular cylinder of 6-in radius with concave side upstream is submerged in water flowing 3 ft/s. Calculate the drag for a cylinder 24 ft long.

5.57 A projectile of form *a*, Fig. 5.25, is 108 mm in diameter and travels at 1 km/s through air, $\rho = 1$ kg/m^3; $c = 300$ m/s. What is its drag?

5.58 On the basis of the discussion of the Mach angle, explain why a supersonic airplane is often seen before it is heard.

5.59 If an airplane 1 mi above the earth passes over an observer and the observer does not hear the plane until it has traveled 1.6 mi further, what is its speed? Sound velocity is 1080 ft/s. What is its Mach angle?

5.60 Give some reason for the discontinuity in the curves of Fig. 5.23 at the angle of attack of 22°.

5.61 What is the ratio of lift to drag for the airfoil section of Fig. 5.23 for an angle of attack of 2°?

5.62 Determine the settling velocity of small metal spheres, sp gr 4.5, diameter 0.1 mm, in crude oil at 25°C.

5.63 A spherical dust particle at an altitude of 80 km is radioactive as a result of an atomic explosion. Determine the time it will take to settle to earth if it falls in accordance with Stokes' law. Its size and sp gr are 25 μm and 2.5. Neglect wind effects. Use isothermal atmosphere at -18°C.

5.64 How large a spherical particle of dust, sp gr 2.5, will settle in atmospheric air at 20°C in obedience to Stokes' law? What is the settling velocity?

5.65 The Chézy coefficient is 127 for flow in a rectangular channel 6 ft wide, 3 ft deep, with bottom slope of 0.0016. What is the discharge?

5.66 A rectangular channel 1 m wide, $\lambda = 0.005$, $S = 0.0064$, carries 1 m^3/s. Determine the velocity.

5.67 What is the value of the Manning roughness factor n in Prob. 5.66?

5.68 A rectangular, brick-lined channel 6 ft wide and 4 ft deep carries 210 cfs. What slope is required for the channel?

5.69 The channel cross section shown in Fig. 5.45 is made of unplaned wood and has a slope of 0.001. What is the discharge?

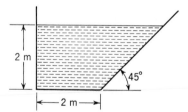

Figure 5.45 Problem 5.69.

5.70 A trapezoidal, unfinished concrete channel carries water at a depth of 2 m. Its bottom width is 3 m and side slopes 1 horizontal to 1½ vertical. For a bottom slope of 0.004 what is the discharge?

5.71 A trapezoidal channel with bottom slope 0.003, bottom width 1.2 m, and side slopes 2 horizontal to 1 vertical carries 6 m^3/s at a depth of 1.2 m. What is the Manning roughness factor?

5.72 A trapezoidal earth canal, bottom width 8 ft and side slopes 2 on 1 (2 horizontal to 1 vertical), is to be constructed to carry 280 cfs. The best velocity for nonscouring is 2.8 ft/s with this material. What is the bottom slope required?

5.73 What diameter is required of a semicircular corrugated-metal channel to carry 2 m^3/s when its slope is 0.006?

5.74 A semicircular corrugated-metal channel 9 ft in diameter has a bottom slope of 0.004. What is its capacity when flowing full?

5.75 Calculate the depth of flow of 60 m³/s in a gravel trapezoidal channel with bottom width of 4 m, side slopes of 3 horizontal to 1 vertical, and bottom slope of 0.0009.

5.76 What is the velocity of flow of 260 cfs in a rectangular channel 12 ft wide? $S = 0.0049; n = 0.014$.

5.77 A trapezoidal channel, brick-lined, is to be constructed to carry 35 m³/s a distance of 8 km with a head loss of 5 m. The bottom width is 4 m, the side slopes 1 on 1. What is the velocity?

5.78 How does the discharge vary with depth in Fig. 5.46?

5.79 How does the velocity vary with depth in Fig. 5.46?

5.80 Determine the depth of flow in Fig. 5.46 for discharge of 12 cfs. The channel is made of riveted steel with bottom slope 0.01.

5.81 Determine the depth y (Fig. 5.47) for maximum velocity for given n and S.

5.82 Determine the depth y (Fig. 5.47) for maximum discharge for given n and S.

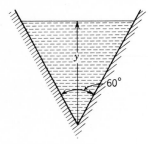

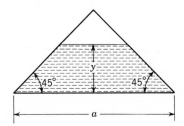

Figure 5.46 Problems 5.78, 5.79, 5.80.   **Figure 5.47** Problems 5.81, 5.82.

5.83 A test on 300-mm-diameter pipe with water showed a gage difference of 280 mm on a mercury-water manometer connected to two piezometer rings 120 m apart. The flow was 0.23 m³/s. What is the friction factor?

5.84 By using the Blasius equation for determination of friction factor, determine the horsepower per mile required to pump 3.0 cfs liquid, $v = 3.3 \times 10^{-4}$ ft²/s, $\gamma = 55$ lb/ft³, through an 18-in pipeline.

5.85 Determine the head loss per kilometer required to maintain a velocity of 3 m/s in a 10-mm-diameter pipe. $v = 4 \times 10^{-5}$ m²/s.

5.86 Fluid flows through a 10-mm-diameter tube at a Reynolds number of 1800. The head loss is 30 m in 100 m of tubing. Calculate the discharge, in liters per minute.

5.87 What size galvanized-iron pipe is needed to be "hydraulically smooth" at $\mathbf{R} = 3.5 \times 10^5$? (A pipe is said to be hydraulically smooth when it has the same losses as a smoother pipe under the same conditions.)

5.88 Above what Reynolds number is the flow through a 3-m-diameter riveted steel pipe, $\epsilon = 3$ mm, independent of the viscosity of the fluid?

5.89 Determine the absolute roughness of a 1-ft-diameter pipe that has a friction factor $f = 0.03$ for $\mathbf{R} = 1,000,000$.

5.90 What diameter clean galvanized-iron pipe has the same friction factor for $\mathbf{R} = 100,000$ as a 300-mm-diameter cast-iron pipe?

5.91 Under what conditions do the losses in an artificially roughened pipe vary as some power of the velocity greater than the second?

5.92 Why does the friction factor increase as the velocity decreases in laminar flow in a pipe?

5.93 By use of Eq. (5.10.13) calculate the friction factor for atmospheric air at 80°F, $V = 50$ ft/s in a 3-ft-diameter galvanized pipe.

5.94 Water at 20°C is to be pumped through a kilometer of 200-mm-diameter wrought-iron pipe at the rate of 60 l/s. Compute the head loss and power required.

5.95 16,000 ft^3/min atmospheric air at 90°F is conveyed 1000 ft through a 4-ft-diameter wrought-iron pipe. What is the head loss in inches of water?

5.96 What power motor for a fan must be purchased to circulate standard air in a wind tunnel at 500 km/h? The tunnel is a closed loop, 60 m long, and it can be assumed to have a constant circular cross section with a 2-m diameter. Assume smooth pipe.

5.97 Must provision be made to cool the air at some section of the tunnel described in Prob. 5.96? To what extent?

5.98 2.0 cfs oil, $\mu = 0.16$ P, $\gamma = 54$ lb/ft^3, is pumped through a 12-in pipeline of cast iron. If each pump produces 80 psi, how far apart may they be placed?

5.99 A 60-mm-diameter smooth pipe 150 m long conveys 10 l/s water at 25°C from a water main, $p = 1.6$ MN/m^2, to the top of a building 25 m above the main. What pressure can be maintained at the top of the building?

5.100 For water at 150°F calculate the discharge for the pipe of Fig. 5.48.

5.101 In Fig. 5.48, how much power would be required to pump 160 gpm from a reservoir at the bottom of the pipe to the reservoir shown?

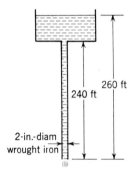

Figure 5.48 Problems 5.100, 5.101.

5.102 A 12-mm-diameter commercial steel pipe 15 m long is used to drain an oil tank. Determine the discharge when the oil level in the tank is 2 m above the exit end of the pipe. $\mu = 0.10$ P; $\gamma = 8$ kN/m^3.

5.103 Two liquid reservoirs are connected by 200 ft of 2-in-diameter smooth tubing. What is the flow rate when the difference in elevation is 50 ft? $\nu = 0.001$ ft^2/s. Work by use of the Moody diagram and by Eq. (5.10.15).

5.104 For a head loss of 80 mm water in a length of 200 m for flow of atmospheric air at 15°C through a 1.25-m-diameter duct, $\epsilon = 1$ mm, calculate the flow in cubic meters per minute. Work by use of Moody diagram and by Eq. (5.10.15).

5.105 A gas of molecular weight 37 flows through a 24-in-diameter galvanized duct at 90 psia and 100°F. The head loss per 100 ft of duct is 2 in H$_2$O. What is the flow in slugs per hour? $\mu = 0.0194$ cP.

5.106 What is the power per kilometer required for a 70 percent efficient blower to maintain the flow of Prob. 5.105?

5.107 The 100 lb$_m$/min air required to ventilate a mine is admitted through 3000 ft of 12-in-diameter galvanized pipe. Neglecting minor losses, what head, in inches of water, does a blower have to produce to furnish this flow? $p = 14$ psia; $t = 90$°F.

5.108 In Fig. 5.44 $H = 20$ m, $L = 150$ m, $D = 50$ mm, $S = 0.85$, $\mu = 4$ cP, $\epsilon = 1$ mm. Find the newtons per second flowing.

5.109 In a process, 10,000 lb/h of distilled water at 70°F is conducted through a smooth tube between two reservoirs having a distance between them of 30 ft and a difference in elevation of 4 ft. What size tubing is needed?

5.110 What size of new cast-iron pipe is needed to transport 400 l/s water at 25°C for 1 km with head loss of 2 m? Work by use of the Moody diagram and Eq. (5.10.18).

5.111 Two types of steel plate, having surface roughnesses of $\epsilon_1 = 0.0003$ ft and $\epsilon_2 = 0.0001$ ft, have a cost differential of 10 percent more for the smoother plate. With an allowable stress in each of 10,000 psi, which plate should be selected to convey 100 cfs water at 200 psi with a head loss of 6 ft/mi?

5.112 An old pipe 2 m in diameter has a roughness of $\epsilon = 30$ mm. A 12-mm-thick lining would reduce the roughness to $\epsilon = 1$ mm. How much in pumping costs would be saved per year per kilometer of pipe for water at 20°C with discharge of 6 m³/s? The pumps and motors are 80 percent efficient, and power costs 2 cents per kilowatthour.

5.113 Calculate the diameter of new wood-stave pipe in excellent condition needed to convey 300 cfs water at 60°F with a head loss of 1 ft per 1000 ft of pipe. Work by use of the Moody diagram and Eq. (5.10.18).

5.114 Two oil reservoirs with difference in elevation of 5 m are connected by 300 m of commercial steel pipe. What size must the pipe be to convey 50 l/s? $\mu = 0.05$ kg/m·s, $\gamma = 8$ kN/m³.

5.115 300 cfs air, $p = 16$ psia, $t = 70$°F, is to be delivered to a mine with a head loss of 3 in H_2O per 1000 ft. What size galvanized pipe is needed?

5.116 Compute the losses, in joules per newton, due to flow of 25 m³/min air, $p = 1$ atm, $t = 20$°C, through a sudden expansion from 300- to 900-mm pipe. How much head would be saved by using a 10° conical diffuser?

5.117 Calculate the value of H in Fig. 5.49 for 125 l/s water at 15°C through commercial steel pipe. Include minor losses.

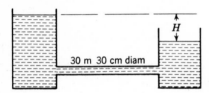

30 m 30 cm diam

Figure 5.49 Problems 5.117, 5.119, 5.120.

5.118 In Prob. 5.28 what would be the discharge if a globe valve were inserted into the line? Assume a smooth pipe and a well-rounded inlet, with $\mu = 1$ cP. Solve by use of the Moody diagram and by the iterative method using Eqs. (5.10.24) to (5.10.27).

5.119 In Fig. 5.49 for $H = 3$ m, calculate the discharge of oil, $S = 0.8$, $\mu = 7$ cP, through smooth pipe. Include minor losses.

5.120 If a valve is placed in the line in Prob. 5.119 and adjusted to reduce the discharge by one-half, what is K for the valve and what is its equivalent length of pipe at this setting?

5.121 A water line connecting two reservoirs at 70°F has 5000 ft of 24-in-diameter steel pipe, three standard elbows, a globe valve, and a re-entrant pipe entrance. What is the difference in reservoir elevations for 20 cfs?

5.122 Determine the discharge in Prob. 5.121 if the difference in elevation is 40 ft.

5.123 What size commercial steel pipe is needed to convey 200 l/s water at 20°C 5 km with a head drop of 4 m? The line connects two reservoirs, has a re-entrant entrance, a submerged outlet, four standard elbows, and a globe valve.

5.124 What is the equivalent length of 2-in-diameter pipe, $f = 0.022$, for (a) a re-entrant pipe entrance, (b) a sudden expansion from 2 to 4 in diameter, (c) a globe valve and a standard tee?

5.125 Find H in Fig. 5.50 for 200 gpm oil flow, $\mu = 0.1$ P, $\gamma = 60$ lb/ft³, for the angle valve wide open.

5.126 Find K for the angle valve in Prob. 5.125 for flow of 10 l/s at the same H.

5.127 What is the discharge through the system of Fig. 5.50 for water at 25°C when $H = 8$ m?

5.128 Compare the smooth-pipe curve on the Moody diagram with Eq. (5.10.4) for $\mathbf{R} = 10^5, 10^6, 10^7$.

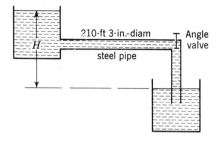

Figure 5.50 Problems 5.125, 5.126, 5.127.

5.129 Check the location of line $\epsilon/D = 0.0002$ on the Moody diagram with Eq. (5.10.7).

5.130 In Eq. (5.10.7) show that, when $\epsilon = 0$, it reduces to Eq. (5.10.4) and that, when **R** is very large, it reduces to Eq. (5.10.6).

5.131 The pumping system shown in Fig. 5.51 has a pump head-discharge curve $H = 40 - 24Q^2$, with head in meters and discharge in cubic meters per second. The pipe lengths include a correction for minor losses. Determine the flow through the system, in liters per second. For efficiency of pumping system of 72 percent determine the power required. The pump requires a suction head of at least one-half atmosphere to avoid cavitation. What is the maximum discharge and power required to reach this maximum?

5.132 In Fig. 5.52 the rocker plate has a width of 1 ft. Calculate (a) the load the bearing will sustain and (b) the drag on the bearing. Assume no flow normal to the paper.

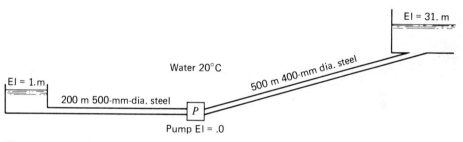

Figure 5.51 Problem 5.131.

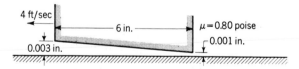

Figure 5.52 Problems 5.132, 5.133, 5.134.

5.133 Find the maximum pressure in the fluid of Prob. 5.132, and determine its location.

5.134 Determine the pressure center for the rocker plate of Prob. 5.132.

5.135 Show that a shaft concentric with a bearing can sustain no load.

SIX
COMPRESSIBLE FLOW

In Chap. 5, viscous incompressible fluid-flow situations were mainly considered. In this chapter on compressible flow, one new variable enters, the density, and one extra equation is available, the equation of state, which relates pressure and density. The other equations—continuity, momentum, and the first and second laws of thermodynamics—are also needed in the analysis of compressible fluid-flow situations. Topics in steady one-dimensional flow of a perfect gas are discussed. The one-dimensional approach is limited to applications in which the velocity and density may be considered constant over any cross section. When density changes are gradual and do not change by more than a few percent, the flow may be treated as incompressible with the use of an average density.

The following topics are treated in this chapter: perfect-gas relations, speed of a sound wave, Mach number, isentropic flow, shock waves, Fanno and Rayleigh lines, adiabatic flow, flow with heat transfer, isothermal flow, and the analogy between shock waves and open-channel waves.

6.1 PERFECT-GAS RELATIONS

In Sec. 1.6 [Eq. (1.6.2)] a perfect gas is defined as a fluid that has constant specific heats and follows the law

$$p = \rho RT \qquad (6.1.1)$$

in which p and T are the absolute pressure and absolute temperature, respectively, ρ is the density, and R the gas constant. In this section specific heats are defined, the specific heat ratio is introduced and related to specific heats and the gas constant, internal energy and enthalpy are related to temperature, entropy relations are established, and the isentropic and reversible polytropic processes are introduced.

In general, the specific heat c_v at constant volume is defined by

$$c_v = \left(\frac{\partial u}{\partial T}\right)_v \tag{6.1.2}$$

in which u is the internal energy† per unit mass. In words, c_v is the amount of internal-energy increase required by a unit mass of gas to increase its temperature by one degree when its volume is held constant. In thermodynamic theory it is proved that u is a function only of temperature for a perfect gas.

The specific heat c_p at constant pressure is defined by

$$c_p = \left(\frac{\partial h}{\partial T}\right)_p \tag{6.1.3}$$

in which h is the enthalpy per unit mass given by $h = u + p/\rho$. Since p/ρ is equal to RT and u is a function only of temperature for a perfect gas, h depends only on temperature. Many of the common gases, such as water vapor, hydrogen, oxygen, carbon monoxide, and air, have a fairly small change in specific heats over the temperature range 500 to 1000°R, and an intermediate value is taken for their use as perfect gases. Table C.3 of Appendix C lists some common gases with values of specific heats at 80°F.

For perfect gases Eq. (6.1.2) becomes

$$du = c_v \, dT \tag{6.1.4}$$

and Eq. (6.1.3) becomes

$$dh = c_p \, dT \tag{6.1.5}$$

Then, from

$$h = u + \frac{p}{\rho} = u + RT$$

differentiating gives

$$dh = du + R \, dT$$

and substitution of Eqs. (6.1.4) and (6.1.5) leads to

$$c_p = c_v + R \tag{6.1.6}$$

which is valid for any gas obeying Eq. (6.1.1) (even when c_p and c_v are changing with temperature). If c_p and c_v are given in heat units per unit mass, i.e., kilocalorie per kilogram per kelvin or Btu per slug per degree Rankine, then R must have the same units. The conversion factor is 1 kcal = 4187 J or 1 Btu = 778 ft·lb.

The *specific heat ratio* k is defined as the ratio

$$k = \frac{c_p}{c_v} \tag{6.1.7}$$

---

† The definitions for c_v and c_p are for equilibrium conditions; hence, the internal energy e of Eq. (3.3.7) is u.

Solving with Eq. (6.1.6) gives

$$c_p = \frac{k}{k-1} R \qquad c_v = \frac{R}{k-1} \qquad (6.1.8)$$

Entropy Relations

The first law of thermodynamics for a system states that the heat added to a system is equal to the work done by the system plus its increase in internal energy [Eq. (3.8.4)]. In terms of the entropy s the equation takes the form

$$T\,ds = du + p\,d\frac{1}{\rho} \qquad (3.8.6)$$

which is a relation between thermodynamic properties and must hold for all pure substances.

The internal-energy change for a perfect gas is

$$u_2 - u_1 = c_v(T_2 - T_1) \qquad (6.1.9)$$

and the enthalpy change is

$$h_2 - h_1 = c_p(T_2 - T_1) \qquad (6.1.10)$$

The change in entropy

$$ds = \frac{du}{T} + \frac{p}{T}d\frac{1}{\rho} = c_v\frac{dT}{T} + R\rho\,d\frac{1}{\rho} \qquad (6.1.11)$$

may be obtained from Eqs. (6.1.4) and (6.1.1). After integrating,

$$s_2 - s_1 = c_v \ln \frac{T_2}{T_1} + R \ln \frac{\rho_1}{\rho_2} \qquad (6.1.12)$$

By use of Eqs. (6.1.8) and (6.1.1), Eq. (6.1.12) becomes

$$s_2 - s_1 = c_v \ln \left[\frac{T_2}{T_1}\left(\frac{\rho_1}{\rho_2}\right)^{k-1}\right] \qquad (6.1.13)$$

or

$$s_2 - s_1 = c_v \ln \left[\frac{p_2}{p_1}\left(\frac{\rho_1}{\rho_2}\right)^{k}\right] \qquad (6.1.14)$$

and

$$s_2 - s_1 = c_v \ln \left[\left(\frac{T_2}{T_1}\right)^{k}\left(\frac{p_2}{p_1}\right)^{1-k}\right] \qquad (6.1.15)$$

These equations are forms of the second law of thermodynamics.

If the process is reversible, $ds = dq_H/T$, or $T\,ds = dq_H$; further, if the process should also be adiabatic, $dq_H = 0$. Thus $ds = 0$ for a reversible, adiabatic process, or $s = $ const; the *reversible, adiabatic* process is therefore *isentropic*. Then, from Eq. (6.1.14) for $s_2 = s_1$,

$$\frac{p_1}{\rho_1^k} = \frac{p_2}{\rho_2^k} \qquad (6.1.16)$$

Equation (6.1.16) combined with the general gas law yields

$$\frac{T_2}{T_1} = \left(\frac{p_2}{p_1}\right)^{(k-1)/k} = \left(\frac{\rho_2}{\rho_1}\right)^{k-1} \tag{6.1.17}$$

The enthalpy change for an isentropic process is

$$h_2 - h_1 = c_p(T_2 - T_1) = c_p T_1\left(\frac{T_2}{T_1} - 1\right) = c_p T_1\left[\left(\frac{p_2}{p_1}\right)^{(k-1)/k} - 1\right] \tag{6.1.18}$$

The *polytropic* process is defined by

$$\frac{p}{\rho^n} = \text{const} \tag{6.1.19}$$

and is an approximation to certain actual processes in which p would plot substantially as a straight line against ρ on log-log paper. This relation is frequently used to calculate the work when the polytropic process is reversible, by substitution into the relation $W = \int p\, d\mathcal{V}$. Heat transfer occurs in a reversible polytropic process except when $n = k$, the isentropic case.

Example 6.1 Express R in kilocalories per kilogram per kelvin for helium. From Table C.3, $R = 2077$ m·N/kg·K; therefore,

$$R = (2077 \text{ m·N/kg·K})\frac{1 \text{ kcal}}{4187 \text{ m·N}} = 0.496 \text{ kcal/kg·K}$$

Example 6.2 Compute the value of R from the values of k and c_p for air and check in Table C.3. From Eq. (6.1.8)

$$R = \frac{k-1}{k}c_p = \frac{1.40 - 1.0}{1.40}(0.240 \text{ Btu/lb}_m\cdot°R) = 0.0686 \text{ Btu/lb}_m\cdot°R$$

By converting from Btu to foot-pounds

$$R = (0.0686 \text{ Btu/lb}_m\cdot°R)(778 \text{ ft·lb/Btu}) = 53.3 \text{ ft·lb/lb}_m\cdot°R$$

which checks with the value in Table C.3.

Example 6.3 Compute the enthalpy change in 5 kg of oxygen when the initial conditions are $p_1 = 130$ kPa abs, $t_1 = 10°C$ and the final conditions are $p_2 = 500$ kPa abs, $t_2 = 95°C$. Enthalpy is a function of temperature only. By Eq. (6.1.10) the enthalpy change for 5 kg oxygen is

$$H_2 - H_1 = 5 \text{ kg} \times c_p(T_2 - T_1)$$
$$= (5 \text{ kg})(0.219 \text{ kcal/kg·K})(95 - 10 \text{ K}) = 93.08 \text{ kcal}$$

Example 6.4 Determine the entropy change in 4.0 slugs of water vapor when the initial conditions are $p_1 = 6$ psia, $t_1 = 110°F$ and the final conditions are $p_2 = 40$ psia, $t_2 = 38°F$. From Eq. (6.1.15) and Table C.3

$$S_2 - S_1 = -(0.271 \text{ Btu/lb}_m\cdot°R)(4.0 \text{ slugs})(32.17 \text{ lb}_m/\text{slug})$$
$$= -34.7 \text{ Btu/°R}$$

Example 6.5 A cylinder containing 2 kg nitrogen at 0.14 MPa abs and 5°C is compressed isentropically to 0.3 MPa abs. Find the final temperature and the work required.
By Eq. (6.1.17)

$$T_2 = T_1 \left(\frac{p_2}{p_1}\right)^{(k-1)/k} = (273 + 5 \text{ K})\left(\frac{0.3}{0.14}\right)^{(1.4-1)/1.4} = 345.6 \text{ K} = 72.6°C$$

From the principle of conservation of energy, the work done on the gas must equal its increase in internal energy, since there is no heat transfer in an isentropic process; i.e.,

$$u_2 - u_1 = c_v(T_2 - T_1) \text{ kcal/kg} = \text{work/kg}$$

or $\quad$ Work $= (2 \text{ kg})(0.177 \text{ kcal/kg}\cdot\text{K})(345.6 - 278 \text{ K}) = 23.93 \text{ kcal}$

Example 6.6 3.0 slugs of air is involved in a reversible polytropic process in which the initial conditions $p_1 = 12$ psia, $t_1 = 60°F$ change to $p_2 = 20$ psia, and volume $\mathcal{V} = 1011$ ft^3. Determine (a) the formula for the process, (b) the work done on the air, (c) the amount of heat transfer, and (d) the entropy change.

(a) $\qquad\qquad \rho_1 = \dfrac{p_1}{RT_1} = \dfrac{12 \times 144}{53.3 \times 32.17(460 + 60)} = 0.00194 \text{ slug/ft}^3$

R was converted to foot-pounds per slug per degree Rankine by multiplying by 32.17. Also,

$$\rho_2 = \tfrac{3}{1011} = 0.002967 \text{ slug/ft}^3$$

From Eq. (6.1.19)

$$\frac{p_1}{\rho_1^n} = \frac{p_2}{\rho_2^n}$$

$$n = \frac{\ln (p_2/p_1)}{\ln (\rho_2/\rho_1)} = \frac{\ln \frac{20}{12}}{\ln (0.002967/0.00194)} = 1.20$$

hence $\qquad\qquad\qquad \dfrac{p}{\rho^{1.2}} = \text{const}$

describes the polytropic process.
(b) Work of expansion is

$$W = \int_{\mathcal{V}_1}^{\mathcal{V}_2} p \, d\mathcal{V}$$

This is the work done by the gas on its surroundings. Since

$$p_1 \mathcal{V}_1^n = p_2 \mathcal{V}_2^n = p\mathcal{V}^n$$

by substituting into the integral,

$$W = p_1 \mathcal{V}_1^n \int_{\mathcal{V}_1}^{\mathcal{V}_2} \frac{d\mathcal{V}}{\mathcal{V}^n} = \frac{p_2 \mathcal{V}_2 - p_1 \mathcal{V}_1}{1 - n} = \frac{mR}{1 - n}(T_2 - T_1)$$

if m is the mass of gas: $\mathcal{V}_2 = 1011$ ft^3 and

$$\mathcal{V}_1 = \mathcal{V}_2 \left(\frac{p_2}{p_1}\right)^{1/n} = 1011\left(\frac{20}{12}\right)^{1/1.2} = 1547 \text{ ft}^3$$

Then $\qquad\qquad W = \dfrac{20 \times 144 \times 1011 - 12 \times 144 \times 1548}{1 - 1.2} = -1,184,000 \text{ ft}\cdot\text{lb}$

Hence, the work done on the gas is 1,184,000 ft·lb.

(c) From the first law of thermodynamics the heat added minus the work done by the gas must equal the increase in internal energy; i.e.,

$$Q_H - W = U_2 - U_1 = c_v m(T_2 - T_1)$$

First

$$T_2 = \frac{p_2}{\rho_2 R} = \frac{20 \times 144}{0.002965 \times 53.3 \times 32.17} = 566°R$$

Then

$$Q_H = -\frac{1,184,000}{778} + 0.171 \times 32.17 \times 3(566 - 520)$$

$$= -761 \text{ Btu}$$

761 Btu was transferred from the mass of air.

(d) From Eq. (6.1.14) the entropy change is computed:

$$s_2 - s_1 = 0.171 \ln\left[\frac{20}{12}\left(\frac{0.00194}{0.002967}\right)^{1.4}\right] = -0.01436 \text{ Btu/lb}_m \cdot °R$$

and

$$S_2 - S_1 = -0.01436 \times 3 \times 32.17 = -1.386 \text{ Btu/°R}$$

A rough check on the heat transfer can be made by using Eq. (3.8.5) and an average temperature $T = (520 + 566)/2 = 543$ and by remembering that the losses are zero in a reversible process.

$$Q_H = T(S_2 - S_1) = 543(-1.386) = -753 \text{ Btu}$$

EXERCISES

6.1.1 Specific heat at constant volume is defined by

(a) kc_p (b) $\left(\frac{\partial u}{\partial T}\right)_p$ (c) $\left(\frac{\partial T}{\partial u}\right)_v$ (d) $\left(\frac{\partial u}{\partial T}\right)_v$ (e) none of these answers

6.1.2 Specific heat at constant pressure, for a perfect gas, is *not* given by

(a) kc_v (b) $\left(\frac{\partial h}{\partial T}\right)_p$ (c) $\frac{h_2 - h_1}{T_2 - T_1}$ (d) $\frac{\Delta u + \Delta(p/\rho)}{\Delta T}$

(e) any of these answers

6.1.3 For a perfect gas, the enthalpy (a) always increases owing to losses; (b) depends upon the pressure only; (c) depends upon the temperature only; (d) may increase while the internal energy decreases; (e) satisfies none of these answers.

6.1.4 The following classes of substances may be considered perfect gases: (a) ideal fluids; (b) saturated steam, water vapor, and air; (c) fluids with a constant bulk modulus of elasticity; (d) water vapor, hydrogen, and nitrogen at low pressure; (e) none of these answers.

6.1.5 c_p and c_v are related by (a) $k = c_p/c_v$; (b) $k = c_p c_v$; (c) $k = c_v/c_p$; (d) $c_p = c_v^k$; (e) none of these answers.

6.1.6 If $c_p = 0.30$ Btu/lb$_m \cdot$°R and $k = 1.66$, in foot-pounds per slug per degree Fahrenheit, c_v equals (a) 0.582; (b) 1452; (c) 4524; (d) 7500; (e) none of these answers.

6.1.7 If $c_p = 0.30$ kcal/kg·K and $k = 1.33$, the gas constant in kilocalories per kilogram per kelvin is (a) 0.075; (b) 0.099; (c) 0.399; (d) 0.699; (e) none of these answers.

6.1.8 $R = 62$ ft·lb/lb$_m$·°R and $c_p = 0.279$ Btu/lb$_m$·°F. The isentropic exponent k is (a) 1.2; (b) 1.33; (c) 1.66; (d) 1.89; (e) none of these answers.

6.1.9 The specific heat ratio is given by

(a) $\dfrac{1}{1 - R/c_p}$ (b) $1 + \dfrac{c_v}{R}$ (c) $\dfrac{c_p}{c_v} + R$ (d) $\dfrac{1}{1 - c_v/R}$

(e) none of these answers

6.1.10 The entropy change for a perfect gas is (a) always positive; (b) a function of temperature only; (c) $(\Delta q_H/T)_{rev}$; (d) a thermodynamic property depending upon temperature and pressure; (e) a function of internal energy only.

6.1.11 An isentropic process is always (a) irreversible and adiabatic; (b) reversible and isothermal; (c) frictionless and adiabatic; (d) frictionless and irreversible; (e) none of these answers.

6.1.12 The relation $p = \text{const } \rho^k$ holds only for those processes that are (a) reversible polytropic; (b) isentropic; (c) frictionless isothermal; (d) adiabatic irreversible; (e) none of these answers.

6.1.13 The reversible polytropic process is (a) adiabatic frictionless; (b) given by $p/\rho = \text{const}$; (c) given by $p\rho^k = \text{const}$; (d) given by $p/\rho^n = \text{const}$; (e) none of these answers.

6.1.14 A reversible polytropic process could be given by

(a) $\dfrac{T_1}{T_2} = \left(\dfrac{\rho_1}{\rho_2}\right)^{n-1}$ (b) $\dfrac{p_1}{p_2} = \left(\dfrac{\rho_2}{\rho_1}\right)^n$ (c) $\dfrac{T_1}{T_2} = \left(\dfrac{p_1}{p_2}\right)^{n-1}$ (d) $\dfrac{T_1}{T_2} = \left(\dfrac{\rho_1}{\rho_2}\right)^{(n-1)/n}$

(e) none of these answers

6.1.15 In a reversible polytropic process (a) some heat transfer occurs; (b) the entropy remains constant; (c) the enthalpy remains constant; (d) the internal energy remains constant; (e) the temperature remains constant.

6.2 SPEED OF A SOUND WAVE; MACH NUMBER

The speed of a small disturbance in a conduit can be determined by application of the momentum equation and the continuity equation. The question is first raised whether a *stationary* small change in velocity, pressure, and density can occur in a channel. By referring to Fig. 6.1, the continuity equation can be written

$$\rho V A = (\rho + d\rho)(V + dV)A$$

in which A is the cross-sectional area of channel. The equation can be reduced to

$$\rho\, dV + V\, d\rho = 0$$

When the momentum equation (3.11.2) is applied to the control volume within the dotted lines,

$$pA - (p + dp)A = \rho V A(V + dV - V)$$

or $$dp = -\rho V\, dV$$

If $\rho\, dV$ is eliminated between the two equations,

$$V^2 = \frac{dp}{d\rho} \tag{6.2.1}$$

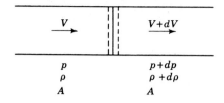

Figure 6.1 Steady flow in prismatic channel with sudden small change in velocity, pressure, and density.

Hence, a small disturbance or sudden change in conditions in steady flow can occur only when the particular velocity $V = \sqrt{dp/d\rho}$ exists in the conduit. Now this problem can be converted to the unsteady flow of a small disturbance through still fluid by superposing on the whole system and its surroundings the velocity V to the left, since this in no way affects the dynamics of the system. This is called the speed of sound c in the medium. The disturbance from a point source would cause a spherical wave to emanate, but at some distance from the source the wavefront would be essentially linear or one-dimensional. Large disturbances, e.g., a bomb explosion, may travel faster than the speed of sound. The equation for speed of sound

$$c = \sqrt{\frac{dp}{d\rho}} \tag{6.2.2}$$

may be expressed in several useful forms. The bulk modulus of elasticity can be introduced:

$$K = -\frac{dp}{d\mathcal{U}/\mathcal{U}}$$

in which $\mathcal{U}$ is the volume of fluid subjected to the pressure change dp. Since

$$\frac{d\mathcal{U}}{\mathcal{U}} = \frac{dv_s}{v_s} = -\frac{d\rho}{\rho}$$

K may be expressed as

$$K = \frac{\rho \, dp}{d\rho}$$

Then, from Eq. (6.2.2),

$$c = \sqrt{\frac{K}{\rho}} \tag{6.2.3}$$

This equation applies to liquids as well as gases.

Example 6.7 Carbon tetrachloride has a bulk modulus of elasticity of 1.124 GPa and a density of 1593 kg/m^3. What is the speed of sound in the medium?

$$c = \sqrt{\frac{K}{\rho}} = \sqrt{\frac{1.124 \times 10^9 \text{ N/m}^2}{1593 \text{ kg/m}^3}} = 840 \text{ m/s}$$

Since the pressure and temperature changes due to passage of a sound wave are extremely small, the process is almost reversible. Also, the relatively rapid process of passage of the wave, together with the minute temperature changes, makes the process almost adiabatic. In the limit, the process may be considered to be isentropic,

$$p\rho^{-k} = \text{const} \qquad \frac{dp}{d\rho} = \frac{kp}{\rho}$$

and

$$c = \sqrt{\frac{kp}{\rho}} \qquad\qquad (6.2.4)$$

or, from the perfect-gas law $p = \rho RT$,

$$c = \sqrt{kRT} \qquad\qquad (6.2.5)$$

which shows that the speed of sound in a perfect gas is a function of absolute temperature only. In flow of gas through a conduit, the speed of sound generally changes from section to section as the temperature is changed by density changes and friction effects. In isothermal flow the speed of sound remains constant.

The Mach number has been defined as the ratio of velocity of a fluid to the local velocity of sound in the medium,

$$\mathbf{M} = \frac{V}{c} \qquad\qquad (6.2.6)$$

Squaring the Mach number produces V^2/c^2, which may be interpreted as the ratio of kinetic energy of the fluid to its thermal energy, since kinetic energy is proportional to V^2 and thermal energy is proportional to T. The Mach number is a measure of the importance of compressibility. In an incompressible fluid K is infinite and $\mathbf{M} = 0$. For perfect gases

$$K = kp \qquad\qquad (6.2.7)$$

when the compression is isentropic.

Example 6.8 What is the speed of sound in dry air at sea level when $t = 68°F$ and in the stratosphere when $t = -67°F$?
At sea level, from Eq. (6.2.5),

$$c = \sqrt{1.4 \times 32.17 \times 53.3(460 + 68)} = 1126 \text{ ft/s}$$

and in the stratosphere

$$c = \sqrt{1.4 \times 32.17 \times 53.3(460 - 67)} = 971 \text{ ft/s}$$

EXERCISES

6.2.1 Select the expression that does *not* give the speed of a sound wave: (a) $\sqrt{kRT}$; (b) $\sqrt{k\rho/p}$; (c) $\sqrt{dp/d\rho}$; (d) $\sqrt{kp/\rho}$; (e) $\sqrt{K/\rho}$.

6.2.2 The speed of sound in water, in feet per second, under ordinary conditions is about (a) 460; (b) 1100; (c) 4600; (d) 11,000; (e) none of these answers.

6.2.3 The speed of sound in an ideal gas varies directly as (a) the density; (b) the absolute pressure; (c) the absolute temperature; (d) the bulk modulus of elasticity; (e) none of these answers.

6.3 ISENTROPIC FLOW

Frictionless adiabatic, or isentropic, flow is an ideal that cannot be reached in the flow of real gases. It is approached, however, in flow through transitions, nozzles, and venturi meters where friction effects are minor, owing to the short distances traveled, and heat transfer is minor because the changes that a particle undergoes are slow enough to keep the velocity and temperature gradients small.† The performance of fluid machines is frequently compared with the performance assuming isentropic flow. In this section one-dimensional steady flow of a perfect gas through converging and converging-diverging ducts is studied.

Some very general results can be obtained by use of Euler's equation (3.5.8), neglecting elevation changes,

$$V \, dV + \frac{dp}{\rho} = 0 \tag{6.3.1}$$

and the continuity equation

$$\rho A V = \text{const} \tag{6.3.2}$$

Differentiating $\rho A V$ and then dividing through by $\rho A V$ gives

$$\frac{d\rho}{\rho} + \frac{dV}{V} + \frac{dA}{A} = 0 \tag{6.3.3}$$

From Eq. (6.2.2) dp is obtained and substituted into Eq. (6.3.1), yielding

$$V \, dV + c^2 \frac{d\rho}{\rho} = 0 \tag{6.3.4}$$

Eliminating $d\rho/\rho$ in the last two equations and rearranging give

$$\frac{dA}{dV} = \frac{A}{V}\left(\frac{V^2}{c^2} - 1\right) = \frac{A}{V}(\mathbf{M}^2 - 1) \tag{6.3.5}$$

The assumptions underlying this equation are that the flow is steady and frictionless. No restrictions as to heat transfer have been imposed. Equation (6.3.5) shows that, for subsonic flow ($\mathbf{M} < 1$), dA/dV is always negative; i.e., the channel area must decrease for increasing velocity. As dA/dV is zero for $\mathbf{M} = 1$ only, the velocity keeps increasing until the minimum section or throat is reached, and that is the only section at which sonic flow may occur. Also, for Mach numbers greater

---

† H. W. Liepmann and A. Roshko, "Elements of Gas Dynamics," p. 51, Wiley, New York, 1957.

than unity (supersonic flow) dA/dV is positive and the area must increase for an increase in velocity. Hence, to obtain supersonic steady flow from a fluid at rest in a reservoir, it must first pass through a converging duct and then a diverging duct.

When the analysis is restricted to isentropic flow, Eq. (6.1.16) may be written

$$p = p_1 \rho_1^{-k} \rho^k \qquad (6.3.6)$$

Differentiating and substituting for dp in Eq. (6.3.1) give

$$V \, dV + k \frac{p_1}{\rho_1^k} \rho^{k-2} \, d\rho = 0$$

Integration yields

$$\frac{V^2}{2} + \frac{k}{k-1} \frac{p_1}{\rho_1^k} \rho^{k-1} = \text{const}$$

or
$$\frac{V_1^2}{2} + \frac{k}{k-1} \frac{p_1}{\rho_1} = \frac{V_2^2}{2} + \frac{k}{k-1} \frac{p_2}{\rho_2} \qquad (6.3.7)$$

Equation (6.3.7) can be derived from Eq. (3.8.2) for adiabatic flow $(dq_H = 0)$ using Eq. (6.1.8). This avoids the restriction to isentropic flow. This equation is useful when expressed in terms of temperature; from $p = \rho RT$

$$\frac{V_1^2}{2} + \frac{k}{k-1} RT_1 = \frac{V_2^2}{2} + \frac{k}{k-1} RT_2 \qquad (6.3.8)$$

For adiabatic flow from a reservoir where conditions are given by p_0, ρ_0, T_0, at any other section

$$\frac{V^2}{2} = \frac{kR}{k-1} (T_0 - T) \qquad (6.3.9)$$

In terms of the local Mach number V/c, with $c^2 = kRT$,

$$\mathbf{M}^2 = \frac{V^2}{c^2} = \frac{2kR(T_0 - T)}{(k-1)kRT} = \frac{2}{k-1} \left(\frac{T_0}{T} - 1 \right)$$

or
$$\frac{T_0}{T} = 1 + \frac{k-1}{2} \mathbf{M}^2 \qquad (6.3.10)$$

From Eqs. (6.3.10) and (6.1.17), which now restrict the following equations to isentropic flow,

$$\frac{p_0}{p} = \left(1 + \frac{k-1}{2} \mathbf{M}^2 \right)^{k/(k-1)} \qquad (6.3.11)$$

and
$$\frac{\rho_0}{\rho} = \left(1 + \frac{k-1}{2} \mathbf{M}^2 \right)^{1/(k-1)} \qquad (6.3.12)$$

Flow conditions are termed critical at the throat section when the velocity there is sonic. Sonic conditions are marked with an asterisk. $\mathbf{M} = 1$; $c^* = V^* = \sqrt{kRT^*}$.

By applying Eqs. (6.3.10) to (6.3.12) to the throat section for critical conditions (for $k = 1.4$ in the numerical portion),

$$\frac{T^*}{T_0} = \frac{2}{k+1} = 0.833 \qquad k = 1.40 \qquad (6.3.13)$$

$$\frac{p^*}{p_0} = \left(\frac{2}{k+1}\right)^{k/(k-1)} = 0.528 \qquad k = 1.40 \qquad (6.3.14)$$

$$\frac{\rho^*}{\rho_0} = \left(\frac{2}{k+1}\right)^{1/(k-1)} = 0.634 \qquad k = 1.40 \qquad (6.3.15)$$

These relations show that, for airflow, the absolute temperature drops about 17 percent from reservoir to throat, the critical pressure is 52.8 percent of the reservoir pressure, and the density is reduced by about 37 percent.

The variation of area with the Mach number for the critical case is obtained by use of the continuity equation and Eqs. (6.3.10) to (6.3.15). First

$$\rho A V = \rho^* A^* V^* \qquad (6.3.16)$$

in which A^* is the minimum, or throat, area. Then

$$\frac{A}{A^*} = \frac{\rho^*}{\rho} \frac{V^*}{V} \qquad (6.3.17)$$

Now, $V^* = c^* = \sqrt{kRT^*}$ and $V = cM = M\sqrt{kRT}$, so that

$$\frac{V^*}{V} = \frac{1}{M}\sqrt{\frac{T^*}{T}} = \frac{1}{M}\sqrt{\frac{T^*}{T_0}}\sqrt{\frac{T_0}{T}} = \frac{1}{M}\left\{\frac{1 + [(k-1)/2]M^2}{(k+1)/2}\right\}^{1/2} \qquad (6.3.18)$$

by use of Eqs. (6.3.13) and (6.3.10). In a similar manner

$$\frac{\rho^*}{\rho} = \frac{\rho^*}{\rho_0}\frac{\rho_0}{\rho} = \left\{\frac{1 + [(k-1)/2]M^2}{(k+1)/2}\right\}^{1/(k-1)} \qquad (6.3.19)$$

By substituting the last two equations into Eq. (6.3.17),

$$\frac{A}{A^*} = \frac{1}{M}\left\{\frac{1 + [(k-1)/2]M^2}{(k+1)/2}\right\}^{(k+1)/2(k-1)} \qquad (6.3.20)$$

which yields the variation of area of duct in terms of Mach number. A/A^* is never less than unity, and for any value greater than unity there will be two values of Mach number, one less than and one greater than unity. For gases with $k = 1.40$, Eq. (6.3.20) reduces to

$$\frac{A}{A^*} = \frac{1}{M}\left(\frac{5 + M^2}{6}\right)^3 \qquad k = 1.40 \qquad (6.3.21)$$

The maximum mass flow rate $\dot{m}_{max}$ can be expressed in terms of the throat area and reservoir conditions:

$$\dot{m}_{max} = \rho^* A^* V^* = \rho_0\left(\frac{2}{k+1}\right)^{1/(k-1)} A^* \sqrt{\frac{kR2T_0}{k+1}}$$

by Eqs. (6.3.15) and (6.3.13). Replacing ρ_0 by p_0/RT_0 gives

$$\dot{m}_{max} = \frac{A^* p_0}{\sqrt{T_0}} \sqrt{\frac{k}{R} \left(\frac{2}{k+1} \right)^{(k+1)/(k-1)}} \tag{6.3.22}$$

For $k = 1.40$ this reduces to

$$\dot{m}_{max} = 0.686 \frac{A^* p_0}{\sqrt{RT_0}} \tag{6.3.23}$$

which shows that the mass flow rate varies linearly as A^* and p_0 and inversely as the square root of the absolute temperature.

For subsonic flow throughout a converging-diverging duct, the velocity at the throat must be less than sonic velocity, or $M_t < 1$ with subscript t indicating the throat section. The mass rate of flow $\dot{m}$ is obtained from

$$\dot{m} = \rho V A = A \sqrt{2 p_0 \rho_0 \frac{k}{k-1} \left(\frac{p}{p_0} \right)^{2/k} \left[1 - \left(\frac{p}{p_0} \right)^{(k-1)/k} \right]} \tag{6.3.24}$$

which is derived from Eqs. (6.3.9) and (6.3.6) and the perfect-gas law. This equation holds for any section and is applicable as long as the velocity at the throat is subsonic. It may be applied to the throat section, and for this section, from Eq. (6.3.14),

$$\frac{p_t}{p_0} \geq \left(\frac{2}{k+1} \right)^{k/(k-1)}$$

where p_t is the throat pressure. When the equals sign is used in the expression, Eq. (6.3.24) reduces to Eq. (6.3.22).

For maximum mass flow rate, the flow downstream from the throat may be either supersonic or subsonic, depending upon the downstream pressure. Substituting Eq. (6.3.22) for $\dot{m}$ in Eq. (6.3.24) and simplifying gives

$$\left(\frac{p}{p_0} \right)^{2/k} \left[1 - \left(\frac{p}{p_0} \right)^{(k-1)/k} \right] = \frac{k-1}{2} \left(\frac{2}{k+1} \right)^{(k+1)/(k-1)} \left(\frac{A^*}{A} \right)^2 \tag{6.3.25}$$

A may be taken as the outlet area and p as the outlet pressure. For a given A^*/A (less than unity) there will be two values of p/p_0 between zero and unity, the upper value for subsonic flow through the diverging duct and the lower value for supersonic flow through the diverging duct. For all other pressure ratios less than the upper value complete isentropic flow is impossible and shock waves form in or just downstream from the diverging duct. They are briefly discussed in the following section.

Appendix Table C.4 is quite helpful in solving isentropic flow problems for $k = 1.4$. Equations (6.3.10), (6.3.11), (6.3.12), and (6.3.21) are presented in tabular form.

Example 6.9 A preliminary design of a wind tunnel to produce Mach number 3.0 at the exit is desired. The mass flow rate is 1 kg/s for $p_0 = 90$ kPa abs, $t_0 = 25°C$. Determine (a) the throat

area, (b) the outlet area, and (c) the velocity, pressure, temperature, and density at the outlet.
(a) The throat area can be determined from Eq. (6.3.23):

$$A^* = \frac{\dot{m}_{max}\sqrt{RT_0}}{0.686p_0} = \frac{(1 \text{ kg/s})\sqrt{(287 \text{ m}\cdot\text{N/kg}\cdot\text{K})(273 + 25 \text{ K})}}{0.686(90,000 \text{ N/m}^2)}$$

$$= 0.00474 \text{ m}^2$$

(b) The area of outlet is determined from Table C.4:

$$\frac{A}{A^*} = 4.23 \qquad A = 4.23(0.00474 \text{ m}^2) = 0.0200 \text{ m}^2$$

(c) From Table C.4

$$\frac{p}{p_0} = 0.027 \qquad \frac{\rho}{\rho_0} = 0.076 \qquad \frac{T}{T_0} = 0.357$$

From the gas law

$$\rho_0 = \frac{p_0}{RT_0} = \frac{90,000 \text{ N/m}^2}{(287 \text{ m}\cdot\text{N/kg}\cdot\text{K})(273 + 25 \text{ K})} = 1.0523 \text{ kg/m}^3$$

hence, at the exit

$$p = 0.027(90,000 \text{ N/m}^3) = 2.43 \text{ kPa abs}$$

$$T = 0.357(273 + 25 \text{ K}) = -166.6°C$$

$$\rho = 0.076(1.0523 \text{ kg/m}^3) = 0.0800 \text{ kg/m}^3$$

From the continuity equation

$$V = \frac{\dot{m}_{max}}{\rho A} = \frac{1 \text{ kg/s}}{(0.08 \text{ kg/m}^3)(0.020 \text{ m}^2)} = 625 \text{ m/s}$$

Example 6.10 A converging-diverging air duct has a throat cross section of 0.40 ft^2 and an exit cross section of 1.0 ft^2. Reservoir pressure is 30 psia, and temperature is 60°F. Determine the range of Mach numbers and the pressure range at the exit for isentropic flow. Find the maximum flow rate.
From Table C.4 [Eq. (6.3.21)] **M** = 2.44 and 0.24. Each of these values of Mach number at the exit is for critical conditions; hence, the Mach number range for isentropic flow is 0 to 0.24 and the one value 2.44.
From Table C.4 [Eq. (6.3.11)] for **M** = 2.44, $p = 30 \times 0.064 = 1.92$ psia, and for **M** = 0.24, $p = 30 \times 0.961 = 28.83$ psia. The downstream pressure range is then from 28.83 to 30 psia, and the isolated point is 1.92 psia.
The maximum mass flow rate is determined from Eq. (6.3.23):

$$\dot{m}_{max} = \frac{0.686 \times 0.40 \times 30 \times 144}{\sqrt{53.3 \times 32.17(460 + 60)}} = 1.255 \text{ slugs/s} = 40.4 \text{ lb}_m/\text{s}$$

Example 6.11 A converging-diverging duct in an air line downstream from a reservoir has a 50-mm-diameter throat. Determine the mass rate of flow when $p_0 = 0.8$ MPa abs, $t_0 = 33°C$, and $p = 0.5$ MPa abs at the throat.

$$\rho_0 = \frac{p_0}{RT} = \frac{800,000 \text{ N/m}^2}{(287 \text{ m}\cdot\text{N/kg}\cdot\text{K})(273 + 33 \text{ K})} = 9.109 \text{ kg/m}^3$$

From Eq. (6.3.24)

$$\dot{m} = \frac{\pi}{4}(0.05 \text{ m})^2 \sqrt{2(800{,}000 \text{ N/m}^2)(9.109 \text{ kg/m}^3)\frac{1.4}{1.4-1}\left(\frac{5}{8}\right)^{2/1.4}}$$

$$\times \sqrt{\left[1 - \left(\frac{5}{8}\right)^{0.4/1.4}\right]} = 3.554 \text{ kg/s}$$

EXERCISES

6.3.1 The differential equation for energy in isentropic flow may take the form

(a) $dp + d(\rho V^2) = 0$ (b) $\dfrac{dV}{V} + \dfrac{d\rho}{\rho} + \dfrac{dA}{A} = 0$ (c) $2V\,dV + \dfrac{dp}{\rho} = 0$

(d) $V\,dV + \dfrac{dp}{\rho} = 0$ (e) none of these answers

6.3.2 In isentropic flow, the temperature (a) cannot exceed the reservoir temperature; (b) cannot drop and then increase again downstream; (c) is independent of the Mach number; (d) is a function of Mach number only; (e) remains constant in duct flow.

6.3.3 The critical pressure ratio for isentropic flow of carbon monoxide is (a) 0.528; (b) 0.634; (c) 0.833; (d) 1.0; (e) none of these answers.

6.3.4 Select the correct statement regarding flow through a converging-diverging tube. (a) When the Mach number at exit is greater than unity, no shock wave has developed in the tube. (b) When the critical pressure ratio is exceeded, the Mach number at the throat is greater than unity. (c) For sonic velocity at the throat, one and only one pressure or velocity can occur at a given downstream location. (d) The Mach number at the throat is always unity. (e) The density increases in the downstream direction throughout the converging portion of the tube.

6.3.5 Select the correct statement regarding frictionless flow: (a) In diverging conduits the velocity always decreases; (b) The velocity is always sonic at the throat of a converging-diverging tube; (c) In supersonic flow the area decreases for increasing velocity; (d) Sonic velocity cannot be exceeded at the throat of a converging-diverging tube; (e) At Mach zero the velocity is sonic.

6.4 SHOCK WAVES

In one-dimensional flow the only type of shock wave that can occur is a normal compression shock wave, as illustrated in Fig. 6.2. For a complete discussion of converging-diverging flow for all downstream pressure ranges,† oblique shock waves must be taken into account as they occur at the exit. In the preceding section isentropic flow was shown to occur throughout a converging-diverging tube for a range of downstream pressures in which the flow was subsonic through-out and for one downstream pressure for supersonic flow through the diffuser (diverging portion). In this section the normal shock wave in a diffuser is studied, with isentropic flow throughout the tube, except for the shock wave surface. The shock wave occurs in supersonic flow and reduces the flow to subsonic flow, as

† H. W. Liepmann and A. Roshko, "Elements of Gas Dynamics," Wiley, New York, 1957.

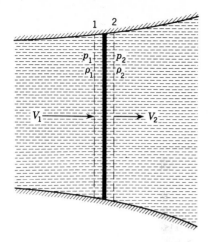

Figure 6.2 Normal compression shock wave.

proved in the following section. It has very little thickness, of the order of the molecular mean free path of the gas. The controlling equations for adiabatic flow are (Fig. 6.2)

Continuity:
$$G = \frac{\dot{m}}{A} = \rho_1 V_1 = \rho_2 V_2 \tag{6.4.1}$$

Energy:
$$\frac{V_1^2}{2} + h_1 = \frac{V_2^2}{2} + h_2 = h_0 = \frac{V^2}{2} + \frac{k}{k-1}\frac{p}{\rho} \tag{6.4.2}$$

which are obtained from Eq. (3.8.1) for no change in elevation, no heat transfer, and no work done. $h = u + p/\rho = c_p T$ is the enthalpy, and h_0 is the value of stagnation enthalpy, i.e., its value in the reservoir or where the fluid is at rest. Equation (6.4.2) holds for real fluids and is valid both upstream and downstream from a shock wave. The momentum equation (3.11.2) for a control volume between sections 1 and 2 becomes

$$(p_1 - p_2)A = \rho_2 A V_2^2 - \rho_1 A V_1^2$$

or
$$p_1 + \rho_1 V_1^2 = p_2 + \rho_2 V_2^2 \tag{6.4.3}$$

For given upstream conditions h_1, p_1, V_1, ρ_1 the three equations are to be solved for p_2, ρ_2, V_2. The equation of state for a perfect gas is also available for use, $p = \rho RT$. The value of p_2 is

$$p_2 = \frac{1}{k+1}[2\rho_1 V_1^2 - (k-1)p_1] \tag{6.4.4}$$

Once p_2 is determined, by combination of the continuity and momentum equations

$$p_1 + \rho_1 V_1^2 = p_2 + \rho_1 V_1 V_2 \tag{6.4.5}$$

V_2 is readily obtained. Finally, ρ_2 is obtained from the continuity equation.

For given upstream conditions, with $M_1 > 1$, the values of p_2, V_2, ρ_2, and $M_2 = V_2/\sqrt{kp_2/\rho_2}$ exist and $M_2 < 1$. By eliminating V_1 and V_2 between Eqs. (6.4.1), (6.4.2), and (6.4.3), the Rankine-Hugoniot equations are obtained:

$$\frac{p_2}{p_1} = \frac{[(k+1)/(k-1)](\rho_2/\rho_1) - 1}{[(k+1)/(k-1)] - \rho_2/\rho_1} \tag{6.4.6}$$

and

$$\frac{\rho_2}{\rho_1} = \frac{1 + [(k+1)/(k-1)]p_2/p_1}{[(k+1)/(k-1)] + p_2/p_1} = \frac{V_1}{V_2} \tag{6.4.7}$$

These equations, relating conditions on either side of the shock wave, take the place of the isentropic relation, Eq. (6.1.16), $p\rho^{-k} = $ const.

From Eq. (6.4.2), the energy equation,

$$\frac{V^2}{2} + \frac{k}{k-1}\frac{p}{\rho} = \frac{c^{*2}}{2} + \frac{c^{*2}}{k-1} = \frac{k+1}{k-1}\frac{c^{*2}}{2} \tag{6.4.8}$$

since the equation holds for all points in adiabatic flow without change in elevation, and $c^* = \sqrt{kp^*/\rho^*}$ is the velocity of sound. Dividing Eq. (6.4.3) by Eq. (6.4.1) gives

$$V_1 - V_2 = \frac{p_2}{\rho_2 V_2} - \frac{p_1}{\rho_1 V_1}$$

and, by eliminating p_2/ρ_2 and p_1/ρ_1 by use of Eq. (6.4.8), leads to

$$V_1 - V_2 = (V_1 - V_2)\left[\frac{c^{*2}(k+1)}{2kV_1 V_2} + \frac{k-1}{2k}\right] \tag{6.4.9}$$

which is satisfied by $V_1 = V_2$ (no shock wave) or by

$$V_1 V_2 = c^{*2} \tag{6.4.10}$$

It may be written

$$\frac{V_1}{c^*}\frac{V_2}{c^*} = 1 \tag{6.4.11}$$

When V_1 is greater than c^*, the upstream Mach number is greater than unity and V_2 is less than c^*, and so the final Mach number is less than unity, and vice versa. It is shown in the following section that the process can occur only from supersonic upstream to subsonic downstream.

By use of Eq. (6.1.14), together with Eqs. (6.4.4), (6.4.6), and (6.4.7), an expression for change of entropy across a normal shock wave may be obtained in terms of M_1 and k. From Eq. (6.4.4)

$$\frac{p_2}{p_1} = \frac{1}{k+1}\left[\frac{2k\rho_1 V_1^2}{kp_1} - (k-1)\right] \tag{6.4.12}$$

Since $c_1^2 = kp_1/\rho_1$ and $\mathbf{M}_1 = V_1/c_1$, from Eq. (6.4.12),

$$\frac{p_2}{p_1} = \frac{2k\mathbf{M}_1^2 - (k-1)}{k+1} \qquad (6.4.13)$$

Placing this value of p_2/p_1 in Eq. (6.4.7) yields

$$\frac{\rho_2}{\rho_1} = \frac{\mathbf{M}_1^2(k+1)}{2 + \mathbf{M}_1^2(k-1)}$$

Substituting these pressure and density ratios into Eq. (6.1.14) gives

$$s_2 - s_1 = c_v \ln\left\{ \frac{2k\mathbf{M}_1^2 - k + 1}{k+1}\left[\frac{2 + \mathbf{M}_1^2(k-1)}{\mathbf{M}_1^2(k+1)}\right]^k \right\} \qquad (6.4.14)$$

By substitution of $\mathbf{M}_1 > 1$ into this equation for the appropriate value of k, the entropy may be shown to increase across the shock wave, indicating that the normal shock may proceed from supersonic flow upstream to subsonic flow downstream. Substitution of values of $\mathbf{M}_1 < 1$ into Eq. (6.4.14) has no meaning, since Eq. (6.4.13) yields a negative value of the ratio p_2/p_1.

Example 6.12 If a normal shock wave occurs in the flow of helium, $p_1 = 1$ psia, $t_1 = 40°F$, $V_1 = 4500$ ft/s, find p_2, ρ_2, V_2, and t_2.
From Table C.3, $R = 386$, $k = 1.66$, and

$$\rho_1 = \frac{p_1}{RT_1} = \frac{1 \times 144}{386 \times 32.17(460 + 40)} = 0.0000232 \text{ slug/ft}^3$$

From Eq. (6.4.4)

$$p_2 = \frac{1}{1.66 + 1}[2 \times 0.0000232 \times 4500^2 - (1.66 - 1) \times 144 \times 1]$$

$$= 317 \text{ lb/ft}^2 \text{ abs}$$

From Eq. (6.4.5)

$$V_2 = V_1 - \frac{p_2 - p_1}{\rho_1 V_1} = 4500 - \frac{317 - 144}{4500 \times 0.0000232} = 2843 \text{ ft/s}$$

From Eq. (6.4.1)

$$\rho_2 = \rho_1 \frac{V_1}{V_2} = 0.0000232 \times \frac{4500}{2843} = 0.0000367 \text{ slug/ft}^3$$

and

$$t_2 = T_2 - 460 = \frac{p_2}{\rho_2 R} - 460 = \frac{317}{0.0000367 \times 32.17 \times 386} - 460 = 236°F$$

EXERCISES

6.4.1 In a normal shock wave in one-dimensional flow the (a) velocity, pressure, and density increase; (b) pressure, density, and temperature increase; (c) velocity, temperature, and density increase; (d) pressure, density, and momentum per unit time increase; (e) entropy remains constant.

6.4.2 A normal shock wave (a) is reversible; (b) may occur in a converging tube; (c) is irreversible; (d) is isentropic; (e) is none of these answers.

6.4.3 Across a normal shock wave in a converging-diverging nozzle for adiabatic flow the following relations are valid: (a) continuity and energy equations, equation of state, isentropic relation; (b) energy and momentum equations, equation of state, isentropic relation; (c) continuity, energy, and momentum equations; equation of state; (d) equation of state, isentropic relation, momentum equation, mass-conservation principle; (e) none of these answers.

6.4.4 Across a normal shock wave there is an increase in (a) p, $\mathbf{M}$, s; (b) p, s; decrease in $\mathbf{M}$; (c) p; decrease in s, $\mathbf{M}$; (d) p, $\mathbf{M}$; no change in s; (e) p, $\mathbf{M}$, T.

6.5 FANNO AND RAYLEIGH LINES

To examine more closely the nature of the flow change in the short distance across a shock wave, where the area may be considered constant, the continuity and energy equations are combined for steady, frictional, adiabatic flow. By considering upstream conditions fixed, that is, p_1, V_1, ρ_1, a plot may be made of all possible conditions at section 2, Fig. 6.2. The lines on such a plot for constant mass flow G are called *Fanno lines*. The most revealing plot is that of enthalpy against entropy, i.e., an hs diagram.

The entropy equation for a perfect gas, Eq. (6.1.14), is

$$s - s_1 = c_v \ln\left[\frac{p}{p_1}\left(\frac{\rho_1}{\rho}\right)^k\right] \tag{6.5.1}$$

The energy equation for adiabatic flow with no change in elevation, from Eq. (6.4.2), is

$$h_0 = h + \frac{V^2}{2} \tag{6.5.2}$$

and the continuity equation for no change in area, from Eq. (6.4.1), is

$$G = \rho V \tag{6.5.3}$$

The equation of state, linking h, p, and ρ, is

$$h = c_p T = \frac{c_p p}{R\rho} \tag{6.5.4}$$

By eliminating p, ρ, and V from the four equations,

$$s = s_1 + c_v \ln\left[\frac{\rho_1^k}{p_1}\frac{R}{c_p}\left(\frac{\sqrt{2}}{G}\right)^{k-1}\right] + c_v \ln\left[h(h_0 - h)^{(k-1)/2}\right] \tag{6.5.5}$$

which is shown on Fig. 6.3 (not to scale). To find the conditions for maximum entropy, Eq. (6.5.5) is differentiated with respect to h and ds/dh is set equal to zero. By indicating by subscript a values at the maximum entropy point,

$$\frac{ds}{dh} = 0 = \frac{1}{h_a} - \frac{k-1}{2}\frac{1}{h_0 - h_a} \qquad \text{or} \qquad h_a = \frac{2}{k+1}h_0$$

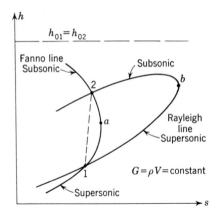

Figure 6.3 Fanno and Rayleigh lines.

After substituting this into Eq. (6.5.2) to find V_a,

$$h_0 = \frac{k+1}{2} h_a = h_a + \frac{V_a^2}{2}$$

and

$$V_a^2 = (k-1)h_a = (k-1)c_p T_a = (k-1)\frac{kR}{k-1} T_a = kRT_a = c_a^2 \qquad (6.5.6)$$

Hence, the maximum entropy at point a is for $\mathbf{M} = 1$, or sonic conditions. For $h > h_a$ the flow is subsonic, and for $h < h_a$ the flow is supersonic. The two conditions, before and after the shock, must lie on the proper Fanno line for the area at which the shock wave occurs. The momentum equation was not used to determine the Fanno line, and so the complete solution is not determined yet.

Rayleigh Line

Conditions before and after the shock must also satisfy the momentum and continuity equations. Assuming constant upstream conditions and constant area, Eqs. (6.5.1), (6.5.3), (6.5.4), and (6.4.1) are used to determine the *Rayleigh line.* Eliminating V in the continuity and momentum equations gives

$$p + \frac{G^2}{\rho} = \text{const} = B \qquad (6.5.7)$$

Next, eliminating p from this equation and the entropy equation gives

$$s = s_1 + c_v \ln \frac{\rho_1^k}{p_1} + c_v \ln \frac{B - G^2/\rho}{\rho^k} \qquad (6.5.8)$$

Enthalpy may be expressed as a function of ρ and upstream conditions, from Eq. (6.5.7):

$$h = c_p T = c_p \frac{p}{R\rho} = \frac{c_p}{R} \frac{1}{\rho}\left(B - \frac{G^2}{\rho}\right) \qquad (6.5.9)$$

The last two equations determine s and h in terms of the parameter ρ and plot on the hs diagram as indicated in Fig. 6.3. This is a *Rayleigh line*. The value of maximum entropy is found by taking $ds/d\rho$ and $dh/d\rho$ from the equations; then by division and equating to zero, using subscript b for maximum point:

$$\frac{ds}{dh} = \frac{c_v}{c_p} R\rho_b \frac{G^2/[\rho_b(B - G^2/\rho_b)] - k}{2G^2/\rho_b - B} = 0$$

To satisfy this equation, the numerator must be zero and the denominator not zero. The numerator set equal to zero yields

$$k = \frac{G^2}{\rho_b(B - G^2/\rho_b)} = \frac{\rho_b^2 V_b^2}{\rho_b p_b} \qquad \text{or} \qquad V_b^2 = \frac{kp_b}{\rho_b} = c_b^2$$

that is, $\mathbf{M} = 1$. For this value the denominator is not zero. Again, as with the Fanno line, sonic conditions occur at the point of maximum entropy. Since the flow conditions must be on both curves, just before and just after the shock wave, it must suddenly change from one point of intersection to the other. The entropy cannot decrease, as no heat is being transferred from the flow, so that the upstream point must be the intersection with least entropy. In all gases investigated the intersection in the subsonic flow has the greater entropy. Thus the shock occurs from supersonic to subsonic.

The Fanno and Rayleigh lines are of value in analyzing flow in constant-area ducts. These are treated in Secs. 6.6 and 6.7.

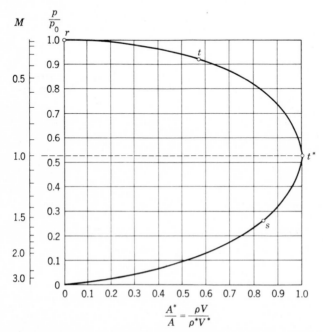

Figure 6.4 Isentropic relations for a converging-diverging nozzle ($k = 1.4$). (*By permission, from H. W. Liepmann and A. Roshko, "Elements of Gas Dynamics," John Wiley & Sons, Inc., New York, 1957.*)

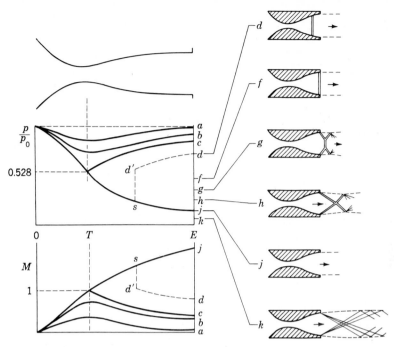

Figure 6.5 Various pressure and Mach number configurations for flow through a nozzle. (*By permission from H. W. Liepmann and A. Roshko, "Elements of Gas Dynamics," John Wiley & Sons, Inc., New York, 1957.*)

Converging-Diverging Nozzle Flow

Following the presentation of Liepmann and Roshko (see references at end of chapter), the various flow situations for converging-diverging nozzles are investigated. Equation (6.3.20) gives the relation between area ratio and Mach number for isentropic flow throughout the nozzle. By use of Eq. (6.3.11) the area ratio is obtained as a function of pressure ratio

$$\frac{A^*}{A} = \frac{\rho V}{\rho^* V^*} = \frac{[1 - (p/p_0)^{(k-1)/k}]^{1/2}(p/p_0)^{1/k}}{[(k-1)/2]^{1/2}[2/(k+1)]^{(k+1)/2(k-1)}} \qquad (6.5.10)$$

Figure 6.4 is a plot of area ratio vs. pressure ratio and **M**, good only for isentropic flow ($k = 1.4$).

By use of the area ratios the distribution of pressure and Mach number along a given converging-diverging nozzle can now be plotted. Figure 6.5 illustrates the various flow conditions that may occur. If the downstream pressure is p_c or greater, isentropic subsonic flow occurs throughout the tube. If the pressure is at j, isentropic flow occurs throughout, with subsonic flow to the throat, sonic flow at the throat, and supersonic flow downstream. For downstream pressure between c and f, a shock wave occurs within the nozzle, as shown for p_d. For pressure at p_f a normal shock wave occurs at the exit; and for pressures between p_f and p_j oblique shock waves at the exit develop.

EXERCISES

6.5.1 A Fanno line is developed from the following equations: (a) momentum and continuity; (b) energy and continuity; (c) momentum and energy; (d) momentum, continuity, and energy; (e) none of these answers.

6.5.2 A Rayleigh line is developed from the following equations: (a) momentum and continuity; (b) energy and continuity; (c) momentum and energy; (d) momentum, continuity, and energy; (e) none of these answers.

6.5.3 Select the correct statement regarding a Fanno or Rayleigh line: (a) Two points having the same value of entropy represent conditions before and after a shock wave. (b) pV is held constant along the line. (c) Mach number always increases with entropy. (d) The subsonic portion of the curve is at higher enthalpy than the supersonic portion. (e) Mach 1 is located at the maximum enthalpy point.

6.6 ADIABATIC FLOW WITH FRICTION IN CONDUITS

Gas flow through a pipe or constant-area duct is analyzed in this section subject to the following assumptions:

1. Perfect gas (constant specific heats).
2. Steady, one-dimensional flow.
3. Adiabatic flow (no heat transfer through walls).
4. Constant friction factor over length of conduit.
5. Effective conduit diameter D is four times hydraulic radius (cross-sectioned area divided by perimeter).
6. Elevation changes are unimportant compared with friction effects.
7. No work added to or extracted from the flow.

The controlling equations are continuity, energy, momentum, and the equation of state. The Fanno line, developed in Sec. 6.5 and shown in Fig. 6.3, was for constant area and used the continuity and energy equations; hence, it applies to adiabatic flow in a duct of constant area. A particle of gas at the upstream end of the duct may be represented by a point on the appropriate Fanno line for proper stagnation enthalpy h_0 and mass flow rate G per unit area. As the particle moves downstream, its properties change, owing to friction or irreversibilities such that the entropy always increases in adiabatic flow. Thus the point representing these properties moves along the Fanno line toward the maximum s point, where $\mathbf{M} = 1$. If the duct is fed by a converging-diverging nozzle, the flow may originally be supersonic; the velocity must then decrease downstream. If the flow is subsonic at the upstream end, the velocity must increase in the downstream direction.

For exactly one length of pipe, depending upon upstream conditions, the flow is just sonic ($\mathbf{M} = 1$) at the downstream end. For shorter lengths of pipe, the flow will not have reached sonic conditions at the outlet, but for longer lengths of pipe, there must be shock waves (and possibly *choking*) if supersonic and choking effects if subsonic. Choking means that the mass flow rate specified cannot take place in this situation and less flow will occur. Table 6.1 indicates the trends in properties

Table 6.1

Property	Subsonic flow	Supersonic flow
Velocity V	Increases	Decreases
Mach number **M**	Increases	Decreases
Pressure p	Decreases	Increases
Temperature T	Decreases	Increases
Density ρ	Decreases	Increases
Stagnation enthalpy	Constant	Constant
Entropy	Increases	Increases

of a gas in adiabatic flow through a constant-area duct, as can be shown from the equations in this section.

The gas cannot change gradually from subsonic to supersonic or vice versa in a constant-area duct.

The momentum equation must now include the effects of wall shear stress and is conveniently written for a segment of duct of length δx (Fig. 6.6):

$$pA - \left(p + \frac{dp}{dx}\,\delta x\right)A - \tau_0 \pi D\,\delta x = \rho V A\left(V + \frac{dV}{dx}\,\delta x - V\right)$$

Upon simplification,

$$dp + \frac{4\tau_0}{D}\,dx + \rho V\,dV = 0 \qquad (6.6.1)$$

By use of Eq. (5.10.2), $\tau_0 = \rho f V^2/8$, in which f is the Darcy-Weisbach friction factor,

$$dp + \frac{f\rho V^2}{2D}\,dx + \rho V\,dV = 0 \qquad (6.6.2)$$

For constant f, or average value over the length of reach, this equation can be transformed into an equation for x as a function of Mach number. By dividing Eq. (6.6.2) by p,

$$\frac{dp}{p} + \frac{f}{2D}\frac{\rho V^2}{p}\,dx + \frac{\rho V}{p}\,dV = 0 \qquad (6.6.3)$$

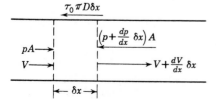

Figure 6.6 Notation for application of momentum equation.

Each term is now developed in terms of $\mathbf{M}$. By definition $V/c = \mathbf{M}$,

$$V^2 = \mathbf{M}^2 \frac{kp}{\rho} \tag{6.6.4}$$

or

$$\frac{\rho V^2}{p} = k\mathbf{M}^2 \tag{6.6.5}$$

for the middle term of the momentum equation. Rearranging Eq. (6.6.4) gives

$$\frac{\rho V}{p} dV = k\mathbf{M}^2 \frac{dV}{V} \tag{6.6.6}$$

Now to express dV/V in terms of $\mathbf{M}$, from the energy equation,

$$h_0 = h + \frac{V^2}{2} = c_p T + \frac{V^2}{2} \tag{6.6.7}$$

Differentiating gives

$$c_p \, dT + V \, dV = 0 \tag{6.6.8}$$

Dividing through by $V^2 = \mathbf{M}^2 kRT$ yields

$$\frac{c_p}{R} \frac{1}{k\mathbf{M}^2} \frac{dT}{T} + \frac{dV}{V} = 0$$

Since $c_p/R = k/(k-1)$,

$$\frac{dT}{T} = -\mathbf{M}^2(k-1) \frac{dV}{V} \tag{6.6.9}$$

Differentiating $V^2 = \mathbf{M}^2 kRT$ and dividing by the equation give

$$2 \frac{dV}{V} = 2 \frac{d\mathbf{M}}{\mathbf{M}} + \frac{dT}{T} \tag{6.6.10}$$

Eliminating dT/T in Eqs. (6.6.9) and (6.6.10) and simplifying lead to

$$\frac{dV}{V} = \frac{d\mathbf{M}/\mathbf{M}}{[(k-1)/2]\mathbf{M}^2 + 1} \tag{6.6.11}$$

which permits elimination of dV/V from Eq. (6.6.6), yielding

$$\frac{\rho V}{p} dV = \frac{k\mathbf{M} \, d\mathbf{M}}{[(k-1)/2]\mathbf{M}^2 + 1} \tag{6.6.12}$$

And finally, to express dp/p in terms of $\mathbf{M}$, from $p = \rho RT$ and $G = \rho V$,

$$pV = GRT \tag{6.6.13}$$

By differentiation

$$\frac{dp}{p} = \frac{dT}{T} - \frac{dV}{V}$$

Equations (6.6.9) and (6.6.11) are used to eliminate dT/T and dV/V:

$$\frac{dp}{p} = -\frac{(k-1)\mathbf{M}^2 + 1}{[(k-1)/2]\mathbf{M}^2 + 1}\frac{d\mathbf{M}}{\mathbf{M}} \tag{6.6.14}$$

Equations (6.6.5), (6.6.12), and (6.6.14) are now substituted into the momentum equation (6.6.3). After rearranging,

$$\frac{f}{D}dx = \frac{2(1-\mathbf{M}^2)}{k\mathbf{M}^3\{[(k-1)/2]\mathbf{M}^2 + 1\}}d\mathbf{M}$$

$$= \frac{2}{k}\frac{d\mathbf{M}}{\mathbf{M}^3} - \frac{k+1}{k}\frac{d\mathbf{M}}{\mathbf{M}\{[(k-1)/2]\mathbf{M}^2 + 1\}} \tag{6.6.15}$$

which can be integrated directly. By using the limits $x = 0$, $\mathbf{M} = \mathbf{M}_0$, $x = l$, $\mathbf{M} = \mathbf{M}$,

$$\frac{fl}{D} = -\frac{1}{k\mathbf{M}^2}\bigg]_{\mathbf{M}_0}^{\mathbf{M}} - \frac{k+1}{2k}\ln\frac{\mathbf{M}^2}{[(k-1)/2]\mathbf{M}^2 + 1}\bigg]_{\mathbf{M}_0}^{\mathbf{M}} \tag{6.6.16}$$

$$= \frac{1}{k}\left(\frac{1}{\mathbf{M}_0^2} - \frac{1}{\mathbf{M}^2}\right) + \frac{k+1}{2k}\ln\left[\left(\frac{\mathbf{M}_0}{\mathbf{M}}\right)^2\frac{(k-1)\mathbf{M}^2 + 2}{(k-1)\mathbf{M}_0^2 + 2}\right] \tag{6.6.17}$$

For $k = 1.4$, this reduces to

$$\frac{fl}{D} = \frac{5}{7}\left(\frac{1}{\mathbf{M}_0^2} - \frac{1}{\mathbf{M}^2}\right) + \frac{6}{7}\ln\left[\left(\frac{\mathbf{M}_0}{\mathbf{M}}\right)^2\frac{\mathbf{M}^2 + 5}{\mathbf{M}_0^2 + 5}\right] \qquad k = 1.4 \tag{6.6.18}$$

If $\mathbf{M}_0$ is greater than 1, $\mathbf{M}$ cannot be less than 1; and if $\mathbf{M}_0$ is less than 1, $\mathbf{M}$ cannot be greater than 1. For the limiting condition $\mathbf{M} = 1$ and $k = 1.4$,

$$\frac{fl_{\text{max}}}{D} = \frac{5}{7}\left(\frac{1}{\mathbf{M}_0^2} - 1\right) + \frac{6}{7}\ln\frac{6\mathbf{M}_0^2}{\mathbf{M}_0^2 + 5} \qquad k = 1.4 \tag{6.6.19}$$

There is some evidence[†] to indicate that friction factors may be smaller in supersonic flow.

Example 6.13 Determine the maximum length of 50-mm-ID pipe, $f = 0.02$ for flow of air, when the Mach number at the entrance to the pipe is 0.30.
From Eq. (6.6.19)

$$\frac{0.02}{0.05}L_{\text{max}} = \frac{5}{7}\left(\frac{1}{0.3^2} - 1\right) + \frac{6}{7}\ln\frac{6 \times 0.30^2}{0.30^2 + 5}$$

from which $L_{\text{max}} = 13.25$ m.

The pressure, velocity, and temperature may also be expressed in integral form in terms of the Mach number. To simplify the equations that follow, they will

---

[†] J. H. Keenan and E. P. Neumann, Measurements of Friction in a Pipe for Subsonic and Supersonic Flow of Air, *J. Appl. Mech.*, vol. 13, no. 2, p. A-91, 1946.

be integrated from upstream conditions to conditions at $\mathbf{M} = 1$, indicated by p^*, V^*, and T^*. From Eq. (6.6.14)

$$\frac{p^*}{p_1} = \mathbf{M}_0 \sqrt{\frac{(k-1)\mathbf{M}_0^2 + 2}{k+1}} \qquad (6.6.20)$$

From Eq. (6.6.11)

$$\frac{V^*}{V_0} = \frac{1}{\mathbf{M}_0} \sqrt{\frac{(k-1)\mathbf{M}_0^2 + 2}{k+1}} \qquad (6.6.21)$$

From Eqs. (6.6.9) and (6.6.11)

$$\frac{dT}{T} = -(k-1)\frac{\mathbf{M}\, d\mathbf{M}}{[(k-1)/2]\mathbf{M}^2 + 1}$$

which, when integrated, yields

$$\frac{T^*}{T_0} = \frac{(k-1)\mathbf{M}_0^2 + 2}{k+1} \qquad (6.6.22)$$

Example 6.14 A 4.0-in-ID pipe, $f = 0.020$, has air at 14.7 psia and at $t = 60°F$ flowing at the upstream end with Mach number 3.0. Determine L_{max}, p^*, V^*, T^*, and values of p_0', V_0, T_0, and L at $\mathbf{M} = 2.0$.
From Eq. (6.6.19)

$$\frac{0.02}{0.333} L_{max} = \frac{5}{7}\left(\frac{1}{9} - 1\right) + \frac{6}{7} \ln \frac{6 \times 3^2}{3^2 + 5}$$

from which $L_{max} = 8.69$ ft. If the flow originated at $\mathbf{M} = 2$, the length L_{max} is given by the same equation:

$$\frac{0.02}{0.333} L_{max} = \frac{5}{7}\left(\frac{1}{4} - 1\right) + \frac{6}{7} \ln \frac{6 \times 2^2}{2^2 + 5}$$

from which $L_{max} = 5.08$ ft.
Hence, the length from the upstream section at $\mathbf{M} = 3$ to the section where $\mathbf{M} = 2$ is $8.69 - 5.08 = 3.61$ ft.
The velocity at the entrance is

$$V = \sqrt{kRT}\,\mathbf{M} = \sqrt{1.4 \times 53.3 \times 32.17(460 + 60)} \times 3 = 3352 \text{ ft/s}$$

From Eqs. (6.6.20) to (6.6.22)

$$\frac{p^*}{14.7} = 3\sqrt{\frac{0.4 \times 3^2 + 2}{2.4}} = 4.583$$

$$\frac{V^*}{3352} = \frac{1}{3}\sqrt{\frac{0.4 \times 3^2 + 2}{2.4}} = 0.509$$

$$\frac{T^*}{520} = \frac{0.4 \times 3^2 + 2}{2.4} = \frac{7}{3}$$

So $p^* = 67.4$ psia, $V^* = 1707$ ft/s, $T^* = 1213°$R. For $\mathbf{M} = 2$ the same equations are now solved for p'_0, V'_0, and T'_0:

$$\frac{67.4}{p'_0} = 2\sqrt{\frac{0.4 \times 2^2 + 2}{2.4}} = 2.45$$

$$\frac{1707}{V'_0} = \frac{1}{2}\sqrt{\frac{0.4 \times 2^2 + 2}{2.4}} = 0.6124$$

$$\frac{1213}{T'_0} = \frac{0.4 \times 2^2 + 2}{2.4} = \frac{3}{2}$$

So $p'_0 = 27.5$ psia, $V'_0 = 2787$ ft/s, and $T'_0 = 809°$R.

EXERCISES

6.6.1 Choking in pipe flow means that (a) a valve is closed in the line; (b) a restriction in flow area occurs; (c) the specified mass flow rate cannot occur; (d) shock waves always occur; (e) supersonic flow occurs somewhere in the line.

6.6.2 In subsonic adiabatic flow with friction in a pipe (a) V, $\mathbf{M}$, s increase; p, T, ρ decrease; (b) p, V, $\mathbf{M}$ increase; T, ρ decrease; (c) p, $\mathbf{M}$, s increase; V, T, ρ decrease; (d) ρ, $\mathbf{M}$, s increase; V, T, p decrease; (e) T, V, s increase; $\mathbf{M}$, p, ρ decrease.

6.6.3 In supersonic adiabatic flow with friction in a pipe (a) V, $\mathbf{M}$, s increase; p, T, ρ decrease; (b) p, T, s increase; ρ, V, $\mathbf{M}$ decrease; (c) p, $\mathbf{M}$, s increase; V, T, ρ decrease; (d) p, T, ρ, s increase; V, $\mathbf{M}$ decrease; (e) p, ρ, s increase; V, $\mathbf{M}$, T decrease.

6.7 FRICTIONLESS FLOW THROUGH DUCTS WITH HEAT TRANSFER

The steady flow of a perfect gas (with constant specific heats) through a constant-area duct is considered in this section. Friction is neglected, and no work is done on or by the flow.

The appropriate equations for analysis of this case are

Continuity: $$G = \frac{\dot{m}}{A} = \rho V \tag{6.7.1}$$

Momentum: $$p + \rho V^2 = \text{const} \tag{6.7.2}$$

Energy: $$q_H = h_2 - h_1 + \frac{V_2^2 - V_1^2}{2}$$

$$= c_p(T_2 - T_1) + \frac{V_2^2 - V_1^2}{2}$$

$$= c_p(T_{02} - T_{01}) \tag{6.7.3}$$

T_{01} and T_{02} are the isentropic stagnation temperatures, i.e., the temperatures produced at a section by bringing the flow isentropically to rest.

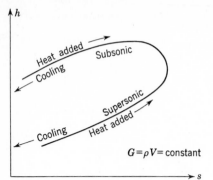

$$G = \rho V = \text{constant}$$

Figure 6.7 Rayleigh line.

The Rayleigh line, obtained from the solution of momentum and continuity for a constant cross section by neglecting friction, is very helpful in examining the flow. First, eliminating V in Eqs. (6.7.1) and (6.7.2) gives

$$p + \frac{G^2}{\rho} = \text{const} \tag{6.7.4}$$

which is Eq. (6.5.7). Equations (6.5.8) and (6.5.9) express the entropy s and enthalpy h in terms of the parameter ρ for the assumptions of this section, as in Fig. 6.7.

Since by Eq. (3.8.4), for no losses, entropy can increase only when heat is added, the properties of the gas must change as indicated in Fig. 6.7, moving toward the maximum entropy point as heat is added. At the maximum s point there is no change in entropy for a small change in h, and isentropic conditions apply to the point. The speed of sound under isentropic conditions is given by $c = \sqrt{dp/d\rho}$ as given by Eq. (6.2.2). From Eq. (6.7.4), by differentiation,

$$\frac{dp}{d\rho} = \frac{G^2}{\rho^2} = V^2$$

using Eq. (6.7.1). Hence at the maximum s point of the Rayleigh line, $V = \sqrt{dp/d\rho}$ also and $\mathbf{M} = 1$, or sonic conditions prevail. The addition of heat to supersonic flow causes the Mach number of the flow to decrease toward $\mathbf{M} = 1$, and if just the proper amount of heat is added, $\mathbf{M}$ becomes 1. If more heat is added, choking results and conditions at the upstream end are altered to reduce the mass rate of flow. The addition of heat to subsonic flow causes an increase in the Mach number toward $\mathbf{M} = 1$, and again, too much heat transfer causes choking with an upstream adjustment of mass flow rate to a smaller value.

From Eq. (6.7.3) it is noted that the increase in isentropic stagnation pressure is a measure of the heat added. From $V^2 = \mathbf{M}^2 kRT$, $p = \rho RT$, and continuity,

$$pV = GRT \qquad \text{and} \qquad \rho V^2 = kp\mathbf{M}^2$$

Now, from the momentum equation,

$$p_1 + kp_1 \mathbf{M}_1^2 = p_2 + kp_2 \mathbf{M}_2^2$$

and
$$\frac{p_1}{p_2} = \frac{1 + k\mathbf{M}_2^2}{1 + k\mathbf{M}_1^2} \tag{6.7.5}$$

Writing this equation for the limiting case $p_2 = p^*$ when $\mathbf{M}_2 = 1$ gives

$$\frac{p}{p^*} = \frac{1 + k}{1 + k\mathbf{M}^2} \tag{6.7.6}$$

with p the pressure at any point in the duct where $\mathbf{M}$ is the corresponding Mach number. For the subsonic case, with $\mathbf{M}$ increasing to the right (Fig. 6.7), p must decrease; and for the supersonic case, as $\mathbf{M}$ decreases toward the right, p must increase.

To develop the other pertinent relations, the energy equation (6.7.3) is used,

$$c_p T_0 = \frac{kR}{k-1} T_0 = \frac{kR}{k-1} T + \frac{V^2}{2}$$

in which T_0 is the isentropic stagnation temperature and T the free-stream temperature at the same section. Applying this to section 1, after dividing through by $kRT_1/(k-1)$, yields

$$\frac{T_{01}}{T_1} = 1 + (k-1)\frac{\mathbf{M}_1^2}{2} \tag{6.7.7}$$

and for section 2

$$\frac{T_{02}}{T_2} = 1 + (k-1)\frac{\mathbf{M}_2^2}{2} \tag{6.7.8}$$

Dividing Eq. (6.7.7) by Eq. (6.7.8) gives

$$\frac{T_{01}}{T_{02}} = \frac{T_1}{T_2}\frac{2 + (k-1)\mathbf{M}_1^2}{2 + (k-1)\mathbf{M}_2^2} \tag{6.7.9}$$

The ratio T_1/T_2 is determined in terms of the Mach numbers as follows. From the perfect-gas law, $p_1 = \rho_1 RT_1$, $p_2 = \rho_2 RT_2$,

$$\frac{T_1}{T_2} = \frac{p_1}{p_2}\frac{\rho_2}{\rho_1} \tag{6.7.10}$$

From continuity, $\rho_2/\rho_1 = V_1/V_2$; and by definition

$$\mathbf{M}_1 = \frac{V_1}{\sqrt{kRT_1}} \qquad \mathbf{M}_2 = \frac{V_2}{\sqrt{kRT_2}}$$

so that
$$\frac{V_1}{V_2} = \frac{\mathbf{M}_1}{\mathbf{M}_2}\sqrt{\frac{T_1}{T_2}}$$

and
$$\frac{\rho_2}{\rho_1} = \frac{\mathbf{M}_1}{\mathbf{M}_2}\sqrt{\frac{T_1}{T_2}} \tag{6.7.11}$$

Now substituting Eqs. (6.7.5) and (6.7.11) into Eq. (6.7.10) and simplifying gives

$$\frac{T_1}{T_2} = \left(\frac{\mathbf{M}_1}{\mathbf{M}_2}\frac{1 + k\mathbf{M}_2^2}{1 + k\mathbf{M}_1^2}\right)^2 \tag{6.7.12}$$

This equation substituted into Eq. (6.7.9) leads to

$$\frac{T_{01}}{T_{02}} = \left(\frac{\mathbf{M}_1}{\mathbf{M}_2}\frac{1 + k\mathbf{M}_2^2}{1 + k\mathbf{M}_1^2}\right)^2\frac{2 + (k-1)\mathbf{M}_1^2}{2 + (k-1)\mathbf{M}_2^2} \tag{6.7.13}$$

When this equation is applied to the downstream section where $T_{02} = T_0^*$ and $\mathbf{M}_2 = 1$ and the subscripts for the upstream section are dropped, the result is

$$\frac{T_0}{T_0^*} = \frac{\mathbf{M}^2(k+1)[2 + (k-1)\mathbf{M}^2]}{(1 + k\mathbf{M}^2)^2} \tag{6.7.14}$$

All the necessary equations for determination of frictionless flow with heat transfer in a constant-area duct are now available. Heat transfer per unit mass is given by $q_H = c_p(T_0^* - T_0)$ for $\mathbf{M} = 1$ at the exit. Use of the equations is illustrated in the following example.

Example 6.15 Air at $V_1 = 300$ ft/s, $p = 40$ psia, $t = 60°F$ flows into a 4.0-in-diameter duct. How much heat transfer per unit mass is needed for sonic conditions at the exit? Determine pressure, temperature, and velocity at the exit and at the section where $\mathbf{M} = 0.70$.

$$\mathbf{M}_1 = \frac{V_1}{\sqrt{kRT_1}} = \frac{300}{\sqrt{1.4 \times 53.3 \times 32.17(460 + 60)}} = 0.268$$

The isentropic stagnation temperature at the entrance, from Eq. (6.7.7), is

$$T_{01} = T_1\left(1 + \frac{k-1}{2}\mathbf{M}_1^2\right) = 520(1 + 0.2 \times 0.268^2) = 527°R$$

The isentropic stagnation temperature at the exit, from Eq. (6.7.14), is

$$T_0^* = \frac{T_0(1 + k\mathbf{M}^2)^2}{(k+1)\mathbf{M}^2[2 + (k-1)\mathbf{M}^2]} = \frac{527(1 + 1.4 \times 0.268^2)^2}{2.4 \times 0.268^2(2 + 0.4 \times 0.268^2)}$$

$$= 1827°R$$

The heat transfer per slug of air flowing is

$$q_H = c_p(T_0^* - T_{01}) = 0.24 \times 32.17(1827 - 527) = 10{,}037 \text{ Btu/slug}$$

The pressure at the exit, Eq. (6.7.6), is

$$p^* = p\frac{1 + k\mathbf{M}^2}{k+1} = \frac{40}{2.4}(1 + 1.4 \times 0.268^2) = 18.34 \text{ psia}$$

and the temperature, from Eq. (6.7.12),

$$T^* = T\left[\frac{1 + k\mathbf{M}^2}{(k+1)\mathbf{M}}\right]^2 = 520\left(\frac{1 + 1.4 \times 0.268^2}{2.4 \times 0.268}\right)^2 = 1522°R$$

At the exit,

$$V^* = c^* = \sqrt{kRT^*} = \sqrt{1.4 \times 53.3 \times 32.17 \times 1522} = 1911 \text{ ft/s}$$

Table 6.2 Trends in flow properties

Property	Heating M > 1	Heating M < 1	Cooling M > 1	Cooling M < 1
Pressure p	Increases	Decreases	Decreases	Increases
Velocity V	Decreases	Increases	Increases	Decreases
Isentropic stagnation temperature T_0	Increases	Increases	Decreases	Decreases
Density ρ	Increases	Decreases	Decreases	Increases
Temperature T	Increases	Increases for $M < 1/\sqrt{k}$ Decreases for $M > 1/\sqrt{k}$	Decreases	Decreases for $M < 1/\sqrt{k}$ Increases for $M > 1/\sqrt{k}$

At the section where $M = 0.7$, from Eq. (6.7.6),

$$p = p^* \frac{k+1}{1 + kM^2} = \frac{18.34 \times 2.4}{1 + 1.4 \times 0.7^2} = 26.1 \text{ psia}$$

From Eq. (6.7.12)

$$T = T^* \left[\frac{(k+1)M}{1 + kM^2} \right]^2 = 1522 \left(\frac{2.4 \times 0.7}{1 + 1.4 \times 0.7^2} \right)^2 = 1511°R$$

and

$$V = M\sqrt{kRT} = 0.7\sqrt{1.4 \times 53.3 \times 32.17 \times 1511} = 1333 \text{ ft/s}$$

The trends in flow properties are shown in Table 6.2.

For curves and tables tabulating the various equations, consult the books by Cambel and Jennings, Keenan and Kaye, and Shapiro listed in the references at the end of this chapter.

EXERCISES

6.7.1 Select the correct statement regarding frictionless duct flow with heat transfer: (a) Adding heat to supersonic flow increases the Mach number. (b) Adding heat to subsonic flow increases the Mach number. (c) Cooling supersonic flow decreases the Mach number. (d) The Fanno line is valuable in analyzing the flow. (e) The isentropic stagnation temperature remains constant along the pipe.

6.7.2 Select the correct trends in flow properties for frictionless duct flow with heat transferred to the pipe $M < 1$: (a) p, V increase; ρ, T, T_0 decrease; (b) V, T_0 increase; p, ρ decrease; (c) p, ρ, T increase; V, T_0 decrease; (d) V, T increase; p, ρ, T_0 decrease; (e) T_0, V, ρ increase; p, T decrease.

6.7.3 Select the correct trends for cooling in frictionless duct flow $M > 1$: (a) V increases; p, ρ, T, T_0 decrease; (b) p, V increase; ρ, T, T_0 decrease; (c) p, ρ, V increase; T, T_0 decrease; (d) p, ρ increase; V, T, T_0 decrease; (e) V, T, T_0 increase; p, ρ decrease.

6.8 STEADY ISOTHERMAL FLOW IN LONG PIPELINES

In the analysis of isothermal flow of a perfect gas through long ducts, neither the Fanno nor the Rayleigh line is applicable, since the Fanno line applies to adiabatic flow and the Rayleigh line to frictionless flow. An analysis somewhat similar to those of the preceding two sections is carried out to show the trend in properties with Mach number.

The appropriate equations are

$$\text{Momentum [Eq. (6.6.3)]:} \quad \frac{dp}{p} + \frac{f}{2D}\frac{\rho V^2}{p}\,dx + \frac{\rho V}{p}\,dV = 0 \tag{6.8.1}$$

$$\text{Equation of state:} \quad \frac{p}{\rho} = \text{const} \qquad \frac{dp}{p} = \frac{d\rho}{\rho} \tag{6.8.2}$$

$$\text{Continuity:} \quad \rho V = \text{const} \qquad \frac{d\rho}{\rho} = -\frac{dV}{V} \tag{6.8.3}$$

$$\text{Energy [Eq. (6.7.7)]:} \quad T_0 = T\left[1 + \frac{(k-1)}{2}\mathbf{M}^2\right] \tag{6.8.4}$$

in which T_0 is the isentropic stagnation temperature at the section where the free-stream static temperature is T and the Mach number is $\mathbf{M}$.

$$\text{Stagnation pressure [Eq. (6.3.11)]:} p_0 = p\left(1 + \frac{k-1}{2}\mathbf{M}^2\right)^{k/(k-1)} \tag{6.8.5}$$

in which p_0 is the pressure (at the section of p and $\mathbf{M}$) obtained by reducing the velocity to zero isentropically.

From definitions and the above equations

$$V = c\mathbf{M} = \sqrt{kRT}\,\mathbf{M} \qquad \frac{dV}{V} = \frac{d\mathbf{M}}{\mathbf{M}} = \frac{d\mathbf{M}^2}{2\mathbf{M}^2}$$

$$\frac{\rho V}{p}\,dV = \frac{V\,dV}{RT} = \frac{c^2}{RT}\mathbf{M}\,d\mathbf{M} = k\mathbf{M}\,d\mathbf{M}$$

$$\frac{\rho V^2}{p} = \frac{c^2\mathbf{M}^2}{RT} = k\mathbf{M}^2$$

Substituting into the momentum equation (6.8.1) yields

$$\frac{dp}{p} = -\frac{f}{2D}k\mathbf{M}^2\,dx - k\mathbf{M}\,d\mathbf{M} = -\frac{d\mathbf{M}}{\mathbf{M}}$$

so

$$\frac{d\mathbf{M}}{\mathbf{M}}(1 - k\mathbf{M}^2) = \frac{f}{2D}k\mathbf{M}^2\,dx$$

and

$$\frac{dp}{p} = \frac{d\rho}{\rho} = -\frac{dV}{V} = -\frac{1}{2}\frac{d\mathbf{M}^2}{\mathbf{M}^2} = -\frac{k\mathbf{M}^2}{1 - k\mathbf{M}^2}\frac{f\,dx}{2D} \tag{6.8.6}$$

The differential dx is positive in the downstream direction, and so one may conclude that the trends in properties vary according as $\mathbf{M}$ is less than or greater than $1/\sqrt{k}$. For $\mathbf{M} < 1/\sqrt{k}$, the pressure and density decrease and velocity and Mach number increase, with the opposite trends for $\mathbf{M} > 1/\sqrt{k}$; hence, the Mach number always approaches $1/\sqrt{k}$, in place of unity for adiabatic flow in pipelines.

To determine the direction of heat transfer, differentiate Eq. (6.8.4) and then divide by it, remembering that T is constant:

$$\frac{dT_0}{T_0} = \frac{k - 1}{2 + (k - 1)\mathbf{M}^2} \, d\mathbf{M}^2 \tag{6.8.7}$$

Eliminating $d\mathbf{M}^2$ in this equation and Eq. (6.8.6) gives

$$\frac{dT_0}{T_0} = \frac{k(k - 1)\mathbf{M}^4}{(1 - k\mathbf{M}^2)[2 + (k - 1)\mathbf{M}^2]} \frac{f \, dx}{D} \tag{6.8.8}$$

which shows that the isentropic stagnation temperature increases for $\mathbf{M} < 1/\sqrt{k}$, indicating that heat is transferred to the fluid. For $\mathbf{M} > 1/\sqrt{k}$ heat transfer is from the fluid.

From Eqs. (6.8.5) and (6.8.6)

$$\frac{dp_0}{p_0} = \frac{2 - (k + 1)\mathbf{M}^2}{2 + (k - 1)\mathbf{M}^2} \frac{k\mathbf{M}^2}{k\mathbf{M}^2 - 1} \frac{f \, dx}{2D} \tag{6.8.9}$$

Table 6.3 shows the trends of fluid properties.

By integration of the various Eqs. (6.8.6) in terms of $\mathbf{M}$, the change with Mach number is found. The last two terms yield

$$\frac{f}{D} \int_0^{L_{\max}} dx = \frac{1}{k} \int_M^{1/\sqrt{k}} \frac{(1 - k\mathbf{M}^2)}{\mathbf{M}^4} \, d\mathbf{M}^2$$

or

$$\frac{f}{D} L_{\max} = \frac{1 - k\mathbf{M}^2}{k\mathbf{M}^2} + \ln \, (k\mathbf{M}^2) \tag{6.8.10}$$

Table 6.3 Trends in fluid properties for isothermal flow

Property	$\mathbf{M} < 1/\sqrt{k}$ subsonic	$\mathbf{M} > 1/\sqrt{k}$ subsonic or supersonic
Pressure p	Decreases	Increases
Density ρ	Decreases	Increases
Velocity V	Increases	Decreases
Mach number $\mathbf{M}$	Increases	Decreases
Stagnation temperature T_0	Increases	Decreases
Stagnation pressure p_0	Decreases	Increases for $\mathbf{M} < \sqrt{2}/(k + 1)$ Decreases for $\mathbf{M} > \sqrt{2}/(k + 1)$

in which L_{max}, as before, represents the maximum length of duct. For greater lengths choking occurs, and the mass rate is decreased. To find the pressure change,

$$\int_p^{p*t} \frac{dp}{p} = -\frac{1}{2} \int_M^{1/\sqrt{k}} \frac{dM^2}{M^2}$$

and

$$\frac{p^{*t}}{p} = \sqrt{k}\,M \qquad\qquad (6.8.11)$$

The superscript $*t$ indicates conditions at $M = 1/\sqrt{k}$, and M and p represent values at any upstream section.

Example 6.16 Helium enters a 100-mm-ID pipe from a converging-diverging nozzle at $M = 1.30$, $p = 14$ kN/m^2 abs, $T = 225$ K. Determine for isothermal flow (a) the maximum length of pipe for no choking, (b) the downstream conditions, and (c) the length from the exit to the section where $M = 1.0$. $f = 0.016$.
(a) From Eq. (6.8.10) for $k = 1.66$

$$\frac{0.016 L_{max}}{0.1 \text{ m}} = \frac{1 - 1.66 \times 1.3^2}{1.66 \times 1.3^2} + \ln\,(1.66 \times 1.3^2)$$

from which $L_{max} = 2.425$ m.
(b) From Eq. (6.8.11)

$$p^{*t} = p\sqrt{k}\,M = 14 \text{ kN/m}^2 \sqrt{1.66}\,1.3 = 23.45 \text{ kN/m}^2 \text{ abs}$$

The Mach number at the exit is $1/\sqrt{1.66} = 0.776$. From Eqs. (6.8.6)

$$\int_V^{V*t} \frac{dV}{V} = \frac{1}{2} \int_M^{1/\sqrt{k}} \frac{dM^2}{M^2} \qquad \text{or} \qquad \frac{V^{*t}}{V} = \frac{1}{\sqrt{k}\,M}$$

At the upstream section

$$V = M\sqrt{kRT} = 1.3\sqrt{1.66 \times 2077 \times 225} = 1145 \text{ m/s}$$

and

$$V^{*t} = \frac{V}{\sqrt{k}\,M} = \frac{1145 \text{ m/s}}{\sqrt{1.66}\,1.3} = 683.6 \text{ m/s}$$

(c) From Eq. (6.8.10) for $M = 1$,

$$\frac{0.016}{0.1 \text{ m}} L'_{max} = \frac{1 - 1.66}{1.66} + \ln\,1.66$$

or $L'_{max} = 0.683$ m. $M = 1$ occurs 0.683 m from the exit.

EXERCISES

6.8.1 In steady, isothermal flow in long pipelines, the significant value of M for determining trends in flow properties is (a) $1/k$; (b) $1/\sqrt{k}$; (c) 1; (d) $\sqrt{k}$; (e) k.
6.8.2 Select the correct trends in fluid properties for isothermal flow in ducts for $M < 0.5$: (a) V increases; M, T_0, p, p_0, ρ decrease; (b) V, M increase; T_0, p, p_0, ρ decrease; (c) V, M, T_0 increase; p, p_0, ρ decrease; (d) V, T_0 increase; M, p, p_0, ρ decrease; (e) V, M, p_0, T_0 increase; p, ρ decrease.

6.9 ANALOGY OF SHOCK WAVES TO OPEN-CHANNEL WAVES

Both the oblique and normal shock waves in a gas have their counterpart in open-channel flow. An elementary surface wave has a speed in still liquid of $\sqrt{gy}$, in which y is the depth in a wide, open channel. When flow in the channel is such that $V = V_c = \sqrt{gy}$, the Froude number is unity and flow is said to be *critical*; i.e., a small disturbance cannot be propagated upstream. This is analogous to sonic flow at the throat of a tube, with Mach number unity. For liquid velocities greater than $V_c = \sqrt{gy}$ the Froude number is greater than unity and the velocity is supercritical, which is analogous to supersonic gas flow. Changes in depth are analogous to changes in density in gas flow.

The continuity equation in an open channel of constant width is

$$Vy = \text{const}$$

and the continuity equation for compressible flow in a tube of constant cross section is

$$V\rho = \text{const}$$

Compressible fluid density ρ and open-channel depth y are analogous.

The same analogy is also present in the energy equation. The energy equation for a horizontal open channel of constant width, neglecting friction, is

$$\frac{V^2}{2g} + y = \text{const}$$

After differentiating,

$$V\,dV + g\,dy = 0$$

By substitution from $V_c = \sqrt{gy}$ to eliminate g,

$$V\,dV + V_c^2\,\frac{dy}{y} = 0$$

which is to be compared with the energy equation for compressible flow [Eq. (6.3.4)]

$$V\,dV + c^2\,\frac{d\rho}{\rho} = 0$$

The two critical velocities V_c and c are analogous, and hence y and ρ are analogous.

By applying the momentum equation to a small depth change in horizontal open-channel flow, and to a sudden density change in compressible flow, the density and the open-channel depth can again be shown to be analogous. In effect, the analogy is between the Froude number and the Mach number.

Analogous to the normal shock wave is the hydraulic jump, which causes a sudden change in velocity and depth and a change in Froude number from greater than unity to less than unity. Analogous to the oblique shock and rarefaction waves in gas flow are oblique liquid waves produced in a channel by changes in the direction of the channel walls or by changes in floor elevation.

A body placed in an open channel with flow at Froude number greater than unity causes waves on the surface that are analogous to shock and rarefaction waves on a similar (two-dimensional) body in a supersonic wind tunnel. Changes to greater depth are analogous to compression shock, and changes to lesser depth to rarefaction waves. Shallow water tanks, called *ripple tanks*, have been used to study supersonic flow situations.

EXERCISES

6.9.1 The speed of a sound wave in a gas is analogous to (a) the speed of flow in an open channel; (b) the speed of an elementary wave in an open channel; (c) the change in depth in an open channel; (d) the speed of a disturbance traveling upstream in moving liquid; (e) none of these answers.

6.9.2 A normal shock wave is analogous to (a) an elementary wave in still liquid; (b) the hydraulic jump; (c) open-channel conditions with $F < 1$; (d) flow of liquid through an expanding nozzle; (e) none of these answers.

PROBLEMS

6.1 3 kg of a perfect gas, molecular weight 36, had its temperature increased 2°C when 6.4 kJ of work was done on it in an insulated constant-volume chamber. Determine c_v and c_p.

6.2 A gas of molecular weight 48 has $c_p = 0.372$. What is c_v for this gas?

6.3 Calculate the specific heat ratio k for Probs. 6.1 and 6.2.

6.4 The enthalpy of a gas is increased by 0.4 Btu/lb$_m$·°R when heat is added at constant pressure, and the internal energy is increased by 0.3 Btu/lb$_m$·°R when the volume is maintained constant and heat is added. Calculate the molecular weight.

6.5 Calculate the enthalpy change of 2 kg carbon monoxide from $p_1 = 14$ kN/m^2 abs, $t_1 = 5°C$ to $p_2 = 30$ kN/m^2 abs, $t_2 = 170°C$.

6.6 Calculate the entropy change in Prob. 6.5.

6.7 From Eq. (6.1.13) and the perfect-gas law, derive the equation of state for isentropic flow.

6.8 Compute the enthalpy change per slug for helium from $t_1 = 0°F$, $p_1 = 15$ psia to $t_2 = 120°F$ in an isentropic process.

6.9 In an isentropic process 1 kg oxygen with a volume of 150 l at 15°C has its absolute pressure doubled. What is the final temperature?

6.10 Work out the expression for density change with temperature for a reversible polytropic process.

6.11 Hydrogen at 50 psia, 30°F, has its temperature increased to 120°F by a reversible polytropic process with $n = 1.20$. Calculate the final pressure.

6.12 A gas has a density decrease of 10 percent in a reversible polytropic process when the temperature decreases from 45 to 5°C. Compute the exponent n for the process.

6.13 A projectile moves through water at 80°F at 2000 ft/s. What is its Mach number?

6.14 If an airplane travels at 1350 km/h at sea level, $p = 101$ kPa abs, $t = 20°C$, and at the same speed in the stratosphere where $t = -55°C$, how much greater is the Mach number in the latter case?

6.15 What is the speed of sound through hydrogen at 80°F?

6.16 Derive the equation for speed of a small liquid wave in an open channel by using the methods of Sec. 6.2 for determination of speed of sound (Fig. 6.8).

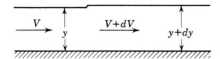

Figure 6.8 Problem 6.16.

6.17 By using the energy equation

$$V\,dV + \frac{dp}{\rho} + d(\text{losses}) = 0$$

the continuity equation $\rho V = $ const, and $c = \sqrt{dp/d\rho}$, show that for subsonic flow in a pipe the velocity must increase in the downstream direction.

6.18 Isentropic flow of air occurs at a section of a pipe where $p = 40$ psia, $t = 90°F$, and $V = 537$ ft/s. An object is immersed in the flow, which brings the velocity to zero. What are the temperature and pressure at the stagnation point?

6.19 What is the Mach number for the flow of Prob. 6.18?

6.20 How do the temperature and pressure at the stagnation point in isentropic flow compare with reservoir conditions?

6.21 Air flows from a reservoir at 90°C, 7 atm. Assuming isentropic flow, calculate the velocity, temperature, pressure, and density at a section where $\mathbf{M} = 0.60$.

6.22 Oxygen flows from a reservoir where $p_0 = 100$ psia, $t_0 = 90°F$, to a 6-in-diameter section where the velocity is 600 ft/s. Calculate the mass rate of flow (isentropic) and the Mach number, pressure, and temperature at the 6-in section.

6.23 Helium discharges from a $\frac{1}{2}$-in-diameter converging nozzle at its maximum rate for reservoir conditions of $p = 4$ atm, $t = 25°C$. What restrictions are placed on the downstream pressure? Calculate the mass flow rate and velocity of the gas at the nozzle.

6.24 Air in a reservoir at 250 psia, $t = 290°F$, flows through a 2-in-diameter throat in a converging-diverging nozzle. For $\mathbf{M} = 1$ at the throat, calculate p, ρ, and T there.

6.25 What must be the velocity, pressure, density, temperature, and diameter at a cross section of the nozzle of Prob. 6.24 where $\mathbf{M} = 2.4$?

6.26 Nitrogen in sonic flow at a 25-mm-diameter throat section has a pressure of 50 kN/m² abs, $t = -20°C$. Determine the mass flow rate.

6.27 What is the Mach number for Prob. 6.26 at a 40-mm-diameter section in supersonic and in subsonic flow?

6.28 What diameter throat section is needed for critical flow of 0.5 lb$_m$/s carbon monoxide from a reservoir where $p = 300$ psia, $t = 100°F$?

6.29 A supersonic nozzle is to be designed for airflow with $\mathbf{M} = 3.5$ at the exit section, which is 200 mm in diameter and has a pressure of 7 kN/m² abs and temperature of $-85°C$. Calculate the throat area and reservoir conditions.

6.30 In Prob. 6.29 calculate the diameter of cross section for $\mathbf{M} = 1.5, 2.0$, and 2.5.

6.31 For reservoir conditions of $p_0 = 180$ psia, $t_0 = 120°F$, air flows through a converging-diverging tube with a 3.0-in-diameter throat with a maximum Mach number of 0.80. Determine the mass rate of flow and the diameter, pressure, velocity, and temperature at the exit where $\mathbf{M} = 0.50$.

6.32 Calculate the exit velocity and the mass rate of flow of nitrogen from a reservoir where $p = 4$ atm, $t = 25°C$, through a converging nozzle of 60 mm diameter discharging to atmosphere.

6.33 Reduce Eq. (6.3.25) to its form for airflow. Plot p/p_0 vs. A^*/A for the range of p/p_0 from 0.98 to 0.02.

6.34 By utilizing the plot of Prob. 6.33, find the two pressure ratios for $A^*/A = 0.50$.

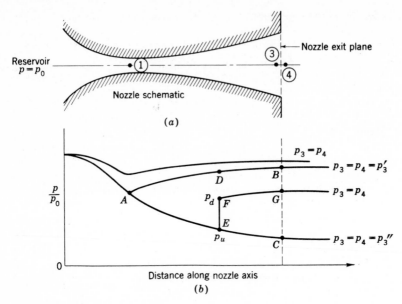

Figure 6.9 Problems 6.40 to 6.47.

6.35 In a converging-diverging duct in supersonic flow of hydrogen, the throat diameter is 50 mm. Determine the pressure ratios p/p_0 in the converging and diverging ducts where the diameter is 57 mm.

6.36 A shock wave occurs in a duct carrying air where the upstream Mach number is 2.0 and upstream temperature and pressure are 15°C and 20 kN/m^2 abs. Calculate the Mach number, pressure, temperature, and velocity after the shock wave.

6.37 Show that entropy has increased across the shock wave of Prob. 6.36.

6.38 Conditions immediately before a normal shock wave in airflow are $p_u = 6$ psia, $t_u = 100°F$, $V_u = 1800$ ft/s. Find M_u, M_d, p_d, and t_d, where the subscript d refers to conditions just downstream from the shock wave.

6.39 For $A = 0.16$ ft^2 in Prob. 6.38, calculate the entropy increase across the shock wave in Btu per second per degree Rankine.

6.40 From Eqs. (6.3.1), (6.3.4), and (6.3.5) deduce that at the throat of a convergent-divergent (De Laval) nozzle, point 1 of Fig. 6.9a, $dp = 0$, $d\rho = 0$ for $M \neq 1$ (cf. Fig. 6.9b). Are these differentials zero for $M = 1$? Explain.

6.41 From Eqs. (6.3.1), (6.3.4), and (6.3.5) justify the slopes of the curves shown in Fig. 6.9b. Do not consider *EFG*.

6.42 For the nozzle described below, plot curves *ADB* and *AEC* (Fig. 6.9b). (*Suggestions:* Determine only one intermediate point. Use section VI.) The reservoir has air at 300 kPa abs and 40°C when sonic conditions are obtained at the throat.

	Section									
	I	II	III	IV	V	VI	VII	VIII	IX	X exit
Distance downstream from throat, mm	5	10	15	20	25	30	35	40	45	50
A/A^* ($A^* = 27$ cm^2)	1.030	1.050	1.100	1.133	1.168	1.200	1.239	1.269	1.310	1.345

6.43 By using the data from Prob. 6.42, determine p_3/p_0 when a normal shock wave occurs at section VI.

6.44 Could a flow discontinuity occur at section VI of Prob. 6.42 so that the flow path would be described by $ADFG$ of Fig. 6.9b? (*Hint:* Determine the entropy changes.)

6.45 What is p_3/p_0 when a normal shock wave occurs just inside the nozzle exit? (*Hint:* $p_d = p_4$ and p_u is p_3 for isentropic flow up to section VI of Prob. 6.42.)

6.46 Suggest what might occur just outside the nozzle if there is a receiver pressure p_4 which is above that for which the gas flows isentropically throughout the nozzle into the receiver, point C of Fig. 6.9b, but below that for which a normal shock is possible at the nozzle exit (cf. Prob. 6.45).

6.47 Speculate on what occurs within and without the nozzle if the receiver pressure is below that corresponding to point C of Fig. 6.9b.

6.48 Show, from the equations of Sec. 6.6, that temperature, pressure, and density decrease in real, adiabatic duct flow for subsonic conditions and increase for supersonic conditions.

6.49 What length of 100-mm-diameter insulated duct, $f = 0.018$, is needed when oxygen enters at $M = 3.0$ and leaves at $M = 2.0$?

6.50 Air enters an insulated pipe at $M = 0.4$ and leaves at $M = 0.6$. What portion of the duct length is required for the flow to occur at $M = 0.5$?

6.51 Determine the maximum length, without choking, for the adiabatic flow of air in a 110-mm-diameter duct, $f = 0.025$, when upstream conditions are $t = 50°C$, $V = 200$ m/s, $p = 2$ atm. What are the pressure and temperature at the exit?

6.52 What minimum size insulated duct is required to transport nitrogen 1000 ft? The upstream temperature is 80°F, and the velocity there is 200 ft/s. $f = 0.020$.

6.53 Find the upstream and downstream pressures in Prob. 6.52 for 3 lb$_m$/s flow.

6.54 What is the maximum mass rate of flow of air from a reservoir, $t = 15°C$, through 6 m of insulated 25-mm-diameter pipe, $f = 0.020$, discharging to atmosphere? $p = 1$ atm.

6.55 In frictionless oxygen flow through a duct the following conditions prevail at inlet and outlet: $V_1 = 300$ ft/s; $t_1 = 80°F$; $M_2 = 0.5$. Find the heat added per slug and the pressure ratio p_1/p_2.

6.56 In frictionless air the flow through a 120-mm-diameter duct 0.15 kg/s enters at $t = 0°C$, $p = 7$ kN/m^2 abs. How much heat, in kilocalories per kilogram, can be added without choking the flow?

6.57 Frictionless flow through a duct with heat transfer causes the Mach number to decrease from 2 to 1.75. $k = 1.4$. Determine the temperature, velocity, pressure, and density ratios.

6.58 In Prob. 6.57 the duct is 2 in square, $p_1 = 15$ psia, and $V_1 = 2000$ ft/s. Calculate the mass rate of flow for air flowing.

6.59 How much heat must be transferred per kilogram to cause the Mach number to increase from 2 to 2.8 in a frictionless duct carrying air? $V_1 = 500$ m/s.

6.60 Oxygen at $V_1 = 525$ m/s, $p = 80$ kN/m^2 abs, $t = -10°C$ flows in a 60-mm-diameter frictionless duct. How much heat transfer per kilogram is needed for sonic conditions at the exit?

6.61 Prove the density, pressure, and velocity trends given in Sec. 6.8 in the table of trends in flow properties.

6.62 Apply the first law of thermodynamics, Eq. (3.8.1), to isothermal flow of a perfect gas in a horizontal pipeline, and develop an expression for the heat added per slug flowing.

6.63 Air is flowing at constant temperature through a 3-in-diameter horizontal pipe, $f = 0.02$. At the entrance $V_1 = 300$ ft/s, $t = 120°F$, $p_1 = 30$ psia. What is the maximum pipe length for this flow, and how much heat is transferred to the air per pound mass?

6.64 Air at 15°C flows through a 25-mm-diameter pipe at constant temperature. At the entrance $V_1 = 60$ m/s, and at the exit $V_2 = 90$ m/s. $f = 0.016$. What is the length of the pipe?

6.65 If the pressure at the entrance of the pipe of Prob. 6.64 is 1.5 atm, what is the pressure at the exit and what is the heat transfer to the pipe per second?

6.66 Hydrogen enters a pipe from a converging nozzle at $M = 1$, $p = 2$ psia, $t = 0°F$. Determine, for isothermal flow, the maximum length of pipe, in diameters, and the pressure change over this length. $f = 0.016$.

6.67 Oxygen flows at constant temperature of 20°C from a pressure tank, $p = 130$ atm, through 10 ft of 3-mm-ID tubing to another tank where $p = 110$ atm. $f = 0.016$. Determine the mass rate of flow.

6.68 In isothermal flow of nitrogen at 80°F, 2 lb_m/s is to be transferred 100 ft from a tank where $p = 200$ psia to a tank where $p = 160$ psia. What minimum size of tubing, $f = 0.016$, is needed?

REFERENCES

Cambel, A. B., and B. H. Jennings: "Gas Dynamics," McGraw-Hill, New York, 1958.

Keenan, J. H., and J. Kaye: "Gas Tables," Wiley, New York, 1948.

Liepmann, H. W., and A. Roshko: "Elements of Gas Dynamics," Wiley, New York, 1957.

Owczarek, J. A.: "Fundamentals of Gas Dynamics," International Textbook, Scranton, Pa., 1964.

Shapiro, A. H.: "The Dynamics and Thermodynamics of Compressible Fluid Flow," vol. 1, Ronald, New York, 1953.

Van Wylen, G. J., and R. E. Sonntag: "Fundamentals of Classical Thermodynamics," SI Version, 2d ed., Wiley, New York, 1976.

SEVEN
IDEAL-FLUID FLOW

In the preceding chapters most of the relations have been developed for one-dimensional flow, i.e., flow in which the average velocity at each cross section is used and variations across the section are neglected. Many design problems in fluid flow, however, require more exact knowledge of velocity and pressure distributions, such as in flow over curved boundaries along an airplane wing, through the passages of a pump or compressor, or over the crest of a dam. An understanding of two- and three-dimensional flow of a nonviscous, incompressible fluid provides the student with a much broader approach to many real fluid-flow situations. There are also analogies that permit the same methods to apply to flow through porous media.

In this chapter the principles of irrotational flow of an ideal fluid are developed and applied to elementary flow cases. After the flow requirements are established, Euler's equation is derived and the velocity potential is defined. Euler's equation is integrated to obtain Bernoulli's equation, and stream functions and boundary conditions are developed. Flow cases are then studied in two dimensions.

7.1 REQUIREMENTS FOR IDEAL-FLUID FLOW

The Prandtl hypothesis, Sec. 5.6, states that, for fluids of low viscosity, the effects of viscosity are appreciable only in a narrow region surrounding the fluid boundaries. For incompressible flow situations in which the boundary layer remains thin, ideal-fluid results may be applied to flow of a real fluid to a satisfactory degree of approximation. Converging or accelerating flow situations generally have thin boundary layers, but decelerating flow may have separation of the boundary layer and development of a large wake that is difficult to predict analytically.

An ideal fluid must satisfy the following requirements:

1. The continuity equation, Sec. 3.4, div $\mathbf{q} = 0$, or

$$\frac{\partial u}{\partial x} + \frac{\partial v}{\partial y} + \frac{\partial w}{\partial z} = 0$$

2. Newton's second law of motion at every point at every instant
3. Neither penetration of fluid into nor gaps between fluid and boundary at any solid boundary

If, in addition to requirements 1, 2, and 3, the assumption of irrotational flow is made, the resulting fluid motion closely resembles real-fluid motion for fluids of low viscosity, outside boundary layers.

Using the above conditions, the application of Newton's second law to a fluid particle leads to the Euler equation, which, together with the assumption of irrotational flow, can be integrated to obtain the Bernoulli equation. The unknowns in a fluid-flow situation with given boundaries are velocity and pressure at every point. Unfortunately, in most cases it is impossible to proceed directly to equations for velocity and pressure distribution from the boundary conditions.

7.2 EULER'S EQUATION OF MOTION

Euler's equation of motion along a streamline (one-dimensional) was developed in Sec. 3.5 by use of the momentum and continuity equations, and by use of Eq. (2.2.5). In this section it is developed from Eq. (2.2.5) for the xyz-coordinate system in any orientation, with the assumption that gravity is the only body force acting. Since Euler's equation is based on a frictionless fluid, the vector equation (2.2.5)

$$\mathbf{f} - \mathbf{j}'\gamma = \rho\mathbf{a} \qquad \mathbf{f} = -\nabla p \qquad (2.2.5)$$

may be reorganized into the proper form. The unit vector $\mathbf{j}'$ is directed vertically upward in the coordinate direction h. The quantity $-\mathbf{j}'\gamma$ is the pull of gravity per unit volume in the h direction. It may be decomposed into its x, y, z components (Fig. 7.1) $-\mathbf{i}\gamma \cos\theta_x$, $-\mathbf{j}\gamma \cos\theta_y$, $-\mathbf{k}\gamma \cos\theta_z$, in which θ_x, θ_y, and θ_z represent the angles between the h axis and the x, y, z axes, respectively. These are the direction cosines of h with respect to the xyz system of coordinates, and they may be written

$$\cos\theta_x = \frac{\partial h}{\partial x} \qquad \cos\theta_y = \frac{\partial h}{\partial y} \qquad \cos\theta_z = \frac{\partial h}{\partial z}$$

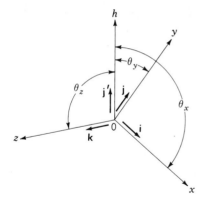

Figure 7.1 Arbitrary orientation of xyz-coordinate system.

For example, $\partial h/\partial x$ is the change in h for unit change in x when y, z, and t are constant. In equation form,

$$\mathbf{j}'\gamma = \mathbf{i}\gamma\,\frac{\partial h}{\partial x} + \mathbf{j}\gamma\,\frac{\partial h}{\partial y} + \mathbf{k}\gamma\,\frac{\partial h}{\partial z} = \nabla\gamma h$$

The operation ∇ applied to the scalar h yields the gradient of h, as in Eq. (2.2.2). Equation (2.2.5) now becomes

$$-\nabla(p + \gamma h) = \rho\mathbf{a} = \rho\left(\mathbf{i}\,\frac{du}{dt} + \mathbf{j}\,\frac{dv}{dt} + \mathbf{k}\,\frac{dw}{dt}\right) \tag{7.2.1}$$

The component equations of Eq. (7.2.1) are

$$-\frac{\partial}{\partial x}(p + \gamma h) = \rho\,\frac{du}{dt}$$

$$-\frac{\partial}{\partial y}(p + \gamma h) = \rho\,\frac{dv}{dt} \tag{7.2.2}$$

$$-\frac{\partial}{\partial z}(p + \gamma h) = \rho\,\frac{dw}{dt}$$

u, v, w are velocity components in the x, y, z directions, respectively, at any point; du/dt is the x component of acceleration of the fluid particle at (x, y, z). Since u is a function of x, y, z, and t and x, y, and z are coordinates of the moving fluid particle, they become functions of t; hence,

$$a_x = \frac{du}{dt} = \frac{\partial u}{\partial x}\frac{dx}{dt} + \frac{\partial u}{\partial y}\frac{dy}{dt} + \frac{\partial u}{\partial z}\frac{dz}{dt} + \frac{\partial u}{\partial t}$$

However, dx/dt, dy/dt, and dz/dt are the velocity components of the particle, so the x component of the particle acceleration, a_x, is

$$\frac{du}{dt} = u\,\frac{\partial u}{\partial x} + v\,\frac{\partial u}{\partial y} + w\,\frac{\partial u}{\partial z} + \frac{\partial u}{\partial t}$$

By treating dv/dt and dw/dt in a similar manner, the Euler equations in three dimensions for a frictionless fluid are

$$-\frac{1}{\rho}\frac{\partial}{\partial x}(p + \gamma h) = u\frac{\partial u}{\partial x} + v\frac{\partial u}{\partial y} + w\frac{\partial u}{\partial z} + \frac{\partial u}{\partial t} \tag{7.2.3}$$

$$-\frac{1}{\rho}\frac{\partial}{\partial y}(p + \gamma h) = u\frac{\partial v}{\partial x} + v\frac{\partial v}{\partial y} + w\frac{\partial v}{\partial z} + \frac{\partial v}{\partial t} \tag{7.2.4}$$

$$-\frac{1}{\rho}\frac{\partial}{\partial z}(p + \gamma h) = u\frac{\partial w}{\partial x} + v\frac{\partial w}{\partial y} + w\frac{\partial w}{\partial z} + \frac{\partial w}{\partial t} \tag{7.2.5}$$

The first three terms on the right-hand sides of the equations are *convective-acceleration* terms, depending upon changes of velocity with space. The last term is the *local acceleration*, depending upon velocity change with time at a point.

Natural Coordinates in Two-Dimensional Flow

Euler's equations in two dimensions are obtained from the general-component equations by setting $w = 0$ and $\partial/\partial z = 0$; thus

$$-\frac{1}{\rho}\frac{\partial}{\partial x}(p + \gamma h) = u\frac{\partial u}{\partial x} + v\frac{\partial u}{\partial y} + \frac{\partial u}{\partial t} \tag{7.2.6}$$

$$-\frac{1}{\rho}\frac{\partial}{\partial y}(p + \gamma h) = u\frac{\partial v}{\partial x} + v\frac{\partial v}{\partial y} + \frac{\partial v}{\partial t} \tag{7.2.7}$$

By taking particular directions for the x and y axes, they can be reduced to a form that makes them easier to understand. If the x axis, called the s axis, is taken parallel to the velocity vector at a point (Fig. 7.2), it is then tangent to the streamline through the point. The y axis, called the n axis, is drawn toward the center of curvature of the streamline. The velocity component u is v_s, and the component v is v_n. As v_n is zero at the point, Eq. (7.2.6) becomes

$$-\frac{1}{\rho}\frac{\partial}{\partial s}(p + \gamma h) = v_s\frac{\partial v_s}{\partial s} + \frac{\partial v_s}{\partial t} \tag{7.2.8}$$

Although v_n is zero at the point (s, n), its rates of change with respect to s and t are not necessarily zero. Equation (7.2.7) becomes

$$-\frac{1}{\rho}\frac{\partial}{\partial n}(p + \gamma h) = v_s\frac{\partial v_n}{\partial s} + \frac{\partial v_n}{\partial t} \tag{7.2.9}$$

When the velocity at s and at $s + \delta s$ along the streamline is considered, v_n changes from zero to δv_n. With r the radius of curvature of the streamline at s, from similar triangles (Fig. 7.2),

$$\frac{\delta s}{r} = \frac{\delta v_n}{v_s} \qquad \text{or} \qquad \frac{\partial v_n}{\partial s} = \frac{v_s}{r}$$

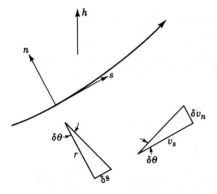

Figure 7.2 Notation for natural coordinates.

Substituting into Eq. (7.2.9) gives

$$-\frac{1}{\rho}\frac{\partial}{\partial n}(p + \gamma h) = \frac{v_s^2}{r} + \frac{\partial v_n}{\partial t} \qquad (7.2.10)$$

For steady flow of an incompressible fluid Eqs. (7.2.6) and (7.2.10) may be written

$$-\frac{1}{\rho}\frac{\partial}{\partial s}(p + \gamma h) = \frac{\partial}{\partial s}\left(\frac{v_s^2}{2}\right) \qquad (7.2.11)$$

and

$$-\frac{1}{\rho}\frac{\partial}{\partial n}(p + \gamma h) = \frac{v_s^2}{r} \qquad (7.2.12)$$

Equation (7.2.11) can be integrated with respect to s to produce Eq. (3.6.1), with the constant of integration varying with n, that is, from one streamline to another. Equation (7.2.12) shows how pressure head varies across streamlines. With v_s and r known functions of n, Eq. (7.2.12) can be integrated.

Example 7.1 A container of liquid is rotated with angular velocity ω about a vertical axis as a solid. Determine the variation of pressure in the liquid.

n is the radial distance, measured inwardly, $dn = -dr$, and $v_s = \omega r$. By integrating Eq. (7.2.12),

$$-\frac{1}{\rho}(p + \gamma h) = -\int \frac{\omega^2 r^2}{r}\,dr + \text{const}$$

or

$$\frac{1}{\rho}(p + \gamma h) = \frac{\omega^2 r^2}{2} + \text{const}$$

To evaluate the constant, if $p = p_0$ when $r = 0$ and $h = 0$,

$$p = p_0 - \gamma h + \rho\frac{\omega^2 r^2}{2}$$

which shows that the pressure is hydrostatic along a vertical line and increases as the square of the radius. Integration of Eq. (7.2.11) shows that the pressure is constant for a given h and v_s, that is, along a streamline. These results are the same as for rotation in relative equilibrium determined in Sec. 2.9.

EXERCISES

7.2.1 The units for Euler's equations of motion are given by (a) force per unit mass; (b) velocity; (c) energy per unit weight; (d) force per unit weight; (e) none of these answers.

7.2.2 Euler's equations of motion are a mathematical statement that at every point (a) rate of mass inflow equals rate of mass outflow; (b) force per unit mass equals acceleration; (c) the energy does not change with the time; (d) Newton's third law of motion holds; (e) the fluid momentum is constant.

7.3 IRROTATIONAL FLOW; VELOCITY POTENTIAL

In this section it is shown that the assumption of irrotational flow leads to the existence of a velocity potential. By use of these relations and the assumption of a conservative body force, the Euler equations can be integrated.

The individual particles of a frictionless incompressible fluid initially at rest cannot be caused to rotate. This can be visualized by considering a small free body of fluid in the shape of a sphere. Surface forces act normal to its surface, since the fluid is frictionless, and therefore act through the center of the sphere. Similarly, the body force acts at the mass center. Hence, no torque can be exerted on the sphere, and it remains without rotation. Likewise, once an ideal fluid has rotation, there is no way of altering it, as no torque can be exerted on an elementary sphere of the fluid.

An analytical expression for fluid rotation of a particle about an axis parallel to the z axis is developed. The rotation component may be defined as the average angular velocity of two infinitesimal linear elements that are mutually perpendicular to each other and to the axis of rotation. The two line elements may conveniently be taken as δx and δy in Fig. 7.3, although any other two perpendicular elements in the plane through the point would yield the same result. The particle is at $P(x, y)$, and it has velocity components u, v in the xy plane. The angular velocities of δx and δy are sought. The angular velocity of δx is

$$\frac{v + (\partial v/\partial x)\, \delta x - v}{\delta x} = \frac{\partial v}{\partial x} \text{ rad/s}$$

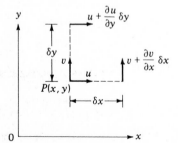

Figure 7.3 Rotation in a fluid.

and the angular velocity of δy is

$$-\frac{u + (\partial u/\partial y)\, \delta y - u}{\delta y} = -\frac{\partial u}{\partial y} \text{ rad/s}$$

if counterclockwise is positive. Hence, by definition, the rotation component ω_z of a fluid particle at (x, y) is

$$\omega_z = \frac{1}{2}\left(\frac{\partial v}{\partial x} - \frac{\partial u}{\partial y}\right) \tag{7.3.1}$$

Similarly, the two other rotation components, ω_x and ω_y, about axes parallel to x and to y are

$$\omega_x = \frac{1}{2}\left(\frac{\partial w}{\partial y} - \frac{\partial v}{\partial z}\right) \qquad \omega_y = \frac{1}{2}\left(\frac{\partial u}{\partial z} - \frac{\partial w}{\partial x}\right) \tag{7.3.2}$$

The rotation vector $\boldsymbol{\omega}$ is

$$\boldsymbol{\omega} = \mathbf{i}\omega_x + \mathbf{j}\omega_y + \mathbf{k}\omega_z \tag{7.3.3}$$

The *vorticity vector*, curl $\mathbf{q} = \nabla \times \mathbf{q}$, is defined as twice the rotation vector. It is given by $2\boldsymbol{\omega}$.

By assuming that the fluid has no rotation, i.e., it is irrotational, curl $\mathbf{q} = 0$, or from Eqs. (7.3.1) and (7.3.2)

$$\frac{\partial v}{\partial x} = \frac{\partial u}{\partial y} \qquad \frac{\partial w}{\partial y} = \frac{\partial v}{\partial z} \qquad \frac{\partial u}{\partial z} = \frac{\partial w}{\partial x} \tag{7.3.4}$$

These restrictions on the velocity must hold at every point (except special singular points or lines). The first equation is the irrotational condition for two-dimensional flow. It is the condition that the differential expression

$$u\, dx + v\, dy$$

is exact, say

$$u\, dx + v\, dy = -d\phi = -\frac{\partial \phi}{\partial x}\, dx - \frac{\partial \phi}{\partial y}\, dy \tag{7.3.5}$$

The minus sign is arbitrary; it is a convention that causes the value of ϕ to decrease in the direction of the velocity. By comparing terms in Eq. (7.3.5), $u = -\partial \phi/\partial x$, $v = -\partial \phi/\partial y$. This proves the existence, in two-dimensional flow, of a function ϕ such that its negative derivative with respect to any direction is the velocity component in that direction. It can also be demonstrated for three-dimensional flow. In vector form,

$$\mathbf{q} = -\text{grad }\phi = -\nabla\phi \tag{7.3.6}$$

is equivalent to

$$u = -\frac{\partial \phi}{\partial x} \qquad v = -\frac{\partial \phi}{\partial y} \qquad w = -\frac{\partial \phi}{\partial z} \tag{7.3.7}$$

The assumption of a velocity potential is equivalent to the assumption of irrotational flow, as

$$\text{curl} \, (-\text{grad} \, \phi) = -\nabla \times \nabla\phi = 0 \qquad (7.3.8)$$

because $\nabla \times \nabla = 0$. This is shown from Eq. (7.3.7) by cross-differentiation:

$$\frac{\partial u}{\partial y} = -\frac{\partial^2\phi}{\partial x \, \partial y} \qquad \frac{\partial v}{\partial x} = -\frac{\partial^2\phi}{\partial y \, \partial x}$$

proving $\partial v/\partial x = \partial u/\partial y$, etc.

Substitution of Eqs. (7.3.7) into the continuity equation

$$\frac{\partial u}{\partial x} + \frac{\partial v}{\partial y} + \frac{\partial w}{\partial z} = 0$$

yields

$$\frac{\partial^2\phi}{\partial x^2} + \frac{\partial^2\phi}{\partial y^2} + \frac{\partial^2\phi}{\partial z^2} = 0 \qquad (7.3.9)$$

In vector form this is

$$\nabla \cdot \mathbf{q} = -\nabla \cdot \nabla\phi = -\nabla^2\phi = 0 \qquad (7.3.10)$$

and is written $\nabla^2\phi = 0$. Equation (7.3.9) or (7.3.10) is the *Laplace equation*. Any function ϕ that satisfies the Laplace equation is a possible irrotational fluid-flow case. As there are an infinite number of solutions to the Laplace equation, each of which satisfies certain flow boundaries, the main problem is the selection of the proper function for the particular flow case.

Because ϕ appears to the first power in each term, Eq. (7.3.9), is a linear equation, and the sum of two solutions also is a solution; e.g., if ϕ_1 and ϕ_2 are solutions of Eq. (7.3.9), then $\phi_1 + \phi_2$ is a solution; thus

$$\nabla^2\phi_1 = 0 \qquad \nabla^2\phi_2 = 0$$

then

$$\nabla^2(\phi_1 + \phi_2) = \nabla^2\phi_1 + \nabla^2\phi_2 = 0$$

Similarly, if ϕ_1 is a solution, $C\phi_1$ is a solution if C is constant.

EXERCISES

7.3.1 Select the value of ϕ that satisfies continuity; (a) $x^2 + y^2$; (b) $\sin x$; (c) $\ln (x + y)$; (d) $x + y$; (e) none of these answers.

7.3.2 In irrotational flow of an ideal fluid (a) a velocity potential exists; (b) all particles must move in straight lines; (c) the motion must be uniform; (d) the flow is always steady; (e) the velocity must be zero at a boundary.

7.3.3 A function ϕ that satisfies the Laplace equation (a) must be linear in x and y; (b) is a possible case of rotational fluid flow; (c) does not necessarily satisfy the continuity equation; (d) is a possible fluid-flow case; (e) is none of these answers.

7.3.4 If both ϕ_1 and ϕ_2 are solutions of the Laplace equation, which of the following is also a solution? (a) $\phi_1 - 2\phi_2$; (b) $\phi_1 \phi_2$; (c) ϕ_1/ϕ_2; (d) ϕ_1^2; (e) none of these answers.

7.3.5 Select the relation that must hold if the flow is irrotational. (a) $\partial u/\partial y + \partial v/\partial x = 0$; (b) $\partial u/\partial x = \partial v/\partial y$; (c) $\partial^2 u/\partial x^2 + \partial^2 v/\partial y^2 = 0$; (d) $\partial u/\partial y = \partial v/\partial x$; (e) none of these answers.

7.4 INTEGRATION OF EULER'S EQUATIONS; BERNOULLI EQUATION

Equation (7.2.3) can be rearranged so that every term contains a partial derivative with respect to x. From Eq. (7.3.4)

$$v \frac{\partial u}{\partial y} = v \frac{\partial v}{\partial x} = \frac{\partial}{\partial x} \frac{v^2}{2} \qquad w \frac{\partial u}{\partial z} = w \frac{\partial w}{\partial x} = \frac{\partial}{\partial x} \frac{w^2}{2}$$

and from Eq. (7.3.7)

$$\frac{\partial u}{\partial t} = -\frac{\partial}{\partial x} \frac{\partial \phi}{\partial t}$$

Making these substitutions into Eq. (7.2.3) and rearranging give

$$\frac{\partial}{\partial x} \left(\frac{p}{\rho} + gh + \frac{u^2}{2} + \frac{v^2}{2} + \frac{w^2}{2} - \frac{\partial \phi}{\partial t} \right) = 0$$

As $u^2 + v^2 + w^2 = q^2$, the square of the speed,

$$\frac{\partial}{\partial x} \left(\frac{p}{\rho} + gh + \frac{q^2}{2} - \frac{\partial \phi}{\partial t} \right) = 0 \qquad (7.4.1)$$

Similarly, for the y and z directions,

$$\frac{\partial}{\partial y} \left(\frac{p}{\rho} + gh + \frac{q^2}{2} - \frac{\partial \phi}{\partial t} \right) = 0 \qquad (7.4.2)$$

$$\frac{\partial}{\partial z} \left(\frac{p}{\rho} + gh + \frac{q^2}{2} - \frac{\partial \phi}{\partial t} \right) = 0 \qquad (7.4.3)$$

The quantities within the parentheses are the same in Eqs. (7.4.1) to (7.4.3). Equation (7.4.1) states that the quantity is not a function of x, since the derivative with respect to x is zero. Similarly, the other equations show that the quantity is not a function of y or z. Therefore, it can be a function of t only, say $F(t)$:

$$\frac{p}{\rho} + gh + \frac{q^2}{2} - \frac{\partial \phi}{\partial t} = F(t) \qquad (7.4.4)$$

In steady flow $\partial \phi / \partial t = 0$ and $F(t)$ becomes a constant E:

$$\frac{p}{\rho} + gh + \frac{q^2}{2} = E \qquad (7.4.5)$$

The available energy is everywhere constant throughout the fluid. This is Bernoulli's equation for an irrotational fluid.

The pressure term can be separated into two parts, the hydrostatic pressure p_s and the dynamic pressure p_d, so that $p = p_s + p_d$. Inserting in Eq. (7.4.5) gives

$$gh + \frac{p_s}{\rho} + \frac{p_d}{\rho} + \frac{q^2}{2} = E$$

The first two terms may be written

$$gh + \frac{p_s}{\rho} = \frac{1}{\rho}(p_s + \gamma h)$$

with h measured vertically upward. The expression is a constant, since it expresses the hydrostatic law of variation of pressure. These two terms may be included in the constant E. After dropping the subscript on the dynamic pressure, there remains

$$\frac{p}{\rho} + \frac{q^2}{2} = E \qquad (7.4.6)$$

This simple equation permits the variation in pressure to be determined if the speed is known or vice versa. Assuming both the speed q_0 and the dynamic pressure p_0 to be known at one point,

$$\frac{p_0}{\rho} + \frac{q_0^2}{2} = \frac{p}{\rho} + \frac{q^2}{2}$$

or

$$p = p_0 + \frac{\rho q_0^2}{2}\left[1 - \left(\frac{q}{q_0}\right)^2\right] \qquad (7.4.7)$$

Example 7.2 A submarine moves through water at a speed of 30 ft/s. At a point A on the submarine 5 ft above the nose, the velocity of the submarine relative to the water is 50 ft/s. Determine the dynamic pressure difference between this point and the nose, and determine the difference in total pressure between the two points.

If the submarine is stationary and the water is moving past it, the velocity at the nose is zero and the velocity at A is 50 ft/s. By selecting the dynamic pressure at infinity as zero, from Eq. (7.4.6)

$$E = 0 + \frac{q_0^2}{2} = \frac{30^2}{2} = 450 \text{ ft·lb/slug}$$

For the nose

$$\frac{p}{\rho} = E = 450 \qquad p = 450 \times 1.935 = 870 \text{ lb/ft}^2$$

For point A

$$\frac{p}{\rho} = E - \frac{q^2}{2} = 450 - \frac{50^2}{2} \qquad \text{and} \qquad p = 1.935\left(\frac{30^2}{2} - \frac{50^2}{2}\right) = -1548 \text{ lb/ft}^2$$

Therefore, the difference in dynamic pressure is

$$-1548 - 870 = -2418 \text{ lb/ft}^2$$

The difference in total pressure can be obtained by applying Eq. (7.4.5) to point A and to the nose n,

$$gh_A + \frac{p_A}{\rho} + \frac{q_A^2}{2} = gh_n + \frac{p_n}{\rho} + \frac{q_n^2}{2}$$

Hence

$$p_A - p_n = \rho\left(gh_n - gh_A + \frac{q_n^2 - q_A^2}{2}\right) = 1.935\left(-5g - \frac{50^2}{2}\right)$$

$$= -2740 \text{ lb/ft}^2$$

It may also be reasoned that the actual pressure difference varies by 5γ from the dynamic pressure difference since A is 5 ft above the nose, or

$$-2418 - 5 \times 62.4 = -2740 \text{ lb/ft}^2$$

EXERCISES

7.4.1 Euler's equations of motion can be integrated when it is assumed that (a) the continuity equation is satisfied; (b) the fluid is incompressible; (c) a velocity potential exists and the density is constant; (d) the flow is rotational and incompressible; (e) the fluid is nonviscous.

7.4.2 The Bernoulli equation in steady ideal-fluid flow states that (a) the velocity is constant along a streamline; (b) the energy is constant along a streamline but may vary across streamlines; (c) when the speed increases, the pressure increases; (d) the energy is constant throughout the fluid; (e) the net flow rate into any small region must be zero.

7.4.3 An unsteady-flow case may be transformed into a steady-flow case (a) regardless of the nature of the problem; (b) when two bodies are moving toward each other in an infinite fluid; (c) when an unsymmetrical body is rotating in an infinite fluid; (d) when a single body translates in an infinite fluid; (e) under no circumstances.

7.5 STREAM FUNCTIONS; BOUNDARY CONDITIONS

Two stream functions are defined: one for two-dimensional flow, where all lines of motion are parallel to a fixed plane, say the xy plane, and the flow is identical in each of these planes, and the other for three-dimensional flow with axial symmetry, i.e., all flow lines are in planes intersecting the same line or axis, and the flow is identical in each of these planes.

Two-Dimensional Stream Function

If A, P represent two points in one of the flow planes, e.g., the xy plane (Fig. 7.4), and if the plane has unit thickness, the rate of flow across any two lines ACP, ABP must be the same if the density is constant and no fluid is created or destroyed within the region, as a consequence of continuity. Now, if A is a fixed point and P a movable point, the flow rate across any line connecting the two points is a function of the position of P. If this function is ψ, and if it is taken as a sign

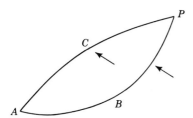

Figure 7.4 Fluid region showing the positive flow direction used in the definition of a stream function.

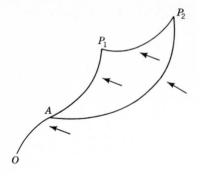

Figure 7.5 Flow between two points in a fluid region.

convention that it denotes the flow rate from right to left as the observer views the line from A looking toward P, then

$$\psi = \psi(x, y)$$

is defined as the stream function.

If ψ_1, ψ_2 represent the values of stream function at points P_1, P_2 (Fig. 7.5), respectively, then $\psi_2 - \psi_1$ is the flow across $P_1 P_2$ and is independent of the location of A. Taking another point O in the place of A changes the values of ψ_1, ψ_2 by the same amount, viz., the flow across OA. Then ψ is indeterminate to the extent of an arbitrary constant.

The velocity components u, v in the x, y directions can be obtained from the stream function. In Fig. 7.6a, the flow $\delta\psi$ across $\overline{AP} = \delta y$, from right to left, is $-u\,\delta y$, or

$$u = -\frac{\delta\psi}{\delta y} = -\frac{\partial\psi}{\partial y} \tag{7.5.1}$$

and similarly

$$v = \frac{\delta\psi}{\delta x} = \frac{\partial\psi}{\partial x} \tag{7.5.2}$$

In words, the partial derivative of the stream function with respect to any direction gives the velocity component $+90°$ (counterclockwise) to that direction. In plane polar coordinates

$$v_r = -\frac{1}{r}\frac{\partial\psi}{\partial\theta} \qquad v_\theta = \frac{\partial\psi}{\partial r}$$

from Fig. 7.6b.

When the two points P_1, P_2 of Fig. 7.5 lie on the same streamline, $\psi_1 - \psi_2 = 0$ as there is no flow across a streamline. Hence, a streamline is given by $\psi = $ const. By comparing Eqs. (7.3.3) with Eqs. (7.5.1) and (7.5.2),

$$\frac{\partial\phi}{\partial x} = \frac{\partial\psi}{\partial y} \qquad \frac{\partial\phi}{\partial y} = -\frac{\partial\psi}{\partial x} \tag{7.5.3}$$

These are the Cauchy-Riemann equations.

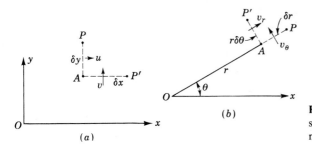

Figure 7.6 Selection of path to show relation of velocity components to stream function.

(a)

(b)

By Eqs. (7.5.3) a stream function can be found for each velocity potential. If the velocity potential satisfies the Laplace equation, the stream function also satisfies it. Hence, the stream function may be considered as velocity potential for another flow case.

Stokes' Stream Function for Axially Symmetric Flow

In any one of the planes through the axis of symmetry select two points A, P such that A is fixed and P is variable. Draw a line connecting AP. The flow through the surface generated by rotating AP about the axis of symmetry is a function of the position of P. Let his function be $2\pi\psi$, and let the axis of symmetry be the x axis of a cartesian system of reference. Then ψ is a function of x and $\hat{\omega}$, where

$$\hat{\omega} = \sqrt{y^2 + z^2}$$

is the distance from P to the x axis. The surfaces $\psi = $ const are stream surfaces.

To find the relation between ψ and the velocity components u, v' parallel to the x axis and the $\hat{\omega}$ axis (perpendicular to the x axis), respectively, a procedure similar to that for two-dimensional flow is employed. Let PP' be an infinitesimal step parallel first to $\hat{\omega}$ and then to x; that is, $PP' = \delta\hat{\omega}$ and then $PP' = \delta x$. The resulting relations between stream function and velocity are given by

$$-2\pi\hat{\omega}\,\delta\hat{\omega}\,u = 2\pi\,\delta\psi \quad \text{and} \quad 2\pi\hat{\omega}\,\delta x\,v' = 2\pi\,\delta\psi$$

Solving for u, v' gives

$$u = -\frac{1}{\hat{\omega}}\frac{\partial\psi}{\partial\hat{\omega}} \qquad v' = \frac{1}{\hat{\omega}}\frac{\partial\psi}{\partial x} \qquad (7.5.4)$$

The same sign convention is used as in the two-dimensional case.

The relations between stream function and potential function are

$$\frac{\partial\phi}{\partial x} = \frac{1}{\hat{\omega}}\frac{\partial\psi}{\partial\hat{\omega}} \qquad \frac{\partial\phi}{\partial\hat{\omega}} = -\frac{1}{\hat{\omega}}\frac{\partial\psi}{\partial x} \qquad (7.5.5)$$

In three-dimensional flow with axial symmetry, ψ has the dimensions $L^3 T^{-1}$, or volume per unit time.

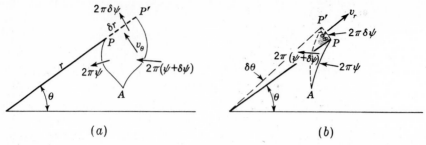

(a) (b)

Figure 7.7 Displacement of P to show the relation between velocity components and Stokes' stream function.

The stream function is used for flow about bodies of revolution that are frequently expressed most readily in spherical polar coordinates. Let r be the distance from the origin and θ be the polar angle; the meridian angle is not needed because of axial symmetry. Referring to Fig. 7.7a and b,

$$2\pi r \sin \theta \; \delta r \; v_\theta = 2\pi \; \delta \psi \qquad -2\pi r \sin \theta \; r \; \delta\theta \; v_r = 2\pi \; \delta\psi$$

from which

$$v_\theta = \frac{1}{r \sin \theta} \frac{\partial \psi}{\partial r} \qquad v_r = -\frac{1}{r^2 \sin \theta} \frac{\partial \psi}{\partial \theta} \qquad (7.5.6)$$

and

$$\frac{1}{\sin \theta} \frac{\partial \psi}{\partial \theta} = r^2 \frac{\partial \phi}{\partial r} \qquad \frac{\partial \psi}{\partial r} = -\sin \theta \frac{\partial \phi}{\partial \theta} \qquad (7.5.7)$$

These expressions are useful in dealing with flow about spheres, ellipsoids, and disks and through apertures.

Boundary Conditions

At a fixed boundary the velocity component normal to the boundary must be zero at every point on the boundary (Fig. 7.8):

$$\mathbf{q} \cdot \mathbf{n}_1 = 0 \qquad (7.5.8)$$

$\mathbf{n}_1$ is a unit vector normal to the boundary. In scalar notation this is easily expressed in terms of the velocity potential

$$\frac{\partial \phi}{\partial n} = 0 \qquad (7.5.9)$$

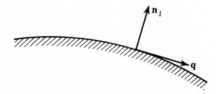

Figure 7.8 Notation for boundary condition at a fixed boundary.

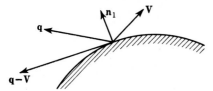

Figure 7.9 Notation for boundary condition at a moving boundary.

at all points on the boundary. For a moving boundary (Fig. 7.9), where the boundary point has the velocity **V**, the fluid velocity component normal to the boundary must equal the velocity of the boundary normal to the boundary; thus

$$\mathbf{q} \cdot \mathbf{n}_1 = \mathbf{V} \cdot \mathbf{n}_1 \qquad (7.5.10)$$

or

$$(\mathbf{q} - \mathbf{V}) \cdot \mathbf{n}_1 = 0 \qquad (7.5.11)$$

For two fluids in contact, a dynamical boundary condition is required; viz., the pressure must be continuous across the interface.

A stream surface in steady flow (fixed boundaries) satisfies the condition for a boundary and may be taken as a solid boundary.

EXERCISES

7.5.1 The Stokes' stream function applies to (*a*) all three-dimensional ideal-fluid-flow cases; (*b*) ideal (nonviscous) fluids only; (*c*) irrotational flow only; (*d*) cases of axial symmetry; (*e*) none of these cases.

7.5.2 The Stokes' stream function has the value $\psi = 1$ at the origin and the value $\psi = 2$ at (1, 1, 1). The discharge through the surface between these points is (*a*) 1; (*b*) π; (*c*) 2π; (*d*) 4; (*e*) none of these answers.

7.5.3 The two-dimensional stream function (*a*) is constant along an equipotential surface; (*b*) is constant along a streamline; (*c*) is defined for irrotational flow only; (*d*) relates velocity and pressure; (*e*) is none of these answers.

7.5.4 In two-dimensional flow $\psi = 4$ ft^2/s at (0, 2) and $\psi = 2$ ft^2/s at (0, 1). The discharge between the two points is (*a*) from left to right; (*b*) 4π cfs/ft; (*c*) 2 cfs/ft; (*d*) $1/\pi$ cfs/ft; (*e*) none of these answers.

7.5.5 The boundary condition for steady flow of an ideal fluid is that the (*a*) velocity is zero at the boundary; (*b*) velocity component normal to the boundary is zero; (*c*) velocity component tangent to the boundary is zero; (*d*) boundary surface must be stationary; (*e*) continuity equation must be satisfied.

7.6 THE FLOW NET

In two-dimensional flow the flow net is of great benefit; it is taken up in this section.

The line given by $\phi(x, y)$ = const is called an *equipotential* line. It is a line along which the value of ϕ (the velocity potential) does not change. Since velocity v_s in any direction s is given by

$$v_s = -\frac{\partial \phi}{\partial s} = -\lim_{\Delta s \to 0} \frac{\Delta \phi}{\Delta s}$$

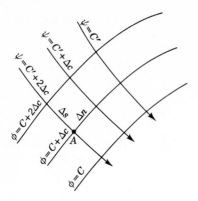

Figure 7.10 Elements of a flow net.

and $\Delta\phi$ is zero for two closely spaced points on an equipotential line, the velocity vector has no component in the direction defined by the line through the two points. In the limit as $\Delta s \to 0$ this proves that there is no velocity component tangent to an equipotential line and, therefore, the velocity vector must be everywhere *normal* to an equipotential line (except at singular points where the velocity is zero or infinite).

The line $\psi(x, y) = \text{const}$ is a streamline and is everywhere tangent to the velocity vector. Streamlines and equipotential lines are therefore *orthogonal*; i.e., they intersect at right angles, except at singular points. A *flow net* is composed of a family of equipotential lines and a corresponding family of streamlines with the constants varying in arithmetical progression. It is customary to let the change in constant between adjacent equipotential lines and adjacent streamlines be the same, for example, Δc. In Fig. 7.10, if the distance between streamlines is Δn and the distance between equipotential lines is Δs at some small region in the flow net, the approximate velocity v_s is then given in terms of the spacing of the equipotential lines [Eq. (7.3.7)],

$$v_s \approx -\frac{\Delta\phi}{\Delta s} = -\frac{-\Delta c}{\Delta s} = \frac{\Delta c}{\Delta s}$$

or in terms of the spacing of streamlines [Eqs. (7.5.1) and (7.5.2)],

$$v_s \approx \frac{\Delta\psi}{\Delta n} = \frac{\Delta c}{\Delta n}$$

These expressions are approximate when Δc is finite, but when Δc becomes very small, the expressions become exact and yield velocity at a point. As both velocities referred to are the same, the equations show that $\Delta s = \Delta n$, or that the flow net consists of an orthogonal grid that reduces to perfect squares in the limit as the grid size approaches zero.

Once a flow net has been found by any means to satisfy the boundary conditions and to form an orthogonal net reducing to perfect squares in the limit as the number of lines is increased, the flow net is the only solution for the particular boundaries, as uniqueness theorems in hydrodynamics prove. In steady flow when the boundaries are stationary, the boundaries themselves become part of the flow net, as they are streamlines. The problem of finding the flow net to satisfy given fixed boundaries may be considered purely as a *graphical exercise*, i.e., the construction of an orthogonal system of lines that compose the boundaries and reduce to perfect squares in the limit as the number of lines increases. This is one of the practical methods employed in two-dimensional-flow analysis, although it usually requires many attempts and much erasing.

Another practical method of obtaining a flow net for a particular set of fixed boundaries is the *electrical analogy*. The boundaries in a model are formed out of strips of nonconducting material mounted on a flat nonconducting surface, and the end equipotential lines are formed out of a conducting strip, e.g., brass or copper. An electrolyte (conducting liquid) is placed at uniform depth in the flow space and a voltage potential is applied to the two end conducting strips. By means of a probe and a voltmeter, lines with constant drop in voltage from one end are mapped out and plotted. These are equipotential lines. By reversing the process and making the flow boundaries out of conducting material and the end equipotential lines from nonconducting material, the streamlines are mapped.

A special conducting paper, called *Teledeltos paper*, may be used in place of a tank with an electrolyte. Silver ink is used to form a conducting strip or line having constant voltage. One cuts the paper to the size and shape needed, places the constant-voltage lines on the paper with a heavy line of silver ink, then marks the intermediate points of constant voltage directly on the paper, using the same circuits as with an electrolyte.

The relaxation method† numerically determines the value of potential function at points throughout the flow, usually located at the intersections of a square grid. The Laplace equation is written as a difference equation, and it is shown that the value of potential function at a grid point is the average of the four values at the neighboring grid points. Near the boundaries special formulas are required. With values known at the boundaries, each grid point is computed based on the assumed values at the neighboring grid points; then these values are improved by repeating the process until the changes are within the desired accuracy. This method is particularly convenient for solution with high-speed digital computers.

Use of the Flow Net

After a flow net for a given boundary configuration has been obtained, it may be used for all irrotational flows with geometrically similar boundaries. It is necessary to know the velocity at a single point and the pressure at one point. Then, by use of the flow net, the velocity can be determined at every other point. Application of the Bernoulli equation [Eq. (7.4.7)] produces the dynamic pressure. If the

---

† C.-S. Yih, Ideal-Fluid Flow, p. 4–67 in V. L. Streeter (ed.), "Handbook of Fluid Dynamics," McGraw-Hill, New York, 1961.

velocity is known, e.g., at A (Fig. 7.10), Δn or Δs can be scaled from the adjacent lines. Then $\Delta c \approx \Delta n v_s \approx \Delta s v_s$. With the constant Δc determined for the whole grid in this manner, measurement of Δs or Δn at any other point permits the velocity to be computed there,

$$v_s \approx \frac{\Delta c}{\Delta s} = \frac{\Delta c}{\Delta n}$$

The concepts underlying the flow net have been developed for irrotational flow of an ideal fluid. Because of the similarity of differential equations describing groundwater flow and irrotational flow, the flow net can also be used to determine streamlines and lines of constant piezometric head $(h + p/\gamma)$ for percolation through homogeneous porous media. The flow cases of the next section may then be interpreted in terms of the very rotational, viscous flow at extremely small velocities through a porous medium.

EXERCISE

7.6.1 An equipotential surface (a) has no velocity component tangent to it; (b) is composed of streamlines; (c) is a stream surface; (d) is a surface of constant dynamic pressure; (e) is none of these answers.

7.7 TWO-DIMENSIONAL FLOW

Two simple flow cases that may be interpreted for flow along straight boundaries are first examined; then the source, vortex, doublet, uniform flow, and flow around a cylinder, with and without circulation, are discussed.

Flow around a Corner

The potential function

$$\phi = A(x^2 - y^2)$$

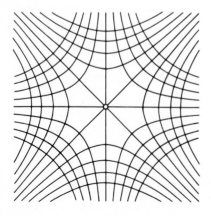

Figure 7.11 Flow net for flow around 90° bend.

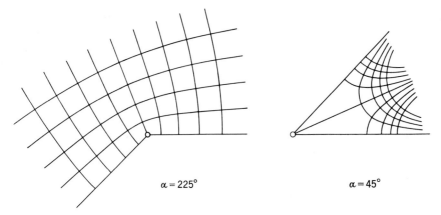

$\alpha = 225°$ $\alpha = 45°$

Figure 7.12 Flow net for flow along two inclined surfaces.

has as its stream function

$$\psi = 2Axy = Ar^2 \sin 2\theta$$

in which r and θ are polar coordinates. It is plotted for equal-increment changes in ϕ and ψ in Fig. 7.11. Conditions at the origin are not defined, as it is a stagnation point. As any of the streamlines may be taken as fixed boundaries, the plus axes may be taken as walls, yielding flow into a 90° corner. The equipotential lines are hyperbolas having axes coincident with the coordinate axes and asymptotes given by $y = \pm x$. The streamlines are rectangular hyperbolas having $y = \pm x$ as axes and the coordinate axes as asymptotes. From the polar form of the stream function it is noted that the two lines $\theta = 0$ and $\theta = \pi/2$ are the streamline $\psi = 0$.

This case may be generalized to yield flow around a corner with angle α. By examining

$$\phi = Ar^{\pi/\alpha} \cos \frac{\pi\theta}{\alpha} \qquad \psi = Ar^{\pi/\alpha} \sin \frac{\pi\theta}{\alpha}$$

it is noted the streamline $\psi = 0$ is now given by $\theta = 0$ and $\theta = \alpha$. Two flow nets are shown in Fig. 7.12, for the cases $\alpha = 225°$ and $\alpha = 45°$.

Source

A line normal to the xy plane, from which fluid is imagined to flow uniformly in all directions *at right angles* to it, is a source. It appears as a point in the customary two-dimensional flow diagram. The total flow per unit time per unit length of line is called the *strength* of the source. As the flow is in radial lines from the source, the velocity a distance r from the source is determined by the strength divided by the flow area of the cylinder, or $2\pi\mu/2\pi r$, in which the strength is $2\pi\mu$. Then, since by

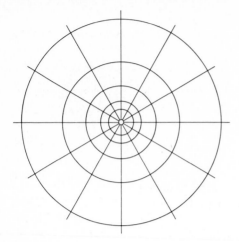

Figure 7.13 Flow net for source or vortex.

Eq. (7.3.7) the velocity in any direction is given by the negative derivative of the velocity potential with respect to the direction,

$$-\frac{\partial \phi}{\partial r} = \frac{\mu}{r} \qquad \frac{\partial \phi}{\partial \theta} = 0$$

and

$$\phi = -\mu \ln r$$

is the velocity potential, in which ln indicates the natural logarithm and r is the distance from the source. This value of ϕ satisfies the Laplace equation in two dimensions.

The streamlines are radial lines from the source, i.e.,

$$\frac{\partial \psi}{\partial r} = 0 \qquad -\frac{1}{r}\frac{\partial \psi}{\partial \theta} = \frac{\mu}{r}$$

From the second equation

$$\psi = -\mu\theta$$

Lines of constant ϕ (equipotential lines) and constant ψ are shown in Fig. 7.13. A *sink* is a negative source, a line into which fluid is flowing.

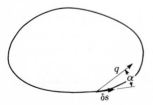

Figure 7.14 Notation for definition of circulation.

Vortex

In examining the flow case given by selecting the stream function for the source as a velocity potential,

$$\phi = -\mu\theta \qquad \psi = \mu \ln r$$

which also satisfies the Laplace equation, it is seen that the equipotential lines are radial lines and the streamlines are circles. The velocity is in a tangential direction only, since $\partial\phi/\partial r = 0$. It is

$$q = -\frac{1}{r}\frac{\partial\phi}{\partial\theta} = \frac{\mu}{r}$$

since $r\,\delta\theta$ is the length element in the tangential direction.

In referring to Fig. 7.14, the *flow along a closed curve* is called the *circulation*. The flow along an element of the curve is defined as the product of the length element δs of the curve and the component of the velocity tangent to the curve, $q \cos \alpha$. Hence, the circulation Γ around a closed path C is

$$\Gamma = \int_C q \cos \alpha \, ds = \int_C \mathbf{q} \cdot d\mathbf{s}$$

The velocity distribution given by the equation $\phi = -\mu\theta$ is for the *vortex* and is such that the circulation around any closed path that contains the vortex is constant. The value of the circulation is the strength of the vortex. By selecting any circular path of radius r to determine the circulation, $\alpha = 0°$, $q = \mu/r$, and $ds = r\,d\theta$; hence,

$$\Gamma = \int_C q \cos \alpha \, ds = \int_0^{2\pi} \frac{\mu}{r} r \, d\theta = 2\pi\mu$$

At the point $r = 0$, $q = \mu/r$ goes to infinity; hence, this point is called a singular point. Figure 7.13 shows the equipotential lines and streamlines for the vortex.

Doublet

The two-dimensional doublet is defined as the limiting case as a source and sink of equal strength approach each other so that the product of their strength and the distance between them remains a constant $2\pi\mu$. μ is called the *strength* of the

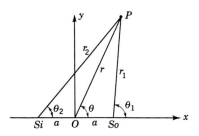

Figure 7.15 Notation for derivation of a two-dimensional doublet.

doublet. The axis of the doublet is from the sink toward the source, i.e., the line along which they approach each other.

In Fig. 7.15 a source is located at $(a, 0)$ and a sink of equal strength at $(-a, 0)$. The velocity potential for both, at some point P, is

$$\phi = -m \ln r_1 + m \ln r_2$$

with r_1, r_2 measured from source and sink, respectively, to the point P. Thus $2\pi m$ is the strength of source and sink. To take the limit as a approaches zero for $2am = \mu$, the form of the expression for ϕ must be altered. The terms r_1 and r_2 may be expressed in terms of the polar coordinates r, θ by the cosine law, as follows:

$$r_1^2 = r^2 + a^2 - 2ar \cos \theta = r^2 \left[1 + \left(\frac{a}{r}\right)^2 - 2\frac{a}{r} \cos \theta \right]$$

$$r_2^2 = r^2 + a^2 + 2ar \cos \theta = r^2 \left[1 + \left(\frac{a}{r}\right)^2 + 2\frac{a}{r} \cos \theta \right]$$

Rewriting the expression for ϕ with these relations gives

$$\phi = -\frac{m}{2} (\ln r_1^2 - \ln r_2^2) = -\frac{m}{2} \left\{ \ln r^2 + \ln \left[1 + \left(\frac{a}{r}\right)^2 - 2\frac{a}{r} \cos \theta \right] \right.$$

$$\left. - \ln r^2 - \ln \left[1 + \left(\frac{a}{r}\right)^2 + 2\frac{a}{r} \cos \theta \right] \right\}$$

By the series expression,

$$\ln (1 + x) = x - \frac{x^2}{2} + \frac{x^3}{3} - \frac{x^4}{4} + \cdots$$

$$\phi = -\frac{m}{2} \left\{ \left(\frac{a}{r}\right)^2 - 2\frac{a}{r} \cos \theta - \frac{1}{2} \left[\left(\frac{a}{r}\right)^2 - 2\frac{a}{r} \cos \theta \right]^2 \right.$$

$$+ \frac{1}{3} \left[\left(\frac{a}{r}\right)^2 - 2\frac{a}{r} \cos \theta \right]^3 - \cdots - \left[\left(\frac{a}{r}\right)^2 + 2\frac{a}{r} \cos \theta \right]$$

$$\left. + \frac{1}{2} \left[\left(\frac{a}{r}\right)^2 + 2\frac{a}{r} \cos \theta \right]^2 - \frac{1}{3} \left[\left(\frac{a}{r}\right)^2 + 2\frac{a}{r} \cos \theta \right]^3 + \cdots \right\}$$

After simplifying,

$$\phi = 2am \left[\frac{\cos \theta}{r} + \left(\frac{a}{r}\right)^2 \frac{\cos \theta}{r} - \left(\frac{a}{r}\right)^4 \frac{\cos \theta}{r} - \frac{4}{3} \left(\frac{a}{r}\right)^2 \frac{\cos^3 \theta}{r} + \cdots \right]$$

Now, if $2am = \mu$ and if the limit is taken as a approaches zero,

$$\phi = \frac{\mu \cos \theta}{r}$$

which is the velocity potential for a two-dimensional doublet at the origin, with axis in the $+x$ direction.

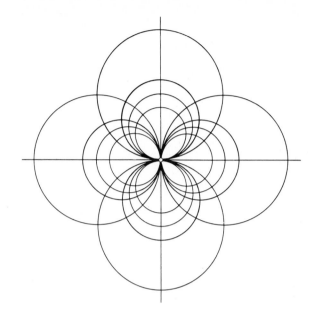

Figure 7.16 Equipotential lines and streamlines for the two-dimensional doublet.

Using the relations

$$v_r = -\frac{\partial \phi}{\partial r} = -\frac{1}{r}\frac{\partial \psi}{\partial \theta} \qquad v_\theta = -\frac{1}{r}\frac{\partial \phi}{\partial \theta} = \frac{\partial \psi}{\partial r}$$

gives for the doublet

$$\frac{\partial \psi}{\partial \theta} = -\frac{\mu \cos \theta}{r} \qquad \frac{\partial \psi}{\partial r} = \frac{\mu}{r^2}\sin \theta$$

After integrating,

$$\psi = -\frac{\mu \sin \theta}{r}$$

is the stream function for the doublet. The equations in cartesian coordinates are

$$\phi = \frac{\mu x}{x^2 + y^2} \qquad \psi = -\frac{\mu y}{x^2 + y^2}$$

Rearranging gives

$$\left(x - \frac{\mu}{2\phi}\right)^2 + y^2 = \frac{\mu^2}{4\phi^2} \qquad x^2 + \left(y + \frac{\mu}{2\psi}\right)^2 = \frac{\mu^2}{4\psi^2}$$

The lines of constant ϕ are circles through the origin with centers on the x axis, and the streamlines are circles through the origin with centers on the y axis, as shown in Fig. 7.16. The origin is a singular point where the velocity goes to infinity.

Uniform Flow

Uniform flow in the $-x$ direction, $u = -U$, is expressed by

$$\phi = Ux \qquad \psi = Uy$$

In polar coordinates,

$$\phi = Ur \cos \theta \qquad \psi = Ur \sin \theta$$

Flow around a Circular Cylinder

The addition of the flow due to a doublet and a uniform flow results in flow around a circular cylinder; thus

$$\phi = Ur \cos \theta + \frac{\mu \cos \theta}{r} \qquad \psi = Ur \sin \theta - \frac{\mu \sin \theta}{r}$$

As a streamline in steady flow is a possible boundary, the streamline $\psi = 0$ is given by

$$0 = \left(Ur - \frac{\mu}{r} \right) \sin \theta$$

which is satisfied by $\theta = 0$, π, or by the value of r that makes

$$Ur - \frac{\mu}{r} = 0$$

If this value is $r = a$, which is a circular cylinder, then

$$\mu = Ua^2$$

and the streamline $\psi = 0$ is the x axis and the circle $r = a$. The potential and stream functions for uniform flow around a circular cylinder of radius a are, by substitution of the value of μ,

$$\phi = U\left(r + \frac{a^2}{r} \right) \cos \theta \qquad \psi = U\left(r - \frac{a^2}{r} \right) \sin \theta$$

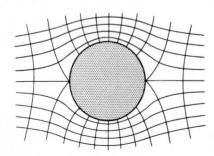

Figure 7.17 Equipotential lines and streamlines for flow around a circular cylinder.

for the uniform flow in the $-x$ direction. The equipotential lines and streamlines for this case are shown in Fig. 7.17.

The velocity at any point in the flow can be obtained from either the velocity potential or the stream function. On the surface of the cylinder the velocity is necessarily tangential and is expressed by $\partial\psi/\partial r$ for $r = a$; thus

$$q\Big|_{r=a} = U\left(1 + \frac{a^2}{r^2}\right)\sin\theta\Big|_{r=a} = 2U\sin\theta$$

The velocity is zero (stagnation point) at $\theta = 0$, π and has maximum values of $2U$ at $\theta = \pi/2$, $3\pi/2$. For the dynamic pressure zero at infinity, with Eq. (7.4.7) for $p_0 = 0$, $q_0 = U$,

$$p = \frac{\rho}{2}U^2\left[1 - \left(\frac{q}{U}\right)^2\right]$$

which holds for any point in the plane except the origin. For points on the cylinder,

$$p = \frac{\rho}{2}U^2(1 - 4\sin^2\theta)$$

The maximum pressure, which occurs at the stagnation points, is $\rho U^2/2$; and the minimum pressure, at $\theta = \pi/2$, $3\pi/2$, is $-3\rho U^2/2$. The points of zero dynamic pressure are given by $\sin\theta = \pm\frac{1}{2}$, or $\theta = \pm\pi/6$, $\pm5\pi/6$. A cylindrical pitot-static tube is made by providing three openings in a cylinder, at 0 and $\pm30°$, as the difference in pressure between 0 and $\pm30°$ is the dynamic pressure $\rho U^2/2$.

The drag on the cylinder is shown to be zero by integration of the x component of the pressure force over the cylinder; thus

$$\text{Drag} = \int_0^{2\pi} pa\cos\theta\,d\theta = \frac{\rho a U^2}{2}\int_0^{2\pi}(1 - 4\sin^2\theta)\cos\theta\,d\theta = 0$$

Similarly, the lift force on the cylinder is zero.

Flow around a Circular Cylinder with Circulation

The addition of a vortex to the doublet and the uniform flow results in flow around a circular cylinder with circulation,

$$\phi = U\left(r + \frac{a^2}{r}\right)\cos\theta - \frac{\Gamma}{2\pi}\theta \qquad \psi = U\left(r - \frac{a^2}{r}\right)\sin\theta + \frac{\Gamma}{2\pi}\ln r$$

The streamline $\psi = (\Gamma/2\pi)\ln a$ is the circular cylinder $r = a$. At great distances from the origin, the velocity remains $u = -U$, showing that flow around a circular cylinder is maintained with addition of the vortex. Some of the streamlines are shown in Fig. 7.18.

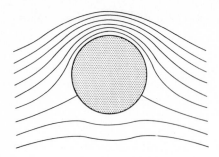

Figure 7.18 Streamlines for flow around a circular cylinder with circulation.

The velocity at the surface of the cylinder, necessarily tangent to the cylinder, is

$$q = \frac{\partial \psi}{\partial r}\bigg|_{r=a} = 2U \sin \theta + \frac{\Gamma}{2\pi a}$$

Stagnation points occur when $q = 0$; that is,

$$\sin \theta = -\frac{\Gamma}{4\pi U a}$$

When the circulation is $4\pi U a$, the two stagnation points coincide at $r = a$, $\theta = -\pi/2$. For larger circulation, the stagnation point moves away from the cylinder.

The pressure at the surface of the cylinder is

$$p = \frac{\rho U^2}{2}\left[1 - \left(2\sin\theta + \frac{\Gamma}{2\pi a U}\right)^2\right]$$

The drag again is zero. The lift, however, becomes

$$\text{Lift} = -\int_0^{2\pi} pa \sin \theta \, d\theta$$

$$= -\frac{\rho a U^2}{2}\int_0^{2\pi}\left[1 - \left(2\sin\theta + \frac{\Gamma}{2\pi a U}\right)^2\right]\sin\theta \, d\theta = \rho U \Gamma$$

showing that the lift is directly proportional to the density of fluid, the approach velocity U, and the circulation Γ. This thrust, which acts at right angles to the approach velocity, is referred to as the *Magnus effect*. The Flettner rotor ship was designed to utilize this principle by mounting circular cylinders with axes vertical on the ship and then mechanically rotating the cylinders to provide circulation. Airflow around the rotors produces the thrust at right angles to the relative wind direction. The close spacing of streamlines along the upper side of Fig. 7.18 indicates that the velocity is high there and that the pressure must then be correspondingly low.

The theoretical flow around a circular cylinder with circulation can be transformed† into flow around an airfoil with the same circulation and the same lift.

The airfoil develops its lift by producing a circulation around it due to its shape. It can be shown‡ that the lift is $\rho U \Gamma$ for any cylinder in two-dimensional flow. The angle of inclination of the airfoil relative to the approach velocity (angle of attack) greatly affects the circulation. For large angles of attack, the flow does not follow the wing profile, and the theory breaks down.

It should be mentioned that all two-dimensional ideal-fluid-flow cases may be conveniently handled by complex-variable theory and by a system of *conformal mapping*, which transforms a flow net from one configuration to another by a suitable complex-variable mapping function.

Example 7.3 A source with strength 0.2 m³/s·m and a vortex with strength 1 m²/s are located at the origin. Determine the equations for velocity potential and stream function. What are the velocity components at $x = 1$ m, $y = 0.5$ m?

The velocity potential for the source is

$$\phi = -\frac{0.2}{2\pi} \ln r \qquad \text{m}^2/\text{s}$$

and the corresponding stream function is

$$\psi = -\frac{0.2}{2\pi} \theta \qquad \text{m}^2/\text{s}$$

The velocity potential for the vortex is

$$\phi = -\frac{1}{2\pi} \theta \qquad \text{m}^2/\text{s}$$

and the corresponding stream function is

$$\psi = \frac{1}{2\pi} \ln r \qquad \text{m}^2/\text{s}$$

Adding the respective functions gives

$$\phi = -\frac{1}{\pi}\left(0.1 \ln r + \frac{\theta}{2}\right) \qquad \text{and} \qquad \psi = -\frac{1}{\pi}\left(0.1\theta - \frac{1}{2} \ln r\right)$$

The radial and tangential velocity components are

$$v_r = -\frac{\partial \phi}{\partial r} = \frac{1}{10\pi r} \qquad v_\theta = -\frac{1}{r}\frac{\partial \phi}{\partial \theta} = \frac{1}{2\pi r}$$

At $(1, 0.5)$, $r = \sqrt{1^2 + 0.5^2} = 1.117$ m, $v_r = 0.0285$ m/s, $v_\theta = 0.143$ m/s.

---

† V. L. Streeter, "Fluid Dynamics," pp. 137–155, McGraw-Hill, New York, 1948.
‡ *Ibid.*

EXERCISES

7.7.1 Select the relation that must hold in two-dimensional, irrotational flow: (a) $\partial\phi/\partial x = \partial\psi/\partial y$; (b) $\partial\phi/\partial x = -\partial\psi/\partial y$; (c) $\partial\phi/\partial y = \partial\psi/\partial x$; (d) $\partial\phi/\partial x = \partial\phi/\partial y$; (e) none of these answers.

7.7.2 A source in two-dimensional flow (a) is a point from which fluid is imagined to flow outward uniformly in all directions; (b) is a line from which fluid is imagined to flow uniformly in all directions at right angles to it; (c) has a strength defined as the speed at unit radius; (d) has streamlines that are concentric circles; (e) has a velocity potential independent of the radius.

7.7.3 The two-dimensional vortex (a) has a strength given by the circulation around a path enclosing the vortex; (b) has radial streamlines; (c) has a zero circulation around it; (d) has a velocity distribution that varies directly as the radial distance from the vortex; (e) creates a velocity distribution that has rotation throughout the fluid.

PROBLEMS

7.1 Compute the gradient of the following two-dimensional scalar functions:

$\quad$ (a) $\phi = -2\ln(x^2 + y^2)$ $\quad$ (b) $\phi = Ux + Vy$ $\quad$ (c) $\phi = 2xy$

7.2 Compute the divergence of the gradients of ϕ found in Prob. 7.1.

7.3 Compute the curl of the gradients of ϕ found in Prob. 7.1.

7.4 For $\mathbf{q} = \mathbf{i}(x + y) + \mathbf{j}(y + z) + \mathbf{k}(x^2 + y^2 + z^2)$ find the components of rotation at (2, 2, 2).

7.5 Derive the equation of continuity for two-dimensional flow in polar coordinates by equating the net efflux from a small polar element to zero (Fig. 7.19). It is

$$\frac{\partial v_r}{\partial r} + \frac{v_r}{r} + \frac{1}{r}\frac{\partial v_\theta}{\partial \theta} = 0$$

7.6 The x component of velocity is $u = x^2 + z^2 + 5$, and the y component is $v = y^2 + z^2$. Find the simplest z component of velocity that satisfies continuity.

7.7 A velocity potential in two-dimensional flow is $\phi = y + x^2 - y^2$. Find the stream function for this flow.

7.8 The two-dimensional stream function for a flow is $\psi = 9 + 6x - 4y + 7xy$. Find the velocity potential.

7.9 Derive the partial differential equations relating ϕ and ψ for two-dimensional flow in plane polar coordinates.

7.10 From the continuity equation in polar coordinates in Prob. 7.5, derive the Laplace equation in the same coordinate system.

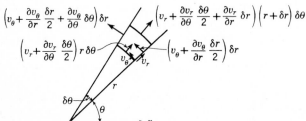

$\left(v_\theta + \dfrac{\partial v_\theta}{\partial r}\dfrac{\delta r}{2} + \dfrac{\partial v_\theta}{\partial \theta}\delta\theta\right)\delta r$ $\qquad$ $\left(v_r + \dfrac{\partial v_r}{\partial \theta}\dfrac{\delta\theta}{2} + \dfrac{\partial v_r}{\partial r}\delta r\right)\left(r + \delta r\right)\delta\theta$

$\left(v_r + \dfrac{\partial v_r}{\partial \theta}\dfrac{\delta\theta}{2}\right)r\,\delta\theta$ $\qquad$ $\left(v_\theta + \dfrac{\partial v_\theta}{\partial r}\dfrac{\delta r}{2}\right)\delta r$

Figure 7.19 Problems 7.5, 7.10.

7.11 Does the function $\phi = 1/r$ satisfy the Laplace equation in two dimensions? In three-dimensional flow is it satisfied?

7.12 By use of the equations developed in Prob. 7.9 find the two-dimensional stream function for $\phi = \ln r$.

7.13 Find the Stokes' stream function for $\phi = 1/r$.

7.14 For the Stokes' stream function $\psi = 9r^2 \sin^2 \theta$, find ϕ in cartesian coordinates.

7.15 In Prob. 7.14 what is the discharge between stream surfaces through the points $r = 1$, $\theta = 0$ and $r = 1$, $\theta = \pi/4$?

7.16 Write the boundary conditions for steady flow around a sphere, of radius a, at its surface and at infinity.

7.17 A circular cylinder of radius a has its center at the origin and is translating with velocity V in the direction. Write the boundary condition in terms of ϕ that is to be satisfied at its surface and at infinity.

7.18 A circular cylinder 8 ft in diameter rotates at 500 rpm. When in an airstream, $\rho = 0.002$ slug/ft^3, moving at 400 ft/s, what is the lift force per foot of cylinder, assuming 90 percent efficiency in developing circulation from the rotation?

7.19 Show that, if two stream functions ψ_1 and ψ_2 satisfy the Laplace equation, $\nabla^2\psi = 0$ for $\psi = \psi_1 + \psi_2$.

7.20 Show that, if u_1, v_1 and u_2, v_2 are the velocity components of two velocity potentials ϕ_1 and ϕ_2 which satisfy the Laplace equation, then for $\phi = \phi_1 + \phi_2$ the velocity components are $u = u_1 + u_2$ and $v = v_1 + v_2$.

7.21 A two-dimensional source is located at $(1, 0)$ and another one of the same strength at $(-1, 0)$. Construct the velocity vector at $(0, 0)$, $(0, 1)$, $(0, -1)$, $(0, -2)$, and $(1, 1)$. (*Hint:* By using the results of Prob. 7.20, draw the velocity components by adding the individual velocity components induced at the point in question by each source, without regard to the other, due to its strength and location.)

7.22 Determine the velocity potential for a source located at $(1, 0)$. Write the equation for the velocity potential for the source system described in Prob. 7.21.

7.23 Draw a set of streamlines for each of the sources described in Prob. 7.21 and from this diagram construct the streamlines for the combined flow. (*Hint:* For each of the sources draw streamlines separated by an angle of $\pi/6$. Finally, combine the intersection points of those rays for which $\psi_1 + \psi_2$ is constant.)

7.24 Does the line $x = 0$ form a line in the flow field described in Prob. 7.21 for which there is no velocity component normal to it? Is this line a streamline? Could this line be the trace of a solid plane lamina that was submerged in the flow? Does the velocity potential determined in Prob. 7.22 describe the flow in the region $x > 0$ for a source located at a distance of unity from a plane wall? Justify your answers.

7.25 Determine the equation for the velocity on the line $x = 0$ for the flow described in Prob. 7.21. Find an equation for the pressure on the surface whose trace is $x = 0$. What is the force on one side of this plane due to the source located at distance of unity from it? Water is the fluid.

7.26 In two-dimensional flow what is the nature of the flow given by $\phi = 7x + 2 \ln r$?

7.27 By using a method similar to that suggested in Prob. 7.23, draw the potential lines for the flow given in Prob. 7.26.

7.28 By using the suggestion in Prob. 7.23, draw a flow net for a flow consisting of a source and a vortex which are located at the origin. Use the same value for μ in both the source and the vortex.

7.29 A source discharging 20 cfs/ft is located at $(-1, 0)$, and a sink of twice the strength is located at $(2, 0)$. For dynamic pressure at the origin of 100 lb/ft^2, $\rho = 1.8$ slugs/ft^3, find the velocity and dynamic pressure at $(0, 1)$ and $(1, 1)$.

7.30 Select the strength of doublet needed to portray a uniform flow of 20 m/s around a cylinder of radius 2 m.

7.31 Develop the equations for flow around a Rankine cylinder formed by a source, an equal sink, and a uniform flow. If $2a$ is the distance between source and sink, their strength is $2\pi\mu$, and U is the uniform velocity, develop an equation for length of the body.

TWO

APPLICATIONS OF FLUID MECHANICS

In Part 1 the fundamental concepts and equations have been developed and illustrated by many examples and simple applications. Fluid resistance, dimensional analysis, compressible flow, and ideal-fluid flow have been presented. In Part 2 several of the important fields of application of fluid mechanics are explored: measurement of flow, turbomachinery, closed-conduit and open-channel flow, and unsteady flow.

EIGHT
FLUID MEASUREMENT

Fluid measurements include the determination of pressure, velocity, discharge, shock waves, density gradients, turbulence, and viscosity. There are many ways these measurements may be taken, e.g., direct, indirect, gravimetric, volumetric, electronic, electromagnetic, and optical. Direct measurements for discharge consist in the determination of the volume or weight of fluid that passes a section in a given time interval. Indirect methods of discharge measurement require the determination of head, difference in pressure, or velocity at several points in a cross section and, with these, computing the discharge. The most precise methods are the gravimetric or volumetric determinations, in which the weight or volume is measured by weigh scales or by a calibrated tank for a time interval that is measured by a stopwatch.

Pressure and velocity measurements are first undertaken in this chapter, followed by positive-displacement meters, rate meters, river-flow measurement, and turbulence and viscosity measurement.

8.1 PRESSURE MEASUREMENT

The measurement of pressure is required in many devices that determine the velocity of a fluid stream or its rate of flow, because of the relation between velocity and pressure given by the energy equation. The static pressure of a fluid in motion is its pressure when the velocity is undisturbed by the measurement.

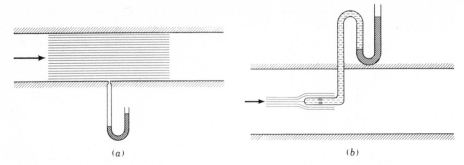

(a)   (b)

Figure 8.1 Static-pressure-measuring devices: (a) piezometer opening; (b) static tube.

Figure 8.1a indicates one method of measuring static pressure, the *piezometer opening*. When the flow is parallel, as indicated, the pressure variation is hydrostatic normal to the streamlines; hence, by measuring the pressure at the wall, the pressure at any other point in the cross section can be determined. The piezometer opening should be small, with length of opening at least twice the diameter, and should be normal to the surface, with no burrs at its edges, because small eddies would form and distort the measurement. A small amount of rounding of the opening is permissible. Any slight misalignment or roughness at the opening may cause errors in measurement; therefore, it is advisable to use several piezometer openings connected together into a *piezometer ring*. When the surface is rough in the vicinity of the opening, the reading is unreliable. For small irregularities it may be possible to smooth the surface around the opening.

For rough surfaces, the *static tube* (Fig. 8.1b) may be used. It consists of a tube that is directed upstream with the end closed. It has radial holes in the cylindrical portion downstream from the nose. The flow is presumed to be moving by the openings as if it were undisturbed. There are disturbances, however, due to both the nose and the right-angled leg that is normal to the flow. The static tube should be calibrated, as it may read too high or too low. If it does not read true static pressure, the discrepancy Δh normally varies as the square of the velocity of flow by the tube; i.e.,

$$\Delta h = C \frac{v^2}{2g}$$

in which C is determined by towing the tube in still fluid where pressure and velocity are known or by inserting it into a smooth pipe that contains a piezometer ring.

Such tubes are relatively insensitive to the Reynolds number and to Mach numbers below unity. Their alignment with the flow is not critical, so that an error of but a few percent is to be expected for a yaw misalignment of 15°.

The piezometric opening may lead to a bourdon gage, a manometer, a micro-manometer, or an electronic transducer. The transducers depend upon very small deformations of a diaphragm due to pressure change to create an electronic signal. The principle may be that of a strain gage and a Wheatstone bridge circuit, or it may rely on motion in a differential transformer, a capacitance chamber, or the piezoelectric behavior of a crystal under stress.

EXERCISES

8.1.1 A piezometer opening is used to measure (*a*) the pressure in a static fluid; (*b*) the velocity in a flowing stream; (*c*) the total pressure; (*d*) the dynamic pressure; (*e*) the undisturbed fluid pressure.

8.1.2 A static tube is used to measure (*a*) the pressure in a static fluid; (*b*) the velocity in a flowing stream; (*c*) the total pressure; (*d*) the dynamic pressure; (*e*) the undisturbed fluid pressure.

8.1.3 The piezoelectric properties of quartz are used to measure (*a*) temperature; (*b*) density; (*c*) velocity; (*d*) pressure; (*e*) none of these.

8.1.4 Water for a pipeline was diverted into a weigh tank for exactly 10 min. The increased weight in the tank was 4765 lb. The average flow rate in gallons per minute was (*a*) 66.1; (*b*) 57.1; (*c*) 7.95; (*d*) 0.13; (*e*) none of these.

8.1.5 A rectangular tank with cross-sectional area of 8 m^2 was filled to a depth of 1.3 m by a steady flow of liquid for 12 min. The rate of flow, in liters per second, was (*a*) 14.44; (*b*) 867; (*c*) 901; (*d*) 6471; (*e*) none of these.

8.2 VELOCITY MEASUREMENT

Since determining velocity at a number of points in a cross section permits evaluating the discharge, velocity measurement is an important phase of measuring flow. Velocity can be found by measuring the time an identifiable particle takes to move a known distance. This is done whenever it is convenient or necessary. This technique has been developed to study flow in regions which are so small that the normal flow would be greatly disturbed and perhaps disappear if an instrument were introduced to measure the velocity. A transparent viewing region must be made available, and by means of a strong light and a powerful microscope the very minute impurities in the fluid can be photographed with a high-speed motion-picture camera. From such motion pictures the velocity of the particles, and therefore the velocity of the fluid in a small region, can be determined.

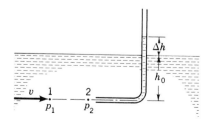

Figure 8.2 Simple pitot tube.

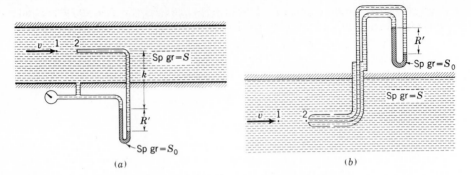

Figure 8.3 Velocity measurement: (a) pitot tube and piezometer opening; (b) pitot-static tube.

Normally, however, the device used does not measure velocity directly, but yields a measurable quantity that can be related to velocity. The *pitot tube* operates on such a principle and is one of the most accurate methods of measuring velocity. In Fig. 8.2 a glass tube or hypodermic needle with a right-angled bend is used to measure the velocity v in an open channel. The tube opening is directed upstream so that the fluid flows into the opening until the pressure builds up in the tube sufficiently to withstand the impact of velocity against it. Directly in front of the opening the fluid is at rest. The streamline through 1 leads to the point 2, called the *stagnation point*, where the fluid is at rest, and there divides and passes around the tube. The pressure at 2 is known from the liquid column within the tube. Bernoulli's equation, applied between points 1 and 2, produces

$$\frac{v^2}{2g} + \frac{p_1}{\gamma} = \frac{p_2}{\gamma} = h_0 + \Delta h$$

since both points are at the same elevation. As $p_1/\gamma = h_0$, the equation reduces to

$$\frac{v^2}{2g} = \Delta h \tag{8.2.1}$$

or
$$v = \sqrt{2g\,\Delta h} \tag{8.2.2}$$

Practically, it is very difficult to read the height Δh from a free surface.

The pitot tube measures the stagnation pressure, which is also referred to as the *total pressure*. The total pressure is composed of two parts, the static pressure h_0 and the dynamic pressure Δh, expressed in length of a column of the flowing fluid (Fig. 8.2). The dynamic pressure is related to velocity head by Eq. (8.2.1).

By combining the static-pressure measurement and the total-pressure measurement, i.e., measuring each and connecting to opposite ends of a differential manometer, the dynamic pressure head is obtained. Figure 8.3a illustrates one arrangement. Bernoulli's equation applied from 1 to 2 is

$$\frac{v^2}{2g} + \frac{p_1}{\gamma} = \frac{p_2}{\gamma} \tag{8.2.3}$$

The equation for the manometer, in units of length of water, is

$$\frac{p_1}{\gamma} S + kS + R'S_0 - (k + R')S = \frac{p_2}{\gamma} S$$

Simplifying gives

$$\frac{p_2 - p_1}{\gamma} = R'\left(\frac{S_0}{S} - 1\right) \tag{8.2.4}$$

Substituting for $(p_2 - p_1)/\gamma$ in Eq. (8.2.3) and solving for v results in

$$v = \sqrt{2gR'\left(\frac{S_0}{S} - 1\right)} \tag{8.2.5}$$

The pitot tube is also insensitive to flow alignment, and an error of only a few percent occurs if the tube has a yaw misalignment of less than 15°.

The static tube and pitot tube may be combined into one instrument, called a *pitot-static tube* (Fig. 8.3b). Analyzing this system in a manner similar to that in Fig. 8.3a shows that the same relations hold; Eq. (8.2.5) expresses the velocity, but the uncertainty in the measurement of static pressure requires a corrective coefficient C to be applied:

$$v = C\sqrt{2gR'\left(\frac{S_0}{S} - 1\right)} \tag{8.2.6}$$

A particular form of pitot-static tube with a blunt nose, the *Prandtl tube*, has been so designed that the disturbances due to nose and leg cancel, leaving $C = 1$ in the equation. For other pitot-static tubes the constant C must be determined by calibration.

The *current meter* (Fig. 8.4a) is used to measure the velocity of liquid flow in open channels. The cups are so shaped that the drag varies with orientation, causing a relatively slow rotation. With an electric circuit and headphones, an audible signal is detected for a fixed number of revolutions. The number of signals in a given time is a function of the velocity. The meters are calibrated by towing them through liquid at known speeds. For measuring high-velocity flow a current meter with a propeller as rotating element is used, as it offers less resistance to the flow.

Air velocities are measured with cup-type or vane (propeller) anemometers

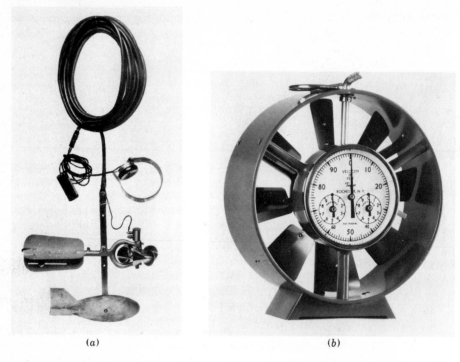

(a) (b)

Figure 8.4 Velocity-measuring devices: (a) Price current meter for liquids (*W. and L. E. Gurley*); (b) air anemometer. (*Taylor Instrument Co.*)

(Fig. 8.4b) which drive generators indicating air velocity directly or drive counters indicating the number of revolutions.

By so designing the vanes that they have very low inertia, employing precision bearings and optical tachometers which effectively take no power to drive them, anemometers can be made to read very low air velocities. They can be sensitive enough to measure the convection air currents which the human body causes by its heat emission to the atmosphere.

Velocity Measurement in Compressible Flow

The pitot-static tube may be used for velocity determinations in compressible flow. In Fig. 8.3b the velocity reduction from free-stream velocity at 1 to zero at 2 takes place very rapidly without significant heat transfer. Friction plays a very small part, so that the compression may be assumed to be isentropic. Applying Eq. (6.3.7) to points 1 and 2 of Fig. 8.3b with $V_2 = 0$ gives

$$\frac{V_1^2}{2} = \frac{k}{k-1}\left(\frac{p_2}{\rho_2} - \frac{p_1}{\rho_1}\right) = \frac{kR}{k-1}(T_2 - T_1) = c_p T_1\left(\frac{T_2}{T_1} - 1\right) \qquad (8.2.7)$$

The substitution of c_p is from Eq. (6.1.8). Equation (6.1.17) then gives

$$\frac{V_1^2}{2} = c_p T_1 \left[\left(\frac{p_2}{p_1} \right)^{(k-1)/k} - 1 \right] = c_p T_2 \left[1 - \left(\frac{p_1}{p_2} \right)^{(k-1)/k} \right] \qquad (8.2.8)$$

The static pressure p_1 may be obtained from the side openings of the pitot tube, and the stagnation pressure may be obtained from the impact opening leading to a simple manometer, or $p_2 - p_1$ may be found from the differential manometer. If the tube is not so designed that true static pressure is measured, it must be calibrated and the true static pressure computed.

Gas velocities may be measured with a *hot-wire anemometer*, which works on the principle that the resistance to the flow of electricity through a fine platinum wire is a function of cooling due to gas flow around it. Cooled film sensors are also used for gas flow and have been adapted to liquid flow.

EXERCISES

8.2.1 The simple pitot tube measures the (a) static pressure; (b) dynamic pressure; (c) total pressure; (d) velocity at the stagnation point; (e) difference in total and dynamic pressure.

8.2.2 A pitot-static tube (C = 1) is used to measure air speeds. With water in the differential manometer and a gage difference of 3 in, the air speed for $\gamma = 0.0624$ lb/ft^3, in feet per second, is (a) 4.01; (b) 15.8; (c) 24.06; (d) 127; (e) none of these.

8.2.3 The pitot-static tube measures (a) static pressure; (b) dynamic pressure; (c) total pressure; (d) difference in static and dynamic pressure; (e) difference in total and dynamic pressure.

8.2.4 The temperature of a known flowing gas can be determined from measurement of (a) static and stagnation pressure only; (b) velocity and stagnation pressure only; (c) velocity and dynamic pressure only; (d) velocity and stagnation temperature only; (e) none of these answers.

8.2.5 The velocity of a known flowing gas may be determined from measurement of (a) static and stagnation pressure only; (b) static pressure and temperature only; (c) static and stagnation temperature only; (d) stagnation temperature and stagnation pressure only; (e) none of these answers.

8.2.6 The hot-wire anemometer is used to measure (a) pressure in gases; (b) pressure in liquids; (c) wind velocities at airports; (d) gas velocities; (e) liquid discharges.

8.3 POSITIVE-DISPLACEMENT METERS

One volumetric meter, the positive-displacement meter, has pistons or partitions which are displaced by the flowing fluid and a counting mechanism that records the number of displacements in any convenient unit, such as liters or cubic feet.

A common meter is the *disk meter*, or *wobble meter* used on many domestic water-distribution systems. The disk oscillates in a passageway, so that a known volume of fluid moves through the meter for each oscillation. A stem normal to the disk operates a gear train which, in turn, operates a counter. In good condition, these meters are accurate to within 1 percent. When they are worn, the error may be very large for small flows, such as those caused by a leaky faucet.

The flow of natural gas at low pressure is usually measured by a volumetric meter with a traveling partition. The partition is displaced by gas inflow to one end of the chamber in which it operates, and then, by a change in valving, it is displaced to the opposite end of the chamber. The oscillations operate a counting mechanism.

Oil flow or high-pressure gas flow in a pipeline is frequently measured by a rotary meter in which cups or vanes move about an annular opening and displace a fixed volume of fluid for each revolution. Radial or axial pistons may be so arranged that the volume of a continuous flow through them is determined by rotations of a shaft.

Positive-displacement meters normally have no timing equipment that measures the rate of flow. The rate of steady flow may be determined with a stopwatch to record the time for displacement of a given volume of fluid.

EXERCISE

8.3.1 A piston-type displacement meter has a volume displacement of 35 cm³ per revolution of its shaft. The discharge, in liters per minute, for 1000 rpm is (a) 1.87; (b) 4.6; (c) 35; (d) 40.34; (e) none of these.

8.4 ORIFICES

A *rate meter* is a device that determines, generally by a single measurement, the quantity (weight or volume) per unit time that passes a given cross section. Included among rate meters are the orifice, nozzle, venturi meter, rotometer, and weir. The orifice is discussed in this section; the venturi meter, nozzle, and some other closed-conduit devices are discussed in Sec. 8.5; and weirs are discussed in Sec. 8.6.

Orifice in a Reservoir

An orifice may be used for measuring the rate of flow out of a reservoir or through a pipe. An orifice in a reservoir or tank may be in the wall or in the bottom. It is an opening, usually round, through which the fluid flows, as in Fig. 8.5. It may be square-edged, as shown, or rounded, as in Fig. 3.12. The area of the orifice is the area of the opening. With the square-edged orifice, the fluid jet contracts during a short distance of about one-half diameter downstream from the opening. The portion of the flow that approaches along the wall cannot make a right-angled turn at the opening and therefore maintains a radial velocity component that reduces the jet area. The cross section where the contraction is greatest is called the *vena contracta*. The streamlines are parallel throughout the jet at this section, and the pressure is atmospheric.

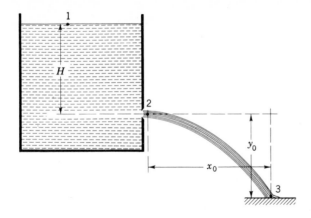

Figure 8.5 Orifice in a reservoir.

The head H on the orifice is measured from the center of the orifice to the free surface. The head is assumed to be held constant. Bernoulli's equation applied from a point 1 on the free surface to the center of the *vena contracta*, point 2, with local atmospheric pressure as datum and point 2 as elevation datum, neglecting losses, is written

$$\frac{V_1^2}{2g} + \frac{p_1}{\gamma} + z_1 = \frac{V_2^2}{2g} + \frac{p_2}{\gamma} + z_2$$

Inserting the values gives

$$0 + 0 + H = \frac{V_2^2}{2g} + 0 + 0$$

or

$$V_2 = \sqrt{2gH} \tag{8.4.1}$$

This is only the *theoretical* velocity, because the losses between the two points were neglected. The ratio of the *actual* velocity V_a to the theoretical velocity V_t is called the *velocity coefficient* C_v; that is,

$$C_v = \frac{V_a}{V_t} \tag{8.4.2}$$

Hence,

$$V_{2a} = C_v\sqrt{2gH} \tag{8.4.3}$$

The actual discharge Q_a from the orifice is the product of the actual velocity at the *vena contracta* and the area of the jet. The ratio of jet area A_2 at *vena contracta* to area of orifice A_0 is symbolized by another coefficient, called the *coefficient of contraction* C_c:

$$C_c = \frac{A_2}{A_0} \tag{8.4.4}$$

The area at the *vena contracta* is $C_c A_0$. The actual discharge is thus

$$Q_a = C_v C_c A_0 \sqrt{2gH} \qquad (8.4.5)$$

It is customary to combine the two coefficients into a *discharge coefficient* C_d,

$$C_d = C_v C_c \qquad (8.4.6)$$

from which

$$Q_a = C_d A_0 \sqrt{2gH} \qquad (8.4.7)$$

There is no way to compute the losses between points 1 and 2; hence, C_v must be determined experimentally. It varies from 0.95 to 0.99 for the square-edged or rounded orifice. For most orifices, such as the square-edged one, the amount of contraction cannot be computed, and test results must be used. There are several methods for obtaining one or more of the coefficients. By measuring area A_0, the head H, and the discharge Q_a (by gravimetric or volumetric means), C_d is obtained from Eq. (8.4.7). Determination of either C_v or C_c then permits determination of the other by Eq. (8.4.6). Several methods follow.

1. Trajectory method. By measuring the position of a point on the trajectory of the free jet downstream from the *vena contracta* (Fig. 8.5) the actual velocity V_a can be determined if air resistance is neglected. The x component of velocity does not change; therefore, $V_a t = x_0$, in which t is the time for a fluid particle to travel from the *vena contracta* to point 3. The time for a particle to drop a distance y_0 under the action of gravity when it has no initial velocity in that direction is expressed by $y_0 = gt^2/2$. After t is eliminated in the two relations,

$$V_a = \frac{x_0}{\sqrt{2y_0/g}}$$

With V_{2t} determined by Eq. (8.4.1), the ratio $V_a/V_t = C_v$ is known.
2. Direct measuring of V_a. With a pitot tube placed at the *vena contracta*, the actual velocity V_a is determined.

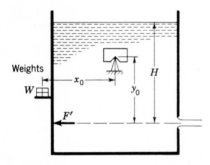

Figure 8.6 Momentum method for determination of C_v and C_c.

3. Direct measuring of jet diameter. With outside calipers, the diameter of jet at the *vena contracta* may be approximated. This is not a precise measurement and in general is less satisfactory than the other methods.

4. Use of momentum equation. When the reservoir is small enough to be suspended on knife-edges, as in Fig. 8.6, it is possible to determine the force F that creates the momentum in the jet. With the orifice opening closed, the tank is leveled by adding or subtracting weights. With the orifice discharging, a force creates the momentum in the jet and an equal and opposite force F' acts against the tank. By addition of sufficient weights W, the tank is again leveled. From the figure, $F' = Wx_0/y_0$. With the momentum equation,

$$\Sigma F_x = \frac{Q\gamma}{g}(V_{x_{\text{out}}} - V_{x_{\text{in}}}) \qquad \text{or} \qquad \frac{Wx_0}{y_0} = \frac{Q_a\gamma V_a}{g}$$

as $V_{x_{\text{in}}}$ is zero and V_a is the final velocity. Since the actual discharge is measured, V_a is the only unknown in the equation.

Losses in Orifice Flow

The head loss in flow through an orifice is determined by applying the energy equation with a loss term for the distance between points 1 and 2 (Fig. 8.5),

$$\frac{V_{1a}^2}{2g} + \frac{p_1}{\gamma} + z_1 = \frac{V_{2a}^2}{2g} + \frac{p_2}{\gamma} + z_2 + \text{losses}$$

Substituting the values for this case gives

$$\text{Losses} = H - \frac{V_{2a}^2}{2g} = H(1 - C_v^2) = \frac{V_{2a}^2}{2g}\left(\frac{1}{C_v^2} - 1\right) \qquad (8.4.8)$$

in which Eq. (8.4.3) has been used to obtain the losses in terms of H and C_v or V_{2a} and C_v.

> **Example 8.1** A 75-mm-diameter orifice under a head of 4.88 m discharges 8900 N water in 32.6 s. The trajectory was determined by measuring $x_0 = 4.76$ m for a drop of 1.22 m. Determine C_v, C_c, C_d, the head loss per unit weight, and the power loss.
> The theoretical velocity V_{2t} is
>
> $$V_{2t} = \sqrt{2gH} = \sqrt{2 \times 9.806 \times 4.88} = 9.783 \text{ m/s}$$
>
> The actual velocity is determined from the trajectory. The time to drop 1.22 m is
>
> $$t = \sqrt{\frac{2y_0}{g}} = \sqrt{\frac{2 \times 1.22}{9.806}} = 0.499 \text{ s}$$

and the velocity is expressed by

$$x_0 = V_{2a}t \qquad V_{2a} = \frac{4.76}{0.499} = 9.539 \text{ m/s}$$

Then

$$C_v = \frac{V_{2a}}{V_{2t}} = \frac{9.539}{9.783} = 0.975$$

The actual discharge Q_a is

$$Q_a = \frac{8900}{9806 \times 32.6} = 0.0278 \text{ m}^3/\text{s}$$

With Eq. (8.4.7)

$$C_d = \frac{Q_a}{A_0\sqrt{2gH}} = \frac{0.0278}{\pi(0.0375^2)\sqrt{2 \times 9.806 \times 4.88}} = 0.643$$

Hence, from Eq. (8.4.6),

$$C_c = \frac{C_d}{C_v} = \frac{0.643}{0.975} = 0.659$$

The head loss, from Eq. (8.4.8), is

$$\text{Loss} = H(1 - C_v^2) = 4.88(1 - 0.975^2) = 0.241 \text{ m} \cdot \text{N/N}$$

The power loss is

$$Q\gamma(\text{loss}) = 0.0278 \times 9806 \times 0.241 = 65.7 \text{ W}$$

The Borda mouthpiece (Fig. 8.7), a short, thin-walled tube about one diameter long that projects into the reservoir (re-entrant), permits application of the momentum equation, which yields one relation between C_v and C_d. The velocity along the wall of the tank is almost zero at all points; hence, the pressure distribution is hydrostatic. With the component of force exerted on the liquid by the tank parallel to the axis of the tube, there is an unbalanced force due to the

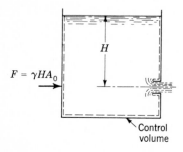

Figure 8.7 The Borda mouthpiece.

opening, which is $\gamma H A_0$. The final velocity is V_{2a}; the initial velocity is zero; and Q_a is the actual discharge. Then

$$\gamma H A_0 = Q_a \frac{\gamma}{g} V_{2a}$$

and
$$Q_a = C_d A_0 \sqrt{2gH} \qquad V_{2a} = C_v \sqrt{2gH}$$

Substituting for Q_a and V_{2a} and simplifying lead to

$$1 = 2C_d C_v = 2C_v^2 C_c$$

Orifice in a Pipe

The square-edged orifice in a pipe (Fig. 8.8) causes a contraction of the jet downstream from the orifice opening. For incompressible flow Bernoulli's equation applied from section 1 to the jet at its *vena contracta*, section 2, is

$$\frac{V_{1t}^2}{2g} + \frac{p_1}{\gamma} = \frac{V_{2t}^2}{2g} + \frac{p_2}{\gamma}$$

The continuity equation relates V_{1t} and V_{2t} with the contraction coefficient $C_c = A_2/A_0$,

$$V_1 \frac{\pi D_1^2}{4} = V_2 C_c \frac{\pi D_0^2}{4} \qquad\qquad (8.4.9)$$

After eliminating V_{1t},

$$\frac{V_{2t}^2}{2g} \left[1 - C_c^2 \left(\frac{D_0}{D_1} \right)^4 \right] = \frac{p_1 - p_2}{\gamma}$$

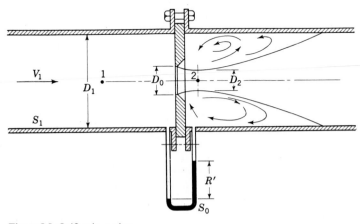

Figure 8.8 Orifice in a pipe.

and by solving for V_{2t} the result is

$$V_{2t} = \sqrt{\frac{2g(p_1 - p_2)/\gamma}{1 - C_c^2(D_0/D_1)^4}}$$

Multiplying by C_v to obtain the actual velocity at the *vena contracta* gives

$$V_{2a} = C_v \sqrt{\frac{2(p_1 - p_2)/\rho}{1 - C_c^2(D_0/D_1)^4}}$$

and, finally multiplying by the area of the jet, $C_c A_0$, produces the actual discharge Q,

$$Q = C_d A_0 \sqrt{\frac{2(p_1 - p_2)/\rho}{1 - C_c^2(D_0/D_1)^4}} \tag{8.4.10}$$

in which $C_d = C_v C_c$. In terms of the gage difference R', Eq. (8.4.10) becomes

$$Q = C_d A_0 \sqrt{\frac{2gR'(S_0/S_1 - 1)}{1 - C_c^2(D_0/D_1)^4}} \tag{8.4.11}$$

Because of the difficulty in determining the two coefficients separately, a simplified formula is generally used,

$$Q = C A_0 \sqrt{\frac{2\,\Delta p}{\rho}} \tag{8.4.12}$$

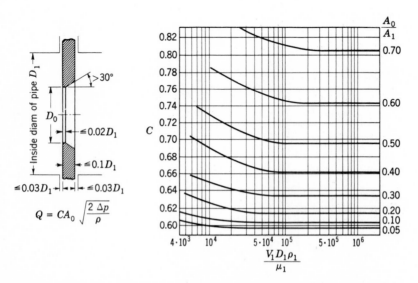

Figure 8.9 VDI orifice and discharge coefficients. (*Ref. 11 in NACA Tech. Mem. 952.*)

or its equivalent,

$$Q = CA_0 \sqrt{2gR'\left(\frac{S_0}{S_1} - 1\right)} \qquad (8.4.13)$$

Values of C are given in Fig. 8.9 for the VDI orifice.

By a procedure explained in the next section, Eq. (8.4.12) may be modified by an expansion factor Y (Fig. 8.13) to yield actual mass rate of compressible (isentropic) flow.

$$\dot{m} = CYA_0 \sqrt{2\rho_1 \, \Delta p} \qquad (8.4.14)$$

The location of the pressure taps is usually so specified that an orifice can be installed in a conduit and used with sufficient accuracy without performing a calibration at the site.

Unsteady Orifice Flow from Reservoirs

In the orifice situations considered, the liquid surface in the reservoir has been assumed to be held constant. An unsteady-flow case of some practical interest is that of determining the time to lower the reservoir surface a given distance. Theoretically, Bernoulli's equation applies only to steady flow, but if the reservoir surface drops slowly enough, the error from using Bernoulli's equation is negligible. The volume discharged from the orifice in time δt is $Q \, \delta t$, which must just equal the reduction in volume in the reservoir in the same time increment (Fig. 8.10), $A_R(-\delta y)$, in which A_R is the area of liquid surface at height y above the orifice. Equating the two expressions gives

$$Q \, \delta t = -A_R \, \delta y$$

Solving for δt and integrating between the limits $y = y_1$, $t = 0$ and $y = y_2$, $t = t$ yields

$$t = \int_0^t dt = -\int_{y_1}^{y_2} \frac{A_R \, dy}{Q}$$

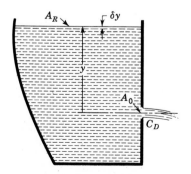

Figure 8.10 Notation for falling head.

The orifice discharge Q is $C_d A_0 \sqrt{2gy}$. After substitution for Q,

$$t = -\frac{1}{C_d A_0 \sqrt{2g}} \int_{y_1}^{y_2} A_R y^{-1/2} \, dy$$

When A_R is known as a function of y, the integral can be evaluated. Consistently with other SI or U.S. customary units, t is in seconds. For the special case of a tank with constant cross section,

$$t = -\frac{A_R}{C_d A_0 \sqrt{2g}} \int_{y_1}^{y_2} y^{-1/2} \, dy = \frac{2 A_R}{C_d A_0 \sqrt{2g}} \left(\sqrt{y_1} - \sqrt{y_2} \right)$$

Example 8.2 A tank has a horizontal cross-sectional area of 2 m² at the elevation of the orifice, and the area varies linearly with elevation so that it is 1 m² at a horizontal cross section 3 m above the orifice. For a 100-mm-diameter orifice, $C_d = 0.65$, compute the time, in seconds, to lower the surface from 2.5 to 1 m above the orifice.

$$A_R = 2 - \frac{y}{3} \quad \text{m}^2$$

and

$$t = -\frac{1}{0.65\pi(0.05^2)\sqrt{2 \times 9.806}} \int_{2.5}^{1} \left(2 - \frac{y}{3} \right) y^{-1/2} \, dy = 73.8 \text{ s}$$

EXERCISES

8.4.1 The actual velocity at the *vena contracta* for flow through an orifice from a reservoir is expressed by (a) $C_v \sqrt{2gH}$; (b) $C_c \sqrt{2gH}$; (c) $C_d \sqrt{2gH}$; (d) $\sqrt{2gH}$; (e) $C_v V_a$.

8.4.2 A fluid jet discharging from a 20-mm-diameter orifice has a diameter 17.5 mm at its *vena contracta*. The coefficient of contraction is (a) 1.31; (b) 1.14; (c) 0.875; (d) 0.766; (e) none of these answers.

8.4.3 The ratio of actual discharge to theoretical discharge through an orifice is (a) $C_c C_v$; (b) $C_c C_d$; (c) $C_v C_d$; (d) C_d/C_v; (e) C_d/C_c.

8.4.4 The losses in orifice flow are

(a) $\dfrac{1}{C_v^2} \left(\dfrac{V_{2a}^2}{2g} - 1 \right)$ (b) $\dfrac{V_{2t}^2}{2g} - \dfrac{V_{2a}^2}{2g}$ (c) $H(C_v^2 - 1)$

(d) $H - \dfrac{V_{2t}^2}{2g}$ (e) none of these answers

8.4.5 For a liquid surface to lower at a constant rate, the area of reservoir A_R must vary with head y on the orifice, as (a) $\sqrt{y}$; (b) y; (c) $1/\sqrt{y}$; (d) $1/y$; (e) none of these answers.

8.4.6 A 50-mm-diameter Borda mouthpiece discharges 7.68 l/s under a head of 3 m. The velocity coefficient is (a) 0.96; (b) 0.97; (c) 0.98; (d) 0.99; (e) none of these answers.

8.5 VENTURI METER, NOZZLE, AND OTHER RATE DEVICES

Venturi Meter

The venturi meter is used to measure the rate of flow in a pipe. It is generally a casting (Fig. 8.11) consisting of an upstream section which is the same size as the pipe, has a bronze liner, and contains a piezometer ring for measuring static pressure; a converging conical section; a cylindrical throat with a bronze liner containing a piezometer ring; and a gradually diverging conical section leading to a cylindrical section the size of the pipe. A differential manometer is attached to the two piezometer rings. The size of a venturi meter is specified by the pipe and throat diameter; e.g., a 6 by 4 in venturi meter fits a 6-in-diameter pipe and has a 4-in-diameter throat. For accurate results the venturi meter should be preceded by at least 10 diameters of straight pipe. In the flow from the pipe to the throat, the velocity is greatly increased and the pressure correspondingly decreased. The amount of discharge in incompressible flow is shown to be a function of the manometer reading.

The pressures at the upstream section and throat are *actual pressures*, and the velocities from Bernoulli's equation are *theoretical velocities*. When losses are considered in the energy equation, the velocities are *actual velocities*. First, with the Bernoulli equation (i.e., without a head-loss term) the theoretical velocity at the throat is obtained. Then, by multiplying this by the velocity coefficient C_v, the actual velocity is obtained. The actual velocity times the actual area of the throat determines the actual discharge. From Fig. 8.11

$$\frac{V_{1t}^2}{2g} + \frac{p_1}{\gamma} + h = \frac{V_{2t}^2}{2g} + \frac{p_2}{\gamma} \qquad (8.5.1)$$

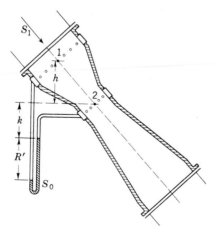

Figure 8.11 Venturi meter.

in which elevation datum is taken through point 2. V_1 and V_2 are average velocities at sections 1 and 2, respectively; hence, α_1, α_2 are assumed to be unity. With the continuity equation $V_1 D_1^2 = V_2 D_2^2$,

$$\frac{V_1^2}{2g} = \frac{V_2^2}{2g}\left(\frac{D_2}{D_1}\right)^4 \tag{8.5.2}$$

which holds for either the actual velocities or the theoretical velocities. Equation (8.5.1) may be solved for V_{2t},

$$\frac{V_{2t}^2}{2g}\left[1 - \left(\frac{D_2}{D_1}\right)^4\right] = \frac{p_1 - p_2}{\gamma} + h$$

and

$$V_{2t} = \sqrt{\frac{2g[h + (p_1 - p_2)/\gamma]}{1 - (D_2/D_1)^4}} \tag{8.5.3}$$

Introducing the velocity coefficient $V_{2a} = C_v V_{2t}$ gives

$$V_{2a} = C_v\sqrt{\frac{2g[h + (p_1 - p_2)/\gamma]}{1 - (D_2/D_1)^4}} \tag{8.5.4}$$

After multiplying by A_2, the actual discharge Q is determined to be

$$Q = C_v A_2\sqrt{\frac{2g[h + (p_1 - p_2)/\gamma]}{1 - (D_2/D_1)^4}} \tag{8.5.5}$$

The gage difference R' may now be related to the pressure difference by writing the equation for the manometer. In units of length of water (S_1 is the specific gravity of flowing fluid and S_0 the specific gravity of manometer liquid),

$$\frac{p_1}{\gamma}S_1 + (h + k + R')S_1 - R'S_0 - kS_1 = \frac{p_2}{\gamma}S_1$$

Simplifying gives

$$h + \frac{p_1 - p_2}{\gamma} = R'\left(\frac{S_0}{S_1} - 1\right) \tag{8.5.6}$$

By substituting into Eq. (8.5.5),

$$Q = C_v A_2\sqrt{\frac{2gR'(S_0/S_1 - 1)}{1 - (D_2/D_1)^4}} \tag{8.5.7}$$

which is the venturi meter equation for incompressible flow. The contraction coefficient is unity; hence, $C_v = C_d$. It should be noted that h has dropped out of the equation. The discharge depends upon the gage difference R' regardless of the orientation of the venturi meter; whether it is horizontal, vertical, or inclined, exactly the same equation holds.

C_v is determined by calibration, i.e., by measuring the discharge and the gage difference and solving for C_v, which is usually plotted against the Reynolds number. Experimental results for venturi meters are given in Fig. 8.12. They are

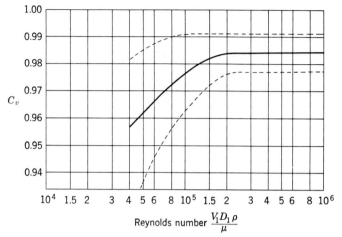

Figure 8.12 Coefficient C_v for venturi meters. ("*Fluid Meters: Their Theory and Application*," *5th ed., American Society of Mechanical Engineers, 1956.*)

applicable to diameter ratios D_2/D_1 from 0.25 to 0.75 within the tolerances shown by the dotted lines. Where feasible, a venturi meter should be so selected that its coefficient is constant over the range of Reynolds numbers for which it is to be used.

The coefficient may be slightly greater than unity for venturi meters that are unusually smooth inside. This does not mean that there are no losses; it results from neglecting the kinetic-energy correction factors α_1, α_2 in the Bernoulli equation. Generally, α_1 is greater than α_2, since the reducing section acts to make the velocity distribution uniform across section 2.

The venturi meter has a low overall loss owing to the gradually expanding conical section, which aids in reconverting the high kinetic energy at the throat into pressure energy. The loss is about 10 to 15 percent of the head change between sections 1 and 2.

Venturi Meter for Compressible Flow

The theoretical flow of a compressible fluid through a venturi meter is substantially isentropic and is obtained from Eqs. (6.3.2), (6.3.6), and (6.3.7). When multiplied by C_v, the velocity coefficient, it yields for mass flow rate

$$\dot{m} = C_v A_2 \sqrt{\frac{[2k/(k-1)]p_1\rho_1(p_2/p_1)^{2/k}[1-(p_2/p_1)^{(k-1)/k}]}{1-(p_2/p_1)^{2/k}(A_2/A_1)^2}} \qquad (8.5.8)$$

The velocity coefficient is the same as for liquid flow. Equation (8.5.5), when reduced to horizontal flow and modified by insertion of an expansion factor, can be applied to compressible flow

$$\dot{m} = C_v Y A_2 \sqrt{\frac{2\rho_1 \, \Delta p}{1-(D_2/D_1)^4}} \qquad (8.5.9)$$

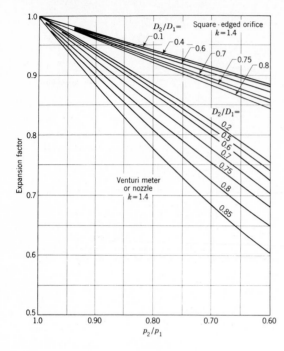

Figure 8.13 Expansion factors.

Y can be found by solving Eqs. (8.5.9) and (8.5.8) and is shown to be a function of k, p_2/p_1, and A_2/A_1. Values of Y are plotted in Fig. 8.13 for $k = 1.40$; hence, by the use of Eq. (8.5.9) and Fig. 8.13 compressible flow can be computed for a venturi meter.

Flow Nozzle

The ISA (Instrument Society of America) flow nozzle (originally the VDI flow nozzle) is shown in Fig. 8.14. It has no contraction of the jet other than that of the nozzle opening; therefore, the contraction coefficient is unity.

Equations (8.5.5) and (8.5.7) hold equally well for the flow nozzle. For a horizontal pipe ($h = 0$), Eq. (8.5.5) may be written

$$Q = CA_2 \sqrt{\frac{2\,\Delta p}{\rho}} \tag{8.5.10}$$

in which

$$C = \frac{C_v}{\sqrt{1 - (D_2/D_1)^4}} \tag{8.5.11}$$

and $\Delta p = p_1 - p_2$. The value of coefficient C in Fig. 8.14 is for use in Eq. (8.5.10). When the coefficient given in the figure is to be used, it is important that the

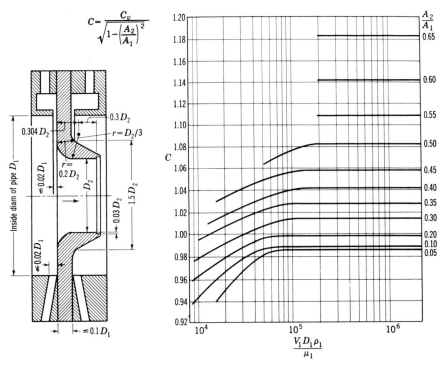

$$C = \frac{C_v}{\sqrt{1 - \left(\frac{A_2}{A_1}\right)^2}}$$

Figure 8.14 ISA (VDI) flow nozzle and discharge coefficients. (*Ref. 11 in NACA Tech. Mem. 952.*)

dimensions shown be closely adhered to, particularly in the location of the piezometer openings (two methods shown) for measuring pressure drop. At least 10 diameters of straight pipe should precede the nozzle.

The flow nozzle is less costly than the venturi meter. It has the disadvantage that the overall losses are much higher because of the lack of guidance of jet downstream from the nozzle opening.

Compressible flow through a nozzle is found by Eq. (8.5.9) and Fig. 8.13 if $k = 1.4$. For other values of specific heat ratio k, Eq. (8.5.8) may be used.

Example 8.3 Determine the flow through a 6-in-diameter waterline that contains a 4-in-diameter flow nozzle. The mercury-water differential manometer has a gage difference of 10 in. Water temperature is 60°F.
From the data given, $S_0 = 13.6$, $S_1 = 1.0$, $R' = \frac{10}{12} = 0.833$ ft, $A_2 = \pi/36 = 0.0873$ ft^2, $\rho = 1.938$ slugs/ft^3, $\mu = 2.359 \times 10^{-5}$ lb·s/ft^2. Substituting Eq. (8.5.11) into Eq. (8.5.7) gives

$$Q = CA_2 \sqrt{2gR'\left(\frac{S_0}{S_1} - 1\right)}$$

From Fig. 8.14, for $A_2/A_1 = (\frac{4}{6})^2 = 0.444$, assume that the horizontal region of the curves applies. Hence, $C = 1.056$; then compute the flow and the Reynolds number.

$$Q = 1.056 \times 0.0873 \sqrt{64.4 \times 0.833\left(\frac{13.6}{1.0} - 1.0\right)} = 2.40 \text{ cfs}$$

Then
$$V_1 = \frac{Q}{A_1} = \frac{2.40}{\pi/16} = 12.21 \text{ ft/s}$$

and
$$\mathbf{R} = \frac{V_1 D_1 \rho}{\mu} = \frac{12.21 \times 1.938}{2 \times 2.359 \times 10^{-5}} = 502,000$$

The chart shows the value of C to be correct; therefore, the discharge is 2.40 cfs.

Elbow Meter

The elbow meter for incompressible flow is one of the simplest flow-rate-measuring devices. Piezometer openings on the inside and on the outside of the elbow are connected to a differential manometer. Because of centrifugal force at the bend, the difference in pressures is related to the discharge. A straight calming length should precede the elbow, and, for accurate results, the meter should be calibrated in place.† As most pipelines have an elbow, it may be used as the meter. After calibration the results are as reliable as with a venturi meter or a flow nozzle.

Rotameter

The rotameter is a variable-area meter that consists of an enlarging transparent tube and a metering "float" (actually heavier than the liquid) that is displaced upward by the upward flow of fluid through the tube. The tube is graduated to read the flow directly. Notches in the float cause it to rotate and thus maintain a central position in the tube. The greater the flow, the higher the position the float assumes.

Electromagnetic Flow Devices

If a magnetic field is set up across a nonconducting tube and a conducting fluid flows through the tube, an induced voltage is produced across the flow which can be measured if electrodes are embedded in the tube walls.‡ The voltage is a linear function of the volume rate passing through the tube. Either an ac or a dc field may be used, with a corresponding signal generated at the electrodes. A disadvantage of the method is the small signal received and the large amount of amplification needed. The device has been used to measure the flow in blood vessels.

† W. M. Lansford, The Use of an Elbow in a Pipe Line for Determining the Rate of Flow in a Pipe, *Univ. Ill. Eng. Exp. Stn. Bull.* 289, December 1936.
‡ H. G. Elrod, Jr., and R. R. Fouse, An Investigation of Electromagnetic Flowmeters, *Trans. ASME*, vol. 74, p. 589, May 1952.

EXERCISES

8.5.1 Which of the following measuring instruments is a rate meter? (*a*) Current meter; (*b*) disk meter; (*c*) hot-wire anemometer; (*d*) pitot tube; (*e*) venturi meter.

8.5.2 The discharge coefficient for a 40 by 20 mm venturi meter at a Reynolds number of 200,000 is (*a*) 0.95; (*b*) 0.96; (*c*) 0.973; (*d*) 0.983; (*e*) 0.992.

8.5.3 Select the correct statement: (*a*) The discharge through a venturi meter depends upon Δp only and is independent of orientation of the meter. (*b*) A venturi meter with a given gage difference R' discharges at a greater rate when the flow is vertically downward through it than when the flow is vertically upward. (*c*) For a given pressure difference the equations show that the discharge of gas is greater through a venturi meter when compressibility is taken into account than when it is neglected. (*d*) The coefficient of contraction of a venturi meter is unity. (*e*) The overall loss is the same in a given pipeline whether a venturi meter or a nozzle with the same D_2 is used.

8.5.4 The expansion factor Y depends upon (*a*) k, p_2/p_1, and A_2/A_1; (*b*) R', p_2/p_1, and A_2/A_1; (*c*) k, R', and p_2/p_1; (*d*) k, R', and A_2/A_1; (*e*) none of these answers.

8.6 WEIRS

Open-channel flow may be measured by a *weir*, which is an obstruction in the channel that causes the liquid to back up behind it and flow over it or through it. By measuring the height of upstream liquid surface, the rate of flow is determined. Weirs constructed from a sheet of metal or other material so that the jet, or *nappe*, springs free as it leaves the upstream face are called *sharp-crested weirs*. Other weirs, such as the *broad-crested weir*, support the flow in a longitudinal direction.

The sharp-crested rectangular weir (Fig. 8.15) has a horizontal crest. The nappe is contracted at top and bottom as shown. An equation for discharge can be derived if the contractions are neglected. Without contractions the flow appears as in Fig. 8.16. The nappe has parallel streamlines with atmospheric pressure throughout.

Bernoulli's equation applied between 1 and 2 is

$$H + 0 + 0 = \frac{v^2}{2g} + H - y + 0$$

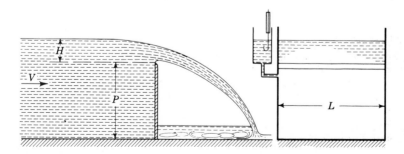

Figure 8.15 Sharp-crested rectangular weir.

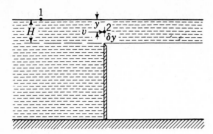

Figure 8.16 Weir nappe without contractions.

in which the velocity head at section 1 is neglected. By solving for v,

$$v = \sqrt{2gy}$$

The theoretical discharge Q_t is

$$Q_t = \int v \, dA = \int_0^H vL \, dy = \sqrt{2g} \, L \int_0^H y^{1/2} \, dy = \tfrac{2}{3}\sqrt{2g} \, LH^{3/2}$$

in which L is the width of weir. Experiment shows that the exponent of H is correct but the coefficient is too great. The contractions and losses reduce the actual discharge to about 62 percent of the theoretical, or

$$Q = \begin{cases} 3.33LH^{3/2} & \text{U.S. customary units} \\ 1.84LH^{3/2} & \text{SI units} \end{cases} \qquad (8.6.1)$$

When the weir does not extend completely across the width of the channel, it has *end contractions*, illustrated in Fig. 8.17a. An empirical correction for the reduction of flow is accomplished by subtracting $0.1H$ from L for each end contraction. The weir in Fig. 8.15 is said to have its end contractions *suppressed*.

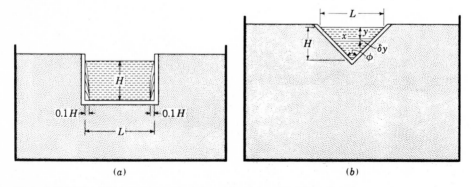

Figure 8.17 Weirs: (a) horizontal with end contractions; (b) V-notch weir.

The head H is measured upstream from the weir a sufficient distance to avoid the surface contraction. A hook gage mounted in a stilling pot connected to a piezometer opening determines the water-surface elevation from which the head is determined.

When the height P of weir (Fig. 8.15) is small, the velocity head at 1 cannot be neglected. A correction may be added to the head,

$$Q = CL\left(H + \alpha \frac{V^2}{2g}\right)^{3/2} \tag{8.6.2}$$

in which V is velocity and α is greater than unity, usually taken as about 1.4, which accounts for the nonuniform velocity distribution. Equation (8.6.2) must be solved for Q by trial, since both Q and V are unknown. As a first trial, the term $\alpha V^2/2g$ may be neglected to approximate Q. With this trial discharge, a value of V is computed, since

$$V = \frac{Q}{L(P + H)}$$

For small discharges the V-notch weir is particularly convenient. The contraction of the nappe is neglected, and the theoretical discharge is computed (Fig. 8.17b) as follows.

The velocity at depth y is $v = \sqrt{2gy}$, and the theoretical discharge is

$$Q_t = \int v\, dA = \int_0^H vx\, dy$$

By similar triangles, x may be related to y,

$$\frac{x}{H - y} = \frac{L}{H}$$

After substituting for v and x,

$$Q_t = \sqrt{2g}\frac{L}{H}\int_0^H y^{1/2}(H - y)\, dy = \frac{4}{15}\sqrt{2g}\frac{L}{H}H^{5/2}$$

Expressing L/H in terms of the angle ϕ of the V notch gives

$$\frac{L}{2H} = \tan \frac{\phi}{2}$$

Hence,

$$Q_t = \frac{8}{15}\sqrt{2g}\tan \frac{\phi}{2}H^{5/2}$$

The exponent in the equation is approximately correct, but the coefficient must be reduced by about 42 percent because of the neglected contractions. An approximate equation for a 90° V-notch weir is

$$Q = \begin{cases} 2.50H^{2.50} & \text{U.S. customary units} \\ 1.38H^{2.50} & \text{SI units} \end{cases} \tag{8.6.3}$$

Experiments show that the coefficient is increased by roughening the upstream side of the weir plate, which causes the boundary layer to grow thicker. The greater amount of slow-moving liquid near the wall is more easily turned, and hence there is less contraction of the nappe.

The broad-crested weir (Fig. 8.18a) supports the nappe so that the pressure variation is hydrostatic at section 2. Bernoulli's equation applied between points 1 and 2 can be used to find the velocity v_2 at height z, neglecting the velocity of approach,

$$H + 0 + 0 = \frac{v_2^2}{2g} + z + (y - z)$$

In solving for v_2,

$$v_2 = \sqrt{2g(H - y)}$$

z drops out; hence, v_2 is constant at section 2. For a weir of width L normal to the plane of the figure, the theoretical discharge is

$$Q = v_2 Ly = Ly\sqrt{2g(H - y)} \tag{8.6.4}$$

A plot of Q as abscissa against the depth y as ordinate, for constant H, is given in Fig. 8.18b. The depth is shown to be that which yields the maximum discharge, by the following reasoning.

A gate or other obstruction placed at section 3 of Fig. 8.18a can completely stop the flow by making $y = H$. Now, if a small flow is permitted to pass section 3

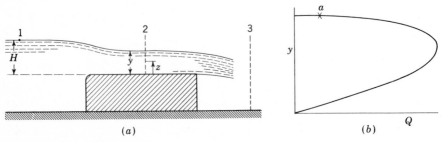

Figure 8.18 Broad-crested weir.

(holding H constant), the depth y becomes a little less than H and the discharge is, for example, as shown by point a on the depth-discharge curve. By further lifting of the gate or obstruction at section 3, the discharge-depth relation follows the upper portion of the curve until the maximum discharge is reached. Any additional removal of downstream obstructions, however, has no effect upon the discharge, because the velocity of flow at section 2 is $\sqrt{gy}$, which is exactly the speed an elementary wave can travel in still liquid of depth y. Hence, the effect of any additional lowering of the downstream surface elevation cannot travel upstream to affect further the value of y, and the discharge occurs at the maximum value. This depth y, called the *critical depth*, is discussed in Sec. 11.5. The speed of an elementary wave is derived in Sec. 12.10.

By taking dQ/dy and with the result set equal to zero, for constant H,

$$\frac{dQ}{dy} = 0 = L\sqrt{2g(H-y)} + Ly\frac{1}{2}\frac{-2g}{\sqrt{2g(H-y)}}$$

and solving for y gives
$$y = \tfrac{2}{3}H$$

Inserting the value of H, that is, $3y/2$, into the equation for velocity v_2 gives

$$v_2 = \sqrt{gy}$$

and substituting the value of y into Eq. (8.6.4) leads to

$$Q_t = \begin{cases} 3.09LH^{3/2} & \text{U.S. customary units} \\ 1.705LH^{3/2} & \text{SI units} \end{cases} \tag{8.6.5}$$

Experiments show that, for a well-rounded upstream edge, the discharge is

$$Q = \begin{cases} 3.03LH^{3/2} & \text{U.S. customary units} \\ 1.67LH^{3/2} & \text{SI units} \end{cases} \tag{8.6.6}$$

which is within 2 percent of the theoretical value. The flow, therefore, adjusts itself to discharge at the maximum rate.

Since viscosity and surface tension have a minor effect on the discharge coefficients of weirs, a weir should be calibrated with the liquid that it will measure.

Example 8.4 Tests on a 60° V-notch weir yield the following values of head H on the weir and discharge Q:

H, ft	0.345	0.356	0.456	0.537	0.568	0.594	0.619	0.635	0.654	0.665
Q, cfs	0.107	0.110	0.205	0.303	0.350	0.400	0.435	0.460	0.490	0.520

By means of the theory of least squares, determine the constants in $Q = CH^m$ for this weir.

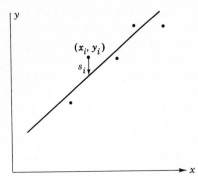

Figure 8.19 Log-log plot of Q vs. H for V-notch weir.

By taking the logarithm of each side of the equation

$$\ln Q = \ln C + m \ln H \qquad \text{or} \qquad y = B + mx$$

it is noted that the best values of B and m are needed for a straight line through the data when plotted on log-log paper.

By the theory of least squares, the best straight line through the data points is the one yielding a minimum value of the sums of the squares of vertical displacements of each point from the line; or, from Fig. 8.19,

$$F = \sum_{i=1}^{i=n} s_i^2 = \Sigma[y_i - (B + mx_i)]^2$$

where n is the number of experimental points. To minimize F, $\partial F/\partial B$ and $\partial F/\partial m$ are taken and set equal to zero, yielding two equations in the two unknowns B and m, as follows:

$$\frac{\partial F}{\partial B} = 0 = 2\Sigma[y_i - (B + mx_i)](-1)$$

from which

$$\Sigma y_i - nB - m\Sigma x_i = 0 \tag{1}$$

and

$$\frac{\partial F}{\partial m} = 0 = 2\Sigma[y_i - (B + mx_i)](-x_i)$$

or

$$\Sigma x_i y_i - B\Sigma x_i - m\Sigma x_i^2 = 0 \tag{2}$$

Solving Eqs. (1) and (2) for m gives

$$m = \frac{\Sigma x_i y_i / \Sigma x_i - \Sigma y_i / n}{\Sigma x_i^2 / \Sigma x_i - \Sigma x_i / n} \qquad B = \frac{\Sigma y_i - m\Sigma x_i}{n}$$

These equations are readily solved by an electronic hand calculator having the Σ key, or a simple program may be written for the digital computer. The answer for the data of this problem is $m = 2.437$, $C = 1.395$.

Measurement of River Flow

Daily records of the discharge of rivers over long periods of time are essential to economic planning for utilization of their water resources or protection against floods. The daily measurement of discharge by determining velocity distribution over a cross section of the river is costly. To avoid the cost and still obtain daily records, *control sections* are established where the river channel is stable, i.e., with little change in bottom or sides of the stream bed. The control section is frequently at a break in slope of the river bottom where it becomes steeper downstream.

A gage rod is mounted at the control section, and the elevation of water surface is determined by reading the waterline on the rod; in some installations float-controlled recording gages keep a continuous record of river elevation. A *gage height-discharge* curve is established by taking current-meter measurements from time to time as the river discharge changes and plotting the resulting discharge against the gage height.

With a stable control section the gage height-discharge curve changes very little, and current-meter measurements are infrequent. For unstable control sections the curve changes continuously, and discharge measurements must be made every few days to maintain an accurate curve.

Daily readings of gage height produce a daily record of the river discharge.

EXERCISES

8.6.1 The discharge through a V-notch weir varies as (a) $H^{-1/2}$; (b) $H^{1/2}$; (c) $H^{3/2}$; (d) $H^{5/2}$; (e) none of these answers.

8.6.2 The discharge of a rectangular sharp-crested weir with end contractions is less than for the same weir with end contractions suppressed by (a) 5%; (b) 10%; (c) 15%; (d) no fixed percent; (e) none of these answers.

8.7 MEASUREMENT OF TURBULENCE

Turbulence is a characteristic of the flow. It affects the calibration of measuring instruments and has an important effect upon heat transfer, evaporation, diffusion, and many other phenomena connected with fluid movement.

Turbulence is generally specified by two quantities, the *size* and the *intensity* of the fluctuations. In steady flow the temporal mean velocity components at a

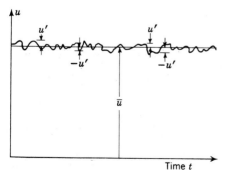

Figure 8.20 Turbulent fluctuations in direction of flow.

point are constant. If these mean values are $\bar{u}, \bar{v}, \bar{w}$ and the velocity components at an instant are u, v, w, the fluctuations are given by u', v', w', in

$$u = \bar{u} + u' \qquad v = \bar{v} + v' \qquad w = \bar{w} + w'$$

The root mean square of measured values of the fluctuations (Fig. 8.20) is a measure of the intensity of the turbulence. These are $\sqrt{\overline{(u')^2}}, \sqrt{\overline{(v')^2}}, \sqrt{\overline{(w')^2}}$.

The size of the fluctuation is an average measure of the size of eddy, or vortex, in the flow. When two velocity-measuring instruments (hot-wire anemometers) are placed adjacent to each other in a flow, the velocity fluctuations are correlated, i.e., they tend to change in unison. Separating these instruments reduces this correlation. The distance between instruments for zero correlation is a measure of the size of the fluctuation. Another method for determining turbulence is discussed in Sec. 5.6.

8.8 MEASUREMENT OF VISCOSITY

The treatment of fluid measurement is concluded with a discussion of methods for determining viscosity. Viscosity may be measured in a number of ways: (1) by use of Newton's law of viscosity, (2) by use of the Hagen-Poiseuille equation, (3) by methods that require calibration with fluids of known viscosity.

By measurement of the velocity gradient du/dy and the shear stress τ, in Newton's law of viscosity [Eq. (1.1.1)],

$$\tau = \mu \, \frac{du}{dy} \tag{8.8.1}$$

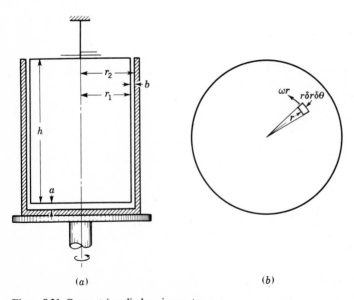

(a) $\qquad\qquad\qquad\qquad\qquad$ (b)

Figure 8.21 Concentric-cylinder viscometer.

the dynamic or absolute viscosity can be computed. This is the most basic method, because it determines all other quantities in the defining equation for viscosity. By means of a cylinder that rotates at a known speed with respect to an inner concentric stationary cylinder, du/dy is determined. By measurement of torque on the stationary cylinder, the shear stress can be computed. The ratio of shear stress to rate of change of velocity expresses the viscosity.

A schematic view of a concentric-cylinder viscometer is shown in Fig. 8.21a. When the speed of rotation is N rpm and the radius is r_2, the fluid velocity at the surface of the outer cylinder is $2\pi r_2 N/60$. With clearance b

$$\frac{du}{dy} = \frac{2\pi r_2 N}{60b}$$

The equation is based on $b \ll r_2$. The torque T_c on the inner cylinder is measured by a torsion wire from which it is suspended. By attaching a disk to the wire, its rotation can be determined by a fixed pointer. If the torque due to fluid below the bottom of the inner cylinder is neglected, the shear stress is

$$\tau = \frac{T_c}{2\pi r_1^2 h}$$

Substituting into Eq. (8.8.1) and solving for the viscosity yields

$$\mu = \frac{15 T_c b}{\pi^2 r_1^2 r_2 h N} \tag{8.8.2}$$

When the clearance a is so small that the torque contribution from the bottom is appreciable, it can be calculated in terms of the viscosity.

Referring to Fig. 8.21b,

$$\delta T = r\tau \; \delta A = r\mu \frac{\omega r}{a} r \; \delta r \; \delta\theta$$

in which the velocity change is ωr in the distance a. Integrating over the circular area of the disk and letting $\omega = 2\pi N/60$ leads to

$$T_d = \frac{\mu}{a}\frac{\pi}{30} N \int_0^{r_1} \int_0^{2\pi} r^3 \; dr \; d\theta = \frac{\mu\pi^2}{a60} N r_1^4 \tag{8.8.3}$$

The torque due to disk and cylinder must equal the torque T in the torsion wire, so that

$$T = \frac{\mu\pi^2 N r_1^4}{a60} + \frac{\mu\pi^2 r_1^2 r_2 h N}{15b} = \frac{\mu\pi^2 N r_1^2}{15}\left(\frac{r_1^2}{4a} + \frac{r_2 h}{b}\right) \tag{8.8.4}$$

in which all quantities except μ are known. The flow between the surfaces must be laminar for Eqs. (8.8.2) to (8.8.4) to be valid.

Often the geometry of the inner cylinder is altered to eliminate the torque which acts on the lower surface. If the bottom surface of the inner cylinder is made concave, a pocket of air will be trapped between the bottom surface of the inner

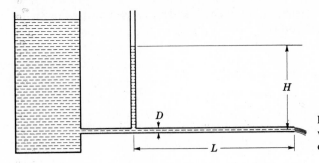

Figure 8.22 Determination of viscosity by flow through a capillary tube.

cylinder and the fluid in the rotating outer cup. A well-designed cup and a careful filling procedure will ensure the condition whereby the torque measured will consist of that produced in the annulus between the two cylinders and a minute amount resulting from the action of the air on the bottom surface. Naturally, the viscometer must be provided with a temperature-controlled bath and a variable-speed drive which can be carefully regulated. Such design refinements are needed in order to obtain the rheological diagrams (cf. Fig. 1.2) for the fluid under test.

The measurement of all quantities in the Hagen-Poiseuille equation, except μ, by a suitable experimental arrangement is another basic method for determination of viscosity. A setup as in Fig. 8.22 may be used. Some distance is required for the fluid to develop its characteristic velocity distribution after it enters the tube; therefore, the head or pressure must be measured by some means at a point along the tube. The volume $\mathcal{V}$ of flow can be measured over a time t where the reservoir

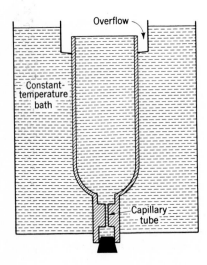

Figure 8.23 Schematic view of Saybolt viscometer.

surface is held at a constant level. This yields Q; and by determining γ, Δp can be computed. Then with L and D known, from Eq. (5.2.10a),

$$\mu = \frac{\Delta p \pi D^4}{128 Q L}$$

An adaptation of the capillary tube for industrial purposes is the *Saybolt viscometer* (Fig. 8.23). A short capillary tube is utilized, and the time is measured for 60 cm^3 of fluid to flow through the tube under a falling head. The time, in seconds, is the Saybolt reading. This device measures kinematic viscosity, evident from a rearrangement of Eq. (5.2.10a). When $\Delta p = \rho g h$, $Q = \mho/t$; and when the terms that are the same regardless of the fluid are separated,

$$\frac{\mu}{\rho t} = \frac{g h \pi D^4}{128 \mho L} = C_1$$

Although the head h varies during the test, it varies over the same range for all liquids; and the terms on the right-hand side may be considered as a constant of the particular instrument. Since $\mu/\rho = v$, the kinematic viscosity is

$$v = C_1 t$$

which shows that the kinematic viscosity varies directly as the time t. The capillary tube is quite short, and so the velocity distribution is not established. The flow tends to enter uniformly, and then, owing to viscous drag at the walls, to slow down there and speed up in the center region. A correction in the above equation is needed, which is of the form C/t; hence,

$$v = C_1 t + \frac{C_2}{t}$$

The approximate relation between viscosity and Saybolt seconds is expressed by

$$v = 0.0022t - \frac{1.80}{t}$$

in which v is in stokes and t in seconds.

For measuring viscosity there are many other industrial methods that generally have to be calibrated for each special case to convert to the absolute units. One consists of several tubes containing "standard" liquids of known graduated viscosities with a steel ball in each of the tubes. The time for the ball to fall the length of the tube depends upon the viscosity of the liquid. By placing the test sample in a similar tube, its viscosity can be approximated by comparison with the other tubes.

The flow of a fluid in a capillary tube is the basis for viscometers of the Oswald-Cannon-Fenske, or Ubbelohde, type. In essence, the viscometer is a U tube one leg of which is a fine capillary tube connected to a reservoir above. The tube is held vertically, and a known quantity of fluid is placed in the reservoir and allowed to flow by gravity through the capillary. The time is recorded for the free surface in the reservoir to fall between two scribed marks. A calibration constant for each instrument takes into account the variation of the capillary's bore from the standard, the bore's uniformity, entrance conditions, and the slight unsteadiness due to the falling head during the 1- to 2-min test. Various bore sizes can be obtained to cover a wide range of viscosities. Exact procedures for carrying out the tests are contained in the standards of the American Society for Testing and Materials.

EXERCISE

8.8.1 A homemade viscometer of the Saybolt type is calibrated by two measurements with liquids of known kinematic viscosity. For $v = 0.461$ St, $t = 97$ s, and for $v = 0.18$ St, $t = 46$ s. The coefficients C_1, C_2 in $v = C_1 t + C_2/t$ are

(a) $C_1 = 0.005$ (b) $C_1 = 0.0044$ (c) $C_1 = 0.0046$ (d) $C_1 = 0.00317$
    $C_2 = -2.3$ $C_2 = 3.6$ $C_2 = 1.55$ $C_2 = 14.95$

(e) none of these answers

PROBLEMS

8.1 A static tube (Fig. 8.1b) indicates a static pressure that is 1 kPa too low when liquid is flowing at 2 m/s. Calculate the correction to be applied to the indicated pressure for the liquid flowing at 5 m/s.

8.2 Four piezometer openings in the same cross section of a cast-iron pipe indicate the following pressures for simultaneous readings: 43, 42.6, 42.4, 37 mm Hg. What value should be taken for the pressure?

8.3 A simple pitot tube (Fig. 8.2) is inserted into a small stream of flowing oil, $\gamma = 55$ lb/ft³, $\mu = 0.65$ P, $\Delta h = 1.5$ in, $h_0 = 5$ in. What is the velocity at point 1?

8.4 A stationary body immersed in a river has a maximum pressure of 69 kPa exerted on it at a distance of 5.4 m below the free surface. Calculate the river velocity at this depth.

8.5 From Fig. 8.3 derive the equation for velocity at 1.

8.6 In Fig. 8.3 air is flowing ($p = 16$ psia, $t = 40°F$) and water is in the manometer. For $R' = 1.2$ in, calculate the velocity of air.

8.7 In Fig. 8.3 air is flowing ($p = 101$ kPa abs, $t_1 = 5°C$) and mercury is in the manometer. For $R' = 200$ mm, calculate the velocity at 1 (a) for isentropic compression of air between 1 and 2 and (b) for air considered incompressible.

8.8 A pitot-static tube directed into a 4 m/s water stream has a gage difference of 37 mm on a water-mercury differential manometer. Determine the coefficient for the tube.

8.9 A pitot-static tube, $C = 1.12$, has a gage difference of 10 mm on a water-mercury manometer when directed into a water stream. Calculate the velocity.

8.10 A pitot-static tube of the Prandtl type has the following value of gage difference R' for the radial distance from center of a 3-ft-diameter pipe:

r, ft	0.0	0.3	0.6	0.9	1.2	1.48
R', in	4.00	3.91	3.76	3.46	3.02	2.40

Water is flowing, and the manometer fluid has a specific gravity of 2.93. Calculate the discharge.

8.11 What would be the gage difference on a water-nitrogen manometer for flow of nitrogen at 200 m/s, using a pitot-static tube? The static pressure is 175 kPa abs, and the corresponding temperature is 25°C. True static pressure is measured by the tube.

8.12 Measurements in an airstream indicate that the stagnation pressure is 15 psia, the static pressure is 10 psia, and the stagnation temperature is 102°F. Determine the temperature and velocity of the airstream.

8.13 0.5 kg/s nitrogen flows through a 50-mm-diameter tube with stagnation temperature of 38°C and undisturbed temperature of 20°C. Find the velocity and static and stagnation pressures.

8.14 A disk meter has a volumetric displacement of 27 cm^3 for one complete oscillation. Calculate the flow, in liters per minute, for 86.5 oscillations per minute.

8.15 A disk water meter with volumetric displacement of 40 cm^3 per oscillation requires 470 oscillations per minute to pass 0.32 l/s and 3840 oscillations per minute to pass 2.57 l/s. Calculate the percent error, or slip, in the meter.

8.16 A volumetric tank 4 ft in diameter and 5 ft high was filled with oil in 16 min 32.4 s. What is the average discharge in gallons per minute?

8.17 A weigh tank receives 75 N liquid, sp gr 0.86, in 14.9 s. What is the flow rate, in liters per minute?

8.18 Determine the equation for trajectory of a jet discharging horizontally from a small orifice with head of 5 m and velocity coefficient of 0.96. Neglect air resistance.

8.19 An orifice of area 30 cm^2 in a vertical plate has a head of 1.1 m of oil, sp gr 0.91. It discharges 6790 N of oil in 79.3 s. Trajectory measurements yield $x_0 = 2.25$ m, $y_0 = 1.23$ m. Determine C_v, C_c, C_d.

8.20 Calculate Y, the maximum rise of a jet from an inclined plate (Fig. 8.24), in terms of H and α. Neglect losses.

8.21 In Fig. 8.24, for $\alpha = 45°$, $Y = 0.48H$. Neglecting air resistance of the jet, find C_v for the orifice.

8.22 Show that the locus of maximum points of the jet of Fig. 8.24 is given by $X^2 = 4Y(H - Y)$ when losses are neglected.

8.23 A 3-in-diameter orifice discharges 64 ft^3 liquid, sp gr 1.07, in 82.2 s under a 9-ft head. The velocity at the *vena contracta* is determined by a pitot-static tube with coefficient 1.0. The manometer liquid is acetylene tetrabromide, sp gr 2.96, and the gage difference is $R' = 3.35$ ft. Determine C_v, C_c, and C_d.

8.24 A 100-mm-diameter orifice discharges 44.6 l/s water under a head of 2.75 m. A flat plate held normal to the jet just downstream from the *vena contracta* requires a force of 320 N to resist impact of the jet. Find C_d, C_v, and C_c.

8.25 Compute the discharge from the tank shown in Fig. 8.25.

8.26 For $C_v = 0.96$ in Fig. 8.25, calculate the losses in meter-newtons per newton and in meter-newtons per second.

8.27 Calculate the discharge through the orifice of Fig. 8.26.

8.28 For $C_v = 0.93$ in Fig. 8.26, determine the losses in joules per newton and in watts.

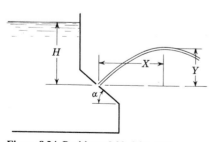

Figure 8.24 Problems 8.20, 8.21, 8.22.

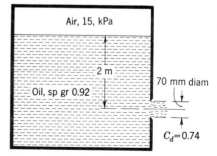

Figure 8.25 Problems 8.25, 8.26.

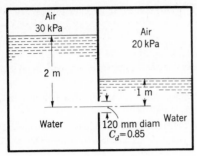

Figure 8.26 Problems 8.27, 8.28.

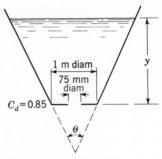

Figure 8.27 Problem 8.35.

8.29 A 4-in-diameter orifice discharges 1.60 cfs liquid under a head of 11.8 ft. The diameter of jet at the *vena contracta* is found by calipering to be 3.47 in. Calculate C_v, C_d, and C_c.

8.30 A Borda mouthpiece 50 mm in diameter has a discharge coefficient of 0.51. What is the diameter of the issuing jet?

8.31 A 75-mm-diameter orifice, $C_d = 0.82$, is placed in the bottom of a vertical tank that has a diameter of 1.5 m. How long does it take to draw the surface down from 3 to 2.5 m?

8.32 Select the size of orifice that permits a tank of horizontal cross section 1.5 m² to have the liquid surface drawn down at the rate of 160 mm/s for 3.35 m head on the orifice. $C_d = 0.63$.

8.33 A 4-in-diameter orifice in the side of a 6-ft-diameter tank draws the surface down from 8 to 4 ft above the orifice in 83.7 s. Calculate the discharge coefficient.

8.34 Select a reservoir of such size and shape that the liquid surface drops 1 m/min over a 3-m distance for flow through a 100-mm-diameter orifice. $C_d = 0.74$.

8.35 In Fig. 8.27 the truncated cone has an angle $\theta = 60°$. How long does it take to draw the liquid surface down from $y = 4$ m to $y = 1$ m?

8.36 Calculate the dimensions of a tank such that the surface velocity varies inversely as the distance from the centerline of an orifice draining the tank. When the head is 300 mm, the velocity of fall of the surface is 30 mm/s; orifice diameter is 12.5 mm, $C_d = 0.66$.

8.37 Determine the time required to raise the right-hand surface of Fig. 8.28 by 2 ft.

8.38 How long does it take to raise the water surface of Fig. 8.29 2 m? The left-hand surface is a large reservoir of constant water-surface elevation.

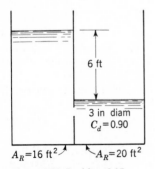

Figure 8.28 Problem 8.37.

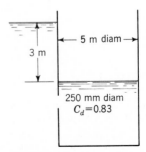

Figure 8.29 Problem 8.38.

8.39 Show that for incompressible flow the losses per unit weight of fluid between the upstream section and throat of a venturi meter are $KV_2^2/2g$ if

$$K = [(1/C_v)^2 - 1][1 - (D_2/D_1)^4]$$

8.40 A 4 by 2 m venturi meter carries water at 25°C. A water-air differential manometer has a gage difference of 60 mm. What is the discharge?

8.41 What is the pressure difference between the upstream section and throat of a 150 by 75 mm horizontal venturi meter carrying 50 l/s water at 48°C?

8.42 A 12 by 6 in venturi meter is mounted in a vertical pipe with the flow upward. 2000 gpm oil, sp gr 0.80, $\mu = 0.1$ P, flows through the pipe. The throat section is 6 in above the upstream section. What is $p_1 - p_2$?

8.43 Air flows through a venturi meter in a 55-mm-diameter pipe having a throat diameter of 30 mm, $C_v = 0.97$. For $p_1 = 830$ kPa abs, $t_1 = 15°C$, $p_2 = 550$ kPa abs, calculate the mass per second flowing.

8.44 Oxygen, $p_1 = 40$ psia, $t_1 = 120°F$, flows through a 1 by $\frac{1}{2}$ in venturi meter with a pressure drop of 6 psi. Find the mass per second flowing and the throat velocity.

8.45 Air flows through a 80-mm-diameter ISA flow nozzle in a 120-mm-diameter pipe. $p_1 = 150$ kPa abs; $t_1 = 5°C$; and a differential manometer with liquid, sp gr 2.93, has a gage difference of 0.8 m when connected between the pressure taps. Calculate the mass rate of flow.

8.46 A 2.5-in-diameter ISA nozzle is used to measure flow of water at 40°F in a 6-in-diameter pipe. What gage difference on a water-mercury manometer is required for 300 gpm?

8.47 Determine the discharge in a 300-mm-diameter line with a 160-mm-diameter VDI orifice for water at 20°C when the gage difference is 300 mm on an acetylene tetrabromide (sp gr 2.94)-water differential manometer.

8.48 A 10-mm-diameter VDI orifice is installed in a 25-mm-diameter pipe carrying nitrogen at $p_1 = 8$ atm, $t_1 = 50°C$. For a pressure drop of 140 kPa across the orifice, calculate the mass flow rate.

8.49 Air at 1 atm, $t = 21°C$ flows through a 1-m-square duct that contains a 500-mm-diameter square-edged orifice. With a head loss of 60 mm H_2O across the orifice, compute the flow, in cubic meters per minute.

8.50 A 6-in-diameter VDI orifice is installed in a 12-in-diameter oil line, $\mu = 6$ cP, $\gamma = 52$ lb/ft³. An oil-air differential manometer is used. For a gage difference of 20 in determine the flow rate, in gallons per minute.

8.51 A rectangular sharp-crested weir 4 m long with end contractions suppressed is 1.3 m high. Determine the discharge when the head is 200 mm.

8.52 In Fig. 8.15, $L = 8$ ft, $P = 1.8$ ft, $H = 0.80$ ft. Estimate the discharge over the weir. $C = 3.33$.

8.53 A rectangular sharp-crested weir with end contractions is 1.5 m long. How high should it be placed in a channel to maintain an upstream depth of 2.25 m for 0.45 m³/s flow?

8.54 Determine the head on a 60° V-notch weir for discharge of 170 l/s.

8.55 Tests on a 90° V-notch weir gave the following results: $H = 180$ mm, $Q = 19.4$ l/s, $H = 410$ mm, $Q = 150$ l/s. Determine the formula for the weir.

8.56 A sharp-crested rectangular weir 3 ft long with end contractions suppressed and a 90° V-notch weir are placed in the same weir box, with the vertex of the 90° V-notch weir 6 in below the rectangular weir crest. Determine the head on the V-notch weir (a) when the discharges are equal and (b) when the rectangular weir discharges its greatest amount above the discharge of the V-notch weir.

8.57 A broad-crested weir 1.6 m high and 3 m long has a well-rounded upstream corner. What head is required for a flow of 2.85 m³/s?

8.58 A circular disk 180 mm in diameter has a clearance of 0.3 mm from a flat plate. What torque is required to rotate the disk 800 rpm when the clearance contains oil, $\mu = 0.8$ P?

8.59 The concentric-cylinder viscometer (Fig. 8.21a) has the following dimensions: $a = 0.012$ in, $b = 0.05$ in, $r_1 = 2.8$ in, $h = 6.0$ in. The torque is 20 lb·in when the speed is 160 rpm. What is the viscosity?

8.60 With the apparatus of Fig. 8.22, $D = 0.5$ mm, $L = 1$ m, $H = 0.75$ m, and 60 cm^3 was discharged in 1 h 30 min. What is the viscosity in poises? $S = 0.83$.

REFERENCES

ASME Symposium on Flow, Its Measurement and Control in Science and Industry, Pittsburgh, May 9–14, 1971.

Dowden, R. Rosemary: "Fluid Flow Measurement: A Bibliography," BHRA Fluid Engineering, Cranfield, Bedford, England, 1972.

"Fluid Meters," 6th ed., American Society of Mechanical Engineers, New York, 1971.

NINE
TURBOMACHINERY

To turn a fluid stream or change the magnitude of its velocity requires that forces be applied. When a moving vane deflects a fluid jet and changes its momentum, forces are exerted between vane and jet and work is done by displacement of the vane. Turbomachines make use of this principle: the axial and centrifugal pumps, blowers, and compressors, by continuously doing work on the fluid, add to its energy; the impulse, Francis, and propeller turbines and steam and gas turbines continuously extract energy from the fluid and convert it into torque on a rotating shaft; the fluid coupling and the torque converter, each consisting of a pump and a turbine built together, make use of the fluid to transmit power smoothly. Designing efficient turbomachines utilizes both theory and experimentation. A good design of given size and speed may be readily adapted to other speeds and other geometrically similar sizes by application of the theory of scaled models, as outlined in Sec. 4.5.

Similarity relations are first discussed in this chapter by consideration of homologous units and specific speed. Elementary cascade theory is next taken up, before the theory of turbomachines is considered. Water reaction turbines are dealt with before pumps and blowers; then the impulse turbine and centrifugal compressor follow. The chapter closes with a discussion of cavitation.

9.1 HOMOLOGOUS UNITS; SPECIFIC SPEED

In utilizing scaled models in designing turbomachines, geometric similitude is required as well as geometrically similar velocity vector diagrams at entrance to, or exit from, the impellers. Viscous effects must, unfortunately, be neglected, as

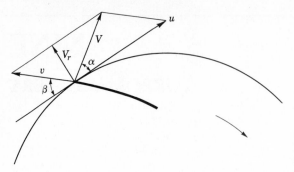

Figure 9.1 Velocity vector diagram for exit from a pump impeller.

it is generally impossible to satisfy the above two conditions and have equal Reynolds numbers in model and prototype. Two geometrically similar units having similar velocity vector diagrams are *homologous*. They will also have geometrically similar streamlines.

The velocity vector diagram in Fig. 9.1 at exit from a pump impeller can be used to formulate the condition for similar streamline patterns. The blade angle is β; u is the peripheral speed of the impeller at the end of the vane or blade; v is the velocity of fluid *relative* to the vane, and V is the absolute velocity leaving the impeller, the vector sum of u and v; V_r is the radial component of V and is proportional to the discharge; α is the angle which the absolute velocity makes with u, the tangential direction. According to geometric similitude, β must be the same for two units, and for similar streamlines α also must be the same in each case.

It is convenient to express the fact that α is to be the same in any of a series of turbomachines, called homologous units, by relating the speed of rotation N, the impeller diameter (or other characteristic dimension) D, and the flow rate Q. For constant α, V_r is proportional to V ($V_r = V \sin \alpha$) and u is proportional to V_r. Hence, the conditions for constant α in a homologous series of units may be expressed as

$$\frac{V_r}{u} = \text{const}$$

The discharge Q is proportional to $V_r D^2$, since any cross-sectional flow area is proportional to D^2. The speed of rotation N is proportional to u/D. When these values are inserted,

$$\frac{Q}{ND^3} = \text{const} \tag{9.1.1}$$

expresses the condition in which geometrically similar units are homologous.

The discharge Q through homologous units can be related to head H and a representative cross-sectional area A by the orifice formula

$$Q = C_d A \sqrt{2gH}$$

in which C_d, the discharge coefficient, varies slightly with the Reynolds number and so causes a small change in efficiency with size in a homologous series. The change in discharge with the Reynolds number is referred to as *scale effect*. The smaller machines, having smaller hydraulic radii of passages, have lower Reynolds numbers and correspondingly higher friction factors; hence, they are less efficient. The change in efficiency from model to prototype may be from 1 to 4 percent. However, in the homologous theory, the scale effect must be neglected, and so an empirical correction for change in efficiency with size is used [see Eq. (9.4.2)]. As $A \sim D^2$, the discharge equation may be

$$\frac{Q}{D^2 \sqrt{H}} = \text{const} \tag{9.1.2}$$

Eliminating Q between Eqs. (9.1.1) and (9.1.2) gives

$$\frac{H}{N^2 D^2} = \text{const} \tag{9.1.3}$$

Equations (9.1.1) and (9.1.3) are most useful in determining performance characteristics for one unit from those of a homologous unit of different size and speed.†

---

† Application of dimensional analysis is illuminating. The variables appearing to be pertinent to the flow relations for similar units would be $F(H,Q,N,D,g) = 0$. There are two dimensions involved, L and T; N and D may be selected as the repeating variables, yielding

$$f\left(\frac{Q}{ND^3}, \frac{H}{D}, \frac{g}{N^2 D}\right) = 0$$

Solving for H gives

$$H = D f_1\left(\frac{Q}{ND^3}, \frac{g}{N^2 D}\right)$$

Experiment shows that the second dimensionless parameter actually occurs to the power -1; hence,

$$H = \frac{N^2 D^2}{g} f_2\left(\frac{Q}{ND^3}\right) \qquad \text{or} \qquad \frac{gH}{N^2 D^2} = f_2\left(\frac{Q}{ND^3}\right)$$

The characteristic curve for a pump in dimensionless form is the plot of Q/ND^3 as abscissa against $H/(N^2 D^2/g)$ as ordinate. This curve, obtained from tests on one unit of the series, then applies to all homologous units, and it may be converted to the usual characteristic curve by selecting desired values of N and D. As power is proportional to γQH, the dimensionless power term is

$$\frac{\gamma}{\gamma} \frac{Q}{ND^3} \frac{H}{N^2 D^2/g} = \frac{\text{power}}{\rho N^3 D^5}$$

Example 9.1 A prototype test of a mixed-flow pump with a 72-in-diameter discharge opening, operating at 225 rpm, resulted in the following characteristics:

H, ft	Q, cfs	e, %	H, ft	Q, cfs	e, %	H, ft	Q, cfs	e, %
60	200	69	47.5	330	87.3	35	411	82
57.5	228	75	45	345	88	32.5	425	79
55	256	80	42.5	362	87.4	30	438	75
52.5	280	83.7	40	382	86.3	27.5	449	71
50	303	86	37.5	396	84.4	25	459	66.5

What size and synchronous speed (60 Hz) of homologous pump should be used to produce 200 cfs at 60 ft head at point of best efficiency? Find the characteristic curves for this case. Subscript 1 refers to the 72-in pump. For best efficiency $H_1 = 45, Q_1 = 345, e = 88$ percent. With Eqs. (9.1.1) and (9.1.3),

$$\frac{H}{N^2 D^2} = \frac{H_1}{N_1^2 D_1^2} \qquad \frac{Q}{ND^3} = \frac{Q_1}{N_1 D_1^3}$$

or

$$\frac{60}{N^2 D^2} = \frac{45}{225^2 \times 72^2} \qquad \frac{200}{ND^3} = \frac{345}{225 \times 72^3}$$

After solving for N and D,

$$N = 366 \text{ rpm} \qquad D = 51.1 \text{ in}$$

The nearest synchronous speed (3600 divided by number of pairs of poles) is 360 rpm. To maintain the desired head of 60 ft, a new D is necessary. Its size can be computed:

$$D = \sqrt{\tfrac{60}{45} \times \tfrac{225}{360}} \times 72 = 52 \text{ in}$$

The discharge at best efficiency is then

$$Q = \frac{Q_1 ND^3}{N_1 D_1^3} = 345 \times \frac{360}{225}\left(\frac{52}{72}\right)^3 = 208 \text{ cfs}$$

which is slightly more capacity than required. With $N = 360$ and $D = 52$, equations for transforming the corresponding values of H and Q for any efficiency can be obtained:

$$H = H_1\left(\frac{ND}{N_1 D_1}\right)^2 = H_1\left(\frac{360}{225} \times \frac{52}{72}\right)^2 = 1.335 H_1$$

and

$$Q = Q_1 \frac{ND^3}{N_1 D_1^3} = Q_1\left(\frac{360}{225}\right)\left(\frac{52}{72}\right)^3 = 0.603 Q_1$$

The characteristics of the new pump are

H, ft	Q, cfs	e, %	H, ft	Q, cfs	e, %	H, ft	Q, cfs	e, %
80	121	69	63.5	200	87.3	46.7	248	82
76.7	138	75	60	208	88	43.4	257	79
73.4	155	80	56.7	219	87.4	40	264	75
70	169	83.7	53.5	231	86.3	36.7	271	71
66.7	183	86	50	239	84.4	33.4	277	66.5

The efficiency of the 52-in pump might be a fraction of a percent less than that of the 72-in pump, as the hydraulic radii of flow passages are smaller, so that the Reynolds number would be less.

Specific Speed

The specific speed of a homologous unit is a constant widely used in selecting the type of unit and in preliminary design. It is usually defined differently for a pump and a turbine.

The specific speed N_s of a homologous series of pumps is defined as the speed of some one unit of the series of such size that it delivers unit discharge at unit head. It is obtained as follows. Eliminating D in Eqs. (9.1.1) and (9.1.3) and rearranging give

$$\frac{N\sqrt{Q}}{H^{3/4}} = \text{const} \tag{9.1.4}$$

By definition of specific speed, the constant is N_s, the speed of a unit for $Q = 1$, $H = 1$;

$$N_s = \frac{N\sqrt{Q}}{H^{3/4}} \tag{9.1.5}$$

The specific speed of a series is usually defined for the point of best efficiency, i.e., for the speed, discharge, and head that is most efficient.

The specific speed of a homologous series of turbines is defined as the speed of a unit of the series of such size that it produces unit power with unit head. Since power P is proportional to QH,

$$\frac{P}{QH} = \text{const} \tag{9.1.6}$$

The terms D and Q may be eliminated from Eqs. (9.1.1), (9.1.3), and (9.1.6) to produce

$$\frac{N\sqrt{P}}{H^{5/4}} = \text{const} \tag{9.1.7}$$

For unit power and unit head the constant of Eq. (9.1.7) becomes the speed, or the specific speed N_s of the series, so that

$$N_s = \frac{N\sqrt{P}}{H^{5/4}} \tag{9.1.8}$$

The specific speed of a unit required for a given discharge and head can be estimated from Eqs. (9.1.5) and (9.1.8). For pumps handling large discharges at low heads a high specific speed is indicated; for a high-head turbine producing relatively low power (small discharge) the specific speed is low. Experience has shown that for best efficiency one particular type of pump or turbine is usually indicated for a given specific speed.

Because Eqs. (9.1.5) and (9.1.8) are not dimensionally correct (γ and g have been included in the constant term), the value of specific speed depends on the units involved. For example, in the United States Q is commonly expressed in gallons per minute, millions of gallons per day, or cubic feet per second when referring to specific speeds of pumps.

Centrifugal pumps have low specific speeds; mixed-flow pumps have medium specific speeds; and axial-flow pumps have high specific speeds. Impulse turbines have low specific speeds; Francis turbines have medium specific speeds; and propeller turbines have high specific speeds.

EXERCISES

9.1.1 Two units are homologous when they are geometrically similar and have (a) similar streamlines; (b) the same Reynolds number; (c) the same efficiency; (d) the same Froude number; (e) none of these answers.

9.1.2 The following two relations are necessary for homologous units: (a) $H/ND^3 = $ const; $Q/N^2D^2 = $ const; (b) $Q/D^2\sqrt{H} = $ const; $H/N^3D = $ const; (c) $P/QH = $ const; $H/N^2D^2 = $ const; (d) $N\sqrt{Q}/H^{3/2} = $ const; $N\sqrt{P}/H^{3/4} = $ const; (e) none of these answers.

9.1.3 The specific speed of a pump is defined as the speed of a unit (a) of unit size with unit discharge at unit head; (b) of such size that it requires unit power for unit head; (c) of such size that it delivers unit discharge at unit head; (d) of such size that it delivers unit discharge at unit power; (e) none of these answers.

9.2 ELEMENTARY CASCADE THEORY

Turbomachines either do work on a fluid or extract work from it in a continuous manner by having it flow through a series of moving (and possibly fixed) vanes. By examination of flow through a series of similar blades or vanes, called a *cascade*, some of the requirements of an efficient system can be developed. Consider, first, flow through the simple fixed cascade system of Fig. 9.2. The velocity vector representing the fluid has been turned through an angle by the presence of the cascade system. A force has been exerted on the fluid, but (neglecting friction

Figure 9.2 Simple cascade system.

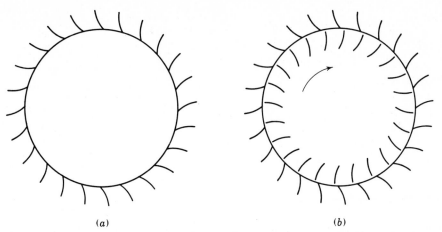

Figure 9.3 Cascades of vanes on periphery of a circular cylinder: (*a*) stationary vanes; (*b*) rotating cascade within a fixed cascade.

effects and turbulence) no work is done on the fluid. Section 3.11 deals with forces on a single vane.

Since turbomachines are rotational devices, the cascade system may be arranged symmetrically around the periphery of a circle, as in Fig. 9.3*a*. If the fluid approaches the fixed cascade in a radial direction, it has its moment of momentum changed from zero to a value dependent upon the mass per unit time flowing, the tangential component of velocity V_t developed, and the radius, from Eq. (3.12.5),

$$T = \rho Q r V_t \tag{9.2.1}$$

Again, no work is done by the fixed-vane system.

Consider now another series of vanes (Fig. 9.3*b*) rotating within the fixed-vane system at a speed ω. For efficient operation of the system it is important that the fluid flow onto the moving vanes with the least disturbance, i.e., in a tangential

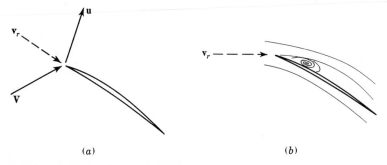

Figure 9.4 Flow onto blades: (*a*) flow tangent to blade; (*b*) flow separation, or shock, with relative velocity not tangent to leading edge.

manner, as illustrated in Fig. 9.4*a*. When the relative velocity is not tangent to the blade at its entrance, separation may occur, as shown in Fig. 9.4*b*. The losses tend to increase rapidly (about as the square) with angle from the tangential and radically impair the efficiency of the machine. Separation also frequently occurs when the approaching relative velocity is tangential to the vane, owing to curvature of the vanes or to expansion of the flow passages, which causes the boundary layer to thicken and come to rest. These losses are called *shock* or *turbulence* losses. When the fluid exits from the moving cascade, it will generally have its velocity altered in both magnitude and direction, thereby changing its moment of momentum and either doing work on the cascade or having work done on it by the moving cascade. In the case of a turbine it is desired to have the fluid leave with no moment of momentum. An old saying in turbine design is " have the fluid enter without shock and leave without velocity."

Turbomachinery design requires the proper arrangement and shaping of passages and vanes so that the purpose of the design can be met most efficiently. The particular design depends upon the purpose of the machine, the amount of work to be done per unit mass of fluid, and the fluid density.

9.3 THEORY OF TURBOMACHINES

Turbines extract useful work from fluid energy; and pumps, blowers, and turbocompressors add energy to fluids by means of a runner consisting of vanes rigidly attached to a shaft. Since the only displacement of the vanes is in the tangential

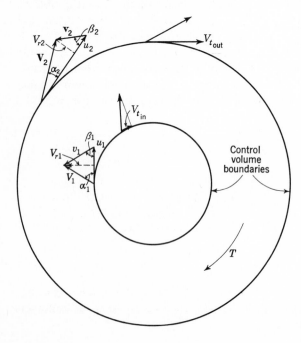

Figure 9.5 Steady flow through control volume with circular symmetry.

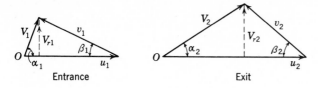

Entrance   Exit

Figure 9.6 Polar vector diagrams.

direction, work is done by the displacement of the tangential components of force on the runner. The radial components of force on the runner have no displacement in a radial direction and hence can do no work.

In turbomachine theory, friction is neglected, and the fluid is assumed to have perfect guidance through the machine, i.e., an infinite number of thin vanes, and so the relative velocity of the fluid is always tangent to the vane. This yields circular symmetry and permits the moment-of-momentum equation, Sec. 3.12, to take the simple form of Eq. (3.12.5), for steady flow,

$$T = \rho Q[(rV_t)_{\text{out}} - (rV_t)_{\text{in}}] \tag{9.3.1}$$

in which T is the torque acting on the fluid within the control volume (Fig. 9.5) and $\rho Q(rV_t)_{\text{out}}$ and $\rho Q(rV_t)_{\text{in}}$ represent the moment of momentum leaving and entering the control volume, respectively.

The polar vector diagram is generally used in studying vane relations (Fig. 9.6), with subscript 1 for entering fluid and subscript 2 for exiting fluid. V is the absolute fluid velocity; u is the peripheral velocity of the runner; and v is the fluid velocity relative to the runner. The absolute velocities V, u are laid off from O, and the relative velocity connects them as shown. V_u is designated as the component of absolute velocity in the tangential direction; α is the angle the absolute velocity V makes with the peripheral velocity u; and β is the angle the relative velocity makes with $-u$, or it is the *blade angle*, as perfect guidance is assumed. V_r is the absolute velocity component normal to the periphery. In this notation Eq. (9.3.1) becomes

$$T = \rho Q(r_2 V_2 \cos \alpha_2 - r_1 V_1 \cos \alpha_1)$$
$$= \rho Q(r_2 V_{u2} - r_1 V_{u1}) = \dot m(r_2 V_{u2} - r_1 V_{u1}) \tag{9.3.2}$$

The mass per unit time flowing is $\dot m = \rho Q = (\rho Q)_{\text{out}} = (\rho Q)_{\text{in}}$. In the form above, when T is positive, the fluid moment of momentum increases through the control volume, as for a pump. For T negative, moment of momentum of the fluid is decreased, as for a turbine runner. When $T = 0$, as in passages where there are no vanes,

$$rV_u = \text{const}$$

This is *free-vortex* motion, with the tangential component of velocity varying inversely with radius. It is discussed in Sec. 7.7 and compared with the forced vortex in Sec. 2.9.

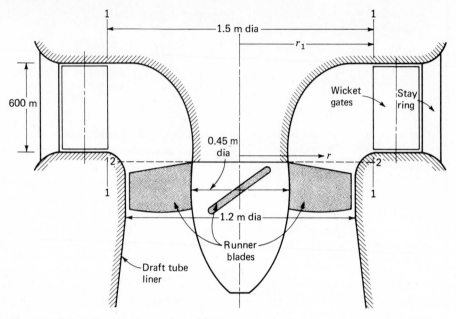

Figure 9.7 Schematic view of propeller turbine.

Example 9.2 The wicket gates of Fig. 9.7 are turned so that the flow makes an angle of 45° with a radial line at section 1, where the speed is 4.005 m/s. Determine the magnitude of tangential velocity component V_u over section 2.

Since no torque is exerted on the flow between sections 1 and 2, the moment of momentum is constant and the motion follows the free-vortex law

$$V_u r = \text{const}$$

At section 1

$$V_{u1} = 4.005 \cos 45° = 2.832 \text{ m/s}$$

Then

$$V_{u1} r_1 = (2.832 \text{ m/s})(0.75 \text{ m}) = 2.124 \text{ m}^2/\text{s}$$

Across section 2

$$V_{u2} = \frac{2.124 \text{ m}^2/\text{s}}{r \text{ m}}$$

At the hub $V_u = 2.124/0.225 = 9.44$ m/s, and at the outer edge $V_u = 2.124/0.6 = 3.54$ m/s.

Head and Energy Relations

By multiplying Eq. (9.3.2) by the rotational speed ω (rad/s) of runner,

$$T\omega = \rho Q(\omega r_2 V_{u2} - \omega r_1 V_{u1}) = \rho Q(u_2 V_{u2} - u_1 V_{u1}) \qquad (9.3.3)$$

For no losses the power available from a turbine is $Q \, \Delta p = Q\gamma H$, in which H is the head on the runner, since $Q\gamma$ is the weight per unit time and H the potential energy per unit weight. Similarly, a pump runner produces work $Q\gamma H$, in which H is the pump head. The power exchange is

$$T\omega = Q\gamma H \tag{9.3.4}$$

Solving for H, using Eq. (9.3.3) to eliminate T, gives

$$H = \frac{u_2 V_{u2} - u_1 V_{u1}}{g} \tag{9.3.5}$$

For turbines the sign is reversed in Eq. (9.3.5).

For pumps the *actual* head H_{a_p} produced is

$$H_{a_p} = e_h H = H - H_L \tag{9.3.6}$$

and for turbines the actual head H_{a_t} is

$$H_{a_t} = \frac{H}{e_h} = H + H_L \tag{9.3.7}$$

in which e_h is the hydraulic efficiency of the machine and H_L represents all the internal fluid losses in the machine. The overall efficiency of the machines is further reduced by bearing friction, by friction caused by fluid between runner and housing, and by leakage or flow that passes around the runner without going through it. These losses do not affect the head relations.

Pumps are generally so designed that the angular momentum of fluid entering the runner (impeller) is zero. Then

$$H = \frac{u_2 V_2 \cos \alpha_2}{g} \tag{9.3.8}$$

Turbines are so designed that the angular momentum is zero at the exit section of the runner for conditions at best efficiency; hence,

$$H = \frac{u_1 V_1 \cos \alpha_1}{g} \tag{9.3.9}$$

In writing the energy equation for a pump, with Eqs. (9.3.5) and (9.3.6),

$$H_p = \left(\frac{V_2^2}{2g} + \frac{p_2}{\gamma} + z_2\right) - \left(\frac{V_1^2}{2g} + \frac{p_1}{\gamma} + z_1\right)$$

$$= \frac{u_2 V_2 \cos \alpha_2 - u_1 V_1 \cos \alpha_1}{g} - H_L \tag{9.3.10}$$

for which it is assumed that all streamlines through the pump have the same total energy. With the relations among the absolute velocity V, the velocity v relative to the runner, and the velocity u of runner, from the vector diagrams (Fig. 9.6) by the law of cosines,

$$u_1^2 + V_1^2 - 2u_1 V_1 \cos \alpha_1 = v_1^2 \qquad u_2^2 + V_2^2 - 2u_2 V_2 \cos \alpha_2 = v_2^2$$

Eliminating the absolute velocities V_1, V_2 in these relations and in Eq. (9.3.10) gives

$$H_L = \frac{u_2^2 - u_1^2}{2g} - \frac{v_2^2 - v_1^2}{2g} - \frac{p_2 - p_1}{\gamma} - (z_2 - z_1) \qquad (9.3.11)$$

or

$$H_L = \frac{u_2^2 - u_1^2}{2g} - \left[\left(\frac{v_2^2}{2g} + \frac{p_2}{\gamma} + z_2\right) - \left(\frac{v_1^2}{2g} + \frac{p_1}{\gamma} + z_1\right)\right] \qquad (9.3.12)$$

The losses are the difference in centrifugal head, $(u_2^2 - u_1^2)/2g$, and in the head change in the relative flow. For no loss, the increase in pressure head, from Eq. (9.3.11), is

$$\frac{p_2 - p_1}{\gamma} + z_2 - z_1 = \frac{u_2^2 - u_1^2}{2g} - \frac{v_2^2 - v_1^2}{2g} \qquad (9.3.13)$$

With no flow through the runner, v_1, v_2 are zero, and the head rise is as expressed in the relative equilibrium relation [Eq. (2.9.6)]. When flow occurs, the head rise is equal to the centrifugal head minus the difference in relative velocity heads.

For the case of a turbine, exactly the same equations result.

Example 9.3 A centrifugal pump with a 700-mm-diameter impeller runs at 1800 rpm. The water enters without whirl, and $\alpha_2 = 60°$. The actual head produced by the pump is 17 m. Find its hydraulic efficiency when $V_2 = 6$ m/s.
From Eq. (9.3.8) the theoretical head is

$$H = \frac{u_2 V_2 \cos \alpha_2}{g} = \frac{1800(2\pi)(0.35)(6)(0.50)}{60(9.806)} = 20.18 \text{ m}$$

The actual head is 17 m; hence, the hydraulic efficiency is

$$e_h = \frac{17}{20.18} = 84.2\%$$

EXERCISES

9.3.1 A shaft transmits 150 kW at 600 rpm. The torque in newton-meters is (a) 26.2; (b) 250; (c) 2390; (d) 4780; (e) none of these answers.

9.3.2 What torque is required to give 100 cfs water a moment of momentum so that it has a tangential velocity of 10 ft/s at a distance of 6 ft from the axis? (a) 116 lb·ft; (b) 1935 lb·ft; (c) 6000 lb·ft; (d) 11,610 lb·ft; (e) none of these answers.

9.3.3 The moment of momentum of water is reduced by 27,100 N·m in flowing through vanes on a shaft turning 400 rpm. The power developed on the shaft is, in kilowatts, (a) 181.5; (b) 1134; (c) 10,800; (d) not determinable; insufficient data; (e) none of these answers.

9.3.4 Liquid moving with constant angular momentum has a tangential velocity of 4.0 ft/s 10 ft from the axis of rotation. The tangential velocity 5 ft from the axis is, in feet per second, (a) 2; (b) 4; (c) 8; (d) 16; (e) none of these answers.

9.4 REACTION TURBINES

In the *reaction* turbine a portion of the energy of the fluid is converted into kinetic energy by the fluid's passing through adjustable wicket gates (Fig. 9.8) before entering the runner, and the remainder of the conversion takes place through the runner. All passages are filled with liquid, including the passage (draft tube) from the runner to the downstream liquid surface. The static fluid pressure occurs on both sides of the vanes and hence does no work. The work done is entirely due to the conversion to kinetic energy.

The reaction turbine is quite different from the impulse turbine, discussed in Sec. 9.6. In an impulse turbine all the available energy of the fluid is converted into kinetic energy by a nozzle that forms a free jet. The energy is then taken from the jet by suitable flow through moving vanes. The vanes are partly filled, with the jet open to the atmosphere throughout its travel through the runner.

In contrast, in the reaction turbine the kinetic energy is appreciable as the fluid leaves the runner and enters the draft tube. The function of the draft tube is to reconvert the kinetic energy to flow energy by a gradual expansion of the flow cross section. Application of the energy equation between the two ends of the draft tube shows that the action of the tube is to reduce the pressure at its upstream end to less than atmospheric pressure, thus increasing the effective head across the runner to the difference in elevation between head water and tail water, less losses.

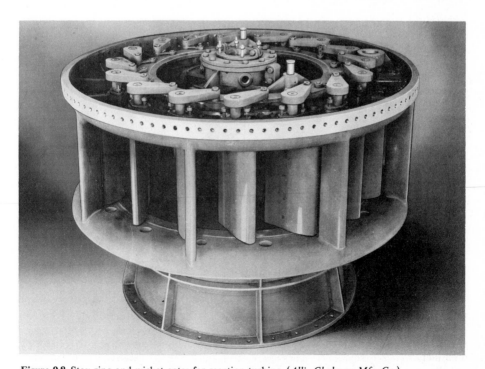

Figure 9.8 Stay ring and wicket gates for reaction turbine. (*Allis-Chalmers Mfg. Co.*)

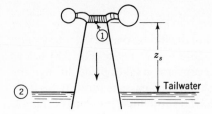

Figure 9.9 Draft tube.

By referring to Fig. 9.9, the energy equation from 1 to 2 yields

$$z_s + \frac{V_1^2}{2g} + \frac{p_1}{\gamma} = 0 + 0 + 0 + \text{losses}$$

The losses include the expansion loss, friction, and velocity head loss at the exit from the draft tube, all of which are quite small; hence,

$$\frac{p_1}{\gamma} = -z_s - \frac{V_1^2}{2g} + \text{losses} \tag{9.4.1}$$

shows that considerable vacuum is produced at section 1, which effectively increases the head across the turbine runner. The turbine setting must not be too high, or cavitation occurs in the runner and draft tube (see Sec. 9.8).

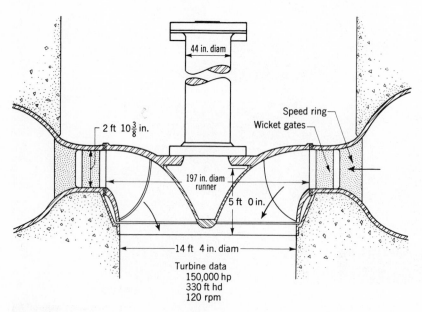

Figure 9.10 Francis turbine for Grand Coulee, Columbia Basin Project. (*Newport News Shipbuilding and Dry Dock Co.*)

Example 9.4 A turbine has a velocity of 6 m/s at the entrance to the draft tube and a velocity of 1.2 m/s at the exit. For friction losses of 0.1 m and a tail water 5 m below the entrance to the draft tube, find the pressure head at the entrance.

From Eq. (9.4.1)

$$\frac{p_1}{\gamma} = -5 - \frac{6^2}{2 \times 9.806} + \frac{1.2^2}{2 \times 9.806} + 0.1 = -6.66 \text{ m}$$

as the kinetic energy at the exit from the draft tube is lost. Hence, a suction head of 6.66 m is produced by the presence of the draft tube.

There are two forms of the reaction turbine in common use, the *Francis* turbine (Fig. 9.10) and the *propeller* (axial-flow) turbine (Fig. 9.11). In both, all passages flow full, and energy is converted to useful work entirely by changing the

Figure 9.11 Field view of installation of runner of 24,500-hp, 100-rpm, 41-ft-head, Kaplan adjustable-runner hydraulic turbine. Box Canyon Project, Public Utility District No. 1 of Pend Oreille County, Washington. Plant placed in operation in 1955. (*Allis-Chalmers Mfg. Co.*)

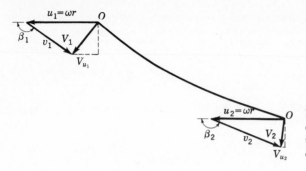

Figure 9.12 Velocity diagrams for entrance and exit of a propeller turbine blade at fixed radial distance.

moment of momentum of the liquid. The flow passes first through the wicket gates, which impart a tangential and a radially inward velocity to the fluid. A space between the wicket gates and the runner permits the flow to close behind the gates and move as a free vortex, without external torque being applied.

In the Francis turbine (Fig. 9.10) the fluid enters the runner so that the relative velocity is tangent to the leading edge of the vanes. The radial component is gradually changed to an axial component, and the tangential component is reduced as the fluid traverses the vane, so that at the runner exit the flow is axial with very little whirl (tangential component) remaining. The pressure has been reduced to less than atmospheric, and most of the remaining kinetic energy is reconverted to flow energy by the time it discharges from the draft tube. The Francis turbine is best suited to medium-head installations from 80 to 600 ft (25 to 180 m) and has an efficiency between 90 and 95 percent for the larger installations. Francis turbines are designed in the specific speed range of 10 to 110 (ft, hp, rpm) or 40 to 420 (m, kW, rpm) with best efficiency in the range 40 to 60 (ft, hp, rpm) or 150 to 230 (m, kW, rpm).

In the propeller turbine (Fig. 9.7), after passing through the wicket gates, the flow moves as a free vortex and has its radial component changed to axial component by guidance from the fixed housing. The moment of momentum is constant, and the tangential component of velocity is increased through the reduction in radius. The blades are few in number, relatively flat, and with very little curvature; they are so placed that the relative flow entering the runner is tangential to the leading edge of the blade. The relative velocity is high, as with the Pelton wheel, and changes slightly in traversing the blade. The velocity diagrams in Fig. 9.12 show how the tangential component of velocity is reduced. Propeller turbines are made with blades that pivot around the hub, thus permitting the blade angle to be adjusted for different gate openings and for changes in head. They are particularly suited for low-head installations, up to 30 m, and have top efficiencies around 94 percent. Axial-flow turbines are designed in the specific speed range of 100 to 210 (ft, hp, rpm) or 380 to 800 (m, kW, rpm) with best efficiency from 120 to 160 (ft, hp, rpm) or 460 to 610 (m, kW, rpm).

The windmill is a form of axial-flow turbine. Since it has no fixed vanes to give an initial tangential component to the airstream, it must impart the tangential

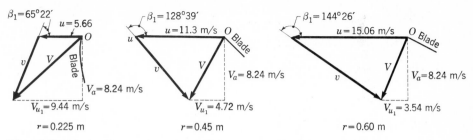

$\beta_1 = 65°22'$ $u = 5.66$ $V_a = 8.24$ m/s $V_{u_1} = 9.44$ m/s $r = 0.225$ m

$\beta_1 = 128°39'$ $u = 11.3$ m/s $V_a = 8.24$ m/s $V_{u_1} = 4.72$ m/s $r = 0.45$ m

$\beta_1 = 144°26'$ $u = 15.06$ m/s $V_a = 8.24$ m/s $V_{u_1} = 3.54$ m/s $r = 0.60$ m

Figure 9.13 Velocity diagrams for angle of leading edge of a propeller turbine blade.

component to the air with the moving vanes. The airstream expands in passing through the vanes with a reduction in its axial velocity.

Example 9.5 Assuming uniform axial velocity over section 2 of Fig. 9.7 and using the data of Example 9.2, determine the angle of the leading edge of the propeller at $r = 0.225, 0.45$, and 0.6 m for a propeller speed of 240 rpm.
At $r = 0.225$ m,

$$u = \tfrac{240}{60}(2\pi)(0.225) = 5.66 \text{ m/s} \qquad V_u = 9.44 \text{ m/s}$$

At $r = 0.45$ m,

$$u = \tfrac{240}{60}(2\pi)(0.45) = 11.3 \text{ m/s} \qquad V_u = 4.72 \text{ m/s}$$

At $r = 0.6$ m,

$$u = \tfrac{240}{60}(2\pi)(0.6) = 15.06 \text{ m/s} \qquad V_u = 3.54 \text{ m/s}$$

The discharge through the turbine is, from section 1,

$$Q = (0.6 \text{ m})(1.5 \text{ m})(\pi)(4.005 \text{ m/s})(\cos 45°) = 8 \text{ m}^3/\text{s}$$

Hence, the axial velocity at section 2 is

$$V_a = \frac{8}{\pi(0.6^2 - 0.225^2)} = 8.24 \text{ m/s}$$

Figure 9.13 shows the initial vane angle for the three positions.

Moody[†] has developed a formula to estimate the efficiency of a unit of a homologous series of turbines when the efficiency of one of the series is known:

$$e = 1 - (1 - e_1)\left(\frac{D_1}{D}\right)^{1/4} \tag{9.4.2}$$

in which e_1 and D_1 are usually the efficiency and the diameter of a model.

† Lewis F. Moody, The Propeller Type Turbine, *Trans. ASCE*, vol. 89, p. 628, 1926.

EXERCISE

9.4.1 A reaction-type turbine discharges 34 m³/s under a head of 7.5 m and with an overall efficiency of 91 percent. The power developed is, in kilowatts, (a) 2750; (b) 2500; (c) 2275; (d) 70.7; (e) none of these answers.

9.5 PUMPS AND BLOWERS

Pumps add energy to liquids and blowers to gases. The procedure for designing them is the same for both, except when the density is appreciably increased. Turbopumps and -blowers are *radial-flow*, *axial-flow*, or a combination of the two, called *mixed-flow*. For high heads the radial (centrifugal) pump, frequently with two or more stages (two or more impellers in series), is best adapted. For large flows under small heads the axial-flow pump or blower (Fig. 9.14*a*) is best suited. The mixed-flow pump (Fig. 9.14*b*) is used for medium head and medium discharge.

The equations developed in Sec. 9.2 apply just as well to pumps and blowers as to turbines. The usual centrifugal pump has a suction, or inlet, pipe leading to the center of the impeller, a radial outward-flow runner, as in Fig. 9.15, and a collection pipe or spiral casing that guides the fluid to the discharge pipe. Ordinarily, no fixed vanes are used, except for multistage units in which the flow is relatively small and the additional fluid friction is less than the additional gain in conversion of kinetic energy to pressure energy upon leaving the impeller.

Figure 9.16 shows a sectional elevation of a large centrifugal pump. For lower heads and greater discharges (relatively) the impellers vary as shown in Fig. 9.17, from high head at left to low head at right with the axial-flow impeller. The

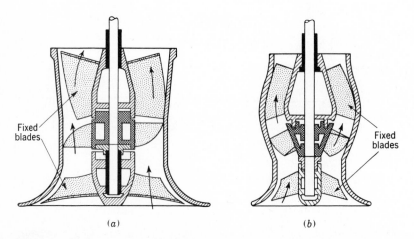

Fixed
blades

Fixed
blades

(*a*)   (*b*)

Figure 9.14 Well-type pumps: (*a*) axial flow; (*b*) mixed flow. (*Ingersoll-Rand Co.*)

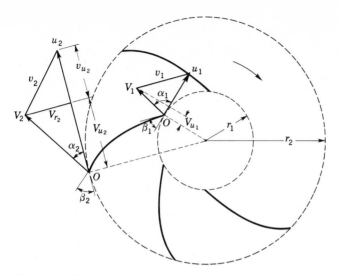

Figure 9.15 Velocity relations for flow through a centrifugal-pump impeller.

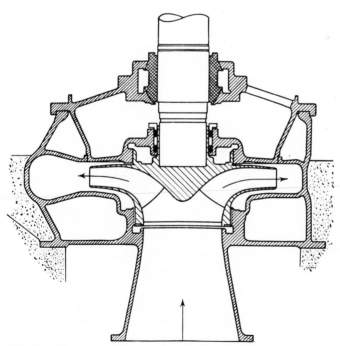

Figure 9.16 Sectional elevation of Eagle Mountain and Hayfield pumps, Colorado River Aqueduct. (*Worthington Corp.*)

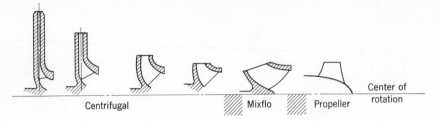

Figure 9.17 Impeller types used in pumps and blowers. (*Worthington Corp.*)

specific speed increases from left to right. A chart for determining the types of pump for best efficiency is given in Fig. 9.18 for water.

Centrifugal and mixed-flow pumps are designed in the specific speed range 500 to 6500 and axial pumps from 5000 to 11,000; speed is expressed in revolutions per minute, discharge in gallons per minute, and head in feet.

Characteristic curves showing head, efficiency, and brake horsepower as a function of discharge for a typical centrifugal pump with backward-curved vanes are given in Fig. 9.19. Pumps are not as efficient as turbines, in general, owing to the inherently high losses that result from conversion of kinetic energy into flow energy.

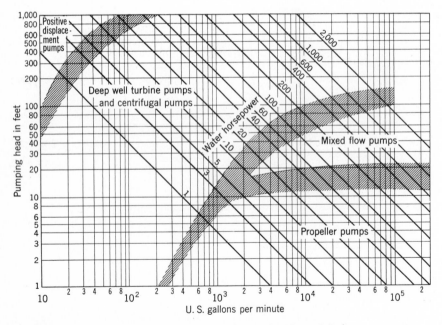

Figure 9.18 Chart for selection of type of pump. (*Fairbanks, Morse & Co.*)

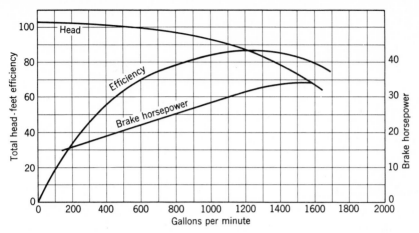

Figure 9.19 Characteristic curves for typical centrifugal pump; 10-in impeller, 1750 rpm. (*Ingersoll-Rand Co.*)

Theoretical Head-Discharge Curve

A theoretical head-discharge curve may be obtained by use of Eq. (9.3.8) and the vector diagrams of Fig. 9.6. From the exit diagram of Fig. 9.6

$$V_2 \cos \alpha_2 = V_{u2} = u_2 - V_{r2} \cot \beta_2$$

From the discharge, if b_2 is the width of the impeller at r_2 and vane thickness is neglected,

$$Q = 2\pi r_2 b_2 V_{r2}$$

Eliminating V_{r2} and substituting the preceding two equations into Eq. (9.3.8) give

$$H = \frac{u_2^2}{g} - \frac{u_2 Q \cot \beta_2}{2\pi r_2 b_2 g} \tag{9.5.1}$$

For a given pump and speed, H varies linearly with Q, as shown in Fig. 9.20. The usual design of centrifugal pump has $\beta_2 < 90°$, which gives decreasing head with

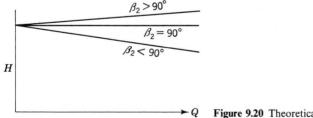

Figure 9.20 Theoretical head-discharge curves.

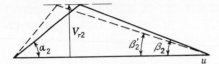

Figure 9.21 Effect of circulatory flow.

increasing discharge. For blades radial at the exit, $\beta_2 = 90°$ and the theoretical head is independent of discharge. For blades curved forward, $\beta_2 > 90°$, and the head rises with discharge.

Actual Head-Discharge Curve

By subtracting head losses from the theoretical head-discharge curve, the actual head-discharge curve is obtained. The most important subtraction is not an actual loss; it is a failure of the finite number of blades to impart the relative velocity with angle β_2 of the blades. Without perfect guidance (infinite number of blades) the fluid actually is discharged as if the blades had an angle β'_2 which is less than β_2 (Fig. 9.21) for the same discharge. This inability of the blades to impart proper guidance reduces V_{u2} and hence decreases the actual head produced. This is called *circulatory flow*, and it is shown in Fig. 9.22. Fluid friction in flow through the fixed and moving passages causes losses that are proportional to the square of the discharge. They are shown in Fig. 9.22. The final head loss to consider is that of turbulence, the loss due to improper relative-velocity angle at the blade inlet. The pump can be designed for one discharge (at a given speed) at which the relative velocity is tangent to the blade at the inlet. This is the point of best efficiency, and shock or turbulence losses are negligible. For other discharges the loss varies about as the square of the discrepancy in approach angle, as shown in Fig. 9.22. The final lower line then represents the actual head-discharge curve. Shutoff head is usually about $u_2^2/2g$, or half of the theoretical shutoff head.

In addition to the head losses and reductions, pumps and blowers have torque losses due to bearing- and packing-friction and disk-friction losses from the fluid between the moving impeller and housing. Internal leakage is also an important

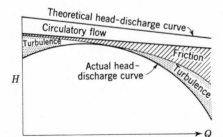

Figure 9.22 Head-discharge relations.

power loss, in that fluid which has passed through the impeller, with its energy increased, escapes through clearances and flows back to the suction side of the impeller.

Example 9.6 A centrifugal water pump has an impeller (Fig. 9.15) with $r_2 = 12$ in, $r_1 = 4$ in, $\beta_1 = 20°$, $\beta_2 = 10°$. The impeller is 2 in wide at $r = r_1$ and $\frac{3}{4}$ in wide at $r = r_2$. For 1800 rpm, neglecting losses and vane thickness, determine (a) the discharge for shockless entrance when $\alpha_1 = 90°$, (b) α_2 and the theoretical head H, (c) the horsepower required, and (d) the pressure rise through the impeller.

(a) The peripheral speeds are

$$u_1 = \tfrac{1800}{60}(2\pi)(\tfrac{1}{3}) = 62.8 \text{ ft/s} \qquad u_2 = 3u_1 = 188.5 \text{ ft/s}$$

The vector diagrams are shown in Fig. 9.23. With u_1 and the angles α_1, β_1 known, the entrance diagram is determined, $V_1 = u_1 \tan 20° = 22.85$ ft/s; hence,

$$Q = 22.85(\pi)(\tfrac{2}{3})(\tfrac{2}{12}) = 7.97 \text{ cfs}$$

(b) At the exit the radial velocity V_{r2} is

$$V_{r2} = \frac{7.97 \times 12}{2\pi \times 0.75} = 20.3 \text{ ft/s}$$

By drawing u_2 (Fig. 9.23) and a parallel line distance V_{r2} from it, the vector triangle is determined when β_2 is laid off. Thus

$$v_{u2} = 20.3 \cot 10° = 115 \text{ ft/s} \qquad V_{u2} = 188.5 - 115 = 73.5 \text{ ft/s}$$

$$\alpha_2 = \tan^{-1} \frac{20.3}{73.5} = 15°26' \qquad V_2 = 20.3 \csc 15°26' = 76.2 \text{ ft/s}$$

From Eq. (9.3.8)

$$H = \frac{u_2 V_2 \cos \alpha_2}{g} = \frac{u_2 V_{u2}}{g} = \frac{188.5 \times 73.5}{32.2} = 430 \text{ ft}$$

(c)
$$hp = \frac{Q\gamma H}{550} = \frac{7.97 \times 62.4 \times 430}{550} = 388$$

(d) By applying the energy equation from entrance to exit of the impeller, including the energy H added (elevation change across impeller is neglected),

$$H + \frac{V_1^2}{2g} + \frac{p_1}{\gamma} = \frac{V_2^2}{2g} + \frac{p_2}{\gamma}$$

and
$$\frac{p_2 - p_1}{\gamma} = 430 + \frac{22.85^2}{64.4} - \frac{76.2^2}{64.4} = 348 \text{ ft}$$

or
$$p_2 - p_1 = 348 \times 0.433 = 151 \text{ psi}$$

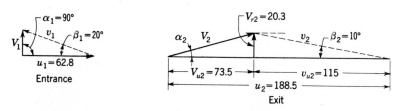

Figure 9.23 Vector diagrams for entrance and exit of pump impeller.

EXERCISES

9.5.1 The head developed by a pump with hydraulic efficiency of 80 percent, for $u_2 = 100$ ft/s, $V_2 = 60$ ft/s, $\alpha_2 = 45°$, $\alpha_1 = 90°$, is (a) 52.6; (b) 105.3; (c) 132; (d) 165; (e) none of these answers.

9.5.2 Select the correct relations for pump vector diagrams. (a) $\alpha_1 = 90°$; $v_1 = u_1 \cot \beta_1$; (b) $V_{u2} = u_2 - V_{r2} \cot \beta_2$; (c) $\omega_2 = r_2/u_2$; (d) $r_1 V_1 = r_2 V_{r2}$; (e) none of these answers.

9.6 IMPULSE TURBINES

The impulse turbine is one in which all available energy of the flow is converted by a nozzle into kinetic energy at atmospheric pressure before the fluid contacts the moving blades. Losses occur in flow from the reservoir through the pressure pipe (penstock) to the base of the nozzle, which may be computed from pipe friction data. At the base of the nozzle the available energy, or total head, is

$$H_a = \frac{p_1}{\gamma} + \frac{V_1^2}{2g} \tag{9.6.1}$$

from Fig. 9.24. With C_v the nozzle coefficient, the jet velocity V_2 is

$$V_2 = C_v\sqrt{2gH_a} = C_v\sqrt{2g\left(\frac{p_1}{\gamma} + \frac{V_1^2}{2g}\right)} \tag{9.6.2}$$

The head lost in the nozzle is

$$H_a - \frac{V_2^2}{2g} = H_a - C_v^2 H_a = H_a(1 - C_v^2) \tag{9.6.3}$$

and the efficiency of the nozzle is

$$\frac{V_2^2/2g}{H_a} = \frac{C_v^2 H_a}{H_a} = C_v^2 \tag{9.6.4}$$

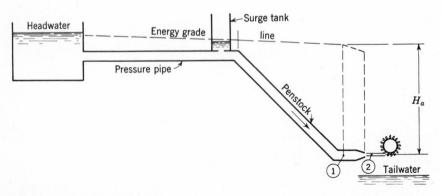

Figure 9.24 Impulse turbine system.

Figure 9.25 Southern California Edison, Big Creek 2A, 1948, $8\frac{1}{2}$-in-diameter jet impulse buckets and disk in process of being reamed; 56,000 hp, 2200 ft head, 300 rpm. (*Allis-Chalmers Mfg. Co.*)

The jet, with velocity V_2, strikes double-cupped buckets (Figs. 9.25 and 9.26) which split the flow and turn the relative velocity through the angle θ (Fig. 9.26). The x component of momentum is changed by (Fig. 9.26)

$$F = \rho Q(v_r - v_r \cos \theta)$$

and the power exerted on the vanes is

$$Fu = \rho Q u v_r (1 - \cos \theta) \tag{9.6.5}$$

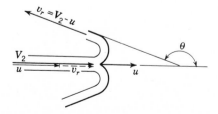

Figure 9.26 Flow-through bucket.

To maximize the power theoretically, two conditions must be met: $\theta = 180°$ and uv_r must be a maximum; that is, $u(V_2 - u)$ must be a maximum. To determine when $(uv_r)_{max}$ occurs, differentiate with respect to u and equate to zero,

$$(V_2 - u) + u(-1) = 0$$

The condition is met when $u = V_2/2$. After making these substitutions into Eq. (9.6.5),

$$Fu = \rho Q \frac{V_2}{2}\left(V_2 - \frac{V_2}{2}\right)[1 - (-1)] = \gamma Q \frac{V_2^2}{2g} \qquad (9.6.6)$$

which accounts for the total kinetic energy of the jet. The velocity diagram for these values shows that the absolute velocity leaving the vanes is zero.

Practically, when vanes are arranged on the periphery of a wheel (Fig. 9.25), the fluid must retain enough velocity to move out of the way of the following bucket. Most of the practical impulse turbines are Pelton wheels. The jet is split in two and turned in a horizontal plane, and half is discharged from each side to avoid any unbalanced thrust on the shaft. There are losses due to the splitter and to friction between jet and bucket surface, which make the most economical speed somewhat less than $V_2/2$. It is expressed in terms of the *speed factor*

$$\phi = \frac{u}{\sqrt{2gH_a}} \qquad (9.6.7)$$

For most efficient turbine operation ϕ has been found to depend upon specific speed as shown in Table 9.1. The angle θ of the bucket is usually 173 to 176°. If the diameter of the jet is d and the diameter of the wheel is D at the centerline of the buckets, it has been found in practice that the diameter ratio D/d should be about $54/N_s$ (ft, hp, rpm), or $206/N_s$ (m, kW, rpm) for maximum efficiency.

Table 9.1 Dependence of ϕ on specific speed*

Specific speed N_s		ϕ
(m, kW, rpm)	(ft, hp, rpm)	
7.62	2	0.47
11.42	3	0.46
15.24	4	0.45
19.05	5	0.44
22.86	6	0.433
26.65	7	0.425

* Modified from J. W. Daily, Hydraulic Machinery, in H. Rouse (ed.), "Engineering Hydraulics," p. 943, Wiley, New York, 1950.

In the majority of installations only one jet is used; it discharges horizontally against the lower periphery of the wheel as shown in Fig. 9.24. The wheel speed is carefully regulated for the generation of electric power. A governor operates a needle valve that controls the jet discharge by changing its area. So V_2 remains practically constant for a wide range of positions of the needle valve.

The efficiency of the power conversion drops off rapidly with change in head (which changes V_2), as is evident when power is plotted against V_2 for constant u in Eq. (9.6.5). The wheel operates in atmospheric air although it is enclosed by a housing. It is therefore essential that the wheel be placed above the maximum floodwater level of the river into which it discharges. The head from nozzle to tail water is wasted. Because of their inefficiency at other than the design head and because of the wasted head, Pelton wheels usually are employed for high heads, e.g., from 200 m to more than 1 km. For high heads, the efficiency of the complete installation, from head water to tail water, may be in the high 80s.

Impulse wheels with a single nozzle are most efficient in the specific speed range of 2 to 6, when P is in horsepower, H is in feet, and N is in revolutions per minute. Multiple nozzle units are designed in the specific speed range of 6 to 12.

Example 9.7 A Pelton wheel is to be selected to drive a generator at 600 rpm. The water jet is 75 mm in diameter and has a velocity of 100 m/s. With the blade angle at 170°, the ratio of vane speed to initial jet speed at 0.47, and neglecting losses, determine (a) diameter of wheel to centerline of buckets (vanes), (b) power developed, and (c) kinetic energy per newton remaining in the fluid.

(a) The peripheral speed of the wheel is

$$u = 0.47 \times 100 = 47 \text{ m/s}$$

Then
$$\frac{600}{60}\left(2\pi \frac{D}{2}\right) = 47 \text{ m/s} \quad \text{or} \quad D = 1.495 \text{ m}$$

(b) From Eq. (9.6.5), the power, in kilowatts, is computed to be

$$(1000 \text{ kg/m}^3) \frac{\pi}{4} (0.075 \text{ m})^2 (100 \text{ m/s})(47 \text{ m/s})(100 - 47 \text{ m/s})$$

$$\times [1 - (-0.9848)] \frac{1 \text{ kW}}{1000 \text{ W}} = 2184 \text{ kW}$$

(c) From Fig. 3.34, the absolute velocity components leaving the vane are

$$V_x = (100 - 47)(-0.9848) + 47 = -5.2 \text{ m/s}$$
$$V_y = (100 - 47)(0.1736) = 9.2 \text{ m/s}$$

The kinetic energy remaining in the jet is

$$\frac{5.2^2 + 9.2^2}{2 \times 9.806} = 5.69 \text{ m} \cdot \text{N/N}$$

Example 9.8 A small impulse wheel is to be used to drive a generator for 60-Hz power. The head is 100 m, and the discharge is 40 l/s. Determine the diameter of the wheel at the centerline of the buckets and the speed of the wheel. $C_v = 0.98$. Assume efficiency of 80 percent. The power is

$$P = \gamma Q H_a e = 9806 \times 0.040 \times 100 \times 0.80 = 31.38 \text{ kW}$$

By taking a trial value of N_s of 15,

$$N = \frac{N_s H_a^{5/4}}{\sqrt{P}} = \frac{15 \times 100^{5/4}}{\sqrt{31.38}} = 847 \text{ rpm}$$

For 60-Hz power the speed must be 3600 divided by the number of pairs of poles in the generator. For five pairs of poles the speed would be $3600/5 = 720$ rpm, and for four pairs of poles it would be $3600/4 = 900$ rpm. The closer speed 900 is selected. Then

$$N_s = \frac{N\sqrt{P}}{H_a^{5/4}} = \frac{900\sqrt{31.38}}{100^{5/4}} = 15.94$$

For $N_s = 15.94$, take $\phi = 0.448$ from Table 9.1,

$$u = \phi\sqrt{2gH_a} = 0.448\sqrt{2 \times 9.806 \times 100} = 19.84 \text{ m/s}$$

and

$$\omega = \frac{900}{60}2\pi = 94.25 \text{ rad/s}$$

The peripheral speed u and D and ω are related:

$$u = \frac{\omega D}{2} \qquad D = \frac{2u}{\omega} = \frac{2 \times 19.84}{94.25} = 421 \text{ mm}$$

The diameter d of the jet is obtained from the jet velocity V_2; thus

$$V_2 = C_v\sqrt{2gH_a} = 0.98\sqrt{2 \times 9.806 \times 100} = 43.4 \text{ m/s}$$

$$a = \frac{Q}{V_2} = \frac{0.040}{43.4} = 9.22 \text{ cm}^2 \qquad d = \sqrt{\frac{4a}{\pi}} = \sqrt{\frac{0.000922}{0.7854}} = 34.3 \text{ mm}$$

where a is the area of jet. Hence, the diameter ratio D/d is

$$\frac{D}{d} = \frac{421}{34.3} = 12.27$$

The desired diameter ratio for best efficiency is

$$\frac{D}{d} = \frac{206}{N_s} = \frac{206}{15.94} = 12.92$$

so the ratio D/d is satisfactory. The wheel diameter is 421 mm, and the speed is 900 rpm.

EXERCISES

9.6.1 An impulse turbine (a) always operates submerged; (b) makes use of a draft tube; (c) is most suited for low-head installations; (d) converts pressure head into velocity head throughout the vanes; (e) operates by initial complete conversion to kinetic energy.

9.6.2 A 24-in diameter Pelton wheel turns at 400 rpm. Select from the following the head, in feet, best suited for this wheel: (a) 7; (b) 30; (c) 120; (d) 170; (e) 480.

9.7 CENTRIFUGAL COMPRESSORS

Centrifugal compressors operate according to the same principles as turbo-machines for liquids. It is important for the fluid to enter the impeller without shock, i.e., with the relative velocity tangent to the blade. Work is done on the gas by rotation of the vanes, the moment-of-momentum equation relating torque to production of tangential velocity. At the impeller exit the high-velocity gas must have its kinetic energy converted in part to flow energy by suitable expanding flow passages. For adiabatic compression (no cooling of the gas) the actual work w_a of compression per unit mass is compared with the work w_{th} per unit mass to compress the gas to the same pressure isentropically. For cooled compressors the work w_{th} is based on the isothermal work of compression to the same pressure as the actual case. Hence,

$$\eta = \frac{w_{th}}{w_a} \tag{9.7.1}$$

is the formula for efficiency of a compressor.

The efficiency formula for compression of a perfect gas is developed for the adiabatic compressor, assuming no internal leakage in the machine, i.e., no short-circuiting of high-pressure fluid back to the low-pressure end of the impeller. Centrifugal compressors are usually multistage, with pressure ratios up to 3 across a single stage. From the moment-of-momentum equation (9.3.2) with inlet absolute velocity radial, $\alpha_1 = 90°$, the theoretical torque T_{th} is

$$T_{th} = \dot{m} V_{u2} r_2 \tag{9.7.2}$$

in which $\dot{m}$ is the mass per unit time being compressed, V_{u2} is the tangential component of the absolute velocity leaving the impeller, and r_2 is the impeller radius at exit. The actual applied torque T_a is greater than the theoretical torque by the torque losses due to bearing and packing friction plus disk friction; hence,

$$T_{th} = T_a \eta_m \tag{9.7.3}$$

if η_m is the mechanical efficiency of the compressor.

In addition to the torque losses, there are irreversibilities due to flow through the machine. The actual work of compression through the adiabatic machine is obtained from the steady-flow energy equation (3.8.1), neglecting elevation changes and replacing $u + p/\rho$ by h,

$$-w_a = \frac{V_{2a}^2 - V_1^2}{2} + h_2 - h_1 \tag{9.7.4}$$

The isentropic work of compression can be obtained from Eq. (3.8.1) in differential form, neglecting the z terms,

$$-dw_{th} = V\,dV + d\frac{p}{\rho} + du = V\,dV + \frac{dp}{\rho} + pd\frac{1}{\rho} + du$$

The last two terms are equal to $T\,ds$ from Eq. (3.8.6), which is zero for isentropic flow, so that

$$-dw_{th} = V\,dV + \frac{dp}{\rho} \tag{9.7.5}$$

By integrating for $p/\rho^k = $ const between sections 1 and 2,

$$-w_{th} = \frac{V_{2th}^2 - V_1^2}{2} + \frac{k}{k-1}\left(\frac{p_2}{\rho_{2th}} - \frac{p_1}{\rho_1}\right)$$

$$= \frac{V_{2th}^2 - V_1^2}{2} + c_p T_1\left[\left(\frac{p_2}{p_1}\right)^{(k-1)/k} - 1\right] \qquad (9.7.6)$$

The efficiency may now be written

$$\eta = \frac{-w_{th}}{-w_a} = \frac{(V_{2th}^2 - V_1^2)/2 + c_p T_1[(p_2/p_1)^{(k-1)/k} - 1]}{(V_{2a}^2 - V_1^2)/2 + c_p(T_{2a} - T_1)} \qquad (9.7.7)$$

since $h = c_p T$. In terms of Eqs. (9.7.2) and (9.7.3)

$$-w_a = \frac{T_a \omega}{\dot{m}} = \frac{T_{th}\omega}{\eta_m \dot{m}} = \frac{V_{u2} r_2 \omega}{\eta_m} = \frac{V_{u2} u_2}{\eta_m} \qquad (9.7.8)$$

then

$$\eta = \frac{\eta_m}{V_{u2} u_2}\left\{c_p T_1\left[\left(\frac{p_2}{p_1}\right)^{(k-1)/k} - 1\right] + \frac{V_{2th}^2 - V_1^2}{2}\right\} \qquad (9.7.9)$$

Use of this equation is made in the following example.

Example 9.9 An adiabatic turbocompressor has blades that are radial at the exit of its 150-mm-diameter impeller. It is compressing 0.5 kg/s air at 98.06 kPa abs, $t = 15°C$, to 294.18 kPa abs. The entrance area is 60 cm², and the exit area is 35 cm²; $\eta = 0.75$; $\eta_m = 0.90$. Determine the rotational speed of the impeller and the actual temperature of air at the exit.
The density at the inlet is

$$\rho_1 = \frac{p_1}{RT_1} = \frac{9.806 \times 10^4 \text{ N/m}^2}{(287 \text{ J/kg·K})(273 + 15 \text{ K})} = 1.186 \text{ kg/m}^3$$

and the velocity at the entrance is

$$V_1 = \frac{\dot{m}}{\rho_1 A_1} = \frac{0.5 \text{ kg/s}}{(1.186 \text{ kg/m}^3)(0.006 \text{ m}^2)} = 70.26 \text{ m/s}$$

The theoretical density at the exit is

$$\rho_{2th} = \rho_1\left(\frac{p_2}{p_1}\right)^{1/k} = 1.186 \times 3^{1/1.4} = 2.60 \text{ kg/m}^3$$

and the theoretical velocity at the exit is

$$V_{2th} = \frac{\dot{m}}{\rho_{2th} A_2} = \frac{0.5}{2.60 \times 0.0035} = 54.945 \text{ m/s}$$

For radial vanes at the exit, $V_{u2} = u_2 = \omega r_2$. From Eq. (9.7.9)

$$u_2^2 = \frac{\eta_m}{\eta}\left\{c_p T_1\left[\left(\frac{p_2}{p_1}\right)^{(k-1)/k} - 1\right] + \frac{V_{2th}^2 - V_1^2}{2}\right\}$$

$$= \frac{0.90}{0.75}\left[(0.24 \times 4187)(273 + 15)(3^{0.4/1.4} - 1) + \frac{54.945^2 - 70.26^2}{2}\right]$$

and $u_2 = 359.56$ m/s. Then

$$\omega = \frac{u_2}{r_2} = \frac{359.56}{0.075} = 4794 \text{ rad/s}$$

and

$$N = \omega \frac{60}{2\pi} = 4794 \frac{60}{2\pi} = 45{,}781 \text{ rpm}$$

The theoretical work w_{th} is the term in the brackets in the expression for u_2^2. It is $-w_{th} = 0.1058 \times 10^6$ m·N/kg. Then, from Eq. (9.7.1),

$$w_a = \frac{w_{th}}{\eta} = -\frac{1.058 \times 10^5}{0.75} = -1.411 \times 10^5 \text{ m·N/kg}$$

Since the kinetic-energy term is small, Eq. (9.7.4) can be solved for $h_2 - h_1$ and a trial solution effected:

$$h_2 - h_1 = c_p(T_{2a} - T_1) = 1.411 \times 10^5 + \frac{70.26^2 - V_{2a}^2}{2}$$

As a first approximation, let $V_{2a} = V_{2th} = 54.945$; then

$$T_{2a} = 288 + \frac{1}{0.24 \times 4187}\left(1.411 \times 10^5 + \frac{70.26^2 - 54.945^2}{2}\right) = 429.4 \text{ K}$$

For this temperature the density at the exit is 2.387 kg/m³, and the velocity is 59.85 m/s. Insertion of this value in place of 54.945 reduces the temperature to $T_{2a} = 429.1$ K.

9.8 CAVITATION

When a liquid flows into a region where its pressure is reduced to vapor pressure, it boils and vapor pockets develop in it. The vapor bubbles are carried along with the liquid until a region of higher pressure is reached, where they suddenly collapse. This process is called *cavitation*. If the vapor bubbles are near to (or in contact with) a solid boundary when they collapse, the forces exerted by the liquid rushing into the cavities create very high localized pressures that cause pitting of the solid surface. The phenomenon is accompanied by noise and vibrations that have been described as similar to gravel going through a centrifugal pump.

In a flowing liquid, the *cavitation parameter σ* is useful in characterizing the susceptibility of the system to cavitate. It is defined by

$$\sigma = \frac{p - p_v}{\rho V^2/2} \tag{9.8.1}$$

in which p is the absolute pressure at the point of interest, p_v is the vapor pressure of the liquid, ρ is the density of the liquid, and V is the undisturbed, or reference, velocity. The cavitation parameter is a form of pressure coefficient. Two geometrically similar systems would be equally likely to cavitate or would have the same degree of cavitation for the same value of σ. When $\sigma = 0$, the pressure is reduced to vapor pressure and boiling should occur.

Tests made on chemically pure liquids show that they will sustain high tensile stresses, of the order of megapascals, which is in contradiction to the concept of cavities forming when pressure is reduced to vapor pressure. Since there is generally spontaneous boiling when vapor pressure is reached with commercial or technical liquids, it is generally accepted that nuclei must be present around which the vapor bubbles form and grow. The nature of the nuclei is not thoroughly understood, but they may be microscopic dust particles or other contaminants, which are widely dispersed through technical liquids.

Cavitation bubbles may form on nuclei, grow, then move into an area of higher pressure and collapse, all in a few thousandths of a second in flow within a turbomachine. In aerated water the bubbles have been photographed as they move through several oscillations, but this phenomenon does not seem to occur in nonaerated liquids. Surface tension of the vapor bubbles appears to be an important property accounting for the high-pressure pulses resulting from collapse of a vapor bubble. Experiments indicate pressures of the order of a gigapascal based on the analysis of strain waves in a photoelastic specimen exposed to cavitation.† Pressures of this magnitude appear to be reasonable, in line with the observed mechanical damage caused by cavitation.

The formation and collapse of great numbers of bubbles on a surface subject that surface to intense local stressing, which appears to damage the surface by fatigue. Some ductile materials withstand battering for a period, called the *incubation period*, before damage is noticeable, whereas brittle materials may lose mass immediately. There may be certain electrochemical, corrosive, and thermal effects which hasten the deterioration of exposed surfaces. Rheingans‡ has collected a series of measurements made by magnetostriction-oscillator tests, showing mass losses of various metals used in hydraulic machines (see Table 9.2).

Protection against cavitation should start with the hydraulic design of the system in order to avoid the low pressures if practicable. Otherwise, use of special cavitation-resistant materials or coatings may be effective. Small amounts of air entrained into water systems have markedly reduced cavitation damage, and recent studies indicate that cathodic protection is helpful.

The formation of vapor cavities decreases the useful channel space for liquid and thus decreases the efficiency of a fluid machine. Cavitation causes three undesirable conditions: lowered efficiency, damage to flow passages, and noise and vibrations. Curved vanes are particularly susceptible to cavitation on their convex sides and may have localized areas where cavitation causes pitting or failure. Since all turbomachinery and ship propellers and many hydraulic structures are subject to cavitation, special attention must be given to it in their design.

A *cavitation index* σ' is useful in the proper selection of turbomachinery and in its location with respect to suction or tail-water elevation. The minimum pressure in a pump or turbine generally occurs along the convex side of vanes near the

† G. W. Sutton, A Photoelastic Study of Strain Waves Caused by Cavitation, *J. Appl. Mech.*, vol. 24, pt. 3, pp. 340–348, 1957.

‡ W. J. Rheingans, Selecting Materials to Avoid Cavitation Damage, *Mater. Des. Eng.*, pp. 102–106, 1958.

Table 9.2 Mass loss in materials used in hydraulic machines

Alloy	Mass loss after 2 h, mg
Rolled stellite†	0.6
Welded aluminum bronze‡	3.2
Cast aluminum bronze§	5.8
Welded stainless steel (two layers, 17% Cr, 7% Ni)	6.0
Hot-rolled stainless steel (26% Cr, 13% Ni)	8.0
Tempered, rolled stainless steel (12% Cr)	9.0
Cast stainless steel (18% Cr, 8% Ni)	13.0
Cast stainless steel (12% Cr)	20.0
Cast manganese bronze	80.0
Welded mild steel	97.0
Plate steel	98.0
Cast steel	105.0
Aluminum	124.0
Brass	156.0
Cast iron	224.0

† This material is not suitable for ordinary use, in spite of its high resistance, because of its high cost and difficulty in machining.
‡ Ampco-Trode 200: 83%Cu, 10.3% Al, 5.8% Fe.
§ Ampco 20: 83.1% Cu, 12.4% Al, 4.1% Fe.

low-pressure side of the impeller. In Fig. 9.27, if e is the point of minimum pressure, Bernoulli's equation applied between e and the downstream liquid surface, neglecting losses between the two points, may be written

$$\frac{p_e}{\gamma} + \frac{V_e^2}{2g} + z_s = \frac{p_a}{\gamma} + 0 + 0$$

in which p_a is the atmospheric pressure and p_e is the absolute pressure. For cavitation to occur at e, the pressure must be equal to or less than p_v, the vapor pressure. If $p_e = p_v$,

$$\sigma' = \frac{V_e^2}{2gH} = \frac{p_a - p_v - \gamma z_s}{\gamma H} \tag{9.8.2}$$

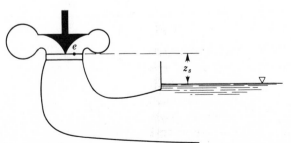

Figure 9.27 Turbine or pump setting.

is the ratio of energy available at e to total energy H across the unit, since the only energy is kinetic energy. The ratio σ' is a cavitation index or number. The critical value σ_c may be determined by a test on a model of the homologous series. For cavitationless performance, the low-pressure setting z_s for an impeller installation must be so fixed that the resulting value of σ' is greater than that of σ_c.

Example 9.10 Tests on a pump model indicate a $\sigma_c = 0.10$. A homologous unit is to be installed at a location where $p_a = 90$ kPa and $p_v = 3.5$ kPa and is to pump water against a head of 25 m. What is the maximum permissible suction head?
Solving Eq. (9.8.2) for z_s and substituting the values of σ_c, H, p_a, and p_v give

$$z_s = \frac{p_a - p_v}{\gamma} - \sigma' H = \frac{90{,}000 - 3500}{9806} - 0.10 \times 25 = 6.32 \text{ m}$$

The less the value of z_s, the greater the value of the plant σ' and the greater the assurance against cavitation.

The *net positive suction head* (NPSH) is frequently used in the specification of minimum suction conditions for a turbomachine. It is

$$\text{NPSH} = \frac{V_e^2}{2g} = \frac{p_a - p_v - \gamma z_s}{\gamma} \tag{9.8.3}$$

A test is run on the machine to determine the maximum value of z_s for operation of the machine with no impairment of efficiency and without objectionable noise or damage. Then, from this test NPSH is calculated from Eq. (9.8.3). Any setting of this machine where the suction lift is less than z_s, as found from Eq. (9.8.3), is then acceptable. Note that z_s is positive when the suction reservoir is below the turbomachine, as in Fig. 9.27.

A *suction specific speed S* for homologous units may be formulated. By elimination of D_e in the two equations,

$$\text{NPSH} = \frac{V_e^2}{2g} \sim \frac{Q^2}{D_e^4} \qquad \frac{Q}{ND_e^3} = \text{const}$$

S is obtained:

$$S = \frac{N\sqrt{Q}}{(\text{NPSH})^{3/4}} \tag{9.8.4}$$

When different units of a series are operating under cavitating conditions, equal values of S indicate a similar degree of cavitation. When cavitation is not present, the equation is not valid.

EXERCISES

9.8.1 The cavitation parameter is defined by

(a) $\dfrac{p_v - p}{\rho V^2/2}$ (b) $\dfrac{p_{\text{atm}} - p_v}{\rho V^2/2}$ (c) $\dfrac{p - p_v}{\gamma V^2/2}$ (d) $\dfrac{p - p_v}{\rho V^2/2}$ (e) none of these answers

9.8.2 Cavitation is caused by (a) high velocity; (b) low barometric pressure; (c) high pressure; (d) low pressure; (e) low velocity.

PROBLEMS

9.1 By use of Eqs. (9.1.1) and (9.1.3), together with $P = \gamma QH$ for power, develop the homologous relation for P in terms of speed and diameter.

9.2 A centrifugal pump is driven by an induction motor that reduces in speed as the pump load increases. A test determines several sets of values of N, Q, H for the pump. How is a characteristic curve of the pump for a constant speed determined from these data?

9.3 What is the specific speed of the pump of Example 9.1 at its point of best efficiency?

9.4 Plot the dimensionless characteristic curve of the pump of Example 9.1. On this same curve plot several points from the characteristics of the new (52-in) pump. Why are they not exactly on the same curve?

9.5 Determine the size and synchronous speed of a pump homologous to the 72-in pump of Example 9.1 that will produce 3 m³/s at 100 m head at its point of best efficiency.

9.6 Develop the characteristic curve for a homologous pump of the series of Example 9.1 for 18-in-diameter discharge and 1800 rpm.

9.7 A pump with a 200-mm-diameter impeller discharges 100 l/s at 1140 rpm and 10 m head at its point of best efficiency. What is its specific speed?

9.8 A hydroelectric site has a head of 100 m and an average discharge of 10 m³/s. For a generator speed of 200 rpm, what specific speed turbine is needed? Assume an efficiency of 92 percent.

9.9 A model turbine, $N_s = 36$, with a 14-in-diameter impeller develops 27 hp at a head of 44 ft and an efficiency of 86 percent. What are the discharge and speed of the model?

9.10 What size and synchronous speed of homologous unit of Prob. 9.9 would be needed to discharge 600 cfs at 260 ft of head?

9.11 22 m³/s water flowing through the fixed vanes of a turbine has a tangential component of 2 m/s at a radius of 1.25 m. The impeller, turning at 180 rpm, discharges in an axial direction. What torque is exerted on the impeller?

9.12 In Prob. 9.11, neglecting losses, what is the head on the turbine?

9.13 A generator with speed $N = 240$ rpm is to be used with a turbine at a site where $H = 120$ m and $Q = 8$ m³/s. Neglecting losses, what tangential component must be given to the water at $r = 1$ m by the fixed vanes? What torque is exerted on the impeller? How much horsepower is produced?

9.14 At what angle should the wicket gates of a turbine be set to extract 9000 kW from a flow of 25 m³/s? The diameter of the opening just inside the wicket gates is 3.5 m, and the height is 1 m. The turbine runs at 200 rpm, and flow leaves the runner in an axial direction.

9.15 For a given setting of wicket gates how does the moment of momentum vary with the discharge?

9.16 Assuming constant axial velocity just above the runner of the propeller turbine of Prob. 9.14, calculate the tangential-velocity components if the hub radius is 300 mm and the outer radius is 900 mm.

9.17 Determine the vane angles β_1, β_2 for entrance and exit from the propeller turbine of Prob. 9.16 so that no angular momentum remains in the flow. (Compute the angles for inner radius, outer radius, and midpoint.)

9.18 Neglecting losses, what is the head on the turbine of Prob. 9.14?

9.19 The hydraulic efficiency of a turbine is 95 percent, and its theoretical head is 80 m. What is the actual head required?

9.20 A turbine model test with 260-mm-diameter impeller showed an efficiency of 90 percent. What efficiency could be expected from a 1.2-m-diameter impeller?

9.21 Construct a theoretical head-discharge curve for the following specifications of a centrifugal pump: $r_1 = 50$ mm, $r_2 = 100$ mm, $b_1 = 25$ mm, $b_2 = 20$ mm, $N = 1200$ rpm, and $\beta_2 = 30°$.

9.22 A centrifugal water pump (Fig. 9.15) has an impeller $r_1 = 2.75$ in, $b_1 = 1\frac{3}{8}$ in, $r_2 = 4.5$ in, $b_2 = \frac{3}{4}$ in, $\beta_1 = 30°$, $\beta_2 = 45°$ (b_1, b_2 are impeller width at r_1 and r_2, respectively). Neglect thickness of vanes. For 1800 rpm, calculate (a) the design discharge for no prerotation of entering fluid, (b) α_2 and

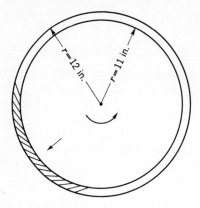

Figure 9.28 Problem 9.28.

the theoretical head at point of best efficiency, and (c) for hydraulic efficiency of 85 percent and overall efficiency of 78 percent, the actual head produced, losses in foot-pounds per pound, and brake horsepower.

9.23 A centrifugal pump has an impeller with dimensions $r_1 = 75$ mm, $r_2 = 160$ mm, $b_1 = 50$ mm, $b_2 = 30$ mm, $\beta_1 = \beta_2 = 30°$. For a discharge of 55 l/s and shockless entry to vanes compute (a) the speed, (b) the head, (c) the torque, (d) the power, and (e) the pressure rise across impeller. Neglect losses. $\alpha_1 = 90°$.

9.24 A centrifugal water pump with impeller dimensions $r_1 = 2$ in, $r_2 = 5$ in, $b_1 = 3$ in, $b_2 = 1.5$ in, $\beta_2 = 60°$ is to pump 5 cfs at 64 ft head. Determine (a) β_1, (b) the speed, (c) the horsepower, and (d) the pressure rise across the impeller. Neglect losses, and assume no shock at the entrance. $\alpha_1 = 90°$.

9.25 Select values of r_1, r_2, β_1, β_2, b_1, and b_2 of a centrifugal impeller to take 30 l/s water from a 100-mm-diameter suction line and increase its energy by 15 m·N/N. $N = 1200$ rpm; $\alpha_1 = 90°$. Neglect losses.

9.26 A pump has blade angles $\beta_1 = \beta_2$; $b_1 = 2b_2 = 25$ mm, $r_1 = r_2/3 = 50$ mm. For a theoretical head of 30 m at a discharge at best efficiency of 30 l/s, determine the blade angles and speed of the pump. Neglect thickness of vanes and assume perfect guidance. (*Hint*: Write down every relation you know connecting β_1, β_2, b_1, b_2, r_1, r_2, u_1, u_2, H_{th}, Q, V_{r2}, V_{u2}, V_1, ω, and N from the two velocity vector diagrams, and by substitution reduce to one unknown.)

9.27 A mercury-water differential manometer, $R' = 700$ mm, is connected from the 100-mm-diameter suction pipe to the 80-mm-diameter discharge pipe of a pump. The centerline of the suction pipe is 300 mm below the discharge pipe. For $Q = 60$ l/s water, calculate the head developed by the pump.

9.28 The impeller for a blower (Fig. 9.28) is 18 in wide. It has straight blades and turns at 1200 rpm. For 10,000 ft³/min air, $\gamma = 0.08$ lb/ft³, calculate (a) entrance and exit blade angles ($\alpha_1 = 90°$), (b) the head produced, in inches of water, and (c) the theoretical horsepower required.

9.29 An air blower is to be designed to produce pressure of 100 mm H_2O when operating at 3600 rpm. $\gamma = 11.5$ N/m³; $r_2 = 1.1r_1$; $\beta_2 = \beta_1$; width of impeller is 100 mm; $\alpha_1 = 90°$. Find r_1.

9.30 In Prob. 9.29, when $\beta_1 = 30°$, calculate the discharge in cubic meters per minute.

9.31 A site for a Pelton wheel has a steady flow of 55 l/s with a nozzle velocity of 75 m/s. With a blade angle of 174° and $C_v = 0.98$, for 60-Hz power, determine (a) the diameter of wheel, (b) the speed, (c) the power, (d) the energy remaining in the water. Neglect losses.

9.32 An impulse wheel is to be used to develop 50-Hz power at a site where $H = 120$ m, and $Q = 75$ l/s. Determine the diameter of the wheel and its speed. $C_v = 0.97$; $e = 82$ percent.

9.33 Develop the equation for efficiency of a cooled compressor,

$$\eta = \frac{\eta_m}{V_{u2}u_2}\left(\frac{V_{2th}^2 - V_1^2}{2} + \frac{p_1}{\rho_1}\ln\frac{p_2}{\rho_1}\right)$$

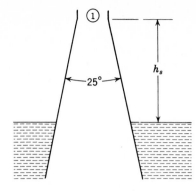

Figure 9.29 Problem 9.35.

9.34 Find the rotational speed in Example 9.9 for a cooled compressor. Use results of Prob. 9.33, with the actual air temperature at exit 15°C.

9.35 A turbine draft tube (Fig. 9.29) expands from 6 to 18 ft diameter. At section 1 the velocity is 30 ft/s for vapor pressure of 1 ft and barometric pressure of 32 ft of water. Determine h_s for incipient cavitation (pressure equal to vapor pressure at section 1).

9.36 What is the cavitation parameter at a point in flowing water where $t = 20°C$, $p = 14$ kPa and the velocity is 12 m/s?

9.37 A turbine with $\sigma_c = 0.08$ is to be installed at a site where $H = 60$ m and a water barometer stands at 8.3 m. What is the maximum permissible impeller setting above tail water?

REFERENCES

Church, A. H.: "Centrifugal Pumps and Blowers," Wiley, New York, 1944.

Daily, J. W.: Hydraulic Machinery, in H. Rouse (ed.), "Engineering Hydraulics," Wiley, New York, 1950.

Eisenberg, P., and M. P. Tulin: Cavitation, sec. 12 in V. L. Streeter (ed.), "Handbook of Fluid Dynamics," McGraw-Hill, New York, 1961.

Moody, L. F.: Hydraulic Machinery, in C. V. Davis (ed.), "Handbook of Applied Hydraulics," 2d ed., McGraw-Hill, New York, 1952.

Norrie, D. H.: "An Introduction to Incompressible Flow Machines," American Elsevier, New York, 1963.

Stepanoff, A. J.: "Centrifugal and Axial Flow Pumps," Wiley, New York, 1948.

Wislicenus, G. F.: "Fluid Mechanics of Turbomachinery," McGraw-Hill, New York, 1947.

TEN

STEADY CLOSED-CONDUIT FLOW

The basic procedures for solving problems in incompressible steady flow in closed conduits are presented in Sec. 5.10, where simple pipe flow situations are discussed, including losses due to change in cross section or direction of flow. The great majority of practical problems deal with turbulent flow, and velocity distributions in turbulent pipe flow are discussed in Sec. 5.4. The Darcy-Weisbach equation is introduced in Chap. 5 to relate frictional losses to flow rate in pipes, with the friction factor determined from the Moody diagram. Exponential friction formulas commonly used in commercial and industrial applications are discussed in this chapter. The use of the hydraulic and energy grade lines in solving problems is reiterated before particular applications are developed. Complex flow problems are investigated, including hydraulic systems that incorporate various different elements such as pumps and piping networks. The use of the digital computer in analysis and design becomes particularly relevant when multielement systems are being investigated. The hand-held electronic programmable calculator is effective for iterative solutions such as the small-network problems.

10.1 EXPONENTIAL PIPE-FRICTION FORMULAS

Industrial pipe-friction formulas are usually empirical, of the form

$$\frac{h_f}{L} = \frac{RQ^n}{D^m} \tag{10.1.1}$$

in which h_f/L is the head loss per unit length of pipe (slope of the energy grade line), Q the discharge, and D the inside pipe diameter. The resistance coefficient R is a function of pipe roughness only. An equation with specified exponents and coefficient R is valid only for the fluid viscosity for which it is developed, and it is normally limited to a range of Reynolds numbers and diameters. In its range of applicability such an equation is convenient, and nomographs are often used to aid problem solution.

The Hazen-Williams[†] formula for flow of water at ordinary temperatures through pipes is of this form, with R given by

$$R = \frac{4.727}{C^n} \qquad \text{U.S. customary units} \tag{10.1.2}$$

$$R = \frac{10.675}{C^n} \qquad \text{SI units} \tag{10.1.3}$$

with $n = 1.852$, $m = 4.8704$, and C dependent upon roughness as follows:

$$C = \begin{cases} 140 & \text{extremely smooth, straight pipes; asbestos-cement} \\ 130 & \text{very smooth pipes; concrete; new cast iron} \\ 120 & \text{wood stave; new welded steel} \\ 110 & \text{vitrified clay; new riveted steel} \\ 100 & \text{cast iron after years of use} \\ 95 & \text{riveted steel after years of use} \\ 60 \text{ to } 80 & \text{old pipes in bad condition} \end{cases}$$

One can develop a special-purpose formula for a particular application by using the Darcy-Weisbach equation and friction factors from the Moody diagram or, alternatively, by using experimental data if available.[‡] Exponential formulas developed from experimental results are generally very useful and handy in the region over which the data were gathered. Extrapolations and applications to other situations must be carried out with caution.

A comparison between the Hazen-Williams equation and the Darcy-Weisbach equation with friction factors from the Moody diagram is presented in Fig. 10.1. It shows equivalent values of f vs. Reynolds number for three typical Hazen-Williams roughness values 70, 100, and 140. The fluid is water at 15°C.

By equating the slope of the hydraulic grade line in the Darcy-Weisbach equation, $h_f/L = fQ^2/2gDA^2$, to Eq. (10.1.1), solving for f, and introducing the Reynolds number to eliminate Q (in SI),

$$f = \frac{1014.2}{C^{1.852}D^{0.0184}} \mathbf{R}^{-0.148} \tag{10.1.4}$$

† E. F. Brater, "Handbook of Hydraulics," 6th ed., McGraw-Hill, New York, 1976, p. 6-17.

‡ V. L. Streeter and E. B. Wylie, "Fluid Mechanics," 6th edition, McGraw-Hill, New York, 1975, pp. 545–547.

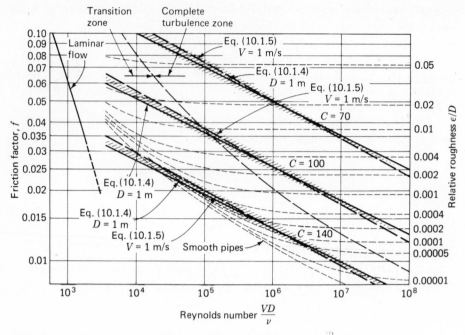

Figure 10.1 Comparison: Hazen-Williams vs. Darcy-Weisbach on Moody diagram.

For a given Hazen-Williams coefficient C and diameter D the friction factor reduces with increasing Reynolds number. A similar solution for f in terms of C, Reynolds number, and V can be developed by combining the same equations and eliminating D,

$$f = \frac{1304.56 V^{0.0184}}{C^{1.852}} \mathbf{R}^{-0.1664} \tag{10.1.5}$$

It may be noted that f is not strongly dependent upon pipe diameter in Eq. (10.1.4). Similarly, the friction factor is not strongly dependent upon the velocity in Eq. (10.1.5).

In Fig. 10.1, at the three selected values of C, Eq. (10.1.4) is shown for a particular diameter of 1 m and Eq. (10.1.5) is shown for a specific velocity of 1 m/s. The shaded region around each of these lines shows the range of practical variation of the variables ($0.025 < D < 6$ m, 0.030 m/s $< V < 30$ m/s).

The two formulations Darcy-Weisbach vs. Hazen-Williams for calculation of losses in a pipeline can be seen to be significantly different. The Darcy-Weisbach equation is probably more rationally based than other empirical exponential formulations and has received wide acceptance. However, when specific experimental data are available and an exponential formula based on the data is developed, it is preferable to the more general Moody diagram approach. The data must be reliable, and the equation can be considered valid only over the range of gathered data.

10.2 HYDRAULIC AND ENERGY GRADE LINES

The concepts of *hydraulic* and *energy grade lines* are useful in analyzing more complex flow problems. If, at each point along a pipe system, the term p/γ is determined and plotted as a vertical distance above the center of the pipe, the locus of points is the hydraulic grade line. More generally, the plot of the two terms

$$\frac{p}{\gamma} + z$$

for the flow, as ordinates, against length along the pipe, as abscissas, produces the hydraulic grade line. The hydraulic grade line, or piezometric head line, is the locus of heights to which liquid would rise in vertical glass tubes connected to piezometer openings in the line. When the pressure in the line is less than atmospheric, p/γ is negative and the hydraulic grade line is below the pipeline.

The energy grade line is a line joining a series of points marking the available energy in meter-newtons per newton for each point along the pipe as ordinate, plotted against distance along the pipe as the abscissa. It consists of the plot of

$$\frac{V^2}{2g} + \frac{p}{\gamma} + z$$

for each point along the line. By definition, the energy grade line is always vertically above the hydraulic grade line a distance of $V^2/2g$, neglecting the kinetic-energy correction factor.

The hydraulic and energy grade lines are shown in Fig. 10.2 for a simple pipeline containing a square-edged entrance, a valve, and a nozzle at the end of the line. To construct these lines when the reservoir surface is given, it is necessary first to apply the energy equation from the reservoir to the exit, including all minor losses as well as pipe friction, and to solve for the velocity head $V^2/2g$. Then, to find the elevation of hydraulic grade line at any point, the energy equation is applied from the reservoir to that point, including all losses between the two

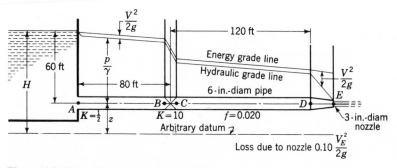

Figure 10.2 Hydraulic and energy grade lines.

points. The equation is solved for $p/\gamma + z$, which is plotted above the arbitrary datum. To find the energy grade line at the same point, the equation is solved for $V^2/2g + p/\gamma + z$, which is plotted above the arbitrary datum.

The reservoir surface is the hydraulic grade line and is also the energy grade line. At the square-edged entrance the energy grade line drops by $0.5V^2/2g$ because of the loss there, and the hydraulic grade line drops $1.5V^2/2g$. This is made obvious by applying the energy equation between the reservoir surface and a point just downstream from the pipe entrance:

$$H + 0 + 0 = \frac{V^2}{2g} + z + \frac{p}{\gamma} + 0.5\frac{V^2}{2g}$$

Solving for $z + p/\gamma$,

$$z + \frac{p}{\gamma} = H - 1.5\frac{V^2}{2g}$$

shows the drop of $1.5V^2/2g$. The head loss due to the sudden entrance does not actually occur at the entrance itself, but over a distance of 10 or more diameters of pipe downstream. It is customary to show it at the fitting.

Example 10.1 Determine the elevation of hydraulic and energy grade lines at points A, B, C, D, and E of Fig. 10.2. $z = 10$ ft.

First, solving for the velocity head is accomplished by applying the energy equation from the reservoir to E,

$$10 + 60 + 0 + 0 = \frac{V_E^2}{2g} + 10 + 0 + \frac{1}{2}\frac{V^2}{2g} + 0.020\frac{200}{0.50}\frac{V^2}{2g} + 10\frac{V^2}{2g} + 0.10\frac{V_E^2}{2g}$$

From the continuity equation, $V_E = 4V$. After simplifying,

$$60 = \frac{V^2}{2g}\left(16 + \frac{1}{2} + 8 + 10 + 16 \times 0.1\right) = 36.1\frac{V^2}{2g}$$

and $V^2/2g = 1.66$ ft. Applying the energy equation for the portion from the reservoir to A gives

$$70 + 0 + 0 = \frac{V^2}{2g} + \frac{p}{\gamma} + z + 0.5\frac{V^2}{2g}$$

Hence, the hydraulic grade line at A is

$$\frac{p}{\gamma} + z\bigg|_A = 70 - 1.5\frac{V^2}{2g} = 70 - 1.5 \times 1.66 = 67.51 \text{ ft}$$

The energy grade line for A is

$$\frac{V^2}{2g} + z + \frac{p}{\gamma} = 67.51 + 1.66 = 69.17 \text{ ft}$$

For B,

$$70 + 0 + 0 = \frac{V^2}{2g} + \frac{p}{\gamma} + z + 0.5\frac{V^2}{2g} + 0.02\frac{80}{0.5}\frac{V^2}{2g}$$

and

$$\frac{p}{\gamma} + z\bigg|_B = 70 - (1.5 + 3.2)1.66 = 62.19 \text{ ft}$$

the energy grade line is at $62.19 + 1.66 = 63.85$ ft.

Across the valve the hydraulic grade line drops by $10V^2/2g$, or 16.6 ft. Hence, at C the energy and hydraulic grade lines are at 47.25 ft and 45.59 ft, respectively.

At point D,

$$70 = \frac{V^2}{2g} + \frac{p}{\gamma} + z + \left(10.5 + 0.02\frac{200}{0.50}\right)\frac{V^2}{2g}$$

and

$$\frac{p}{\gamma} + z\bigg|_D = 70 - 19.5 \times 1.66 = 37.6 \text{ ft}$$

with the energy grade line at $37.6 + 1.66 = 39.26$ ft.

At point E the hydraulic grade line is 10 ft, and the energy grade line is

$$z + \frac{V_E^2}{2g} = 10 + 16\frac{V^2}{2g} = 10 + 16 \times 1.66 = 36.6 \text{ ft}$$

The *hydraulic gradient* is the slope of the hydraulic grade line if the conduit is horizontal; otherwise, it is

$$\frac{d(z + p/\gamma)}{dL}$$

The *energy gradient* is the slope of the energy grade line if the conduit is horizontal; otherwise, it is

$$\frac{d(z + p/\gamma + V^2/2g)}{dL}$$

In many situations involving long pipelines the minor losses may be neglected (when less than 5 percent of the pipe friction losses), or they may be included as equivalent lengths of pipe which are added to actual length in solving the problem. For these situations the value of the velocity head $V^2/2g$ is small compared with $f(L/D)V^2/2g$ and is neglected.

In this special but very common case, when minor effects are neglected, the energy and hydraulic grade lines are superposed. The single grade line, shown in Fig. 10.3, is commonly referred to as the *hydraulic grade line*. No change in hydraulic grade line is shown for minor losses. For these situations with long pipelines the hydraulic gradient becomes h_f/L, with h_f given by the Darcy-Weisbach equation

$$h_f = f\frac{L}{D}\frac{V^2}{2g} \tag{10.2.1}$$

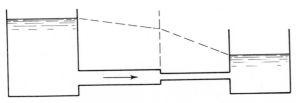

Figure 10.3 Hydraulic grade line for long pipelines where minor losses are neglected or included as equivalent lengths of pipe.

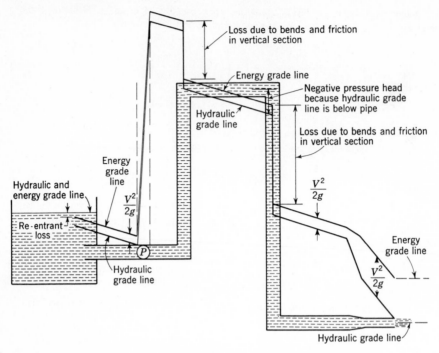

Figure 10.4 Hydraulic and energy grade lines for a system with pump and siphon.

or by Eq. (10.1.1). Flow (except through a pump) is always in the direction of decreasing energy grade line.

Pumps add energy to the flow, a fact which may be expressed in the energy equation either by including a *negative loss* or by stating the energy per unit weight added as a positive term on the upstream side of the equation. The hydraulic grade line rises sharply at a pump. Figure 10.4 shows the hydraulic and energy grade lines for a system with a pump and a siphon. The true slope of the grade lines can be shown only for horizontal lines.

Example 10.2 A pump with a shaft input of 7.5 kW and an efficiency of 70 percent is connected in a waterline carrying 0.1 m³/s. The pump has a 150-mm-diameter suction line and a 120-mm-diameter discharge line. The suction line enters the pump 1 m below the discharge line. For a suction pressure of 70 kN/m², calculate the pressure at the discharge flange and the rise in the hydraulic grade line across the pump.

If the energy added in meter-newtons per newton is symbolized by E, the fluid power added is

$$Q\gamma E = 7500 \times 0.70 \quad \text{or} \quad E = \frac{7500 \times 0.7}{0.1 \times 9806} = 5.354 \text{ m}$$

Applying the energy equation from suction flange to discharge flange gives

$$\frac{V_s^2}{2g} + \frac{p_s}{\gamma} + 0 + 5.354 = \frac{V_d^2}{2g} + \frac{p_d}{\gamma} + 1$$

in which the subscripts s and d refer to the suction and discharge conditions, respectively. From the continuity equation

$$V_s = \frac{0.1 \times 4}{0.15^2 \pi} = 5.66 \text{ m/s} \qquad V_d = \frac{0.1 \times 4}{0.12^2 \pi} = 8.84 \text{ m/s}$$

Solving for p_d gives

$$\frac{p_d}{\gamma} = \frac{5.66^2}{2 \times 9.806} + \frac{70,000}{9806} + 5.354 - \frac{8.84^2}{2 \times 9.806} - 1 = 9.141 \text{ m}$$

and $p_d = 89.6 \text{ kN/m}^2$. The rise in hydraulic grade line is

$$\left(\frac{p_d}{\gamma} + 1\right) - \frac{p_s}{\gamma} = 9.141 + 1 - \frac{70,000}{9806} = 3.002 \text{ m}$$

In this example much of the energy was added in the form of kinetic energy, and the hydraulic grade line rises only 3.002 m for a rise of energy grade line of 5.354 m.

A turbine takes energy from the flow and causes a sharp drop in both the energy and the hydraulic grade lines. The energy removed per unit weight of fluid may be treated as a loss in computing grade lines.

10.3 THE SIPHON

A closed conduit, arranged as in Fig. 10.5, which lifts the liquid to an elevation higher than its free surface and then discharges it at a lower elevation is a *siphon*. It has certain limitations in its performance due to the low pressures that occur near the summit s.

Assuming that the siphon flows full, with a continuous liquid column throughout it, the application of the energy equation for the portion from 1 to 2 produces the equation

$$H = \frac{V^2}{2g} + K\frac{V^2}{2g} + f\frac{L}{D}\frac{V^2}{2g}$$

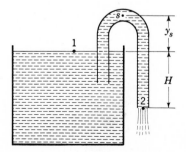

Figure 10.5 Siphon.

in which K is the sum of all the minor-loss coefficients. Factoring out the velocity head gives

$$H = \frac{V^2}{2g}\left(1 + K + \frac{fL}{D}\right) \tag{10.3.1}$$

which is solved in the same fashion as the simple pipe problems of the first or second type. With the discharge known, the solution for H is straightforward, but the solution for velocity with H given is a trial solution started by assuming an f.

The pressure at the summit s is found by applying the energy equation for the portion between 1 and s after Eq. (10.3.1) is solved. It is

$$0 = \frac{V^2}{2g} + \frac{p_s}{\gamma} + y_s + K'\frac{V^2}{2g} + f\frac{L'}{D}\frac{V^2}{2g}$$

in which K' is the sum of the minor-loss coefficients between the two points and L' is the length of conduit upstream from s. Solving for the pressure gives

$$\frac{p_s}{\gamma} = -y_s - \frac{V^2}{2g}\left(1 + K' + \frac{fL'}{D}\right) \tag{10.3.2}$$

which shows that the pressure is negative and that it decreases with y_s and $V^2/2g$. If the solution of the equation should be a value of p_s/γ equal to or less than the vapor pressure† of the liquid, then Eq. (10.3.1) is not valid because the vaporization of portions of the fluid column invalidates the incompressibility assumption used in deriving the energy equation.

Although Eq. (10.3.1) is not valid for this case, theoretically there will be a discharge so long as y_s plus the vapor pressure is less than local atmospheric pressure expressed in length of the fluid column. When Eq. (10.3.2) yields a pressure less than vapor pressure at s, the pressure at s may be taken as vapor pressure. Then, with this pressure known, Eq. (10.3.2) is solved for $V^2/2g$, and the discharge is obtained therefrom. It is assumed that air does not enter the siphon at 2 and break at s the vacuum that produces the flow.

Practically, a siphon does not work satisfactorily when the pressure intensity at the summit is close to vapor pressure. Air and other gases come out of solution at the low pressures and collect at the summit, thus reducing the length of the right-hand column of liquid that produces the low pressure at the summit. Large siphons that operate continuously have vacuum pumps to remove the gases at the summits.

The lowest pressure may not occur at the summit but somewhere downstream from that point, because friction and minor losses may reduce the pressure more than the decrease in elevation increases pressure.

---

† A liquid boils when its pressure is reduced to its vapor pressure. The vapor pressure is a function of temperature for a particular liquid. Water has a vapor pressure of 0.0619 m (0.203 ft) H_2O abs at 0°C (32°F) and 10.33 m (33.91 ft) H_2O abs at 100°C (212°F). See Appendix C.

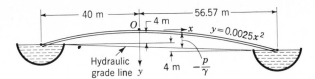

Figure 10.6 Siphon connecting two reservoirs.

Example 10.3 Neglecting minor losses and considering the length of pipe equal to its horizontal distance, determine the point of minimum pressure in the siphon of Fig. 10.6.

When minor losses are neglected, the kinetic-energy term $V^2/2g$ is usually neglected also. Then the hydraulic grade line is a straight line connecting the two liquid surfaces. Coordinates of two points on the line are

$$x = -40 \text{ m}, \ y = 4 \text{ m} \qquad \text{and} \qquad x = 56.57 \text{ m}, \ y = 8 \text{ m}$$

The equation of the line is, by substitution into $y = mx + b$,

$$y = 0.0414x + 5.656 \text{ m}$$

The minimum pressure occurs where the distance between hydraulic grade line and pipe is a maximum,

$$\frac{p}{\gamma} = 0.0025x^2 - 0.0414x - 5.656$$

To find minimum p/γ, set $d(p/\gamma)/dx = 0$, which yields $x = 8.28$, and $p/\gamma = -5.827$ m of fluid flowing. The minimum point occurs where the slopes of the pipe and of the hydraulic grade line are equal.

10.4 PIPES IN SERIES

When two pipes of different sizes or roughnesses are so connected that fluid flows through one pipe and then through the other, they are said to be connected in series. A typical series-pipe problem, in which the head H may be desired for a given discharge or the discharge wanted for a given H, is illustrated in Fig. 10.7. Applying the energy equation from A to B, including all losses, gives

$$H + 0 + 0 = 0 + 0 + 0 + K_e \frac{V_1^2}{2g} + f_1 \frac{L_1}{D_1} \frac{V_1^2}{2g} + \frac{(V_1 - V_2)^2}{2g} + f_2 \frac{L_2}{D_2} \frac{V_2^2}{2g} + \frac{V_2^2}{2g}$$

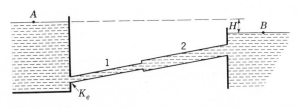

Figure 10.7 Pipes connected in series.

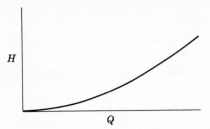

H

Q

Figure 10.8 Plot of calculated H for selected values of Q.

in which the subscripts refer to the two pipes. The last item is the head loss at exit from pipe 2. With the continuity equation

$$V_1 D_1^2 = V_2 D_2^2$$

V_2 is eliminated from the equations, so that

$$H = \frac{V_1^2}{2g}\left\{ K_e + \frac{f_1 L_1}{D_1} + \left[1 - \left(\frac{D_1}{D_2}\right)^2\right]^2 + \frac{f_2 L_2}{D_2}\left(\frac{D_1}{D_2}\right)^4 + \left(\frac{D_1}{D_2}\right)^4 \right\}$$

For known lengths and sizes of pipes this reduces to

$$H = \frac{V_1^2}{2g}(C_1 + C_2 f_1 + C_3 f_2) \tag{10.4.1}$$

in which C_1, C_2, C_3 are known. With the discharge given, the Reynolds number is readily computed, and the f's may be looked up in the Moody diagram. Then H is found by direct substitution. With H given, V_1, f_1, f_2 are unknowns in Eq. (10.4.1). By assuming values of f_1 and f_2 (they may be assumed equal), a trial V_1 is found from which trial Reynolds numbers are determined and values of f_1, f_2 looked up. With these new values, a better V_1 is computed from Eq. (10.4.1). Since f varies so slightly with the Reynolds number, the trial solution converges very rapidly. The same procedures apply for more than two pipes in series.

In place of the assumption of f_1 and f_2 when H is given, a graphical solution may be utilized in which several values of Q are assumed in turn, and the corresponding values of H are calculated and plotted against Q, as in Fig. 10.8. By connecting the points with a smooth curve, it is easy to read off the proper Q for the given value of H.

Example 10.4 In Fig. 10.7, $K_e = 0.5$, $L_1 = 300$ m, $D_1 = 600$ mm, $\epsilon_1 = 2$ mm, $L_2 = 240$ m, $D_2 = 1$ m, $\epsilon_2 = 0.3$ mm, $\nu = 3 \times 10^{-6}$ m²/s, and $H = 6$ m. Determine the discharge through the system.

From the energy equation

$$6 = \frac{V_1^2}{2g}\left[0.5 + f_1\frac{300}{0.6} + (1 - 0.6^2)^2 + f_2\frac{240}{1.0}0.6^4 + 0.6^4\right]$$

After simplifying,

$$6 = \frac{V_1^2}{2g}(1.0392 + 500f_1 + 31.104f_2)$$

From $\epsilon_1/D_1 = 0.0033$, $\epsilon_2/D_2 = 0.0003$, and Fig. 5.32, values of f are assumed for the fully turbulent range:

$$f_1 = 0.026 \qquad f_2 = 0.015$$

By solving for V_1 with these values, $V_1 = 2.848$ m/s, $V_2 = 1.025$ m/s,

$$\mathbf{R}_1 = \frac{2.848 \times 0.6}{3 \times 10^{-6}} = 569,600 \qquad \mathbf{R}_2 = \frac{1.025 \times 1.0}{3 \times 10^{-6}} = 341,667$$

From Fig. 5.32, $f_1 = 0.0265$, $f_2 = 0.0168$. By solving again for V_1, $V_1 = 2.819$ m/s, and $Q = 0.797$ m^3/s.

Equivalent Pipes

Series pipes can be solved by the method of equivalent lengths. Two pipe systems are said to be equivalent when the same head loss produces the same discharge in both systems. From Eq. (10.2.1)

$$h_{f_1} = f_1 \frac{L_1}{D_1}\frac{Q_1^2}{(D_1^2\pi/4)^2 2g} = \frac{f_1 L_1}{D_1^5}\frac{8Q_1^2}{\pi^2 g}$$

and for a second pipe

$$h_{f_2} = \frac{f_2 L_2}{D_2^5}\frac{8Q_2^2}{\pi^2 g}$$

For the two pipes to be equivalent,

$$h_{f_1} = h_{f_2} \qquad Q_1 = Q_2$$

After equating $h_{f_1} = h_{f_2}$ and simplifying,

$$\frac{f_1 L_1}{D_1^5} = \frac{f_2 L_2}{D_2^5}$$

Solving for L_2 gives

$$L_2 = L_1\frac{f_1}{f_2}\left(\frac{D_2}{D_1}\right)^5 \tag{10.4.2}$$

which determines the length of a second pipe to be equivalent to that of the first pipe. For example, to replace 300 m of 250-mm pipe with an equivalent length of 150-mm pipe, the values of f_1 and f_2 must be approximated by selecting a discharge within the range intended for the pipes. Say $f_1 = 0.020, f_2 = 0.018$, then

$$L_2 = 300\frac{0.020}{0.018}\left(\frac{150}{250}\right)^5 = 25.9 \text{ m}$$

For these assumed conditions 25.9 m of 150-mm pipe is equivalent to 300 m of 250-mm pipe.

Hypothetically, two or more pipes composing a system may also be replaced by a pipe which has the same discharge for the same overall head loss.

Example 10.5 Solve Example 10.4 by means of equivalent pipes.
First, by expressing the minor losses in terms of equivalent lengths, for pipe 1

$$K_1 = 0.5 + (1 - 0.6^2)^2 = 0.91 \qquad L_{e_1} = \frac{K_1 D_1}{f_1} = \frac{0.91 \times 0.6}{0.026} = 21 \text{ m}$$

and for pipe 2

$$K_2 = 1 \qquad L_{e_2} = \frac{K_2 D_2}{f_2} = \frac{1 \times 1}{0.015} = 66.7 \text{ m}$$

The values of f_1, f_2 are selected for the fully turbulent range as an approximation. The problem is now reduced to 321 m of 600-mm pipe and 306.7 m of 1-m pipe. By expressing the 1-m pipe in terms of an equivalent length of 600-mm pipe, by Eq. (10.4.2)

$$L_e = \frac{f_2}{f_1} L_2 \left(\frac{D_1}{D_2}\right)^5 = 306.7 \frac{0.015}{0.026} \left(\frac{0.6}{1.0}\right)^5 = 13.76 \text{ m}$$

By adding to the 600-mm pipe, the problem is reduced to finding the discharge through 334.76 m of 600-mm pipe, $\epsilon_1 = 2$ mm, $H = 6$ m,

$$6 = f \frac{334.76}{0.6} \frac{V^2}{2g}$$

With $f = 0.026$, $V = 2.848$ m/s, and $\mathbf{R} = 2.848 \times 0.6/(3 \times 10^{-6}) = 569{,}600$.
For $\epsilon/D = 0.0033$, $f = 0.0265$, $V = 2.821$, and $Q = \pi(0.3^2)(2.821) = 0.798$ m³/s. By use of Eq. (5.10.15), $Q = 0.781$ m³/s.

10.5 PIPES IN PARALLEL

A combination of two or more pipes connected as in Fig. 10.9, so that the flow is divided among the pipes and then is joined again, is a *parallel-pipe* system. In series pipes the same fluid flows through all the pipes and the head losses are cumulative, but in parallel pipes the head losses are the same in any of the lines and the discharges are cumulative.

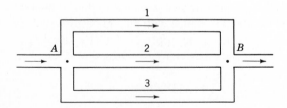

Figure 10.9 Parallel-pipe system.

In analyzing parallel-pipe systems, it is assumed that the minor losses are added into the lengths of each pipe as equivalent lengths. From Fig. 10.9 the conditions to be satisfied are

$$h_{f_1} = h_{f_2} = h_{f_3} = \frac{p_A}{\gamma} + z_A - \left(\frac{p_B}{\gamma} + z_B\right) \qquad (10.5.1)$$

$$Q = Q_1 + Q_2 + Q_3$$

in which z_A, z_B are elevations of points A and B, and Q is the discharge through the approach pipe or the exit pipe.

Two types of problems occur: (1) with elevation of hydraulic grade line at A and B known, to find the discharge Q; (2) with Q known, to find the distribution of flow and the head loss. Sizes of pipe, fluid properties, and roughnesses are assumed to be known.

The first type is, in effect, the solution of simple pipe problems for discharge, since the head loss is the drop in hydraulic grade line. These discharges are added to determine the total discharge.

The second type of problem is more complex, as neither the head loss nor the discharge for any one pipe is known. The recommended procedure is as follows:

1. Assume a discharge Q_1' through pipe 1.
2. Solve for h_{f_1}', using the assumed discharge.
3. Using h_{f_1}', find Q_2', Q_3'.
4. With the three discharges for a common head loss, now assume that the given Q is split up among the pipes in the same proportion as Q_1', Q_2', Q_3'; thus

$$Q_1 = \frac{Q_1'}{\Sigma Q'} Q \qquad Q_2 = \frac{Q_2'}{\Sigma Q'} Q \qquad Q_3 = \frac{Q_3'}{\Sigma Q'} Q \qquad (10.5.2)$$

5. Check the correctness of these discharges by computing h_{f_1}, h_{f_2}, h_{f_3} for the computed Q_1, Q_2, Q_3.

This procedure works for any number of pipes. By judicious choice of Q_1', obtained by estimating the percent of the total flow through the system that should pass through pipe 1 (based on diameter, length, and roughness), Eq. (10.5.2) produces values that check within a few percent, which is well within the range of accuracy of the friction factors.

Example 10.6 In Fig. 10.9, $L_1 = 3000$ ft, $D_1 = 1$ ft, $\epsilon_1 = 0.001$ ft; $L_2 = 2000$ ft, $D_2 = 8$ in, $\epsilon_2 = 0.0001$ ft; $L_3 = 4000$ ft, $D_3 = 16$ in, $\epsilon_3 = 0.0008$ ft; $\rho = 2.00$ slugs/ft³, $\nu = 0.00003$ ft²/s, $p_A = 80$ psi, $z_A = 100$ ft, $z_B = 80$ ft. For a total flow of 12 cfs, determine flow through each pipe and the pressure at B.

Assume $Q_1' = 3$ cfs; then $V_1' = 3.82$, $\mathbf{R}_1' = 3.82 \times 1/0.00003 = 127{,}000$, $\epsilon_1/D_1 = 0.001$, $f_1' = 0.022$, and

$$h_{f_1}' = 0.022 \frac{3000}{1.0} \frac{3.82^2}{64.4} = 14.97 \text{ ft}$$

For pipe 2

$$14.97 = f'_2 \frac{2000}{0.667} \frac{V'^2_2}{2g}$$

Then $\epsilon_2/D_2 = 0.00015$. Assume $f'_2 = 0.020$; then $V'_2 = 4.01$ ft/s, $\mathbf{R}'_2 = 4.01 \times \frac{2}{3} \times 1/0.00003 = 89,000$, $f'_2 = 0.019$, $V'_2 = 4.11$ ft/s, $Q'_2 = 1.44$ cfs.

For pipe 3

$$14.97 = f'_3 \frac{4000}{1.333} \frac{V'^2_3}{2g}$$

Then $\epsilon_3/D_3 = 0.0006$. Assume $f'_3 = 0.020$; then $V'_3 = 4.01$ ft/s, $\mathbf{R}'_3 = 4.01 \times 1.333/0.00003 = 178,000$, $f'_3 = 0.020$, $Q'_3 = 5.60$ cfs.

The total discharge for the assumed conditions is

$$\Sigma Q' = 3.00 + 1.44 + 5.60 = 10.04 \text{ cfs}$$

Hence

$$Q_1 = \frac{3.00}{10.04} 12 = 3.58 \text{ cfs} \qquad Q_2 = \frac{1.44}{10.04} 12 = 1.72 \text{ cfs}$$

$$Q_3 = \frac{5.60}{10.04} 12 = 6.70 \text{ cfs}$$

Check the values of h_1, h_2, h_3:

$$V_1 = \frac{3.58}{\pi/4} = 4.56 \qquad \mathbf{R}_1 = 152,000 \qquad f_1 = 0.021 \qquad h_{f_1} = 20.4 \text{ ft}$$

$$V_2 = \frac{1.72}{\pi/9} = 4.93 \qquad \mathbf{R}_2 = 109,200 \qquad f_2 = 0.019 \qquad h_{f_2} = 21.6 \text{ ft}$$

$$V_3 = \frac{6.70}{4\pi/9} = 4.80 \qquad \mathbf{R}_3 = 213,000 \qquad f_3 = 0.019 \qquad h_{f_3} = 20.4 \text{ ft}$$

f_2 is about midway between 0.018 and 0.019. If 0.018 had been selected, h_2 would be 20.4 ft.

To find p_B,

$$\frac{p_A}{\gamma} + z_A = \frac{p_B}{\gamma} + z_B + h_f$$

or

$$\frac{p_B}{\gamma} = \frac{80 \times 144}{64.4} + 100 - 80 - 20.8 = 178.1$$

in which the average head loss was taken. Then

$$p_B = \frac{178.1 \times 64.4}{144} = 79.6 \text{ psi}$$

10.6 BRANCHING PIPES

A simple *branching-pipe* system is shown in Fig. 10.10. In this situation the flow through each pipe is wanted when the reservoir elevations are given. The sizes and types of pipes and fluid properties are assumed known. The Darcy-Weisbach

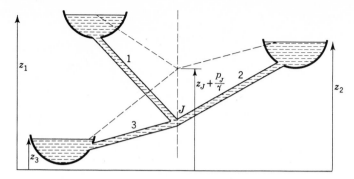

Figure 10.10 Three interconnected reservoirs.

equation must be satisfied for each pipe, and the continuity equation must be satisfied. It takes the form that the flow into the junction J must just equal the flow out of the junction. Flow must be out of the highest reservoir and into the lowest; hence, the continuity equation may be either

$$Q_1 = Q_2 + Q_3 \quad\text{or}\quad Q_1 + Q_2 = Q_3$$

If the elevation of hydraulic grade line at the junction is above the elevation of the intermediate reservoir, flow is *into* it; but if the elevation of hydraulic grade line at J is below the intermediate reservoir, the flow is *out* of it. Minor losses may be expressed as equivalent lengths and added to the actual lengths of pipe.

The solution is effected by first assuming an elevation of hydraulic grade line at the junction and then computing Q_1, Q_2, Q_3 and substituting into the continuity equation. If the flow into the junction is too great, a higher grade-line elevation, which will reduce the inflow and increase the outflow, is assumed.

Example 10.7 In Fig. 10.10, find the discharges for water at 20°C with the following pipe data and reservoir elevations: $L_1 = 3000$ m, $D_1 = 1$ m, $\epsilon_1/D_1 = 0.0002$; $L_2 = 600$ m, $D_2 = 0.45$ m, $\epsilon_2/D_2 = 0.002$; $L_3 = 1000$ m, $D_3 = 0.6$ m, $\epsilon_3/D_3 = 0.001$; $z_1 = 30$ m, $z_2 = 18$ m, $z_3 = 9$ m.
Assume $z_J + p_J/\gamma = 23$ m. Then

$$7 = f_1 \frac{3000}{1} \frac{V_1^2}{2g} \qquad f_1 = 0.014 \qquad V_1 = 1.75 \text{ m/s} \qquad Q_1 = 1.380 \text{ m}^3/\text{s}$$

$$5 = f_2 \frac{600}{0.45} \frac{V_2^2}{2g} \qquad f_2 = 0.024 \qquad V_2 = 1.75 \text{ m/s} \qquad Q_2 = 0.278 \text{ m}^3/\text{s}$$

$$14 = f_3 \frac{1000}{0.60} \frac{V_3^2}{2g} \qquad f_3 = 0.020 \qquad V_3 = 2.87 \text{ m/s} \qquad Q_3 = 0.811 \text{ m}^3/\text{s}$$

so that the inflow is greater than the outflow by

$$1.380 - 0.278 - 0.811 = 0.291 \text{ m}^3/\text{s}$$

Assume $z_J + p_J/\gamma = 24.6$ m. Then

$$5.4 = f_1 \frac{3000}{1} \frac{V_1^2}{2g} \qquad f_1 = 0.015 \qquad V_1 = 1.534 \text{ m/s} \qquad Q_1 = 1.205 \text{ m}^3/\text{s}$$

$$6.6 = f_2 \frac{600}{0.45} \frac{V_2^2}{2g} \qquad f_2 = 0.024 \qquad V_2 = 2.011 \text{ m/s} \qquad Q_2 = 0.320 \text{ m}^3/\text{s}$$

$$15.6 = f_3 \frac{1000}{0.60} \frac{V_3^2}{2g} \qquad f_3 = 0.020 \qquad V_3 = 3.029 \text{ m/s} \qquad Q_3 = 0.856 \text{ m}^3/\text{s}$$

The inflow is still greater by 0.029 m³/s. By extrapolating linearly, $z_J + p_J/\gamma = 24.8$ m, $Q_1 = 1.183$, $Q_2 = 0.325$, $Q_3 = 0.862$ m³/s.

In pumping from one reservoir to two or more other reservoirs, as in Fig. 10.11, the characteristics of the pump must be known. Assuming that the pump runs at constant speed, its head depends upon the discharge. A suitable procedure is as follows:

1. Assume a discharge through the pump.
2. Compute the hydraulic-grade-line elevation at the suction side of the pump.
3. From the pump characteristic curve find the head produced and add it to suction hydraulic grade line.
4. Compute drop in hydraulic grade line to the junction J and determine elevation of hydraulic grade line there.
5. For this elevation, compute flow into reservoirs 2 and 3.
6. If flow into J equals flow out of J, the problem is solved. If flow into J is too great, assume less flow through the pump and repeat the procedure.

This procedure is easily plotted on a graph, so that the intersection of two elevations vs. flow curves yields the answer.

More complex branching-pipe problems may be solved with a similar approach by beginning with a trial solution. However, the network-analysis procedure in Sec. 10.8 is recommended for multibranch systems as well as for multi-parallel-loop systems. Such problems are most easily handled with a digital computer.

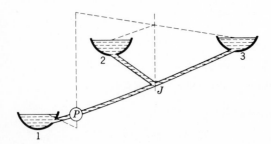

Figure 10.11 Pumping from one reservoir to two other reservoirs.

10.7 NETWORKS OF PIPES

Interconnected pipes through which the flow to a given outlet may come from several circuits are called a *network of pipes*, in many ways analogous to flow through electric networks. Problems on these in general are complicated and require trial solutions in which the elementary circuits are balanced in turn until all conditions for the flow are satisfied.

The following conditions must be satisfied in a network of pipes:

1. The algebraic sum of the pressure drops around each circuit must be zero.
2. Flow into each junction must equal flow out of the junction.
3. The Darcy-Weisbach equation, or equivalent exponential friction formula, must be satisfied for each pipe; i.e., the proper relation between head loss and discharge must be maintained for each pipe.

The first condition states that the pressure drop between any two points in the circuit, for example, A and G (Fig. 10.12), must be the same whether through the pipe AG or through $AFEDG$. The second condition is the continuity equation.

Since it is impractical to solve network problems analytically, methods of successive approximations are utilized. The Hardy Cross method[†] is one in which flows are assumed for each pipe so that continuity is satisfied at every junction. A correction to the flow in each circuit is then computed in turn and applied to bring the circuits into closer balance.

Minor losses are included as equivalent lengths in each pipe. Exponential equations are commonly used, in the form $h_f = rQ^n$, where $r = RL/D^m$ in Eq. (10.1.1). The value of r is a constant in each pipeline (unless the Darcy-Weisbach equation is used) and is determined in advance of the loop-balancing procedure. The corrective term is obtained as follows.

[†] Hardy Cross, Analysis of Flow in Networks of Conduits or Conductors, *Univ. Ill. Bull. 286,* November 1936.

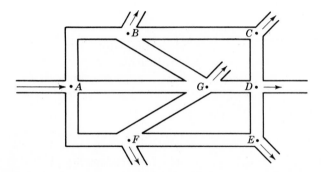

Figure 10.12 Pipe network.

For any pipe in which Q_0 is an assumed initial discharge

$$Q = Q_0 + \Delta Q \tag{10.7.1}$$

where Q is the correct discharge and ΔQ is the correction. Then for each pipe,

$$h_f = rQ^n = r(Q_0 + \Delta Q)^n = r(Q_0^n + nQ_0^{n-1}\,\Delta Q + \cdots)$$

If ΔQ is small compared with Q_0, all terms of the series after the second may be dropped. Now for a circuit,

$$\Sigma h_f = \Sigma rQ\,|Q|^{n-1} = \Sigma rQ_0\,|Q_0|^{n-1} + \Delta Q\Sigma rn\,|Q_0|^{n-1} = 0$$

in which ΔQ has been taken out of the summation because it is the same for all pipes in the circuit and absolute-value signs have been added to account for the direction of summation around the circuit. The last equation is solved for ΔQ in each circuit in the network

$$\Delta Q = -\frac{\Sigma rQ_0\,|Q_0|^{n-1}}{\Sigma rn\,|Q_0|^{n-1}} \tag{10.7.2}$$

When ΔQ is applied to each pipe in a circuit in accordance with Eq. (10.7.1), the directional sense is important; i.e., it adds to flows in the clockwise direction and subtracts from flows in the counterclockwise direction.

Steps in an arithmetic procedure may be itemized as follows:

1. Assume the best distribution of flows that satisfies continuity by careful examination of the network.
2. For each pipe in an elementary circuit, calculate and sum the net head loss $\Sigma h_f = \Sigma rQ^n$. Also calculate $\Sigma rn\,|Q|^{n-1}$ for the circuit. The negative ratio, by Eq. (10.7.2) yields the correction, which is then added algebraically to each flow in the circuit to correct it.
3. Proceed to another elementary circuit and repeat the correction process of 2. Continue for all elementary circuits.
4. Repeat 2 and 3 as many times as needed until the corrections (ΔQ's) are arbitrarily small.

The values of r occur in both numerator and denominator; hence, values proportional to the actual r may be used to find the distribution. Similarly, the apportionment of flows may be expressed as a percent of the actual flows. To find a particular head loss, the actual values of r and Q must be used after the distribution has been determined.

Example 10.8 The distribution of flow through the network of Fig. 10.13 is desired for the inflows and outflows as given. For simplicity n has been given the value 2.0.

The assumed distribution is shown in diagram a. At the upper left the term $\Sigma rQ_0\,|Q_0|^{n-1}$ is computed for the lower circuit number 1. Next to the diagram on the left is the computation of

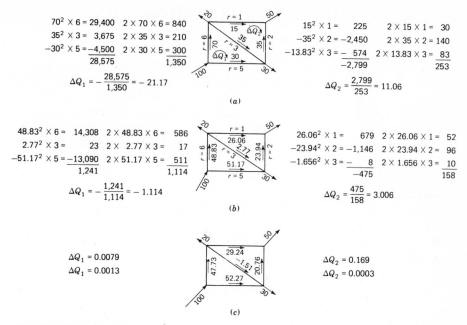

$70^2 \times 6 = 29,400 \quad 2 \times 70 \times 6 = 840$
$35^2 \times 3 = 3,675 \quad 2 \times 35 \times 3 = 210$
$-30^2 \times 5 = \underline{-4,500} \quad 2 \times 30 \times 5 = \underline{300}$
$ 28,575 1,350$

$$\Delta Q_1 = -\frac{28,575}{1,350} = -21.17$$

$15^2 \times 1 = 225 \quad 2 \times 15 \times 1 = 30$
$-35^2 \times 2 = -2,450 \quad 2 \times 35 \times 2 = 140$
$-13.83^2 \times 3 = \underline{-574} \quad 2 \times 13.83 \times 3 = \underline{83}$
$ -2,799 253$

$$\Delta Q_2 = \frac{2,799}{253} = 11.06$$

(a)

$48.83^2 \times 6 = 14,308 \quad 2 \times 48.83 \times 6 = 586$
$2.77^2 \times 3 = 23 \quad 2 \times 2.77 \times 3 = 17$
$-51.17^2 \times 5 = \underline{-13,090} \quad 2 \times 51.17 \times 5 = \underline{511}$
$ 1,241 1,114$

$$\Delta Q_1 = -\frac{1,241}{1,114} = -1.114$$

$26.06^2 \times 1 = 679 \quad 2 \times 26.06 \times 1 = 52$
$-23.94^2 \times 2 = -1,146 \quad 2 \times 23.94 \times 2 = 96$
$-1.656^2 \times 3 = \underline{-8} \quad 2 \times 1.656 \times 3 = \underline{10}$
$ -475 158$

$$\Delta Q_2 = \frac{475}{158} = 3.006$$

(b)

$\Delta Q_1 = 0.0079$
$\Delta Q_1 = 0.0013$

$\Delta Q_2 = 0.169$
$\Delta Q_2 = 0.0003$

(c)

Figure 10.13 Solution for flow in a simple network.

$\Sigma nr |Q_0|^{n-1}$ for the same circuit. The same format is used for the second circuit in the upper right of the figure. The corrected flow after the first step for the top horizontal pipe is determined as $15 + 11.06 = 26.06$ and for the diagonal as $35 + (-21.17) + (-11.06) = 2.77$. Diagram *(b)* shows the flows after one correction. Diagram *(c)* shows the values after four corrections.

Very simple networks, such as the one shown in Fig. 10.13, may be solved with the hand-held programmable calculator if it has memory storage of about 15 and about 100 program steps. For networks larger than the previous example or for networks that contain multiple reservoirs, supply pumps, or booster pumps, the Hardy Cross loop-balancing method may be programmed for numerical solution on a digital computer. Such a program is provided in the next section.

A number of more general methods†,‡,§ are available, primarily based upon the Hardy Cross loop-balancing or node-balancing schemes. In the more general methods the system is normally modeled with a set of simultaneous equations which are solved by the Newton-Raphson method. Some programmed solutions‡,§ are very useful as design tools, since pipe sizes or roughnesses may be treated as unknowns in addition to junction pressures and flows.

---

† R. Epp and A. G. Fowler, Efficient Code for Steady-State Flows in Networks, *J. Hydraul. Div. ASCE*, vol. 96, no. HY1, pp. 43–56, January 1970.

‡ Uri Shamir and C. D. D. Howard, Water Distribution Systems Analysis, *J. Hydraul. Div., ASCE*, vol. 94, no. HY1, pp. 219–234, January 1968.

§ Michael A. Stoner, A New Way to Design Natural Gas Systems, *Pipe Line Ind.*, vol. 32, no. 2, pp. 38–42, 1970.

10.8 COMPUTER PROGRAM FOR STEADY-STATE HYDRAULIC SYSTEMS

Hydraulic systems that contain components different from pipelines can be handled by replacing the component with an equivalent length of pipeline. When the additional component is a pump, special consideration is needed. Also, in systems that contain more than one fixed hydraulic-grade-line elevation, a special artifice must be introduced.

For systems with multiple fixed-pressure-head elevations, Fig. 10.14, *pseudo loops* are created to account for the unknown outflows and inflows at the reservoirs and to satisfy continuity conditions during balancing. A pseudo loop is created by using an imaginary pipeline that interconnects each pair of fixed pressure levels. These imaginary pipelines carry no flow but maintain a fixed drop in the hydraulic grade line equal to the difference in elevation of the reservoirs. If head drop is considered positive in an assumed positive direction in the imaginary pipe, then the correction in loop 3, Fig. 10.14, is

$$\Delta Q_3 = -\frac{150 - 135 - r_4 Q_4 |Q_4|^{n-1} - r_1 Q_1 |Q_1|^{n-1}}{n r_4 |Q_4|^{n-1} + n r_1 |Q_1|^{n-1}} \tag{10.8.1}$$

This correction is applied to pipes 1 and 4 only. If additional real pipelines existed in a pseudo loop, each would be adjusted accordingly during each loop-balancing iteration. The terms in Eq. (10.8.1) may be identified easily by relating to Eq. (10.7.2). Alternatively, the same equation may be generated by application of Newton's method (Appendix B).

A pump in a system may be considered as a flow element with a negative head loss equal to the head rise that corresponds to the flow through the unit. The pump-head-discharge curve, element 8 in Fig. 10.14, may be expressed by a cubic equation

$$H = A_0 + A_1 Q_8 + A_2 Q_8^2 + A_3 Q_8^3$$

where A_0 is the shutoff head of the pump. The correction in loop 4 is

$$\Delta Q_4 = -\frac{135 - 117 - (A_0 + A_1 Q_8 + A_2 Q_8^2 + A_3 Q_8^3) + r_5 Q_5 |Q_5|^{n-1}}{n r_5 |Q_5|^{n-1} - (A_1 + 2A_2 Q_8 + 3A_3 Q_8^2)} \tag{10.8.2}$$

This correction is applied to pipe 5 and to pump 8 in the loop. Equation (10.8.2) is developed by application of Newton's method to the loop. For satisfactory balancing of networks with pumping stations, the slope of the head-discharge curve should always be less than or equal to zero.

The FORTRAN IV program (Fig. 10.15) may be used to analyze a wide variety of liquid steady-state pipe flow problems. The Hardy Cross loop-balancing

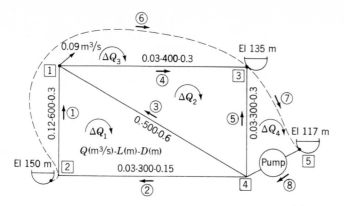

Figure 10.14 Sample network.

method is used. Pipeline flows described by the Hazen-Williams equation or laminar or turbulent flows analyzed with the Darcy-Weisbach equation can be handled; multiple reservoirs or fixed pressure levels, as in a sprinkler system, can be analyzed; and systems with booster pumps or supply pumps can be treated. Either U.S. customary or SI units may be used by proper specification of input data.

A network is visualized as a combination of elements that are interconnected at junctions. The elements may include pipelines, pumps, and imaginary elements which are used to create pseudo loops in multiple-reservoir systems. All minor losses are handled by estimating equivalent lengths and adding them onto the actual pipe lengths. Each element in the system is numbered up to a maximum of 100, without duplication and not necessarily consecutively. A positive flow direction is assigned to each element, and, as in the arithmetic solution, an estimated flow is assigned to each element such that continuity is satisfied at each junction. The assigned positive flow direction in a pump must be in the intended direction of normal pump operation. Any solution with backward flow through a pump is invalid. The flow direction in the imaginary element that creates a pseudo loop indicates only the direction of fixed positive head drop, since the flow must be zero in this element. Each junction, which may represent the termination of a single element or the intersection of many elements, is numbered up to a maximum of 100, without duplication and not necessarily sequentially. An outflow or inflow at a junction is defined during the assignment of initial element flows.

The operation of the program is best visualized in two major parts: the first performs the balancing of each loop in the system successively and then repeats in an iterative manner until the sum of all loop flow corrections is less than a specified tolerance. At the end of this balancing process the element flows are computed and printed. The second part of an analysis involves the computation of the hydraulic-grade-line elevations at junctions in the system. Each of these parts requires a special indexing of the system configuration in the input data. The

```
C HARDY CROSS LOOP BALANCING INCLUDING MULTIPLE RESERVOIRS & PUMPS(PU)
C HAZEN-WILLIAMS(HW) OR DARCY-WEISBACH(DW) MAY BE USED FOR PIPES
C ENGLISH(EN) OR SI UNITS(SI) MAY BE USED, POSITIVE DH IN ELEMENT IS HEAD DROP
      DIMENSION ITY(4),ID(2),ITYPE(100),ELEM(500),IND(500),Q(100),H(100)
     2,S(20),IX(240)
      DATA ITY/'HW','DW','PS','PU'/,IE/'&&'/,ID/'EN','SI'/
   10 DO 12 J=1,500
      IF (J.LE.100)ITYPE(J)=5
      IF (J.LE.100) H(J)=-1000.
      IF (J.LE.240)IX(J)=0
   12 IND(J)=0
C READ PARAMETERS FOR PROBLEM, AND ELEMENT DATA
      READ (5,15,END=99) NT,KK,TOL,VNU,DEF
   15 FORMAT (A2,I8,F10.4,F10.7,F10.5)
      IF (NT.EQ.ID(2))GO TO 20
      WRITE (6,18) VNU
   18 FORMAT (' ENGLISH UNITS SPECIFIED, VISCOSITY IN FT**2/SEC=',F10.7)
      UNITS=4.727
      G=32.174
      GO TO 22
   20 WRITE (6,21) VNU
   21 FORMAT (' SI UNITS SPECIFIED, VISCOSITY IN M**2/SEC=',F10.7)
      UNITS=10.674
      G=9.806
   22 WRITE (6,24) TOL,KK
   24 FORMAT (' DESIRED FLOW TOLERANCE=',F5.3,' NO. OF ITERATIONS=',I5//
     2' PIPE    Q(CFS OR M**3/S)  L(FT OR M)  D(FT OR M)  HW C OR EPS')
   26 READ (5,30) NT,I,QQ,X1,X2,X3,X4,X5
   30 FORMAT (A2,3X,I5,3F10.3,F10.5,2F10.3)
      IF (NT.EQ.IE) GO TO 68
      Q(I)=QQ
      DO 32 NTY=1,4
      IF (NT.EQ.ITY(NTY))GO TO 33
   32 CONTINUE
   33 ITYPE(I)=NTY
      KP=4*(I-1)+1
      GO TO (41,42,53,64),NTY
   41 IF (X3.EQ.0.)X3=DEF
      ELEM(KP)=UNITS*X1/(X3**1.852*X2**4.8704)
      EX=1.852
      GO TO 43
   42 IF (X3.EQ.0.)X3=DEF
      EX=2.
      ELEM(KP)=X1/(2.*G*X2**5*.7854*.7854)
      ELEM(KP+1)=1./(.7854*X2*VNU)
      ELEM(KP+2)=X3/(3.7*X2)
   43 WRITE (6,45)I,Q(I),X1,X2,X3
   45 FORMAT (I5,F18.3,F12.1,F12.3,F14.5)
      EN=EX-1.
      GO TO 26
   53 ELEM(KP)=X1
      WRITE (6,55) I,X1
   55 FORMAT (I5,' RESERVOIR ELEV DIFFERENCE=',F10.2)
      GO TO 26
   64 ELEM(KP)=X2
      ELEM(KP+3)=(X5-3.*(X4-X3)-X2)/(6.*X1**3)
      ELEM(KP+2)=(X4-2.*X3+X2)/(2.*X1**2)-ELEM(KP+3)*3.*X1
      ELEM(KP+1)=(X3-X2)/X1-ELEM(KP+2)*X1-ELEM(KP+3)*X1*X1
      WRITE (6,66) I,X1,X2,X3,X4,X5,(ELEM(KP+J-1),J=1,4)
   66 FORMAT (I5,' PUMP CURVE, DQ=',F7.3,' H=',4F8.1/5X,
     2' COEF IN PUMP EQ=',4F11.3)
      GO TO 26
C READ LOOP INDEXING DATA, IND=NO. PIPES,PIPE,PIPE,ETC. CLOCKWISE+,CC-
   68 I1=1
   70 I2=I1+14
      READ (5,75) NT,(IND(I),I=I1,I2)
   75 FORMAT (A2,2X,15I4)
      IF (NT.EQ.IE) GO TO 78
      I1=I2+1
      GO TO 70
   78 IF (I1.EQ.1) GO TO 140
      WRITE (6,79) (IND(I),I=1,I1)
   79 FORMAT (' IND='/(15I4))
```

```
C BALANCE ALL LOOPS
      DO 130 K=1,KK
      DDQ=0.
      IP=1
   80 I1=IND(IP)
      IF(I1.LE.0) GO TO 124
      DH=0.
      HDQ=0.
      DO 110 J=1,I1
      I=IND(IP+J)
      IF (I) 81,110,82
   81 S(J)=-1.
      I=-I
      GO TO 83
   82 S(J)=1.
   83 NTY=ITYPE(I)
      KP=4*(I-1)+1
      GO TO (91,92,103,104),NTY
   91 R=ELEM(KP)
      GO TO 95
   92 REY=ELEM(KP+1)*ABS(Q(I))
      IF(REY.LT.1.) REY=1.
      IF (REY-2000.) 93,94,94
   93 R=ELEM(KP)*64./REY
      GO TO 95
   94 R=ELEM(KP)*1.325/(ALOG(ELEM(KP+2)+5.74/REY**.9))**2
   95 DH=DH+S(J)*R*Q(I)*ABS(Q(I))**EN
      HDQ=HDQ+EX*R*ABS(Q(I))**EN
      GO TO 110
  103 DH=DH+S(J)*ELEM(KP)
      GO TO 110
  104 DH=DH+S(J)*(ELEM(KP)+Q(I)*(ELEM(KP+1)+Q(I)*(ELEM(KP+2)+Q(I)*
     2ELEM(KP+3))))
      HDQ=HDQ-(ELEM(KP+1)+2.*ELEM(KP+2)*Q(I)+3.*ELEM(KP+3)*Q(I)**2)
  110 CONTINUE
      IF (ABS(HDQ).LT..0001) HDQ=1.
      DQ=-DH/HDQ
      DDQ=DDQ+ABS(DQ)
      DO 120 J=1,I1
      I=IABS(IND(IP+J))
      IF (ITYPE(I).EQ.3) GO TO 120
      Q(I)=Q(I)+S(J)*DQ
  120 CONTINUE
      IP=IP+I1+1
      GO TO 80
  124 WRITE (6,125) K,DDQ
  125 FORMAT (' ITERATION NO.',I4,' SUM OF FLOW CORRECTIONS=',F10.4)
      IF (DDQ.LT.TOL) GO TO 140
  130 CONTINUE
  140 WRITE (6,141)
  141 FORMAT (' ELEMENT    FLOW')
      DO 150 I=1,100
      NTY=ITYPE(I)
      GO TO (142,142,150,142,150),NTY
  142 WRITE (6,143) I,Q(I)
  143 FORMAT (I5,F10.3)
  150 CONTINUE
C READ DATA FOR FGL COMPUTATION, IX=JUNC,ELEMENT,JUNC,ELEM,JUNC,ETC.
  152 READ (5,155) NT,K,HH
  155 FORMAT (A2,I8,F10.3)
      IF (NT.EQ.IE) GO TO 160
      H(K)=HH
      GO TO 152
  160 I1=1
  162 I2=I1+14
      READ (5,75)NT,(IX(K),K=I1,I2)
      IF (NT.EQ.IE) GO TO 170
      I1=I2+1
      GO TO 162
  170 WRITE (6,171) (IX(I),I=1,I1)
  171 FORMAT (' IX='/(15I4))
      IP=1
```

```
180 DO 200 J=1,238,2
    IF (J.EQ.1) I1=IX(IP)
    I=IX(IP+J)
    N=IX(IP+J+1)
    IF(I) 181,199,182
181 SS=-1.
    I=-I
    GO TO 183
182 SS=1.
183 NTY=ITYPE(I)
    KP=4*(I-1)+1
    GO TO (184,185,189,190,199),NTY
184 R=ELEM(KP)
    GO TO 188
185 REY=ELEM(KP+1)*ABS(Q(I))
    IF (REY.LT.1.) REY=1.
    IF (REY-2000.) 186,187,187
186 R=ELEM(KP)*64./REY
    GO TO 188
187 R=ELEM(KP)*1.325/(ALOG(ELEM(KP+2)+5.74/REY**.9))**2
188 H(N)=H(I1)-SS*R*Q(I)*ABS(Q(I))**EN
    GO TO 199
189 H(N)=H(I1)-SS*ELEM(KP)
    GO TO 199
190 H(N)=H(I1)+SS*(ELEM(KP)+Q(I)*(ELEM(KP+1)+Q(I)*(ELEM(KP+2)+Q(I)*
    2ELEM(KP+3))))
199 IF (IX(J+IP+3).EQ.0) GO TO 210
    IF (IX(J+IP+2).EQ.0) GO TO 205
200 I1=N
205 IP=IP+J+3
    GO TO 180
210 WRITE (6,215)
215 FORMAT (' JUNCTION  HEAD')
    DO 220 N=1,100
    IF (H(N).EQ.-1000.) GO TO 220
    WRITE (6,143) N,H(N)
220 CONTINUE
    GO TO 10
99  STOP
    END
```

Figure 10.15 FORTRAN program for hydraulic systems.

indexing of the system loops for balancing is placed in the vector IND. A series of integer values identifies each loop sequentially by the number of elements in the loop followed by the element number of each element in the loop. The directional sense of flow in each element is identified by a positive element number for the clockwise direction and a negative element number for counterclockwise. The second part of the program requires an identification of one or more junctions with known heads. Then a series of junction and element numbers indexes a continuous path through the system to all junctions where the hydraulic grade line is wanted. The path may be broken at any point by an integer zero followed by a new junction where the head is known. These data are stored in the vector IX by a junction number where the head is known followed by a contiguous element number and junction number. Again the positive element number is used in the assigned flow direction, and the negative element number is used when tracing a path against the assigned element flow direction. Any continuous path may be broken by inserting a zero; then a new path is begun with a new initial junction, an element, and a node, etc. All junction hydraulic-grade-line elevations that are computed are printed.

As shown below, the type of each element is identified in the input data, and each element is identified in the program by the assignment of a unique numerical value in the vector ITYPE.

Element	Data	Program
Hazen-Williams pipeline	HW	1
Darcy-Weisbach pipeline	DW	2
Pseudo element	PS	3
Pump	PU	4

The physical data associated with each element are entered on separate cards. In the program the physical data that describe all elements in the system are stored in the vector ELEM, with five locations reserved for each element. As an example of the position of storage of element information, the data pertaining to element number 13 are located in positions 61 to 65 in ELEM.

Data preparation for the program is best visualized in four steps, as shown in Fig. 10.16 and described below. Formated input is used, as shown in Fig. 10.16 and the program.

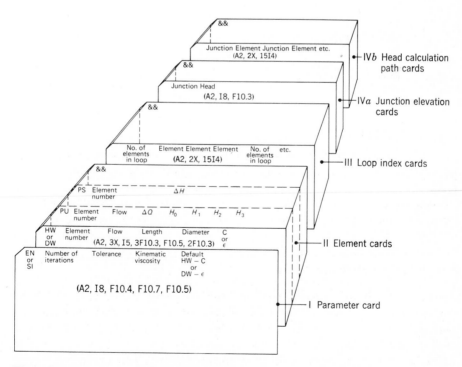

Figure 10.16 Data cards for Hardy Cross program.

Step 1: Parameter Description Card

The type of unit to be used in the analysis is defined by the characters EN for English, or U.S. customary, units or SI for the SI units. An integer defines the maximum number of iterations to be allowed during the balancing scheme. An acceptable tolerance is set for the sum of the absolute values of the corrections in each loop during each iteration. The liquid kinematic viscosity must be specified if the Darcy-Weisbach equation is used for pipeline losses. If the Hazen-Williams equation is used, a default value for the coefficient C may be defined, or if the Darcy-Weisbach equation is used, a default value for absolute pipe roughness may be defined. If the default value is used on the parameter card, it need not be placed on the element cards; however, if it is, the element data override the default value.

Step 2: Element Cards

Each element in the system requires a separate card. Pipeline elements require either HW or DW to indicate the equation for the problem solution, the element number, the estimated flow, the length, the inside diameter, and (if the default value is not used) either the Hazen-Williams coefficient or the pipe roughness for the Darcy-Weisbach equation. Pump elements require PU to indicate the element type, the element number, the estimated flow, a flow increment ΔQ at which values of pump head are specified, and four values of head from the pump-characteristic curve beginning at shutoff head and at equal flow intervals of ΔQ. The pseudo element for the pseudo loop requires PS to indicate the type, the element number, a zero or blank for the flow, and a difference in elevation between the interconnected fixed-pressure-head levels with head drop positive. The end of the element data is indicated by a card with "&&" in the first two columns.

Step 3: Loop Index Cards

These data are supplied with 15 integer numbers per card in the following order: the number of elements in a loop (maximum of 20) followed by the element number of each element in the loop with a negative sign to indicate counterclockwise flow direction. This information is repeated until all loops are defined. The end of step 3 data is indicated by a card with "&&" in columns 1 and 2.

Step 4: Head Calculation Cards

Junctions with fixed elevations are identified on separate cards by giving the junction number and the hydraulic-grade-line elevation. There must be one or more of these cards followed by a card with "&&" in the first two columns to indicate the end of this type of data.

The path to be followed in computing the hydraulic-grade-line elevations is specified by supplying 15 integer values per card in the following order: junction

```
SI        30      .001      .000001      100.
HW         1      .12        600.          .3
HW         2      .03        300.          .15
HW         3      .0         500.          .6
HW         4      .03        400.          .3
HW         5      .03        300.          .3
PS         6                  15.
PS         7                  18.
PU         8      .06         .03        30.        29.        26.        20.
&&
           3   2   1  -3   3   4  -5   3   3   6  -4  -1   3   5   7
           8
&&
           5    117.
&&
           5   8   4   2   2   1   1   4   3
&&
SI UNITS SPECIFIED, VISCOSITY IN M**2/SEC= 0.0000010
DESIRED FLOW TOLERANCE=0.001 NO. OF ITERATIONS=    30
PIPE       Q(CFS OR M**3/S)   L(FT OR M)   D(FT OR M)   HW C OR EPS
   1             0.120          600.0        0.300       100.00000
   2             0.030          300.0        0.150       100.00000
   3             0.000          500.0        0.600       100.00000
   4             0.030          400.0        0.300       100.00000
   5             0.030          300.0        0.300       100.00000
   6 RESERVOIR ELEV DIFFERENCE=       15.00
   7 RESERVOIR ELEV DIFFERENCE=       18.00
   8 PUMP CURVE, DQ= 0.030  H=     30.0      29.0      26.0      20.0
     COEF IN PUMP EQ=      30.000    -11.111    -555.556   -6172.836
IND=
   3   2   1  -3   3   4  -5   3   3   6  -4  -1   3   5   7
   8   0   0   0   0   0   0   0   0   0   0   0   0   0   0
   0
ITERATION NO.   1 SUM OF FLOW CORRECTIONS=      0.1385
ITERATION NO.   2 SUM OF FLOW CORRECTIONS=      0.1040
ITERATION NO.   3 SUM OF FLOW CORRECTIONS=      0.0372
ITERATION NO.   4 SUM OF FLOW CORRECTIONS=      0.0034
ITERATION NO.   5 SUM OF FLOW CORRECTIONS=      0.0006
ELEMENT    FLOW
   1       0.143
   2      -0.034
   3       0.027
   4       0.080
   5       0.094
   8       0.087
 IX=
   5   8   4   2   2   1   1   4   3   0   0   0   0   0   0
   0
JUNCTION  HEAD
   1      137.811
   2      150.044
   3      135.044
   4      137.797
   5      117.000
```

Figure 10.17 Program input and output for Example 10.9.

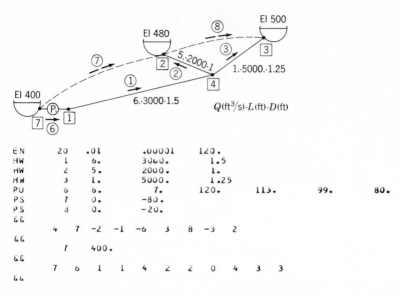

```
EN      20     .01        .00001     120.
HW       1     6.         3000.        1.5
HW       2     5.         2000.        1.
HW       3     1.         5000.        1.25
PU       6     6.            7.       120.      113.      99.     80.
PS       7     0.          -80.
PS       3     0.          -20.
&&
         4     7    -2   -1   -6    3    8   -3    2
&&
         1    400.
&&
         7     6     1    1    4    2    2    0    4    3    3
&&
```

Figure 10.18 Input data for branching-pipe system in U.S. customary units with Hazen-Williams formula.

number where the head is known, element number (with a negative sign to indicate a path opposite to the assumed flow direction), junction number, etc. If one wants a new path to begin at a junction different from the last listed junction, a single zero is added, followed by a junction where the head is known, element number, junction number, etc. The end of step 4 data is indicated by a card with "&&" in columns 1 and 2.

Example 10.9 The program in Fig. 10.15 is used to solve the network problem displayed in Fig. 10.14. The pump data are as follows:

Q, m^3/s	0	0.03	0.06	0.09
H, m	30	29	26	20

The Hazen-Williams pipeline coefficient for all pipes is 100. Figure 10.17 displays the input data and the computer output for this problem.

Figures 10.18 to 10.20 give input data for three systems which can be solved with this program.

10.9 CONDUITS WITH NONCIRCULAR CROSS SECTIONS

In this chapter so far, only circular pipes have been considered. For cross sections that are noncircular, the Darcy-Weisbach equation may be applied if the term D can be interpreted in terms of the section. The concept of the *hydraulic radius R*

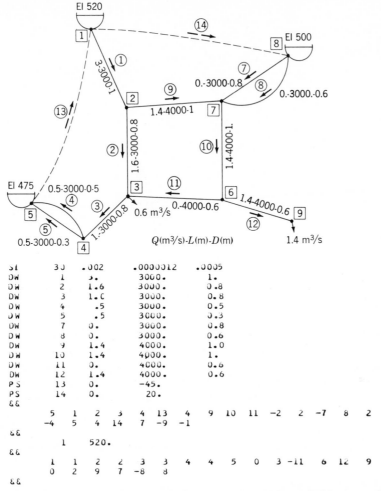

Figure 10.19 Input data for hydraulic system. SI units and Darcy-Weisbach equation.

permits circular and noncircular sections to be treated in the same manner. The hydraulic radius is defined as the cross-sectional area divided by the *wetted perimeter*. Hence, for a circular section,

$$R = \frac{\text{area}}{\text{perimeter}} = \frac{\pi D^2/4}{\pi D} = \frac{D}{4} \qquad (10.9.1)$$

and the diameter is equivalent to $4R$. Assuming that the diameter may be replaced by $4R$ in the Darcy-Weisbach equation, in the Reynolds number, and in the relative roughness,

$$h_f = f\,\frac{L}{4R}\,\frac{V^2}{2g} \qquad R = \frac{V4R\rho}{\mu} \qquad \frac{\epsilon}{D} = \frac{\epsilon}{4R} \qquad (10.9.2)$$

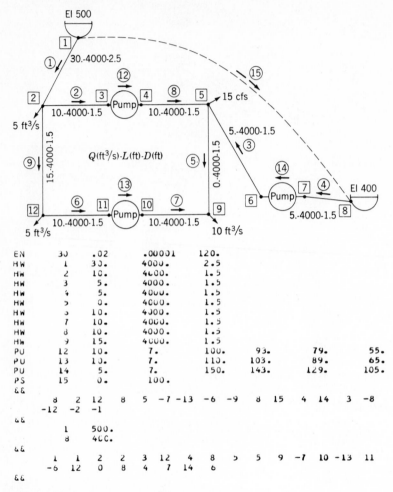

Figure 10.20 Input for booster-pump system.

Noncircular sections may be handled in a similar manner. The Moody diagram applies as before. The assumptions in Eqs. (10.9.2) cannot be expected to hold for odd-shaped sections but should give reasonable values for square, oval, triangular, and similar types of sections.

Example 10.10 Determine the head loss, in millimeters of water, required for flow of 300 m³/min of air at 20°C and 100 kPa through a rectangular galvanized-iron section 700 mm wide, 350 mm high, and 70 m long.

$$R = \frac{A}{P} = \frac{0.7 \times 0.35}{2(0.7 + 0.35)} = 0.117 \text{ m} \qquad \frac{\epsilon}{4R} = \frac{0.00015}{4 \times 0.117} = 0.00032$$

$$V = \frac{300}{60(0.7)(0.35)} = 20.41 \text{ m/s} \qquad \mu = 2.2 \times 10^{-5} \text{ N·s/m}^2$$

$$\rho = \frac{p}{R'T} = \frac{100{,}000}{287(273 + 20)} = 1.189 \text{ kg/m}^3$$

$$\mathbf{R} = \frac{VD\rho}{\mu} = \frac{V4R\rho}{\mu} = \frac{20.41 \times 4 \times 0.117 \times 1.189}{2.2 \times 10^{-5}} = 516{,}200$$

From Fig. 5.32, $f = 0.0165$

$$h_f = f \frac{L}{4R} \frac{V^2}{2g} = 0.0165 \frac{70}{4 \times 0.117} \frac{\overline{20.41^2}}{2 \times 9.806} = 52.42 \text{ m}$$

The specific weight of air is $\rho g = 1.189 \times 9.806 = 11.66 \text{ N/m}^3$. In millimeters of water,

$$\frac{52.42 \times 11.66 \times 1000}{9806} = 62.33 \text{ mm}$$

10.10 AGING OF PIPES

The Moody diagram, with the values of absolute roughness shown there, is for new, clean pipe. With use, pipes become rougher, owing to corrosion, incrustations, and deposition of material on the pipe walls. The speed with which the friction factor changes with time depends greatly on the fluid being handled. Colebrook and White[†] found that the absolute roughness ϵ increases linearly with time,

$$\epsilon = \epsilon_0 + \alpha t \tag{10.10.1}$$

in which ϵ_0 is the absolute roughness of the new surface. Tests on a pipe are required to determine α.

The time variation of the Hazen-Williams coefficient has been summarized graphically[‡] for water-distribution systems in seven major U.S. cities. Although it is not a linear variation, the range of values for the average rate of decline in C may typically be between 0.5 and 2 per year, with the larger values generally applicable in the first years following installation. The only sure way that accurate coefficients can be obtained for older water mains is through field tests.

[†] C. F. Colebrook and C. M. White, The Reduction of Carrying Capacity of Pipes with Age, *J. Inst. Civ. Eng. Lond.*, 1937.

[‡] W. D. Hudson, Computerized Pipeline Design, *Transp. Eng. J. ASCE*, vol. 99, no. TE1, 1973.

PROBLEMS

10.1 Sketch the hydraulic and energy grade lines for Fig. 10.21. $H = 8$ m.

10.2 Calculate the value of K for the valve of Fig. 10.21 so that the discharge of Prob. 10.1 is reduced by one-half. Sketch the hydraulic and energy grade lines.

10.3 Compute the discharge of the system in Fig. 10.22. Draw the hydraulic and energy grade lines.

10.4 What head is needed in Fig. 10.22 to produce a discharge of 0.3 m³/s?

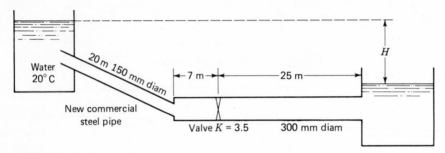

Figure 10.21 Problems 10.1, 10.2.

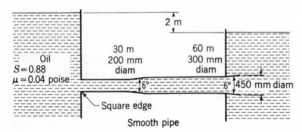

Figure 10.22 Problems 10.3, 10.4.

10.5 Calculate the discharge through the siphon of Fig. 10.23 with the conical diffuser removed. $H = 4$ ft.

10.6 Calculate the discharge in the siphon of Fig. 10.23 for $H = 8$ ft. What is the minimum pressure in the system?

10.7 Find the discharge through the siphon of Fig. 10.24. What is the pressure at A? Estimate the minimum pressure in the system.

10.8 Neglecting minor losses other than the valve, sketch the hydraulic grade line for Fig. 10.25. The globe valve has a loss coefficient $K = 4.5$.

10.9 What is the maximum height of point A (Fig. 10.25) for no cavitation? Barometer reading is 29.5 in Hg.

10.10 Two reservoirs are connected by three clean cast-iron pipes in series; $L_1 = 300$ m, $D_1 = 200$ mm; $L_2 = 360$ m, $D_2 = 300$ mm; $L_3 = 1200$ m, $D_3 = 450$ mm. When $Q = 0.1$ m³/s water at 20°C, determine the difference in elevation of the reservoirs.

10.11 Solve Prob. 10.10 by the method of equivalent lengths.

10.12 For a difference in elevation of 10 m in Prob. 10.10, find the discharge by use of the Hazen-Williams equation.

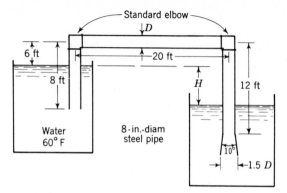

Figure 10.23 Problems 10.5, 10.6.

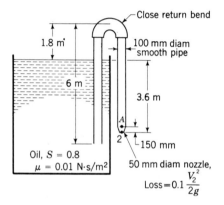

Figure 10.24 Problem 10.7.

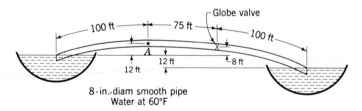

Figure 10.25 Problems 10.8, 10.9.

10.13 Determine the discrepancy between the head loss calculated with the Darcy-Weisbach equation and with the Hazen-Williams equation for 15°C water flowing in a 1-m-diameter welded steel pipeline. $C = 120$, $\epsilon = 0.2$ mm. Graph the discrepancy as a function of Reynolds number, $10^4 < \mathbf{R} < 10^7$.

10.14 What diameter smooth pipe is required to convey 8 l/s kerosene at 32°C 150 m with a head of 5 m? There are a valve and other minor losses with total K of 7.6.

10.15 Air at atmospheric pressure and 60°F is carried through two horizontal pipes ($\epsilon = 0.06$) in series.

The upstream pipe is 400 ft of 24 in diameter, and the downstream pipe is 100 ft of 36 in diameter. Estimate the equivalent length of 18-in ($\epsilon = 0.003$) pipe. Neglect minor losses.

10.16 What pressure drop, in inches of water, is required for flow of 6000 cfm in Prob. 10.15? Include losses due to sudden expansion.

10.17 Two pipes are connected in parallel between two reservoirs; $L_1 = 2500$ m, $D_1 = 1.2$-m-diameter old cast-iron pipe, $C = 100$; $L_2 = 2500$ m, $D_2 = 1$ m, $C = 90$. For a difference in elevation of 3.6 m determine the total flow of water at 20°C.

10.18 For 4.5 m³/s flow in the system of Prob. 10.17, determine the difference in elevation of reservoir surfaces.

10.19 Three smooth tubes are connected in parallel: $L_1 = 40$ ft, $D_1 = \frac{1}{2}$ in; $L_2 = 60$ ft, $D_2 = 1$ in; $L_3 = 50$ ft, $D_3 = \frac{3}{4}$ in. For total flow of 30 gpm oil, $\gamma = 55$ lb/ft³, $\mu = 0.65$ P, what is the drop in hydraulic grade line between junctions?

10.20 Determine the discharge of the system of Fig. 10.26 for $L = 600$ m, $D = 500$ mm, $\epsilon = 0.5$ mm, and $H = 8$ m, with the pump A characteristics given.

10.21 Determine the discharge through the system of Fig. 10.26 for $L = 4000$ ft, $D = 24$-in smooth pipe, $H = 40$ ft, with pump B characteristics.

10.22 Construct a head-discharge-efficiency table for pumps A and B (Fig. 10.26) connected in series. (SI units).

10.23 Construct a head–discharge-efficiency table for pumps A and B (Fig. 10.26) connected in parallel. (U.S. customary units.)

10.24 Find the discharge through the system of Fig. 10.26 for pumps A and B in series; 1600 m of 300-mm clean cast-iron pipe, $H = 30$ m.

10.25 Determine the power needed to drive pumps A and B in Prob. 10.24.

10.26 Find the discharge through the system of Fig. 10.26 for pumps A and B in parallel; 2000 m of 500-mm steel pipe, $H = 10$ m.

10.27 Determine the power needed to drive the pumps in Prob. 10.26.

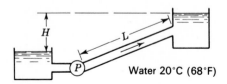

Water 20°C (68°F)

Pump A

Hp		Qp		e
m	ft	l/s	cfs	%
21.3	70	0	0	0
18.3	60	56.6	2.00	59
16.8	55	72.5	2.56	70
15.2	50	85.8	3.03	76
13.7	45	97.7	3.45	78
12.2	40	108	3.82	76.3
10.7	35	116	4.11	72
9.1	30	127	4.48	65
7.6	25	130	4.59	56.5
6.1	20	134	4.73	42

Pump B

Hp		Qp		e
m	ft	l/s	cfs	%
24.4	80	0	0	0
21.3	70	74	2.60	54
18.3	60	112	3.94	70
15.2	50	140	4.96	80
12.2	40	161	5.70	73
9.1	30	174	6.14	60
6.1	20	177	6.24	40

Figure 10.26 Problems 10.20 to 10.27, 10.39.

10.28 For $H = 12$ m in Fig. 10.27, find the discharge through each pipe, $\mu = 8$ cP; sp gr $= 0.9$.

10.29 Find H in Fig. 10.27 for 0.03 m³/s flowing. $\mu = 5$ cP; sp gr $= 0.9$.

10.30 Find the equivalent length of 300-mm-diameter clean cast-iron pipe to replace the system of Fig. 10.28. For $H = 10$ m, what is the discharge?

10.31 With velocity of 1 m/s in the 200-mm-diameter pipe of Fig. 10.28 calculate the flow through the system and the head H required.

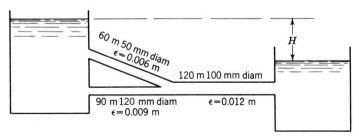

60 m 50 mm diam
$\epsilon = 0.006$ m

120 m 100 mm diam

90 m 120 mm diam $\epsilon = 0.012$ m
$\epsilon = 0.009$ m

H

Figure 10.27 Problems 10.28, 10.29.

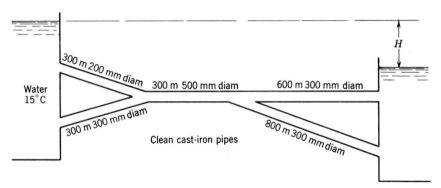

Water 15°C

300 m 200 mm diam   300 m 500 mm diam   600 m 300 mm diam

300 m 300 mm diam

Clean cast-iron pipes

800 m 300 mm diam

H

Figure 10.28 Problems 10.30, 10.31.

10.32 In Fig. 10.29 find the flow through the system when the pump is removed.

10.33 If the pump of Fig. 10.29 is delivering 80 l/s toward J, find the flow into A and B and the elevation of the hydraulic grade line at J.

10.34 The pump is adding 7500 W fluid power to the flow (toward J) in Fig. 10.29. Find Q_A and Q_B.

10.35 With pump A of Fig. 10.26 in the system of Fig. 10.29, find Q_A, Q_B, and the elevation of the hydraulic grade line at J.

10.36 With pump B of Fig. 10.26 in the system of Fig. 10.29, find the flow into B and the elevation of the hydraulic grade line at J.

10.37 For flow of 30 l/s into B of Fig. 10.29, what head is produced by the pump? For pump efficiency of 70 percent, how much power is required?

10.38 Find the flow through the system of Fig. 10.30 for no pump in the system.

10.39 (a) With pumps A and B of Fig. 10.26 in parallel in the system of Fig. 10.30, find the flow into B, C, and D and the elevation of the hydraulic grade line at J_1 and J_2. (b) Assume all the pipes in Fig. 10.30 are cast iron. Prepare the data for solution to this problem by use of the program in Fig. 10.15.

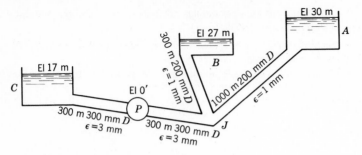

Figure 10.29 Problems 10.32 to 10.37.

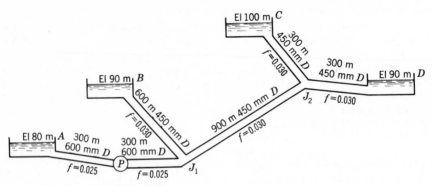

Figure 10.30 Problems 10.38, 10.39.

10.40 Calculate the flow through each of the pipes of the network shown in Fig. 10.31. $n = 2$.

10.41 Determine the flow through each line of Fig. 10.32. $n = 2$.

10.42 By use of the program in Fig. 10.15, solve Prob. 10.35.

10.43 By use of the program in Fig. 10.15 solve the problems given in (*a*) Fig. 10.18, (*b*) Fig. 10.19, (*c*) Fig. 10.20.

10.44 Determine the slope of the hydraulic grade line for flow of atmospheric air at 80°F through a rectangular 18 by 6 in galvanized-iron conduit. $V = 30$ ft/s.

10.45 What size square conduit is needed to convey 300 l/s water at 15°C with slope of hydraulic grade line of 0.001? $\epsilon = 1$ mm.

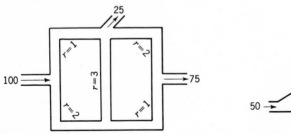

Figure 10.31 Problem 10.40.

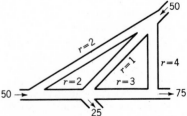

Figure 10.32 Problem 10.41.

10.46 Calculate the discharge of oil, $S = 0.85$, $\mu = 4$ cP, through 30 m of 50 by 120 mm sheet-metal conduit when the head loss is 600 mm. $\epsilon = 0.00015$ m.

10.47 A duct, with cross section an equilateral triangle 1 ft on a side, conveys 6 cfs water at 60°F. $\epsilon = 0.003$ ft. Calculate the slope of the hydraulic grade line.

10.48 A clean 700-mm-diameter cast-iron water pipe has its absolute roughness doubled in 5 years of service. Estimate the head loss per 1000 m for a flow of 400 l/s when the pipe is 25 years old.

10.49 An 18-in-diameter pipe has an f of 0.020 when new for 5 ft/s water flow of 60°F. In 10 years $f = 0.029$ for $V = 3$ ft/s. Find f for 4 ft/s at end of 20 years.

10.50 The hydraulic grade line in a system is
  (a) Always above the energy grade line
  (b) Always above the closed conduit
  (c) Always sloping downward in the direction of flow
  (d) The velocity head below the energy grade line
  (e) Upward in direction of flow when pipe is inclined downward

10.51 In solving a series-pipe problem for discharge, the energy equation is used along with the continuity equation to obtain an expression that contains a $V^2/2g$ and f_1, f_2, etc. The next step in the solution is to assume (a) Q; (b) V; (c) $\mathbf{R}$; (d) $f_1, f_2, \ldots$; (e) none of these quantities.

10.52 One pipe system is said to be equivalent to another pipe system when the following two quantities are the same: (a) h, Q; (b) L, Q; (c) L, D; (d) f, D; (e) V, D.

10.53 In parallel-pipe problems
  (a) The head losses through each pipe are added to obtain the total head loss.
  (b) The discharge is the same through all the pipes.
  (c) The head loss is the same through each pipe.
  (d) A direct solution gives the flow through each pipe when the total flow is known.
  (e) A trial solution is not needed.

10.54 Branching-pipe problems are solved
  (a) Analytically by using as many equations as unknowns
  (b) By assuming the head loss is the same through each pipe
  (c) By equivalent lengths
  (d) By assuming a distribution which satisfies continuity and computing a correction
  (e) By assuming the elevation of hydraulic grade line at the junction point and trying to satisfy continuity

10.55 In networks of pipes
  (a) The head loss around each elementary circuit must be zero.
  (b) The (horsepower) loss in all circuits is the same.
  (c) The elevation of hydraulic grade line is assumed for each junction.
  (d) Elementary circuits are replaced by equivalent pipes.
  (e) Friction factors are assumed for each pipe.

10.56 The following quantities are computed by using $4R$ in place of diameter for noncircular sections:
  (a) Velocity, relative roughness
  (b) Velocity, head loss
  (c) Reynolds number, relative roughness, head loss
  (d) Velocity, Reynolds number, friction factor
  (e) None of these answers

10.57 Experiments show that in the aging of pipes
  (a) The friction factor increases linearly with time.
  (b) A pipe becomes smoother with use.
  (c) The absolute roughness increases linearly with time.
  (d) No appreciable trends can be found.
  (e) The absolute roughness decreases with time.

ELEVEN

STEADY FLOW IN OPEN CHANNELS

A broad coverage of topics in open-channel flow has been selected for this chapter. Steady uniform flow was discussed in Sec. 5.9, and application of the momentum equation to the hydraulic jump in Sec. 3.11. Weirs were introduced in Sec. 8.6. In this chapter open-channel flow is first classified and then the *shape* of optimum canal cross sections is discussed, followed by a section on flow through a floodway. The hydraulic jump and its application to stilling basins is then treated, followed by a discussion of specific energy and critical depth which leads into gradually varied flow. Water-surface profiles are classified and related to channel control sections. Transitions are next discussed, with one special application to the critical-depth meter.

The mechanics of flow in open channels is more complicated than closed-conduit flow owing to the presence of a free surface. The hydraulic grade line coincides with the free surface, and, in general, its position is unknown.

For laminar flow to occur, the cross section must be extremely small, the velocity very small, or the kinematic viscosity extremely high. One example of laminar flow is given by a thin film of liquid flowing down an inclined or vertical plane. This case is treated by the methods developed in Chap. 5 (see Prob. 5.10). Pipe flow has a lower critical Reynolds number of 2000, and this same value may be applied to an open channel when the diameter D is replaced by $4R$. R is the hydraulic radius, which is defined as the cross-sectional area of the channel divided by the wetted perimeter. In the range of Reynolds number, based on R in place of D, $\mathbf{R} = VR/\nu < 500$ flow is laminar, $500 < \mathbf{R} < 2000$ flow is *transitional* and may be either laminar or turbulent, and $\mathbf{R} > 2000$ flow is generally turbulent.

Most open-channel flows are turbulent, usually with water as the liquid. The methods for analyzing open-channel flow are not developed to the extent of those for closed conduits. The equations in use assume complete turbulence, with the head loss proportional to the square of the velocity. Although practically all data on open-channel flow have been obtained from experiments on the flow of water, the equations should yield reasonable values for other liquids of low viscosity. The material in this chapter applies to turbulent flow only.

11.1 CLASSIFICATION OF FLOW

Open-channel flow occurs in a large variety of forms, from flow of water over the surface of a plowed field during a hard rain to the flow at constant depth through a large prismatic channel. It may be classified as steady or unsteady, uniform or nonuniform. *Steady uniform flow* occurs in very long inclined channels of constant cross section in those regions where *terminal velocity* has been reached, i.e., where the head loss due to turbulent flow is exactly supplied by the reduction in potential energy due to the uniform decrease in elevation of the bottom of the channel. The depth for steady uniform flow is called the *normal depth*. In steady uniform flow the discharge is constant and the depth is everywhere constant along the length of the channel. Several equations are in common use for determining the relations between the average velocity, the shape of the cross section, its size and roughness, and the slope, or inclination, of the channel bottom (Sec. 5.9).

Steady nonuniform flow occurs in any irregular channel in which the discharge does not change with the time; it also occurs in regular channels when the flow depth and hence the average velocity change from one cross section to another. For gradual changes in depth or section, called *gradually varied flow*, methods are available, by numerical integration or step-by-step means, for computing flow depths for known discharge, channel dimensions and roughness, and given conditions at one cross section. For those reaches of a channel where pronounced changes in velocity and depth occur in a short distance, as in a transition from one cross section to another, model studies are frequently made. The *hydraulic jump* is one example of steady nonuniform flow; it is discussed in Secs. 3.11 and 11.4.

Unsteady uniform flow rarely occurs in open-channel flow. *Unsteady nonuniform flow* is common but is difficult to analyze. Wave motion is an example of this type of flow, and its analysis is complex when friction is taken into account. Positive and negative *surge waves* in a rectangular channel are analyzed, neglecting effects of friction, in Secs. 12.9 and 12.10.

Flow is also classified as *tranquil* or *rapid*. When flow occurs at low velocities so that a small disturbance can travel upstream and thus change upstream conditions, it is said to be tranquil flow† ($\mathbf{F} < 1$). Conditions upstream are affected by downstream conditions, and the flow is controlled by the downstream conditions.

---

† See Sec. 4.4 for definition and discussion of the Froude number $\mathbf{F}$.

When flow occurs at such high velocities that a small disturbance, such as an elementary wave is swept downstream, the flow is described as *shooting* or *rapid* ($\mathbf{F} > 1$). Small changes in downstream conditions do not effect any change in upstream conditions; hence, the flow is controlled by upstream conditions. When flow is such that its velocity is just equal to the velocity of an elementary wave, the flow is said to be critical ($\mathbf{F} = 1$).

The terms "subcritical" and "supercritical" are also used to classify flow velocities. *Subcritical* refers to tranquil flow at velocities less than critical, and *supercritical* corresponds to rapid flows when velocities are greater than critical.

Velocity Distribution

The velocity at a solid boundary must be zero, and in open-channel flow it generally increases with distance from the boundaries. The maximum velocity does not occur at the free surface but is usually below the free surface a distance of 0.05 to 0.25 of the depth. The average velocity along a vertical line is sometimes determined by measuring the velocity at 0.6 of the depth, but a more reliable method is to take the average of the velocities at 0.2 and 0.8 of the depth, according to measurements of the U.S. Geological Survey.

EXERCISES

11.1.1 In open-channel flow (a) the hydraulic grade line is always parallel to the energy grade line; (b) the energy grade line coincides with the free surface; (c) the energy and hydraulic grade lines coincide; (d) the hydraulic grade line can never rise; (e) the hydraulic grade line and free surface coincide.

11.1.2 Tranquil flow must always occur (a) above normal depth; (b) below normal depth; (c) above critical depth; (d) below critical depth; (e) on adverse slopes.

11.2 BEST HYDRAULIC CHANNEL CROSS SECTIONS

Some channel cross sections are more efficient than others in that they provide more area for a given wetted perimeter. In general, when a channel is constructed, the excavation, and possibly the lining, must be paid for. From the Manning formula it is shown that when the area of cross section is a minimum, the wetted perimeter is also a minimum, and so both lining and excavation approach their minimum value for the same dimensions of channel. The *best hydraulic section* is one that has the least wetted perimeter or its equivalent, the least area for the type of section. The Manning formula is

$$Q = \frac{C_m}{n} A R^{2/3} S^{1/2} \tag{11.2.1}$$

in which Q is the discharge (L^3/T), A the cross-sectional flow area, R (area divided by wetted perimeter P) the hydraulic radius, S the slope of energy grade line, n the

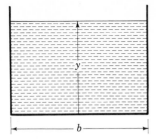

Figure 11.1 Rectangular cross section.

Manning roughness factor (Table 5.2), C_m an empirical constant $(L^{1/3}/T)$ equal to 1.49 in U.S. customary units and to 1.0 in SI units. With Q, n, and S known, Eq. (11.2.1) may be written

$$A = cP^{2/5} \qquad (11.2.2)$$

in which c is known. This equation shows that P is a minimum when A is a minimum. To find the best hydraulic section for a *rectangular* channel (Fig. 11.1) $P = b + 2y$ and $A = by$. Then

$$A = (P - 2y)y = cP^{2/5}$$

by elimination of b. The value of y is sought for which P is a minimum. Differentiating with respect to y gives

$$\left(\frac{dP}{dy} - 2\right)y + P - 2y = \frac{2}{5}cP^{-3/5}\,\frac{dP}{dy}$$

Setting $dP/dy = 0$ gives $P = 4y$, or since $P = b + 2y$,

$$b = 2y \qquad (11.2.3)$$

Therefore, the depth is one-half the bottom width, independent of the size of rectangular section.

To find the best hydraulic *trapezoidal* section (Fig. 11.2) $A = by + my^2$, $P = b + 2y\sqrt{1 + m^2}$. After eliminating b and A in these equations and Eq. (11.2.2),

$$A = by + my^2 = (P - 2y\sqrt{1 + m^2})y + my^2 = cP^{2/5} \qquad (11.2.4)$$

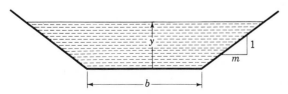

Figure 11.2 Trapezoidal cross section.

By holding m constant and by differentiating with respect to y, $\partial P/\partial y$ is set equal to zero; thus

$$P = 4y\sqrt{1 + m^2} - 2my \qquad (11.2.5)$$

Again, by holding y constant, Eq. (11.2.4) is differentiated with respect to m, and $\partial P/\partial m$ is set equal to zero, producing

$$\frac{2m}{\sqrt{1 + m^2}} = 1$$

After solving for m,

$$m = \frac{\sqrt{3}}{3}$$

and after substituting for m in Eq. (11.2.5),

$$P = 2\sqrt{3}\,y \qquad b = 2\frac{\sqrt{3}}{3}\,y \qquad A = \sqrt{3}\,y^2 \qquad (11.2.6)$$

which shows that $b = P/3$ and hence the sloping sides have the same length as the bottom. As $\tan^{-1} m = 30°$, the best hydraulic section is one-half a hexagon. For trapezoidal sections with m specified (maximum slope at which wet earth will stand) Eq. (11.2.5) is used to find the best bottom-width-to-depth ratio.

The semicircle is the best hydraulic section of all possible open-channel cross sections.

Example 11.1 Determine the dimensions of the most economical trapezoidal brick-lined channel to carry 200 m³/s with a slope of 0.0004.
With Eq. (11.2.6),

$$R = \frac{A}{P} = \frac{y}{2}$$

and by substituting into Eq. (11.2.1),

$$200 = \frac{1.00}{0.016}\sqrt{3}\,y^2\left(\frac{y}{2}\right)^{2/3}\sqrt{0.0004}$$

or $\qquad\qquad y^{8/3} = 146.64 \qquad y = 6.492$ m

and from Eq. (11.2.6), $b = 7.5$ m.

EXERCISES

11.2.1 The best hydraulic rectangular cross section occurs when ($b =$ bottom width, $y =$ depth) (a) $y = 2b$; (b) $y = b$; (c) $y = b/2$; (d) $y = b^2$; (e) $y = b/5$.

11.2.2 The best hydraulic canal cross section is defined as (a) the least expensive canal cross section; (b) the section with minimum roughness coefficient; (c) the section that has a maximum area for a given flow; (d) the one that has a minimum perimeter; (e) none of these answers.

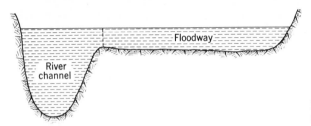

Figure 11.3 Floodway cross section.

11.3 STEADY UNIFORM FLOW IN A FLOODWAY

A practical open-channel problem of importance is the computation of discharge through a floodway (Fig. 11.3). In general, the floodway is much rougher than the river channel and its depth (and hydraulic radius) is much less. The slope of energy grade line must be the same for both portions. The discharge for each portion is determined separately, using the dashed line of Fig. 11.3 as the separation line for the two sections (but not as solid boundary), and then the discharges are added to determine the total capacity of the system.

Since both portions have the same slope, the discharge may be expressed as

$$Q_1 = K_1\sqrt{S} \qquad Q_2 = K_2\sqrt{S}$$

or
$$Q = (K_1 + K_2)\sqrt{S} \qquad (11.3.1)$$

in which the value of K is

$$K = \frac{C_m}{n} AR^{2/3}$$

from Manning's formula and is a function of depth only for a given channel with fixed roughness. By computing K_1 and K_2 for different elevations of water surface, their sum may be taken and plotted against elevation. From this plot it is easy to determine the slope of energy grade line for a given depth and discharge from Eq. (11.3.1).

11.4 HYDRAULIC JUMP; STILLING BASINS

The relations among the variables V_1, y_1, V_2, y_2 for a hydraulic jump to occur in a horizontal rectangular channel are developed in Sec. 3.11. Another way of determining the conjugate depths for a given discharge is the $F + M$ method. The momentum equation applied to the free body of liquid between y_1 and y_2 (Fig. 11.4) is, for unit width ($V_1 y_1 = V_2 y_2 = q$),

$$\frac{\gamma y_1^2}{2} - \frac{\gamma y_2^2}{2} = \rho q(V_2 - V_1) = \rho V_2^2 y_2 - \rho V_1^2 y_1$$

Figure 11.4 Hydraulic jump in horizontal rectangular channel.

Rearranging gives

$$\frac{\gamma y_1^2}{2} + \rho V_1^2 y_1 = \frac{\gamma y_2^2}{2} + \rho V_2^2 y_2 \qquad (11.4.1)$$

or

$$F_1 + M_1 = F_2 + M_2 \qquad (11.4.2)$$

in which F is the hydrostatic force at the section and M is the momentum per second passing the section. By writing $F + M$ for a given discharge q per unit width

$$F + M = \frac{\gamma y^2}{2} + \frac{\rho q^2}{y} \qquad (11.4.3)$$

a plot is made of $F + M$ as abscissa against y as ordinate, Fig. 11.5, for $q = 10$ cfs/ft. Any vertical line intersecting the curve cuts it at two points having the same value of $F + M$; hence, they are conjugate depths. The value of y for minimum $F + M$ [by differentiation of Eq. (11.4.3) with respect to y and setting $d(F + M)/dy$ equal to zero] is

$$y_c = \left(\frac{q^2}{g}\right)^{1/3} \qquad (11.4.4)$$

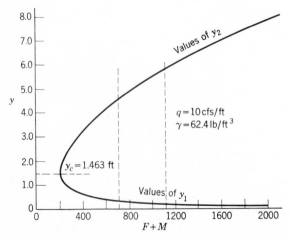

Figure 11.5 $F + M$ curve for hydraulic jump.

The jump must always occur from a depth less than this value to a depth greater than this value. This depth is the *critical depth*, which is shown in the following section to be the depth of minimum energy. Therefore, the jump always occurs from rapid flow to tranquil flow. The fact that available energy is lost in the jump prevents any possibility of its suddenly changing from the higher conjugate depth to the lower conjugate depth.

The conjugate depths are directly related to the Froude numbers before and after the jump,

$$\mathbf{F}_1 = \frac{V_1}{\sqrt{gy_1}} \qquad \mathbf{F}_2 = \frac{V_2}{\sqrt{gy_2}} \qquad\qquad (11.4.5)$$

From the continuity equation

$$V_1^2 y_1^2 = gy_1^3 \mathbf{F}_1^2 = V_2^2 y_2^2 = gy_2^3 \mathbf{F}_2^2$$

or
$$\mathbf{F}_1^2 y_1^3 = \mathbf{F}_2^2 y_2^3 \qquad\qquad (11.4.6)$$

From Eq. (11.4.1)

$$y_1^2 \left(1 + 2\frac{V_1^2}{gy_1}\right) = y_2^2 \left(1 + 2\frac{V_2^2}{gy_2}\right)$$

Substituting from Eqs. (11.4.5) and (11.4.6) gives

$$(1 + 2\mathbf{F}_1^2)\mathbf{F}_1^{-4/3} = (1 + 2\mathbf{F}_2^2)\mathbf{F}_2^{-4/3} \qquad\qquad (11.4.7)$$

The value of $\mathbf{F}_2$ in terms of $\mathbf{F}_1$ is obtained from the hydraulic jump equation (3.11.23)

$$y_2 = \frac{-y_1}{2} + \sqrt{\left(\frac{y_1}{2}\right)^2 + 2\frac{V_1^2 y_1}{g}} \qquad \text{or} \qquad 2\frac{y_2}{y_1} = -1 + \sqrt{1 + 8\frac{V_1^2}{gy_1}}$$

By Eqs. (11.4.5) and (11.4.6)

$$\mathbf{F}_2 = \frac{2\sqrt{2}\,\mathbf{F}_1}{(\sqrt{1 + 8\mathbf{F}_1^2} - 1)^{3/2}} \qquad\qquad (11.4.8)$$

These equations apply only to a rectangular section.

The Froude number before the jump is always greater than unity, and after the jump it is always less than unity.

Stilling Basins

A stilling basin is a structure for dissipating available energy of flow below a spillway, outlet works, chute, or canal structure. In the majority of existing installations a hydraulic jump is housed within the stilling basin and used as the energy dissipator. This discussion is limited to rectangular basins with horizontal floors although sloping floors are used in some cases to save excavation. An authoritative and comprehensive work† by personnel of the Bureau of Reclamation classified the hydraulic jump as an effective energy dissipator in terms of the Froude number $F_1(V_1/\sqrt{gy_1})$ entering the basin as follows:

At $F_1 = 1$ to 1.7. Standing wave. There is only a slight difference in conjugate depths. Near $F_1 = 1.7$ a series of small rollers develops.

At $F_1 = 1.7$ to 2.5. Pre-jump. The water surface is quite smooth; the velocity is fairly uniform; and the head loss is low. No baffles required if proper length of pool is provided.

At $F_1 = 2.5$ to 4.5. Transition. Oscillating action of entering jet, from bottom of basin to surface. Each oscillation produces a large wave of irregular period that can travel downstream for miles and damage earth banks and riprap. If possible, it is advantageous to avoid this range of Froude numbers in stilling basin design.

At $F_1 = 4.5$ to 9. Range of good jumps. The jump is well balanced, and the action is at its best. Energy absorption (irreversibilities) ranges from 45 to 70 percent. Baffles and sills may be utilized to reduce length of basin.

At $F_1 = 9$ upward. Effective but rough. Energy dissipation up to 85 percent. Other types of stilling basins may be more economical.

Baffle blocks are frequently used at the entrance to a basin to corrugate the flow. They are usually regularly spaced with gaps about equal to block widths. Sills, either triangular or dentated, are frequently employed at the downstream end of a basin to aid in holding the jump within the basin and to permit some shortening of the basin.

The basin should be paved with high-quality concrete to prevent erosion and cavitation damage. No irregularities in floor or training walls should be permitted. The length of the jump, about $6y_2$, should be within the paved basin, with good riprap downstream if the material is easily eroded.

Example 11.2 A hydraulic jump occurs downstream from a 15-m-wide sluice gate. The depth is 1.5 m, and the velocity is 20 m/s. Determine (a) the Froude number and the Froude number corresponding to the conjugate depth, (b) the depth and velocity after the jump, and (c) the power dissipated by the jump.

(a)
$$F_1 = \frac{V_1}{\sqrt{gy_1}} = \frac{20}{\sqrt{9.806 \times 1.5}} = 5.215$$

From Eq. (11.4.8)

$$F_2 = \frac{2\sqrt{2} \times 5.215}{(\sqrt{1 + 8 \times 5.215^2} - 1)^{3/2}} = 0.2882$$

(b)
$$F_2 = \frac{V_2}{\sqrt{gy_2}} \qquad V_2 y_2 = V_1 y_1 = 1.5 \times 20 = 30 \text{ m}^2/\text{s}$$

---

† Research Study on Stilling Basins, Energy Dissipators, and Associated Appurtenances, Progress Report II, *U.S. Bur. Reclam. Hydraul. Lab. Rep.* Hyd-399, Denver, June 1, 1955. In this report the Froude number was defined as $V/\sqrt{gy}$.

Then

$$V_2^2 = \mathbf{F}_2^2 g y_2 = \mathbf{F}_2^2 g \frac{30}{V_2}$$

and $\qquad V_2 = (0.2882^2 \times 9.806 \times 30)^{1/3} = 2.90 \text{ m/s}$

$\qquad\qquad y_2 = 10.34 \text{ m}$

(c) From Eq. (3.11.24), the head loss h_j in the jump is

$$h_j = \frac{(y_2 - y_1)^3}{4 y_1 y_2} = \frac{(10.34 - 1.50)^3}{4 \times 1.5 \times 10.34} = 11.13 \text{ m} \cdot \text{N/N}$$

The power dissipated is

$$\text{Power} = \gamma Q h_j = 9806 \times 15 \times 30 \times 11.13 = 49.1 \text{ MW}$$

EXERCISE

11.4.1 Supercritical flow can never occur (a) directly after a hydraulic jump; (b) in a mild channel; (c) in an adverse channel; (d) in a horizontal channel; (e) in a steep channel.

11.5 SPECIFIC ENERGY; CRITICAL DEPTH

The energy per unit weight E_s with elevation datum taken as the bottom of the channel is called the *specific energy*. It is a convenient quantity to use in studying open-channel flow and was introduced by Bakhmeteff in 1911. It is plotted vertically above the channel floor,

$$E_s = y + \frac{V^2}{2g} \tag{11.5.1}$$

A plot of specific energy for a particular case is shown in Fig. 11.6. In a rectangular channel, in which q is the discharge per unit width, with $Vy = q$,

$$E_s = y + \frac{q^2}{2gy^2} \tag{11.5.2}$$

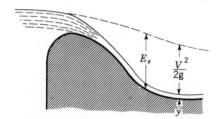

Figure 11.6 Example of specific energy.

It is of interest to note how the specific energy varies with the depth for a constant discharge (Fig. 11.7). For small values of y the curve goes to infinity along the E_s axis, while for large values of y the velocity head term is negligible and the curve approaches the 45° line $E_s = y$ asymptotically. The specific energy has a minimum value below which the given q cannot occur. The value of y for minimum E_s is obtained by setting dE_s/dy equal to zero, from Eq. (11.5.2), holding q constant,

$$\frac{dE_s}{dy} = 0 = 1 - \frac{q^2}{gy^3}$$

or

$$y_c = \left(\frac{q^2}{g}\right)^{1/3} \tag{11.5.3}$$

The depth for minimum energy y_c is called *critical depth*. Eliminating q^2 in Eqs. (11.5.2) and (11.5.3) gives

$$E_{s_{min}} = \tfrac{3}{2}y_c \tag{11.5.4}$$

showing that the critical depth is two-thirds of the specific energy. Eliminating E_s in Eqs. (11.5.1) and (11.5.4) gives

$$V_c = \sqrt{gy_c} \tag{11.5.5}$$

The velocity of flow at critical condition V_c is $\sqrt{gy_c}$, which was used in Sec. 8.6 in connection with the broad-crested weir. Another method of arriving at the critical condition is to determine the maximum discharge q that could occur for a given specific energy. The resulting equations are the same as Eqs. (11.5.3) to (11.5.5).

For nonrectangular cross sections, as illustrated in Fig. 11.8, the specific-energy equation takes the form

$$E_s = y + \frac{Q^2}{2gA^2} \tag{11.5.6}$$

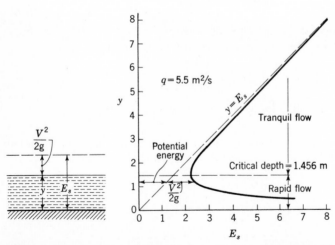

Figure 11.7 Specific energy required for flow of a given discharge at various depths.

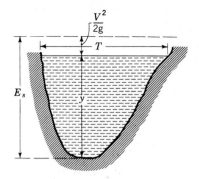

Figure 11.8 Specific energy for a nonrectangular section.

in which A is the cross-sectional area. To find the critical depth,

$$\frac{dE_s}{dy} = 0 = 1 - \frac{Q^2}{gA^3}\frac{dA}{dy}$$

From Fig. 11.8, the relation between dA and dy is expressed by

$$dA = T\,dy$$

in which T is the width of the cross section at the liquid surface. With this relation,

$$\frac{Q^2}{gA_c^3}T_c = 1 \qquad\qquad (11.5.7)$$

The critical depth must satisfy this equation. Eliminating Q in Eqs. (11.5.6) and (11.5.7) gives

$$E_s = y_c + \frac{A_c}{2T_c} \qquad\qquad (11.5.8)$$

This equation shows that the minimum energy occurs when the velocity head is one-half the average depth A/T. Equation (11.5.7) may be solved by trial for irregular sections by plotting

$$f(y) = \frac{Q^2 T}{gA^3}$$

Critical depth occurs for that value of y which makes $f(y) = 1$.

Example 11.3 Determine the critical depth for 10 m³/s flowing in a trapezoidal channel with bottom width 3 m and side slopes 1 horizontal to 2 vertical (1 on 2).

$$A = 3y + \frac{y^2}{2} \qquad T = 3 + y$$

Hence,

$$f(y) = \frac{10^2(3 + y)}{9.806(3y + y^2/2)^3} = \frac{10.198(3 + y)}{(3y + 0.5y^2)^3} = 1.0$$

By trial

y	2.0	1.2	0.8	1.0	0.99	0.98	0.985	0.984
$f(y)$	0.1	0.53	1.92	0.95	0.982	1.014	0.998	1.0014

The critical depth is 0.984 m. This trial solution is easily carried out by programmable calculator.

In uniform flow in an open channel, the energy grade line slopes downward parallel to the bottom of the channel, thus showing a steady decrease in available energy. The specific energy, however, remains constant along the channel, since $y + V^2/2g$ does not change. In nonuniform steady flow, the energy grade line always slopes downward, or the available energy is decreased. The specific energy may either increase or decrease, depending upon the slope of the channel bottom, the discharge, the depth of flow, properties of the cross section, and channel roughness. In Fig. 11.6 the specific energy increases during flow down the steep portion of the channel and decreases along the horizontal channel floor.

The specific-energy and critical-depth relations are essential in studying gradually varied flow and in determining control sections in open-channel flow.

The head loss in a hydraulic jump is easily displayed by drawing the $F + M$ curve (Fig. 11.5) and the specific-energy curve (Fig. 11.7) to the same vertical scale for the same discharge. Conjugate depths exist where any given vertical line intersects the $F + M$ curve. The specific energy at the upper depth may be observed to be always less than the specific energy at the corresponding lower conjugate depth.

EXERCISES

11.5.1 Flow at critical depth occurs when (a) changes in upstream resistance alter downstream conditions; (b) the specific energy is a maximum for a given discharge; (c) any change in depth requires more specific energy; (d) the normal depth and critical depth coincide for a channel; (e) the velocity is given by $\sqrt{2gy}$.

11.5.2 Critical depth in a rectangular channel is expressed by (a) $\sqrt{Vy}$; (b) $\sqrt{2gy}$; (c) $\sqrt{gy}$; (d) $\sqrt{q/g}$; (e) $(q^2/g)^{1/3}$.

11.5.3 Critical depth in a nonrectangular channel is expressed by (a) $Q^2T/gA^3 = 1$; (b) $QT^2/gA^2 = 1$; (c) $Q^2A^3/gT^2 = 1$; (d) $Q^2/gA^3 = 1$; (e) none of these answers.

11.5.4 The specific energy for the flow expressed by $V = 4.43$ m/s $y = 1$ m, in meter-newtons per newton, is (a) 2; (b) 3; (c) 5.43; (d) 9.86; (e) none of these answers.

11.5.5 The minimum possible specific energy for a flow is 2.475 ft·lb/lb. The discharge per foot of width, in cubic feet per second, is (a) 4.26; (b) 12.02; (c) 17; (d) 22.15; (e) none of these answers.

11.6 GRADUALLY VARIED FLOW

Gradually varied flow is steady nonuniform flow of a special class. The depth, area, roughness, bottom slope, and hydraulic radius change very slowly (if at all) along the channel. The basic assumption required is that the head-loss rate at a given section is given by the Manning formula for the same depth and discharge, regardless of trends in the depth. Solving Eq. (11.2.1) for the head loss per unit length of channel produces

$$S = -\frac{\Delta E}{\Delta L} = \left(\frac{nQ}{C_m A R^{2/3}}\right)^2 \tag{11.6.1}$$

in which S is now the slope of the energy grade line or, more specifically, the sine of the angle the energy grade line makes with the horizontal. In gradually varied flow the slopes of energy grade line, hydraulic grade line, and bottom are all different. Computations of gradually varied flow may be carried out either by the *standard-step method* or by *numerical integration*. Horizontal channels of great width are treated as a special case that may be integrated.

Standard-Step Method

Applying the energy equation between two sections a finite distance ΔL apart, Fig. 11.9, including the loss term, gives

$$\frac{V_1^2}{2g} + S_0 \Delta L + y_1 = \frac{V_2^2}{2g} + y_2 + S \Delta L \tag{11.6.2}$$

Solving for the length of reach gives

$$\Delta L = \frac{(V_1^2 - V_2^2)/2g + y_1 - y_2}{S - S_0} \tag{11.6.3}$$

If conditions are known at one section, e.g., section 1, and the depth y_2 is wanted a distance ΔL away, a trial solution is required. The procedure is as follows:

1. Assume a depth y_2; then compute A_2, V_2.
2. For the assumed y_2 find an average y, P, and A for the reach [for prismatic channels $y = (y_1 + y_2)/2$ with A and R computed for this depth] and compute S.
3. Substitute in Eq. (11.6.3) to compute ΔL.
4. If ΔL is not correct, assume a new y_2 and repeat the procedure.

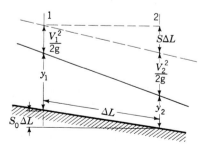

Figure 11.9 Gradually varied flow.

Example 11.4 At section 1 of a canal the cross section is trapezoidal, $b_1 = 10$ m, $m_1 = 2$, $y_1 = 7$ m, and at section 2, downstream 200 m, the bottom is 0.08 m higher than at section 1, $b_2 = 15$ m, and $m_2 = 3$. $Q = 200$ m^3/s, $n = 0.035$. Determine the depth of water at section 2.

$$A_1 = b_1 y_1 + m_1 y_1^2 = 10 \times 7 + 2 \times 7^2 = 168 \text{ m}^2 \qquad V_1 = \tfrac{200}{168} = 1.19 \text{ m/s}$$

$$P_1 = b_1 + 2y_1\sqrt{m_1^2 + 1} = 10 + 2 \times 7\sqrt{2^2 + 1} = 41.3 \text{ m}$$

$$S_0 = -\frac{0.08}{200} = -0.0004$$

Since the bottom has an adverse slope, i.e., it is rising in the downstream direction, and since section 2 is larger than section 1, y_2 is probably less than y_1. Assume $y_2 = 6.9$ m; then

$$A_2 = 15 \times 6.9 + 3 \times 6.9^2 = 246 \text{ m}^2 \qquad V_2 = \tfrac{200}{246} = 0.813 \text{ m/s}$$

and $\qquad P_2 = 15 + 2 \times 6.9\sqrt{10} = 58.6$ m

The average $A = 207$ and average wetted perimeter $P = 50.0$ are used to find an average hydraulic radius for the reach, $R = 4.14$ m. Then

$$S = \left(\frac{nQ}{C_m A R^{2/3}}\right)^2 = \left(\frac{0.035 \times 200}{1.0 \times 207 \times 4.14^{2/3}}\right)^2 = 0.000172$$

Substituting into Eq. (11.6.3) gives

$$\Delta L = \frac{(1.19^2 - 0.813^2)/(2 \times 9.806) + 7 - 6.9}{0.000172 + 0.0004} = 242 \text{ m}$$

A larger y_2, for example, 6.92 m, would bring the computed value of length closer to the actual length.

The standard-step method is easily followed with the programmable calculator if about 20 memory storage spaces and about 100 program steps are available. In the first trial y_2 is used to evaluate ΔL_{new}. Then a linear proportion yields a new trial $y_{2_{new}}$ for the next step; thus

$$\frac{y_1 - y_2}{\Delta L_{new}} = \frac{y_1 - y_{2_{new}}}{\Delta L_{given}}$$

or

$$y_{2_{new}} = y_1 + (y_2 - y_1)\frac{\Delta L_{given}}{\Delta L_{new}}$$

A few iterations yield complete information on section 2.

Numerical Integration Method

A more satisfactory procedure, particularly for flow through channels having a constant shape of cross section and constant bottom slope, is to obtain a differential equation in terms of y and L and then perform the integration numerically. When ΔL is considered as an infinitesimal in Fig. 11.9, the rate of change of available energy is equal to the rate of head loss $-\Delta E/\Delta L$ given by Eq. (11.6.1), or

$$\frac{d}{dL}\left(\frac{V^2}{2g} + z_0 - S_0 L + y\right) = -\left(\frac{nQ}{C_m A R^{2/3}}\right)^2 \qquad (11.6.4)$$

in which $z_0 - S_0 L$ is the elevation of bottom of channel at L, z_0 is the elevation of bottom at $L = 0$, and L is measured positive in the downstream direction. After performing the differentiation,

$$-\frac{V}{g}\frac{dV}{dL} + S_0 - \frac{dy}{dL} = \left(\frac{nQ}{C_m AR^{2/3}}\right)^2 \qquad (11.6.5)$$

Using the continuity equation $VA = Q$ leads to

$$\frac{dV}{dL} A + V \frac{dA}{dL} = 0$$

and expressing $dA = T \, dy$, in which T is the liquid-surface width of the cross section, gives

$$\frac{dV}{dL} = -\frac{VT}{A}\frac{dy}{dL} = -\frac{QT}{A^2}\frac{dy}{dL}$$

Substituting for V in Eq. (11.6.5) yields

$$\frac{Q^2}{gA^3} T \frac{dy}{dL} + S_0 - \frac{dy}{dL} = \left(\frac{nQ}{C_m AR^{2/3}}\right)^2$$

and solving for dL gives

$$dL = \frac{1 - Q^2 T/gA^3}{S_0 - (nQ/C_m AR^{2/3})^2} \, dy \qquad (11.6.6)$$

After integrating,

$$L = \int_{y_1}^{y_2} \frac{1 - Q^2 T/gA^3}{S_0 - (nQ/C_m AR^{2/3})^2} \, dy \qquad (11.6.7)$$

in which L is the distance between the two sections having depths y_1 and y_2.

When the numerator of the integrand is zero, critical flow prevails; there is no change in L for a change in y (neglecting curvature of the flow and nonhydrostatic pressure distribution at this section). Since this is not a case of gradual change in depth, the equations are not accurate near critical depth. When the denominator of the integrand is zero, uniform flow prevails and there is no change in depth along the channel. The flow is at *normal depth*.

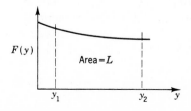

Figure 11.10 Numerical integration of equation for gradually varied flow.

For a channel of prismatic cross section, constant n and S_0, the integrand becomes a function of y only,

$$F(y) = \frac{1 - Q^2 T/gA^3}{S_0 - (nQ/C_m AR^{2/3})^2}$$

and the equation can be integrated numerically by plotting $F(y)$ as ordinate against y as abscissa. The area under the curve (Fig. 11.10) between two values of y is the length L between the sections, since

$$L = \int_{y_1}^{y_2} F(y)\, dy$$

Example 11.5 A trapezoidal channel, $b = 3$ m, $m = 1$, $n = 0.014$, $S_0 = 0.001$, carries 28 m³/s. If the depth is 3 m at section 1, determine the water-surface profile for the next 700 m downstream. To determine whether the depth increases or decreases, the slope of the energy grade line at section 1 is computed [using Eq. (11.6.1)]

$$A = by + my^2 = 3 \times 3 + 1 \times 3^2 = 18 \text{ m}^2$$

$$P = b + 2y\sqrt{m^2 + 1} = 11.485 \text{ m}$$

and

$$R = \frac{18}{11.485} = 1.567 \text{ m}$$

Then

$$S = \left(\frac{0.014 \times 28}{18 \times 1.567^{2/3}}\right)^2 = 0.00026$$

By substituting into Eq. (11.5.7) the values A, Q, and $T = 9$ m, $Q^2 T/gA^3 = 0.12$, showing that the depth is above critical. With the depth greater than critical and the energy grade line less steep than the bottom of the channel, the specific energy is increasing. When the specific energy increases above critical, the depth of flow increases. Δy is then positive. Substituting into Eq. (11.6.7) yields

$$L = \int_{3}^{y} \frac{1 - 79.95T/A^3}{0.001 - 0.1537/(A^2 R^{4/3})}\, dy$$

The following table evaluates the terms of the integrand.

y	A	P	R	T	Numerator	$10^6 \times$ Denominator	$F(y)$	L
3	18	11.48	1.57	9	0.8766	739	1185	0
3.2	19.84	12.05	1.65	9.4	0.9038	799	1131	231.6
3.4	21.76	12.62	1.72	9.8	0.9240	843	1096	454.3
3.6	23.76	13.18	1.80	10.2	0.9392	876	1072	671.1
3.8	25.84	13.75	1.88	10.6	0.9509	901	1056	883.9

The integral $\int F(y)\, dy$ can be evaluated by plotting the curve and taking the area under it between $y = 3$ and the following values of y. As $F(y)$ does not vary greatly in this example, the average of

$F(y)$ may be used for each reach (the trapezoidal rule); and when it is multiplied by Δy, the length of reach is obtained. Between $y = 3$ and $y = 3.2$

$$\frac{1185 + 1131}{2}\,0.2 = 231.6$$

Between $y = 3.2$ and $y = 3.4$

$$\frac{1131 + 1096}{2}\,0.2 = 222.7$$

and so on. Five points on it are known, so the water surface can be plotted. A more accurate way of summing $F(y)$ to obtain L is by use of Simpson's rule, Appendix B.1. The procedure used is equivalent to a Runge-Kutta second-order solution of a differential equation, Appendix B.5. A programmable calculator (about 20 memory storage spaces and 75 program steps) was used to carry out this solution. By taking $\Delta y = 0.1$ m in place of 0.2 m, the length to $y = 3.6$ m is 0.6 m less.

Horizontal Channels of Great Width

For channels of great width the hydraulic radius equals the depth; and for horizontal channel floors, $S_0 = 0$; hence, Eq. (11.6.7) can be simplified. The width may be considered as unity; that is, $T = 1$, $Q = q$ and $A = y$, $R = y$; thus

$$L = -\int_{y_1}^{y} \frac{1 - q^2/gy^3}{n^2 q^2/C_m^2 y^{10/3}}\, dy \tag{11.6.8}$$

or, after performing the integration,

$$L = -\frac{3}{13}\left(\frac{C_m}{nq}\right)^2 (y^{13/3} - y_1^{13/3}) + \frac{3}{4g}\left(\frac{C_m}{n}\right)^2 (y^{4/3} - y_1^{4/3}) \tag{11.6.9}$$

Example 11.6 After contracting below a sluice gate water flows onto a wide horizontal floor with a velocity of 15 m/s and a depth of 0.7 m. Find the equation for the water-surface profile, $n = 0.015$.
From Eq. (11.6.9), with x replacing L as distance from section 1, where $y_1 = 0.7$, and with $q = 0.7 \times 15 = 10.5$ m²/s,

$$x = -\frac{3}{13}\left(\frac{1}{0.015 \times 10.5}\right)^2 (y^{13/3} - 0.7^{13/3}) + \frac{3}{4 \times 9.806}\left(\frac{1}{0.015}\right)^2 (y^{4/3} - 0.7^{4/3})$$

$$= -209.3 - 9.30y^{13/3} + 340y^{4/3}$$

Critical depth occurs [Eq. (11.5.3)] at

$$y_c = \left(\frac{q^2}{g}\right)^{1/3} = \left(\frac{10.5^2}{9.806}\right)^{1/3} = 2.24 \text{ m}$$

The depth must increase downstream, since the specific energy decreases, and the depth must move toward the critical value for less specific energy. The equation does not hold near the critical depth because of vertical accelerations that have been neglected in the derivation of gradually varied flow. If the channel is long enough for critical depth to be attained before the end of the channel, the high-velocity flow downstream from the gate may be drowned or a jump may occur. The water-surface calculation for the subcritical flow must begin with critical depth at the downstream end of the channel.

The computation of water-surface profiles with the aid of a digital computer is discussed after the various types of gradually varied flow profiles are classified.

EXERCISE

11.6.1 Gradually varied flow is (*a*) steady uniform flow; (*b*) steady nonuniform flow; (*c*) unsteady uniform flow; (*d*) unsteady nonuniform flow; (*e*) none of these answers.

11.7 CLASSIFICATION OF SURFACE PROFILES

A study of Eq. (11.6.7) reveals many types of surface profiles, each of which has its definite characteristics. The bottom slope is classified as *adverse, horizontal, mild, critical,* and *steep;* and, in general, the flow can be above the normal depth or below the normal depth, and it can be above critical depth or below critical depth.

The various profiles are plotted in Fig. 11.11; the procedures used are discussed for the various classifications in the following paragraphs. A very wide channel is assumed in the reduced equations which follow, with $R = y$.

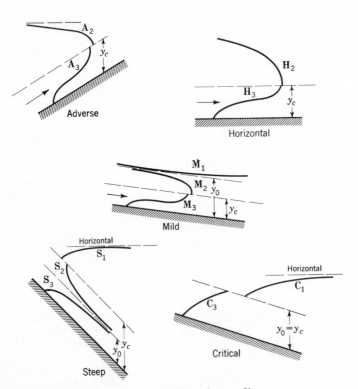

Figure 11.11 The various typical liquid-surface profiles.

Adverse Slope Profiles

When the channel bottom rises in the direction of flow (S_0 is negative), the resulting surface profiles are said to be adverse. There is no normal depth, but the flow may be either below critical depth or above critical depth. Below critical depth the numerator is negative, and Eq. (11.6.6) has the form

$$dL = \frac{1 - C_1/y^3}{S_0 - C_2/y^{10/3}}\, dy$$

where C_1 and C_2 are positive constants. Here $F(y)$ is positive and the depth increases downstream. This curve is labeled $\mathbf{A_3}$ and shown in Fig. 11.11. For depths greater than critical depth, the numerator is positive, and $F(y)$ is negative; i.e., the depth decreases in the downstream direction. For y very large, $dL/dy = 1/S_0$, which is a horizontal asymptote for the curve. At $y = y_c$, dL/dy is 0, and the curve is perpendicular to the critical-depth line. This curve is labeled $\mathbf{A_2}$.

Horizontal Slope Profiles

For a horizontal channel $S_0 = 0$, the normal depth is infinite and flow may be either below critical depth or above critical depth. The equation has the form

$$dL = -Cy^{1/3}(y^3 - C_1)\, dy$$

For y less than critical, dL/dy is positive, and the depth increases downstream. It is labeled $\mathbf{H_3}$. For y greater than critical ($\mathbf{H_2}$ curve), dL/dy is negative, and the depth decreases downstream. These equations are integrable analytically for very wide channels.

Mild Slope Profiles

A mild slope is one on which the normal flow is tranquil, i.e., where normal depth y_0 is greater than critical depth. Three profiles may occur, $\mathbf{M_1}, \mathbf{M_2}, \mathbf{M_3}$, for depth above normal, below normal and above critical, and below critical, respectively. For the $\mathbf{M_1}$ curve, dL/dy is positive and approaches $1/S_0$ for very large y; hence, the $\mathbf{M_1}$ curve has a horizontal asymptote downstream. As the denominator approaches zero as y approaches y_0, the normal depth is an asymptote at the upstream end of the curve. Thus, dL/dy is negative for the $\mathbf{M_2}$ curve, with the upstream asymptote the normal depth, and $dL/dy = 0$ at critical. The $\mathbf{M_3}$ curve has an increasing depth downstream, as shown.

Critical Slope Profiles

When the normal depth and the critical depth are equal, the resulting profiles are labeled $\mathbf{C_1}$ and $\mathbf{C_3}$ for depth above and below critical, respectively. The equation has the form

$$dL = \frac{1}{S_0} \frac{1 - b/y^3}{1 - b_1/y^{10/3}}\, dy$$

with both numerator and denominator positive for C_1 and negative for C_3. Therefore, the depth increases downstream for both. For large y, dL/dy approaches $1/S_0$; hence, a horizontal line is an asymptote. The value of dL/dy at critical depth is $0.9/S_0$; hence, curve C_1 is convex upward. Curve C_3 also is convex upward, as shown.

Steep Slope Profiles

When the normal flow is rapid in a channel (normal depth less than critical depth), the resulting profiles S_1, S_2, S_3 are referred to as steep profiles: S_1 is above the normal and critical, S_2 between critical and normal, and S_3 below normal depth. For curve S_1 both numerator and denominator are positive, and the depth increases downstream approaching a horizontal asymptote. For curve S_2 the numerator is negative and the denominator positive but approaching zero at $y = y_0$. The curve approaches the normal depth asymptotically. The S_3 curve has a positive dL/dy as both numerator and denominator are negative. It plots as shown on Fig. 11.11.

It should be noted that a given channel may be classified as mild for one discharge, critical for another discharge, and steep for a third discharge, since normal depth and critical depth depend upon different functions of the discharge. The use of the various surface profiles is discussed in the next section.

EXERCISE

11.7.1 The hydraulic jump always occurs from (*a*) an M_3 curve to an M_1 curve; (*b*) an H_3 curve to an H_2 curve; (*c*) an S_3 curve to an S_1 curve; (*d*) below normal depth to above normal depth; (*e*) below critical depth to above critical depth.

11.8 CONTROL SECTIONS

A small change in downstream conditions cannot be relayed upstream when the depth is critical or less than critical; hence, downstream conditions do not control the flow. All rapid flows are controlled by upstream conditions, and computations of surface profiles must be started at the upstream end of a channel.

Tranquil flows are affected by small changes in downstream conditions and therefore are controlled by them. Tranquil-flow computations must start at the downstream end of a reach and be carried upstream.

Control sections occur at entrances and exits to channels and at changes in channel slopes, under certain conditions. A gate in a channel can be a control for both the upstream and downstream reaches. Three control sections are illustrated in Fig. 11.12. In *a* the flow passes through critical at the entrance to a channel, and depth can be computed there for a given discharge. The channel is steep; therefore, computations proceed downstream. In *b* a change in channel slope from mild to

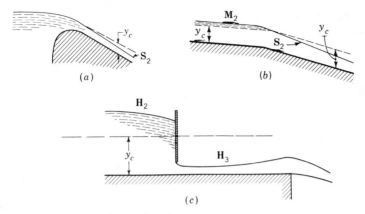

Figure 11.12 Channel control sections.

steep causes the flow to pass through critical at the break in grade. Computations proceed both upstream and downstream from the control section at the break in grade. In *c* a gate in a horizontal channel provides control both upstream and downstream from it. The various curves are labeled according to the classification in Fig. 11.11.

The hydraulic jump occurs whenever the conditions required by the momentum equation are satisfied. In Fig. 11.13, liquid issues from under a gate in rapid flow along a horizontal channel. If the channel were short enough, the flow could discharge over the end of the channel as an H_3 curve. With a longer channel, however, the jump occurs, and the resulting profile consists of pieces of H_3 and H_2 curves with the jump in between. In computing these profiles for a known discharge, the H_3 curve is computed, starting at the gate (contraction coefficient must be known) and proceeding downstream until it is clear that the depth will reach critical before the end of the channel is reached. Then the H_2 curve is computed, starting with critical depth at the end of the channel and proceeding upstream. The depths conjugate to those along H_3 are computed and plotted as shown. The intersection of the conjugate-depth curve and the H_2 curve locates the position of the jump. The channel may be so long that the H_2 curve is everywhere

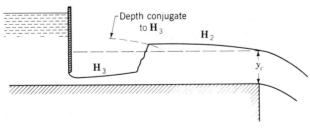

Figure 11.13 Hydraulic jump between two control sections.

```
C WATER SURFACE PROFILE IN RECT, TRAPEZOIDAL OR TRIANGULAR CHANNEL.
C XL=LENGTH, B=BOT WIDTH, Z=SIDE SLOPE, RN=MANNING N, SO=BOT SLOPE, Q=FLOW.
C YCONT=CONTROL DEPTH. IF YCONT=0. IN DATA, YCONT IS SET EQUAL TO YC.
C IN SUBCRITICAL FLOW, CONTROL IS DOWNSTREAM AND DISTANCES ARE MEASURED
C IN THE UPSTREAM DIRECTION
C IN SUPERCRITICAL FLOW, CONTROL IS U.S. AND DISTANCES ARE MEASURED D.S.
      DATA ISI/'SI'/
      AREA(YY)=YY*(B+Z*YY)
      PER(YY)=B+2.*YY*SQRT(1.+Z*Z)
      YCRIT(YY)=1.-Q*Q*(B+2.*Z*YY)/(G*AREA(YY)**3)
      YNORM(YY)=1.-Q*Q*CON/(AREA(YY)**3.333/PER(YY)**1.333)
      DL(YY)=YCRIT(YY)/(YNORM(YY)*SO)
      FPM(YY)=GAM*(YY*YY*(B*.5+Z*YY/3.)+Q*Q/(G*AREA(YY)))
      ENERGY(YY)=YY+Q*Q/(2.*G*AREA(YY)**2)
5     READ (5,7,END=99) IUNIT,XL,B,Z,RN,SO,Q,YCONT
7     FORMAT (A2/F10.1,3F10.4,F10.6,2F10.3)
      IF (IUNIT.EQ.ISI) GO TO 10
      GAM=62.4
      G=32.2
      CON=(RN/1.486)**2/SO
      WRITE (6,9)
9     FORMAT (' ENGLISH UNITS')
      GO TO 12
10    G=9.806
      GAM=9802.
      CON=RN**2/SO
      WRITE (6,11)
11    FORMAT (' SI UNITS')
12    WRITE (6,13) XL,Q,B,Z,RN,SO
13    FORMAT (/' CHANNEL LENGTH=',F10.1,' DISCHARGE =',F12.3/
     2' B=',F8.2,' Z=',F6.3,' RN=',F6.4,' SO=',F9.6)
      NN=30
C DETERMINATION OF CRITICAL AND NORMAL DEPTHS
      UP=30.
      DN=0.
      YC=15.
      DO 20 I=1,15
      IF (YCRIT(YC)) 14,21,15
14    DN=YC
      GO TO 20
15    UP=YC
20    YC=(UP+DN)*.5
21    IF (YCONT.EQ.0.) YCONT=YC
      IF (SO.LE.0.) GO TO 33
      UP=40.
      DN=0.
      YN=20.
      DO 30 I=1,15
      IF (YNORM(YN)) 23,31,24
23    DN=YN
      GO TO 30
24    UP=YN
30    YN=(UP+DN)*.5
31    WRITE (6,32) YN,YC
32    FORMAT (/' NORMAL DEPTH=',F7.3,' CRITICAL DEPTH=',F7.3)
      GO TO 35
33    YN=3.*YC
34    FORMAT (/' CRITICAL DEPTH=',F7.3)
      WRITE (6,34) YC
35    IF (YN.LT.YC) GO TO 50
C MILD,ADVERSE, OR HORIZONTAL CHANNEL, YN. GT. YC
      IF (YCONT .LT.YC) GO TO 45
C SUBCRITICAL FLOW, YCONT .GE. YC
      SIGN=-1.
      DY=(YCONT-YN)*.998/NN
      WRITE (6,44) YCONT
44    FORMAT (/' CONTROL IS DOWNSTREAM, DEPTH=',F7.3)
      GO TO 60
C SUPERCRITICAL FLOW
45    SIGN=1.
      DY=(YC-YCONT)/NN
      WRITE (6,46) YCONT
46    FORMAT (/' CONTROL IS UPSTREAM, DEPTH=',F7.3)
      GO TO 60
```

```
C STEEP CHANNEL, YN .LT. YC
  50 IF (YCONT.LE.YC) GO TO 55
C SUBCRITICAL FLOW, YCONT .GT.YC
     SIGN=-1.
     DY=(YCONT-YC)/NN
     WRITE (6,44) YCONT
     GO TO 60
C SUPERCRITICAL FLOW, YCONT .LE. YC
  55 SIGN=1.
     NN=NN*2
     DY=(YN-YCONT)*.998/NN
     WRITE (6,46) YCONT
  60 SL=0.
     Y=YCONT
     E=ENERGY(Y)
     FM=FPM(Y)
     WRITE (6,62)
  62 FORMAT (//'        DISTANCE        DEPTH      ENERGY          F+M ')
     WRITE (6,64) SL,Y, E, FM
  64 FORMAT (5X,F10.1,2F12.3,F13.0)
C WATER SURFACE PROFILE CALCULATION
     DO 80 I=1,NN,2
     Y2=YCONT+SIGN*DY*(I+1)
     DX=DY*(DL(Y)+DL(Y2)+4.*DL(YCONT+SIGN*I*DY))/3.
     SL=SL+DX
     IF(SL.GT.XL) GO TO 82
     Y=Y2
     E=ENERGY(Y)
     FM=FPM(Y)
     IF (I.EQ.NN-1.AND.SL.LT.0.) SL=XL
  80 WRITE (6,64)SL,Y,E,FM
     GO TO 5
  82 Y=Y2-SIGN*2.*DY*(SL-XL)/DX
     E=ENERGY(Y)
     FM=FPM(Y)
     WRITE (6,64)XL,Y,E,FM
     GO TO 5
  99 STOP
     END
SI UNITS, EXAMPLE 11.7
   200.      2.5      0.8      0.012     0.025      25.      0.0
SI UNITS
   600.      2.5      0.8      0.012     0.0002     25.      0.907
SI UNITS
   600.      2.5      0.8      0.012     0.0002     25.      2.0
```

Figure 11.14 FORTRAN program for water-surface profiles.

greater than the depth conjugate to H_3. A *drowned jump* then occurs, with H_2 extending to the gate.

All sketches are drawn to a greatly exaggerated vertical scale, since usual channels have small bottom slopes.

11.9 COMPUTER CALCULATION OF GRADUALLY VARIED FLOW

In Sec. 11.6 the standard-step and numerical-integration methods of computing water-surface profiles were introduced. The repetitious calculation in the latter method is easily handled by digital computer. The program, listed in Fig. 11.14, calculates the steady gradually varied water-surface profile in any prismatic rectangular, symmetric trapezoidal or triangular channel. The concepts of physical control sections in a channel must be understood in order to use the program successfully.

Input data include the specification of the system of units (SI or ENGLISH) in the first columns of the first data card, followed by the channel dimensions, discharge, and water-surface control depth on the second card. If the control depth is left blank or set to zero in data, it is automatically assumed to be the critical depth in the program. For subcritical flow the control is downstream, and distances are measured in the upstream direction. For supercritical flow the control depth is upstream, and distances are measured in the downstream direction.

The program begins with several line functions to compute the various variables and functions in the problem. After the necessary data input, critical depth is computed, followed by the normal-depth calculation if normal depth exists. The bisection method, Appendix B.3, is used in these calculations. The type of profile is then categorized, and finally the water-surface profile, specific energy, and $F + M$ are calculated and printed. Simpson's rule is used in the integration for the water-surface profile, Appendix B.1.

The program can be utilized for other channel sections, such as circular or parabolic, by simply changing the line functions at the beginning.

Example 11.7 A trapezoidal channel, $B = 2.5$ m, side slope $= 0.8$, has two bottom slopes. The upstream portion is 200 m long, $S_0 = 0.025$, and the downstream portion, 600 m long, $S_0 = 0.0002$, $n = 0.012$. A discharge of 25 m³/s enters at critical depth from a reservoir at the upstream end, and at the downstream end of the system the water depth is 2 m. Determine the water-surface profiles throughout the system, including jump location.

Three separate sets of data, shown in Fig. 11.14, are needed to obtain the results used to plot the solution as shown in Fig. 11.15. The first set for the steep upstream channel has a control depth equal to zero since it will be automatically assumed critical depth in the program. The second set is for the supercritical flow in the mild channel. It begins at a control depth equal to the end depth from the upstream channel and computes the water surface downstream to the critical depth. The third set of data uses the 2-m downstream depth as the control depth and computes in the upstream direction. Figure 11.16 shows the computer output from the last two data sets. The jump is located by finding the position of equal $F + M$ from the output of the last two data sets.

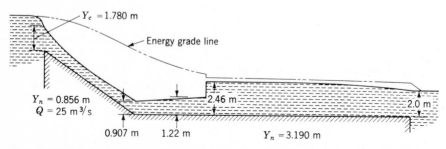

Figure 11.15 Solution to Example 11.7 as obtained from computer.

```
SI UNITS
CHANNEL LENGTH=      600.0 DISCHARGE =      25.000
B=     2.50 Z= 0.800 RN=0.0120 SO= 0.000200
NORMAL DEPTH=   3.190 CRITICAL DEPTH=   1.780
CONTROL IS UPSTREAM, DEPTH=   0.907
        DISTANCE        DEPTH       ENERGY              F+M
            0.0         0.907       4.630           225573.
           25.8         0.965       4.160           211572.
           51.2         1.023       3.786           199573.
           76.0         1.082       3.487           189272.
          100.1         1.140       3.247           180431.
          123.3         1.198       3.054           172859.
          145.5         1.256       2.900           166399.
          166.5         1.315       2.777           160926.
          186.0         1.373       2.679           156334.
          204.0         1.431       2.603           152534.
          220.0         1.489       2.544           149455.
          233.9         1.547       2.500           147035.
          245.4         1.606       2.469           145221.
          254.1         1.664       2.448           143970.
          259.6         1.722       2.437           143242.
          261.5         1.780       2.433           143007.
SI UNITS
CHANNEL LENGTH=      600.0 DISCHARGE =      25.000
B=     2.50 Z= 0.800 RN=0.0120 SO= 0.000200
NORMAL DEPTH=   3.190 CRITICAL DEPTH=   1.780
CONTROL IS DOWNSTREAM, DEPTH=   2.000
        DISTANCE        DEPTH       ENERGY              F+M
            0.0         2.000       2.474           146109.
           33.0         2.079       2.504           148634.
           80.5         2.158       2.541           151844.
          145.4         2.238       2.583           155711.
          231.3         2.317       2.630           160211.
          343.0         2.396       2.681           165326.
          486.6         2.475       2.734           171040.
          600.0         2.524       2.769           174857.
```

Figure 11.16 Computer output, Example 11.7.

11.10 TRANSITIONS

At entrances to channels and at changes in cross section and bottom slope, the structure that conducts the liquid from the upstream section to the new section is a *transition*. Its purpose is to change the shape of flow and surface profile in such a manner that minimum losses result. A transition for tranquil flow from a rectangular channel to a trapezoidal channel is illustrated in Fig. 11.17. Applying the energy equation from section 1 to section 2 gives

$$\frac{V_1^2}{2g} + y_1 = \frac{V_2^2}{2g} + y_2 + z + E_1 \qquad (11.10.1)$$

In general, the sections and depths are determined by other considerations, and z must be determined for the expected available energy loss E_1. By good design, i.e., with slowly tapering walls and flooring with no sudden changes in cross-sectional area, the losses can be held to about one-tenth the difference between velocity heads for accelerated flow and to about three-tenths the difference between velocity heads for retarded flow. For rapid flow, wave mechanics is required in designing the transitions.†

† A. T. Ippen, Channel Transitions and Controls, in H. Rouse (ed.), "Engineering Hydraulics," Wiley, New York, 1950.

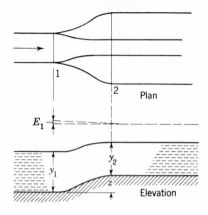

Figure 11.17 Transition from rectangular channel to trapezoidal channel for tranquil flow.

Example 11.8 In Fig. 11.17, 400 cfs flows through the transition; the rectangular section is 8 ft wide; and $y_1 = 8$ ft. The trapezoidal section is 6 ft wide at the bottom with side slopes $1:1$, and $y_2 = 7.5$ ft. Determine the rise z in the bottom through the transition.

$$V_1 = \frac{400}{64} = 6.25 \qquad \frac{V_1^2}{2g} = 0.61 \qquad A_2 = 101.25 \text{ ft}^2$$

$$V_2 = \frac{400}{101.25} = 3.95 \qquad \frac{V_2^2}{2g} = 0.24 \qquad E_1 = 0.3\left(\frac{V_1^2}{2g} - \frac{V_2^2}{2g}\right) = 0.11$$

After substituting into Eq. (11.10.1),

$$z = 0.61 + 8 - 0.24 - 7.5 - 0.11 = 0.76 \text{ ft}$$

The *critical-depth meter*† is an excellent device for measuring discharge in an open channel. The relations for determination of discharge are worked out for a rectangular channel of constant width, Fig. 11.18, with a raised floor over a reach of channel about $3y_c$ long. The raised floor is of such height that the restricted section becomes a control section with critical velocity occurring over it. By measuring only the upstream depth y_1, the discharge per foot of width is accurately determined. Applying the energy equation from section 1 to the critical section (exact location unimportant), including the transition-loss term, gives

$$\frac{V_1^2}{2g} + y_1 = z + y_c + \frac{V_c^2}{2g} + \frac{1}{10}\left(\frac{V_c^2}{2g} - \frac{V_1^2}{2g}\right)$$

Since

$$y_c + \frac{V_c^2}{2g} = E_c \qquad \frac{V_c^2}{2g} = \frac{E_c}{3}$$

† E. F. Brater and H. W. King, "Handbook of Hydraulics," 6th ed., pp. **8**-14 to **8**-16, McGraw-Hill, New York, 1976.

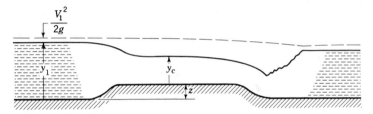

Figure 11.18 Critical-depth meter.

in which E_c is the specific energy at critical depth,

$$y_1 + 1.1 \frac{V_1^2}{2g} = z + 1.033 E_c \tag{11.10.2}$$

From Eq. (11.5.3)

$$y_c = \frac{2}{3} E_c = \left(\frac{q^2}{g}\right)^{1/3} \tag{11.10.3}$$

In Eqs. (11.10.2) and (11.10.3) E_c is eliminated and the resulting equation solved for q,

$$q = 0.517 g^{1/2} \left(y_1 - z + 1.1 \frac{V_1^2}{2g}\right)^{3/2}$$

Since $q = V_1 y_1$, V_1 can be eliminated,

$$q = 0.517 g^{1/2} \left(y_1 - z + \frac{0.55}{g} \frac{q^2}{y_1^2}\right)^{3/2} \tag{11.10.4}$$

The equation is solved by trial. As y_1 and z are known and the right-hand term containing q is small, it may first be neglected for an approximate q. A value a little larger than the approximate q may be substituted on the right-hand side. When the two q's are the same the equation is solved. Once z and the width of channel are known, a chart or table may be prepared yielding Q for any y_1. Experiments indicate that accuracy within 2 to 3 percent may be expected.

With tranquil flow a jump occurs downstream from the meter, and with rapid flow a jump occurs upstream from the meter.

Example 11.9 In a critical-depth meter 2 m wide with $z = 0.3$ m the depth y_1 is measured to be 0.75 m. Find the discharge.

$$q = 0.517(9.806^{1/2})(0.45^{3/2}) = 0.489 \ \text{m}^2/\text{s}$$

As a second approximation let q be 0.50,

$$q = 0.517(9.806^{1/2})\left(0.45 + \frac{0.55}{9.806} \times 0.5^2\right)^{3/2} = 0.512 \ \text{m}^2/\text{s}$$

and as a third approximation, 0.513,

$$q = 0.517(9.806^{1/2})\left(0.45 + \frac{0.55}{9.806} \times 0.513^2\right)^{3/2} = 0.513 \text{ m}^2/\text{s}$$

Then

$$Q = 2 \times 0.513 = 1.026 \text{ m}^3/\text{s}$$

EXERCISES

11.10.1 The loss through a diverging transition is about

(a) $0.1\dfrac{(V_1 - V_2)^2}{2g}$ (b) $0.1\dfrac{(V_1^2 - V_2^2)}{2g}$ (c) $0.3\dfrac{(V_1 - V_2)^2}{2g}$ (d) $0.3\dfrac{V_1^2 - V_2^2}{2g}$

(e) none of these answers

11.10.2 A critical-depth meter (a) measures the depth at the critical section; (b) is always preceded by a hydraulic jump; (c) must have a tranquil flow immediately upstream; (d) always has a hydraulic jump downstream; (e) always has a hydraulic jump associated with it.

PROBLEMS

11.1 Show that, for laminar flow to be insured down an inclined surface, the discharge per unit width cannot be greater than 500ν. (See Prob. 5.10.)

11.2 Calculate the depth of laminar flow of water at 20°C down a plane surface making an angle of 30° with the horizontal for the lower critical Reynolds number. (See Prob. 5.10.)

11.3 Calculate the depth of turbulent flow at $R = VR/\nu = 500$ for flow of water at 20°C down a plane surface making an angle θ of 30° with the horizontal. Use Manning's formula. $n = 0.01$; $S = \sin \theta$.

11.4 A rectangular channel is to carry 1.2 m³/s at a slope of 0.009. If the channel is lined with galvanized iron, $n = 0.011$, what is the minimum number of square meters of metal needed for each 100 m of channel? Neglect freeboard.

11.5 A trapezoidal channel, with side slopes 2 on 1 (2 horizontal to 1 vertical), is to carry 20 m³/s with a bottom slope of 0.0009. Determine the bottom width, depth, and velocity for the best hydraulic section. $n = 0.025$.

11.6 A trapezoidal channel made out of brick, with bottom width 6 ft and with bottom slope 0.001, is to carry 600 cfs. What should the side slopes and depth of channel be for the least number of bricks?

11.7 What radius semicircular corrugated-metal channel is needed to convey 2.5 m³/s a distance of 1 km with a head loss of 2 m? Can you find another cross section that requires less perimeter?

11.8 Determine the best hydraulic trapezoidal section to convey 85 m³/s with a bottom slope of 0.002. The lining is finished concrete.

11.9 Calculate the discharge through the channel and floodway of Fig. 11.19 for steady uniform flow, with $S = 0.0009$ and $y = 8$ ft.

11.10 For 200 m³/s flow in the section of Fig. 11.19 when the depth over the floodway is 1.2 m, calculate the energy gradient.

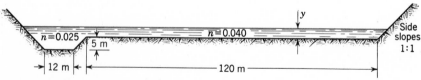

Figure 11.19 Problems 11.9, 11.10, 11.11.

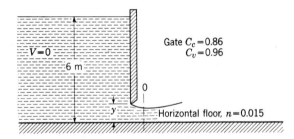

Gate $C_c = 0.86$
$C_v = 0.96$

$V = 0$

6 m

0

y

Horizontal floor, $n = 0.015$

Figure 11.20 Problems 11.22 to 11.26.

11.11 For 25,000 cfs flow through the section of Fig. 11.19, find the depth of flow in the floodway when the slope of the energy grade line is 0.0004.

11.12 Draw an $F + M$ curve for 2.5 m³/s per meter of width.

11.13 Draw the specific-energy curve for 2.5 m³/s per meter of width on the same chart as Prob. 11.12. What is the energy loss in a jump whose upstream depth is 0.5 m?

11.14 Prepare a plot of Eq. (11.4.7).

11.15 With $q = 100$ cfs/ft and $\mathbf{F}_1 = 3.5$, determine v_1, y_1, and the conjugate depth y_2.

11.16 Determine the two depths having a specific energy of 2 m for 1 m³/s per meter of width.

11.17 What is the critical depth for flow of 1.5 m³/s per meter of width?

11.18 What is the critical depth for flow of 0.3 m³/s through the cross section of Fig. 5.47?

11.19 Determine the critical depth for flow of 8.5 m³/s through a trapezoidal channel with a bottom width of 2.5 m and side slopes of 1 on 1.

11.20 An unfinished concrete rectangular channel 12 ft wide has a slope of 0.0009. It carries 480 cfs and has a depth of 7 ft at one section. By using the step method and taking one step only, compute the depth 1000 ft downstream.

11.21 Solve Prob. 11.20 by taking two equal steps. What is the classification of this water-surface profile?

11.22 A very wide gate (Fig. 11.20) admits water to a horizontal channel. Considering the pressure distribution hydrostatic at section 0, compute the depth at section 0 and the discharge per meter of width when $y = 1.0$ m.

11.23 If the depth at section 0 of Fig. 11.20 is 600 mm and the discharge per meter of width is 6 m²/s, compute the water-surface curve downstream from the gate.

11.24 Draw the curve of conjugate depths for the surface profile of Prob. 11.23.

11.25 If the very wide channel in Fig. 11.20 extends downstream 700 m and then has a sudden dropoff, compute the flow profile upstream from the end of the channel for $q = 6$ m²/s by integrating the equation for gradually varied flow.

11.26 Using the results of Probs. 11.24 and 11.25, determine the position of a hydraulic jump in the channel.

11.27 (a) In Fig. 11.21 the depth downstream from the gate is 0.6 m and the velocity is 12 m/s. For a very wide channel, compute the depth at the downstream end of the adverse slope. (b) Solve part (a) by use of the computer program given in Fig. 11.14, or write a similar program to obtain the solution.

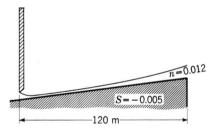

$n = 0.012$

$S = -0.005$

120 m

Figure 11.21 Problem 11.27.

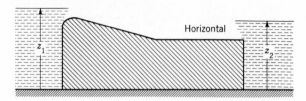

Figure 11.22 Problems 11.28 to 11.30.

11.28 Sketch (without computation) and label all the liquid-surface profiles that can be obtained from Fig. 11.22 by varying z_1, z_2, and the lengths of the channels for $z_2 < z_1$, with a steep, inclined channel.

11.29 In Fig. 11.22 determine the possible combinations of control sections for various values of z_1, z_2 and various channel lengths for $z_1 > z_2$, with the inclined channel always steep.

11.30 Sketch the various liquid-surface profiles and control sections for Fig. 11.22 obtained by varying channel length for $z_2 > z_1$.

11.31 Show an example of a channel that is mild for one discharge and steep for another discharge. What discharge is required for it to be critical?

11.32 Use the computer program in Fig. 11.14 or a similar program you have written to locate the hydraulic jump in a 90° triangular channel, 0.5 km long, that carries a flow of 1 m³/s, $n = 0.015$, $S_0 = 0.001$. The upstream depth is 0.2 m, and the downstream depth is 0.8 m.

11.33 Design a transition from a trapezoidal section, 8 ft bottom width and side slopes 1 on 1, depth 4 ft, to a rectangular section, 6 ft wide and 6 ft deep, for a flow of 250 cfs. The transition is to be 20 ft long, and the loss is one-tenth the difference between velocity heads. Show the bottom profile, and do not make any sudden changes in cross-sectional area.

11.34 A transition from a rectangular channel, 2.6 m wide and 2 m deep, to a trapezoidal channel, bottom width 4 m and side slopes 2 on 1, with depth 1.3 m has a loss four-tenths the difference between velocity heads. The discharge is 5.6 m³/s. Determine the difference between elevations of channel bottoms.

11.35 A critical-depth meter 16 ft wide has a rise in bottom of 2.0 ft. For an upstream depth of 3.52 ft determine the flow through the meter.

11.36 With flow approaching a critical-depth meter site at 6 m/s and a Froude number of 3, what is the minimum amount the floor must be raised?

REFERENCES

Bakhmeteff, B. A.: "Hydraulics of Open Channels," McGraw-Hill, New York, 1932.
Chow, V. T.: "Open-Channel Hydraulics," McGraw-Hill, New York, 1959.
Henderson, F. M.: "Open Channel Flow," Macmillan, New York, 1966.

TWELVE

UNSTEADY FLOW

Up to this point practically all flow cases examined have been for steady flow or have been reducible to a steady-flow situation. As technology advances and larger equipment is constructed or higher speeds employed, the problems of hydraulic transients become increasingly important. The hydraulic transients not only cause dangerously high pressures but produce excessive noise, fatigue, pitting due to cavitation, and disruption of normal control of circuits. Owing to the inherent period of certain systems of pipes, resonant vibrations may be incurred which can be destructive.

Hydraulic-transient analysis deals with the calculation of pressures and velocities during an unsteady-state mode of operation of a system. This may be caused by adjustment of a valve in a piping system, stopping a pump, or innumerable other possible changes in system operation.

The analysis of unsteady flow is much more complex than that of steady flow. Another independent variable enters, time, and equations may be partial differential equations rather than ordinary differential equations. The digital computer is ideally suited to the solution of such problems because of its large storage capacity and its ability to operate at very high computing rates. This chapter is in two parts: the first part dealing with closed-conduit transients, and the second part with frictionless open-channel unsteady flow.

First, oscillation of a U tube is studied, followed by its application to pipelines and reservoirs, the use of surge tanks, and the establishment of flow in a system. Equations are next developed for cases with more severe changes in velocity that require consideration of liquid compressibility and pipe-wall elasticity (usually called waterhammer). The open-channel cases are the positive and negative surge waves in a frictionless prismatic channel with instantaneous gate changes.

A   FLOW IN CLOSED CONDUITS

The unsteady-flow cases in closed conduits are treated as one-dimensional distributed-parameter problems. The equation of motion or the unsteady linear-momentum equation is used, and the unsteady continuity equation takes special forms. With nonlinear resistance terms for friction and other effects, the differential equations are frequently solved by numerical methods using the programmable calculator or the digital computer.

12.1 OSCILLATION OF LIQUID IN A U TUBE

Three cases of oscillations of liquid in a simple U tube are of interest: (1) frictionless liquid, (2) laminar resistance, and (3) turbulent resistance.

Frictionless Liquid

For the frictionless case, Euler's equation of motion in unsteady form [Eq. (3.5.6)] may be applied. It is

$$\frac{1}{\rho}\frac{\partial p}{\partial s} + g\frac{\partial z}{\partial s} + v\frac{\partial v}{\partial s} + \frac{\partial v}{\partial t} = 0$$

When sections 1 and 2 are designated (Fig. 12.1) and the equation is integrated from 1 to 2, for incompressible flow

$$\frac{p_2 - p_1}{\rho} + g(z_2 - z_1) + \frac{v_2^2 - v_1^2}{2} + \int_1^2 \frac{\partial v}{\partial t}\,ds = 0 \qquad (12.1.1)$$

But $p_1 = p_2$ and $v_1 = v_2$; also, $\partial v/\partial t$ is independent of s, hence,

$$g(z_2 - z_1) = -L\frac{\partial v}{\partial t} \qquad (12.1.2)$$

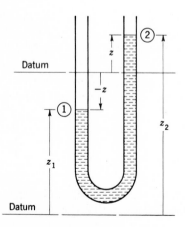

Figure 12.1 Oscillation of liquid in a U tube.

in which L is the length of liquid column. By changing the elevation datum to the equilibrium position through the menisci, $g(z_2 - z_1) = 2gz$; since v is a function of t only, $\partial v/\partial t$ may be written dv/dt, or d^2z/dt^2,

$$\frac{d^2z}{dt^2} = \frac{dv}{dt} = -\frac{2g}{L}z \qquad (12.1.3)$$

The general solution of this equation is

$$z = C_1 \cos \sqrt{\frac{2g}{L}}\, t + C_2 \sin \sqrt{\frac{2g}{L}}\, t$$

in which C_1 and C_2 are arbitrary constants of integration. The solution is readily checked by differentiating twice and substituting into the differential equation. To evaluate the constants, if $z = Z$ and $dz/dt = 0$ when $t = 0$, then $C_1 = Z$ and $C_2 = 0$, or

$$z = Z \cos \sqrt{\frac{2g}{L}}\, t \qquad (12.1.4)$$

This equation defines a simple harmonic motion of a meniscus, with a period for a complete oscillation of $2\pi\sqrt{L/2g}$. Velocity of the column may be obtained by differentiating z with respect to t.

Example 12.1 A frictionless fluid column 2.18 m long has a speed of 2 m/s when $z = 0.5$ m. Find (a) the maximum value of z, (b) the maximum speed, and (c) the period.
(a) Differentiating Eq. (12.1.4), after substituting for L, gives

$$\frac{dz}{dt} = -3Z \sin 3t$$

If t_1 is the time when $z = 0.5$ and $dz/dt = 2$,

$$0.5 = Z \cos 3t_1 \qquad -2 = -3Z \sin 3t_1$$

Dividing the second equation by the first equation gives

$$\tan 3t_1 = \tfrac{4}{3}$$

or $3t_1 = 0.927$ rad, $t_1 = 0.309$ s, $\sin 3t_1 = 0.8$, and $\cos 3t_1 = 0.6$. Then $Z = 0.5/\cos 3t_1 = 0.5/0.6 = 0.833$ m, the maximum value of z.
(b) The maximum speed occurs when $\sin 3t = 1$, or

$$3Z = 3 \times 0.833 = 2.499 \text{ m/s}$$

(c) The period is

$$2\pi \sqrt{\frac{L}{2g}} = 2.094 \text{ s}$$

Laminar Resistance

When a shear stress τ_0 at the wall of the tube resists motion of the liquid column, it may be introduced into the Euler equation of motion along a streamline (Fig. 3.8).

The resistance in length δs is $\tau_0\, \pi D\, \delta s$. After dividing through by the mass of the particle $\rho A\, \delta s$, it is $4\tau_0/\rho D$, and Eq. (12.1.1) becomes

$$\frac{1}{\rho}\frac{\partial p}{\partial s} + g\,\frac{\partial z}{\partial s} + v\,\frac{\partial v}{\partial s} + \frac{\partial v}{\partial t} + \frac{4\tau_0}{\rho D} = 0 \qquad (12.1.5)$$

This equation is good for either laminar or turbulent resistance. The assumption is made that the frictional resistance in unsteady flow is the same as for the steady flow at the same velocity. From the Poiseuille equation the shear stress at the wall of a tube is

$$\tau_0 = \frac{8\mu v}{D} \qquad (12.1.6)$$

After making the substitution for τ_0 in Eq. (12.1.5) and integrating with respect to s as before,

$$g(z_2 - z_1) + L\,\frac{\partial v}{\partial t} + \frac{32 v v L}{D^2} = 0$$

Setting $2gz = g(z_2 - z_1)$, changing to total derivatives, and replacing v by dz/dt give

$$\frac{d^2 z}{dt^2} + \frac{32v}{D^2}\frac{dz}{dt} + \frac{2g}{L}z = 0 \qquad (12.1.7)$$

In effect, the column is assumed to have the average velocity dz/dt at any cross section.

By substitution

$$z = C_1 e^{at} + C_2 e^{bt}$$

can be shown to be the general solution of Eq. (12.1.7), provided that

$$a^2 + \frac{32v}{D^2}a + \frac{2g}{L} = 0 \qquad \text{and} \qquad b^2 + \frac{32v}{D^2}b + \frac{2g}{L} = 0$$

C_1 and C_2 are arbitrary constants of integration that are determined by given values of z and dz/dt at a given time. To keep a and b distinct, since the equations defining them are identical, they are taken with opposite signs before the radical term in solution of the quadratics; thus

$$a = -\frac{16v}{D^2} + \sqrt{\left(\frac{16v}{D^2}\right)^2 - \frac{2g}{L}} \qquad b = -\frac{16v}{D^2} - \sqrt{\left(\frac{16v}{D^2}\right)^2 - \frac{2g}{L}}$$

To simplify the formulas, if

$$m = \frac{16v}{D^2} \qquad n = \sqrt{\left(\frac{16v}{D^2}\right)^2 - \frac{2g}{L}}$$

then

$$z = C_1 e^{-mt+nt} + C_2 e^{-mt-nt}$$

When the initial condition is taken that $t = 0$, $z = 0$, $dz/dt = V_0$, then by substitution $C_1 = -C_2$, and

$$z = C_1 e^{-mt}(e^{nt} - e^{-nt}) \qquad (12.1.8)$$

Since

$$\frac{e^{nt} - e^{-nt}}{2} = \sinh nt$$

Eq. (12.1.8) becomes

$$z = 2C_1 e^{-mt} \sinh nt$$

By differentiating with respect to t,

$$\frac{dz}{dt} = 2C_1(-me^{-mt} \sinh nt + ne^{-mt} \cosh nt)$$

and setting $dz/dt = V_0$ for $t = 0$ gives

$$V_0 = 2C_1 n$$

since $\sinh 0 = 0$ and $\cosh 0 = 1$. Then

$$z = \frac{V_0}{n} e^{-mt} \sinh nt \qquad (12.1.9)$$

This equation gives the displacement z of one meniscus of the column as a function of time, starting with the meniscus at $z = 0$ when $t = 0$, and rising with velocity V_0.

Two principal cases† are to be considered. When

$$\frac{16v}{D^2} > \sqrt{\frac{2g}{L}}$$

n is a real number and the viscosity is so great that the motion is damped out in a partial cycle with z never becoming negative, Fig. 12.2 ($m/n = 2$). The time t_0 for maximum z to occur is found by differentiating z [Eq. (12.1.9)] with respect to t and equating to zero,

$$\frac{dz}{dt} = 0 = \frac{V_0}{n}(-me^{-mt} \sinh nt + ne^{-mt} \cosh nt)$$

or $\quad \tanh nt_0 = \dfrac{n}{m} \qquad (12.1.10)$

Substitution of this value of t into Eq. (12.1.9) yields the maximum displacement Z

$$Z = \frac{V_0}{\sqrt{m^2 - n^2}}\left(\frac{m - n}{m + n}\right)^{m/2n} = V_0 \sqrt{\frac{L}{2g}}\left(\frac{m - n}{m + n}\right)^{m/2n} \qquad (12.1.11)$$

---

† A third case, $16v/D^2 = \sqrt{2g/L}$, must be treated separately, yielding $z = V_0 t e^{-mt}$. The resulting oscillation is for a partial cycle only and is a limiting case of $16v/D^2 > \sqrt{2g/L}$.

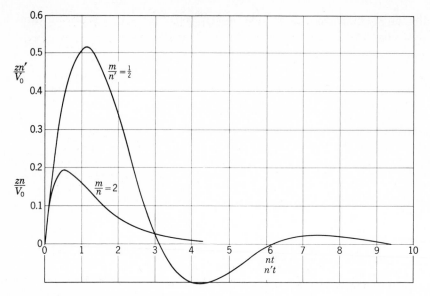

Figure 12.2 Position of meniscus as a function of time for oscillation of liquid in a U tube with laminar resistance.

The second case, when

$$\frac{16v}{D^2} < \sqrt{\frac{2g}{L}}$$

results in a negative term, within the radical,

$$n = \sqrt{-1\left[\frac{2g}{L} - \left(\frac{16v}{D^2}\right)^2\right]} = i\sqrt{\frac{2g}{L} - \left(\frac{16v}{D^2}\right)^2} = in'$$

in which $i = \sqrt{-1}$ and n' is a real number. Replacing n by in' in Eq. (12.1.9) produces the real function

$$z = \frac{V_0}{in'}e^{-mt}\sinh in't = \frac{V_0}{n'}e^{-mt}\sin n't \qquad (12.1.12)$$

since

$$\sin n't = \frac{1}{i}\sinh in't$$

The resulting motion of z is an oscillation about $z = 0$ with decreasing amplitude, as shown in Fig. 12.2 for the case $m/n' = \frac{1}{2}$. The time t_0 of maximum or minimum displacement is obtained from Eq. (12.1.12) by equating $dz/dt = 0$, producing

$$\tan n't_0 = \frac{n'}{m} \qquad (12.1.13)$$

There are an indefinite number of values of t_0 satisfying this expression, corresponding with all the maximum and minimum positions of a meniscus. By substitution of t_0 into Eq. (12.1.12)

$$Z = \frac{V_0}{\sqrt{n'^2 + m^2}} \exp\left(-\frac{m}{n'} \tan^{-1} \frac{n'}{m}\right) = V_0 \sqrt{\frac{L}{2g}} \exp\left(-\frac{m}{n'} \tan^{-1} \frac{n'}{m}\right)$$
$$(12.1.14)$$

Example 12.2 A 1.0-in-diameter U tube contains oil, $v = 1 \times 10^{-4}$ ft^2/s, with a total column length of 120 in. Applying air pressure to one of the tubes makes the gage difference 16 in. By quickly releasing the air pressure, the oil column is free to oscillate. Find the maximum velocity, the maximum Reynolds number, and the equation for position of one meniscus z, in terms of time.

The assumption is made that the flow is laminar, and the Reynolds number will be computed on this basis. The constants m and n are

$$m = \frac{16v}{D^2} = \frac{16 \times 10^{-4}}{(\frac{1}{12})^2} = 0.2302$$

$$n = \sqrt{\left(\frac{16v}{D^2}\right)^2 - \frac{2g}{L}} = \sqrt{0.2302^2 - \frac{2 \times 32.2}{10}} = \sqrt{-6.387} = i2.527$$

or $$n' = 2.527$$

Equations (12.1.12) to (12.1.14) apply to this case, as the liquid will oscillate above and below $z = 0$. The oscillation starts from the maximum position, that is, $Z = 0.667$ ft. By use of Eq. (12.1.14) the velocity (fictitious) when $z = 0$ at time t_0 before the maximum is determined to be

$$V_0 = Z \sqrt{\frac{2g}{L}} \exp\left(\frac{m}{n'} \tan^{-1} \frac{n'}{m}\right) = 0.667 \sqrt{\frac{64.4}{10}} \exp\left(\frac{0.2302}{2.527} \tan^{-1} \frac{2.527}{0.2302}\right)$$

$$= 1.935 \text{ ft/s}$$

and $$\tan n't_0 = \frac{n'}{m} \qquad t_0 = \frac{1}{2.527} \tan^{-1} \frac{2.527}{0.2302} = 0.586 \text{ s}$$

Hence, by substitution into Eq. (12.1.12),

$$z = 0.766 \exp\left[-0.2302(t + 0.586)\right] \sin 2.527(t + 0.586)$$

in which $z = Z$ at $t = 0$. The maximum velocity (actual) occurs for $t > 0$. Differentiating with respect to t to obtain the expression for velocity,

$$V = \frac{dz}{dt} = -0.1763 \exp\left[-0.2302(t + 0.586)\right] \sin 2.57(t + 0.586)$$

$$+ 1.935 \exp\left[-0.2302(t + 0.586)\right] \cos 2.527(t + 0.586)$$

Differentiating again with respect to t and equating to zero to obtain maximum V produces

$$\tan 2.527(t + 0.586) = -0.1837$$

The solution in the second quadrant should produce the desired maximum, $t = 0.584$ s. Substituting this time into the expression for V produces $V = -1.48$ ft/s. The corresponding Reynolds number is

$$\mathbf{R} = \frac{VD}{v} = 1.48\left(\frac{1}{12} \times 10^4\right) = 1234$$

hence the assumption of laminar resistance is justified.

Turbulent Resistance

In the majority of practical cases of oscillation, or surge, in pipe systems there is turbulent resistance. With large pipes and tunnels the Reynolds number is large except for those time periods when the velocity is very near to zero. The assumption of fluid resistance proportional to the square of the average velocity is made (constant f). It closely approximates true conditions, although it yields too small a resistance for slow motions, in which case resistance is almost negligible. The equations will be developed for $f =$ const for oscillation within a simple U tube. This case will then be extended to include oscillation of flow within a pipe or tunnel between two reservoirs, taking into account the minor losses. The assumption is again made that resistance in unsteady flow is given by steady-flow resistance at the same velocity.

Using Eq. (5.10.2) to substitute for τ_0 in Eq. (12.1.5) leads to

$$\frac{1}{\rho}\frac{\partial p}{\partial s} + g\frac{\partial z}{\partial s} + v\frac{\partial v}{\partial s} + \frac{\partial v}{\partial t} + \frac{fv^2}{2D} = 0 \qquad (12.1.15)$$

When this equation is integrated from section 1 to section 2 (Fig. 12.1), the first term drops out as the limits are $p = 0$ in each case, the third term drops out as $\partial v/\partial s \equiv 0$, and the fourth and fifth terms are independent of s; hence,

$$g(z_2 - z_1) + \frac{\partial v}{\partial t}L + \frac{fv^2}{2D}L = 0$$

Since v is a function of t only, the partial may be replaced with the total derivative

$$\frac{dv}{dt} + \frac{f}{2D}v|v| + \frac{2g}{L}z = 0 \qquad (12.1.16)$$

The absolute-value sign on the velocity term is needed so that the resistance opposes the velocity, whether positive or negative. By expressing $v = dz/dt$,

$$\frac{d^2z}{dt^2} + \frac{f}{2D}\frac{dz}{dt}\left|\frac{dz}{dt}\right| + \frac{2g}{L}z = 0 \qquad (12.1.17)$$

This is a nonlinear differential equation because of the v-squared term. It can be integrated once with respect to t, but no closed solution is known for the second integration. It is easily handled by the Runge-Kutta methods (Appendix B.5) with the digital computer when initial conditions are known: $t = t_0, z = z_0, dz/dt = 0$. Much can be learned from Eq. (12.1.17), however, by restricting the motion to the $-z$ direction; thus

$$\frac{d^2z}{dt^2} - \frac{f}{2D}\left(\frac{dz}{dt}\right)^2 + \frac{2g}{L}z = 0 \qquad (12.1.18)$$

The equation may be integrated once,† producing

$$\left(\frac{dz}{dt}\right)^2 = \frac{4gD^2}{f^2L}\left(1 + \frac{fz}{D}\right) + Ce^{fz/D} \qquad (12.1.19)$$

in which C is the constant of integration. To evaluate the constant, if $z = z_m$ for $dz/dt = 0$,

$$C = -\frac{4gD^2}{f^2L}\left(1 + \frac{fz_m}{D}\right)e^{-fz_m/D}$$

and

$$\left(\frac{dz}{dt}\right)^2 = \frac{4gD^2}{f^2L}\left[1 + \frac{fz}{D} - \left(1 + \frac{fz_m}{D}\right)e^{f(z - z_m)/D}\right] \qquad (12.1.20)$$

Although this equation cannot be integrated again, numerical integration of particular situations yields z as a function of t. The equation, however, can be used to determine the magnitude of successive oscillations. At the instants of maximum or minimum z, say z_m and z_{m+1}, respectively, $dz/dt = 0$, and Eq. (12.1.20) simplifies to

$$\left(1 + \frac{fz_m}{D}\right)e^{-fz_m/D} = \left(1 + \frac{fz_{m+1}}{D}\right)e^{-fz_{m+1}/D} \qquad (12.1.21)$$

Since Eq. (12.1.18), the original equation, holds only for decreasing z, z_m must be positive and z_{m+1} negative. To find z_{m+2}, the other meniscus could be considered and z_{m+1} as a positive number substituted into the left-hand side of the equation to determine a minus z_{m+2} in place of z_{m+1} on the right-hand side of the equation.

Example 12.3 A U tube consisting of 500-mm-diameter pipe with $f = 0.03$ has a maximum oscillation (Fig. 12.1) of $z_m = 6$ m. Find the minimum position of the surface and the following maximum.
From Eq. (12.1.21)

$$\left(1 + \frac{0.03 \times 6}{0.5}\right)e^{-0.03(6)/0.5} = (1 + 0.06z_{m+1})e^{-0.06z_{m+1}}$$

or

$$(1 + 0.06z_{m+1})e^{-0.06z_{m+1}} = 0.9488$$

which is satisfied by $z_{m+1} = -4.84$ m. Using $z_m = 4.84$ m in Eq. (12.1.21),

$$(1 + 0.06z_{m+1})e^{-0.06z_{m+1}} = (1 + 0.06 \times 4.84)e^{-0.06(4.84)} = 0.9651$$

which is satisfied by $z_{m+1} = -4.05$ m. Hence, the minimum water surface is $z = -4.84$ m and the next maximum is $z = 4.05$ m.

---

† By substitution of

$$p = \frac{dz}{dt} \qquad \frac{d^2z}{dt^2} = \frac{dp}{dt} = \frac{dp}{dz}\frac{dz}{dt} = p\frac{dp}{dz}$$

then

$$p\frac{dp}{dz} - \frac{f}{2D}p^2 + \frac{2gz}{L} = 0$$

This equation can be made exact by multiplying by the integrating factor $e^{-fz/D}$. For the detailed method see Earl D. Rainville, "Elementary Differential Equations," 3d ed., Macmillan, New York, 1964.

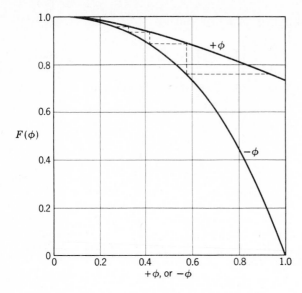

Figure 12.3 Graphical solution of $F(\phi) = (1 + \phi)e^{-\phi}$, yielding successive maximum and minimum displacements.

Equation (12.1.21) can be solved graphically. If $\phi = fz/D$, then

$$F(\phi) = (1 + \phi)e^{-\phi} \qquad (12.1.22)$$

which is conveniently plotted with $F(\phi)$ as ordinate and both $-\phi$ and $+\phi$ on the same abscissa scale (Fig. 12.3). Successive values of ϕ are found as indicated by the dashed stepped line.

Although z cannot be found as a function of t from Eq. (12.1.20), V is given as a function of z, since $V = dz/dt$. The maximum value of V is found by equating $dV^2/dz = 0$ to find its position z'; thus

$$\frac{dV^2}{dz} = 0 = \frac{f}{D} - \left(1 + \frac{fz_m}{D}\right)\left[\exp\frac{f(z' - z_m)}{D}\right]\frac{f}{D}$$

After solving for z',

$$z' = z_m - \frac{D}{f} \ln\left(1 + \frac{fz_m}{D}\right)$$

and after substituting back into Eq. (12.1.20),

$$V_m^2 = \frac{4gD^2}{f^2 L}\left[\frac{fz_m}{D} - \ln\left(1 + \frac{fz_m}{D}\right)\right] \qquad (12.1.23)$$

Oscillation of Two Reservoirs

The equation for oscillation of two reservoirs connected by a pipeline is the same as that for oscillation of a U tube, except for value of constant terms. If z_1 and z_2 represent displacements of the reservoir surfaces from their equilibrium positions

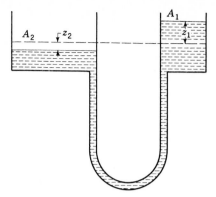

Figure 12.4 Oscillation of two reservoirs.

(Fig. 12.4), and if z represents displacement of a water particle within the connecting pipe from its equilibrium position,

$$zA = z_1 A_1 = z_2 A_2$$

in which A_1 and A_2 are the reservoir areas, assumed to be constant in this derivation. Taking into account minor losses in the system by using the equivalent length L_e of pipe and fittings plus other minor losses, the Euler equation with resistance included is

$$-\gamma A(z_1 + z_2) + \frac{\gamma A f L_e}{2gD}\left(\frac{dz}{dt}\right)^2 = \frac{\gamma A L}{g}\frac{d^2z}{dt^2}$$

for z decreasing. After simplifying

$$\frac{d^2z}{dt^2} - \frac{f}{2D}\frac{L_e}{L}\left(\frac{dz}{dt}\right)^2 + \frac{gA}{L}\left(\frac{1}{A_1} + \frac{1}{A_2}\right)z = 0 \qquad (12.1.24)$$

After comparing with Eq. (12.1.18), f is replaced by $f L_e/L$, and $2g/L$ by $gA(1/A_1 + 1/A_2)/L$. In Eq. (12.1.22)

$$\phi = f\,\frac{L_e}{L}\frac{z}{D}$$

Example 12.4 In Fig. 12.4 a valve is opened suddenly in the pipeline when $z_1 = 40$ ft. $L = 2000$ ft, $A_1 = 200$ ft^2, $A_2 = 300$ ft^2, $D = 3.0$ ft, $f = 0.024$, and minor losses are $3.50V^2/2g$. Determine the subsequent maximum negative and positive surges in the reservoir A_1. The equivalent length of minor losses is

$$\frac{KD}{f} = \frac{3.5 \times 3}{0.024} = 438 \text{ ft}$$

Then $L_e = 2000 + 438 = 2438$ and

$$z_m = \frac{z_1 A_1}{A} = \frac{40 \times 200}{2.25\pi} = 1132 \text{ ft}$$

The corresponding ϕ is

$$\phi = f \frac{L_e}{L} \frac{z_m}{D} = 0.024 \left(\frac{2438}{2000} \right) \left(\frac{1132}{3} \right) = 11.04$$

and

$$F(\phi) = (1 + \phi) e^{-\phi} = (1 + 11.04) e^{-11.04} = 0.000193$$

which is satisfied by $\phi \approx -1.0$. Then

$$F(\phi) = (1 + 1) e^{-1} = 0.736 = (1 + \phi) e^{-\phi}$$

which is satisfied by $\phi = -0.593$. The values of z_m are, for $\phi = -1$,

$$z_m = \frac{\phi L D}{f L_e} = \frac{-1 \times 2000 \times 3}{0.024 \times 2438} = -102.6$$

and for $\phi = 0.593$,

$$z_m = \frac{0.593 \times 2000 \times 3}{0.024 \times 2438} = 60.9$$

The corresponding values of z_1 are

$$z_1 = z_m \frac{A}{A_1} = -102.6 \frac{2.25\pi}{200} = -3.63 \text{ ft}$$

and

$$z_1 = 60.9 \frac{2.25\pi}{200} = 2.15 \text{ ft}$$

EXERCISES

12.1.1 Neglecting friction, the maximum difference in elevation of the two menisci of an oscillating U tube is 1.0 ft, $L = 3.0$ ft. The period of oscillation is, in seconds, (a) 0.52; (b) 1.92; (c) 3.27; (d) 20.6; (e) none of these answers.

12.1.2 The maximum speed of the liquid column in Exercise 12.1.1, in feet per second, is (a) 0.15; (b) 0.31; (c) 1.64; (d) 3.28; (e) none of these answers.

12.1.3 In frictionless oscillation of a U tube, $L = 2.179$ m, $z = 0$, $V = 6$ m/s. The maximum value of z is, in meters, (a) 1.57; (b) 2.179; (c) 2.00; (d) 13.074; (e) none of these answers.

12.1.4 In analyzing the oscillation of a U tube with laminar resistance, the assumption is made that the (a) motion is steady; (b) resistance is constant; (c) Darcy-Weisbach equation applies; (d) resistance is a linear function of the displacement; (e) resistance is the same at any instant as if the motion were steady.

12.1.5 When $16v/D^2 = 5$ and $2g/L = 12$ in oscillation of a U tube with laminar resistance, (a) the resistance is so small that it may be neglected; (b) the menisci oscillate about the $z = 0$ axis; (c) the velocity is a maximum when $z = 0$; (d) the velocity is zero when $z = 0$; (e) the speed of column is a linear function of z.

12.1.6 In laminar resistance to oscillation in a U tube, $m = 1$, $n = \frac{1}{2}$, $V_0 = 3$ ft/s when $t = 0$ and $z = 0$. The time of maximum displacement of meniscus is, in seconds, (a) 0.46; (b) 0.55; (c) 0.93; (d) 1.1; (e) none of these answers.

12.1.7 In Exercise 12.1.6 the maximum displacement, in feet, is (a) 0.53; (b) 1.06; (c) 1.16; (d) 6.80; (e) none of these answers.

12.1.8 In analyzing the oscillation of a U tube with turbulent resistance, the assumption is made that (a) the Darcy-Weisbach equation applies; (b) the Hagen-Poiseuille equation applies; (c) the motion is steady; (d) the resistance is a linear function of velocity; (e) the resistance varies as the square of the displacement.

12.1.9 The maximum displacement is $z_m = 20$ ft for $f = 0.020$, $D = 1.0$ ft in oscillation of a U tube with turbulent flow. The minimum displacement $(-z_{m+1})$ of the same fluid column is (a) -13.3; (b) -15.8; (c) -16.5; (d) -20; (e) none of these answers.

12.2 ESTABLISHMENT OF FLOW

The problem of determination of time for flow to become established in a pipeline when a valve is suddenly opened is easily handled when friction and minor losses are taken into account. After a valve is opened (Fig. 12.5), the head H is available to accelerate the flow in the first instants, but as the velocity increases, the accelerating head is reduced by friction and minor losses. If L_e is the equivalent length of the pipe system, the final velocity V_0 is given by application of the energy equation

$$H = f \frac{L_e}{D} \frac{V_0^2}{2g} \qquad (12.2.1)$$

The equation of motion is

$$\gamma A \left(H - f \frac{L_e}{D} \frac{V^2}{2g} \right) = \frac{\gamma A L}{g} \frac{dV}{dt}$$

Solving for dt and rearranging, with Eq. (12.2.1), give

$$\int_0^t dt = \frac{LV_0^2}{gH} \int_0^V \frac{dV}{V_0^2 - V^2}$$

After integration,

$$t = \frac{LV_0}{2gH} \ln \frac{V_0 + V}{V_0 - V} \qquad (12.2.2)$$

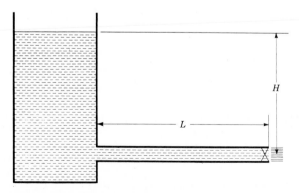

Figure 12.5 Notation for establishment of flow.

The velocity V approaches V_0 asymptotically; i.e., mathematically it takes infinite time for V to attain the value V_0. Practically, for V to reach $0.99\, V_0$ takes

$$t = \frac{LV_0}{gH}\frac{1}{2}\ln\frac{1.99}{0.01} = 2.646\frac{LV_0}{gH}$$

V_0 must be determined by taking minor losses into account, but Eq. (12.2.2) does not contain L_e.

Example 12.5 In Fig. 12.5 the minor losses are $16V^2/2g$, $f = 0.030$, $L = 3000$ m, $D = 2.4$ m, and $H = 20$ m. Determine the time, after the sudden opening of a valve, for velocity to attain nine-tenths the final velocity.

$$L_e = 3000 + \frac{16 \times 2.4}{0.03} = 4280 \text{ m}$$

From Eq. (12.2.1)

$$V_0 = \sqrt{\frac{2gHD}{fL_e}} = \sqrt{\frac{19.612 \times 20 \times 2.4}{0.030 \times 4280}} = 2.708 \text{ m/s}$$

Substituting $V = 0.9V_0$ into Eq. (12.2.2) gives

$$t = \frac{3000 \times 2.708}{19.612 \times 20}\ln\frac{1.90}{0.10} = 60.98 \text{ s}$$

EXERCISE

12.2.1 When a valve is suddenly opened at the downstream end of a long pipe connected at its upstream end with a water reservoir, (a) the velocity attains its final value instantaneously if friction is neglected; (b) the time to attain nine-tenths of its final velocity is less with friction than without friction; (c) the value of f does not affect the time to acquire a given velocity; (d) the velocity increases exponentially with time; (e) the final velocity is attained in a finite time.

12.3 SURGE CONTROL

The oscillation of flow in pipelines, when compressibility effects are not important, is referred to as *surge*. For sudden deceleration of flow due to closure of the flow passage, compressibility of the liquid and elasticity of the pipe walls must be considered; this phenomenon, known as *waterhammer*, is discussed in Secs. 12.4 to 12.8. Oscillations in a U tube are special cases of surge. As one means of eliminating waterhammer, provision is made to permit the liquid to surge into a tank (Fig. 12.6). The valve at the end of a pipeline may be controlled by a turbine governor, and may rapidly stop the flow if the generator loses its load. To destroy all momentum in the long pipe system quickly would require high pressure, which in turn would require a very costly pipeline. With a surge tank as near the valve as feasible, although surge will occur between the reservoir and surge tank, development of high pressure in this reach is prevented. It is still necessary to design the pipeline between surge tank and valve to withstand waterhammer.

Surge tanks may be classified as *simple*, *orifice*, and *differential*. The simple

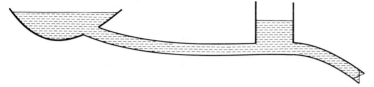

Figure 12.6 Surge tank on a long pipeline.

surge tank has an unrestricted opening into it and must be large enough not to overflow (unless a spillway is provided) or not to be emptied, allowing air to enter the pipeline. It must also be of a size that will not fluctuate in resonance with the governor action on the valve. The period of oscillation of a simple surge tank is relatively long.

The orifice surge tank has a restricted opening, or orifice, between pipeline and tank and hence allows more rapid pressure changes in the pipeline than the simple surge tank. The more rapid pressure change causes a more rapid adjustment of flow to the new valve setting, and losses through the orifice aid in dissipating excess available energy resulting from valve closure.

The differential surge tank (Fig. 12.7) is in effect a combination of an orifice surge tank and a simple surge tank of small cross-sectional area. In case of rapid valve opening a limited amount of liquid is directly available from the central riser, and flow from the large tank supplements this flow. For sudden valve closures the central riser may be so designed that it overflows into the outside tank.

Surge tanks operating under air pressure are utilized in certain circumstances, e.g., after a reciprocating pump. They are generally uneconomical for large pipelines.

Detailed analysis of surge tanks entails a numerical integration of the equation of motion for the liquid in the pipeline, taking into account the particular rate

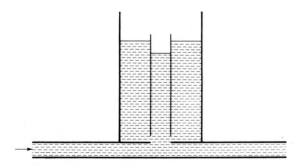

Figure 12.7 Differential surge tank.

of valve closure, together with the continuity equation. The particular type of surge tank to be selected for a given situation depends upon a detailed study of the economics of the pipeline system. High-speed digital computers are most helpful in their design.

Another means of controlling surge and waterhammer is to supply a quick-opening bypass valve that opens when the control valve closes. The quick-opening valve has a controlled slow closure at such a rate that excessive pressure is not developed in the line. The bypass valve wastes liquid, however, and does not provide relief from surge due to opening of the control valve or starting of a pump.

The following sections on closed-conduit flow take into account compressibility of the liquid and elasticity of the pipe walls. Waterhammer calculations may be accomplished in several ways; the characteristics method, recommended for general use in computer solutions, is presented here.

EXERCISE

12.3.1 Surge may be differentiated from waterhammer by (a) the time for a pressure wave to traverse the pipe; (b) the presence of a reservoir at one end of the pipe; (c) the rate of deceleration of flow; (d) the relative compressibility of liquid to expansion of pipe walls; (e) the length-diameter ratio of pipe.

12.4 DESCRIPTION OF THE WATERHAMMER PHENOMENON

Waterhammer may occur in a closed conduit flowing full when there is either a retardation or acceleration of the flow, such as with the change in opening of a valve in the line. If the changes are gradual, the calculations may be carried out by surge methods, considering the liquid incompressible and the conduit rigid. When a valve is rapidly closed in a pipeline during flow, the flow through the valve is reduced. This increases the head on the upstream side of the valve and causes a pulse of high pressure to be propagated upstream at the sonic wave speed a. The action of this pressure pulse is to decrease the velocity of flow. On the downstream side of the valve the pressure is reduced, and a wave of lowered pressure travels downstream at wave speed a, which also reduces the velocity. If the closure is rapid enough and the steady pressure low enough, a vapor pocket may be formed downstream from the valve. When this occurs, the cavity will eventually collapse and produce a high-pressure wave downstream.

Before undertaking the derivation of equations for solution of waterhammer, a description of the sequence of events following sudden closure of a valve at the downstream end of a pipe leading from a reservoir (Fig. 12.8) is given. Friction is neglected in this case. At the instant of valve closure ($t = 0$) the fluid nearest the valve is compressed and brought to rest, and the pipe wall is stretched (Fig. 12.8a). As soon as the first layer is compressed, the process is repeated for the next layer. The fluid upstream from the valve continues to move downstream with undiminished speed until successive layers have been compressed back to the source.

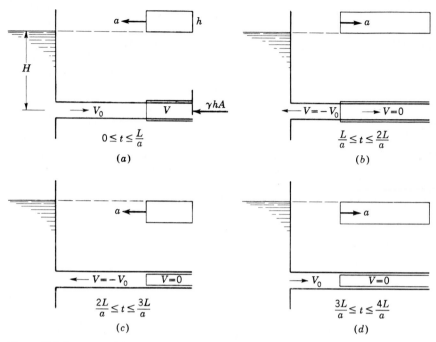

Figure 12.8 Sequence of events for one cycle of sudden closure of a valve.

The high pressure moves upstream as a wave, bringing the fluid to rest as it passes, compressing it, and expanding the pipe. When the wave reaches the upstream end of the pipe ($t = L/a$ s), all the fluid is under the extra head h, all the momentum has been lost, and all the kinetic energy has been converted into elastic energy.

There is an unbalanced condition at the upstream (reservoir) end at the instant of arrival of the pressure wave, as the reservoir pressure is unchanged. The fluid starts to flow backward, beginning at the upstream end. This flow returns the pressure to the value which was normal before closure, the pipe wall returns to normal, and the fluid has a velocity V_0 in the backward sense. This process of conversion travels downstream toward the valve at the speed of sound a in the pipe. At the instant $2L/a$ the wave arrives at the valve, pressures are back to normal along the pipe, and the velocity is everywhere V_0 in the backward direction.

Since the valve is closed, no fluid is available to maintain the flow at the valve and a low pressure develops $(-h)$ such that the fluid is brought to rest. This low-pressure wave travels upstream at speed a and everywhere brings the fluid to rest, causes it to expand because of the lower pressure, and allows the pipe walls to contract. (If the static pressure in the pipe is not sufficiently high to sustain head $-h$ above vapor pressure, the liquid vaporizes in part and continues to move backward over a longer period of time.)

At the instant the negative pressure wave arrives at the upstream end of the pipe, $3L/a$ s after closure, the fluid is at rest but uniformly at head $-h$ less than before closure. This leaves an unbalanced condition at the reservoir, and fluid flows into the pipe, acquiring a velocity V_0 forward and returning the pipe and fluid to normal conditions as the wave progresses downstream at speed a. At the instant this wave reaches the valve, conditions are exactly the same as at the instant of closure, $4L/a$ s earlier.

This process is then repeated every $4L/a$ s. The action of fluid friction and imperfect elasticity of fluid and pipe wall, neglected heretofore, is to damp out the vibration and eventually cause the fluid to come permanently to rest. Closure of a valve in less than $2L/a$ is called *rapid closure*; *slow closure* refers to times of closure greater than $2L/a$.

The sequence of events taking place in a pipe may be compared with the sudden stopping of a freight train when the engine hits an immovable object. The car behind the engine compresses the spring in its forward coupling and stops as it exerts a force against the engine, and each car in turn keeps moving at its original speed until the preceding one suddenly comes to rest. When the caboose is at rest, all the energy is stored in compressing the coupling springs (neglecting losses). The caboose has an unbalanced force exerted on it, and starts to move backward, which in turn causes an unbalanced force on the next car, setting it in backward motion. This action proceeds as a wave toward the engine, causing each car to move at its original speed in a backward direction. If the engine is immovable, the car next to it is stopped by a tensile force in the coupling between it and the engine, analogous to the low-pressure wave in waterhammer. The process repeats itself car by car until the train is again at rest, with all couplings in tension. The caboose is then acted upon by the unbalanced tensile force in its coupling and is set into forward motion, followed in turn by the rest of the cars. When this wave reaches the engine, all cars are in motion as before the original impact. Then the whole cycle is repeated again. Friction acts to reduce the energy to zero in a very few cycles.

EXERCISES

12.4.1 Waterhammer occurs only when (a) $2L/a > 1$; (b) $V_0 > a$; (c) $2L/a = 1$; (d) $2L/a < 1$; (e) compressibility effects are important.

12.4.2 Valve closure is *rapid* only when (t_c = time of closure) (a) $2L/a \geq t_c$; (b) $L/a \geq t_c$; (c) $L/2a \geq t_c$; (d) $t_c = 0$; (e) none of these answers.

12.5 DIFFERENTIAL EQUATIONS FOR CALCULATION OF WATERHAMMER

Two basic mechanics equations are applied to a short segment of fluid in a pipe to obtain the differential equations for transient flow: Newton's second law of motion and the continuity equation. The dependent variables are pressure p and

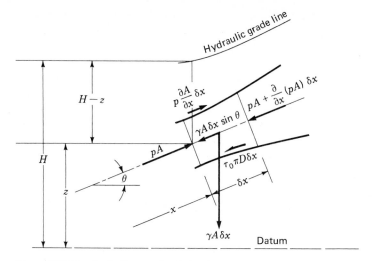

Figure 12.9 Free-body diagram for derivation of equation of motion.

the average velocity V at a cross section. The independent variables are distance x along the pipe measured from the upstream end and time t; hence, $p = p(x, t)$, $V = V(x, t)$. Poisson's ratio effect is not taken into account in this derivation. For pipelines with expansion joints it does not enter into the derivation. Friction is considered to be proportional to the square of the velocity.

Equation of Motion

The fluid element between two parallel planes δx apart, normal to the pipe axis, is taken as a free body for application of Newton's second law of motion in the axial direction (Fig. 12.9). In equation form

$$pA - \left[pA + \frac{\partial}{\partial x}(pA)\,\delta x \right] + p\,\frac{\partial A}{\partial x}\,\delta x - \gamma A\,\delta x \sin \theta - \tau_0 \pi D\,\delta x = \rho A\,\delta x\,\frac{dV}{dt}$$

Dividing through by the mass of the element $\rho A\,\delta x$ and simplifying give

$$-\frac{1}{\rho}\frac{\partial p}{\partial x} - g \sin \theta - \frac{4\tau_0}{\rho D} = \frac{dV}{dt} \tag{12.5.1}$$

For steady turbulent flow, $\tau_0 = \rho f V^2/8$ [Eq. (5.10.2)]. The assumption is made that the friction factor in unsteady flow is the same as in steady flow. Hence, the equation of motion becomes

$$\frac{dV}{dt} + \frac{1}{\rho}\frac{\partial p}{\partial x} + g \sin \theta + \frac{f V|V|}{2D} = 0 \tag{12.5.2}$$

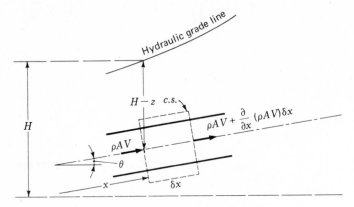

Figure 12.10 Control volume for derivation of continuity equation.

Since friction must oppose the motion, V^2 has been written as $V|V|$. By expanding the acceleration term,

$$\frac{dV}{dt} = V\,\frac{\partial V}{\partial x} + \frac{\partial V}{\partial t}$$

In waterhammer applications the term $V\,\partial V/\partial x$ is generally much smaller than $\partial V/\partial t$; hence, it will be omitted, leaving

$$L_1 = \frac{\partial V}{\partial t} + \frac{1}{\rho}\,\frac{\partial p}{\partial x} + g\sin\theta + \frac{fV|V|}{2D} = 0 \qquad (12.5.3)$$

The equation is indicated by L_1 to distinguish it from the equation of continuity L_2, which is next derived.

Equation of Continuity

The unsteady continuity equation (3.3.1) is applied to the control volume of Fig. 12.10,

$$-\frac{\partial}{\partial x}(\rho AV)\,\delta x = \frac{\partial}{\partial t}(\rho A\,\delta x) \qquad (12.5.4)$$

in which δx is not a function of t. Expanding the equation and dividing through by the mass $\rho A\,\delta x$ give

$$\frac{V}{A}\,\frac{\partial A}{\partial x} + \frac{1}{A}\,\frac{\partial A}{\partial t} + \frac{V}{\rho}\,\frac{\partial\rho}{\partial x} + \frac{1}{\rho}\,\frac{\partial\rho}{\partial t} + \frac{\partial V}{\partial x} = 0 \qquad (12.5.5)$$

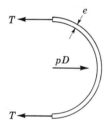

Figure 12.11 Tensile force in pipe wall.

The first two terms are the total derivative $(1/A) \, dA/dt$, and the next two terms are the total derivative $(1/\rho) \, d\rho/dt$, yielding

$$\frac{1}{A}\frac{dA}{dt} + \frac{1}{\rho}\frac{d\rho}{dt} + \frac{\partial V}{\partial x} = 0 \qquad (12.5.6)$$

The first term deals with the elasticity of the pipe wall and its rate of deformation with pressure; the second term takes into account the compressibility of the liquid. For the wall elasticity the rate of change of tensile force per unit length (Fig. 12.11) is $(D/2) \, dp/dt$; when divided by the wall thickness e, it is the rate of change of unit stress $(D/2e) \, dp/dt$; when this is divided by Young's modulus of elasticity for the wall material, the rate of increase of unit strain is obtained, $(D/2eE) \, dp/dt$. After multiplying this by the radius $D/2$, the rate of radial extension is obtained; finally, by multiplying by the perimeter πD, the rate of area increase is obtained:

$$\frac{dA}{dt} = \frac{D}{2eE}\frac{dp}{dt}\frac{D}{2}\pi D$$

and hence

$$\frac{1}{A}\frac{dA}{dt} = \frac{D}{eE}\frac{dp}{dt} \qquad (12.5.7)$$

From the definition of bulk modulus of elasticity of fluid (Chap. 1),

$$K = -\frac{dp}{d\mathcal{V}/\mathcal{V}} = \frac{dp}{d\rho/\rho}$$

and the rate of change of density divided by density yields

$$\frac{1}{\rho}\frac{d\rho}{dt} = \frac{1}{K}\frac{dp}{dt} \qquad (12.5.8)$$

By Eqs. (12.5.7) and (12.5.8), Eq. (12.5.6) becomes

$$\frac{1}{K}\frac{dp}{dt}\left(1 + \frac{K}{E}\frac{D}{e}\right) + \frac{\partial V}{\partial x} = 0 \qquad (12.5.9)$$

It is convenient to express the constants in this equation in the form

$$a^2 = \frac{K/\rho}{1 + (K/E)(D/e)c_1} \qquad (12.5.10)$$

in which c_1 is unity for the pipeline with expansion joints. Equation (12.5.9) now becomes

$$\frac{1}{\rho}\frac{dp}{dt} + a^2 \frac{\partial V}{\partial x} = 0 \tag{12.5.11}$$

By expanding dp/dt,

$$\frac{dp}{dt} = V \frac{\partial p}{\partial x} + \frac{\partial p}{\partial t}$$

Again, for waterhammer applications the term $V\,\partial p/\partial x$ is usually much smaller than $\partial p/\partial t$ and is neglected, yielding

$$L_2 = \frac{\partial p}{\partial t} + \rho a^2 \frac{\partial V}{\partial x} = 0 \tag{12.5.12}$$

which is the continuity equation for a compressible liquid in an elastic pipe. L_1 and L_2 provide two nonlinear partial differential equations in V and p in terms of the independent variables x and t. No general solution to these equations is known, but they can be solved by the method of characteristics for a convenient finite-difference solution with the digital computer.

12.6 THE METHOD-OF-CHARACTERISTICS SOLUTION

Equations L_1 and L_2 in the preceding section contain two unknowns. These equations may be combined with an unknown multiplier as $L = L_1 + \lambda L_2$. Any two real, distinct values of λ yield two equations in V and p that contain all the physics of the original two equations L_1 and L_2 and may replace them in any solution. It may happen that great simplification will result if two particular values of λ are found. L_1 and L_2 are substituted into the equation for L, with some rearrangement.

$$L = \left(\frac{\partial V}{\partial x}\lambda\rho a^2 + \frac{\partial V}{\partial t}\right) + \lambda\left(\frac{\partial p}{\partial x}\frac{1}{\rho\lambda} + \frac{\partial p}{\partial t}\right) + g\sin\theta + \frac{fV|V|}{2D} = 0$$

The first term in parentheses is the total derivative dV/dt if $\lambda\rho a^2 = dx/dt$, from the calculus

$$\frac{dV}{dt} = \frac{\partial V}{\partial x}\frac{dx}{dt} + \frac{\partial V}{\partial t} = \frac{\partial V}{\partial x}\lambda\rho a^2 + \frac{\partial V}{\partial t}$$

Similarly, the second term in parentheses is the total derivative dp/dt if $1/\rho\lambda = dx/dt$. For both statements to be correct, dx/dt must have the same value

$$\frac{dx}{dt} = \lambda\rho a^2 = \frac{1}{\lambda\rho} \quad \text{or} \quad \lambda = \pm\frac{1}{\rho a} \tag{12.6.1}$$

Then
$$\frac{dx}{dt} = \pm a \tag{12.6.2}$$

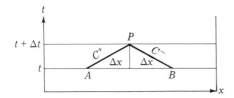

Figure 12.12 xt plot of characteristics along which solution is obtained.

The equation for L now becomes

$$L = \frac{dV}{dt} \pm \frac{1}{\rho a}\frac{dp}{dt} + g\sin\theta + \frac{fV|V|}{2D} = 0 \qquad (12.6.3)$$

subject to the conditions of Eq. (12.6.2). Therefore, two real distinct values of λ have been found that convert the two partial differential equations into the pair of total differential equations (12.6.3), subject to Eqs. (12.6.2).

Since Eq. (12.6.3) is valid only when Eq. (12.6.2) is satisfied, it is convenient to visualize the solution on a plot of x against t, as in Fig. 12.12. The pipe may be considered to be from the origin O upstream to L at its downstream end. Hence, x locates a point in the pipeline and t the time at which the dependent variables V and p are to be determined. First consider that conditions at A are known (that is, V_A, p_A, x_A, and t_A). Then Eq. (12.6.3) with the $+$ sign, called the C^+ equation, is valid along the line AP or an extension of the line. The slope of line AP is $dt/dx = 1/a$, in which a is the speed of an acoustical wave through the pipe. By multiplying Eq. (12.6.3) through by $\rho a\,dt$, and integrating from A to P,

$$\rho a\int_A^P dV + \int_A^P dp + \int_A^P \rho a g \sin\theta\,dt + \int_A^P \rho a\,dt\,\frac{fV|V|}{2D} = 0$$

Since $a\,dt = dx$, the equation may be written in finite-difference form,

$$\rho a(V_P - V_A) + p_P - p_A + \rho g \sin\theta\,\Delta x + \frac{\rho\,\Delta x f|V_A|V_A}{2D} = 0 \qquad (12.6.4)$$

This form of the equation assumes θ constant from A to P along the pipe and evaluates the last term (the friction term, which is usually small) by taking V to be constant at the known value at A. The corresponding C^- equation, in a similar manner, becomes

$$\rho a(V_P - V_B) - p_P + p_B + \rho g \sin\theta\,\Delta x + \frac{\rho\,\Delta x f|V_B|V_B}{2D} = 0 \qquad (12.6.5)$$

since $a\,dt = -dx$. Δx is a positive number, the reach length. These two equations may be solved simultaneously to determine p_P and V_P.

In solving piping problems it is convenient to work with hydraulic grade line H and discharge Q in place of p and V. From Fig. 12.9

$$p_P = \rho g(H_P - z_P) \qquad p_A = \rho g(H_A - z_A)$$

and
$$p_P - p_A = \rho g(H_P - H_A) - \rho g(z_P - z_A)$$

$$= \rho g(H_P - H_A) - \rho g\,\Delta x \sin \theta \tag{12.6.6}$$

Substitution in Eq. (12.6.4), with $V = Q/A$, yields

C^+:
$$H_P = H_A - \frac{a}{gA}(Q_P - Q_A) - \frac{\Delta x f Q_A |Q_A|}{2gDA^2} \tag{12.6.7}$$

C^-:
$$H_P = H_B + \frac{a}{gA}(Q_P - Q_B) + \frac{\Delta x f Q_B |Q_B|}{2gDA^2} \tag{12.6.8}$$

The C^- equation from B to P is obtained in a manner similar to the C^+ equation. To simplify the equations, let

$$B = \frac{a}{gA} \qquad R = \frac{f\,\Delta x}{2gDA^2}$$

Then

C^+:
$$H_P = H_A - B(Q_P - Q_A) - RQ_A|Q_A| \tag{12.6.9}$$

C^-:
$$H_P = H_B + B(Q_P - Q_B) + RQ_B|Q_B| \tag{12.6.10}$$

Example 12.6 The two waterhammer equations are to be used to solve for the head at the dead end of the system of Fig. 12.13. $f = 0.018$, $a = 1200$ m/s. The wave frequency is π rad/s, the natural frequency of the system, that is, $2\pi/(4L/a)$.
The hydraulic grade line at the reservoir is always known

$$H_{P_A} = 100 + 3 \sin \pi t$$

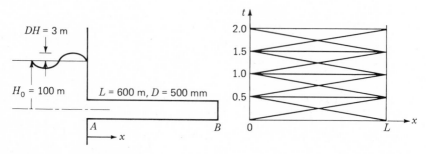

Figure 12.13 Example 12.6.

and the discharge at the downstream end is $Q_B = 0$ at all times, as it is a dead end. By using only one reach, of length L, $dt = L/a = 0.5$ s.

$$B = \frac{a}{gA} = \frac{1200}{9.806(\pi/16)} = 623.25$$

$$R = \frac{f \, \Delta x}{2gDA^2} = \frac{0.018 \times 600}{2 \times 9.806 \times 0.5 \times (\pi/16)^2} = 28.57$$

The two equations (12.6.9) and (12.6.10) become

C^+: $\quad H_{P_B} = H_A - 623.25(Q_{P_B} - Q_A) - 28.57Q_A|Q_A|$

C^-: $\quad H_{P_A} = 100 + 3 \sin \pi t = H_B + 623.25(Q_{P_A} - Q_B) + 28.57Q_B|Q_B|$

At $t = 0$, $H_A = H_B = 100$ m, $Q_A = 0$, $Q_B \equiv 0$. It must be remembered that the procedure is to start with Q_A and H_B known, then solve for H_{P_B} and Q_{P_A}. Next, H_B takes the value of H_{P_B} and Q_A takes the value Q_{P_A}, the time is incremented by Δt, and the procedure is repeated. The simplified equations are

$$H_{P_B} = H_A + 623.25Q_A - 28.57Q_A|Q_A|$$

$$Q_{P_A} = \frac{100 + 3 \sin \pi t - H_B}{623.25}$$

with $H_A = H_B = 100$ for $t = 0$. Results of the calculation, using a programmable calculator are

t	H_{P_A}	Q_{P_A}	H_{P_B}
0	100	0	100
0.5	103	0.00481	100
1	100	0	106
1.5	97	-0.0144	100
2	100	0	88
2.5	103	0.0241	100
3	100	0	117.98
3.5	97	-0.0337	100
4	100	0	76.06
4.5	103	0.0432	100
5	100	0	129.89
5.5	97	-0.0528	100
6	100	0	64.19
6.5	103	0.0623	100
7	100	0	141.70
7.5	97	-0.0717	100
8	100	0	52.45

By exciting the pipe at its natural period, $4L/a$, it is to be noted that a resonance condition is building in the line. The fluctuation will continue to grow until friction forces bring it into a steady oscillatory motion or until vapor pressure is reached. In this example vapor pressure at the dead end would be reached. The equations are not valid then.[†]

† E. B. Wylie and V. L. Streeter, "Fluid Transients," McGraw-Hill, New York, 1978.

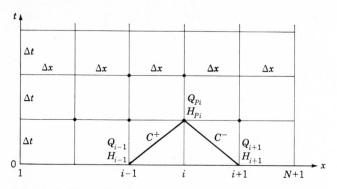

Figure 12.14 Rectangular grid for solution of characteristics equations.

Digital Computer Solution

For the digital computer solution of the transient equations (12.6.9) and (12.6.10) for a single pipe, N reaches of equal length are selected so that $\Delta x = L/N$ and $\Delta t = \Delta x/a$. The C^+ and C^- lines are then the diagonals of the rectangular grid, Fig. 12.14. A subscripted notation is convenient, as shown in the figure. In applying the equations to solve for an internal section, where H_{P_i} and Q_{P_i} are desired, conditions at the earlier time are known, that is, Q_{i-1}, H_{i-1}, Q_{i+1}, H_{i+1}. By collecting the known terms of Eq. (12.6.9) into one constant C_P

$$C_P = H_{i-1} + Q_{i-1}(B - R|Q_{i-1}|) \tag{12.6.11}$$

and for Eq. (12.6.10) into the constant C_M

$$C_M = H_{i+1} - Q_{i+1}(B - R|Q_{i+1}|) \tag{12.6.12}$$

Now Eqs. (12.6.9) and (12.6.10) become

C^+:
$$H_{P_i} = C_P - BQ_{P_i} \tag{12.6.13}$$

C^-:
$$H_{P_i} = C_M + BQ_{P_i} \tag{12.6.14}$$

With C_P and C_M known, solution of Eqs. (12.6.13) and (12.6.14) yields

$$H_{P_i} = \tfrac{1}{2}(C_P + C_M)$$

$$Q_{P_i} = \frac{C_P - H_{P_i}}{B}$$

All interior sections may be calculated in this manner.

Concept of Boundary Conditions

At the upstream end of a pipe, Eq. (12.6.14) for the C^- characteristic provides one equation in the two unknowns Q_{P_1} and H_{P_1} (Fig. 12.15a). One condition is needed exterior to the pipe to relate the pipeline response to the boundary-condition behavior. This condition may be a constant value of one of the variables, such as a

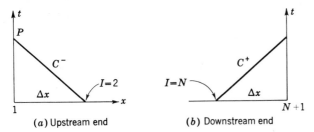

Figure 12.15 Boundary conditions.

constant-head reservoir, a specified variation of one of the variables as a function of time, an algebraic relation between the two variables, or a relation in the form of a differential equation. Some boundary conditions may involve additional variables, e.g., unknown pump speed in the case of a centrifugal pump connected at the upstream end of the line. In this case two independent equations must be available to combine with Eq. (12.6.14) to solve for the three unknowns at each time step. The simplest boundary condition is one in which one of the variables is given as a function of time or held constant. A direct solution of Eq. (12.6.14) for the other variable at each time step provides a complete solution of the interaction of the fluid in the pipeline and the particular boundary. This includes the appropriate reflection and transmission of transient pressure and flow waves that arrive at the pipe end.

At the downstream end of the pipe (Fig. 12.15b), Eq. (12.6.13) for the C^+ characteristic provides one equation in the two variables $H_{P_{N+1}}$ and $Q_{P_{N+1}}$. One external condition is needed that either specifies one of the variables to be constant or a known function of time or provides a relation between the variables in algebraic or differential equation form. The simplest end condition is one in which a variable is held constant, as at a dead end where $Q_{P_{N+1}}$ is zero. Then Eq. (12.6.13) provides a direct solution for $H_{P_{N+1}}$ at each time step.

A computer program is provided in Sec. 12.8; it involves simple boundary conditions on a single pipeline. Complex systems can be visualized as a combination of single pipelines that are handled as described above, with boundary conditions at the pipe ends to transfer the transient response from one pipeline to another and to provide interaction with the system terminal conditions. Thus it may be noted that a complicated system can be treated by a combination of a common solution procedure for the interior of each pipeline, together with a systematic coverage of each terminal and interconnection point in the system. The primary focus in the treatment of a variety of transient liquid-flow problems is on the handling of boundary conditions, which are discussed in the next section.

EXERCISES

12.6.1 The head rise at a valve due to sudden closure is (a) $a^2/2g$; (b) $V_0 a/g$; (c) $V_0 a/2g$; (d) $V_0^2/2g$; (e) none of these answers.

12.6.2 The speed of a pressure wave through a pipe depends upon (a) the length of pipe; (b) the original head at the valve; (c) the viscosity of fluid; (d) the initial velocity; (e) none of these answers.

12.6.3 When the velocity in a pipe is suddenly reduced from 3 to 2 m/s by downstream valve closure, for $a = 980$ m/s the head rise, in meters, is (a) 100; (b) 200; (c) 300; (d) 980; (e) none of these answers.

12.6.4 When $t_c = L/2a$, the proportion of pipe length subjected to maximum heads is, in percent, (a) 25; (b) 50; (c) 75; (d) 100; (e) none of these answers.

12.6.5 When the steady-state value of head at a valve is 120 ft, the valve is given a sudden partial closure such that $\Delta h = 80$ ft. The head at the valve at the instant this reflected wave returns is (a) -80; (b) 40; (c) 80; (d) 200; (e) none of these answers.

12.7 BOUNDARY CONDITIONS

The term *boundary condition* refers to the end condition on each pipeline. It may be a system terminal at a reservoir, valve, etc., or it may be a pipeline connection to another pipeline or a different type of element, e.g., a pump or a storage volume. In each of the many options at the downstream end of the pipe the equation along the C^+ characteristic is used to interface with the particular end condition. A few common boundary conditions follow, and in each case either Eq. (12.6.13) or (12.6.14) is used to represent the pipeline response.

Valve at Downstream End

For steady-state flow through the valve, considered as an orifice,

$$Q_0 = (C_d A_v)_0 \sqrt{2gH_0}$$

with Q_0 the steady-state flow, H_0 the head across the valve, and $(C_d A_v)_0$ the area of the opening times the discharge coefficient. For another opening, in general,

$$Q_P = C_d A_v \sqrt{2gH_P} \qquad (12.7.1)$$

The solution of Eq. (12.7.1) and Eq. (12.6.13) yields

$$Q_{P_{NS}} = -gB(C_d A_v)^2 + \sqrt{[gB(C_d A_v)^2]^2 + (C_d A_v)^2 2gC_P} \qquad (12.7.2)$$

and

$$H_{P_{NS}} = C_P - BQ_{P_{NS}} \qquad (12.7.3)$$

in which the subscript $NS = N + 1$ refers to the downstream section in the pipe.

Junction of Two or More Pipes

At a connection of pipelines of different properties, the continuity equation must be satisfied at each instant of time, and a common hydraulic-grade-line elevation may be assumed at the end of each pipe. These statements implicitly assume that there is no storage at the junction, and they also neglect all minor effects. In multipipe systems it is necessary either to use double-subscript notation, the first subscript referring to the pipe number and the second to the pipe section number, or to use continuous sectioning in the entire system. If the former scheme is used to handle the three-pipe junction in Fig. 12.16 and if Eqs. (12.6.13) and (12.6.14)

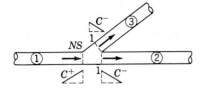

Figure 12.16 Pipeline junction.

are written in the following form, a summation provides a simple solution for the common hydraulic grade line.

$$Q_{P1, NS} = -\frac{H_{P1, NS}}{B_1} + \frac{C_{P1}}{B_1}$$

$$-Q_{P2, 1} = -\frac{H_{P1, NS}}{B_2} + \frac{C_{M2}}{B_2}$$

$$-Q_{P3, 1} = -\frac{H_{P1, NS}}{B_3} + \frac{C_{M3}}{B_3}$$

$$\Sigma Q_P = 0 = -H_{P1, NS}\Sigma\frac{1}{B_i} + \frac{C_{P1}}{B_1} + \frac{C_{M2}}{B_2} + \frac{C_{M3}}{B_3}$$

or

$$H_{P1, NS} = \frac{C_{P1}/B_1 + C_{M2}/B_2 + C_{M3}/B_3}{\Sigma(1/B_i)} \qquad (12.7.4)$$

With the common hydraulic grade line computed, the equations above can be used to determine the flow in each pipe at the junction.

Valve in Line

A valve or orifice between two different pipelines or within a given line must be treated simultaneously with the contiguous end sections of the pipelines. It is assumed that the orifice equation (12.7.1) is valid at any instant for the control volume shown in Fig. 12.17. This assumption neglects any inertia effects in accelerating or decelerating flow through the valve opening and also implies that there is a constant volume of fluid within the indicated control volume. At any instant the flow rate in the end sections are equal, $Q_{P1, NS} = Q_{P2, 1}$, and the orifice

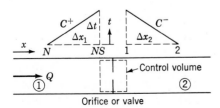

Figure 12.17 Valve in line.

equation for positive flow, written with the subscript notation to identify the pipeline as well as the section, becomes

$$Q_{P2,1} = Q_{P1,NS} = C_d A_v \sqrt{2g(H_{P1,NS} - H_{P2,1})} \qquad Q > 0 \qquad (12.7.5)$$

By use of Eqs. (12.6.13) and (12.6.14)

$$H_{P1,NS} = C_P - B_1 Q_{P1,NS}$$

$$H_{P2,1} = C_M + B_2 Q_{P2,1}$$

elimination of $H_{P1,NS}$ and $H_{P2,1}$ yields

$$Q_{P2,1} = -g(B_1 + B_2)(C_d A_v)^2$$
$$+ \sqrt{[g(B_1 + B_2)(C_d A_v)^2]^2 + 2g(C_d A_v)^2 (C_P - C_M)} \qquad (12.7.6)$$

The plus sign has been taken before the radical, since a positive Q_P must be found. Also, C_P must be greater than C_M for a positive Q_P. In a similar manner starting with

$$Q_{P2,1} = Q_{P1,NS} = -C_d A_v \sqrt{2g(H_{P2,1} - H_{P1,NS})} \qquad Q < 0 \qquad (12.7.7)$$

for negative flow only,

$$Q_{P2,1} = g(B_1 + B_2)(C_d A_v)^2$$
$$- \sqrt{[g(B_1 + B_2)(C_d A_v)^2]^2 + 2g(C_d A_v)^2 (C_M - C_P)} \qquad (12.7.8)$$

The minus sign was taken before the radical to obtain a negative Q_P. Also, C_M must be greater than C_P for a negative solution.

The procedure is to calculate C_P and C_M. If $C_P \geq C_M$, use Eq. (12.7.6). If $C_M > C_P$, use Eq. (12.7.8).

Centrifugal Pump with Speed Known

If a pump is operating at a constant speed, or if the unit is started and the pump and motor come up to speed in a known manner, the interaction of the pump and the fluid in the connecting pipelines can be handled by a fairly simple boundary condition. The homologous conditions, Eqs. (9.1.1) and (9.1.3), when the transient behavior of a given pump is investigated, may be written

$$\frac{H}{N^2} = \text{const} \qquad \frac{Q}{N} = \text{const} \qquad (12.7.9)$$

where H is the head rise across the pump and N is the speed. If the pump characteristic curve is expressed by a parabola, then in the homologous form it may be written

$$\frac{H}{N^2} = C_1 + C_2 \frac{Q}{N} + C_3 \left(\frac{Q}{N}\right)^2 \qquad (12.7.10)$$

The values C_1, C_2, C_3 are found to approximate the pump curve. The compressibility of the fluid in the pump is assumed to be negligible compared with the rest of

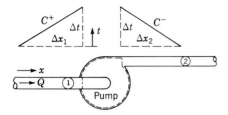

Figure 12.18 Centrifugal pump.

the system. When Eq. (12.7.10) is applied to the pump in Fig. 12.18, it takes the form

$$H_{P2,1} - H_{P1,NS} = C_1 N^2 + C_2 N Q_{P1,NS} + C_3 Q^2_{P1,NS} \qquad (12.7.11)$$

When this equation is combined with Eqs. (12.6.13) and (12.6.14), the discharge is determined as

$$Q_{P1,NS} = \frac{B_1 + B_2 - C_2 N}{2C_3} \left\{ 1 - \left[1 - \frac{4C_3(N^2 C_1 + C_{P1} - C_{M2})}{(B_1 + B_2 - C_2 N)^2} \right]^{1/2} \right\}$$
$$(12.7.12)$$

During a pump start-up, a linear speed rise is often assumed for the speed varia-tion. If the speed of the pump is constant, N may be combined with the pump curve constants. Also, if the pump is operating directly from a suction reservoir, the equation may be simplified by the elimination of the equation along the C^+ characteristic in the suction pipeline.

Example 12.7 Develop the necessary boundary-condition equations for the pump in Fig. 12.19. The pump is to be started with a linear speed rise to N_R in t_0 s. A check valve exists in the discharge pipe. The initial no-flow steady-state head on the downstream side of the check valve is H_C. For a steady flow of Q_0 there is a loss of ΔH_0 across the open check valve. Assume the check valve opens when the pump has developed enough head to exceed H_C.

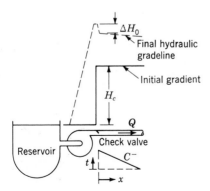

Figure 12.19 Example 12.7.

The equation for the hydraulic grade line downstream from the pump and check valve (after the check valve is open) is

$$H_P = C_1 N^2 + C_2 N Q_P + C_3 Q_P^2 - Q_P^2 \frac{\Delta H_0}{Q_0^2}$$

$$Q_P = \frac{B - C_2 N}{2(C_3 - \Delta H_0 / Q_0^2)} \left\{ 1 - \left[1 + \frac{4(C_3 - \Delta H_0 / Q_0^2)^2 (C_M - C_1 N^2)}{(B - C_2 N)^2} \right]^{1/2} \right\}$$

The equations for the boundary condition are

$$B = \frac{a}{gA}$$

$$C_M = H_2 - Q_2 \left(B - \frac{f \Delta x}{2gDA^2} |Q_2| \right)$$

$$N = \begin{cases} N_R \dfrac{t}{t_0} & t \le t_0 \\ N_R & t > t_0 \end{cases}$$

If $C_1 N^2 > H_C$, Q_P is defined by the above equation for the quadratic, and

$$H_P = C_M + B Q_P$$

If $C_1 N^2 < H_C$, $Q_P = 0$ and $H_P = H_C$.

Accumulator

Many different types of air or gas accumulators are used to help reduce pressure transients in liquid systems. In an analysis of the behavior of the accumulator shown in Fig. 12.20 the pressure is visualized as being the same throughout the

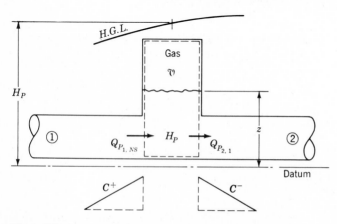

Figure 12.20 Simple accumulator.

indicated control volume at any instant. It is assumed to be frictionless and inertialess. The gas is assumed to follow the reversible polytropic relation

$$H_A \mathcal{V}^n = C \qquad (12.7.13)$$

where H_A is the absolute head equal to the gage plus barometric pressure heads, $\mathcal{V}$ is the gas volume, n is the polytropic exponent, and C is a constant.

The following equations express the relations needed for solution of the simple accumulator shown in Fig. 12.20:
Eq. (12.7.13):

$$H_A (\mathcal{V}_0 + \Delta \mathcal{V})^n = C$$

in which

$$H_A = H_P - z + H_{\text{bar}}$$

the absolute head at end of time $t + \Delta t$. $\mathcal{V}_0$ is the known volume at beginning of Δt, and $\Delta \mathcal{V}$ is the volume change during Δt.

$$\Delta \mathcal{V} = 0.5 \, \Delta t (Q_{P_{2,1}} + Q_{2,1} - Q_{P_{1,NS}} - Q_{1,NS})$$

By use of Eqs. (12.6.13) and (12.6.14) to eliminate $Q_{P_{2,1}}$ and $Q_{P_{1,NS}}$, a single equation in one unknown, H_P, is obtained

$$(H_P - z + H_{\text{bar}}) \left[\mathcal{V}_0 + 0.5 \, \Delta t \left(\frac{H_P - C_M}{B_2} + Q_{2,1} + \frac{H_P - C_P}{B_1} - Q_{1,NS} \right) \right]^n = C$$
$$(12.7.14)$$

z is assumed to hold its value during Δt. It may then be adjusted to its new value by $z = z + \Delta z$, where $\Delta z = -\Delta \mathcal{V} / A_A$ with A_A the cross-sectional area of the accumulator. Equation (12.7.14) may be solved by the Newton-Raphson root-finding technique described in Appendix B.4.

12.8 BASIC WATERHAMMER PROGRAM

In this section a FORTRAN IV program is presented; it calculates transients in a single pipeline with a reservoir upstream and a valve downstream (Fig. 12.21).

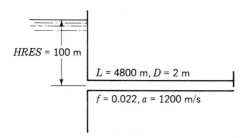

$HRES$ = 100 m

L = 4800 m, D = 2 m

f = 0.022, a = 1200 m/s

Figure 12.21 System for basic waterhammer program.

```
C  BASIC WATERHAMMER PROGRAM FOR UPSTREAM RESERVOIR, SINGLE PIPE,AND
C  DOWNSTREAM VALVE. ELEV DATUM AT VALVE. DARCY WEISBACH FRICTION.
C  F GIVEN. A=WAVE SPEED. CVA=PRODUCT OF DISCHARGE COEF. AND AREA OF
C  VALVE OPENING, WITH VALUES GIVEN FOR INTERVALS DCV.
      DIMENSION Q(6),QP(6),H(6),HP(6),CVA(11)
      NAMELIST/DIN/F,XL,A,N,D,HRES,G,JPR,DCV,TMAX,CVA
   10 READ(5,DIN,END=99)
      WRITE(6,DIN)
      AR=.7854*D*D
      B=A/(G*AR)
      NS=N+1
      DT=XL/(A*N)
      CV=CVA(1)
      HP(1)=HRES
      J=0
      T=.0
      Q0=SQRT(HRES/(F*XL/(2.*G*D*AR*AR)+1./(2.*G*CV*CV)))
      H0=(Q0/CV)**2/(2.*G)
      R=(HRES-H0)/(Q0**2*N)
      DO 11 I=1,NS
      Q(I)=Q0
   11 H(I)=HRES-(I-1)*R*Q0**2
      WRITE(6,1)
      WRITE(6,2)
    1 FORMAT('1      PIEZOMETRIC HEADS AND FLOWS ALONG THE PIPE')
    2 FORMAT('0   TIME    CV    X/L=     .0     .25     .50     .75
     11.0')
   13 WRITE(6,3) T,CV,(H(I),I=1,NS),(Q(I),I=1,NS)
    3 FORMAT('0',F7.3,1X,F7.5,5H   H=,5F8.2/19X,'Q=',5F8.2)
   14 T=T+DT
      J=J+1
      IF(T.GT.TMAX)GO TO 10
      K=T/DCV+1
      CV=CVA(K)+(T-(K-1)*DCV)*(CVA(K+1)-CVA(K))/DCV
C  DOWNSTREAM BOUNDARY CONDITION
      CP=H(N)+Q(N)*(B-R*ABS(Q(N)))
      QP(NS)=-G*B*CV*CV+SQRT((G*B*CV*CV)**2+2.*G*CV*CV*CP)
      HP(NS)=CP-B*QP(NS)
C  UPSTREAM BOUNDARY CONDITION
      QP(1)=Q(2)+(HRES-H(2)-R*Q(2)*ABS(Q(2)))/B
C  INTERIOR SECTIONS
      DO 15 I=2,N
      CP=H(I-1)+Q(I-1)*(B-R*ABS(Q(I-1)))
      CM=H(I+1)-Q(I+1)*(B-R*ABS(Q(I+1)))
      HP(I)=.5*(CP+CM)
   15 QP(I)=(HP(I)-CM)/B
      DO 16 I=1,NS
      H(I)=HP(I)
   16 Q(I)=QP(I)
      IF(J/JPR*JPR.EQ.J)GO TO 13
      GO TO 14
   99 STOP
      END
$DATA
 &DIN F=.022,XL=4800.,A=1200.,N=4,D=2.,HRES=100.,G=9.805,JPR=1,DCV=5.,
 TMAX=40.,CVA=.06,.03,.01,.003,.001,.0005,.0002,4*.0,
 &END
```

Figure 12.22 Basic waterhammer program.

The valve data are given in terms of values of $C_D A$ in the orifice formula

$$QP(NS) = C_D A\sqrt{2g\,HP(NS)} \qquad (12.8.1)$$

The values of $C_D A$, called CVA in the program, are listed as an array, with CVA(1) the initial value at time zero and each following value at time intervals of DCV seconds. The datum for hydraulic grade line must be taken through the centerline of the valve, and no provision is made for backflow through the valve.

```
PIEZOMETRIC HEADS AND FLOWS ALONG THE PIPE
```

TIME	CV	X/L=	.0	.25	.50	.75	1.0
0.000	0.06000	H =	100.00	99.53	99.06	98.58	98.11
		Q =	2.63	2.63	2.63	2.63	2.63
1.000	0.05400	H =	100.00	99.53	99.06	98.58	105.12
		Q =	2.63	2.63	2.63	2.63	2.45
2.000	0.04800	H =	100.00	99.53	99.06	105.56	112.72
		Q =	2.63	2.63	2.63	2.45	2.26
3.000	0.04200	H =	100.00	99.53	106.00	113.13	121.00
		Q =	2.63	2.63	2.45	2.26	2.05
4.000	0.03600	H =	100.00	106.45	113.54	121.38	129.96
		Q =	2.63	2.45	2.26	2.05	1.82
5.000	0.03000	H =	100.00	113.95	121.75	130.31	139.71
		Q =	2.28	2.26	2.05	1.82	1.57
6.000	0.02600	H =	100.00	115.27	130.66	140.03	146.68
		Q =	1.89	1.88	1.82	1.57	1.39
7.000	0.02200	H =	100.00	116.70	133.51	146.98	154.07
		Q =	1.48	1.46	1.40	1.40	1.21
8.000	0.01800	H =	100.00	118.23	133.02	147.54	161.84
		Q =	1.03	1.01	1.04	1.04	1.01
9.000	0.01400	H =	100.00	116.32	132.26	147.87	157.62
		Q =	0.54	0.61	0.64	0.65	0.78
10.000	0.01000	H =	100.00	114.03	131.18	142.35	152.03
		Q =	0.19	0.18	0.22	0.39	0.55
11.000	0.00860	H =	100.00	114.86	124.12	135.35	139.80
		Q =	-0.18	-0.19	-0.08	0.12	0.45
12.000	0.00720	H =	100.00	110.09	119.03	121.59	125.97
		Q =	-0.58	-0.44	-0.30	-0.02	0.36
13.000	0.00580	H =	100.00	104.18	107.56	109.66	110.39
		Q =	-0.70	-0.68	-0.38	-0.06	0.27
14.000	0.00440	H =	100.00	97.48	94.83	96.37	99.70
		Q =	-0.79	-0.64	-0.44	-0.09	0.19
15.000	0.00300	H =	100.00	90.66	86.30	84.88	88.00
		Q =	-0.57	-0.55	-0.35	-0.19	0.12
16.000	0.00260	H =	100.00	88.83	80.72	77.94	73.78
		Q =	-0.31	-0.28	-0.29	-0.13	0.10
17.000	0.00220	H =	100.00	90.06	80.47	69.63	69.57
		Q =	0.00	-0.05	-0.07	-0.01	0.08
18.000	0.00180	H =	100.00	91.64	78.97	72.10	66.78
		Q =	0.20	0.22	0.23	0.15	0.07
28.000	0.00032	H =	100.00	108.63	114.41	113.89	114.56
		Q =	-0.71	-0.59	-0.50	-0.28	0.02
29.000	0.00025	H =	100.00	103.13	105.33	104.50	102.50
		Q =	-0.81	-0.79	-0.50	-0.24	0.01
30.000	0.00020	H =	100.00	96.71	93.24	93.95	94.69
		Q =	-0.87	-0.72	-0.53	-0.21	0.01
31.000	0.00016	H =	100.00	90.12	85.35	83.44	85.60
		Q =	-0.64	-0.61	-0.43	-0.28	0.01
32.000	0.00012	H =	100.00	88.64	80.34	77.01	72.36
		Q =	-0.36	-0.34	-0.36	-0.21	0.00
33.000	0.00008	H =	100.00	90.21	80.30	69.26	68.55
		Q =	-0.05	-0.11	-0.13	-0.08	0.00
34.000	0.00004	H =	100.00	91.66	79.13	71.85	66.27
		Q =	0.15	0.16	0.18	0.09	0.00
35.000	0.00000	H =	100.00	88.92	83.21	76.15	75.26
		Q =	0.38	0.43	0.38	0.26	0.00
36.000	0.00000	H =	100.00	91.55	85.93	86.61	86.08
		Q =	0.71	0.59	0.51	0.29	0.00

Figure 12.23 Output for Example 12.8.

The value of $C_D A$ is found by linear interpolation (called CV). Then

$$QP(NS) = CV\sqrt{2gHP(NS)} \tag{12.8.2}$$

and the C^+ equation

$$HP(NS) = CP - B \cdot QP(NS) \tag{12.8.3}$$

are solved simultaneously for QP(NS) and HP(NS).

Figure 12.22 lists the program. All transient characteristic programs for pipe flow tend to have the same format: Read-in of data; calculation of area, AR, and of B, NS, and DT; then steady-state conditions. The energy equation from the reservoir to the valve (neglecting minor losses) yields, Fig. 12.21,

$$\text{HRES} - f \frac{L}{D} \frac{Q_0^2}{2gA^2} = \frac{Q_0^2}{2gCV^2}$$

from which steady-state flow Q_0 is determined. The initial head at the valve is H_0, and the resistance coefficient R per reach is

$$R = \frac{\text{HRES} - \text{HO}}{NQ_0^2}$$

The head at each section is then calculated and stored along with $Q_i = Q_{0i}$.

Headings for an output table are organized and steady-state conditions are printed out. The time is now incremented, and the program calculates CV, downstream boundary condition, upstream boundary condition, internal sections. QP_i and HP_i are substituted for Q_i and H_i; then results are printed out every JPRth time. The procedure is repeated until the specified time TMAX is reached.

Example 12.8 The system of Fig. 12.21 initially has a valve opening $C_D A = 0.06$ m². At intervals of 5 s, $C_D A$ takes on the values 0.03, 0.01, 0.003, 0.001, 0.0005, 0.0002, 0.0, and remains closed. Calculate the transients of the system for 40 s after the valve starts to close.

The data for the problem are shown at the end of Fig. 12.22, and some of the output is given in Fig. 12.23. The number of reaches was selected as $N = 4$. Hydraulic-grade-line elevations are given in meters and discharges in meters cubed per second.

B OPEN-CHANNEL FLOW

In general, open-channel transients are more complex to handle than closed-conduit transients. Surface-wave motion is an example of open-channel and unsteady flow. The subject is too vast to attempt to cover as part of a chapter. Two special topics are discussed: frictionless positive and negative surge waves.

12.9 FRICTIONLESS POSITIVE SURGE WAVE IN A RECTANGULAR CHANNEL

In this section the surge wave resulting from a sudden change in flow (due to a gate or other mechanism) that increases the depth is studied. A rectangular channel is assumed, and friction is neglected. Such a situation is shown in Fig. 12.24 shortly after a sudden, partial closure of a gate. The problem is analyzed by reducing it to a steady-state problem, as in Fig. 12.25. The continuity equation yields, per unit width,

$$(V_1 + c)y_1 = (V_2 + c)y_2 \tag{12.9.1}$$

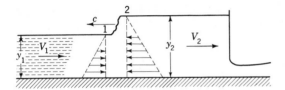

Figure 12.24 Positive surge wave in a rectangular channel.

and the momentum equation for the control volume $1 - 2$, neglecting shear stress on the floor, per unit width, is

$$\frac{\gamma}{2}(y_1^2 - y_2^2) = \frac{\gamma}{g}y_1(V_1 + c)(V_2 + c - V_1 - c) \qquad (12.9.2)$$

By elimination of V_2 in the last two equations,

$$V_1 + c = \sqrt{gy_1}\left[\frac{y_2}{2y_1}\left(1 + \frac{y_2}{y_1}\right)\right]^{1/2} \qquad (12.9.3)$$

In this form the speed of an elementary wave is obtained by letting y_2 approach y_1, yielding

$$V_1 + c = \sqrt{gy} \qquad (12.9.4)$$

For propagation through still liquid $V_1 \to 0$, and the wave speed is $c = \sqrt{gy}$ when the problem is converted back to the unsteady form by superposition of $V = -c$.

In general, Eqs. (12.9.1) and (12.9.2) have to be solved by trial. The hydraulic-jump formula results from setting $c = 0$ in the two equations [see Eq. (3.11.23)].

Example 12.9 A rectangular channel 3 m wide and 2 m deep, discharging 18 m^3/s, suddenly has the discharge reduced to 12 m^3/s at the downstream end. Compute the height and speed of the surge wave.

$V_1 = 3$, $y_1 = 2$, $V_2 y_2 = 4$. With Eqs. (12.9.1) and (12.9.2),

$$6 = 4 + c(y_2 - 2) \qquad \text{and} \qquad y_2^2 - 4 = \frac{2 \times 2}{9.806}(c + 3)(3 - V_2)$$

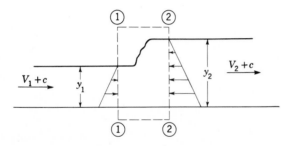

Figure 12.25 Surge problem reduced to a steady-state problem by superposition of surge velocity.

Eliminating c and V_2 gives

$$y_2^2 - 4 = \frac{4}{9.806}\left(\frac{2}{y_2 - 2} + 3\right)\left(3 - \frac{4}{y_2}\right)$$

or

$$\left(\frac{y_2 - 2}{3y_2 - 4}\right)^2 (y_2 + 2)y_2 = \frac{4}{9.806} = 0.407$$

After solving for y_2 by trial, $y_2 = 2.75$ m. Hence, $V_2 = 4/2.75 = 1.455$ m/s. The height of surge wave is 0.75 m, and the speed of the wave is

$$c = \frac{2}{y_2 - 2} = \frac{2}{0.75} = 2.667 \text{ m/s}$$

EXERCISE

12.9.1 An elementary wave can travel upstream in a channel, $y = 4$ ft, $V = 8$ ft/s, with a velocity of (a) 3.35 ft/s; (b) 11.35 ft/s; (c) 16.04 ft/s; (d) 19.35 ft/s; (e) none of these answers.

12.10 FRICTIONLESS NEGATIVE SURGE WAVE IN A RECTANGULAR CHANNEL

The negative surge wave appears as a gradual flattening and lowering of a liquid surface. It occurs, for example, in a channel downstream from a gate that is being closed or upstream from a gate that is being opened. Its propagation is accomplished by a series of elementary negative waves superposed on the existing velocity, each wave traveling at less speed than the one at next greater depth. Application of the momentum equation and the continuity equation to a small depth change produces simple differential expressions relating wave speed c, velocity V, and depth y. Integration of the equations yields liquid-surface profile as a function of time, and velocity as a function of depth or as a function of position along the channel and time (x and t). The fluid is assumed to be frictionless, and vertical accelerations are neglected.

In Fig. 12.26a an elementary disturbance is indicated in which the flow upstream has been slightly reduced. For application of the momentum and continuity equations it is convenient to reduce the motion to a steady one, as in Fig. 12.26b, by imposing a uniform velocity c to the left. The continuity equation is

$$(V - \delta V - c)(y - \delta y) = (V - c)y$$

or, by neglecting the product of small quantities,

$$(c - V)\,\delta y = y\,\delta V \qquad (12.10.1)$$

The momentum equation produces

$$\frac{\gamma}{2}(y - \delta y)^2 - \frac{\gamma}{2}y^2 = \frac{\gamma}{g}(V - c)y[V - c - (V - \delta V - c)]$$

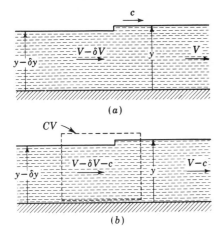

Figure 12.26 Elementary wave.

After simplifying,

$$\delta y = \frac{c - V}{g} \, \delta V \tag{12.10.2}$$

Equating $\delta V / \delta y$ in Eqs. (12.10.1) and (12.10.2) gives

$$c - V = \pm \sqrt{gy} \tag{12.10.3}$$

or

$$c = V \pm \sqrt{gy}$$

The speed of an elementary wave in still liquid at depth y is $\sqrt{gy}$ and with flow the wave travels at the speed $\sqrt{gy}$ *relative* to the flowing liquid.

Eliminating c from Eqs. (12.10.1) and (12.10.2) gives

$$\frac{dV}{dy} = \pm \sqrt{\frac{g}{y}}$$

and integrating leads to

$$V = \pm 2 \sqrt{gy} + \text{const}$$

For a negative wave forming downstream from a gate, Fig. 12.27, by using the plus sign, after an instantaneous partial closure, $V = V_0$ when $y = y_0$, and

$$V_0 = 2 \sqrt{gy_0} + \text{const}$$

After eliminating the constant,

$$V = V_0 - 2 \sqrt{g} \left(\sqrt{y_0} - \sqrt{y} \right) \tag{12.10.4}$$

The wave travels in the $+x$ direction, so that

$$c = V + \sqrt{gy} = V_0 - 2 \sqrt{gy_0} + 3 \sqrt{gy} \tag{12.10.5}$$

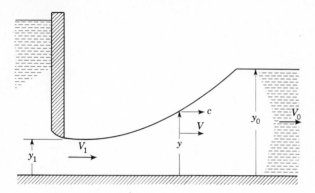

Figure 12.27 Negative wave after gate closure.

If the gate motion occurs at $t = 0$, the liquid-surface position is expressed by $x = ct$, or

$$x = (V_0 - 2\sqrt{gy_0} + 3\sqrt{gy})t \qquad (12.10.6)$$

Eliminating y from Eqs. (12.10.5) and (12.10.6) gives

$$V = \frac{V_0}{3} + \frac{2}{3}\frac{x}{t} - \frac{2}{3}\sqrt{gy_0} \qquad (12.10.7)$$

which is the velocity in terms of x and t.

Example 12.10 In Fig. 12.27 find the Froude number of the undisturbed flow such that the depth y_1 at the gate is just zero when the gate is suddenly closed. For $V_0 = 20$ ft/s, find the liquid-surface equation.

It is required that $V_1 = 0$ when $y_1 = 0$ at $x = 0$ for any time after $t = 0$. In Eq. (12.10.4), with $V = 0$, $y = 0$,

$$V_0 = 2\sqrt{gy_0} \qquad \text{or} \qquad \mathbf{F}_0 = \frac{V_0}{\sqrt{gy_0}} = 2$$

For $V_0 = 20$,

$$y_0 = \frac{V_0^2}{4g} = \frac{20^2}{4g} = 3.11 \text{ ft}$$

By Eq. (12.10.6)

$$x = (20 - 2\sqrt{32.2 \times 3.11} + 3\sqrt{32.2y})t = 17.04\sqrt{y}\,t$$

The liquid surface is a parabola with vertex at the origin and surface concave upward.

Example 12.11 In Fig. 12.27 the gate is partially closed at the instant $t = 0$ so that the discharge is reduced by 50 percent. $V_0 = 6$ m/s, $y_0 = 3$ m. Find V_1, y_1, and the surface profile. The new discharge is

$$q = \frac{6 \times 3}{2} = 9 = V_1 y_1$$

By Eq. (12.10.4)

$$V_1 = 6 - 2\sqrt{9.806}\,(\sqrt{3} - \sqrt{y_1})$$

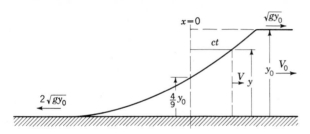

Figure 12.28 Dam-break profile.

Then V_1 and y_1 are found by trial from the last two equations, $V_1 = 4.24$ m/s, $y_1 = 2.12$ m. The liquid-surface equation, from Eq. (12.10.6), is

$$x = (6 - 2\sqrt{3g} + 3\sqrt{gy})t \qquad \text{or} \qquad x = (9.39\sqrt{y} - 4.84)t$$

which holds for the range of values of y between 2.12 and 3 m.

Dam Break

An idealized dam-break water-surface profile, Fig. 12.28 can be obtained from Eqs. (12.10.4) to (12.10.7). From a frictionless, horizontal channel with depth of water y_0 on one side of a gate and no water on the other side of the gate, the gate is suddenly removed. Vertical accelerations are neglected. $V_0 = 0$ in the equations, and y varies from y_0 to 0. The velocity at any section, from Eq. (12.10.4), is

$$V = -2\sqrt{g}(\sqrt{y_0} - \sqrt{y}) \qquad (12.10.8)$$

always in the downstream direction. The water-surface profile is, from Eq. (12.10.6),

$$x = (3\sqrt{gy} - 2\sqrt{gy_0})t \qquad (12.10.9)$$

At $x = 0$, $y = 4y_0/9$, the depth remains constant and the velocity past the section $x = 0$ is, from Eq. (12.10.8),

$$V = -\tfrac{2}{3}\sqrt{gy_0}$$

also independent of time. The leading edge of the wave feathers out to zero height and moves downstream at $V = c = -2\sqrt{gy_0}$. The water surface is a parabola with vertex at the leading edge, concave upward.

With an actual dam break, ground roughness causes a positive surge, or wall of water, to move downstream; i.e., the feathered edge is retarded by friction.

EXERCISES

12.10.1 The speed of an elementary wave in a still liquid is given by (a) $(gy^2)^{1/3}$; (b) $2y/3$; (c) $\sqrt{2gy}$; (d) $\sqrt{gy}$; (e) none of these answers.

12.10.2 A negative surge wave (a) is a positive surge wave moving backward; (b) is an inverted positive surge wave; (c) can never travel upstream; (d) can never travel downstream; (e) is none of the above.

PROBLEMS

12.1 Determine the period of oscillation of a U tube containing $\frac{1}{2}$ l of water. The cross-sectional area is 2.4 cm². Neglect friction.

12.2 A U tube containing alcohol is oscillating with maximum displacement from equilibrium position of 120 mm. The total column length is 1 m. Determine the maximum fluid velocity and the period of oscillation. Neglect friction.

12.3 A liquid, $v = 0.002$ ft²/s, is in a 0.50-in-diameter U tube. The total liquid column is 70 in long. If one meniscus is 15 in above the other meniscus when the column is at rest, determine the time for one meniscus to move to within 1.0 in of its equilibrium position.

12.4 Develop the equations for motion of a liquid in a U tube for laminar resistance when $16v/D^2 = \sqrt{2g/L}$. *Suggestion:* Try $z = e^{-mt}(c_1 + c_2 t)$.

12.5 A U tube contains liquid oscillating with a velocity 2 m/s at the instant the menisci are at the same elevation. Find the time to the instant the menisci are next at the same elevation and determine the velocity then. $v = 1 \times 10^{-5}$ m²/s, $D = 6$ mm, $L = 750$ mm.

12.6 A 10-ft-diameter horizontal tunnel has 10-ft-diameter vertical shafts spaced 1 mi apart. When valves are closed isolating this reach of tunnel, the water surges to a depth of 50 ft in one shaft when it is 20 ft in the other shaft. For $f = 0.022$ find the height of the next two surges.

12.7 Two standpipes 6 m in diameter are connected by 900 m of 2.5-m-diameter pipe; $f = 0.020$, and minor losses are 4.5 velocity heads. One reservoir level is 9 m above the other one when a valve is rapidly opened in the pipeline. Find the maximum fluctuation in water level in the standpipe.

12.8 A valve is quickly opened in a pipe 1200 m long, $D = 0.6$ m, with a 0.3-m-diameter nozzle on the downstream end. Minor losses are $4V^2/2g$, with V the velocity in the pipe, $f = 0.024$, $H = 9$ m. Find the time to attain 95 percent of the steady-state discharge.

12.9 A globe valve ($K = 10$) at the end of a pipe 2000 ft long is rapidly opened. $D = 3.0$ ft, $f = 0.018$, minor losses are $2V^2/2g$, and $H = 75$ ft. How long does it take for the discharge to attain 80 percent of its steady-state value?

12.10 A steel pipeline with expansion joints is 1 m in diameter and has a 10-mm wall thickness. When it is carrying water, determine the speed of a pressure wave.

12.11 Benzene ($K = 150,000$ psi, $S = 0.88$) flows through $\frac{3}{4}$-in-ID steel tubing with $\frac{1}{8}$-in wall thickness. Determine the speed of a pressure wave.

12.12 Determine the maximum time for rapid valve closure on a pipeline: $L = 1000$ m, $D = 1.3$ m, $e = 12$-mm steel pipe, $V_0 = 3$ m/s, water flowing.

12.13 A valve is closed in 5 s at the downstream end of a 3000-m pipeline carrying water at 2 m/s. $a = 1000$ m/s. What is the peak pressure developed by the closure?

12.14 Determine the length of pipe in Prob. 12.13 subjected to the peak pressure.

12.15 A valve is closed at the downstream end of a pipeline in such a manner that only one-third of the line is subjected to maximum pressure. During what proportion of the time $2L/a$ is it closed?

12.16 A pipe, $L = 2000$ m, $a = 1000$ m/s, has a valve on its downstream end, $V_0 = 2.5$ m/s and $H_0 = 20$ m. It closes in three increments, spaced 1 s apart, each area reduction being one-third of the original opening. Find the pressure at the gate and at the midpoint of the pipeline at 1-s intervals for 5 s after initial closure.

12.17 A pipeline, $L = 600$ m, $a = 1200$ m/s, has a valve at its downstream end, $V_0 = 2$ m/s and $H_0 = 30$ m. Determine the pressure at the valve for closure ($C_d = 0.6$):

A_v/A_{v0}	1	0.75	0.60	0.45	0.30	0.15	0
t, s	0	0.5	1.0	1.5	2.0	2.5	3.0

12.18 In Prob. 12.17 determine the peak pressure at the valve for uniform area reduction in 3.0 s.

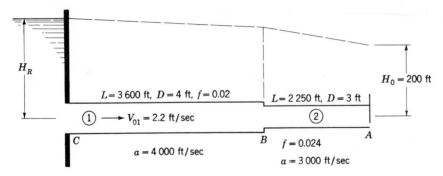

Figure 12.29 Problems 12.25, 12.26.

12.19 Find the maximum area reduction for $\frac{1}{2}$-s intervals for the pipeline of Prob. 12.17 when the maximum head at the valve is not to exceed 50 m. Increase head linearly to 50 m in 1 s, then hold constant.

12.20 Derive the characteristics-method solution for waterhammer with the pressure p and the discharge Q as dependent variables.

12.21 Work Example 12.6 with one-third of the wave frequency. Do resonance conditions still exist?

12.22 Develop a single-pipeline waterhammer program to handle a valve closure at the downstream end of the pipe with a reservoir at the upstream end. The valve closure is given by $C_d A_v/(C_d A_v)_0 = (1 - t/t_c)^m$, where t_c is the time of closure and is 6.2 s, and $m = 3.2$; $L = 5743.5$ ft, $a = 3927$ ft/s, $D = 4$ ft, $f = 0.019$, $V_0 = 3.6$ ft/s, and $H_0 = 300$ ft.

12.23 In Prob. 12.22 place a wave on the reservoir at a period of 1.95 s and obtain a solution with the aid of a computer.

12.24 Develop the computer program to solve Example 12.7 using your own particular input data.

12.25 The valve closure data for the series system shown in Fig. 12.29 is

$C_d A_v/(C_d A_v)_0$	1	0.73	0.5	0.31	0.16	0.05	0.0
t, s	0	1	2	3	4	5	6

Develop a characteristics-method computer program to determine the pressure head and flow at the valve and at the series pipeline connection.

12.26 In Prob. 12.25 place the valve at the connection between the two pipelines and an orifice at the downstream end with a steady-state head drop of 100 ft. Develop a program to analyze the valve motion given in Prob. 12.25.

12.27 In Example 12.6 reduce H_0 to 20 m, let $\omega = \pi/2$, and work the problem until $t = 3$ s.

12.28 A rectangular channel is discharging 50 cfs per foot of width at a depth of 10 ft when the discharge upstream is suddenly increased to 70 cfs/ft. Determine the speed and height of the surge wave.

12.29 In a rectangular channel with velocity 2 m/s flowing at a depth of 2 m, a surge wave 0.3 m high travels upstream. What is the speed of the wave, and how much is the discharge reduced per meter of width?

12.30 A rectangular channel 3 m wide and 2 m deep discharges 28 m³/s when the flow is completely stopped downstream by closure of a gate. Compute the height and speed of the resulting positive surge wave.

12.31 Determine the depth downstream from the gate of Prob. 12.30 after it closes.

12.32 Find the downstream water surface of Prob. 12.30 3 s after closure.

12.33 Determine the water surface 2 s after an ideal dam breaks. Original depth is 30 m.

REFERENCES

Bergeron, L.: "Water Hammer in Hydraulics and Wave Surges in Electricity," translated under the sponsorship of the ASME, Wiley, New York, 1961.

Parmakian, J.: "Waterhammer Analysis," Prentice-Hall, Englewood Cliffs, N.J., 1955 (also Dover, New York, 1963).

Streeter, V. L.: Unsteady Flow Calculations by Numerical Methods, *Trans. ASME J. Basic Eng.*, June 1972.

Wylie, E. B., and V. L. Streeter: "Fluid Transients," McGraw-Hill, New York, 1978.

FORCE SYSTEMS, MOMENTS, AND CENTROIDS

The material in this appendix has been assembled to aid in working with force systems. Simple force systems are briefly reviewed, and first and second moments, including the product of inertia, are discussed. Centroids and centroidal axes are defined.

SIMPLE FORCE SYSTEMS

A free-body diagram for an object or portion of an object shows the action of all other bodies on it. The action of the earth on the object, called a *body force*, is proportional to the mass of the object. In addition, forces and couples may act on the object by contact with its surface. When the free body is at rest or is moving in a straight line with uniform speed, it is said to be in *equilibrium*. By Newton's second law of motion, since there is no acceleration of the free body, the summation of all force components in any direction must be zero and the summation of all moments about any axis must be zero.

Two force systems are equivalent if they have the same value for summation of forces in every direction and the same value for summation of moments about every axis. The simplest equivalent force system is called the *resultant* of the force system. Equivalent force systems always cause the same motion (or lack of motion) of a free body.

In coplanar force systems the resultant is either a force or a couple. In noncoplanar parallel force systems the resultant is either a force or a couple. In general noncoplanar systems the resultant may be a force, a couple, or a force and a couple.

The action of a fluid on any surface may be replaced by the resultant force system that causes the same external motion or reaction as the distributed fluid-force system. In this situation the fluid may be considered to be completely removed, the resultant acting in its place.

FIRST AND SECOND MOMENTS; CENTROIDS

The moment of an area, volume, weight, or mass may be determined in a manner analogous to that of determining the moments of a force about an axis.

First Moments

The moment of an area A about the y axis (Fig. A.1) is expressed by

$$\int_A x \, dA$$

in which the integration is carried out over the area. To determine the moment about a parallel axis, for example, $x = k$, the moment becomes

$$\int_A (x - k) \, dA = \int_A x \, dA - kA \tag{A.1}$$

which shows that there will always be a parallel axis $x = k = \bar{x}$, about which the moment is zero. This axis, called a *centroidal axis*, is obtained from Eq. (A.1) by setting it equal to zero and solving for $\bar{x}$,

$$\bar{x} = \frac{1}{A} \int_A x \, dA \tag{A.2}$$

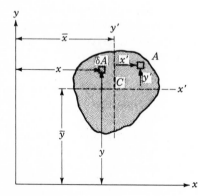

Figure A.1 Notation for first and second moments.

Another centroidal axis may be determined parallel to the x axis,

$$\bar{y} = \frac{1}{A} \int_A y \, dA \tag{A.3}$$

The point of intersection of centroidal axes is called the *centroid* of the area. It may easily be shown, by rotation of axes, that the first moment of the area is zero about any axis through the centroid. When an area has an axis of symmetry, it is a centroidal axis because the moments of corresponding area elements on each side of the axis are equal in magnitude and opposite in sign. When location of the centroid is known, the first moment for any axis may be obtained without integration by taking the product of area and distance from centroid to the axis,

$$\int_A z \, dA = \bar{z}A \tag{A.4}$$

The centroidal axis of a triangle, parallel to one side, is one-third the altitude from that side; the centroid of a semicircle of radius a is $4a/3\pi$ from the diameter.

By taking the first moment of a volume $\mathcal{V}$ about a plane, say the yz plane, the distance to its centroid is similarly determined,

$$\bar{x} = \frac{1}{\mathcal{V}} \int_{\mathcal{V}} x \, d\mathcal{V} \tag{A.5}$$

The mass center of a body is determined by the same procedure,

$$x_m = \frac{1}{M} \int_M x \, dm \tag{A.6}$$

in which dm is an element of mass and M is the total mass of the body. For practical engineering purposes the *center of gravity* of a body is at its mass center.

Second Moments

The second moment of an area A (Fig. A.1) about the y axis is

$$I_y = \int_A x^2 \, dA \tag{A.7}$$

It is called the *moment of inertia* of the area, and it is always positive, since dA is always considered positive. After transferring the axis to a parallel axis through the centroid C of the area,

$$I_c = \int_A (x - \bar{x})^2 \, dA = \int_A x^2 \, dA - 2\bar{x} \int_A x \, dA + \bar{x}^2 \int_A dA$$

Since

$$\int_A x \, dA = \bar{x}A \qquad \int_A x^2 \, dA = I_y \qquad \int_A dA = A$$

therefore,

$$I_c = I_y - \bar{x}^2 A \qquad \text{or} \qquad I_y = I_c + \bar{x}^2 A \tag{A.8}$$

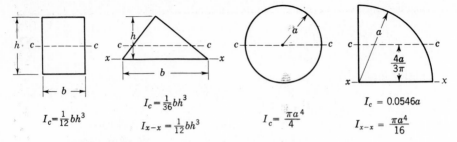

$$I_c = \frac{1}{12}bh^3$$

$$I_c = \frac{1}{36}bh^3$$

$$I_{x-x} = \frac{1}{12}bh^3$$

$$I_c = \frac{\pi a^4}{4}$$

$$I_c = 0.0546a$$

$$I_{x-x} = \frac{\pi a^4}{16}$$

Figure A.2 Moments of inertia of simple areas about centroidal axes.

In words, the moment of inertia of an area about any axis is the sum of the moment of inertia about a parallel axis through the centroid and the product of the area and square of distance between axes. Figure A.2 shows moments of inertia for four simple areas.

The *product of inertia* I_{xy} of an area is expressed by

$$I_{xy} = \int_A xy \, dA \tag{A.9}$$

with the notation of Fig. A.1. It may be positive or negative. By writing the expression for product of inertia about the xy axes, in terms of $\bar{x}$ and $\bar{y}$, Fig. A.1,

$$I_{xy} = \int_A (\bar{x} + x')(\bar{y} + y') \, dA = \bar{x}\bar{y}A + \int_A x'y' \, dA$$

$$+ \bar{x} \int_A y' \, dA + \bar{y} \int_A x' \, dA = \bar{x}\bar{y}A + \bar{I}_{xy}$$

$\bar{I}_{xy}$ is the product of inertia about centroidal axes parallel to the xy axes. Whenever either axis is an axis of symmetry of the area, the product of inertia is zero.

The product of inertia I_{xy} of a triangle having sides b and h along the positive coordinate axes is $b^2h^2/24$.

COMPUTER PROGRAMMING AIDS

FORTRAN IV COMPILER

The programs listed in the text are in FORTRAN IV compiler language (G level). It is assumed that the reader is acquainted with FORTRAN IV. Several useful programming techniques for engineering calculations are not given in first computing courses. These include quadratures, numerical integration using Simpson's rule, parabolic interpolation, solution of algebraic and transcendental equations by the bisection and Newton-Raphson methods, and the Runge-Kutta methods of solving systems of ordinary simultaneous differential equations. These techniques are discussed in this appendix.

B.1 QUADRATURES; NUMERICAL INTEGRATION BY SIMPSON'S RULE

The integral $V = \int_{y_0}^{y_1} F(y)\, dy$ is to be evaluated between the known limits y_0 and y_1 for the known, finite, continuous function $F(y)$. By dividing the interval between y_0 and y_1 into N equal reaches (N even) (Fig. B.1), Simpson's rule may be applied

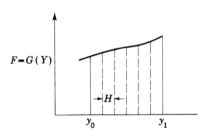

Figure B.1 Determining area under a curve by Simpson's rule.

to find the area under the curve. A statement sequence to find V when we let $G(Y) = F(Y)$ is

```
    H = (Y1 − Y0)/N
    V = G(Y0) + G(Y1)
    DO 2   I = 1,N,2
2   V = V + 4.*G(Y0 + I*H)
    N1 = N − 1
    DO 3   I = 2,N1,2
3   V = V + 2.*G(Y0 + I*H)
    V = V*H/3.
    WRITE(6,1) V
1   FORMAT ('0 V = ', F 20.4)
    If intermediate values are desired:
    V = .0
    N2 = N − 1
    DO 4   J = 1, N2, 2
    I = J − 1
    V = V + H*(G(Y0 + I*H) + G(Y0 + (I + 2)*H)
  1 + 4.*G(Y0 + (I + 1)*H))/3.
    Y = Y0 + (I + 2)*H
4   WRITE(6,5) Y,V
5   FORMAT ('0 Y = ', F 20.4, 5X, 'V = ', F20.4)
```

B.2 PARABOLIC INTERPOLATION

It is frequently desirable to use experimental data in computer programs. For example, consider that A, the area of a reservoir, is known for 10-ft intervals of elevation z, starting with z_0. Then for any elevation within the range of data, the area of reservoir is desired. More generally (Fig. B.2), values of y are known for equal increments in x. If the value of y is desired for the x shown, a parabola through the three points, with axis vertical, is first found by transferring the origin to x', y'.

$$y' = ax'^2 + bx'$$

Let $x_{n+1} - x_n = x_n - x_{n-1} = h$; then

$$y_{n+1} - y_n = ah^2 + bh \qquad y_{n-1} - y_n = ah^2 - bh$$

from which

$$a = \frac{1}{2h^2} (y_{n+1} + y_{n-1} - 2y_n) \qquad b = \frac{1}{2h} (y_{n+1} - y_{n-1})$$

and

$$y = y_n + \frac{\theta^2}{2} (y_{n+1} + y_{n-1} - 2y_n) + \frac{\theta}{2} (y_{n+1} - y_{n-1})$$

in which θh has been substituted for x'.

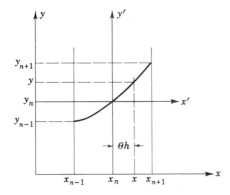

Figure B.2 Parabolic interpolation.

Example B.1 The area of reservoir given for each 10 ft of elevation above $z_0 = 6320$ is

A	348	692	1217
z	6320	6330	6340

and so forth. The program is, for $dz = 10$:

```
    I = (Z − Z0)/DZ + 2
    TH = (Z − Z0 − (I − 1)*DZ)/DZ
    AREA = A(I) + .5*TH*(A(I + 1) − A(I − 1) + TH*(A(I + 1)
    2 + A(I − 1) − 2.*A(I)))
    WRITE (6,1)Z, AREA
  1 FORMAT('0 Z = ', F10.2, 5X, 'AREA = ',F10.2)
```

In this program TH takes the place of θ, and the interpolation is taken so that TH is always negative.

B.3 SOLUTION OF ALBEBRAIC OR TRANSCENDENTAL EQUATIONS BY THE BISECTION METHOD

In the algebraic expression $F(x) = 0$, when a range of values of x is known that contains only one root, the *bisection* method is a practical way to obtain it. It is best shown by an example. The critical depth in a trapezoidal channel is wanted for given flow Q and channel dimensions (Fig. B.3). The formula

$$GG = 1 − \frac{Q^2 T}{gA^3} = 0$$

must be satisfied by some positive depth YCR greater than 0 and less than, say, 100 ft. T is the top width ($B + 2.*M*YCR$). The interval is bisected and this value of YCR tried. If the value of GG is positive, as with the solid line in Fig. B.3b, then the root is less than the midpoint and the upper limit is moved to the midpoint and the remaining half bisected, etc.

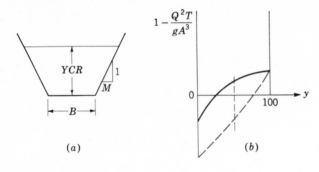

Figure B.3 Bisection method.

In program form

```
REAL M
F(YY) = 1.  −Q**2*(B + 2.*M*YY)/((YY*(B + M*YY))**3*G)
YMAX = 100.
YMIN = .0
YCR = .5*(YMAX + YMIN)
DO  2  J = 1,14
X = F(YCR)
IF(X.GT..0) YMAX = YCR
IF(X.LE..0) YMIN = YCR
2   YCR = .5*(YMAX + YMIN)
WRITE (6,1) YCR
```

The 14 iterations reduce the interval within which the root must be to about 0.01 ft.

B.4 SOLUTION OF TRANSCENDENTAL OR ALGEBRAIC EQUATIONS BY THE NEWTON-RAPHSON METHOD

The Newton-Raphson method† is particularly convenient for solving easily differentiable equations when the value of the desired root is known approximately. Let $y = F(x)$ be the equation, Fig. B.4, with $x = x_0$ an approximate value of the root. The root is at B; that is, for $x = B$, $F(x) = 0$. Starting at $x = x_0$ and drawing the tangent to the curve at A give

$$F'(x_0) = \frac{F(x_0)}{x_0 - x_1} \quad \text{or} \quad x_1 = x_0 - \frac{F(x_0)}{F'(x_0)}$$

It is evident that x_1 is a better approximation to the root than x_0 if there is no point of inflection between A and B and if the slope of the curve does not become zero. If one were to apply these procedures starting at point C (with no inflection

† L. A. Pipes, "Applied Mathematics for Engineers and Physicists," 2d ed., pp. 115–118, McGraw-Hill, New York, 1958.

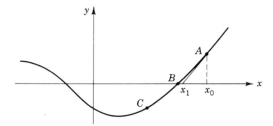

Figure B.4 Newton-Raphson root-finding method.

points or zero slopes), the first application would yield an x_1 to the right of B. The procedure may be repeated three or four times to obtain an accurate value of the root.

Example B.2 It is desired to find the root near $x = x_0$ in the equation $y = a_0 + a_1 x + a_2 x^2 + a_3 \sin \omega x$.

The programming sequence for this solution follows (ω = OMEGA):

```
NAMELIST/DIN/A0,A1,A2,A3,OMEGA,X0,X
READ (5,DIN)
X = X0
DO 1  I = 1, 4
1  X = X − (A0 + A1*X + A2*X*X + A3*SIN(OMEGA*X))/
1  (A1 + 2.*A2*X + A3*OMEGA*COS(OMEGA*X))
WRITE (6,DIN)
END
&DIN A0 = − 1.55,A1 = 1.,A2 = − .5,A3 = 1.,OMEGA = 1.5708,
X0 = .95  &END
```

B.5 RUNGE-KUTTA SOLUTION OF DIFFERENTIAL EQUATIONS

The family of Runge-Kutta solutions is for various orders of accuracy, but they are alike in that the differential equation has its solution extended forward from known conditions by an increment of the independent variable without using information outside this increment.

First Order

In the equation

$$\frac{dy}{dt} = F(y, t)$$

$y = y_n$ when $t = t_n$ and y_{n+1} is desired when $t = t_n + h$. In Fig. B.5,

$$u_1 = hF(y_n, t_n) \qquad y_{n+1} = y_n + u_1 \qquad t_{n+1} = t_n + h$$

the equation is evaluated at the initial known conditions and the extension is taken as the tangent to the curve at this point.

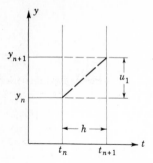

Figure B.5 First-order Runge-Kutta method.

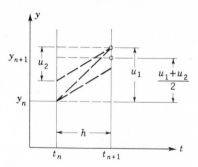

Figure B.6 Second-order Runge-Kutta method.

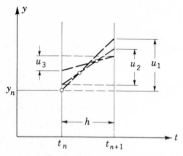

Figure B.7 Third-order Runge-Kutta method.

Second Order

The equation is evaluated at the end points of the interval h, as shown in Fig. B.6:

$$u_1 = hF(y_n, t_n) \qquad u_2 = hF(y_n + u_1, t_n + h)$$
$$y_{n+1} = y_n + \tfrac{1}{2}(u_1 + u_2) \qquad t_{n+1} = t_n + h$$

u_2 is evaluated for the point found by the first-order method.

Third Order

The slope of the curve is evaluated at the initial point, the one-third point, and the two-thirds point as follows (Fig. B.7):

$$u_1 = hF(y_n, t_n)$$

$$u_2 = hF\left(y_n + \frac{u_1}{3}, t_n + \frac{h}{3}\right)$$

$$u_3 = hF\left(y_n + \frac{2u_2}{3}, t_n + \frac{2h}{3}\right)$$

$$y_{n+1} = y_n + \frac{u_1}{4} + \frac{3u_3}{4}$$

$$t_{n+1} = t_n + h$$

Differential equations of higher order and degree may be simplified by expressing them as simultaneous first-order differential equations.

Example B.3 Put Eq. (12.1.17) into suitable form for solution by the third-order Runge-Kutta method.

In Eq. (12.1.17)

$$\frac{d^2z}{dt^2} + \frac{f}{2D}\frac{dz}{dt}\left|\frac{dz}{dt}\right| + \frac{2g}{L}z = 0$$

Let $y = dz/dt$, then $dy/dt = d^2z/dt^2$, and

$$\frac{dy}{dt} = -\frac{f}{2D}y|y| - \frac{2g}{L}z = F_1(y, z, t)$$

$$\frac{dz}{dt} = y = F_2(y, z, t)$$

The two equations are solved simultaneously, from known initial conditions y_n, z_n, t_n:

$$u_{11} = hF_1(y_n, z_n, t_n) = h\left(-\frac{f}{2D}y_n|y_n| - \frac{2g}{L}z_n\right)$$

$$u_{12} = hF_2(y_n, z_n, t_n) = hy_n$$

$$u_{21} = hF_1\left(y_n + \frac{u_{11}}{3}, z_n + \frac{u_{12}}{3}, t_n + \frac{h}{3}\right)$$

$$= h\left[-\frac{f}{2D}\left(y_n + \frac{u_{11}}{3}\right)\left|y_n + \frac{u_{11}}{3}\right| - \frac{2g}{L}\left(z_n + \frac{u_{12}}{3}\right)\right]$$

$$u_{22} = hF_2\left(y_n + \frac{u_{11}}{3}, z_n + \frac{u_{12}}{3}, t_n + \frac{h}{3}\right) = h\left(y_n + \frac{u_{11}}{3}\right)$$

$$u_{31} = hF_1(y_n + \tfrac{2}{3}u_{21}, z_n + \tfrac{2}{3}u_{22}, t_n + \tfrac{2}{3}h)$$

$$= h\left[-\frac{f}{2D}\left(y_n + \frac{2}{3}u_{21}\right)\left|y_n + \frac{2}{3}u_{21}\right| - \frac{2g}{L}\left(z_n + \frac{2}{3}u_{22}\right)\right]$$

$$u_{32} = hF_2(y_n + \tfrac{2}{3}u_{21}, z_n + \tfrac{2}{3}u_{22}, t_n + \tfrac{2}{3}h) = h(y_n + \tfrac{2}{3}u_{21})$$

$$y_{n+1} = y_n + \frac{u_{11}}{4} + \frac{3}{4}u_{31}$$

$$z_{n+1} = z_n + \frac{u_{12}}{4} + \frac{3}{4}u_{32}$$

$$t_{n+1} = t_n + h$$

The equations for simultaneous solution have been written for a general case as well as for the specific case of solution of Eq. (12.1.17).

APPENDIX
C
PHYSICAL PROPERTIES OF FLUIDS

Table C.1 Physical properties of water in SI units

Temp, °C	Specific weight γ, N/m^3	Density ρ, kg/m^3	Viscosity μ, N·s/m^2 $10^3 \mu =$	Kinematic viscosity ν, m^2/s $10^6 \nu =$	Surface tension σ, N/m $100\,\sigma =$	Vapor-pressure head p_v/γ, m	Bulk modulus of elasticity K, N/m^2 $10^{-7} K =$
0	9805	999.9	1.792	1.792	7.62	0.06	204
5	9806	1000.0	1.519	1.519	7.54	0.09	206
10	9803	999.7	1.308	1.308	7.48	0.12	211
15	9798	999.1	1.140	1.141	7.41	0.17	214
20	9789	998.2	1.005	1.007	7.36	0.25	220
25	9779	997.1	0.894	0.897	7.26	0.33	222
30	9767	995.7	0.801	0.804	7.18	0.44	223
35	9752	994.1	0.723	0.727	7.10	0.58	224
40	9737	992.2	0.656	0.661	7.01	0.76	227
45	9720	990.2	0.599	0.605	6.92	0.98	229
50	9697	988.1	0.549	0.556	6.82	1.26	230
55	9679	985.7	0.506	0.513	6.74	1.61	231
60	9658	983.2	0.469	0.477	6.68	2.03	228
65	9635	980.6	0.436	0.444	6.58	2.56	226
70	9600	977.8	0.406	0.415	6.50	3.20	225
75	9589	974.9	0.380	0.390	6.40	3.96	223
80	9557	971.8	0.357	0.367	6.30	4.86	221
85	9529	968.6	0.336	0.347	6.20	5.93	217
90	9499	965.3	0.317	0.328	6.12	7.18	216
95	9469	961.9	0.299	0.311	6.02	8.62	211
100	9438	958.4	0.284	0.296	5.94	10.33	207

Table C.2 Physical properties of water in English units†

Temp, °F	Specific weight γ, lb/ft^3	Density ρ, $slugs/ft^3$	Viscosity μ, $lb \cdot s/ft^2$ $10^5 \mu =$	Kine-matic viscosity v, ft^2/s $10^5 v =$	Surface tension σ, lb/ft $100 \sigma =$	Vapor-pressure head p_v/γ, ft	Bulk modulus of elasticity K, lb/in^2 $10^{-3} K =$
32	62.42	1.940	3.746	1.931	0.518	0.20	293
40	62.43	1.940	3.229	1.664	0.514	0.28	294
50	62.41	1.940	2.735	1.410	0.509	0.41	305
60	62.37	1.938	2.359	1.217	0.504	0.59	311
70	62.30	1.936	2.050	1.059	0.500	0.84	320
80	62.22	1.934	1.799	0.930	0.492	1.17	322
90	62.11	1.931	1.595	0.826	0.486	1.61	323
100	62.00	1.927	1.424	0.739	0.480	2.19	327
110	61.86	1.923	1.284	0.667	0.473	2.95	331
120	61.71	1.918	1.168	0.609	0.465	3.91	333
130	61.55	1.913	1.069	0.558	0.460	5.13	334
140	61.38	1.908	0.981	0.514	0.454	6.67	330
150	61.20	1.902	0.905	0.476	0.447	8.58	328
160	61.00	1.896	0.838	0.442	0.441	10.95	326
170	60.80	1.890	0.780	0.413	0.433	13.83	322
180	60.58	1.883	0.726	0.385	0.426	17.33	313
190	60.36	1.876	0.678	0.362	0.419	21.55	313
200	60.12	1.868	0.637	0.341	0.412	26.59	308
212	59.83	1.860	0.593	0.319	0.404	33.90	300

† This table was compiled primarily from Hydraulic Models, *ASCE Man. Eng. Pract.* 25, 1942.

Table C.3 Properties of gases at low pressures and 80°F (26.67°C)

Gas	Chemical formula	Molec-ular weight	Gas constant R $m \cdot N/kg \cdot K$	Gas constant R $ft \cdot lb/lb_m \cdot °R$	Specific heat, Btu/$lb_m \cdot °R$ or kcal/kg$\cdot$K c_p	Specific heat, Btu/$lb_m \cdot °R$ or kcal/kg$\cdot$K c_v	Specific heat ratio k
Air	—	29.0	287	53.3	0.240	0.171	1.40
Carbon monoxide	CO	28.0	297	55.2	0.249	0.178	1.40
Helium	He	4.00	2077	386	1.25	0.753	1.66
Hydrogen	H_2	2.02	4121	766	3.43	2.44	1.40
Nitrogen	N_2	28.0	297	55.2	0.248	0.177	1.40
Oxygen	O_2	32.0	260	48.3	0.219	0.157	1.40
Water vapor	H_2O	18.0	462	85.8	0.445	0.335	1.33

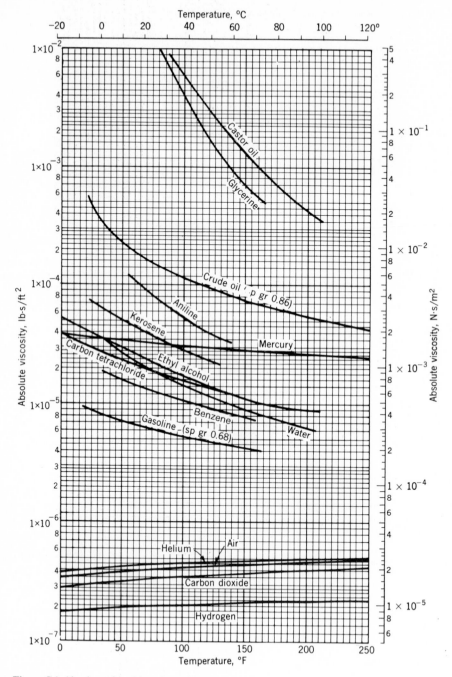

Figure C.1 Absolute viscosities of certain gases and liquids.

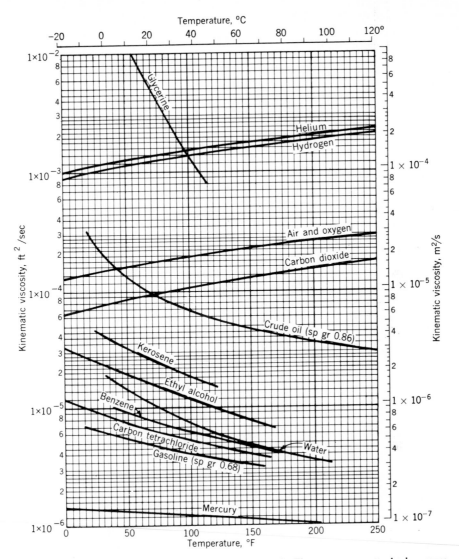

Figure C.2 Kinematic viscosities of certain gases and liquids. The gases are at standard pressure.

Table C.4 One-dimensional isentropic relations (for a perfect gas with constant specific heat; $k = 1.4$)†

M	A/A^*	p/p_0	ρ/ρ_0	T/T_0	M	A/A^*	p/p_0	ρ/ρ_0	T/T_0
0.00	.....	1.000	1.000	1.000	0.78	1.05	0.669	0.750	0.891
0.01	57.87	0.9999	0.9999	0.9999	0.80	1.04	0.656	0.740	0.886
0.02	28.94	0.9997	0.9999	0.9999	0.82	1.03	0.643	0.729	0.881
0.04	14.48	0.999	0.999	0.9996	0.84	1.02	0.630	0.719	0.876
0.06	9.67	0.997	0.998	0.999	0.86	1.02	0.617	0.708	0.871
0.08	7.26	0.996	0.997	0.999	0.88	1.01	0.604	0.698	0.865
0.10	5.82	0.993	0.995	0.998	0.90	1.01	0.591	0.687	0.860
0.12	4.86	0.990	9.993	0.997	0.92	1.01	0.578	0.676	0.855
0.14	4.18	0.986	0.990	0.996	0.94	1.00	0.566	0.666	0.850
0.16	3.67	0.982	0.987	0.995	0.96	1.00	0.553	0.655	0.844
0.18	3.28	0.978	0.984	0.994	0.98	1.00	0.541	0.645	0.839
0.20	2.96	0.973	0.980	0.992	1.00	1.00	0.528	0.632	0.833
0.22	2.71	0.967	0.976	0.990	1.02	1.00	0.516	0.623	0.828
0.24	2.50	0.961	0.972	0.989	1.04	1.00	0.504	0.613	0.822
0.26	2.32	0.954	0.967	0.987	1.06	1.00	0.492	0.602	0.817
0.28	2.17	0.947	0.962	0.985	1.08	1.01	0.480	0.592	0.810
0.30	2.04	0.939	0.956	0.982	1.10	1.01	0.468	0.582	0.805
0.32	1.92	0.932	0.951	0.980	1.12	1.01	0.457	0.571	0.799
0.34	1.82	0.923	0.944	0.977	1.14	1.02	0.445	0.561	0.794
0.36	1.74	0.914	0.938	0.975	1.16	1.02	0.434	0.551	0.788
0.38	1.66	0.905	0.931	0.972	1.18	1.02	0.423	0.541	0.782
0.40	1.59	0.896	0.924	0.969	1.20	1.03	0.412	0.531	0.776
0.42	1.53	0.886	0.917	0.966	1.22	1.04	0.402	0.521	0.771
0.44	1.47	0.876	0.909	0.963	1.24	1.04	0.391	0.512	0.765
0.46	1.42	0.865	0.902	0.959	1.26	1.05	0.381	0.502	0.759
0.48	1.38	0.854	0.893	0.956	1.28	1.06	0.371	0.492	0.753
0.50	1.34	0.843	0.885	0.952	1.30	1.07	0.361	0.483	0.747
0.52	1.30	0.832	0.877	0.949	1.32	1.08	0.351	0.474	0.742
0.54	1.27	0.820	0.868	0.945	1.34	1.08	0.342	0.464	0.736
0.56	1.24	0.808	0.859	0.941	1.36	1.09	0.332	0.455	0.730
0.58	1.21	0.796	0.850	0.937	1.38	1.10	0.323	0.446	0.724
0.60	1.19	0.784	0.840	0.933	1.40	1.11	0.314	0.437	0.718
0.62	1.17	0.772	0.831	0.929	1.42	1.13	0.305	0.429	0.713
0.64	1.16	0.759	0.821	0.924	1.44	1.14	0.297	0.420	0.707
0.66	1.13	0.747	0.812	0.920	1.46	1.15	0.289	0.412	0.701
0.68	1.12	0.734	0.802	0.915	1.48	1.16	0.280	0.403	0.695
0.70	1.09	0.721	0.792	0.911	1.50	1.18	0.272	0.395	0.690
0.72	1.08	0.708	0.781	0.906	1.52	1.19	0.265	0.387	0.684
0.74	1.07	0.695	0.771	0.901	1.54	1.20	0.257	0.379	0.678
0.76	1.06	0.682	0.761	0.896	1.56	1.22	0.250	0.371	0.672

Table C.4 One-dimensional isentropic relations (*continued*)

M	A/A^*	p/p_0	ρ/ρ_0	T/T_0	M	A/A^*	p/p_0	ρ/ρ_0	T/T_0
1.58	1.23	0.242	0.363	0.667	2.30	2.19	0.080	0.165	0.486
1.60	1.25	0.235	0.356	0.661	2.32	2.23	0.078	0.161	0.482
1.62	1.27	0.228	0.348	0.656	2.34	2.27	0.075	0.157	0.477
1.64	1.28	0.222	0.341	0.650	2.36	2.32	0.073	0.154	0.473
1.66	1.30	0.215	0.334	0.645	2.38	2.36	0.071	0.150	0.469
1.68	1.32	0.209	0.327	0.639	2.40	2.40	0.068	0.147	0.465
1.70	1.34	0.203	0.320	0.634	2.42	2.45	0.066	0.144	0.461
1.72	1.36	0.197	0.313	0.628	2.44	2.49	0.064	0.141	0.456
1.74	1.38	0.191	0.306	0.623	2.46	2.54	0.062	0.138	0.452
1.76	1.40	0.185	0.300	0.617	2.48	2.59	0.060	0.135	0.448
1.78	1.42	0.179	0.293	0.612	2.50	2.64	0.059	0.132	0.444
1.80	1.44	0.174	0.287	0.607	2.52	2.69	0.057	0.129	0.441
1.82	1.46	0.169	0.281	0.602	2.54	2.74	0.055	0.126	0.437
1.84	1.48	0.164	0.275	0.596	2.56	2.79	0.053	0.123	0.433
1.86	1.51	0.159	0.269	0.591	2.58	2.84	0.052	0.121	0.429
1.88	1.53	0.154	0.263	0.586	2.60	2.90	0.050	0.118	0.425
1.90	1.56	0.149	0.257	0.581	2.62	2.95	0.049	0.115	0.421
1.92	1.58	0.145	0.251	0.576	2.64	3.01	0.047	0.113	0.418
1.94	1.61	0.140	0.246	0.571	2.66	3.06	0.046	0.110	0.414
1.96	1.63	0.136	0.240	0.566	2.68	3.12	0.044	0.108	0.410
1.98	1.66	0.132	0.235	0.561	2.70	3.18	0.043	0.106	0.407
2.00	1.69	0.128	0.230	0.556	2.72	3.24	0.042	0.103	0.403
2.02	1.72	0.124	0.225	0.551	2.74	3.31	0.040	0.101	0.400
2.04	1.75	0.120	0.220	0.546	2.76	3.37	0.039	0.099	0.396
2.06	1.78	0.116	0.215	0.541	2.78	3.43	0.038	0.097	0.393
2.08	1.81	0.113	0.210	0.536	2.80	3.50	0.037	0.095	0.389
2.10	1.84	0.109	0.206	0.531	2.82	3.57	0.036	0.093	0.386
2.12	1.87	0.106	0.201	0.526	2.84	3.64	0.035	0.091	0.383
2.14	1.90	0.103	0.197	0.522	2.86	3.71	0.034	0.089	0.379
2.16	1.94	0.100	0.192	0.517	2.88	3.78	0.033	0.087	0.376
2.18	1.97	0.097	0.188	0.513	2.90	3.85	0.032	0.085	0.373
2.20	2.01	0.094	0.184	0.508	2.92	3.92	0.031	0.083	0.370
2.22	2.04	0.091	0.180	0.504	2.94	4.00	0.030	0.081	0.366
2.24	2.08	0.088	0.176	0.499	2.96	4.08	0.029	0.080	0.363
2.26	2.12	0.085	0.172	0.495	2.98	4.15	0.028	0.078	0.360
2.28	2.15	0.083	0.168	0.490	3.00	4.23	0.027	0.076	0.357

Table C.5 One-dimensional normal-shock relations (for a perfect gas with $k = 1.4$)†

M_1	M_2	$\frac{p_2}{p_1}$	$\frac{T_2}{T_1}$	$\frac{(p_0)_2}{(p_0)_1}$	M_1	M_2	$\frac{p_2}{p_1}$	$\frac{T_2}{T_1}$	$\frac{(p_0)_2}{(p_0)_1}$
1.00	1.000	1.000	1.000	1.000	1.72	0.635	3.285	1.473	0.847
1.02	0.980	1.047	1.013	1.000	1.74	0.631	3.366	1.487	0.839
1.04	0.962	1.095	1.026	1.000	1.76	0.626	3.447	1.502	0.830
1.06	0.944	1.144	1.039	1.000	1.78	0.621	3.530	1.517	0.821
1.08	0.928	1.194	1.052	0.999	1.80	0.617	3.613	1.532	0.813
1.10	0.912	1.245	1.065	0.999	1.82	0.612	3.698	1.547	0.804
1.12	0.896	1.297	1.078	0.998	1.84	0.608	3.783	1.562	0.795
1.14	0.882	1.350	1.090	0.997	1.86	0.604	3.869	1.577	0.786
1.16	0.868	1.403	1.103	0.996	1.88	0.600	3.957	1.592	0.777
1.18	0.855	1.458	1.115	0.995	1.90	0.596	4.045	1.608	0.767
1.20	0.842	1.513	1.128	0.993	1.92	0.592	4.134	1.624	0.758
1.22	0.830	1.570	1.140	0.991	1.94	0.588	4.224	1.639	0.749
1.24	0.818	1.627	1.153	0.988	1.96	0.584	4.315	1.655	0.740
1.26	0.807	1.686	1.166	0.986	1.98	0.581	4.407	1.671	0.730
1.28	0.796	1.745	1.178	0.983	2.00	0.577	4.500	1.688	0.721
1.30	0.786	1.805	1.191	0.979	2.02	0.574	4.594	1.704	0.711
1.32	0.776	1.866	1.204	0.976	2.04	0.571	4.689	1.720	0.702
1.34	0.766	1.928	1.216	0.972	2.06	0.567	4.784	1.737	0.693
1.36	0.757	1.991	1.229	0.968	2.08	0.564	4.881	1.754	0.683
1.38	0.748	2.055	1.242	0.963	2.10	0.561	4.978	1.770	0.674
1.40	0.740	2.120	1.255	0.958	2.12	0.558	5.077	1.787	0.665
1.42	0.731	2.186	1.268	0.953	2.14	0.555	5.176	1.805	0.656
1.44	0.723	2.253	1.281	0.948	2.16	0.553	5.277	1.822	0.646
1.46	0.716	2.320	1.294	0.942	2.18	0.550	5.378	1.839	0.637
1.48	0.708	2.389	1.307	0.936	2.20	0.547	5.480	1.857	0.628
1.50	0.701	2.458	1.320	0.930	2.22	0.544	5.583	1.875	0.619
1.52	0.694	2.529	1.334	0.923	2.24	0.542	5.687	1.892	0.610
1.54	0.687	2.600	1.347	0.917	2.26	0.539	5.792	1.910	0.601
1.56	0.681	2.673	1.361	0.910	2.28	0.537	5.898	1.929	0.592
1.58	0.675	2.746	1.374	0.903	2.30	0.534	6.005	1.947	0.583
1.60	0.668	2.820	1.388	0.895	2.32	0.532	6.113	1.965	0.575
1.62	0.663	2.895	1.402	0.888	2.34	0.530	6.222	1.984	0.566
1.64	0.657	2.971	1.416	0.880	2.36	0.527	6.331	2.003	0.557
1.66	0.651	3.048	1.430	0.872	2.38	0.525	6.442	2.021	0.549
1.68	0.646	3.126	1.444	0.864	2.40	0.523	6.553	2.040	0.540
1.70	0.641	3.205	1.458	0.856	2.42	0.521	6.666	2.060	0.532

† From Irving Shames, "Mechanics of Fluids," copyright © 1962 by McGraw-Hill, Inc. Used with permission of McGraw-Hill Book Company.

Table C.5 One-dimensional normal-shock relations (*continued*)

M_1	M_2	$\dfrac{p_2}{p_1}$	$\dfrac{T_2}{T_1}$	$\dfrac{(p_0)_2}{(p_0)_1}$	M_1	M_2	$\dfrac{p_2}{p_1}$	$\dfrac{T_2}{T_1}$	$\dfrac{(p_0)_2}{(p_0)_1}$
2.44	0.519	6.779	2.079	0.523	2.76	0.491	8.721	2.407	0.403
2.46	0.517	6.894	2.098	0.515	2.78	0.490	8.850	2.429	0.396
2.48	0.515	7.009	2.118	0.507	2.80	0.488	8.980	2.451	0.389
2.50	0.513	7.125	2.138	0.499	2.82	0.487	9.111	2.473	0.383
2.52	0.511	7.242	2.157	0.491	2.84	0.485	9.243	2.496	0.376
2.54	0.509	7.360	2.177	0.483	2.86	0.484	9.376	2.518	0.370
2.56	0.507	7.479	2.198	0.475	2.88	0.483	9.510	2.541	0.364
2.58	0.506	7.599	2.218	0.468	2.90	0.481	9.645	2.563	0.358
2.60	0.504	7.720	2.238	0.460	2.92	0.480	9.781	2.586	0.352
2.62	0.502	7.842	2.260	0.453	2.94	0.479	9.918	2.609	0.346
2.64	0.500	7.965	2.280	0.445	2.96	0.478	10.055	2.632	0.340
2.66	0.499	8.088	2.301	0.438	2.98	0.476	10.194	2.656	0.334
2.68	0.497	8.213	2.322	0.431	3.00	0.475	10.333	2.679	0.328
2.70	0.496	8.338	2.343	0.424					
2.72	0.494	8.465	2.364	0.417					
2.74	0.493	8.592	2.396	0.410					

Symbol	Quantity	Unit		Dimensions (MLT)
		SI	English	
a	Constant, pulse wave speed	m/s	ft/s	LT^{-1}
a	Acceleration	m/s^2	ft/s^2	LT^{-2}
$\mathbf{a}$	Acceleration vector	m/s^2	ft/s^2	LT^{-2}
a^*	Velocity	m/s	ft/s	LT^{-1}
A	Area	m^2	ft^2	L^2
$\mathbf{A}$	Adverse slope	none	none	
b	Distance	m	ft	L
b	Constant			
c	Speed of surge wave	m/s	ft/s	LT^{-1}
c	Speed of sound	m/s	ft/s	LT^{-1}
c_p	Specific heat, constant pressure	J/kg·K	ft·lb/slug·°R	
c_v	Specific heat, constant volume	J/kg·K	ft·lb/slug·°R	
C	Concentration	No./m^3	No./ft^3	L^{-3}
C	Concentration of marker per unit volume	m^{-3}	ft^{-3}	L^{-3}
C	Coefficient		none	
C	Stress	Pa	lb/ft^2	$ML^{-1}T^{-2}$
C_m	Empirical constant	$m^{1/3}$/s	ft$^{1/3}$/s	$L^{1/3}T^{-1}$
$\mathbf{C}$	Critical slope		none	
D'	Volumetric displacement	m^3	ft^3	L^3
D	Diameter	m	ft	L
D_d	Coefficient of dispersion	m^2/s	ft^2/s	L^2T^{-1}
D_m	Coefficient of molecular diffusion	m^2/s	ft^2/s	L^2T^{-1}
D_t	Coefficient of turbulent diffusion	m^2/s	ft^2/s	L^2T^{-1}

Symbol	Quantity	SI	English	Dimensions (MLT)
e	Efficiency		none	
e	Internal energy per unit mass	J/kg	ft·lb/slug	L^2T^{-2}
E	Internal energy	J	ft·lb	ML^2T^{-2}
E	Specific energy	m·N/N	ft·lb/lb	L
E	Losses per unit weight	m·N/N	ft·lb/lb	L
E	Modulus of elasticity	Pa	lb/ft^2	$ML^{-1}T^{-2}$
f	Friction factor		none	
F	Force	N	lb	MLT^{-2}
F	Force vector	N	lb	MLT^{-2}
F	Froude number		none	
F_B	Buoyant force	N	lb	MLT^{-2}
g	Acceleration of gravity	m/s^2	ft/s^2	LT^{-2}
g_0	Gravitation constant	kg·m/N·s^2	lb$_m$·ft/lb·s^2	
G	Mass flow rate per unit area	kg/s·m^2	slug/s·ft^2	$ML^{-2}T^{-1}$
h	Head, vertical distance	m	ft	L
h	Enthalpy per unit mass	J/kg	ft·lb/slug	L^2T^{-2}
H	Head, elevation of hydraulic grade line	m	ft	L
H	Horizontal slope	none	none	
I	Moment of inertia	m^4	ft^4	L^4
J	Junction point	none	none	
k	Specific heat ratio	none	none	
K	Bulk modulus of elasticity	Pa	lb/ft^2	$ML^{-1}T^{-2}$
K	Minor loss coefficient	none	none	
L	Length	m	ft	L
L	Lift	N	lb	MLT^{-2}
l	Length, mixing length	m	ft	L
ln	Natural logarithm	none	none	
m	Mass	kg	slug	M
m	Form factor, constant	none	none	
m	Strength of source	m^3/s	ft^3/s	L^3T^{-1}
$\dot{m}$	Mass per unit time	kg/s	slug/s	MT^{-1}
M	Molecular weight			
M	Momentum per unit time	N	lb	MLT^{-2}
M	Mild slope	none	none	
M	Mach number	none	none	
$\overline{MG}$	Metacentric height	m	ft	L
n	Exponent, constant	none	none	
n	Normal direction			
n	Manning roughness factor			
n	Number of moles			
$\mathbf{n}_1$	Normal unit vector			
N	Rotation speed	1/s	1/s	T^{-1}
$NPSH$	Net positive suction head	m	ft	L
p	Pressure	Pa	lb/ft^2	$ML^{-1}T^{-2}$
p	Force	N	lb	MLT^{-2}
P	Height of weir	m	ft	L
P	Wetted perimeter	m	ft	L
q	Discharge per unit width	m^2/s	ft^2/s	L^2T^{-1}
q	Velocity	m/s	ft/s	LT^{-1}

Symbol	Quantity	SI	English	Dimensions (MLT)
q	Velocity vector	m/s	ft/s	LT^{-1}
q_H	Heat transfer per unit mass	J/kg	ft·lb/slug	L^2T^{-2}
Q	Discharge	m³/s	ft³/s	L^3T^{-1}
Q_H	Heat transfer per unit time	J/s	ft·lb/s	ML^2T^{-3}
r	Coefficient			
r	Radial distance	m	ft	L
$\mathbf{r}$	Position vector	m	ft	L
R	Hydraulic radius	m	ft	L
R	Gas constant	J/kg·K	ft·lb/slug·°R	
R, R'	Gage difference	m	ft	L
$\mathbf{R}$	Reynolds number	none	none	
s	Distance	m	ft	L
s	Entropy per unit mass	J/Kg·K	ft·lb/slug·°R	
s	Slip	none	none	
S	Entropy	J/K	ft·lb/°R	
S	Specific gravity, slope	none	none	
$\mathbf{s}$	Steep slope	none	none	
t	Time	s	s	T
t	Temperature	°C	°F	
t, t'	Distance, thickness	m	ft	L
T	Temperature	K	°R	
T	Torque	N·m	lb·ft	ML^2T^{-2}
T	Tensile force/ft	N/m	lb/ft	MT^{-2}
T	Top width	m	ft	L
u	Velocity, velocity component	m/s	ft/s	LT^{-1}
u	Peripheral speed	m/s	ft/s	LT^{-1}
u	Intrinsic energy	J/kg	ft·lb/slug	L^2T^{-2}
u_*	Shear stress velocity	m/s	ft/s	LT^{-1}
U	Velocity	m/s	ft/s	LT^{-1}
v	Velocity, velocity component	m/s	ft/s	LT^{-1}
v_s	Specific volume	m³/kg	ft³/slug	$M^{-1}L^3$
$\mathcal{V}$	Volume	m³	ft³	L^3
$\mathbf{V}$	Velocity vector	m/s	ft/s	LT^{-1}
V	Velocity	m/s	ft/s	LT^{-1}
w	Velocity component	m/s	ft/s	LT^{-1}
w	Work per unit mass	J/kg	ft·lb/slug	L^2T^{-2}
W	Work per unit time	J/s	ft·lb/s	ML^2T^{-3}
W	Work of expansion	m·N	ft·lb	ML^2T^{-2}
W_s	Shaft work	m·N	ft·lb	ML^2T^{-2}
W	Weight	N	lb	MLT^{-2}
$\mathbf{W}$	Weber number	none	none	
x	Distance	m	ft	L
x_p	Distance to pressure center	m	ft	L
X	Body-force component per unit mass	N/kg	lb/slug	LT^{-2}
y	Distance, depth	m	ft	L
y_p	Distance to pressure center	m	ft	L
Y	Expansion factor	none	none	
Y	Body-force component per unit mass	N/kg	lb/slug	LT^{-2}
z	Vertical distance	m	ft	L

Symbol	Quantity	Unit		Dimensions (MLT)
		SI	English	
Z	Vertical distance	m	ft	L
Z	Body-force component per unit mass	N/kg	lb/slug	LT^{-2}
α	Kinetic-energy correction factor	none	none	
α	Angle, coefficient	none	none	
β	Momentum correction factor	none	none	
β	Blade angle	none	none	
Γ	Circulation	m^2/s	ft^2/s	L^2T^{-1}
$\mathbf{V}$	Vector operator	1/m	1/ft	L^{-1}
γ	Specific weight	N/m^3	lb/ft^3	$ML^{-2}T^{-2}$
δ	Boundary-layer thickness	m	ft	L
ϵ	Kinematic eddy viscosity	m^2/s	ft^2/s	L^2T^{-1}
ϵ	Roughness height	m	ft	L
η	Eddy viscosity	N·s/m^2	lb·s/ft^2	$ML^{-1}T^{-1}$
η	Head ratio	none	none	
η	Efficiency			
θ	Angle	none	none	
κ	Universal constant	none	none	
λ	Scale ratio, undetermined multiplier	none	none	
μ	Viscosity	N·s/m^2	lb·s/ft^2	$ML^{-1}T^{-1}$
μ	Constant			
v	Kinematic viscosity	m^2/s	ft^2/s	L^2T^{-1}
ϕ	Velocity potential	m^2/s	ft^2/s	L^2T^{-1}
ϕ	Function			
π	Constant	none	none	
Π	Dimensionless parameter	none	none	
ρ	Density	kg/m^3	slug/ft^3	ML^{-3}
σ	Surface tension	N/m	lb/ft	MT^{-2}
σ	Cavitation index	none	none	
τ	Shear stress	Pa	lb/ft^2	$ML^{-1}T^{-2}$
ψ	Stream function, two dimensions	m/s	ft/s	L^2T^{-1}
ψ	Stokes' stream function	m^3/s	ft^3/s	L^3T^{-1}
ω	Angular velocity	rad/s	rad/s	T^{-1}

APPENDIX

E

ANSWERS TO EVEN-NUMBERED PROBLEMS

Chapter 1

1.2	(*a*) Non-Newtonian;
	(*b*) non-Newtonian; (*c*) Newtonian
1.4	Thixotropic
1.6	95.1 lb
1.8	1000 slug·ft/kip·s^2
1.10	8.5 m/s^2
1.12	184.6 slugs/ft·s
1.14	0.004 N·s/m^2
1.16	1.461 N·s/m^2
1.18	88.7 %
1.20	8.059(10)$^{-5}$ ft^2/s, 7.488(10)$^{-2}$ St
1.22	2.51(10)$^{-3}$ in
1.24	6.69(10)$^{-3}$ St

1.26	0.02136 ft^3/lb$_m$; 0.687 ft^3/slug
1.28	(*a*) 2.94;
	(*b*) 3.4014(10)$^{-4}$ m^3/kg, 2.8812(10)^4 N/m^3
1.30	3.2504 kg/m^3, 0.13 kg
1.32	2.255 kg/m^3
1.34	6.39 MN/m^2
1.36	Exponentially
1.38	11 MPa
1.40	0.4 MPa abs
1.42	15.535 psia
1.44	0.14 in
1.46	0.045 in

Chapter 2

2.2	$p_A = 249.6$ psf $= 11,950$ Pa,
	$p_B = p_C = -62.4$ psf $= -2.99$ kPa,
	$p_D = -374.4$ psf $= -17.93$ kPa
2.4	-5.88 kPa, 5.88 kPa,
	5.88 kPa, 22.65 kPa
2.8	68.7 kPa, 0.876 kg/m^3
2.10	(*a*) 16.3 in Hg; (*b*) 18.476 ft H$_2$O;
	(*c*) 6.284 ft; (*d*) 55.16 kPa
2.12	20.843 m
2.14	2.721 m H$_2$O, 3.28 m kerosene,
	0.926 m acetylene tetrabromide

2.16	3.687 m
2.18	5.34 mm Hg, 9.868 m H$_2$O abs
2.20	-454 mm H$_2$O
2.22	(*a*) 11.32 psi; (*b*) 3.344 m H$_2$O gage
2.24	383.43 mm
2.26	50 kPa
2.28	110.53 mm
2.30	36 MN
2.32	-441 N
2.34	(*a*) 51.08 kN; (*b*) 58.9 kN
2.36	$\gamma bh^2/3$

2.38 $1125b^2h^2$ N·m

2.40   0.774 ft below AB

2.42   46.21 kN

2.44   5990.4 lb, 3.0 ft

2.46   602.37 kN

2.48   0.4506 ft

2.50 $x_p = \frac{3}{8}b$, $y_p = \frac{3}{4}h$

2.52   7.429 ft

2.54 $0.433b$

2.56   0.7143 m

2.60   0.58 ft

2.62   0.3334 m

2.66   (a) 11.588 m; (b) 5.12γ, 69.436γ

2.68   9984 lb·ft

2.70   4.227 ft

2.72   1.8 in

2.74   18.75 mm

2.76   2.3575 kN

2.78   (a) 156.896 kN, 4.0833 m, 1.0833 m;
          (b) 179.285 kN, 0.948 m; (c) 0; (d) 0

2.80 -199.68 lb

2.82   9.6 kN, 0.24 m

2.84 $(100 - \frac{20}{3}\rho)r$ lb, 0 lb

2.86   26.304 kN·m

2.88   30.806 kN

2.90   (a) $\gamma r^2/2$, $y_p = 2r/3$ above BD;
          (b) $-\gamma\pi r^2/4$, $x_p = 4r/3\pi$ to left of AC

2.92   16.53 mm

2.94   13.3 m

2.96   1.334 ft

2.98   100 mm, 980.6 N

2.100   126.96 lb

2.102   4.769 m

2.104   No

2.106   (a) $1.348(\sin^2 \theta \cos \theta)^{1/3}$ m;
           (b) $2.19° \leq \theta \leq 54.74°$

2.108   23.256 m/s²

2.110 $p_A = 0$, $p_B = 0.347$ psi, $p_C = 0.069$ psi,
           $p_D = 1.109$ psi, $p_E = 0.693$ psi

2.112 $p_B = -0.52$ psi, $p_C = -0.26$ psi,
           $p_D = 1.30$ psi, $p_E = 1.04$ psi

2.114 $p_A = 0$, $p_B = 22.8$ kPa,
           $p_C = 16.42$ kPa

2.116 $a_x = 0$, $a_y = -g$

2.118 $a_x = 2.0394$ m/s²,
           $a_y = -1.178$ m/s², $\theta = 13.3°$

2.122   31.52 rpm

2.124   0.5 ft left of A, 54.16 rpm

2.126 $2\sqrt{gh_0/r_0}$

2.128 $\dfrac{2(p - p_0)}{\rho\omega^2} + \dfrac{g^2}{\omega^4} = x^2 + \left(y - \dfrac{g}{\omega^2}\right)^2$

2.132 $p = \left[p_0^{(n-1)/n} + \dfrac{n-1}{n}\dfrac{\rho_0\,\omega^2 r^2}{2p_0^{1/n}}\right]^{n/(n-1)}$

2.134   220.5 rpm

2.136   2.3907 kN

2.138 $132.381(1 + 0.0102\omega^2)$

Chapter 3

3.2 $n = 1.85$, turbulent

3.4 $x\sqrt{y} = 2$, $5 - z = 2x$

3.6   22.22 m/s, 54.04 kg/s

3.8 $\dfrac{1.273}{(0.8 - 0.5x/L)^2}$

3.10   Yes

3.12   Yes

3.16 $\dfrac{\partial V_r}{\partial r} + \dfrac{V_r}{r} + \dfrac{1}{r}\dfrac{\partial V_\theta}{\partial \theta} = 0$, yes

3.18 $23.5(10)^6$ ft·lb

3.20   253.3 m·N/s

3.22   19.4 ft, 35.35 ft/s

3.24 $-\frac{5}{3}a$, 46.43 ρa to the left

3.26   2.155 ft, 15.75 ft

3.28   3.83 ft, 15.26 ft

3.30   53.21 cfs/ft, 50.73 cfs/ft

3.32 $r = \dfrac{1}{4(1 + y/H)^{0.25}}$

3.34 $Q = A_2\left[\dfrac{2gR(\rho_m/\rho - 1)}{1 - (D_2/D_1)^4}\right]^{1/2}$

3.36 $R = 0$ for all H

3.38   68.1%, 28.14 ft·lb/lb, 88.14 ft·lb/lb

3.40   60%

3.42   1.037

3.44   0.82 m

3.46 A to B

3.48   3.94 cfs

3.50   6.11

3.52   0.0136 m³/s, 2.731 m

3.54   91.66 m³/s, -77.09 kPa

3.56   101 l/s, -24.73 kN/m², 9.59 kN/m²

3.58   8.75 hp

3.60   18,152 hp

3.62   568.6 gpm

3.64 $0.0667\ H$

3.66 80.68 kPa

3.68 $-0.0033\ \mathrm{m\cdot N/kg\cdot K}$

3.70 $V_{max}\dfrac{2n^2}{(n+1)(2n+1)},\ \dfrac{(2n+1)^2(n+1)}{4n^2(n+2)}$

3.72 12.5ρ

3.74 $(50+g)\rho$

3.76 191.42, 30.62 lb

3.78 157.7 lb

3.80 No

3.82 25.5 MN, 8.449 MN

3.84 18.71 kN, 39.96 kN

3.86 13.06 lb

3.88 340,800 lb, no, $F=\displaystyle\int_A^B p(y)w\,dy$

3.90 59.3%

3.92 106.317 kN, 48.5%, 2434.6 kW

3.94 10.89 kN

3.96 0.579

3.98 $1.617\ \mathrm{m^3/s}$

3.100 92.79 ft/s, 795.7 ft

3.102 0.923

3.104 5.393 MN

3.106 34.79 km

3.108 0

3.110 1865.1 lb, 60°

3.112 $\rho\,\dfrac{u}{V_0}\,q_0(V_0+u)^2\sin^2\theta$

3.114 -8640 N, 1663.2 N

3.116 66.18 N, 20.82 N

3.118 $V_0/3$, 180°

3.122 43.19°, 62.86°

3.124 500 N

3.126 $0.354\ \mathrm{ft\cdot lb/lb}_m$

3.128 10.24 kW

3.130 1.916 m, 1.304 m/s, 1.838 m·N/N, 270.35 kW, 0.0043 kcal/kg

3.134 3.4 m

3.136 23.546 m

3.138 131.9 lb

3.140 96 N·m

3.142 107.1 m

3.144 9.07 N·m

3.146 96.9 rpm

Chapter 4

4.2 (a) $\rho V^2/\Delta p$; (b) $Fg^2/(\rho V^6)$; (c) $t\,\Delta p/\mu$

4.4 $86.4(10)^6\ \mathrm{lb\cdot s^2/ft}$

4.6 Dimensionless, T^{-1}, FLT^{-1}, FL, FL, FLT

4.8 $f\!\left(\dfrac{\Delta h}{l},\dfrac{\mu D}{Q\rho},\dfrac{Q^3\rho^5 g}{\mu^5}\right)=0$

4.10 $\Delta p=c'\gamma\,\Delta z$

4.12 $F_B=c'\mho\rho g$

4.16 $\mathbf{M}=f\!\left(\dfrac{V}{\sqrt{p/\rho}},k\right)$

4.18 3.162 m/s

4.20 $\gamma H^4 f\!\left(\dfrac{\omega H^3}{Q},e\right)$

4.22 $\rho V^2 D^2 f'(\mathbf{R},\mathbf{M})$

4.24 0.185

4.26 Choose model size one seventy-fifth or less of prototype size;

$$\mathrm{loss}_p=\mathrm{loss}_m\!\left(\dfrac{D_m v_p}{D_p v_m}\right)$$

4.28 g, d, yes

4.30 $\omega D^{3/2}\rho_f^{1/2}/\sigma^{1/2}$

4.32 36.98 m/s, 18.59 m³/s; losses the same when expressed in velocity heads

Chapter 5

5.2 $2\mu U/a^2$, $Ua/3$

5.4 0.59 N, $1.636(10)^{-8}\ \mathrm{m^3/s}$

5.6 $Q=\dfrac{p_2-p_1}{2\mu l}\dfrac{2}{3}a^3-Va+\dfrac{U+V}{2}a$

5.8 $\alpha=1.543$, $\beta=1.20$

5.12 $-0.972\ \mathrm{lb/ft^2}$

5.14 Eff = 66.67%

5.16 $2\rho U^2 a/15$, $0.02857\rho U^3 a$

5.18 $\tfrac{4}{3}$

5.20 $0.707 r_0$

5.24 $0.0275\ \mathrm{lb/ft^2}$

5.26 $4.21(10)^{-3}\ \mathrm{ft^3/s}$, 122.6

5.28 $0.00152\ \mathrm{ft^3/s}$

5.30 8 m

5.32 29,200

5.34 $\dfrac{7}{6}\kappa\dfrac{y}{r_0}$

5.36 $2.1026(10)^{-3} \exp\left[-\dfrac{(x-720)^2}{288,000}\right]$

5.38 16.24
5.40 0.223
5.42 8.372 kN
5.44 $\delta = 4.80x/\sqrt{\mathbf{R}_x}$, $\tau_0 = 0.327\sqrt{\mu U^3 \rho/x}$
5.46 $\delta = 0.287x/\mathbf{R}^{1/6}$
5.48 0.905°
5.50 $V = 1.29$ m/s
5.52 1
5.54 3.6 kN
5.56 480.6 lb
5.62 0.00285 m/s
5.64 61 μm, 0.26 m/s
5.66 2.39 m/s
5.68 0.0043
5.70 41.974 m³/s
5.72 0.000482
5.74 234 cfs
5.76 10.11 ft/s
5.78 $Q \sim y^{8/3}$
5.80 1.997 ft
5.82 $0.438a$

5.84 1.594 hp
5.86 6.06 l/s
5.88 1.3×10^6
5.90 181 mm
5.94 14.88 m, 8.75 kW
5.96 1.096 MW
5.98 13 mi
5.100 0.539 cfs
5.102 0.054 l/s
5.104 1456 m³/min
5.106 86.2 kW/km
5.108 26.2 N/s
5.110 0.654 m
5.112 $31,100/yr
5.114 0.215 m ·
5.116 1.4 m, 1.19 m
5.118 0.0021 cfs
5.120 9.2, 145 m
5.122 3.5 cfs
5.124 7.6 ft, 4.26 ft, 89.4 ft
5.126 21.7
5.132 35,560 lb, 27.92 lb
5.134 0.304 ft

Chapter 6

6.2 0.331 kcal/kg·K
6.4 19.86
6.6 0.1237 kcal/K
6.8 4826 Btu
6.10 $T_1/T_2 = (\rho_1/\rho_2)^{n-1}$
6.12 2.276
6.14 15.9%
6.16 $v = \sqrt{gy}$
6.18 114°F, 46.45 psia
6.20 They are the same
6.22 1.7 slugs/s, 11,591 psfa, 516.9°R
6.24 132 psia, 0.0178 slug/ft³, 624.75°R
6.26 0.1059 kg/s
6.28 0.31 m
6.30 0.083 m, 0.1 m, 0.125 m

6.32 321.26 m/s, 2.638 kg/s
6.34 0.095, 0.94
6.36 0.577, 90 kPa abs, 486 K, 254.9 m/s
6.38 1.552, 0.683, 15.9 psia, 299.4°F
6.44 No
6.50 68.2%
6.52 0.98 ft
6.54 0.222 kg/s
6.56 43.42 kcal/kg
6.58 0.168 slug/s
6.60 14.76 kcal/kg
6.62 $q_H = (V_2^2 - V_1^2)/2$
6.64 18.62 m
6.66 3.172, 0.366 psia
6.68 0.150 ft

Chapter 7

7.2 0, 0, 0
7.4 $\omega_x = 1.5$, $\omega_y = -2$, $\omega_z = -0.5$
7.6 $\omega = -2z(x + y)$
7.8 $\phi = -4x + \frac{7}{2}(x^2 - y^2) - 6y + c$
7.12 $\psi = \theta + c$
7.14 $\phi = 18x + c$

7.16 $\left.\dfrac{\partial\phi}{\partial r}\right|_{r=a} = 0$, $-\left.\dfrac{\partial\phi}{\partial r}\right|_{r=\infty} = V_r$,

$-\dfrac{1}{r}\left.\dfrac{\partial\phi}{\partial\theta}\right|_{r=\infty} = V_\theta$

7.18 3790 lb

7.22 $\phi = -\dfrac{\mu}{2} \times$

$\ln \{[(x-1)^2 + y^2][(x+1)^2 + y^2]\}$

7.26 Half body

7.30 80 m^3/s

Chapter 8

8.2	42.67 mm	**8.34**	$D = 1.402y^{1/4}$
8.4	5.67 m/s	**8.36**	$D = 0.2253y^{3/4}$
8.6	68.1 ft/s	**8.38**	159.33 s
8.10	40.54 ft^3/s	**8.40**	3.46 m^3/s
8.12	40.52°F, 859.44 ft/s	**8.42**	3.18 psi
8.14	2.336 l/min	**8.44**	0.00419 slug/s, 540 ft/s
8.16	28.42 gpm	**8.46**	5.75 in
8.18	$y = 0.0543x^2$	**8.48**	0.07002 kg/s
8.20	$y = H \cos^2 \alpha$	**8.50**	5750 gpm
8.24	$C_d = 0.773$, $C_v = 0.977$, $C_c = 0.791$	**8.52**	19.77 cfs
8.26	0.287 m·N/N, 62.4 N·m/s	**8.54**	0.541 m
8.28	0.273 J/N, 163.3 W	**8.56**	3.055 ft, 2.092 ft
8.30	36.1 mm	**8.58**	2.30 N·m
8.32	252.31 mm	**8.60**	8.427×10^{-3} P

Chapter 9

9.2	$Q_c = (Q/N)n$, $H_c = (H/N^2)n^2$, c = corrected, n = const speed	**9.22**	4.12 cfs, 33.17°, 93.93 ft, 79.84 ft, 14.09 ft·lb/lb, 47.85 hp
9.4	Synchronization not exact	**9.24**	43.68°, 1146 rpm, 36.3 hp, 17.15 psi
9.6	$Q = 0.125Q_1$, $H = 4H_1$	**9.26**	30.51°, 1238.2 rpm
9.8	1900 (based on watts)	**9.28**	$\beta_1 = 9.51°$, $\beta_2 = 25.31°$, 5.305 in H$_2$O, 8.36 hp
9.10	89 in, 300 rpm		
9.12	4.81 m	**9.30**	229.87 m^3/min
9.14	13.04°	**9.32**	553 mm, 750 rpm
9.16	19.1 m/s, 57.3 m/s	**9.34**	35,240 rpm
9.18	36.71 m	**9.36**	0.1604
9.20	93.18%		

Chapter 10

10.2	$K = 190$	**10.30**	357.8 m, 159.3 l/s
10.4	14.36 m	**10.32**	$Q_A = 31.69$ l/s, $Q_B = 46.23$ l/s, $Q_C = 77.84$ l/s
10.6	3.99 cfs, −6.49 psi		
10.10	19.66 m	**10.34**	$Q_A = 9.81$ l/s, $Q_B = 39.12$ l/s
10.12	74.4 l/s	**10.36**	$Q_B = 56.15$ l/s, $H_j = 34.82$ m
10.14	82.2 mm	**10.38**	$Q_A = 515$ l/s, $Q_B = -267$ l/s, $Q_C = -454$ l/s, $Q_D = 205$ l/s
10.16	2.823 in H$_2$O		
10.18	15.5 m	**10.40**	58.51, 41.49, 2.36, 31.15, 43.85
10.20	128.1 l/s	**10.44**	0.392
10.24	79.5 l/s	**10.46**	5.59 l/s
10.26	$Q_A = 84.8$ l/s, $Q_B = 139$ l/s	**10.48**	2.044 m
10.28	$Q_1 = 8.701$ l/s, $Q_2 = 192.45$ l/s, $Q_3 = 201.15$ l/s		

Chapter 11

11.2	0.677 mm
11.4	175.1
11.6	$m = 0.686$, $y = 7.775$ ft
11.8	$b = 3.6103$ m, $m = \sqrt{3}/3$
11.10	0.000173
11.16	1.987, 1.667 m

11.18	0.56 m
11.20	7.39 ft
11.22	0.86 m, 8.289 m^2/s
11.34	0.7224 m
11.36	0.3635 m

Chapter 12

12.2	0.5314 m/s, 1.419 s
12.4	$z = V_0\, t e^{-mt}$
12.6	14.682 ft, 14.371 ft
12.8	40.12 s
12.10	1017.4 m/s
12.12	2.007 s

12.14	500 m
12.16	At $t = 3$ s, $H = 274.9$ m at gate, $H = 93.01$ m at midpoint
12.18	1.098 MPa
12.28	24.1 ft/s, 0.83 ft
12.30	4.474 m, 3.772 m/s

Gas properties, 535
Gibson, A. H., 244
Goldstein, S., 4
Gradient, 26, 413–417
Gradually varied flow, 460–466
  computer calculation, 471–473
  integration method, 462–465
  standard step method, 461, 462
Gravity, specific, 12
Gravity dam, 45, 46

Hagen, G. W., 192, 193
Hagen-Poiseuille equation, 161, 192–194,
  238, 365–368
Hardy Cross method, 427–438
Hazen-Williams formula, 411, 412
Head and energy relationships, 382–384
Heat, specific, 15, 263, 264, 535
Heat transfer, 289–293
Hele-Shaw flow, 198
Henderson, F. M., 478
Holley, E. R., 209n.
Holt, M., 181
Homologous units, 373–378
Hot-wire anemometer, 341
Howard, C. D. D., 429n.
Hudson, W. D., 441n.
Hunsaker, J. C., 168, 181
Hydraulic cross sections, best, 450–452
Hydraulic efficiency, 383
Hydraulic grade line, 113, 413–417, 497
Hydraulic gradient, 413–417
Hydraulic jump, 135, 136, 453–457, 469, 472
Hydraulic machinery, 176, 373–409
Hydraulic models, 175–178, 181
Hydraulic radius, 228
Hydraulic structures, 175
Hydrodynamic lubrication, 249–253
Hydrometer, 54, 55
Hydrostatics, 23–67

Ideal fluid, 5, 116
Ideal-fluid flow, 303–331
Ideal plastic, 5
Imaginary free surface, 49, 64–66
Impulse turbines, 396–400

Inertia:
  moment of, 524–526
  product of, 524–526
Internal energy, 93
International System (SI) units, 6
Ippen, A. T., 473n.
Ipsen, D. C., 181
Irreversibility, 105, 106
Irrotational flow, 303–331
Isentropic flow, 83, 271–279, 538, 539
  through nozzles, 271–284, 353, 354
Isentropic process, 264
Isothermal flow, 294–296

Jain, A. K., 239n.
Jennings, B. H., 293, 302
Jet propulsion, 123–129
Jets, fluid action of, 123–129
Journal bearing, 250

Kaplan turbine, 387
Kaye, J., 293, 302
Keenan, J. H., 287, 293, 302
Kinematic eddy viscosity, 200
Kinematic energy, 102
  correction factor, 111, 112
Kinematic viscosity, 10, 534
  of water, 534, 535, 537
King, H. W., 411n., 474n.
Kline, S. J., 181

Laminar flow, 83, 182–193, 448
  through annulus, 190–192
  losses in, 186–188
  between parallel plates, 184–189
  through tubes, 161, 232–249
Langhaar, H. L., 181, 193
Lansford, W. M., 356n.
Laplace equation, 310
Least squares, 361, 362
Liepmann, H., 271, 276, 282, 283, 302
Lift, 223, 224, 328
Lindsey, W. F., 223
Linear momentum, 93, 115–137
  unsteady, 93, 117, 137
Losses, 105, 106

Trattoria

PATRICIA WELLS'
TRATTORIA

HEALTHY, SIMPLE, ROBUST FARE

INSPIRED BY THE SMALL

FAMILY RESTAURANTS OF ITALY

PHOTOGRAPHY BY STEVEN ROTHFELD

WILLIAM MORROW AND COMPANY, INC.
NEW YORK

It is the policy of William Morrow and Company, Inc., and its imprints and affiliates, recognizing the importance of preserving what has been written, to print the books we publish on acid-free paper, and we exert out best efforts to that end.

Library of Congress Cataloging-in-Publication Data

Wells, Patricia.
    Patricia Wells' trattoria / Patricia Wells.
        p.   cm.
    ISBN 0-688-10532-7
    1. Cookery, Italian.   I. Title.
TX723.W45   1993
641.5945—dc20   93-16679
                                 CIP

Printed in the United States of America

First Edition

1 2 3 4 5 6 7 8 9 10

BOOK DESIGN BY BARBARA BACHMAN

This book is dedicated to:

The memory of my maternal grandfather, Felix Ricci, who as a young man left his farm in Ateleta in the Abruzzi region of Italy for a new life in America. Arriving at Ellis Island in 1910, he worked his way across the country, building railroads and saving money to purchase a plot of land. Here he is, in 1934, on his dairy farm in Comstock, Wisconsin.

My mother, Vera Catherine Ricci Kleiber, and my sister, Judith Frances Kleiber Jones, with gratitude for their support all along the way. Here they are, in 1945, the year before they added my name to the Ricci family tree.

Acknowledgments

What are you working on?" is the question posed almost daily by friends, colleagues, readers, acquaintances. During the past several years, of course, the response was *Trattoria!* and, almost universally, that single word managed to inspire eager smiles of recognition and anticipation. For we instantly identify "trattoria" with a simple, generous, full-flavored style of food that is so appealing today.

This book is a testament to the many people—friends and strangers, colleagues and family—who joined me physically and spiritually along the trattoria trail, as I collected the recipes, anecdotes, quotes, tips, and suggestions contained here.

Years ago, when I was a fledgling food journalist, a variety of cooking instructors shared great insights into the Italian way of cooking and of life. Most important among them were Marcella and Victor Hazan—both extraordinary ambassadors of Italian cooking. I will always be grateful to them for their rare spirit of sharing and openness. I also want to thank Giuliano Buglialli for his friendship and the discerning classes he has offered in New York and in Florence.

Many editors have consistently offered support to this and other writing projects, and I specifically want to thank John Vinocur, Pamela Fiori, David Breul, Donna Warner, Ila Stanger, Barbara Peck, Carole Lalli, Mary Simons, and Malachy Duffy for their past, present, and future encouragement.

As I have traveled within Italy, dozens of cooks, chefs, and restaurateurs have kindly allowed me into their kitchens and have shared recipes, tips, and techniques. I particularly want to acknowledge some of them here, including all of the staff at La Frateria di Padre Eligio in Cetona; Ugo and Gigi Salis at Trattoria da Graziella in Fiesole; Maria and Vittorio Becarria and Bruno Galaverna of Osteria Barbabuc in Novello; Angelo Maionchi of Del Cambio in Turin; Carlo Citerrio at Locanda dell'Amorosa in Sinalunga; Elio, Francesco, and Ninetta Mariani at Checchino dal 1887 in Rome; Diana and

Cesare Benelli of Al Covo in Venice; and in Florence, Piero Giannacci at Quattro Stagioni, Francesco Masiero at Il Cammillo, and Fabio and Benedetta Picchi at Cibrèo; and in Milan, Roberto Fontana at Trattoria Casa Fontana and Ezia Calatti at Antica Trattoria della Pesa.

Thank you to Bianca Vetrino in Turin, Enrico and Patricia Jacchia in Rome, Maria Manetti Farrow in Florence, Johanne Killeen and George Germon in Providence, Maggie and Al Shapiro in Normandy, Judith Symonds in Paris, and Rita and Yale Kramer in New York, all friends who made my days in Italy that much more special. Thank you Carlo Scipione Ferrero and Giovanna Bologna for helping with research on the history of the trattoria, and Maria Sanminiatelli for assistance in organizing many of my excursions. I am also grateful to Calvin Trillin for first sending me to Da Giulio in Lucca, where my curiosity about their recipe for "smashed" chicken—pollo al mattone—inspired this book.

Thank you Judy Jones for the care and attention you gave to each of my recipes, and to Steven Rothfeld for your friendship, your shared devotion to the details of life, and, of course, your masterly photos. I am also grateful to Alexandra Guarnaschelli for her expert assistance in the kitchen during our photography sessions. At home, I thank Laura Washburn for her fidelity and last-minute assistance in editing the final manuscript. And thank you Kyle Cathie, for believing in me.

At William Morrow, I am grateful to Barbara M. Bachman.

For almost fifteen years, Susan and Robert Lescher have acted as my literary agents, advisers, and friends, and I thank them for their continued support in helping shape and direct my career.

But my deepest gratitude is saved for my editor and dear friend, Maria Guarnaschelli, who gave one thousand percent to this project. Her insight, encouragement, and vision are unparalleled, not to mention her terrific sense of humor and ability to transform long, intense work sessions into memorable good times.

As ever, this book and I owe everything to Walter Newton Wells, my dear husband and partner in this wonderful life.

▪ ▪ ▪

Contents

Take Me to a Trattoria

Homey, unpretentious, honest, and homemade, that's the heart and soul of Italian trattoria cooking. Robust food—served without frills or fuss—makes up the body and the substance of small family restaurant fare all over Italy.

Several years ago, I began my quest for the quintessential trattoria, along quiet roads framed by rolling vineyards in the Piedmont, by the blue waters of the Ligurian resort of Santa Margherita, down stone alleyways in Siena, through the doors of bustling vineyard eateries in Tuscany, into the back rooms of loud, lively trattorias in Rome, and inside a quirky family spot on the edge of Milan, where they serve more than twenty different versions of risotto. In brief, a joyous land of bright lights, loud voices, full flavors.

Italian food needs no introduction, and even those who neither speak nor read Italian are probably on a first-name basis with many of the dishes found on a typical trattoria menu. Is there anyone who needs a translation of spaghetti, scampi, risotto, or tagliatelle?

But that's only the beginning, and a rather limited one at that. Authentic trattoria cooking—as found in small and large establishments from one end of the country to the other—is as rich, diverse, and pleasurable as any cuisine in the world.

The Italians are masters at roasting, at frying, at composing dishes of pasta, rice, meats, and poultry and salads of infinite variety and surprise. And no culture better understands simplicity and the magic of culinary understatement: Only the Italians can manage to slice a few tomatoes, drizzle them with olive oil, sprinkle them with sea salt and basil, and convince you to make a meal of it.

Along the trattoria trail I've been enraptured by the perfection of a crusty-roasted lamb in Rome, mesmerized by the feather-light delicacy of fried artichokes in Florence, and, in the Tuscan countryside, been brought to my knees by the subtleties of a dish no less noble than spaghetti in tomato sauce.

Whether it's found in the heart of the city or along a dirt road in the country, a trattoria is a essentially a small, informal, family-run restaurant, with home cooking and a homey atmosphere. And because trattorias are generally family ventures, no one is exactly like another.

The word *trattoria*—first appearing in print in 1859—probably comes from the French word *traiteur*, or caterer. But others suggest the word stems from the term *littarae tractoriae*, letters that in Latin times were carried by royalty as they traveled, and were given to messengers as a means of obtaining food and board on their journeys.

Many trattorias—including many featured here—have been in the same family for generations. And though they may have originally served as makeshift gathering spots for card playing and an afternoon sip of wine, they're now full-fledged restaurants, with a true family presence and a commitment to tradition. Other establishments take on a more modern posture, where the owners are young couples who work in tandem, around the clock, updating and re-creating the regional dishes of their childhoods.

When it comes to design and decoration, there is a definite trattoria "look." In the city, in particular, you're almost certain to find glass-paneled doors covered with simple and elegant embroidered white linen curtains; the chairs are almost always bentwood, the china thick and white, the tableware of ordinary stainless steel. If you order the house wine in "caraffa," you're likely to drink it out of small squat juice glasses. Order a bottle of wine from the list, and you'll be upgraded to simple stemmed wine glasses.

Traditionally, the walls are decorated with paintings, drawings, or even frescoes, offered by artists in exchange for food, wine, and atmosphere. Likewise, many establishments are adorned with all manner of thick glazed pottery or with majolica, resting on shelves or covering the walls. While today, little ashtrays, bowls, or plates bearing the name of the establishment may be offered to customers as pleasant souvenirs, they were once sold or swapped for different commodities, mostly wine and pork products.

Often, the kitchens open directly into the dining room, so diners can see the happenings in the kitchen, and the cooks can keep an eye on diners. You may or may not share tables with other diners, and in most cases, you may order one course at a time, adding courses until you say *basta*, or "enough." (For this reason, portions are rarely gargantuan: It is not uncommon to order soup as a first course, pasta as a second course, and meat, fish, or poultry as a main course. But there is nothing to stop one from ordering two pastas in succession, as I've seen Italian businessmen do on many an occasion.)

In the countryside, a trattoria may be little more than a small kitchen and a few outdoor tables, generally canopied by a flourishing grape arbor. Many vineyards boast their own trattoria, where the house wines can be shown off to best advantage.

While regional specialties abound, Italians are highly mobile within their own country, so it's not uncommon to find a Sardinian restaurant in Rome, a Tuscan trattoria in Milan, or a Piemontese eatery in Florence.

Seasonality is the key to all good cooking, and the roster of month-by-month trattoria specialties is no exception. In late August in Tuscany, restaurants offer the prized ovoli mushroom, and come fall you can't pass through the Piedmont without encountering the fragrant white truffle. And if it's December, you're sure to find puntarelle (a sort of wild chicory) on every trattoria menu in Rome.

Wine is hardly an afterthought in trattoria dining, though one's unlikely to find a sommelier in such down-home establishments. The budget-minded (meaning a healthy percentage of the diners) will order the house red or white by the carafe, often (but not always) a wise choice.

Likewise, desserts do not take a place of honor in the trattoria repertoire: Fruit tarts may be purchased from a pastry shop in the neighborhood, or prepackaged ice creams may be served. And always there are bowls of fresh fruit, ranging from slices of plump ripe pineapple to bowls of succulent, rosy-fleshed blood oranges.

While this book represents a decade's worth of expeditions throughout Italy in search of the authentic flavors of this rich land, it is also an attempt to offer—in the one hundred and fifty or so recipes gathered here—a faithful portrait of the colors, aromas, and tastes of Italian cooking today. Intentionally, I have mixed both traditional and modern fare, in hopes that your kitchen might also serve as a model of that rich and generous table.

Patricia Wells
1993

A Trip to
Antipasto Heaven

·

Nothing makes most diners happier than a generous selection of salads, vegetables, and spreads, little bits and bites of this and that. While the litany of dishes may appear intimidating, each can be prepared in advance. Offer, as well, a generous choice of wines, such as a white Vernaccia de San Gimignano from Tuscany, a red Barbera from the Piedmont.

Anchovies Marinated in Lemon Juice, Oil, and Thyme

Grilled Zucchini with Fresh Thyme

Sardinian Parchment Bread

Marinated Baby Artichokes Preserved in Oil

Salt-Cured Black Olives

Seared and Roasted Tomatoes

Individual Eggplant Parmesans

Red Bean and Onion Salad

Silky Sautéed Red Peppers

Aunt Flora's Olive Salad

Lemon-and-Oregano-Seasoned Tuna Mousse

White Bean Salad with Fresh Sage and Thyme

Black Olive Spread

Goat Cheese and Garlic Spread

ANTIPASTI,
STARTERS,
AND SALADS

• • •
Goat Cheese and Garlic Spread
Crema Formaggio all'Olio

"There are two Italies—one composed of the green earth and trans-parent sea, and the mighty ruins of ancient time, and aerial moun-tains, and the warm and radiant atmosphere which is interfused through all things. The other consists of the Italians of the present day, their works and ways. The one is the most sublime and lovely contem-plation that can be conceived by the imagination of man; the other is the most degraded, disgusting, and odious. What do you think? Young women of rank actually eat—you will never guess what—garlick!"

PERCY BYSSHE SHELLEY, LETTER TO LEIGH HUNT,
22 DECEMBER 1818

One spring evening in the Piedmont, more than a dozen of us gathered around a long table at the Osteria Barbabuc, in the hamlet of Novello, for a multi-course feast enlivened with plenty of the local Dolcetto d'Alba and Barbaresco. Little ramekins of this pure white cheese were waiting for us when we were seated, and we quickly devoured this garlic-rich, appetite-stimulating spread. The osteria's cook, Bruno Galaverna, prepares the cheese spread with the local robiola cheese, which is made with either cow's milk, goat's milk, or sheep's milk. When robiola is fresh—no more than three days old—it has a soft, delicate, almost buttery taste. In the Piedmont, one finds many variations of this cheese; sometimes herbs, lemon juice, and freshly ground black pepper are added. I prefer the simplicity of this version, nothing but good fresh cheese, a touch of oil, and a hit of garlic. Be certain to use fresh garlic—stale garlic will make for a bitter spread. Robiola can be found in some specialty shops, but top-quality domestic goat cheese is a worthy substitute.

8 ounces (250 g) robiola or mild fresh goat cheese

2 teaspoons extra-virgin olive oil

2 plump fresh garlic cloves, degermed and minced

In the bowl of a food processor, combine all the ingredients and blend until smooth and silky. Spoon into a ramekin and smooth out with a spatula. Serve immediately, as an appetizer or as part of a cheese course, with breadsticks or slices of toasted Italian bread. (The cheese can be stored, covered and refrigerated, for up to 3 days.)

■ Yield: 1 cup (250 ml) ■
cheese spread

WINE SUGGESTION: We sampled this with the dry, red, and graceful Dolcetto d'Alba, and you might, too.

VARIATION: On a visit to the trattoria Cibrèo in Florence, I sampled a thyme and olive oil version of this spread: To prepare it, simply combine the robiola or goat cheese, oil, and 2 teaspoons fresh thyme leaves, omitting the garlic. Blend in the food processor until smooth and fluffy, and serve as a dip or spread for fresh toast, for an appetizer.

■

The Germ Question

Take a clove of garlic, halve it lengthwise, and examine it. Garlic that is exquisitely fresh, plump, and moist will have a pale, barely visible green sprout running through the center. As garlic gets older that sprout grows, turns darker green, and gives the garlic a more intense flavor. The germ can also add a touch of bitterness to foods when garlic is added in its raw state. As a matter of course, I always remove the germ when garlic will be used raw, leave it in when garlic will be cooked.

• • •
Silky Sautéed Red Peppers

Peperoni in Aceto

With an agreeable flavor balance of both sweet and acid, these smooth, shiny, sautéed red peppers make a welcome first course, can star as part of a antipasto buffet, or are superb as a side dish to a summer meal of cold meats or poultry. The recipe comes from George Germon, chef and owner of Providence, Rhode Island's fine Italian restaurant, Al Forno. He prepared the peppers one day in my kitchen, and kindly allowed me to include them here. Rather than sautéeing the peppers in oil—the more traditional method—he cooked them in vinegar, tossing in a bit of oil at the end. I have also successfully prepared the peppers with balsamic vinegar, making for a sweeter, but no less appealing, dish.

6 red bell peppers (about 2 pounds; 1 kg)

½ cup (125 ml) plus 3 tablespoons best-quality red wine vinegar or balsamic vinegar

1 teaspoon sea salt, or to taste

¼ cup (60 ml) extra-virgin olive oil

1. Wash the peppers, quarter them lengthwise, and remove and discard the seeds and membranes. Place in a very large skillet and toss with the ½ cup (125 ml) vinegar and the salt. Cover and cook over low heat until soft and tender, about 25 minutes. Toss from time to time, adjusting the heat so the peppers cook slowly. By the end of the cooking time, most of the liquid will have evaporated. Do not increase the heat to speed up cooking, or the peppers will scorch and toughen.

2. Once the peppers have softened, transfer them to a large platter. Return the skillet to the heat, increase the heat to medium, and deglaze with the final 3 tablespoons vinegar, using a metal spatula to completely remove any flavorful bits that may have stuck to the pan. Add the oil, and heat just until warmed through, less than 1 minute. Pour the liquid over the peppers, toss, and taste for seasoning. Cool for at least 30 minutes before serving at room temperature: As they cool, the peppers will continue to soften and to absorb the oil and vinegar, giving them a complex, subtle flavor.

Yield: 6 to 8 servings
as an appetizer

Oven-Roasted Peppers

Peperoni al Forno

Whenever I have a batch of these delicious red and green roasted peppers ready at hand, I feel secure, as though my larder were somehow complete. These peppers can wear many hats: as a quick lunch with a slice of grilled bread; as a sauce tossed with warm pasta; as a member of a lovely antipasto table. I've sampled these at trattorias all over Italy. Sometimes they're roasted and served as is with just a touch of oil and salt, and sometimes they have a nice, vinegary tang. Take your choice. Be sure to watch the peppers as they bake: The goal here is peppers that are soft and fully cooked, with most of the skin still attached. If they scorch, or bake at too high a temperature, the skins fall away and the peppers become unpleasantly dry or rubbery, and sometimes bitter. Should you also find yellow or orange peppers in the market, try all four varieties for a festival of color. Or, if you're partial to red peppers, stick with a single hue. Even those who profess a dislike for peppers will be surprised by the sweetness of this dish.

4 red bell peppers (about 1½ pounds, 750 g)

4 green bell peppers (about 1½ pounds, 750 g)

¼ cup (60 ml) extra-virgin olive oil

Fine sea salt to taste

1 tablespoon best-quality red wine vinegar (optional)

1. Preheat the oven to 350°F (175°C; gas mark 4/5).

2. Wash the peppers, quarter them, and remove and discard the seeds and membranes. Place in a covered baking dish large enough to hold them comfortably. Toss with the oil and season lightly with salt.

3. Cover and place in the center of the oven. Bake for 1 to 1½ hours, turning the peppers from time to time so they do not scorch. Remove from the oven and, if desired, toss with the vinegar. Taste for seasoning. Serve warm or at room temperature.

Yield: 8 to 10 servings
as an appetizer

Lemon-and-Oregano-Seasoned Tuna Mousse

Mousse di Tonno

Spread it on toast, place it atop a bed of crisp, dressed greens, or serve it alongside a salad of green beans or steamed beets. This rich and vibrantly flavored tuna appetizer is a standby from Florence's up-to-date trattoria Cibrèo, a place where creativity and imagination are thoroughly uninhibited. This is a great dish to know about when the cupboard is ostensibly bare: Open the pantry, then the refrigerator, and, wow, you've a great snack, a quick lunch, a simple appetizer.

One 6½-ounce (190-g) can imported tuna packed in olive oil (do not drain; see Note)

4 tablespoons (2 ounces; 60 g) unsalted butter, softened

Grated zest (yellow peel) of 1 lemon

2 tablespoons freshly squeezed lemon juice

2 tablespoons extra-virgin olive oil

½ teaspoon dried leaf oregano

1 plump fresh garlic clove, degermed and minced

With a fork, flake the tuna in the can. Transfer, oil and all, to the bowl of a food processor. Add the remaining ingredients and process until smooth and creamy. Taste for seasoning. Transfer to a medium-size bowl and serve at room temperature. (The mousse can be stored, covered and refrigerated, for up to 3 days.)

■ Yield: 1 cup (250 ml) mousse ■

NOTE: If tuna packed in olive oil is unavailable, use top-quality white tuna packed in water. Drain the tuna, discard the water, and add an additional tablespoon of extra-virgin olive oil when preparing the mousse.

∙ ∙ ∎

Caponata

"Plato, when he visited Sicily, was so much struck with the luxury of the Agrigentum, both in their houses and their tables, that a saying of his is still recorded; that they built as if they were never to die, and eat as if they had not an hour to live."

PATRICK BRYDONE, A TOUR THROUGH SICILY AND MALTA, *1773*

Caponata—a smooth, jam-like mixture of eggplant, red peppers, onions, and celery, punctuated by a touch of sugar and vinegar and accented by green olives and capers—is one of my favorite trattoria dishes. I love to make it and adore eating it, especially when the market is filled with glistening purple eggplant. A close cousin of the French ratatouille, caponata is generally chunkier, accented with the crunch and flavor of celery, and laced with the saltiness supplied by the olives and capers. I've sampled many versions of this Sicilian dish throughout Italy, and even tasted one version—north of Venice—that included huge chunks of potatoes. You've succeeded at caponata if each vegetable manages to maintain its own integrity and texture. The vegetables should remain slightly firm, almost crunchy, and should in no way turn to mush. This is achieved by carefully cooking the vegetables separately, then folding them together near the end. Note also, it's important to season this dish lightly as you cook, so in the end, no additional seasoning is necessary. The added step of blanching the olives makes the dish more sophisticated, for unblanched olives can add an aggressive edge. Caponata can be served warm or at room temperature, as part of an antipasto assortment, or as an accompaniment to roast meats or poultry.

(continued)

2 medium onions

2 red bell peppers

1 cup (250 ml) extra-virgin olive
    oil

Fine sea salt to taste

One 16-ounce (480-g) can
    imported Italian plum
    tomatoes in juice or one
    16-ounce (480-g) can crushed
    tomatoes in purée

Several parsley stems, celery
    leaves, and sprigs of thyme,
    tied in a bundle with cotton
    twine

4 plump fresh garlic cloves,
    thinly sliced

8 ribs celery hearts with leaves,
    diced

2 teaspoons fresh thyme leaves

1 firm medium eggplant (about
    1 pound; 500 g), cubed (do not
    peel)

2 tablespoons sugar

½ cup (125 ml) best-quality red
    wine vinegar

1 cup (5 ounces; 150 g) drained
    pitted green olives

¼ cup (60 ml) drained capers,
    rinsed

1. Peel the onions, trim the ends, and cut in half lengthwise. Place each half, cut side down, on a cutting board, and cut crosswise into very thin slices. Set aside. Cut the peppers into thin vertical strips, then halve each strip crosswise. Set aside.

2. In a deep 12-inch (30-cm) skillet, combine the onions, ¼ cup (60 ml) of the oil, and a pinch of salt, and stir to coat the onions with oil. Cook over low heat until soft and translucent, about 5 minutes. Add the peppers and a pinch of salt. Cover and continue cooking for about 5 minutes more. If using whole canned tomatoes, place a food mill over the skillet and purée the tomatoes directly into it. Crushed tomatoes can be added directly from the can. Continue cooking for another 5 minutes. Add the herb bundle and garlic and taste for seasoning. Cover and simmer gently for about 20 minutes, stirring from time to time. Do not overcook: The vegetables should be cooked through but still firm, not mushy. Remove and discard the herb bundle. Remove from the heat and set aside.

3. Meanwhile, in another 12-inch (30-cm) skillet, heat ¼ cup (60 ml) of the oil over moderate heat. Add the celery and cook until it is lightly colored and beginning to turn soft and translucent, 7 to 10 minutes. Transfer to a bowl, and season lightly with salt. Add the thyme, and set aside.

4. In the skillet in which the celery was cooked, heat the remaining oil over moderate heat. When hot, add the eggplant and cook until lightly colored, about 5 minutes. (The eggplant will soak up the oil immediately, but allow it to cook without added oil, and keep the pan moving to avoid scorching.) The eggplant should remain firm.

5. Transfer the eggplant and the celery to the tomato mixture. Taste for seasoning. Cover and simmer gently over low heat until the mixture takes on a soft, jam-like consistency, about 20 minutes.

6. Meanwhile, in a small bowl, combine the sugar and vinegar, and stir to dissolve. Set aside.

7. In a medium saucepan, bring 1 quart (1 l) of water to a boil over high heat. Add the olives and blanch for 2 minutes. Drain and rinse under cold running water. Taste an olive: If it is still very salty, repeat the blanching.

8. Add the sugar-vinegar mixture, the blanched olives, and the capers to the vegetable mixture, and simmer over low heat for 1 to 2 minutes to allow the flavors to blend. Taste for seasoning. Transfer to a large serving bowl to cool. Serve warm or at room temperature, but not chilled.

Yield: 8 to 12 servings
as an appetizer

■ ■ ■

Seasoned Raw Beef

Insalata di Carne Cruda

C arne cruda, or lightly seasoned chopped, raw beef, is Italy's answer to the French steak tartare. This recipe is a specialty of the Piedmont. I've come to prefer this pure, unadulterated version, and particularly love it as the first course of a leisurely Sunday lunch. Use the finest olive oil you can afford, and serve this with plenty of crusty homemade bread. And select the leanest beef available, mixing in the seasoning as lightly as possible so the meat is not packed. Some people prefer seasoning the beef ahead of time and refrigerating it for about an hour before serving so the beef absorbs the seasoning. I prefer to serve it right away, making for a fresh, vibrant flavor.

1 pound (500 g) lean top-round beef or trimmed filet mignon or very lean trimmed sirloin, well chilled, and finely chopped

½ cup (125 ml) celery leaves, snipped with a scissors

½ cup (125 ml) fresh flat-leaf parsley leaves, snipped with a scissors

½ cup (125 ml) extra-virgin olive oil

1 tablespoon freshly squeezed lemon juice

Fine sea salt and freshly ground black pepper to taste

1 lemon, thinly sliced, for garnish

1. In a bowl, combine the beef, half the celery and half the parsley, the oil, and the lemon juice. Toss gently with a fork. Season to taste with salt and pepper.

2. Mound the beef on four chilled salad plates, showering with the remaining parsley and celery leaves. Garnish with slices of lemon.

■ Y i e l d : 4 s e r v i n g s ■

WINE SUGGESTION: Bring out a good red from the Piedmont, a Dolcetto or Nebbiolo.

...

Seared and Roasted Tomatoes

Pomodori al Forno

Few dishes are simpler or more welcoming than a giant platter of fragrant seared and roasted tomatoes. These often take a place of importance in a huge antipasto feast, for they're easy to make and just as easy to love. I've prepared these with both round and oval tomatoes, served warm or at room temperature, with equal success. Just be sure to use a large well-seasoned pan (I use an old, very thin black tin frying pan) that will allow the tomatoes to sear over rather high heat before you roast them in the oven. When served warm, they are a great accompaniment to chicken, meat, or fish.

¼ cup (60 ml) extra-virgin olive oil

8 firm medium-size tomatoes, cored and halved crosswise

Fine sea salt to taste

2 teaspoons fresh thyme leaves

1. Preheat the oven to 400°F (200°C; gas mark 6/7).

2. In a very large skillet, heat the oil over moderately high heat. When hot, place as many tomatoes as will easily fit in the pan, cut side down. Sear, without moving the tomatoes, until they are dark and almost caramelized, 5 to 6 minutes. Transfer the tomatoes, cooked side up, to a baking dish just large enough to hold the tomatoes comfortably. Continue until all the tomatoes are seared. Pour the cooking juices in the skillet over the tomatoes. Season lightly with salt and the thyme.

3. Place the baking dish in the center of the oven and bake, uncovered, until the tomatoes are browned and sizzling, about 30 minutes (or about 15 minutes for oval Roma tomatoes). Serve immediately, or at room temperature.

▪ Yield: 8 servings ▪

Individual Eggplant Parmesans

Parmigiani di Melanzane

I like to think of these delicious individual eggplant "parmesans" as crustless pizzas. I sampled them one August evening at a trattoria on Lake Garda, where the waterside eatery offered some twenty-five or thirty different items from the self-service antipasto table. They make a quick main dish for a light meal, or a luncheon dish served with a green salad. Although trattorias often serve these at room temperature, I think they lose much of their charm. So, for best results, serve hot from the oven. And although this is a simple dish, it requires careful timing and no distractions!

1 firm medium eggplant (about 1 pound; 500 g) (do not peel)

3 to 4 tablespoons extra-virgin olive oil

Fine sea salt to taste

Fresh flat-leaf parsley leaves, fresh thyme leaves, fresh basil leaves, and/or dried leaf oregano to taste (optional)

½ cup (125 ml) Tomato Sauce (page 256)

4 ounces (125 g) fresh whole-milk mozzarella, thinly sliced

¼ cup (1 ounce; 30 g) freshly grated Italian Parmigiano-Reggiano cheese

Dried leaf oregano, for garnish

1. Preheat the oven broiler.

2. Cover a broiling pan with aluminum foil, for easier cleanup. Slice the eggplant crosswise into ½-inch (1-cm) slices. Place the slices on the broiling pan, lightly brush the top of each slice with oil, and season with salt.

3. Place the broiling pan about 5 inches (12.5 cm) from the heat. Broil until the slices are thoroughly and evenly browned, about 5 minutes. Remove the broiling pan from

the oven and turn each slice of eggplant. Lightly brush the uncooked sides with oil. Season with salt, and the herbs, if using. Return to the oven and broil until brown, about 3 minutes.

4. Remove the broiling pan from the oven and top each slice with a spoonful of tomato sauce, a thin slice of mozzarella, and a light sprinkling of Parmesan and of oregano. Return to the oven and broil until the sauce is hot and the cheese is bubbly, about 2 minutes more. The eggplant should be soft when pierced. Serve immediately.

Yield: 4 to 6 servings
as an appetizer

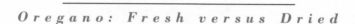

Oregano: Fresh versus Dried

Oregano—known as *origano* in Italian—is perhaps the only herb that is better dried than fresh, for dried leaf oregano has a sweeter, more pungent, more distinctive flavor. It is, in fact, the only dried herb allowed in my spice drawer, and dried oregano finds its way into tuna mousse, salads, and tomato sauces, it's sprinkled on pizzas, and it marries well with both green and black olives. When shopping for the dried herb, be sure to distinguish between ground oregano (which can taste like dirt) and the preferred leaf oregano. Replenish stocks of dried oregano every six months.

Oregano is closely related to marjoram, which has a more delicate, flowery flavor. Wild marjoram—a tall perennial herb—is also sometimes called oregano. To profit from its sweet and pungent flavor, always add oregano near the end of cooking.

· · ·

Grilled Zucchini with Fresh Thyme

Zucchini alla Griglia

Once you've secured firm, small zucchini and have good olive oil and sea salt on hand, there's little more for you to do than heat up the broiler or grill. The pure, sweet, golden freshness of young zucchini should star here, so just brush—don't douse—with oil and seasonings. I like to serve this as part of an antipasto buffet or as a side dish to meats, fish, or poultry. While instructions are given here for broiling in the oven, the zucchini can also be grilled, following the same basic procedure. Just be certain to slice the zucchini very thin, so that it cooks quickly and evenly, and carefully stem the fresh thyme leaves.

5 fresh firm small zucchini (about 1¾ pounds; 875 g), cut into thin lengthwise slices

About 3 tablespoons extra-virgin olive oil

Sea salt to taste

1 tablespoon fresh thyme leaves

1. Preheat the oven broiler.

2. Cover a broiling pan with aluminum foil, for easier cleanup. Place the zucchini slices side by side on the foil-covered broiling pan. Lightly brush with oil, and season with salt.

3. Place the broiling pan about 5 inches (12.5 cm) from the heat. Broil until the slices are golden brown, 2 to 3 minutes. Remove the broiling pan from the oven and, using tongs, turn each slice of zucchini. Lightly brush the uncooked sides with oil. Season with salt. Return to the oven and broil until golden brown, 2 to 3 minutes.

4. Remove the zucchini from the oven and transfer to a shallow dish, layering in overlapping slices. Drizzle lightly with oil and sprinkle with the thyme leaves. Serve warm or at room temperature. The zucchini is best consumed the same day. Refrigeration will alter the vegetable's fresh, delicate flavor, so store at room temperature.

■ Yield: 4 servings ■

Tips for Better Broiling

Broiled—or "grilled"—vegetables are greatly in vogue, but, alas, all too often we are served vegetables that are burnt on the outside and raw on the inside, or were cooked so far in advance they loose all their fresh appeal. Proper oven broiling is simple: Just think of it as baking at intense heat. For best results, vegetables should be cooked quickly at high heat so they do not dry out. Be careful not to broil too close to the heat, or the vegetables will simply burn, not cook.

HERE ARE SOME TIPS:

■ Preheat the broiler for a good fifteen minutes, to ensure a very high and even temperature.

■ Slice the vegetables thin, to expose a maximum of surface area.

■ Broil about 5 inches (12.5 cm) from the heat, so the vegetables cook quickly, but thoroughly, and do not burn.

...

Marinated Baby Artichokes Preserved in Oil

Carciofi sott'Olio

I consider my cupboard bare if I don't have at least one jar of these delicious pickled artichokes on hand. I love them as a snack, as an accompaniment to warm or cold meats or poultry, as a quick topping for pizzas or pasta, or layered in a bowl as part of a large antipasto buffet. I've experimented with many versions of this simple-to-prepare condiment, and find I prefer a very weak pickling solution prepared with white wine vinegar. The light solution and the quick cooking makes for a pickle that's not overly acidic. The hot artichokes are then covered with extra-virgin olive oil, and they take on a special, rich flavor.

In the United States, "baby" artichokes are simply small but fully mature secondary-growth globe artichokes from the "green globe" variety. An average full-sized artichoke weighs in at about two pounds (1 kg). The baby globe artichokes are simply volunteers that appear after the main crop has been picked. The tiny artichokes are sweet and more tender than the larger version, and since the feathery choke is so small and delicate, the heart and choke are both completely edible. In Europe, however, tiny violet-tinged artichokes (in Italy, a favorite variety is "Violetto"—usually weighing about three ounces [90 g] each) are varieties unto themselves, prized for eating raw or for pickling like this.

2 cups (500 ml) Champagne vinegar

2 cups (500 ml) water

2 teaspoons fine sea salt

8 bay leaves, preferably fresh

2½ pounds (1.25 kg) small artichokes (about 25) (see Mail Order Sources, page 325)

1½ cups (375 ml) extra-virgin olive oil, or as needed

4 plump fresh garlic cloves, degermed

2 tablespoons fresh flat-leaf parsley leaves, snipped with a scissors

Antipasto table, left to right, from top: Anchovies Marinated in Lemon Juice, Olive Oil, and Thyme *(PAGE 20)*; Grilled Zucchini with Fresh Thyme *(PAGE 16)*; Sardinian Parchment Bread *(PAGE 184)*; Marinated Baby Artichokes Preserved in Oil *(PAGE 18)*; Salt-Cured Black Olives *(PAGE 280)*; Seared and Roasted Tomatoes *(PAGE 13)*; Individual Eggplant Parmesans *(PAGE 14)*; Red Bean and Onion Salad *(PAGE 40)*; Silky Sautéed Red Peppers *(PAGE 4)*; Aunt Flora's Olive Salad *(PAGE 36)*; Lemon-and-Oregano-Seasoned Tuna Mousse *(PAGE 8)*; White Bean Salad with Fresh Sage and Thyme *(PAGE 42)*; Black Olive Spread *(PAGE 277)*; Goat Cheese and Garlic Spread *(PAGE 2)*

FRESH FAVA BEAN AND PECORINO SALAD *(PAGE 38)*;
BACKGROUND: WALNUT AND PECORINO SALAD *(PAGE 48)*;
 SARDINIAN PARCHMENT BREAD *(PAGE 184)*; GOAT CHEESE AND
FRESH THYME SPREAD *(PAGE 3)*

 Tuscan Five-Bean Soup *(PAGE 70)*;
Sourdough Bread *(PAGE 186)*

 Saffron Butterflies *(PAGE 108)*

1. In a large stainless steel saucepan, combine the vinegar, water, salt, and 4 of the bay leaves.

2. Prepare the artichokes: Rinse the artichokes under cold running water. Using a stainless steel knife to minimize discoloration, trim the stem of an artichoke to about ½ inch (1 cm) from the base. Carefully trim off and discard the stem's fibrous exterior. Bend back the tough outer green leaves one at a time, letting them snap off naturally at the base. Continue snapping off the leaves until only the central cone of yellow leaves with pale green tips remains. Lightly trim the top of the cone of leaves to just below the green tips. Trim any dark green areas from the stem. Depending upon its size, leave whole or halve or quarter the artichoke lengthwise, and transfer immediately to the vinegar mixture to prevent discoloration. Continue until all the artichokes are trimmed.

3. Bring the liquid and artichokes to a simmer over medium-high heat. Simmer, reducing the heat if necessary, until the artichokes are tender but still offer a bit of resistance when pierced with a knife, 7 to 10 minutes. (Cooking time will vary according to the size of the artichokes.) Drain the artichokes, discarding the vinegar mixture and bay leaves.

4. While the artichokes are still warm, marinate them: If the artichokes will be consumed the same day, arrange them in a single layer in a shallow dish and partially cover with oil. To preserve them, arrange in layers in a sterilized 1-quart (1-l) canning jar, or several smaller sterilized jars: Layer the artichokes, garlic, parsley, and the remaining 4 bay leaves, finishing with a layer of artichokes. Completely cover the artichokes with oil. Set aside, uncovered, until the artichokes are completely cool. If necessary, add additional oil to come to the top. Cover securely. Refrigerate, covered, for up to 2 months, making sure that the artichokes are always covered in oil. The artichokes preserved in this manner can be consumed as soon as they are cool, but will have greater depth of flavor if allowed to marinate for at least 24 hours.

■ Y i e l d : 1 q u a r t (1 l i t e r) ■

Anchovies Marinated in Lemon Juice, Olive Oil, and Thyme

Acciughe al Limone

Tiny bites that precede a meal—such as those included in the huge selection of varied antipasti found at many trattorias—are often among the most satisfying. They open the palate and refresh the spirit and the body. One of my favorites is a healthy serving of fresh anchovies that have simply been filleted and marinated ever so quickly in a touch of lemon juice, then anointed with extra-virgin olive oil and a touch of seasoning. Although most of us don't come across fresh anchovies on a regular basis, should you stumble upon them (or beg them off a fisherman or fishmonger), go for it!

1 pound (500 g) fresh anchovies (about 24)

3 tablespoons freshly squeezed lemon juice

2 tablespoons extra-virgin olive oil

Fine sea salt to taste

4 bay leaves, preferably fresh

2 teaspoons fresh thyme leaves

1 shallot, finely minced

1. Lightly rinse (don't wash or soak) the anchovies. Individually head and gut the fish by holding each anchovy firmly just beneath the head and gently pull off the head and attached entrails. Discard the head and entrails. Run your thumb down the center of the fish, and it will easily spread open. Remove and discard the central bone. Carefully pull the anchovy apart into 2 fillets. Lightly run your fingers along each fillet to remove any pieces of bone or entrails. Pat dry, and place the fillets, skin side up, side by side on a platter. Drizzle with the lemon juice, cover with plastic wrap, and refrigerate for 10 minutes.

2. Remove the fish from the refrigerator. Drain off and discard the lemon juice. Drizzle another platter with the olive oil, sprinkle lightly with salt, and cover with the bay leaves. Arrange the anchovy fillets side by side on the platter. Sprinkle with the thyme and minced shallot. Season very lightly with salt. Cover with plastic wrap and refrigerate for 10 minutes. Serve as an appetizer, with freshly grilled toast.

Yield: 4 servings
as an appetizer

Fresh Artichoke Omelet

Tortino di Carciofi

Green and golden, this dish reminds me of springtime and daffodils. Although a tortino is generally a vegetable pie without the pastry, this Florentine specialty is more a cross between an omelet and scrambled eggs. I love artichokes so much that on one trip to Florence I sampled this dish two days in a row: once at the charming little trattoria Sostanza, the next day at the tiny family eatery Il Francescano, near Santa Croce. In this version, the baby artichokes are sliced very thin, then quickly sautéed in olive oil. The eggs are then cooked much like an omelet, but the omelet is not folded: Rather, the cooked artichokes are placed in the center of the cooked eggs, which are gently folded over toward the center. You'd never think of eggs and artichokes as a marriage made in heaven, but they are!

1 lemon

4 baby artichokes (see Mail Order Sources, page 325)

2 tablespoons extra-virgin olive oil

Fine sea salt and freshly ground black pepper to taste

8 extra-fresh large eggs, at room temperature

4 tablespoons (2 ounces; 60 g) unsalted butter, at room temperature

1. Prepare the artichokes: Halve the lemon, squeeze the juice, and place the halved lemon and juice in a bowl filled with cold water. Rinse the artichokes under cold running water. Using a stainless steel knife to minimize discoloration, trim the stem of an artichoke to about ½ inch (1 cm) from the base. Carefully trim off and discard the stem's fibrous exterior. Bend back the tough outer green leaves one at a time, letting them snap off naturally at the base. Continue snapping off the leaves until only the central cone of yellow leaves with pale green tips remains. Lightly trim the top of the cone of leaves to just below the green tips. Trim any dark green areas from the stem. Slice the artichoke lengthwise into very thin slices and transfer immediately to the lemon juice mixture to prevent their discoloring. Continue until all artichokes are trimmed and sliced.

2. Drain the artichokes and pat them dry. In a seasoned or nonstick 10-inch (25-cm) skillet, heat the oil over moderate heat. When the oil is hot but not smoking, add the artichokes and sauté until lightly browned, 3 to 4 minutes. Drain in a sieve set over a small bowl, and season to taste with salt and pepper. Set aside.

3. Prepare the omelet: Break the eggs into a bowl. Cut 2 tablespoons of the butter into small cubes, and add to the eggs. Season with salt and pepper. (Do not beat the eggs at this point.)

4. Warm the seasoned or nonstick pan for several seconds over high heat. Add the remaining 2 tablespoons butter. While the butter melts, beat the eggs lightly with a fork. Tilt the pan to coat the bottom with melted butter. When the butter begins to foam, but before it begins to brown, pour in the beaten eggs. Stir the egg mixture by gently passing a fork through the eggs several times to expose as much of the mixture as possible to the heat of the pan. (Be careful not to scrape the tines of the fork over the bottom of the pan, or the surface will scratch, making the omelet stick.)

5. When the underside of the omelet begins to set, use the fork to lift the edges of the omelet. At the same time, tilt the pan so that any uncooked egg from the top will run under the cooked egg and set. Continue lifting and tilting the pan until all of the egg has set. When the edges of the omelet are firm, but the top is still moist, quickly spoon the artichokes into the center. Fold the edges over about 1 inch (3 cm) all around, slightly enveloping the artichokes. Slip the omelet, right side up, onto a warmed round platter, and season generously with salt and pepper. Slice and transfer to warmed dinner plates.

■ Yield: 4 servings as a first course, ■
2 servings as a main course

Tips for a Fluffy, Moist Omelet

■ Begin with fresh eggs at room temperature.

■ Beat the eggs lightly, and be careful not to overbeat. The yolks and whites should be beaten just to combine them without forming air bubbles that might make the omelet dry out.

■ Heat the pan thoroughly before adding the eggs.

■ Cook any bulky ingredients beforehand—such as the artichokes here—and add them to the omelet at the last minute.

■ Use a well-seasoned (or nonstick) pan that is wide enough to allow the eggs to cook quickly. An omelet that is cooked too slowly will tend to dry out.

• • •

Raw Vegetables Dipped in Olive Oil

Pinzimonio

ITALIAN PROVERB:

"Everything is wholesome to the healthy."

Pinzimonio—a mix of raw vegetables that are dipped in olive oil seasoned with a touch of salt and perhaps fresh black pepper—is an ideal summertime dish, a communal affair that gets everyone eating with their hands, sitting elbow to elbow at long, festive tables. Or, on the more serious side, pinzimonio is the ideal way to "test" the flavors of extra-virgin olive oils. When oils are sampled in their pure state, with just a hint of salt, you can truly judge the flavors, and select a favorite. So, next time you have four or five oils together, make a visit to your greengrocer, invite over a group of friends, and chomp away! The following is simply a suggestion of the vegetables you might include. Arrange them attractively in a large bowl of ice water, to keep them cool. I like to supply each diner with a several small bowls, so they can season their oils (or not) to taste. Also make sure everyone has a good sharp paring knife, for slicing and for trimming the artichokes.

Tender ribs celery with leaves, washed

Baby artichokes

Cherry tomatoes

Fresh baby fava beans in their pods

Fennel bulbs, quartered lengthwise

Red and green bell peppers, quartered lengthwise

Carrots, peeled and quartered lengthwise

Fresh baby onions or scallions

Small cucumbers, quartered lengthwise

Extra-virgin olive oil (several brands if desired)

Fine sea salt and freshly ground black pepper to taste

Fresh country bread

...

Spinach-Parmesan Frittata

Frittata Fredda alla Rustica

I once spent a week in Florence, visiting a different café each morning for breakfast and the morning paper. Since I don't have much of a sweet tooth, I'd opt for an assortment of tiny sandwiches, or panini, ideal Italian finger food. Often, the filling was made up of varied frittate—highly seasoned cooked egg and herb mixtures—perfect for waking up morning tastebuds.

A frittata is a close relative of the omelet, but it is served flat, not rolled, more like a Spanish tortilla, and at room temperature. A frittata is cooked very slowly over low heat, then placed under the broiler until firm. This version is the most classic, and one of my favorites. Try to use the smallest, freshest spinach leaves you can find. When sliced into thin wedges, or used as a sandwich filling, this frittata—always cooled to room temperature before eating—makes perfect snack, luncheon, or picnic fare.

6 large eggs, at room temperature

Fine sea salt and freshly ground black pepper to taste

Freshly grated nutmeg to taste

4 cups (3 ounces; 90 g) loosely packed fresh spinach leaves, rinsed, dried, and finely chopped

1 cup (4 ounces; 125 g) freshly grated Italian Parmigiano-Reggiano cheese

1 tablespoon extra-virgin olive oil

1. Preheat the oven broiler.

2. Crack the eggs into a large bowl and beat lightly with a fork. Add the salt and pepper, nutmeg, spinach, and half the cheese, and beat lightly to combine the ingredients.

3. In a 9-inch (23-cm) ovenproof omelet pan or skillet, heat the oil over moderate heat, swirling the pan to coat the bottom and sides evenly. When the oil is hot but not smoking, add the frittata mixture. Reduce the heat to low and cook slowly, stirring the top two-thirds of the mixture (leaving the bottom part to set, so it doesn't stick) until the eggs have formed small curds and the frittata is brown on the bottom and almost firm in the center, about 4 minutes. The top should still be very soft. With a spatula, lightly loosen the frittata from the edges of the pan, to prevent sticking later on. Sprinkle with the remaining cheese.

4. Transfer the pan to the broiler, placing it about 5 inches (12.5 cm) from the heat, so that the frittata cooks without burning. Broil until the frittata browns lightly on top and becomes puffy and firm, about 2 minutes. (Watch carefully: A minute can make the difference between a golden-brown frittata and one that's overcooked.) Remove the frittata from the broiler and let cool in the pan for 2 minutes. Place a large flat plate over the top of the pan and invert the frittata onto it. Let the frittata cool to room temperature.

5. To serve, cut into wedges and serve with a salad or as a sandwich filling.

■ Yield: 4 to 6 servings ■

• • •

Individual Gorgonzola Soufflés

Tortino Gorgonzola

These rich and memorable individual Gorgonzola soufflés were served as part of a multi-course feast one spring Saturday evening at the Osteria Barbabuc, in the hamlet of Novello in the Piedmont. With my first taste of the warming soufflé, all I could think of was presenting this to guests at home, with a sparkling green arugula salad alongside. That's the way I love to serve this soufflé, for it's a perfect, elegant, and satisfying luncheon dish. If all the ingredients are measured out and ready beforehand, it takes little last-minute preparation. The recipe was shared with me by the osteria's chef, Bruno Galaverna. A tortino is traditionally a simple, rustic vegetable pie, prepared without pastry. Although the quantities of cream and flour here take this beyond the realm of a classic soufflé, the little tortinos puff up beautifully, as if defying tradition.

Butter for preparing the ramekins

2 cups (500 ml) heavy cream

¼ teaspoon fine sea salt

Freshly ground black pepper to taste

¾ cup (100 g) unbleached all-purpose flour

5 large eggs, separated

5 ounces (150 g) imported Gorgonzola cheese, at room temperature, crumbled

1. Preheat the oven to 425°F (220°C; gas mark 8).

2. Thoroughly butter the bottoms and sides of eight ½-cup (125-ml) ramekins.

3. In a large saucepan, combine the cream, salt, and pepper, and scald over moderately high heat, bringing the mixture just to the boiling point. Reduce the heat to low, and add the flour all at once, whisking constantly to prevent lumps from forming. The sauce

will thicken almost immediately. Remove from the heat, and stir in the egg yolks one by one. Then add the Gorgonzola, and stir until the cheese melts into the cream mixture. Set aside.

4. In the bowl of an electric mixer fitted with a whisk, beat the egg whites until stiff but not dry. Whisk one-third of the egg whites into the soufflé mixture and combine thoroughly. (Do not be concerned about deflating the egg whites at this point.) With a large rubber spatula, gently fold in the remaining whites. Do this slowly and patiently. Do not overmix, but be sure that the mixture is well blended and almost no streaks of white remain.

5. Spoon the mixture into the prepared ramekins, filling them three-quarters full, and smoothing out the tops with a spatula. Place the ramekins on a heavy-duty baking sheet and place in the center of the oven. Bake until the soufflés are well risen and the tops are browned, about 15 minutes. Carefully remove from the oven and place each ramekin on a small salad plate. Serve immediately.

■ Yield: 8 servings ■

WINE SUGGESTION: With this soufflé, I love a nice Barbera d'Alba or a Dolcetto d'Alba, both from the Piedmont.

For Easier Separating, and Better Whipped Whites

Eggs are easier to separate when they are cold, but whites whip better at room temperature. So when preparing dishes such as this one, separate the eggs immediately after removing them from the refrigerator, then allow them to come to room temperature before you begin cooking.

··· ·

Herb-Infused Savory Custards

Tartra all'Antica

Creamy, fragrant, and infused with the scents and flavors of fresh rosemary and bay leaf, this elegant and unusual first course is, as chef Angelo Maionchi of Turin's Del Cambio says, "like pasta." That is, it's an appealing backdrop for all sorts of sauces, which can be paraded out to celebrate each season in turn. While a cheese sauce is classic, the chef also serves a cream and asparagus sauce in the spring, and a mushroom and tomato sauce (page 257) in the fall. Any sauce you love for pasta would go well as an accompaniment to this shimmering, molded starter. I love it best with a tomato-based sauce, for I enjoy the play of textures and flavors, as the soft, rich, and refined custard is contrasted with the mildly acidic sauce. Eating this dish is, in fact, a little like having dessert as a first course! I always prepare this in advance, earlier in the day, so there is little last-minute work.

Butter for preparing the ramekins

1 ½ cups (375 ml) whole milk

2 ½ cups (625 ml) heavy cream

   3 tablespoons minced fresh rosemary leaves, measured after mincing

   4 bay leaves, preferably fresh

   4 large eggs

   2 large egg yolks

¼ cup (1 ounce; 30 g) freshly grated Italian Parmigiano-Reggiano cheese

¼ teaspoon freshly grated nutmeg

¼ teaspoon fine sea salt

Freshly ground black pepper to taste

   2 cups (500 ml) warm Tomato-Mushroom Sauce (page 257)

1. Preheat the oven to 350°F (175°C; gas mark 4/5). Butter six 1-cup (250-ml) ramekins.

2. Prepare a large kettle of boiling water for the water bath; set aside.

3. Cut 3 slits in a piece of waxed paper, and use it to line a roasting pan large enough to hold the ramekins. Place the ramekins in the pan, on top of the paper, and set aside. (The paper will prevent the water added to the pan from boiling and splashing up on the custards.)

4. In a medium-size saucepan, combine the milk and cream, and scald over high heat, bringing the mixture just to the boiling point. Remove from the heat, add the rosemary and bay leaves, cover, and set aside to infuse for 10 minutes. Strain the liquid through a fine-mesh sieve into a large bowl. Discard the herbs. Set aside to cool.

5. In a small bowl, blend the eggs and egg yolks lightly with a fork, but do not let the mixture become foamy or frothy, or the custard will be filled with bubbles.

6. When the milk mixture has cooled, add the eggs and stir to blend. Stir in the cheese, nutmeg, salt, and pepper. Taste for seasoning.

7. Divide the custard evenly among the prepared ramekins. Add enough hot water to the roasting pan to reach to about half the depth of the ramekins. Place in the center of the oven and bake until the custards are just set at the edges but still trembling in the center, 50 to 55 minutes.

8. Remove from the oven and carefully remove the ramekins from the water. (The custards can be baked earlier in the day and reheated in a warm water bath for about 10 minutes before serving.) To serve, invert the molds onto warmed salad plates, spooning the warm sauce attractively around the molds.

■ Y i e l d :   6   s e r v i n g s ■

WINE SUGGESTION: A smooth, balanced red is right here, such as a Dolcetto d'Alba from the Piedmont.

Swiss Chard and Parmesan Torte

Torta di Biete

This savory snack is a favorite Mediterranean specialty. Vegetable pies such as this one—prepared with a quick and wholesome olive oil pastry—can be found at pastry and snack shops in many parts of Italy. There are versions flecked with pine nuts and raisins as well, but I prefer the simplicity of this torte. I often prepare it early in the day and serve it with a glass of white wine as an appetizer before dinner. If you have a garden, raise Swiss chard, for it grows like wild and has a sturdy, sensible personality in the kitchen. Short of chard, fresh spinach is a totally worthy substitute.

PASTRY

- 1 cup (135 g) unbleached all-purpose flour
- ¼ teaspoon fine sea salt
- ¼ cup (60 ml) water
- ¼ cup (60 ml) extra-virgin olive oil

FILLING

- 1 pound (500 g) Swiss chard leaves (or substitute fresh spinach leaves)
- Fine sea salt and freshly ground black pepper to taste
- 3 large eggs
- 1 cup (4 ounces; 125 g) freshly grated Italian Parmigiano-Reggiano cheese

1. Preheat the oven to 375°F (190°C; gas mark 5).

2. Prepare the pastry: In a medium-size bowl, combine the flour and salt. Stir in the water, then the oil, mixing until thoroughly blended. Knead briefly. The dough will be

very moist, much like a cookie dough. Press the dough into the bottom of 10½-inch (27-cm) tart tin with a removable bottom. (You do not need to cover the sides of the tin.)

3. Prepare the filling: Wash and dry the green leafy portions of the chard or spinach, trimming and discarding the center stems. Tear the leaves and chop, in several batches, in a food processor.

4. Place the chard in a large shallow frying pan and season with salt and pepper. Wilt the chard or spinach over low heat, and cook until most of the liquid has evaporated, 2 to 3 minutes. Remove from the heat.

5. Combine the eggs and cheese in a medium-size bowl and stir to blend. Stir in the chard or spinach, mix well, and taste for seasoning. Spoon the mixture into the prepared tart tin.

6. Place in the center of the oven and bake until the filling is lightly browned and firm to the touch, about 45 minutes. Remove to a baking rack to cool. Serve at room temperature, cutting into thin wedges. (Do not refrigerate, or the filling will become tough.)

■ Yield: 8 to 12 servings ■

WINE SUGGESTION: I love a bubbly white with this, such as a Prosecco *frizzante*, or a dry white such as a Soave or Orvieto Secco.

. . .

Celery Salad with Anchovy Dressing

Insalata di Puntarelle

One Saturday morning I was wandering through the Campo dei Fiori food market in Rome and noticed that each salad merchant displayed little buckets of water filled with what appeared to be "celery flowers." I asked one vendor, who explained that it was called puntarelle, a local form of wild chicory that is traditionally trimmed into flowers. I sampled the salad the next day at Trattoria Piperno, and wasn't at all surprised to find that puntarelle taste remarkably like celery. The salad can be found in most Roman trattorias, where it is regularly served with this garlic-rich anchovy dressing. Since puntarelle are not easily found outside of Italy, I've taken the liberty of substituting celery, a full-flavored, readily available vegetable we shamefully take for granted. This salad must be served immediately after it is tossed: If left to linger, the dressing wilts the celery and turns it soggy. Note that even though the dressing is prepared in the food processor, the garlic should be minced by hand.

1 head of celery (about 1½ pounds; 790 g), ribs separated, rinsed, and trimmed into 3-inch (8-cm) lengths

ANCHOVY DRESSING
6 tablespoons extra-virgin olive oil

Two 2-ounce (60-g) cans flat anchovy fillets in olive oil

3 plump fresh garlic cloves (or to taste), degermed and minced

Freshly ground black pepper to taste

1. Prepare the celery: Prepare a large bowl of ice water. With a small, sharp knife, cut a celery flower from each piece of celery: Make several lengthwise cuts about one-third the length of each rib, so that the end of each piece fans out into a flower. Place them in the ice water and refrigerate. They will fan out as they chill. (The celery fans are best prepared several hours in advance.)

2. Prepare the dressing: In the bowl of a food processor or a blender, combine the oil, the anchovies and their oil, and the minced garlic, and process or blend into a fairly smooth dressing. Transfer to a small bowl and set aside.

3. At serving time, carefully drain and dry the celery with a clean towel. Place in a large salad bowl. Toss with just enough dressing to lightly coat the celery. Serve immediately, for once tossed, the celery quickly loses crispness and zest. Serve as a salad, with knife and fork, and pass the pepper mill.

■ Yield: 6 to 8 servings ■

Aunt Flora's Olive Salad

Insalata di Olive

As a child I always loved the family gatherings at which my mother's sister, Flora DeAngelo, would prepare her multi-course Italian feasts. This crunchy, fiery, green olive salad was a favored dish in her repertoire. The salad can be served as part of an antipasto platter, or offer it as an accompaniment to cheese, sausages, and crusty bread. I also like it as a spicy condiment for a simple meal, such as Chicken Cooked Under Bricks (page 218), served with Sautéed Spinach (page 62) and Roasted Rosemary Potatoes (page 54). This is an instance in which I depart from my general disdain for dried herbs: Dried leaf oregano does have a special flavor, one that is perfectly natural here.

1 cup (5 ounces; 150 g) drained pimento-stuffed green olives, quartered crosswise

3 ribs tender celery hearts with leaves, diced

1 teaspoon best-quality red wine vinegar

1 tablespoon extra-virgin olive oil

4 plump fresh garlic cloves, degermed and minced

¼ teaspoon dried leaf oregano

¼ teaspoon crushed red peppers (hot red pepper flakes), or to taste

In a small bowl, combine all the ingredients and toss to blend. Cover and refrigerate for at least 2 hours and up to 2 days, tossing occasionally. Serve at room temperature, as a condiment.

■ Yield: 6 to 8 servings ■

Green Olive, Tuna, Celery, and Red Pepper Salad

Insalata di Olive Verde, Tonno, Sedano, e Peperoni

Whenever I prepare a large antipasto buffet, this spunky, colorful, zesty condiment is always on the table. I've seen this dish served at many trattorias, in varied versions. Some cooks add cooked white beans, a pleasant touch that makes for a filling and hearty condiment. I like to serve this with a platter of sliced red tomatoes and good crusty country bread, as a luncheon dish.

1 cup (5 ounces; 150 g) drained pimento-stuffed green olives, halved crosswise

4 to 5 ribs celery hearts with leaves, thinly sliced

1 6½-ounce (190-g) can imported tuna in olive oil (do not drain; see Note)

1 red bell pepper, minced

3 tablespoons extra-virgin olive oil

1 teaspoon best-quality red wine vinegar

Fine sea salt to taste

With a fork, flake the tuna in the can and transfer, oil and all, into a small bowl. Add all the remaining ingredients, and toss to blend. Taste for seasoning. Serve immediately, or cover and refrigerate for up to 8 hours. Serve at room temperature.

▪ Yield: 6 to 8 servings ▪

NOTE: If tuna packed in olive oil is unavailable, use top-quality white tuna packed in water. Drain the tuna, discarding the water, and add an additional tablespoon of extra-virgin olive oil when preparing the salad.

Fresh Fava Bean and Pecorino Salad

Baccelli al Pecorino

As you step inside Cibrèo in Florence—one of my favorite restaurants in all of Italy—your eyes instantly fall on the welcoming sideboard laden with varied appetizers and desserts. The assortment might well include a giant white porcelain bowl of garden-fresh fava beans marinating with cubes of pecorino cheese, oil, lemon juice, and a touch of herbs. While most of us may not be very familiar with these deliciously nutty little beans, they make a vibrant palate-teaser—more like a snack or appetizer than a whole course. When you prepare this at home, don't make the mistake of serving large portions of this filling starter—just a bite or two will do. For this recipe, scour your farmers' markets, and use the smallest and youngest fava beans or broad beans you can find: Once they mature, the skin covering the beans becomes coarse and tough and must be removed before eating. I like the tinge of bitterness in the raw, uncooked beans: It's a great contrast to the almost-sweet young pecorino. If the only sheep's milk cheese you can find is the hard grating variety, use any good-quality soft fresh goat's milk cheese that can easily be cut into cubes.

2 *pounds (1 kg) fresh unshelled fava beans or broad beans (about 2 cups [500 ml] shelled beans)*

3 *tablespoons extra-virgin olive oil*

1 *tablespoon freshly squeezed lemon juice*

1 *teaspoon dried leaf oregano*

3 *tablespoons fresh flat-leaf parsley leaves, snipped with a scissors*

⅛ *teaspoon crushed red peppers (hot red pepper flakes), or to taste*

8 *ounces (250 g) soft sheep's milk cheese (pecorino) or goat's milk cheese, cut in cubes the size of a fava bean*

Fine sea salt and freshly ground black pepper to taste

1. Shell the beans. You should have about 2 cups (500 ml) of beans. Taste one; if the beans are tender and not too bitter, they can be served raw, without removing the skin that covers each bean. If they are tough, blanch the beans in boiling salted water for 30 seconds, then slip off the outer skin, revealing two smaller beans. (This is a very tedious operation, but worth the effort.)

2. In a medium-size bowl, combine the beans and all the remaining ingredients, and toss to blend. Taste for seasoning. Serve immediately, in small portions, as a snack, an appetizer, or as part of a series of antipasto.

■ Yield: 8 to 12 servings ■

NOTE: Should there be any leftovers, sauté the beans and cheese with a touch of oil in a small skillet: They are fragrant and equally delicious as a warm appetizer.

Knowing Beans About Beans

Fresh fava beans—long and round, with velvety pods—are one of the world's great gastronomic treasures. Tender, only faintly bitter, and a brilliant spring green, fava beans—known as *fave* in Italian—are so nourishing they are called the meat of the poor. Fava beans, also known as broad beans, are similar to but not the same as kidney-shaped lima beans, which can become quite tough. When very young and fresh, lima beans—sometimes called wax or butter beans—may be eaten raw. They can be found in farmers' markets, in the pod, or in some supermarkets, already shelled.

Red Bean and Onion Salad

Insalata di Borlotti

This is a favorite antipasto and one that can easily be made in advance. Red-and-white-speckled borlotti, or cranberry (Roman) beans, have a delightfully earthy flavor and always make me feel very healthy and wholesome. I sampled this dish one Sunday evening at a little trattoria on Lake Garda, in the Veneto. Before serving, always retaste the beans, adding additional oil, vinegar, and seasoning as desired. A handful of minced flat-leaf parsley leaves tossed in at the last minute adds a fine touch of color and flavor.

2½ cups (1 pound; 500 g) dried cranberry (borlotti) beans

1 small onion, halved

1 medium carrot, peeled

3 bay leaves, preferably fresh

4 plump fresh garlic cloves, crushed

1 large rib celery

Generous sprig of fresh sage

6 tablespoons extra-virgin olive oil, or to taste

2 small red onions (about 8 ounces; 250 g), minced

¼ cup (60 ml) best-quality red wine vinegar, or to taste

Fine sea salt and freshly ground black pepper to taste

Minced fresh flat-leaf parsley leaves (optional)

1. Rinse the beans, picking them over to remove any pebbles. Place the beans in a large bowl, add boiling water to cover, and set aside for 1 hour. Drain the beans, discarding the water.

2. Place the onion, carrot, bay leaves, garlic, celery, sage, and 2 tablespoons of the oil in a large covered saucepan and add cold water to cover by 1 inch (2.5 cm). Bring just to a simmer over moderate heat, and cook for 15 minutes. Add the drained beans and return to a simmer. Cover and continue cooking just until tender, about 30 minutes to 1 hour more. Check the beans from time to time: They should be slightly firm but not mushy when fully cooked. If necessary, add additional water to keep the beans from drying out. (Cooking time will vary depending upon the freshness of the beans; fresh beans cook more quickly than old ones.)

3. Meanwhile, in a small bowl, toss the minced onions with 2 tablespoons of the oil. (The oil will serve to soften the harshness of the raw onions.) Set aside.

4. Once the beans are cooked, drain them, discarding the herbs and vegetables. Transfer the beans to a large bowl. While still warm, toss with the minced onions and the remaining 2 tablespoons of oil, the vinegar, and the salt and pepper. Taste for seasoning. (The proportions of vinegar and oil given here are only a suggestion: You may prefer more or less.) The beans may be served warm, but are generally served at room temperature, as part of an antipasto table or salad buffet, as a luncheon side dish, or as a simple condiment. The beans remain fresh-flavored for 2 to 3 days, refrigerated.

■ Y i e l d : 8 t o 1 0 s e r v i n g s ■

Bean Lore

The ancient Romans used beans for balloting, both in the courts and in elections. Black beans signified opposition or guilt, white stood for agreement or innocence.

White Bean Salad with Fresh Sage and Thyme

Fagioli all'Olio

This fragrant marriage of white beans and oil, enriched with a few herbs and seasonings, is one of Italy's purest treats. Tuscan trattorias offer these beans year-round, serving them warm as a first course, or at room temperature as part of an antipasto buffet. Sage and thyme add a special depth of flavor.

2½ cups (1 pound; 500 g) dried
   small white beans (navy,
   cannellini, or toscanelli)

1 small onion, halved

1 medium carrot, peeled

3 bay leaves, preferably fresh

4 plump fresh garlic cloves,
   crushed

1 large rib celery

1 generous sprig each of fresh
   sage and thyme

About ½ cup (125 ml) extra-virgin
   olive oil, to taste

Sea salt and freshly ground black
   pepper

3 tablespoons fresh thyme
   leaves, for garnish

1. Rinse the beans, picking them over to remove any pebbles. Place the beans in a large bowl, add boiling water to cover, and set aside for 1 hour. Drain the beans, discarding the water.

2. Place the drained beans in a large covered saucepan and add cold water to cover by 1 inch (2.5 cm). Add the onion, carrot, bay leaves, garlic, celery, sage and thyme sprigs, and 2 tablespoons of the oil. Bring just to a simmer over moderate heat, and simmer for 30 minutes. Season with salt and simmer just until tender, about 30 minutes more. Check the beans from time to time: They should be slightly firm but not mushy when fully cooked. If necessary, add additional water to keep the beans from drying out. (Cooking time will vary depending upon the freshness of the beans; fresh beans cook more quickly than old ones.)

3. Once the beans are cooked, drain them, discarding the herbs and vegetables. Transfer the beans to a large bowl. While still warm, toss with the thyme leaves, season with salt and pepper, and add the remaining 6 tablespoons (95 ml) olive oil. The beans may be served warm, but are generally served at room temperature, as part of an antipasto table or salad buffet. The beans remain fresh-flavored for 2 to 3 days, refrigerated.

▪ Y i e l d :   8   t o   1 0   s e r v i n g s ▪

Tomato and Bread Salad

Panzanella

This is one of those rustic, peasant-style salads that makes you want to pack up a wicker picnic basket and take off for a sunny hill with a beautiful view, a group of friends, and a jug of red wine. Panzanella is a traditional Tuscan salad found in many variations at trattorias throughout the region. No doubt, tomato and bread salad was created out of a need to do *something* with that leftover country bread, whether whole wheat or sourdough. I first sampled panzanella at Il Latini, a boisterous and popular Florentine trattoria where the waiter patiently explained their method of making the salad. The traditional recipe always includes chunks of slightly dried out bread and cubed ripe tomatoes, along with cucumbers, onion, celery, vinegar, and oil. Some cooks trim the crusts from the bread (I don't) and others peel the tomatoes (I don't). The bread can be cubed with a knife or simply ripped into chunks with your hands. I've seen versions with and without garlic, and some enjoy the flavors of green olives or tuna. This is the version I like best, with a pungent hit from both the garlic and the green olives. While some recipes suggest soaking the bread in water first to soften it, I don't unless I am using rock-hard cubes of bread. Soaking makes the bread get soppier faster, and I like to keep the crunch going as long as possible. I prefer to let the natural vegetable juices, vinegar, and oil do the job. Serve this as a luncheon main dish, along with sausages and cheese, and, of course, a simple but fruity red wine.

4 cups (about 8 ounces; 250 g) slightly dried out country bread cubed or torn into pieces

About 1½ pounds (750 g) ripe tomatoes, cored and coarsely chopped

1 red onion, thinly sliced

½ hothouse cucumber, peeled and cut into small cubes

2 ribs celery hearts, cut into small cubes

½ cup loosely packed fresh basil leaves

3 plump fresh garlic cloves, degermed and minced (optional)

1 cup (5 ounces; 150 g) drained pitted green olives, halved crosswise (optional)

1 to 2 tablespoons best-quality red wine vinegar

Fine sea salt to taste

3 to 4 tablespoons extra-virgin olive oil

Freshly ground black pepper to taste

1. Place the bread in a large bowl. Add the tomatoes, onion, cucumber, celery, and basil, and toss gently to blend. If using, add the garlic and green olives. Sprinkle with the vinegar and salt, and toss gently to blend. Spoon the oil over the salad, sprinkle with freshly ground black pepper, and toss once more. Set aside for 30 minutes to allow the bread to absorb the dressing and the flavors to blend.

2. To serve, use a slotted spoon to transfer portions to large dinner plates.

■ Yield: 6 to 8 servings ■

Arugula, Pine Nut, and Parmesan Salad

Insalata di Rughetta, Pignoli, e Parmi

Peppery, vibrant arugula—also known as rughetta or rucola, in Italian—is a remarkable salad green. Its energy can perk up tired palates and spirits like nothing else I know. Here the fresh green is combined with lightly toasted pine nuts and fresh shavings of Parmesan cheese. It's a variation on a dish I sampled at the traditional Florentine trattoria Il Cammillo. There they serve a first course of layered pine nuts, chopped arugula, and shaved Parmesan. I was frustrated by the "tightness" of the dish when I sampled it, and have taken the liberty of turning the combination into a full-fledged salad. Toss it very lightly with a mixture of top-quality vinegar and oil. I serve this often at dinner parties, to welcome raves.

½ cup (2 ounces; 60 g) pine nuts

3 ounces (90 g) stemmed arugula leaves, washed and dried

One 2-ounce (60-g) chunk of Italian Parmigiano-Reggiano cheese

Fine sea salt to taste

About 1 tablespoon best-quality red wine vinegar

Freshly ground black pepper to taste

About 2 tablespoons extra-virgin olive oil

1. Preheat the oven to 350°F (175°C; gas mark 4/5).

2. Spread the nuts loosely on a baking sheet. Toast until lightly browned, about 10 minutes. Check every few minutes to avoid burning the nuts. Remove from the oven, and turn out onto a large plate to cool. (The nuts can be toasted several hours in advance.)

3. In a large shallow salad bowl, combine the arugula and toasted pine nuts. Using a vegetable peeler, shave the Parmesan cheese into long thick strips directly into the bowl. (If the chunk of cheese becomes too small to peel, grate the remaining cheese and add to the bowl.) Season the salad with fine sea salt and toss gently to blend.

4. Drizzle 1 tablespoon vinegar over the salad, season with pepper, and toss to blend. Add just enough oil to very lightly coat the arugula, and toss to blend. Serve immediately, arranging on large serving plates.

■ Yield: 4 servings ■

Walnut and Pecorino Salad

Insalata di Noci e Pecorino

On my last visit to Florence's Cibrèo, a giant white bowl of marinated walnuts and pecorino (sheep's milk cheese) was sitting on a sideboard, welcoming hungry guests as they took their places for the multi-course feast to follow. This simple preparation has since become a family favorite; I serve it sometimes as a before-dinner snack, or, more often, alongside a salad of baby spinach leaves tossed only with fine sea salt and extra-virgin olive oil. The combination is a great substitute for a full cheese course, and yet the wholesome crunch of the toasted walnuts and the soft saltiness of the cheese on the palate make for a fabulous course all its own. The varied ingredients can be measured out in advance, but toss the salad at the last moment, for a fresher, more distinct flavor. If you cannot find top-quality sheep's milk cheese, substitute a good domestic or imported goat's milk cheese.

2 cups (8 ounces; 250 g) freshly cracked walnut halves, toasted and cooled

3 tablespoons extra-virgin olive oil

1 tablespoon freshly squeezed lemon juice

1 teaspoon dried leaf oregano

3 tablespoons fresh flat-leaf parsley leaves, snipped with a scissors

8 ounces (250 g) soft sheep's milk cheese (pecorino) or goat's milk cheese, cubed

Fine sea salt and freshly ground black pepper to taste

In a medium-size bowl, combine all the ingredients and toss to blend. Season to taste. Serve immediately, in portions, as an appetizer or as part of a series of appetizers.

■ Yield: 4 to 6 servings ■

VEGETABLES

Asparagus with Butter and Parmesan

Asparagi alla Parmigiana

This is one of the finest and most flavorful ways I know of preparing asparagus. First the slim and tender green stalks are cooked in boiling salted water, and immediately plunged into ice water to stop the cooking and to keep the delicious spears firm and green. They are then sautéed in a blend of butter and olive oil, and sprinkled with a touch of Parmigiano-Reggiano cheese just before serving. This is a perfect simple first course, which can be followed by a more complicated main course, such as Braised Oxtail with Tomatoes, Onions, and Celery (page 251) or Beef Braised in Barolo Wine (page 247). I sampled this dish one sunny Sunday afternoon in May, at the Trattoria del Castello in Grinzane Cavour in the Piedmont, when everyone had been impatiently awaiting the season's first asparagus, and the true arrival of spring.

Coarse sea salt to taste

1 pound (500 g) small green asparagus stalks, trimmed

2 tablespoons (1 ounce; 30 g) unsalted butter

1 tablespoon extra-virgin olive oil

½ cup (2 ounces; 60 g) freshly grated Italian Parmigiano-Reggiano cheese

1. Prepare a large bowl of ice water and set aside. Bring a large pot of water to a boil, and add 1 tablespoon salt for each quart (liter) of water. Add the asparagus and cook until crisp-tender, about 8 minutes. Remove the asparagus with a slotted spoon and immediately plunge it into the ice water, to cool it down as quickly as possible. (Do not let the asparagus sit in the cold water too long, or it will lose its crispness and fresh flavor.) As soon as the asparagus is cool, drain, and transfer to a thick towel to dry. (The asparagus can be cooked up to 2 hours in advance.)

2. When ready to serve, combine the butter and oil in a large skillet over moderately high heat. When hot, add the asparagus, stir to coat with the fat, and sauté just until warmed through, 2 to 3 minutes. Transfer the asparagus to 4 warmed dinner plates, sprinkle with the cheese, and serve immediately.

■ Yield: 4 servings ■

WINE SUGGESTION: Asparagus is a difficult flavor for wine: A safe choice would be a good Chardonnay.

• • •

Panfried Potatoes with Black Olives

Patate alle Olive

I first sampled this simple dish at a small restaurant in Santa Margherita, along the Ligurian coast near Genoa. I love the play of textures, colors, and flavors in this dish: the softness of the olives contrasted with the crunch of the potatoes; the black against white; and the saltiness of the olives paired with the delicately flavored potatoes. I particularly like this as a side dish to fish or poultry.

1½ pounds (750 g) small red potatoes

3 tablespoons extra-virgin olive oil

½ cup (125 ml) brine-cured black olives, such as Italian Ligurian or French Niçoise olives, drained, pitted, and halved (see page 282)

Fine sea salt and freshly ground black pepper to taste

1. Peel the potatoes and cut them lengthwise into quarters. Rinse in several changes of water and dry thoroughly with a thick, absorbent towel.

2. In a very large heavy skillet, heat the oil over moderately high heat until hot but not smoking. Add the potatoes in a single layer. Reduce the heat to moderate and brown the potatoes thoroughly on one side before turning. Cook the potatoes until they are browned on all sides and tender when pierced with a fork, 15 to 20 minutes total. Add the olives to the pan and cook until warmed through, tossing from time to time, about 1 minute more. Transfer to a warmed serving bowl. Season to taste with salt and pepper, toss, and serve immediately.

■ Yield: 4 servings ■

Tips for Crispy Panfried Potatoes

■ Be sure the potatoes are well rinsed and completely dried before frying them.

■ Do not salt the potatoes until the end of the cooking time. Salting beforehand will encourage them to give up their liquid, making them limp.

■ Do not use a nonstick pan: The potatoes will not brown properly.

■ Heat the oil first, then add the potatoes to the pan.

■ Resist the urge to turn the potatoes too often. Allow them to brown on one side before turning.

■ Season with salt and pepper immediately after removing the potatoes from the frying pan.

Roasted Rosemary Potatoes

Patate al Forno

T he Romans are great at serving meats roasted in a good hot oven, and these attractive, delicious, mouth-watering potatoes are a fine accompaniment. I serve them with just about everything, but love them best with roasted chicken or pork. This is the ideal recipe for someone who insists he doesn't know how to cook, for the preparation takes little skill and little time. I've sampled these using various varieties of both roasting and boiling potatoes with equally good results. The key here is to use small, firm potatoes. The smaller the potatoes are cut, the more quickly they'll cook. Sometimes as a variation, I cut the potatoes crosswise into thin slices for maximum crispiness. These are also delicious served with a dab of Garlic Mayonnaise (page 275).

1½ pounds (750 g) small, firm potatoes

1 tablespoon minced fresh rosemary leaves

3 tablespoons extra-virgin olive oil

Fine sea salt and freshly ground black pepper to taste

1. Preheat the oven to 450°F (230°C; gas mark 9).

2. Peel and quarter the potatoes. Rinse them well and pat thoroughly dry with a thick towel. (To achieve the desired crispy skin when baking, the potatoes must be completely dry.)

3. On a baking sheet, combine the fully dried potatoes, rosemary, and oil, and, with your fingers, toss until the potatoes are well coated and the rosemary is well distributed. Spread the potatoes out in a single layer.

4. Place the baking sheet in the oven, and roast until the potatoes are golden brown and tender when tested with a fork, 20 to 25 minutes. Shake the baking sheet from time to time to redistribute the potatoes. When cooked, season with salt and pepper. Serve immediately.

▪ Yield: 4 to 6 servings ▪

Deep-Fried Zucchini and Zucchini Blossoms

Fritto di Zucca e Fiori di Zucca

I talians are masters at frying and have proven that fried foods need not be fatty, heavy, soggy, or indigestible. Properly fried foods are fragrant, golden, and taste of themselves, for the frying process fixes the flavor of the ingredient. My favorite fried food is the delicate, sunshine-yellow zucchini blossom, which often can be found at farmer's markets in season. If you have a garden, pick the blossoms early in the morning, when they are still tightly closed. While traveling about Italy, I've seen these blossoms served on their own, in tandem with slices of zucchini, or alongside sliced raw baby artichokes. The batter used for frying varies from chef to chef. My favorite version comes from Cesare Benilli at Al Covo in Venice. Beer batter produces a crunchy crust with a slightly—and pleasantly—bitter tang.

BEER BATTER

- ¼ cup (60 ml) water
- ¼ cup (60 ml) beer, at room temperature
- ½ cup (70 g) superfine flour, such as Wondra
- 3 large egg whites

About 2 quarts (2 l) peanut or safflower oil, for deep-frying

16 very fresh zucchini blossoms

About 1 pound (500 g) large fresh zucchini, scrubbed and sliced diagonally into ¼-inch (½-cm) slices (or substitute 1 pound [500 g] fresh baby artichokes, trimmed and quartered)

Fine sea salt to taste

1. Prepare the batter: In a medium-size bowl, whisk together the water and beer. Slowly whisk in the flour, whisking until smooth. The batter will be fairly thick. Set aside for 1 hour, to allow the flour to absorb the liquid.

2. Preheat the oven to 200°F (100°C; gas mark 1) to keep the first batches of fried ingredients warm, as you prepare the others.

3. Pour the oil into a wide 6-quart (6-l) saucepan, or use a deep-fat fryer. The oil should be at least 2 inches (5 cm) deep. Place a deep-fry thermometer in the oil and heat the oil to 375°F (190°C).

4. In the bowl of an electric mixer fitted with a whisk, beat the egg whites until stiff but not dry. Whisk the batter one more time, and, with a spatula, thoroughly fold the egg whites into the batter.

5. With your fingers, dip the ingredients to be fried, a few at a time, into the batter, rolling them to coat evenly with batter. Shake off the excess batter, letting it drip back into the bowl. Carefully lower the blossoms or vegetables, a few at a time, into the hot oil. Fry until the blossoms or vegetables are golden on all sides, turning once, for a total cooking time of about 2 minutes. With a wire skimmer, lift from the oil, drain, and transfer to paper towels. Immediately season with sea salt. Place the first batch in the oven—with the door slightly ajar—to keep warm. Continue frying until all of the ingredients are cooked, allowing the oil to return to 375°F (190°C) before adding each new batch. Serve immediately.

■ Y i e l d : 4 s e r v i n g s ■

WINE SUGGESTION: Sample these with a light white Sauvignon Blanc, perhaps one from Italy's northeast corner, Friuli-Venezia-Giulia.

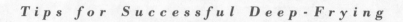

Tips for Successful Deep-Frying

■ Make sure all the ingredients are dry (moist ingredients will splatter).

■ Use fresh oil and plenty of it. Since vegetable oil is fairly inexpensive, I use fresh oil each time I fry. I prefer peanut oil, but safflower or corn oil can also be used. Olive oil is a luxury, and I prefer to save it for other cooking and for salads.

■ Make sure there is at least 2 inches (5 cm) of oil in the pan, and use a pan large enough to contain any splattering. (I use a large, wide enameled steel pan.)

■ Use proper heat and even heat. Invest in a deep-fry thermometer, and leave it in the pan while you cook.

■ Don't crowd the pan. When you add ingredients to hot fat, the temperature is instantly reduced. Always allow the oil to return to the proper temperature before frying another batch.

■ When frying in batter, use superfine flour: It won't cake up and it results in a more delicate batter.

■ Unless otherwise noted, when frying in batter, dip ingredients in the batter at the very last minute, to prevent sogginess.

■ Preheat the oven to keep the first batches of cooked food warm while you continue frying.

■ Drain the ingredients immediately after frying, placing on absorbent paper towels or clean cloth towels. Season with fine sea salt immediately after draining.

■ Serve fried foods hot.

Eggplant Parmesan

Parmigiana di Melanzane

One finds versions of this rich, earthy, and familiar vegetable dish everywhere in Italy. Like so many popular dishes, eggplant Parmesan has been banalized, and cooks often don't take the time to do it right. There are a few keys to a perfect version: Carefully deep-fry the eggplant in oil (if done properly, it won't absorb any more oil than you would use in grilling or baking the slices). Use a good homemade tomato sauce and the freshest mozzarella and Parmesan you can find. And do not prepare it in advance: part of the earthy charm of the dish is its very freshness. If top quality tomatoes cannot be found, substitute 2 cups (500 ml) tomato sauce made from canned tomatoes (page 256). No matter how you make it, you'll be amazed at the fresh look of this dish as it comes from the oven, with a touch of green basil, the golden tones of the lightly browned cheese, and the red tomatoes peeking through—yummm.

(continued)

TOMATO SAUCE

¼ cup (60 ml) extra-virgin olive oil

1 small red or yellow onion, minced

2 plump fresh garlic cloves, minced

Fine sea salt to taste

8 to 10 (about 2 pounds; 1 kg) small ripe tomatoes, peeled, cored, seeded, and chopped

2 firm medium eggplant (about 2 pounds; 1 kg)

2 quarts (2 l) peanut or sunflower oil, for deep-frying

Fine sea salt to taste

1 pound (500 g) fresh whole-milk mozzarella, thinly sliced

1 cup (4 ounces; 125 g) freshly grated Italian Parmigiano-Reggiano cheese

½ cup (125 ml) loosely packed fresh basil leaves, snipped with a scissors

1. Preheat the oven to 400°F (200°C; gas mark 6/7).

2. Prepare the tomato sauce: In a large skillet, heat the oil, onion, garlic, and salt over moderate heat and cook just until the onion is soft and translucent, 3 to 4 minutes. Add the tomatoes, stir to blend, and simmer gently, uncovered, until the sauce begins to thicken, about 15 minutes. Set aside.

3. Rinse and trim the eggplant and cut lengthwise into very thin slices. (Do not peel the eggplant and do not salt it. See box.)

4. Pour the oil into a heavy 3-quart (3-l) saucepan, or use a deep-fat fryer. Heat the oil to 360°F (180°C). Fry the eggplant in batches (2 or 3 slices at a time), until deep golden, 3 to 4 minutes per batch. With a flat wire mesh skimmer, transfer the eggplant slices to paper towels to drain. Season immediately with salt.

5. Spoon several tablespoons of the tomato sauce into a rectangular gratin dish measuring about 9 by 13 inches (23 × 33 cm). Place one-third of the fried eggplant side by side over the sauce. Spoon a thin layer of sauce over the eggplant. Cover with slices of mozzarella, using about half the cheese. Continue with another layer of eggplant, a layer of tomato sauce, and about half the Parmigiano-Reggiano cheese. Continue with the remaining eggplant, tomato sauce, mozzarella, and the final half of the Parmigiano-Reggiano cheese. Sprinkle with the fresh basil.

6. Place the gratin dish in the center of the oven and bake until the cheese is melted and the dish is fragrant and bubbling, about 40 minutes. Remove from the oven and serve warm or at room temperature, but not cold. (Since this dish tends to release a fair amount of liquid, it is best to serve with a slotted spoon.) The dish can be served the next day, but refrigeration tends to change the texture, so it is best to store at room temperature.

▪ Y i e l d : 6 t o 8 s e r v i n g s ▪

P l e a s e   D o n ' t   S a l t   o r   P e e l   t h e   E g g p l a n t

I do not agree with the common practice of salting eggplant as a method of removing bitterness from the vegetable. Truly fresh, firm eggplant is never bitter, so does not require advance salting to extract unwanted flavors. Likewise, the skin of the eggplant imparts a rich, deep flavor, so please do not peel it.

Sautéed Spinach with Garlic, Lemon, and Oil

Spinaci Saltati

Verdant green spinach that's been blanched and rinsed with cold water to drain, then tossed in a pan with oil and the scent of garlic, is one of my favorite vegetables. I serve it alongside roast meats or eat it all on its own. And I love it as I had it served to me at the Osteria del Cinghiale Bianco in Florence, mounded on toast that had been rubbed with garlic, for a classic fettunta, the Florentine version of the antipasto more commonly known as bruschetta.

3 tablespoons coarse sea salt

2 pounds (1 kg) fresh spinach, washed and stemmed, and dried

4 plump fresh garlic cloves, halved

2 tablespoons extra-virgin olive oil

Fine sea salt and freshly ground black pepper to taste

2 tablespoons freshly squeezed lemon juice

1. In a large pot, bring 6 quarts (6 l) of water to a rolling boil over high heat. Add the salt, then add the spinach, stirring to evenly wilt the spinach. Cook just until wilted through, 2 to 3 minutes. Drain the spinach, and rinse thoroughly with cold water to stop the cooking and set the bright green color. Drain again.

2. With a stainless steel knife, coarsely chop the spinach. Place in a fine-mesh sieve set over a bowl and set aside to drain once more, pressing as much liquid as possible from the spinach.

3. In a large skillet, combine the garlic and oil, and cook over moderate heat just until the garlic turns golden, but does not brown, 2 to 3 minutes. Remove and discard the garlic. Add the chopped spinach and cook, tossing gently with the tines of a fork, until warmed through, 2 to 3 minutes. Season with salt and pepper and the lemon juice, and serve immediately.

■ Yield: 4 to 6 servings ■

ITALIAN ADAGE:

"Salt the salad quite a lot,
then generous oil put in the pot,
and vinegar, but just a jot."

Sautéed Baby Artichokes

Carciofi Saltati

If you have a favorite, top-quality extra-virgin olive oil, bring it out for this dish. Nothing makes my palate happier than to bite into these fresh-flavored little artichokes, bathed with the perfume of a great olive oil. The artichokes can be served as a vegetable course or alongside roast pork or chicken. They're also delicious the next day, tossed into an omelet or a salad.

3 tablespoons freshly squeezed lemon juice

1 quart (1 l) water

Sea salt to taste

10 baby artichokes (about 1 pound; 500 g)
(see Mail Order Sources, page 325)

3 tablespoons extra-virgin olive oil

Freshly ground black pepper to taste

1. In a large stainless steel saucepan, combine the lemon juice, water, and ½ teaspoon salt. Set aside.

2. Prepare the artichokes: Rinse the artichokes under cold running water. Using a stainless steel knife to minimize discoloration, trim the stem of an artichoke to about ½ inch (1 cm) from the base. Carefully trim off and discard the stem's fibrous exterior. Bend back the tough outer green leaves one at a time, letting them snap off naturally at the base. Continue snapping off the leaves until only the central cone of yellow leaves with pale green tips remains. Lightly trim the top of the cone of leaves to just below the green tips. Trim any dark green areas from the stem. Depending upon its size, halve or quarter the artichoke, and transfer immediately to the acidulated water to prevent discoloration. Continue until all the artichokes are trimmed.

3. Bring the liquid and artichokes to a simmer over medium-high heat. Simmer, reducing the heat if necessary, until the artichokes are just tender but still offer a bit of resistance when pierced with a knife, 3 to 4 minutes. (Cooking time will vary according to the size of the artichokes.) Thoroughly drain the artichokes, discarding the cooking liquid. (The artichokes may be cooked up to 3 hours in advance. Just be sure to drain them. If they are allowed to cool in the cooking liquid, they will turn mushy.)

4. In a large skillet, heat the olive oil over moderately high heat until hot but not smoking. Add the drained artichokes and sauté just until crisp and evenly browned, 3 to 4 minutes. Season with salt and pepper and serve immediately. The artichokes are also good leftover, served at room temperature.

■ Yield: 4 to 6 servings ■

On Artichokes

In Italy, no soil goes idle. One delight of driving through the Italian countryside is the host of vegetable gardens found tucked into the most unforeseen spots. Gardens are crammed along the side of the road, inserted next to the railroad line, tucked at the edge of a parking lot, stuffed between a ravine and the road's shoulder. Almost always, giant, thistle-like flowering artichokes—*carciofi* in Italian—form the garden's centerpiece.

Artichokes are prized for their rich, sweet, almost grassy flavor, and serve to stimulate salivation, which is why they often appear at the beginning of a meal.

When purchasing artichokes, look for those with flat, tightly closed leaves and bright-green coloring. They should be heavy for their size. (Lightness is a sign they have dried out, and are certain to be tough.) Artichokes thrive on moisture: If they will not be used immediately, lightly trim the stems, wrap in dampened paper towels, and refrigerate.

Braised Artichokes with Garlic and Parsley

Carciofi alla Romana

A marvelously fragrant, vibrant, multipurpose dish. I first sampled this version of the popular Roman-style artichokes at La Frateria di Padre Eligio in Cetona, in Tuscany. I spent a thoroughly educational morning with Chef Walter Tripodi in the restaurant/monastery kitchen, following him about as he prepared the day's luncheon. This was one of that day's dishes, an appealing blend of artichokes, parsley, mint, garlic, and a touch of hot pepper, all braised in white wine and extra-virgin olive oil. Serve the artichokes warm as a first course, or toss them with strands of thick pasta—linguine is great—as a main course. Although Chef Tripodi used tiny, violet-hued choke-less artichokes, I adapted the recipe for the more common globe artichokes.

4 globe artichokes

1 lemon

1 cup (250 ml) loosely packed fresh flat-leaf parsley leaves

1 cup (250 ml) loosely packed fresh mint leaves

8 plump fresh garlic cloves, halved

½ teaspoon crushed red peppers (hot red pepper flakes), or to taste

Fine sea salt to taste

½ cup (125 ml) extra-virgin olive oil

2 cups (500 ml) dry white wine, such as a Chardonnay or Pinot Grigio

1. **Prepare the artichokes:** Prepare a large bowl of cold water. Halve the lemon, squeeze the juice, and add the juice, plus the lemon halves, to the water. Rinse the artichokes under cold running water. Using a stainless steel knife to minimize discoloration, trim the stem of an artichoke to about 1½ inches (4 cm) from the base. Carefully trim off and discard the stem's fibrous exterior. Bend back the tough outer green leaves one at a time, and snap them off at the base. Continue snapping off the leaves until only the central cone of yellow leaves with pale green tips remains. Lightly trim the top of the cone of leaves to just below the green tips. Trim any dark green areas from the base. Halve the artichoke lengthwise. With a small spoon, scrape out, and discard, the hairy choke. Cut each trimmed artichoke half lengthwise into 8 even wedges. Place the wedges in the acidulated water to prevent discoloration. Repeat with the remaining 3 artichokes. Set aside.

2. **Prepare the cooking liquid:** With a large chef's knife, finely chop together the parsley, mint, garlic, crushed red peppers, and salt. Transfer to a 1-quart (1-l) nonreactive heavy saucepan, such as a stainless steel or enameled steel pan. Add the oil and wine. Thoroughly drain the artichoke slices and add to the saucepan. Cover and bring just to a simmer over moderate heat. Reduce the heat to low and simmer very gently until the artichokes are soft and offer no resistance when pierced with a knife, about 45 minutes. (They should still be swimming in liquid.)

3. **To serve warm as a first course or vegetable side dish:** Transfer to warmed shallow soup bowls, spooning the sauce over the artichokes. Pass plenty of crusty bread, for sopping up the delicious sauce.

To serve with pasta: Toss the artichokes with cooked and drained dried linguine, using all the liquid as the sauce. The pasta does not require cheese.

■ Y i e l d : 4 t o 6 s e r v i n g s ■

WINE SUGGESTION: Serve the same simple white wine used in cooking the artichokes, preferably a Chardonnay or Pinot Grigio.

On the Banks

of the Arno

∎

After a diet of Florentine beans and grilled or roasted meats and poultry, Florence always manages to offer up some fine fare from the sea. Here's a Florentine fish menu, to enjoy with a glass or two of Pinot Grigio or Sauvignon Blanc.

Spicy Tomato-Mussel Soup
Sea Bass with Potatoes and Tomatoes in Parchment
Lemon-Rice Tea Cakes

SOUPS

Tuscan Five-Bean Soup

Minestrone alla Toscana

"A first-rate soup is more creative than a second-rate painting."

ABRAHAM MASLOW

Throughout the streets of Florence, one finds an abundance of food and specialty shops, some of them devoted solely to the legume family—dried beans, peas, and lentils—a testament to the Florentine's love for one of nature's healthiest gifts. Many shops even sell customized bean soup "kits," tiny sacks that contain a colorful mix of tiny dried legumes, sometimes with a few grains added for good measure. The mixtures vary, but inevitably include cranberry (borlotti) beans, navy (cannellini) beans, red and green lentils, split peas, and black-eyed peas. You could add tiny adjuki beans, small kidney beans, baby lima beans, barley, or brown lentils to the mix. Just be sure to keep the beans small, and of rather uniform size, so they cook fairly evenly, and estimate about 1 pound (500 g) of beans for every 3 quarts (3 l) of water. I've purchased packages with as many as ten different beans, peas, and grains. This recipe is simply a blueprint: Combine the best beans you can find in your market, and don't be concerned if you have an abundance of dried ingredients remaining. Simply put together your own personalized bean soup kits for your pantry, or create gift packages, with the recipe included, for your friends. If you use fairly fresh beans, and keep them small, you can have dinner on the table in just one hour. The end result is what I think of as a "light" bean soup, one with a complexity of flavors and textures, a dish that's delightfully filling, but not overwhelmingly so. Be sure to serve it with plenty of crusty homemade bread. Wine is not generally served with soups such as this.

½ cup (3 ounces; 90 g) dried cranberry (borlotti) beans

½ cup (3 ounces; 90 g) dried red lentils

½ cup (3 ounces; 90 g) dried green lentils

½ cup (3 ounces; 90 g) dried green split peas

½ cup (3 ounces; 90 g) dried small white (navy) beans (cannellini or toscanelli)

½ cup (3 ounces; 90 g) pearl barley

3 tablespoons extra-virgin olive oil

1 medium onion, diced

1 medium carrot, diced

1 rib celery, thinly sliced

2 plump fresh garlic cloves, minced

Several sprigs of fresh thyme, bay leaves, fresh sage leaves, and celery leaves, tied in a bundle with cotton twine

3 quarts (3 l) water

Sea salt to taste

Extra-virgin olive oil, for the table

Freshly ground black pepper to taste

1. In a colander, combine the legumes and pearl barley. Rinse thoroughly under cold running water. Drain and set aside.

2. In a 6-quart (6-l) heavy-bottomed stockpot, combine the olive oil, onion, carrot, celery, garlic, and herb bundle, and stir to coat with the oil. Cook over moderate heat until the vegetables are fragrant and soft, about 5 minutes. Add the legumes and pearl barley, stir to coat with oil, and cook for 1 minute more. Add the 3 quarts (3 l) water and stir. Cover, bring to a gentle simmer over moderate heat, and cook until the outer shells of the largest beans are tender, about 45 minutes. Add the salt and cook until tender, 15 to 45 minutes more. Stir from time to time to make sure the beans are not sticking to the bottom of the stockpot. (Cooking time will vary according to the size and freshness of the beans.)

3. To serve, remove the herb bundle and ladle the soup—piping hot—into warmed shallow soup bowls. Pass a cruet of extra-virgin olive oil, drizzling directly into each bowl. Pass the pepper mill. (The soup, of course, may be reheated several times over a period of several days. It will thicken. Simply thin with water each time you reheat.)

■ Yield: 6 to 8 servings ■

• • •

Spicy Tomato-Mussel Soup

Zuppa di Cozze

A lthough the Italians traditionally call this a *zuppa*, or soup, this dish is really a cross between a soup and mussels coated with thick, garlic-rich tomato sauce. I sampled this version in Florence, at the family fish restaurant La Capannina di Sante. Chef and owner Sante Collestano served the mussels still in their shells, topped with the fragrant sauce. He tucked rustic slices of toasted Tuscan garlic bread into the bowls before serving so that they began soaking up the marvelous sauce. For a more elegant presentation, remove the mussels from the shells and serve floating in the fragrant tomato sauce. Be sure to use the freshest, smallest, most tender mussels you can find. This is a hearty dish, which can be served as a main course.

THE SAUCE

6 tablespoons extra-virgin olive oil

12 plump fresh garlic cloves, peeled

¾ teaspoon crushed red peppers (hot red pepper flakes), or to taste

One 28-ounce (765-g) can imported Italian plum tomatoes in juice or one 28-ounce (765-g) can crushed tomatoes in purée

Sea salt to taste

THE MUSSELS

4 pounds (2 kg) small fresh mussels, in their shells

3 tablespoons extra-virgin olive oil

1 small onion, minced

A handful of fresh flat-leaf parsley stems, tied in a bundle with cotton twine

1 cup (250 ml) dry white wine, such as a Chardonnay

Freshly ground black pepper to taste

8 thick slices country bread, toasted and rubbed with fresh garlic

A handful of fresh flat-leaf parsley leaves, snipped with a scissors

1. Prepare the sauce: In a heavy saucepan, combine the oil, garlic, and crushed red peppers over moderate heat. Sauté just until the garlic becomes fragrant and the pepper begins to color the oil, 2 to 3 minutes. If using whole canned tomatoes, place a food mill over the saucepan and purée the tomatoes directly into it. Crushed tomatoes can be added directly from the can. Season with salt, and simmer, uncovered, just until the sauce begins to thicken, 10 to 12 minutes. Set aside and keep warm.

2. Meanwhile, thoroughly scrub the mussels and rinse them in several changes of water. If an open mussel closes when you press on it, it is good. If it stays open, the mussel should be discarded. Note that in some markets, mussels are pre-prepared, in that the small black beard that hangs from the mussel has been clipped but not entirely removed. These mussels need only be rinsed before cooking. If the beards have not been clipped, beard the mussels by gently pulling, removing, and discarding the stringy black beard. Do not beard the mussels in advance or they will die and spoil.

3. In a large covered skillet, combine the oil, onion, parsley stems, and wine, and bring to a boil over high heat. Boil for 2 minutes. Add the prepared mussels, sprinkle generously with pepper, and stir. Cook, covered, just until the mussels open, about 5 minutes. Remove the mussels as they open. Do not overcook. Discard any mussels that do not open.

4. Place 2 slices each of toasted garlic bread at an angle at the edge of 4 warmed shallow soup bowls. With a large slotted spoon, transfer the mussels to the soup bowls. Line a fine-mesh sieve with moistened cheesecloth, place the sieve over the saucepan with the tomato sauce, and strain the mussel-cooking liquid through the cheesecloth and into the saucepan. Simmer for 1 to 2 minutes to blend the flavors. Taste for seasoning. Spoon the sauce over the mussels, and sprinkle each serving with snipped parsley leaves. Serve immediately.

▪ Yield: 4 servings ▪

WINE SUGGESTION: Any young, crisp, chilled white would be fine here: Try a Tuscan white—bianco di Toscana—from Antinori, Brolio, or Castello di Volpaia.

Knowing Your Beans

Dried beans can be made tender and flavorful by following just a few simple rules:

■ Consider freshness. Beans may be dried, but that does not make them immortal. When stored in a cool, dark, dry spot in an airtight container, dried beans remain in good form for up to one year. Beyond that, they dry out and toughen. There is no way to make a tough old bean tender. So buy from a reputable shop with quick turnover, and put your own "freshness" date on the package.

■ Salt added to dried beans at the beginning of the cooking period retards tenderness. Once the outer shell is tender—about halfway through the cooking time—salt can no longer inhibit the bean's cooking and softening, so add it then. If you wait until the beans are thoroughly cooked, no amount of salt will season them properly.

■ Acids—such as tomatoes and vinegar—affect bean tenderness. So add acidic ingredients only once the outer shell is tender, about halfway through the cooking time.

■ Hard, mineral-rich water can interfere with the texture of beans. Soft rain water, or filtered water, is ideal for cooking beans.

Pasta and Bean Soup

Pasta e Fagioli

There are about as many versions of pasta and bean soup as there are cooks. This soup—found in almost every region of Italy—varies from a broth-like bean soup to a creamy bean purée gently punctuated with whole beans and tiny pasta to a thick bean paste. I prefer the creamy variety, a satisfying soup that is neither too thin nor too thick, enhanced by a fragrant drizzle of extra-virgin olive oil at serving time. The flavor of the beans is quite delicate and benefits from the company of pancetta, olive oil, herbs, and vegetables. Almost any kind of small dried beans can be used here, such as dried cranberry (borlotti) beans or white (cannellini) beans. I'd advise against using dark red kidney beans: The authenticity would certainly be questionable. Likewise, any tiny pasta—broken bits of vermicelli or spaghetti, tiny elbow macaroni, or tiny stars—is suitable. When adding the pasta, be sure to stir the soup well, for it has a tendency to burn at this point. Note that the soup can be heated and reheated, and like most soups such as this, tastes even better the next day.

I enjoy this soup as a meal unto itself: All one need add is some good, crusty bread, perhaps a bit of cheese afterwards, and a dependable red wine.

(continued)

2½ cups (1 pound; 500 g) dried
  cranberry beans

3 tablespoons extra-virgin olive
  oil

½ cup (2 ounces; 60 g) minced
  pancetta (see Note)

1 medium onion, diced

1 medium carrot, diced

1 large rib celery with leaves,
  diced

4 plump fresh garlic cloves,
  minced

Several sprigs of fresh parsley,
  bay leaves, and celery leaves,
  tied in a bundle with cotton
  twine

3 quarts (3 l) water

Sea salt to taste

½ cup (3 ounces; 90 g) tiny dried
  Italian pasta, such as ditalini,
  broken spaghetti, or tiny
  elbow macaroni

Freshly ground black pepper to
  taste

Extra-virgin olive oil, for the table

1. Rinse the beans, picking them over to remove any pebbles. Place the beans in a large bowl, add boiling water to cover, and set aside for 1 hour. Drain and rinse the beans, discarding the water. Set aside.

2. In a 6-quart (6-l) heavy-bottomed stockpot, combine the oil, pancetta, onion, carrot, celery, garlic, and herb bundle, and stir to coat with the oil. Cook over moderate heat until the vegetables are fragrant and soft, about 5 minutes. Add the drained beans and the 3 quarts (3 l) water. Cover, bring to a simmer over moderate heat, and simmer gently for 30 minutes. Add the salt and continue simmering until the beans are softened, 30 to 60 minutes more. Stir from time to time to make sure the beans do not stick to the bottom of the stockpot. (Cooking time will vary according to the freshness of the beans.)

3. Remove the stockpot from the heat. With a slotted spoon, remove a large ladleful of the beans and vegetables and set aside. Remove and discard the herb bundle. Using a hand blender or immersion mixer, roughly purée the remaining soup directly in the stockpot. (Alternatively, pass the soup through the coarse blade of a food mill and return it to the stockpot.) The soup should have a creamy, but not totally smooth, consistency. Return the reserved beans and vegetables to the stockpot. Return to the heat and bring the soup back to a simmer over moderate heat. Add the pasta, stir, and cook just until the pasta is tender, about 10 minutes more. Stir from time to time to make sure the pasta does not stick to the bottom of the stockpot. Taste for seasoning.

4. To serve, ladle the soup—piping hot—into warmed shallow soup bowls. Pass a cruet of extra-virgin olive oil, drizzling a swirl of oil directly into each bowl of soup. (The soup may be reheated several times over a period of several days. It will thicken. Simply thin with hot water each time you reheat the soup.)

■ Yield: 6 to 8 servings ■

NOTE: If unsmoked Italian pancetta is unavailable, use a lean, top-quality bacon. Blanch it for 1 minute in boiling water, then drain thoroughly. Blanching will remove the smoked flavor from the bacon without cooking it.

That Handy Little Blender

One of the most efficient and practical of modern gadgets is the hand blender (also called an immersion mixer). The hand blender—which resembles an electric hand mixer with a long, thin wand and a rotary blade at the end—is particularly useful for anyone who prepares soups regularly. For best results:

■ Note that a hand blender does its work in seconds, not minutes. Avoid over-blending to maintain the texture of the food.

■ Take the hand blender to the food, not the food to the blender.

■ To prevent splashing, first immerse the blender in the liquid, then turn it on.

Tuscan Bean and Wheat Berry Soup

Gran Farro

F arro is a delicious variety of soft Tuscan wheat berry, a rustic, mottled grain found in tones of wheat and cream. (The grain is also sometimes called spelt.) I first sampled Le Gran Farro—a fragrant soup of beans and wheat berries—at the popular Tuscan trattoria La Mora, in Ponte a Moriano, north of Lucca, a restaurant devoted to continuing regional gastronomic traditions in Italy. The soup is thinner than the traditional bean soup served all over Tuscany, and thus a bit more elegant. It's a favorite in my house: My husband even takes it in his "lunch box" for late nights at the office. With it, serve a bit of red wine, thick slices of homemade bread, and friendship.

1 cup (6 ounces; 180 g) dried small white (navy) beans (cannellini or toscanelli)

1 cup (6 ounces; 180 g) soft whole wheat berries (spelt)

3 tablespoons extra-virgin olive oil

1 medium onion, minced

1 medium carrot, diced

1 rib celery, diced

2 plump fresh garlic cloves, minced

Several sprigs of fresh thyme, bay leaves, and celery leaves, tied in a bundle with cotton twine

2 quarts (2 l) water

Sea salt to taste

Several tablespoons extra-virgin olive oil, for the table

1. Rinse and drain the beans, picking them over to remove any pebbles. Place the beans in a large bowl, add boiling water to cover, and set aside for 1 hour. Drain and rinse the beans, discarding the water. Set aside.

2. Rinse and drain the wheat berries, place them in a small bowl, and add boiling water to cover. Set aside until ready to add to the soup in Step 5 (for a total soaking time of about 1½ hours).

3. In a 6-quart (6-l) heavy-bottomed stockpot, combine the oil, onion, carrot, celery, garlic, and herb bundle, and stir to coat with the oil. Cook over moderate heat until the vegetables are fragrant and soft, about 5 minutes. Add the drained beans, stir to coat with oil, and cook for 1 minute. Add the 2 quarts (2 l) water and stir. Cover, bring to a simmer over moderate heat, and simmer for 30 minutes. Add the salt and continue simmering until the beans are tender, 30 to 60 minutes more, stirring from time to time to make sure the beans do not stick to the bottom of the stockpot. (Cooking time will vary according to the freshness of the beans.)

4. Remove the stockpot from the heat. With a slotted spoon, remove a large ladleful of the beans and vegetables and set aside. Remove and discard the herb bundle. Using a hand blender or immersion mixer, purée the remaining soup directly in the stockpot. (Alternatively, pass the soup through the coarse blade of a food mill and return it to the stockpot.) The soup should have a broth-like, and almost smooth, consistency.

5. Return the reserved beans and vegetables to the stockpot. Return to the heat and bring the soup back to a simmer over moderate heat. Drain and rinse the wheat berries, discarding the water, and add to the soup. Cook just until the wheat berries are swollen and tender, about 1½ hours more. (Cooking time will vary according to the freshness of the wheat berries.) Stir from time to time to make sure the wheat berries do not stick to the bottom of the stockpot. If the soup appears too thick, add a little lukewarm water to thin it out. Taste for seasoning.

6. To serve, ladle the soup—piping hot—into warmed shallow soup bowls. Pass a cruet of extra-virgin olive oil, drizzling a swirl of oil directly into each bowl of soup. (The soup may be reheated several times over a period of several days. It will thicken. Simply thin with lukewarm water each time you reheat the soup.)

■ Y i e l d :   6   t o   8   s e r v i n g s ■

WINE SUGGESTION: A simple red is ideal here, such as a young Chianti Classico or Valpolicella Classico Superiore.

...

Milanese Vegetable Soup

Minestrone alla Milanese

A minestrone is simply a full-flavored vegetable soup, with few limits to variations and interpretations. What all recipes share are an abundance of fresh vegetables, some type of dried beans, and a starch, either pasta or rice, depending upon the region. This version, sampled at the Antica Trattoria della Pesa in Milan, includes the famed Arborio rice of the region. With soups such as this, wine is not recommended. Since the success of the soup does depend upon proper proportions—for perfection of flavor and color—I have included weights for the ingredients in this recipe.

1 cup (6 ounces; 180 g) dried small white (navy) beans (cannellini or toscanelli)

4 tablespoons (2 ounces; 60 g) unsalted butter

½ cup (2 ounces; 60 g) minced pancetta (see Note)

2 medium onions, minced

Sea salt to taste

2 medium carrots, diced

5 to 6 ribs celery hearts with leaves, cut into thin crosswise pieces

1 cup (4 ounces; 125 g) trimmed and diced green beans

½ small white cabbage (12 ounces; 375 g), shredded

2 medium boiling potatoes (12 ounces; 375 g), peeled and diced

2 quarts (2 l) water

One 16-ounce (480-g) can imported Italian plum tomatoes, with their juice

Freshly ground black pepper to taste

1 cup (6 ounces; 180 g) Italian Arborio rice

About 1 cup (4 ounces; 125 g) freshly grated Italian Parmigiano-Reggiano cheese

1. Rinse the beans, picking them over to remove any pebbles. Place the beans in a large bowl, add boiling water to cover, and set aside for 1 hour. Drain the beans, discarding the water. Set aside.

2. In a 6-quart (6-l) heavy-bottomed stockpot, melt the butter over low heat. Add the pancetta, onions, and salt, and stir to coat with the fat. Cook until the onions are soft and translucent, 3 to 4 minutes. Add the carrots, celery, and drained white beans, stir to coat with the fat, and cook for 5 minutes more. Add the green beans, cabbage, potatoes, and the 2 quarts (2 l) water. Place a food mill over the stockpot and purée the tomatoes directly into it. Cover, bring to a simmer over moderate heat, and simmer for 30 minutes. Season to taste with salt and pepper, and continue simmering until the beans are softened and tender, about 30 to 60 minutes more. (Cooking time will vary according to the freshness of the beans.) Add the rice, and simmer just until the rice is tender but still firm to the bite, about 20 minutes more. Taste for seasoning.

3. To serve, stir several tablespoons of the cheese into the soup. Ladle the soup—piping hot—into warmed shallow soup bowls, and pass the remaining cheese separately. (The soup, of course, may be reheated several times over a period of several days. The soup will thicken. Simply add additional water each time you reheat the soup.)

■ Y i e l d : 4 t o 6 s e r v i n g s ■

NOTE: If unsmoked Italian pancetta is not available, use a lean top-quality bacon. Blanch it for 1 minute in boiling water, then drain thoroughly. Blanching will remove the smoked flavor from the bacon without cooking it.

Pasta and Chick Pea Soup

Pasta e Ceci

T he Italians are great soup eaters, and one of the truly classic trattoria soups is this simple blend of chick peas—also known as garbanzo beans—simmered in an aromatic broth, punctuated by bits of pasta, and seasoned at table with best-quality olive oil. The soup should be thick and porridge-like, almost thick enough to hold a spoon upright! Since it's so rich, serve it in small portions, accompanied, at most, by a green salad or simple grilled poultry or fish. It's also a great treat when preceded by a platter of raw vegetables dipped in olive oil, just as I sampled one spring evening at Trattoria Omero, a lively spot with a marvelous view of the hills of Florence. Some foods are simply an excuse for eating something else, and I often think of this golden, harvest-like soup as an excuse for garlic and oil, two favorite foods that always put me in a happy frame of mind.

3 cups (1 pound; 500 g) dried chick peas (garbanzo beans)

3 tablespoons extra-virgin olive oil

1 medium onion, diced

1 medium carrot, diced

1 rib celery, thinly sliced

4 plump fresh garlic cloves, crushed

Several sprigs of fresh parsley, sprigs of sage, bay leaves, and celery leaves, tied in a bundle with cotton twine

2 to 3 quarts (2 to 3 l) cold water

Fine sea salt to taste

½ cup (3 ounces; 90 g) tiny dried Italian pasta, such as ditalini, broken spaghetti, or tiny elbow macaroni

Extra-virgin olive oil, for the table

1. Rinse and drain the chick peas, picking them over to remove any pebbles. Place the chick peas in a large bowl, add boiling water to cover, and set aside for 1 hour. Drain and rinse the chick peas, discarding the water. Set aside.

2. In a 6-quart (6-l) heavy-bottomed stockpot, combine the olive oil, onion, carrot, celery, garlic, and the herb bundle, and stir to coat with the oil. Cook over moderate heat until the vegetables are fragrant and soft, about 5 minutes. Add the chick peas, stir to coat with oil, and cook for 1 minute more. Add 2 quarts (2 l) water and stir. Cover, bring to a simmer over moderate heat, and simmer for 1 hour. Add the salt and continue simmering until the chick peas are tender, about 1 hour more, stirring from time to time to make sure they are not sticking to the bottom of the stockpot. Add additional water if the soup becomes too thick. (Cooking time will vary according to the freshness of the chick peas.)

3. Remove and discard the herb bundle. Using an immersion mixer, roughly purée the soup directly in the stockpot. (Alternatively, pass the soup through the coarse blade of a food mill or purée in batches in a food processor, and return it to the stockpot.) The soup should have a creamy, but not totally smooth, consistency. It should be very thick, almost porridge-like. Season with salt to taste. Add the pasta, stir, and cook just until the pasta is tender, about 10 minutes more, stirring frequently to keep the pasta from sticking. Taste for seasoning.

4. To serve, ladle the soup—piping hot—into warmed shallow soup bowls. Pass a cruet of extra-virgin olive oil, drizzling a swirl of oil directly into each bowl of soup. (The soup, of course, may be reheated several times over a period of several days. If it thickens, simply thin with water each time you reheat the soup.)

■ Y i e l d :   8   t o   1 0   s e r v i n g s ■

Roasted Yellow Pepper Soup

Passato di Peperoni

Sunset-golden peppers in a delicate broth of vegetables and poultry, steaming hot in pristine white soup bowls. Drizzle with the finest extra-virgin olive oil, grill a few slices of thick and crusty homemade bread, and you've got it made. This is the signature soup of Cibrèo, a modern-day Florentine trattoria, where Fabio and Benedetta Picchi serve up full-flavored, imaginative fare based on the country food of their Tuscan youth. Serve this in small portions as a first course, followed by a more substantial main course, such as roast chicken, grilled lamb chops, or osso bucco. If yellow peppers are not to be found, bright red peppers are a far from shabby substitute. Note that cooking the roasted strips of pepper in the oil with the vegetables gives them a richer flavor. And don't skimp when you drizzle on the olive oil: That's what gives this soup its rich and unctuous flavor.

2 tablespoons extra-virgin olive oil

1 large carrot, minced

1 rib celery, minced

1 medium onion, minced

6 (about 2 pounds; 1 kg) yellow bell peppers, roasted and sliced (see box)

Sea salt to taste

2 medium-size potatoes, peeled and diced

1 quart (1 l) water

2 cups (500 ml) chicken stock, preferably homemade (page 272)

Extra-virgin olive oil, for the table

1. In a large stockpot, combine the olive oil, carrot, celery, and onion, and cook over moderate heat until the vegetables are soft and fragrant, about 10 minutes. Add the sliced peppers and cook for 3 to 4 minutes more, for greater flavor intensity. Season with salt. Add the potatoes, water, and chicken stock, cover, and cook over moderate heat until the potatoes are soft, about 20 minutes.

2. Purée in batches in the food processor or blender, or with an immersion mixer. Taste for seasoning. Serve in warmed shallow soup bowls, drizzling each portion with a generous amount of the best olive oil you can find.

▪ Y i e l d :   6   t o   8   s e r v i n g s   ▪

For Better Roasted Peppers

Grilled and roasted peppers are understandably popular today, but like many popular items, they easily soon become the subject of abuse or misunderstanding. Grilled and roasted peppers should be just that, not charred shadows of their former selves. Too often peppers are burnt to a crisp, losing all their fragrant and flavorful essence.

Here are some tips:

■ Select thick-fleshed, thick-skinned peppers. They have more flavor and will better withstand the heat.

■ The best way to grill peppers is to place them at least 3 inches (8 cm) from the heat of the broiler, so they do not come in direct contact with the intense heat and both roast and steam at the same time, making for more moist and tender peppers. Peppers can also be roasted on a grill, over a gas flame, or in a very hot (500°F; 260°C; gas mark 9) oven.

■ Do not pierce the peppers. You want to save that beautifully oily liquid within.

■ Watch the peppers carefully as they cook. Turn them often, using tongs that won't puncture the flesh. The skin should blister, but not burn. (If the skin turns black and charred long before it begins to pull away from the pepper, the heat is too intense.)

■ Once the skin shrinks and peels away from the peppers on all sides, remove the peppers from the heat and seal them in a paper bag or place in a bowl and cover the bowl with plastic wrap. Allow them to cool thoroughly. Remove them from the bag or bowl, being careful not to lose any of the juices. Remove the charred skin from the peppers, carefully remove the seeds, and slice the peppers lengthwise into strips. Do not rinse or wash the peppers once they are peeled, or you will lose the flavorful juices.

DRIED PASTA

···

Penne with Spicy Tomato Sauce

Penne all'Arrabbiata

How can one simple dish give so much pleasure? Sometimes I think I could eat this every day, with the hint of spice from the red peppers, the bite of the garlic, and the flicker of green from the chopped parsley to add a note of freshness. Like many pasta dishes, this one reminds me of the Italian flag, with its proud red, green, and white. *Arrabbiata,* by the way, means "furious" or "angry," which describes the character of the spicy sauce. Traditionally, cheese is not served with this dish. If you begin boiling the water and preparing the sauce at the same time, the dish will take less than thirty minutes to prepare.

¼ cup (60 ml) extra-virgin olive oil

6 plump fresh garlic cloves, minced

½ teaspoon crushed red peppers (hot red pepper flakes), or to taste

Sea salt

One 28-ounce (765-g) can peeled Italian plum tomatoes in juice or one 28-ounce (765-g) can crushed tomatoes in purée

1 pound (500 g) dried Italian tubular pasta, such as penne

1 cup (250 ml) fresh flat-leaf parsley leaves, snipped with a scissors

1. In an unheated skillet large enough to hold the pasta later on, combine the oil, garlic, crushed red peppers, and a pinch of salt, stirring to coat with the oil. Cook over moderate heat just until the garlic turns golden but does not brown, 2 to 3 minutes. If using whole canned tomatoes, place a food mill over the skillet and purée the tomatoes directly into it. Crushed tomatoes can be added directly from the can. Stir to blend, and simmer, uncovered, until the sauce begins to thicken, about 15 minutes. Taste for seasoning.

2. Meanwhile, in a large pot, bring 6 quarts (6 l) of water to a rolling boil. Add 3 tablespoons salt and the penne, stirring to prevent the pasta from sticking. Cook until tender but firm to the bite. Drain thoroughly.

3. Add the drained pasta to the skillet with the tomato sauce. Toss, cover, and let rest over low heat for 1 to 2 minutes to allow the pasta to absorb the sauce. Add the parsley and toss again. Transfer to warmed shallow soup bowls and serve immediately.

■ Y i e l d : 6 s e r v i n g s ■

WINE SUGGESTION: A young Italian red table wine such as a Castelli Romani, from the area just southeast of Rome.

Getting the Most out of Parsley

Parsley is an ingredient in its own right, much more than a simple garnish added for a touch of green. Here, as in many Italian dishes, parsley is essential. To get the most flavor from fresh parsley, stem it first, leaving only the leaves. Place the leaves in a large glass or a deep bowl, and snip the leaves with a sharp scissors. Snipped in this manner, the parsley will be coarsely chopped but won't turn to mush as is often the case when chopped with a knife or in a food processor. Note that whenever one measures the volume of minced, chopped, or snipped herbs, they should be loosely packed.

Pay Attention to Salt

I can't argue enough for the use of top-quality sea salt (not what's known as "kosher salt") in the kitchen. Although sea salt may cost a bit more than traditional table salt, it is well worth the investment in flavor.

Sea salt—both fine and coarse—has a bright, clean, distinct flavor and imparts a truly subtle flavor to foods. Common table salt masks flavors: Rather than make foods taste seasoned, it only makes them taste salty.

I use coarse salt for most cooking (when seasoning sauces, soups, or water for cooking pasta). Fine sea salt, or coarse salt that is ground in a salt mill, is preferable for baking, for the table, and for the last-minute seasoning of a dish.

Always remember to cook and season as you go. Many dishes—such as pastas and beans—cannot successfully be seasoned at the end.

Spaghetti with Red Pesto Sauce

Spaghetti con Pesto Rosso

If you've ever thought of sun-dried tomatoes as a cliché that's passé, try this sauce out on your family and friends and your mind will be changed forever. Sauce the spaghetti very lightly—½ cup (125 ml) of sauce per pound (500 g) of pasta will do—pass the Parmigiano-Reggiano and a bottle of red wine, and live it up!

1 pound (500 g) dried Italian spaghetti

Sea salt

About ½ cup (125 ml) Red Pesto Sauce (page 264)

¼ cup (60 ml) fresh flat-leaf parsley leaves, snipped with a scissors

Freshly grated Italian Parmigiano-Reggiano cheese, for the table (optional)

1. In a large pot, bring 6 quarts (6 l) of water to a rolling boil. As the water is heating, place a large serving bowl over the pot to warm the bowl. When the water is boiling, add 3 tablespoons salt and the spaghetti, stirring to prevent the pasta from sticking. Cook until tender but firm to the bite. Carefully drain the pasta, leaving a few drops of water clinging to the spaghetti so that the sauce will adhere.

2. Add the pasta to the warmed bowl, and toss with the red pesto sauce to blend. Add the parsley and toss again. Transfer to warmed shallow soup bowls and pass the cheese, if desired.

▪ Yield: 6 servings ▪

WINE SUGGESTION: An everyday red is ideal here, such as a Chianti Classico.

• • •

Penne with Vodka and
Spicy Tomato-Cream Sauce

Penne alla Bettola

This is my husband's favorite pasta dish. I know that if I want to make him happy, I just say "vodka pasta," and a broad smile fills his face. There is something wonderfully satisfying about thick tubes of pasta, such as penne, sauced with a mixture of lightly spiced tomatoes and cream. The addition of vodka makes for a very intriguing dish. I'm sure only one out of a thousand people would guess that vodka is the secret ingredient. The recipe comes from La Vecchia Betolla, a lively, elbows-on-the-table trattoria in Florence, where you squeeze onto rough wooden benches and ultimately share in conversation with your neighbors, always close at hand.

¼ cup (60 ml) extra-virgin olive oil

4 plump fresh garlic cloves, minced

½ teaspoon crushed red peppers (hot red pepper flakes), or to taste

Sea salt

One 28-ounce (765-g) can peeled Italian plum tomatoes in juice or one 28-ounce (765-g) can crushed tomatoes in purée

1 pound (500 g) dried Italian tubular pasta, such as penne

2 tablespoons vodka

1 cup (250 ml) heavy cream

¼ cup (60 ml) fresh flat-leaf parsley leaves, snipped with a scissors

1. In an unheated skillet large enough to hold the pasta later on, combine the oil, garlic, crushed red peppers, and a pinch of salt, stirring to coat with the oil. Cook over moderate heat just until the garlic turns golden but does not brown, 2 to 3 minutes. If using whole canned tomatoes, place a food mill over the skillet and purée the tomatoes directly into it. Crushed tomatoes can be added directly from the can. Stir to blend, and simmer, uncovered, until the sauce begins to thicken, about 15 minutes. Taste for seasoning.

2. Meanwhile, in a large pot, bring 6 quarts (6 l) of water to a rolling boil. Add 3 tablespoons salt and the penne, stirring to prevent the pasta from sticking. Cook until tender but firm to the bite. Drain thoroughly.

3. Add the drained pasta to the skillet with the tomato sauce. Toss. Add the vodka, toss again, then add the cream and toss. Cover, reduce the heat to low, and let rest for 1 to 2 minutes to allow the pasta to absorb the sauce. Add the parsley and toss again. Transfer to warmed shallow soup bowls and serve immediately. (Traditionally, cheese is not served with this dish.)

▪ Yield: 6 to 8 servings ▪

WINE SUGGESTION: A good-quality red that can stand up to the spice and cream is ideal here, such as a 2- or 3-year-old Chianti Classico from Tuscany or a California Zinfandel.

. . .

Spaghetti with Capers, Olives, Tomatoes, and Hot Peppers

Spaghetti alla Puttanesca

"Spaghetti can be eaten successfully if you inhale it like a vacuum cleaner."

SOPHIA LOREN

Spaghetti alla puttanesca, or "whore's pasta," is said to have originally been a favorite dish of Italian prostitutes, who could prepare it quickly when they had precious few moments to spend in the kitchen. Today, it's found in trattorias throughout Italy, although it is particularly popular in Rome, where I sampled a pleasantly spiced version at La Campana, a simple restaurant that attracts a glittery and hungry crowd. This is an ideal pasta dish for those who have little in the pantry and little time to spare. When preparing this pasta, don't skimp on the fresh parsley, for it's what adds a great fresh flavor, not to mention a lively color.

¼ cup (60 ml) extra-virgin olive oil

4 flat anchovy fillets cured in salt (see page 279) or in olive oil, drained (if in oil) and minced

3 plump fresh garlic cloves, minced

½ teaspoon crushed red peppers (hot red pepper flakes), or to taste

Sea salt

One 28-ounce (765-g) can peeled Italian plum tomatoes in juice or one 28-ounce (765-g) can crushed tomatoes in purée

15 salt-cured black olives, such as Italian Gaeta or French Nyons olives (see page 280), pitted and halved

2 tablespoons capers, drained and rinsed

1 pound (500 g) dried Italian spaghetti

1 cup (250 ml) fresh flat-leaf parsley leaves, snipped with a scissors

1. In an unheated skillet large enough to hold the pasta later on, combine the oil, anchovies, garlic, crushed red peppers, and a pinch of salt, stirring to coat with the oil. Cook over moderate heat just until the garlic turns golden but does not brown, 2 to 3 minutes. If using whole canned tomatoes, place a food mill over the skillet and purée the tomatoes directly into it. Crushed tomatoes can be added directly from the can. Add the olives and capers. Stir to blend, and simmer, uncovered, until the sauce begins to thicken, about 15 minutes. Taste for seasoning.

2. Meanwhile, in a large pot, bring 6 quarts (6 l) of water to a rolling boil. Add 3 tablespoons salt and the spaghetti, stirring to prevent the pasta from sticking. Cook until tender but firm to the bite. Drain thoroughly.

3. Add the drained pasta to the skillet with the sauce. Toss, cover, and let rest off the heat for 1 to 2 minutes to allow the pasta to absorb the sauce. Add the parsley and toss again. Transfer to warmed shallow soup bowls and serve immediately. (Traditionally, cheese is not served with this dish.)

■ Yield: 6 servings ■

WINE SUGGESTION: A dependable Chianti, such as one from the Antinori or Ricasoli estates.

Spaghetti with Garlic, Oil, and Hot Peppers

Spaghetti con Aglio, Olio, e Peperoncini

"Everything you see I owe to spaghetti."

SOPHIA LOREN

Spaghetti coated with oil, a profusion of garlic, and a hit of pepper is one of the Italy's most universally popular dishes. You'll find variations served in just about every part of the country, sometimes without the hot peppers or without the usual parsley; at other times the herb of choice might be basil, mint, oregano, or rosemary. I adore this version, with a healthy amount of both garlic and hot peppers. It always seems to put me and my guests in a cheery, energetic mood. When preparing this dish at home, be sure to watch the garlic carefully as it cooks. Burnt garlic becomes bitter at once, and turns a sublime dish into one that's thoroughly indigestible. The trick here is to combine the oil, garlic, and hot pepper in an unheated pan, and then heat them together so the garlic does not have a chance to burn. Likewise, adding a touch of oil at the end helps make for a pasta that is evenly and smoothly coated with sauce. Although you will sometimes see this dish served with cheese, that's a mistake: The pasta and sauce are already rich enough, and the cheese is just a lot of lily-gilding!

Sea salt

1 pound (500 g) dried Italian spaghetti

½ cup (125 ml) plus 2 tablespoons extra-virgin olive oil

6 plump fresh garlic cloves, minced

½ teaspoon crushed red peppers (hot red pepper flakes), or to taste

½ cup (125 ml) fresh flat-leaf parsley leaves, snipped with a scissors

1. In a large pot, bring 6 quarts (6 l) of water to a rolling boil over high heat. Add 3 tablespoons salt and the spaghetti, stirring to prevent the pasta from sticking. Cook until tender but firm to the bite. Drain thoroughly.

2. Meanwhile, in an unheated skillet large enough to hold the pasta later on, combine ½ cup (125 ml) of the oil, the garlic, the crushed red peppers, and a pinch of salt. Toss to thoroughly coat the garlic and pepper flakes, and cook over moderate heat just until the garlic turns golden but does not brown, 2 to 3 minutes.

3. Add the drained pasta to the skillet with the sauce. Toss, add the remaining 2 tablespoons of oil, toss thoroughly, and cover. Let rest off the heat for 1 to 2 minutes to allow the pasta to absorb the sauce. Add the parsley and toss again. Transfer to warmed shallow soup bowls and serve immediately. (Traditionally, cheese is not served with this dish.)

■ Yield: 6 servings ■

WINE SUGGESTION: I like a good "daily-drinking" Italian red with this assertive dish, such as a Chianti Classico.

Penne with Zucchini and Spicy Pizza Sauce

Penne con Zucchine alla Pizzaiola

When I'm sad, crabby, or in need of a bit of cheer, this is the pasta dish I turn to for solace. One bite, and I'm in a better mood. The success of this dish depends upon securing very fresh, tender, firm zucchini. Pizzaiola, by the way, is a tomato sauce that tastes a bit like a traditional pizza topping—that is, made with tomatoes, oregano, and garlic. I sampled this version at a little side-street trattoria in Siena. The addition of balsamic vinegar is my own pick-me-up touch, inspired by a recipe from Marcella Hazan. Season with hot pepper according to your taste.

7 tablespoons extra-virgin olive oil

3 tablespoons fresh rosemary leaves, minced

½ teaspoon crushed red peppers (hot red pepper flakes), or to taste

10 plump fresh garlic cloves, slivered

Sea salt

One 28-ounce (765-g) can peeled Italian plum tomatoes in juice or one 28-ounce (765-g) can crushed tomatoes in purée

7 ounces (210 g) firm, fresh zucchini, scrubbed, trimmed, and thinly sliced (do not peel)

½ teaspoon dried leaf oregano

1 pound (500 g) dried Italian tubular pasta, such as penne

2 tablespoons balsamic vinegar

1. In an unheated skillet large enough to hold the pasta later on, heat 6 tablespoons of the oil, the rosemary, crushed red peppers, garlic, and a pinch of salt. Cook over moderate heat just until the garlic turns golden but does not brown, 2 to 3 minutes. If using whole canned tomatoes, place a food mill over the skillet and purée the tomatoes directly into it. Crushed tomatoes can be added directly from the can. Stir to blend, and simmer, uncovered, until the sauce begins to thicken, about 15 minutes.

2. While the sauce is simmering, prepare the zucchini: In a large nonstick skillet, heat the remaining 1 tablespoon oil over moderately high heat. When the oil is hot but not smoking, add the zucchini and sauté just until golden, 2 to 3 minutes. Transfer to a colander to drain any excess oil, season with salt, and toss with the oregano. Set aside.

3. Meanwhile, in a large pot, bring 6 quarts (6 l) of water to a rolling boil. Add 3 tablespoons salt and the penne, stirring to prevent the pasta from sticking. Cook until tender but firm to the bite. Drain thoroughly.

4. Add the drained pasta to the skillet with the tomato sauce. Add the balsamic vinegar and toss. Add the zucchini and toss again. Cover, and let rest over low heat for 1 to 2 minutes to allow the pasta to absorb the sauce. Toss again, transfer to warmed shallow soup bowls, and serve immediately. (Traditionally, cheese is not served with this dish.)

■ Yield: 6 servings ■

WINE SUGGESTION: Either a young Italian red, such as a Barbera d'Alba, or a white, such as a Pinot Grigio.

Gemelli with Eggplant, Tomatoes, and Mozzarella

Gemelli alla Siciliana

On a cool fall day, there are few pastas that warm the soul as this one does. While eggplants are fresh, firm, shiny, and still at their peak, take advantage of this versatile vegetable, which blends so well with the heartiness of the pasta and the richness of a full-flavored tomato sauce. It's in this sort of dish that eggplant takes a starring role, tasting so much like meat you can't believe it isn't. (And whatever you do, don't peel the eggplant: The skin is the source of tremendous flavor.) Be sure to cut the eggplant and cheese into small, even cubes. When you reach into the bowl of pasta, "build" the pasta, sauce, and mozzarella on your fork for a harmonious blending of flavors and textures. A variety of dried pastas might be used here: Although rigatoni is traditional, I find it too bulky and cumbersome and prefer gemelli, ziti, fusilli, or the old standby, penne. While this dish is Sicilian in origin, I've seen it served throughout Italy.

¾ cup (185 ml) extra-virgin olive oil

1 small onion, minced

2 plump fresh garlic cloves, minced

Sea salt

One 28-ounce (765-g) can peeled Italian plum tomatoes in juice or one 28-ounce (765-g) can crushed tomatoes in purée

1 firm medium eggplant (1 pound; 500 g), cubed (do not peel)

1 pound (500 g) dried Italian tubular pasta, such as gemelli, ziti, fusilli, or penne

2 cups (10 ounces; 300 g) cubed whole-milk mozzarella

1. In an unheated skillet large enough to hold the pasta later on, combine ¼ cup (60 ml) of the oil, the onion, garlic, and a pinch of salt, stirring to coat with the oil. Cook over moderate heat just until the garlic turns golden but does not brown, 2 to 3 minutes. If using whole canned tomatoes, place a food mill over the skillet and purée the tomatoes directly into it. Crushed tomatoes can be added directly from the can. Stir to blend, and simmer, uncovered, until the sauce begins to thicken, about 15 minutes. Taste for seasoning.

2. While the sauce is simmering, cook the eggplant: In a large deep skillet, heat the remaining ½ cup (125 ml) oil over moderately high heat. When the oil is hot but not smoking, add the eggplant and cook until lightly colored, about 5 minutes. (The eggplant will soak up the oil immediately, but allow it to cook without added oil, keeping the pan moving to avoid scorching.) Season generously with salt.

3. Add the eggplant to the tomato sauce and keep warm over very low heat. (Neither the sauce nor the eggplant needs additional cooking, but the eggplant should have time to absorb some of the tomato sauce.)

4. Meanwhile, in a large pot, bring 6 quarts (6 l) of water to a rolling boil. Add 3 tablespoons salt and the pasta, stirring to prevent the pasta from sticking. Cook until tender but firm to the bite. Drain thoroughly.

5. Add the drained pasta to the skillet with the tomato sauce. Toss to blend. Cover and let rest off the heat for 1 to 2 minutes to allow the pasta to absorb the sauce. Transfer the pasta to warmed shallow soup bowls and sprinkle each serving with the cubed mozzarella. Serve immediately.

■ Yield: 6 servings ■

WINE SUGGESTION: With this, try a full-bodied red, such as a Montepulciano d'Abruzzo.

• • •
Spaghetti with Pecorino and Pepper

Spaghetti al Cacio e Pepe

T he Romans are great eaters of spaghetti as well as of that deliciously piquant sheep's milk cheese known as pecorino. This simple but sublime dish combines the two, as well as a healthy dose of coarse, freshly ground black pepper. Prepare it for one or for many, serving a sturdy Italian red alongside.

Sea salt

1 pound (500 g) dried Italian spaghetti

¼ cup (60 ml) extra-virgin olive oil

2 cups (8 ounces; 250 g) freshly grated Italian sheep's milk cheese (pecorino), such as Romano

Coarse freshly ground black pepper to taste

1. In a large pot, bring 6 quarts (6 l) of water to a rolling boil. As the water is heating, place a large serving bowl over the pot to warm the bowl. When the water is boiling, add 3 tablespoons salt and the spaghetti, stirring to prevent the pasta from sticking. Cook until tender but firm to the bite. Carefully drain the pasta, leaving a few drops of water clinging to the spaghetti so that the sauce will adhere.

2. Pour the oil into the warmed bowl. Add the pasta, cheese, and plenty of coarse black pepper. Using 2 forks, toss to blend. Transfer to warmed shallow soup bowls and serve immediately.

▪ Y i e l d :   4   t o   6   s e r v i n g s ▪

WINE SUGGESTION: I enjoy this with a sturdy, full-bodied red, such as Barbaresco, Barbera, or a Badia a Coltibuono Riserva.

Tips for Better Pasta

Pasta is a universally loved ingredient, and deserves the best treatment we can give it. Respect your pasta and reward your guests by following these simple rules:

- Cook pasta in plenty of rapidly boiling water—at least 1 quart (1 l) of water for every 3½ ounces (100 g) of pasta.

- Give the water plenty of salt. Use a minimum of 1 teaspoon per quart (liter) of water. Pasta cooked in unsalted water will taste flat and lifeless, no matter how well you salt and flavor the sauce.

- Stir the pasta as it cooks.

- Don't overcook it. Unless you have X-ray vision, "al dente" is not something you can see. You must taste! Both dried pasta and risotto should be cooked al dente, meaning firm to the bite, with no chalkiness in the center. Fresh pasta cooks very quickly, and should be cooked just until tender. Again, taste!

- Drain pasta as soon as it is cooked, but don't overdrain it.

- Sauce pasta as soon as it is drained. Time your cooking so that the sauce is ready when the pasta is. The hotter the pasta, the better it will absorb the sauce.

- Don't oversauce it. When you're done, there should be no sauce left in the bowl.

- Do think about marrying the right sauce to the right pasta.

- Keep it simple. Less is almost always more where pasta is concerned. Let the pasta be the star.

Bucatini with Pancetta, Pecorino, and Black Pepper

Bucatini alla Gricia

A typically Roman preparation, this simple and zesty dish marries the tang of an assertive, aged sheep's milk cheese, the bite of freshly ground black pepper, and the mildly cured flavor of fresh, high-quality pancetta. Traditionally, bucatini alla gricia is made with guanciale, a Roman-style bacon prepared with pork jowls, but a lean, high-quality pancetta is a practical and worthy substitute. Like many Italian pasta sauces, this is more a coating or moistener than a sauce that drowns the pasta, and it can be prepared in the time it takes to get the pasta cooking and uncork a chilled bottle of white wine. The recipe comes from Ninetta Ceccacci Mariani, owner and cook of Rome's family-run Trattoria Checchino dal 1887. Bucatini is a thin, hollow, tubular pasta that's just slightly thicker than traditional spaghetti but thinner than macaroni. I've also successfully prepared this dish with percialetti, a small pierced macaroni that's just slightly thicker than bucatini. Since this is a hearty, substantial dish, a small serving should be sufficient. The recipe can easily be doubled, to serve four.

Sea salt

8 ounces (250 g) dried Italian bucatini, or thick spaghetti

½ cup (2 ounces; 60 g) pancetta, cut into matchsticks (see box)

1 tablespoon extra-virgin olive oil

1 cup (4 ounces; 125 g) freshly grated Italian sheep's milk cheese (pecorino), such as Romano

Coarse freshly ground black pepper to taste

1. In a large pot, bring 3 quarts (3 l) of water to a rolling boil. Add 1½ tablespoons salt and the bucatini, stirring to prevent the pasta from sticking. Cook until tender but firm to the bite. Carefully drain the pasta, leaving a few drops of water clinging to the bucatini so that the sauce will adhere.

2. Meanwhile, in a skillet large enough to hold the pasta later on, combine the pancetta and oil over moderate heat. Sauté until the pancetta is a rosy color but not crisp, 2 to 3 minutes.

3. Add the drained pasta to the skillet and, using 2 forks, toss quickly and thoroughly with the pancetta. Add the cheese, sprinkle generously with coarse black pepper, and toss again. Cover and let rest over low heat for 1 minute to allow the pasta to absorb the sauce. Transfer to warmed shallow soup bowls and serve immediately, passing the pepper mill.

Yield: 2 servings as a main course,
4 servings as a first course

WINE SUGGESTION: A good white table wine would be nice here. If you can find it, sample one from the Latium region of Rome, such as a Castelli Romani or Colli Albani. Or try a young Orvieto Classico from a reputable bottler, such as Antinori or Ruffino.

What, You Have No Pancetta?

There is no American equivalent for pancetta, the unsmoked Italian bacon that is cured with salt and mild spices and rolled. Pancetta is prized for its subtle, delicate flavor and can be found in most Italian specialty shops. If you cannot find it, substitute a very lean, top-quality bacon. Blanch it for 1 minute in boiling water, then drain thoroughly. Blanching will remove the smoked flavor from the bacon without cooking it.

• • •

Rigatoni with Pecorino and Two Peppers

Rigatoni della Casa

T his lusty, quick, and satisfying dish is one of what I call the "Summer House Repertoire." You know, those times when you're in a vacation house kitchen equipped with one or two pots, maybe a bowl or none at all—the bare minimalist *cucina*. The only equipment you'll need for this dish is a pot for boiling pasta and a bowl in which to toss it. To make things even more efficient, place the pasta bowl on top of the pasta pot as the rigatoni cooks, so the bowl is nice and warm at serving time. If you're making this for a crowd, and it's going to be served as a main-dish pasta, count on about 4 ounces (125 g) of pasta and 2 ounces (60 g) of freshly grated sheep's milk cheese, or pecorino, per person. The addition of both red pepper and black pepper makes for a great contrast of flavors. Go easy on the hot red peppers, though, for the fire here should be subtle. Save this recipe for a good Italian sheep's milk cheese, one that is not too dried out nor overly pungent.

Sea salt

1 pound (500 g) dried Italian tubular pasta, such as rigatoni

2 cups (8 ounces; 250 g) freshly grated Italian sheep's milk cheese (pecorino), such as Romano, plus additional for the table

½ teaspoon crushed red peppers (hot red pepper flakes), or to taste

Freshly ground black pepper to taste

1. In a large pot, bring 6 quarts (6 l) of water to a rolling boil. As the water is heating, place a large serving bowl over the pot to warm the bowl. When the water is boiling, add 3 tablespoons salt and the rigatoni, stirring to prevent the pasta from sticking. Cook until tender but firm to the bite. Drain thoroughly.

2. Transfer the drained pasta to the warmed bowl. Add the cheese, crushed red peppers, and a generous amount of black pepper, and toss. Cover and let rest for 1 minute to allow the pasta to absorb the cheese and seasoning. Transfer to warmed shallow soup bowls and serve immediately, with additional grated cheese and red and black pepper, if desired.

■ Yield: 6 servings ■

WINE SUGGESTION: Serve this with a sturdy Italian red, such as a Barbera or Brunello di Montalcino.

Toss, Toss, and Toss Again

How often have you been presented a potentially fabulous pasta dish, only to find the sauce dumped carelessly on top? When this happens, the pasta invariably sticks together in dry clumps, and the sauce becomes little more than an afterthought. In the ideal world, pasta and its accompanying sauce should create a marriage of flavors, not serve simply as neighbors on the same block! And, much like a great salad, a great pasta should be tossed, tossed, and tossed again, so the sauce is thoroughly absorbed into the pasta. Tossing (and a moment's resting time to allow for absorption) is particularly important for thick, dense, tubular pastas—such as penne, gemelli, ziti, or rigatoni.

• • •

Saffron Butterflies

Farfalle allo Zafferano

"Italy, the paradise of earth and the epicure's heaven."

THOMAS NASHE,
THE UNFORTUNATE TRAVELLER, *1594*

I don't know which I like best: looking at the brilliant, happy colors of this dish, savoring the luxurious aroma of the saffron and butter sauce, or devouring it by the mouthful. Farfalle, or little pasta shaped like a butterfly, seems to be tailor-made to the elegance of the sauce. Yes, saffron is expensive, but please do use pure Spanish saffron. You'll only need a teaspoon, and you're worth it! (I enjoy using a mix of saffron here: saffron threads for flecks of elegance, powdered saffron for rich color.) Since the sauce is very rich, serve the saffron butterflies as a first-course pasta, followed by a light main course, such as a whole Baked Sea Bass with Artichokes (page 210). A dusting of freshly grated cheese here is nice, but go easy—you don't want to mask the saffron's distinctive flavor.

Sea salt

8 ounces (250 g) dried Italian farfalle, or butterfly-shaped pasta

2 tablespoons (1 ounce; 30 g) unsalted butter, at room temperature

½ cup (125 ml) heavy cream

1 teaspoon saffron threads (see Mail Order Sources, page 325)

Freshly grated Italian Parmigiano-Reggiano cheese, for the table (optional)

1. In a large pot, bring 3 quarts (3 l) of water to a rolling boil. As the water is heating, place a large serving bowl over the pot to warm the bowl. When the water is boiling, add 1½ tablespoons salt and the farfalle, stirring to prevent the pasta from sticking. Cook the pasta until tender but firm to the bite. Drain thoroughly.

2. Transfer the drained pasta to the warmed bowl, add the butter, cream, and saffron, and toss. Cover and let rest for 1 minute to allow the pasta to absorb the sauce. Transfer to warmed shallow soup bowls and serve immediately, passing cheese if desired.

■ Yield: 4 servings ■

WINE SUGGESTION: A Gattinara or Chianti Riserva remain favorites with this dish.

On Saffron

Saffron—with the powerful name of *zafferano*—has been grown in Italy since the fifteenth century, near Aquila in the Abruzzi. Then, its price equaled that of silver. Today, it's as expensive, as rare, and as fragrantly delicious as truffles. You understand, once you realize it takes 224,000 handpicked stigmas of a crocus flower to make 1 pound (500 grams). In the international market, the most readily available top-quality saffron comes from Spain.

Fusilli with Walnut and Garlic Sauce

Fusilli Salsa di Noci

Rich and fragrant, and elegant as an "all white" dish that reminds me of wintry evenings, this delicious pasta is perfect as a very hearty first course or a main course served with a light salad and fruit gelato to follow. I first sampled this one memorable evening on the terrace of a waterside trattoria in the village of Lazise, along the southern edge of Lake Garda, near Verona in northern Italy. Be sure to toast the walnuts, to bring out their flavor. And go easy on the garlic: The hint should be faint but not overpowering. The little corkscrew pasta know as fusilli is perfect for this sauce, for the tiny flecks of walnuts cling to the pasta, giving you even amounts of sauce and pasta with each bite.

2 plump fresh garlic cloves, degermed and minced

Sea salt

1 cup (4 ounces; 125 g) walnut halves, toasted and cooled

1 cup (250 ml) heavy cream

1 pound (500 g) dried Italian pasta, such as fusilli

½ cup (2 ounces; 60 g) freshly grated Italian Parmigiano-Reggiano cheese

Freshly ground black pepper to taste

1. In a food processor, combine the garlic, a pinch of salt, and the nuts, and process just to coarsely chop the nuts. Add the cream, and process to a fairly smooth sauce. Taste for seasoning. Transfer to a large serving bowl.

2. In a large pot, bring 6 quarts (6 l) of water to a rolling boil. As the water is heating, place the serving bowl over the pot to warm the bowl. When the water is boiling, add 3 tablespoons salt and the fusilli, stirring to prevent the pasta from sticking. Cook the pasta until tender but firm to the bite. Drain thoroughly.

3. Transfer the drained pasta to the warmed bowl, and toss to blend thoroughly. Add the cheese and toss to blend. Season with salt and pepper. Transfer to warmed shallow soup bowls and serve immediately, passing the pepper mill.

▪ Y i e l d : 4 t o 6 s e r v i n g s ▪

WINE SUGGESTION: I like a distinctive white with this, such as Teruzzi & Puthod's Vernaccia di San Gimignano.

Pasta Etiquette

Proper Italian pasta-eating etiquette suggests eating pastas with a fork, not a fork and a spoon. Ideally, all pastas and rice should be served in warm, shallow bowls, rather than on plates. Warm food served in warmed bowls stays hot longer, for the sides of the bowl serve to preserve the heat. The sides of the bowl also come in handy when eating spaghetti or other long pastas: The curve of the bowl serves as a pivoting point for twirling spaghetti on the fork.

...

Rigatoni with Meat and Celery Sauce

Rigatoni Strasciati

"Everything about Florence seems to be colored with a mild violet, like diluted wine."

HENRY JAMES, LETTER TO HENRY JAMES, SR., 26 OCTOBER 1869

Lean meat sauces lightly bathing spirals of pasta are among my favorite trattoria memories: images of chunky white china bowls mounded with fragrant and steaming pasta tossed quickly and ever-so-lightly with a hearty vermilion-tinged sauce. This particular version is found all over Florence, where several trattorias—including the legendary Antico Fattore and the smaller, less well known Quattro Stagioni—make it one of their regular specials. The secret of this sauce, sampled on Wednesdays at Antico Fattore, is lots of celery, good-quality meat (you can use a variety, including bits of prosciutto or pancetta for added intensity), and just the right amount of cooking (about thirty minutes total). *Strasciati* means "to drag," which is what you do with the thick, stubby pasta, dragging it through the sauce to coat it. While rigatoni is classic here, you might also use penne or large pasta shells.

¼ cup (60 ml) extra-virgin olive
   oil

1 small onion, minced

1 cup (250 ml) minced celery

¼ cup (60 ml) minced fresh
   flat-leaf parsley leaves

Sea salt

8 ounces (250 g) beef ground
   round or a mix of lean
   chopped beef, pork, and/or
   veal or prosciutto or pancetta

One 28-ounce (765-g) can peeled
   Italian plum tomatoes in juice
   or one 28-ounce (765-g) can
   crushed tomatoes in purée

Several sprigs of fresh parsley, bay
   leaves, and celery leaves, tied in
   a bundle with cotton twine

1 pound (500 g) dried Italian
   rigatoni, penne, or large shell
   pasta

Freshly grated Italian
   Parmigiano-Reggiano cheese,
   for the table

1. In an unheated skillet large enough to hold the pasta later on, combine the oil, onion, celery, parsley, and a pinch of salt, stirring to coat with the oil. Cook over moderate heat until the mixture is soft and fragrant, 4 to 5 minutes. Add the meat, and toss to blend. Reduce the heat to low and cook, making sure to break the meat up into small bits with a spatula, until the meat changes color, about 5 minutes more. If using whole canned tomatoes, place a food mill over the skillet and purée the tomatoes directly into it. Crushed tomatoes can be added directly from the can. Stir to blend. Add the herb bundle and cook, uncovered, until the sauce begins to thicken, about 20 minutes more. Taste for seasoning. Remove and discard the herb bundle.

2. Meanwhile, in a large pot, bring 6 quarts (6 l) of water to a rolling boil. Add 3 tablespoons salt and the pasta, stirring to prevent the pasta from sticking. Cook until tender but firm to the bite. Drain thoroughly.

3. Add the drained pasta to the skillet with the sauce and toss to blend. Cover and let rest for 1 to 2 minutes, off the heat, to allow the pasta to thoroughly absorb the sauce. Transfer to warmed shallow soup bowls and serve immediately, passing freshly grated cheese.

▪ Y i e l d : 4 t o 6 s e r v i n g s ▪

WINE SUGGESTION: I enjoy this with a light Tuscan red, so why not Chianti Classico?

Spaghetti with Spicy Meat Sauce

Spaghetti alla Giannetto

"It is axiomatic in spaghetti cookery that the pasta must boil freely and loosely in plenty of water so that the surface covering of starch may be washed away. Where this is not done, the result is a thick, sticky, starchy mess that no amount of sauce can redeem."

ANGELO PELLEGRINI, THE UNPREJUDICED PALATE

What could seemingly be more banal than spaghetti and tomato sauce? But once the cheery waitress arrives with a heavy skillet fresh from the kitchen, sets it down at your elbow, and begins to dish out a piping hot portion of spaghetti, you know you're onto something special. This was the scene one cool spring afternoon at Da Giannetto, the bustling trattoria set on the Badia a Coltibuono wine-growing estate in Tuscany.

Beyond the Italian borders, the pasta is often an afterthought, and the sauce is the real star. But at Da Giannetto, the tomato and meat sauce appeared as an almost imperceptible veil, just a whisper of moisture laced with hot pepper, weeping with a suggestion of oil and tiny tidbits of meat. For this simple spaghetti dish, I like to prepare the meat sauce with plenty of red pepper. At the final moment, add a touch of parsley, then freshly grated Parmesan for embellishment. Need one ask for more?

Sea salt

1 pound (500 g) dried Italian spaghetti

2 cups (500 ml) Meat Sauce (page 268)

½ cup (125 ml) fresh flat-leaf parsley leaves, snipped with a scissors

Freshly grated Italian Parmigiano-Reggiano cheese, for the table

A TRIO OF PASTAS, CLOCKWISE FROM TOP LEFT:
TONNARELLI WITH ARUGULA, TOMATOES, AND SHAVED PARMESAN
(PAGE 128);
ANGEL'S HAIR WITH PUNGENT PARSLEY SAUCE *(PAGE 124)*;
SPEEDY LASAGNE *(PAGE 122)*

 TAGLIARINI WITH LEMON SAUCE *(PAGE 130)*

PENNE WITH VODKA AND SPICY TOMATO-CREAM SAUCE *(PAGE 92)*

 **BAKED RISOTTO WITH ASPARAGUS,
SPINACH, AND PARMESAN** *(PAGE 166)*

1. In a large pot, bring 6 quarts (6 l) of water to a rolling boil. Add 3 tablespoons salt and the spaghetti, stirring to prevent the pasta from sticking. Cook until tender but firm to the bite. Drain thoroughly.

2. Meanwhile, in a skillet large enough to hold the pasta later on, warm the meat sauce.

3. Add the drained spaghetti to the skillet, and use 2 large forks to toss to coat the pasta. Add the parsley and toss again. Cover and let rest for 1 to 2 minutes, off the heat, to allow the pasta to thoroughly absorb the sauce. Transfer the pasta to warmed shallow soup bowls and serve immediately, passing freshly grated cheese.

■ Yield: 4 to 6 servings ■

WINE SUGGESTION: It would only be fitting to serve a Chianti Classico from the Badia a Coltibuono estate.

For Cleaner Clams

Since most clams grow in sandy areas, they tend to be sandy themselves. Nothing is more unpleasant than biting into a tender little clam to find a mouthful of grit. To test if your clams are sandy, steam a few open and taste. If they're gritty, purge them in salt water: Thoroughly dissolve 1 tablespoon salt per quart (liter) of cold water needed. Scrub the shells under cold running water, then purge in the salt water for three hours at room temperature. Scoop up the clams with your fingers, leaving behind the grit and sand. You will be amazed at the amount of sand they give up.

Spaghetti with Shrimp, Clams, and Mussels in Tomato Sauce

Spaghetti alla Sante

R obust as well as beautiful, this rustic main dish combines spaghetti and a lusty tomato sauce laced with plump whole cloves of garlic and a hint of hot peppers, all topped off with generous portions of steamed clams, mussels, and shrimp. I sampled this one cool August night in Florence, at the family-run fish restaurant La Capannina di Sante, a pleasant, pine-paneled spot set right along the banks of the Arno, southeast of the center of town. The menu is rudimentary (some might even say limited), but of the half-dozen dishes we sampled that evening, everything was delightfully fresh and energetically seasoned.

The antipasto misto was made up of an almost endless procession of creations, including a full-flavored combination of squid, steamed mussels and clams, and rosy shrimp, all seasoned with lemon juice, basil, and parsley, as well as a platter of breaded and panfried whole anchovies, set on a bed of whole, fresh bay leaves.

But my favorite of all was this spaghetti dish. At La Capannina di Sante the mussels and clams are served in the shell, for a dish that's both pretty to look at and made for casual, hands-on eating. For a more formal presentation, remove the mussels and clams from the shell, and toss the shellfish in with the pasta at the very last minute.

(continued)

¼ cup (60 ml) extra-virgin olive oil

12 plump fresh garlic cloves, peeled

½ teaspoon crushed red peppers (hot red pepper flakes), or to taste

Sea salt

One 28-ounce (765-g) can peeled Italian plum tomatoes in juice or one 28-ounce (765-g) can crushed tomatoes in purée

1 pound (500 g) fresh mussels

1 cup (250 ml) dry white wine, such as a Pinot Grigio

Handful of fresh flat-leaf parsley stems, tied in a bundle with cotton twine

Freshly ground black pepper to taste

1 pound (500 g) fresh littleneck or Manila clams, purged if necessary (see box)

8 ounces (250 g) fresh medium shrimp, shelled

8 ounces (250 g) dried Italian spaghetti

Handful of fresh flat-leaf parsley leaves, snipped with a scissors

1. In a large unheated skillet, combine the oil, garlic, crushed red peppers, and a pinch of salt, stirring to coat with the oil. Cook over moderate heat just until the garlic begins to turn golden but does not brown, 2 to 3 minutes. If using whole canned tomatoes, place a food mill over the skillet and purée the tomatoes directly into it. Crushed tomatoes can be added directly from the can. Stir to blend, and simmer, uncovered, just until the sauce thickens and the garlic offers no resistance when pierced with a knife, about 20 minutes.

2. Meanwhile, thoroughly scrub the mussels, and rinse with several changes of water. If an open mussel closes when you press on it, it is good; if it stays open, the mussel should be discarded. Beard the mussels. (Do not beard the mussels in advance or they will die and spoil. Note that in some markets mussels are pre-prepared, in that the small black beard that hangs from the mussel has been clipped, but not entirely removed. These mussels do not need further attention.)

3. In a very large skillet, combine the wine, parsley stems, and mussels. Sprinkle generously with pepper, cover, and cook just until the mussels open, about 5 minutes. Remove the mussels as they open. Do not overcook. Discard any that do not open.

4. Transfer the mussels, still in their shells, to a large warmed serving bowl. Cover loosely with foil. Set aside and keep warm. Line a sieve with moistened cheesecloth and strain the mussel liquor into the simmering tomato sauce.

5. Using a stiff brush, scrub the clams thoroughly under cold running water. Discard any with broken shells or shells that do not close when tapped.

6. Prepare a large steamer: Fill the steamer pot with 1 cup of water and bring to a boil. Place the clams in the steamer basket, and season generously with freshly ground pepper. Cover and steam the clams over high heat, removing the clams as they open and adding them to the bowl of mussels. (The entire process should take less than 10 minutes.) Discard any shells that do not open. Line a sieve with moistened cheesecloth and strain the clam liquor into the tomato sauce.

7. Taste the tomato sauce. If it is thin, reduce it slightly over low heat. The flavors of the shellfish broths should be distinctive. Toss the cooked clams and mussels, as well as the shrimp, with half of the tomato sauce and keep warm, covered, over low heat. (The shrimp will cook as the pasta is prepared.)

8. Cook the pasta: In a large pot, bring 3 quarts (3 l) of water to a rolling boil. Add 1½ tablespoons salt and the spaghetti, stirring to prevent the pasta from sticking. Cook until tender but firm. Drain thoroughly.

9. Arrange the pasta on a large warmed serving platter and toss with the remaining tomato sauce. Spoon the shellfish and sauce over the pasta, sprinkle with the parsley, and serve immediately. Serve this dish with finger bowls, as well as a large bowl for the mussel and clam shells.

■ Y i e l d :   4   s e r v i n g s ■

WINE SUGGESTION: A crisp white Pinot Grigio is my choice here.

Spaghetti with Marinated Baby Artichokes and Parmesan

Pasta Bianca

"Adapt your dish of spaghetti to circumstances and your state of mind."

GIUSEPPE MAROTTA (NEAPOLITAN WRITER)

One evening, with nothing but hunger on my mind, I created this simple and satisfying spaghetti dish, using what I had on hand at the moment: a box of spaghetti, a jar of homemade marinated baby artichokes, and a touch of parsley, red peppers, and Parmesan. It's a recipe that can accommodate a single diner or a crowd. The dish might be prepared with purchased marinated artichokes, but the flavor will be less fresh and less intense. I like the pale ivory tones of the dish, flecked with a touch of green and of red, and so I call it white pasta, or *pasta bianca*.

2 cups (500 ml) Marinated Baby Artichokes Preserved in Oil (page 18), drained and thinly sliced, marinating oil reserved

¼ teaspoon crushed red peppers (hot red pepper flakes), or to taste

Sea salt

1 pound (500 g) dried Italian spaghetti

¼ cup (60 ml) fresh flat-leaf parsley leaves, snipped with a scissors

1 cup (4 ounces; 125 g) freshly grated Italian Parmigiano-Reggiano cheese

1. In a skillet large enough to hold the pasta later on, combine the sliced artichokes, the reserved oil, and the crushed red peppers. Set aside.

2. In a large pot, bring 6 quarts (6 l) of water to a rolling boil. Add 3 tablespoons salt and the spaghetti, stirring to prevent the pasta from sticking. Cook until tender but firm. Drain thoroughly.

3. While the pasta is cooking, gently warm the artichokes over low heat.

4. Add the drained pasta to the artichokes and toss to blend. Toss in the parsley and about half of the cheese. Transfer to warmed shallow soup bowls and serve immediately, passing the remaining cheese.

■ Yield: 4 to 6 servings ■

WINE SUGGESTION: A fresh white Sauvignon Blanc nicely complements this dish.

Speedy Lasagne

Lasagne Rapide

One evening—having tested various fresh pasta recipes the previous day—I found myself with a few sheets of leftover pasta that I had cut into odd-shaped triangles and left to dry. I had a hunger for the flavors of lasagne, as my palate yearned for that beloved Italian trinity: pasta, tomatoes, and cheese. Time was short, so I improvised, and came up with a light dish that quickly satisfied my craving. I now call it "speedy lasagne," for it cooks quickly atop the stove and requires no special, hard-to-find ingredients. I particularly relish the crunch and character that the onions add to the sauce, while the ricotta adds the sensation of the richness of cream and of cheese, without making the dish the least bit heavy. I like to prepare this with an egg pasta, just slightly extravagant and a bit more satisfying than the eggless variety. I prefer factory-made pappardelle, fettuccine, or tagliatelle.

2 medium onions (about 10 ounces; 300 g), peeled

¼ cup (60 ml) extra-virgin olive oil

¼ teaspoon crushed red peppers (hot red pepper flakes), or to taste

Sea salt

Several sprigs of fresh parsley, bay leaves, sprigs of fresh rosemary, and celery leaves, tied in a bundle with cotton twine

One 28-ounce (765-g) can peeled Italian plum tomatoes in juice or one 28-ounce (765-g) can crushed tomatoes in purée

1 pound (500 g) dried Italian egg noodles, such as pappardelle, fettuccine, or tagliatelle

10 ounces (300 g) whole-milk ricotta, drained

3 tablespoons fresh flat-leaf parsley leaves, snipped with a scissors

1. Slice 1 of the onions in half lengthwise. Place each half, cut side down, on a cutting board, and cut crosswise into very thin slices. Slice the remaining onion in this manner, for about 2 cups (500 ml) sliced onions.

2. In a skillet large enough to hold the pasta later on, combine the sliced onions, oil, crushed red peppers, a pinch of salt, and the herb bundle, stirring to coat with the oil. Cook, uncovered, over very low heat, stirring from time to time, until the onions are very soft and glazed, about 10 minutes.

3. If using whole canned tomatoes, place a food mill over the skillet and purée the tomatoes directly into it. Crushed tomatoes can be added directly from the can. Stir to blend, and simmer, uncovered, until the sauce begins to thicken, about 15 minutes. Taste for seasoning. Remove the herb bundle.

4. Meanwhile, in a large pot, bring 6 quarts (6 l) of water to a rolling boil. Add 3 tablespoons sea salt and the pasta, stirring to prevent the pasta from sticking. Cook until tender but firm to the bite. Drain thoroughly.

5. Add the drained pasta to the skillet with the sauce. Toss. Add about three-quarters of the ricotta, in small spoonfuls, toss again, and cover. Let rest off the heat for 1 to 2 minutes to allow the pasta to absorb the sauce. Toss again and transfer to warmed shallow soup bowls. Garnish with spoonfuls of the remaining ricotta and the parsley. Serve immediately.

■ Yield: 4 to 6 servings ■

WINE SUGGESTION: Serve with a dependable, daily-drinking red wine, such as a Chianti Classico.

Angel's Hair Pasta with Pungent Parsley Sauce

Capellini d'Angelo al Prezzemolo

Delicate angel's hair pasta—capellini—tossed with a pungent blend of parsley, garlic, anchovies, and lemon juice is one of my favorite no-fuss, quick-fix dishes. The sauce knows no season, for parsley can be found in markets year-round. Be sure to prepare the sauce at the last minute, for the anchovy has a tendency to dominate the sauce as it "matures."

Sea salt

1 pound (500 g) dried Italian pasta,
such as capellini or other thin spaghetti

1 recipe freshly prepared Pungent Parsley Sauce (page 260)

1 to 2 tablespoons extra-virgin olive oil

1. In a large pot, bring 6 quarts (6 l) of water to a rolling boil. As the water is heating, place a large serving bowl over the pot to warm the bowl. When the water is boiling, add 3 tablespoons salt and the capellini, stirring to prevent the pasta from sticking. Cook until tender but firm to the bite. Drain thoroughly.

2. Transfer the drained pasta to the warmed bowl, add about three-quarters of the parsley sauce, and toss to evenly coat the pasta with the sauce. Cover for 1 minute, to allow the pasta to absorb the sauce. Toss again, and add the oil, to additionally moisten the pasta and to reinforce the fresh flavor of the sauce. Transfer the pasta to warmed shallow soup bowls, adding a dollop of the remaining parsley sauce to each bowl. Serve immediately.

■ Yield: 4 to 6 servings ■

WINE SUGGESTION: With this peppy sauce, drink a nicely chilled crisp white, such as a Pinot Grigio.

FRESH PASTA

...

Fresh Egg Pasta

Pasta all'Uovo

Because fresh pasta is so readily available almost everywhere, many of us, myself included, have all but abandoned a once-favorite pastime of making our own homemade pasta. But there are many dishes—lasagne for one—and many sauces—tangy lemon sauce for example—that cry out for the very best you can give. Here is a simple, traditional recipe that should bring you back into the kitchen (if you ever left!).

2 cups (265 g) unbleached all-purpose flour

3 large eggs, at room temperature, lightly beaten

BY HAND: Sift the flour onto a clean work surface and make a well in the center. Pour the beaten eggs into the well. With a fork, mix the flour and eggs together until the dough is soft and begins to stick together, about 3 minutes. When the dough forms a mass, transfer it to a lightly floured clean surface and knead until satiny and resilient, 10 to 15 minutes. Cover with a clean cloth and set aside for 1 hour.

IN A FOOD PROCESSOR: Sift the flour into the bowl of a food processor. Pulse the machine, slowly adding the beaten eggs (you may not need all the eggs), until the mixture forms clumps the size of small peas. Do not let the dough form a ball. Turn out onto a lightly floured surface and knead until satiny and resilient, about 10 minutes. Cover with a clean cloth and set aside for 1 hour.

Knead the pasta dough in the pasta machine: Divide the dough into 4 equal portions, covering the unused portions with a clean cloth. Set the rollers of the pasta machine at the widest setting. Flour one-quarter of the dough very lightly, and pass it once through the rollers. Fold the dough in thirds, like a business letter. Press the dough down with your fingertips to fuse the layers and push any air from between them. Turn the dough so that an open end feeds into the roller, and repeat the rolling and folding process (lightly flouring only when necessary) 8 more times.

Reset the rollers for the next thinnest setting. Lightly flour the dough but do not fold it. Pass the dough through the machine again, repeating the process on each remaining setting until the dough is as thin as desired (generally the next-to-last or the last setting on most machines).

Repeat the entire process with the remaining pieces of dough. Let the dough rest on towels until it is taut but not dry, about 15 minutes. Cut into desired lengths by machine or by hand.

■ Yield: 1 pound (500 g) dough ■

Tonnarelli with Arugula, Tomatoes, and Shaved Parmesan

Tonnarelli alla Rughetta

There is no way you can go wrong with such a winning quartet: fresh homemade pasta, leaves of pungent arugula, chunks of sun-kissed ripe tomatoes, and generous shavings of Parmigiano-Reggiano. Just make certain that everything is of top, top quality, as usual! This recipe comes from Arancio d'Oro, a popular trattoria near Rome's Via Condotti, a spot that's a favorite with journalists and neighborhood office workers. Even though this uncooked "sauce" seems like typical summer fare, I sampled it on a blustery day in December, and it warmed me to the tip of my toes. At Arancio d'Oro the dish was prepared with tonnarelli, the square-shaped homemade spaghetti better known as maccheroni alla chitarra. Use any fresh pasta noodle, such as the more finely cut tagliarini, the larger tagliatelle, or ribbons of fettuccine. If arugula cannot be found in the market, substitute very fresh leaves of watercress.

One 2-ounce (60-g) chunk of Italian Parmigiano-Reggiano cheese

4 cups (2 bunches, or about 3 ounces; 100 g) stemmed arugula leaves, washed and dried, coarsely chopped or torn

¼ cup (60 ml) extra-virgin olive oil

4 ripe plum tomatoes (8 ounces; 125 g), cored and coarsely chopped

3 tablespoons coarse sea salt

1 pound (500 g) fresh pasta, such as tonnarelli, fettuccine, or tagliatelle (page 126)

Fine sea salt and freshly ground black pepper to taste

1. In a large pot, bring 6 quarts (6 l) of water to a rolling boil. As the water is heating, place a large serving bowl over the pot to warm the bowl.

2. Using a vegetable peeler, shave the cheese into long, thick strips. (If the chunk of cheese become too small to shave, grate the remaining cheese and add to the bowl.) Place half of the cheese directly into the bowl. Add the arugula, oil, and tomatoes, and toss to blend.

3. Add the coarse sea salt to the boiling water, then add the pasta, stirring to prevent the pasta from sticking. Cook until tender. Drain thoroughly.

4. Add the drained pasta to the bowl, toss, and season generously with fine sea salt and pepper. Divide the pasta among warmed shallow soup bowls. Top with the remaining cheese. Serve immediately, passing the pepper mill.

■ Yield: 4 to 6 servings ■

WINE SUGGESTION: With this pasta, we drank a young Dolcetto d'Alba, from the Piedmont. Its smoothness provided a perfect match for a dish that is quite delicate, despite the peppery flavor of the arugula.

...

Tagliarini with Lemon Sauce

Tagliarini al Limone

"The point of drinking wine is to get in touch with one of the major influences of Western civilization, to taste the sunlight trapped in a bottle, to remember some stony slope in Tuscany. . . ."

JOHN MORTIMER, RUMPOLE AND THE BLIND TASTING

All golden yellow with flecks of green, this dish always reminds me of sunshine and springtime. I first sampled it at the very aristocratic table of the Grand Hotel e la Pace in Montecatini Terme, the old-fashioned spa hotel in the hills of northern Tuscany. Since then, I've ordered it many times in restaurants and trattorias all over Italy. At home, I often make a single portion for myself as a quick, uplifting lunch or dinner, especially on a gray winter's day when it seems that one will never again see spring. The addition of parsley is essential, while cheese is optional. This is one pasta dish that demands good, fresh noodles, such as tagliarini, long homemade pasta that is slightly thinner than tagliatelle. Even the best-quality dried pasta won't do justice to the simple, elegant sauce.

4 tablespoons (2 ounces; 60 g) unsalted butter, at room temperature

1 cup (250 ml) heavy cream

¼ cup (60 ml) freshly squeezed lemon juice

Sea salt

1 pound (500 g) fresh tagliarini, tagliatelle, or fettuccine (page 126)

Grated zest (yellow peel) of 3 lemons

3 tablespoons fresh flat-leaf parsley leaves, snipped with a scissors

Freshly grated Italian Parmigiano-Reggiano cheese, for the table (optional)

1. In a skillet large enough to hold the pasta later on, combine the butter, cream, and lemon juice over low heat. As soon as the butter is melted, remove the skillet from the heat, cover, and set aside.

2. Meanwhile, in a large pot, bring 6 quarts (6 l) of water to a rolling boil. Add 3 tablespoons salt and the pasta, stirring to prevent the pasta from sticking. Cook until tender. Drain, leaving a few drops of water clinging to the pasta so that the sauce will adhere.

3. Transfer the pasta to the skillet, off the heat, and toss to blend. Add the lemon zest, and toss once more. Cover and let rest for 1 to 2 minutes to allow the pasta to thoroughly absorb the sauce. Transfer to warmed shallow soup bowls, shower with the parsley leaves, and serve immediately. Pass freshly grated cheese if desired.

■ Yield: 4 to 6 servings ■

WINE SUGGESTION: Although lemon is the dominant flavor here, it is a fresh flavor, and one that won't fight wine. Either white (a Sauvignon Blanc or an Orvieto) or red (Nebbiolo or Chianti) would be fine with this dish.

The Zest of Life

Zest is the dimpled paper-thin outer rind of any citrus fruit—lemon, orange, grapefruit, or lime—and contains the fruit's essential oils. Distinct from the thick white bitter peel that separates the zest from the fruit, zest is one of the world's most refreshing and versatile flavorings.

Lemons are the Italians' favorite citrus, and the bright, sparkling lemon flavor turns up regularly in pastas, desserts, and ices. While the majority of lemons come from the sunny southern climates, such as Sicily, lemons can be found as far north as Lake Garda, not far from the Swiss border. The appropriately named Lake Garda village of Limone boasts of terraced citrus groves stretched along the lake's shores, where the lemons grow under huge glass structures.

There are several ways to remove the zest of a citrus fruit, and each requires attentiveness, to avoid including the bitter white portion of the rind. Add zest to your cooking by:

■ scrubbing the whole fruit against the tiny holes of a small hand grater;

■ paring the outer rind from the fruit with a vegetable peeler and cutting the rind into very thin strips; or

■ using a zester, a tiny hand-held gadget the size of a vegetable peeler. A zester—my preference—shaves the outer rind in extremely thin, fine strips that result in an even, delicate flavoring.

Citrus should be zested just before it is added to other ingredients, for the flavorful oils will dry out as the zest comes in contact with the air.

▪ ▪ ▪

Lasagne with Basil, Garlic, and Tomato Sauce

Lasagne al Pesto

L iguria is the home of one of Italy's most famous sauces, pesto, that vibrant blend dominated by basil, garlic, and oil. And one of the sauce's most traditional uses is in this dish of lasagne, a far cry from the heavy, layered, baked version most of us are used to. Here, palm-sized rectangles of fresh or dried pasta are simply interlayered with the pungent sauce, making for a dish that's heavenly and oh so simple. I sampled this version one evening in a harborside trattoria in Santa Margherita.

Sea salt

6 ounces (180 g) fresh lasagne (page 126) (or substitute dried Italian lasagne—about 6 sheets)

1 recipe Basil, Garlic, and Tomato Sauce (page 262)

1. In a large pot, bring 6 quarts (6 l) of water to a rolling boil. Add 3 tablespoons salt and the lasagne, stirring gently to prevent the pasta from sticking. Cook fresh pasta until tender, dried pasta until tender but firm to the bite. (Dried lasagne will take 10 to 15 minutes cooking time.) Drain thoroughly but carefully, so the lasagne noodles do not break.

2. Place the drained lasagne on a cutting board and cut each piece in half crosswise. Place a rectangle of pasta in a warmed shallow soup bowl. Whisk the pesto to blend. Top the lasagne with a spoonful of the pesto and smooth out the sauce with the back the spoon. Add 2 more layers of pasta, topping each layer with a spoonful of pesto. Repeat for additional servings, until all the pasta and pesto have been used. Serve immediately.

▪ Yield: 6 servings ▪

WINE SUGGESTION: With pesto, I enjoy a simple Italian white, such as a Pinot Grigio, or, for a change of pace, a Tyrolean Riesling.

Tajarin with Rosemary-Infused Butter

Tajarin al Burro Aromatizzato

Endlessly soothing, this is a pasta I dream about when I'm really, really hungry. My husband and I sampled this one sunny day in May, at the Tre Gallini (Three Hens) trattoria in Turin. *Tajarin* is the Piemontese name for the delicate, tagliatelle-like strands of homemade pasta found all over the region. The sauce here is so simple and tastes so rich, it is totally deceiving. You really can't imagine that you are eating a sauce of nothing but a bit of butter and a pleasant dose of fresh rosemary. At Tre Gallini, they infuse the butter with rosemary and then strain the butter, making for a pasta that looks as though it has no sauce at all. After experimenting with several variations, I decided I liked the look and ease of the unstrained sauce. This is one recipe designed to stimulate your own creative juices: Try saucing the pasta with other fresh herbs, such as fresh summer savory or sage. Whichever herbs you use, mince them by hand: Machine-minced herbs never have the same fresh, vibrant flavor as those minced carefully by hand. Since the fresh pasta is the star here, be sure that it is top quality, be sure to cook it right, and don't overdrain it! Cheese is optional: Sometimes I prefer the sheer simplicity of the pasta, butter, and herbs. Other times, a light touch of freshly grated Parmesan is totally welcome. (Proportions can easily be adjusted for a dish that serves from one to eight people.)

5 tablespoons (2½ ounces; 75 g) unsalted butter

3 to 4 tablespoons minced fresh rosemary leaves (to taste)

Sea salt

1 pound (500 g) fresh tagliatelle or fettuccine (page 126)

Freshly grated Italian Parmigiano-Reggiano cheese, for the table (optional)

1. In a skillet large enough to hold the pasta later on, combine the butter and rosemary over low heat. As soon as the butter is melted, remove the skillet from the heat and cover. Set aside to infuse for 5 minutes.

2. Meanwhile, in a large pot, bring 6 quarts (6 l) of water to a rolling boil. Add 3 tablespoons salt and the pasta, stirring to prevent the pasta from sticking. Cook until tender. Drain, leaving a few drops of water clinging to the pasta so that the sauce will adhere.

3. Transfer the pasta to the skillet, off the heat, and toss to blend. Cover and let rest for 1 to 2 minutes to allow the pasta to thoroughly absorb the sauce. Transfer to warmed shallow soup bowls and serve immediately, with cheese if desired.

■ Yield: 4 to 6 servings ■

WINE SUGGESTION: A dry white Gavi di Gavi or a smooth, young red Dolcetto d'Alba both are excellent matches for this elegant pasta dish.

Tagliatelle with Tricolor Peppers and Basil

Tagliatelle con Peperoni e Basilico

"Italy is so tender, like cooked macaroni—yards and yards of soft tenderness—ravelled round everything."

D. H. LAWRENCE, SEA AND SARDINIA, *1923*

I sampled this dish one sunny Sunday July afternoon in Venice, at the charming neighborhood trattoria Antica Besseta. The small dining room was filled with a large family tucking into platters of fish and pasta, and the room was overflowing with sounds of happy times. I love the purity and simplicity of this colorful dish: golden pasta tossed with a festival of peppers, red, yellow, and green. The spiciness of the hot peppers adds a nice surprise. While the assortment of three colors of peppers is ideal, a mix of green and red peppers will do if yellow ones are not in the market. Whatever you use, the proportions should be half peppers, half pasta, with a gentle nudge of spiciness. This is a terrific first course for a dinner that might include roast poultry or meat.

2 *red bell peppers*

2 *green bell peppers*

2 *yellow bell peppers*

6 *tablespoons extra-virgin olive oil*

½ *teaspoon crushed red peppers (hot red pepper flakes), or to taste*

Sea salt

12 *ounces (375 g) fresh tagliatelle or fettuccine (page 126)*

¼ *cup (60 ml) loosely packed fresh basil leaves*

Freshly grated Italian Parmigiano-Reggiano cheese, for the table (optional)

1. Prepare the sauce: Wash the peppers, core them, halve them lengthwise, and remove the seeds and membranes. Cut each half lengthwise into pencil-thin slices. If the slices are extra long, halve them. Pour the oil into a covered deep 12-inch (30-cm) skillet. Place the pepper slices and crushed red peppers in the skillet, toss with the oil, and season lightly with salt. Cook, covered, over very low heat until very soft, about 40 minutes, stirring from time to time. You may need to cook these over diffused heat. The peppers should not burn or toughen. Be careful not to allow the juices to cook away: You want to retain as much cooking liquid as possible, for an unctuous sauce. (The peppers can be prepared several hours in advance and reheated at serving time.)

2. Meanwhile, in a large pot, bring 6 quarts (6 l) of water to a rolling boil. Add 3 tablespoons salt and the pasta, stirring to prevent the pasta from sticking. Cook until tender. Drain thoroughly.

3. Transfer the pasta to the skillet, off the heat, and toss to blend. Cover and let rest for 1 to 2 minutes to allow the pasta to thoroughly absorb the sauce. Transfer to warmed shallow soup bowls. Snip the basil with a scissors, sprinkling it over the pasta. Serve immediately. Pass freshly grated cheese if desired.

■ Yield: 4 to 6 servings ■

WINE SUGGESTION: At Antica Besseta we drank a local Venetian white. At home, try a good Soave Classico.

Tagliatelle with Tomato Sauce and Butter

Tagliatelle al Pomodoro e Burro

What could be simpler—or more universally appealing—than fresh pasta in tomato sauce? I sampled this soothing version one evening at Milan's small family restaurant Antica Trattoria Della Pesa. The dish leaves the realm of the ordinary by way of the very high quality of the fresh ingredients used. Guests are invited to give the pasta a last toss, allowing the sweet butter to melt into the tomato sauce, enriching the flavor and the fragrance. Since butter is the final touch here, make sure that it is fresh and hasn't been sitting around in the refrigerator absorbing assorted odors.

¼ *cup (60 ml) extra-virgin olive oil*

2 *plump fresh garlic cloves, minced*

Sea salt

One 28-ounce (765-g) can peeled Italian plum tomatoes in juice or one 28-ounce (765-g) can crushed tomatoes in purée

1 *pound (500 g) fresh tagliatelle or fettuccine (page 126)*

4 *tablespoons (2 ounces; 60 g) unsalted butter, at room temperature*

¼ *cup (60 ml) fresh flat-leaf parsley leaves, snipped with a scissors*

Freshly grated Italian Parmigiano-Reggiano cheese, for the table (optional)

1. Prepare the sauce: In an unheated skillet large enough to hold the pasta later on, combine the oil, garlic, and a pinch of salt, stirring to coat with the oil. Cook over moderate heat just until the garlic turns golden but does not brown, 2 to 3 minutes. If using whole canned tomatoes, place a food mill over the skillet and purée the tomatoes directly into it. Crushed tomatoes can be added directly from the can. Stir to blend, and simmer, uncovered, until the sauce begins to thicken, about 15 minutes.

2. Meanwhile, in a large pot, bring 6 quarts (6 l) of water to a rolling boil. Add 3 tablespoons salt and the pasta, stirring to prevent the pasta from sticking. Cook until tender. Drain thoroughly.

3. Transfer the pasta to the skillet, off the heat, and toss to blend. Cover and let rest for 1 to 2 minutes to allow the pasta to thoroughly absorb the sauce. Transfer to warmed shallow soup bowls, placing 1 tablespoon of the butter on top of each serving. Shower with the parsley and serve immediately. Pass freshly grated cheese if desired.

■ Yield: 4 to 6 servings ■

WINE SUGGESTION: Serve a full-flavored red that will enjoy the company of the tomato sauce, a good Chianti or Montepulciano d'Abruzzo.

• • •
Tagliatelle with Prosciutto and Artichokes

Tagliatelle al Prosciutto con Carciofi

A quick, substantial fresh pasta dish, this luxurious mixture of top-quality egg pasta, great ham and cheese, and just a hint of artichokes is one that always seems to take me out of a rut of serving the same, traditional pasta sauces. As with many pasta dishes, the sauce here is almost a background note to the delicious fresh egg pasta, homemade or store-bought. We sampled this one warm evening in May at the fourteenth-century Certosa di Maggiano, just outside of Siena.

2 tablespoons extra-virgin olive oil

1 medium tomato, peeled, cored, quartered, and cut into thin strips

Sea salt

2½ ounces (75 g) thinly sliced prosciutto, cut into matchsticks

2 Marinated Baby Artichokes Preserved in Oil (page 18), drained and cut into matchsticks

8 ounces (250 g) fresh tagliatelle or fettuccine (page 126)

About ¼ cup (1 ounce; 25 g) freshly grated Italian Parmigiano-Reggiano cheese, plus additional for the table

Freshly ground black pepper to taste

¼ cup (60 ml) fresh flat-leaf parsley leaves, snipped with a scissors

1. In a skillet large enough to hold the pasta later on, heat the oil over moderately high heat. When hot, add the tomato and a pinch of salt and cook until most of the liquid has evaporated, about 2 minutes. Add the prosciutto and artichoke hearts and cook just until the prosciutto begins to brown, about 2 minutes more.

2. Meanwhile, in a large pot, bring 3 quarts (3 l) of water to a rolling boil. Add 1½ tablespoons salt and the pasta, stirring to prevent the pasta from sticking. Cook until tender. Drain thoroughly.

3. Transfer the drained pasta to the skillet, tossing quickly and gently with 2 forks. Add the cheese, season generously with pepper, and toss again. Shower with the parsley and toss once more. Transfer to warmed shallow soup bowls and serve immediately, with additional cheese and freshly ground pepper.

Yield: Serves 4 as a first course,
2 as a main course

WINE SUGGESTION: Any good all-purpose red table wine—such as a Chianti Classico from Tuscany—would be good here.

Tagliatelle with Zucchini and Fresh Parsley

Tagliatelle con Zucchini al Prezzemolo

Golden egg noodles, set off with the contrasting color of bright, spring green zucchini and parsley—this dish makes me want to don white linen slacks and a big straw hat, in anticipation of a sparkling sunny day. This is one preparation that demands your complete last-minute attention—so that the pasta is perfectly cooked and drained, then immediately tossed with the zucchini so the zucchini doesn't turn soggy on the way to the dinner table. That said, the dish is a snap, and a marvelously quick first course, one that would go well with grilled baby lamb chops or a simple roast chicken. I sampled this one spring day at a back-street trattoria in Siena, La Vecchia Taverna di Bacco, where it was served with golden, fresh pasta.

*1½ pounds (750 g) firm, fresh zucchini,
scrubbed and trimmed (do not peel)*

⅓ cup (80 ml) extra-virgin olive oil

Sea salt

Freshly ground black pepper to taste

12 ounces (375 g) fresh tagliatelle or fettuccine (page 126)

1 cup (250 ml) fresh flat-leaf parsley leaves, snipped with a scissors

*Freshly grated Italian Parmigiano-Reggiano cheese, for the table
(optional)*

1. Quarter the zucchini lengthwise, then cut into thin slices.

2. In a 12-inch (30-cm) skillet, heat the oil over moderately high heat. When the oil is hot but not smoking, add the zucchini and sauté, shaking the pan vigorously to toss the zucchini, until the slices are lightly golden, 3 to 4 minutes. The zucchini should remain firm and crisp and should not turn soggy. Season generously with salt and pepper, and keep warm.

3. Meanwhile, in a large pot, bring 6 quarts (6 l) of water to a rolling boil. Add 3 tablespoons salt and the pasta, stirring to prevent the pasta from sticking. Cook until tender. Drain thoroughly.

4. Transfer the drained pasta to the skillet with the zucchini, and, using 2 forks, thoroughly toss the pasta to blend. Add the parsley and gently toss again. Taste for seasoning, transfer to warmed shallow soup bowls, and serve immediately. Pass freshly grated cheese if desired.

■ Yield: 4 to 6 servings ■

WINE SUGGESTION: A light white, such as a Sauvignon Blanc, a Soave, or a Verdicchio.

Tagliatelle with Fresh Crabmeat

Tagliatelle al Granchio

Rich, elegant, and sophisticated, this ivory-toned dish is hardly what one thinks of as typical trattoria fare. But when you consider I sampled it in Venice, the most elegant of Italian cities, it makes sense. This is a specialty of Nereo Volpe, owner of the Antica Besseta, a tiny, old-fashioned family trattoria filled with local folks with hearty appetites. I am amazed at the simplicity of this preparation, which requires just three main ingredients: great crabmeat, great pasta, great cream! It's exceptionally rich, so I find that two ounces of pasta satisfies, particularly served as a first course or a delicate main course.

Sea salt

8 ounces (250 g) fresh tagliatelle or fettuccine (page 126)

1 cup (250 ml) heavy cream

*8 ounces (250 g) fresh lump crabmeat, drained, picked over, and flaked
into generous bite-size pieces*

*¼ cup (60 ml) fresh flat-leaf parsley leaves or ¼ cup fresh basil leaves,
snipped with a scissors*

Freshly ground black pepper to taste

1. In a large pot, bring 3 quarts (3 l) of water to a rolling boil. Add 1½ tablespoons salt and the pasta, stirring to prevent the pasta from sticking. Cook until tender but firm. Drain thoroughly.

2. Meanwhile, in a saucepan large enough to hold the pasta later on, warm the cream over low heat. Add the crabmeat and stir gently. Heat just until the crabmeat is warmed through, about 1 minute.

3. Add the drained pasta to the saucepan. With 2 forks, toss the pasta gently over low heat to coat with the sauce. Transfer to warmed shallow soup bowls, sprinkle with the parsley or basil, and serve immediately, passing the pepper mill. (Traditionally, cheese is not served with this pasta.)

■ Yield: 4 servings ■

WINE SUGGESTION: This dish calls for a soft and golden copper-colored white from the Veneto, such as a Pinot Grigio, or a good Chardonnay.

In Search of the Perfect Crab

"Fresh" prepared crabmeat—as opposed to canned or frozen—comes from hard-shell crabs that have usually been steamed whole, in the shell. The meat is then picked from the shell and packed in containers to be sold as refrigerated fresh crabmeat. The prized morsel of crab is the backfin lump meat, often sold simply as lump crabmeat. While crab is generally an expensive delicacy, it's sweet in flavor, and about as "instant" a fresh food as you're likely to find. Many supermarkets and specialty shops now offer pasteurized fresh lump crabmeat that has an amazing six-month shelf life, assuming it's kept carefully refrigerated. Do not opt for canned crabmeat, which is almost as expensive as the fresh or fresh-pasteurized version, but usually rougher in texture and blander in flavor.

Tagliatelle with Arugula and Garlic Sauce

Tagliatelle con Rughetta

One warm summer evening in August, I wandered all over the little Lake Garda village of Lazise in search of the perfect spot for dinner. It became a challenge, sort of a test to prove whether or not I could divine a good restaurant simply by examining each spot from the exterior. I narrowed the choice down to three restaurants, and finally chose one along the harbor, where our group of four trenchermen made short work of the menu, devouring the dozens of antipasto specialties laid out before us, following up with varied platters of pasta. This was my choice, and one that has remained a favorite at home. Be sure to stem the arugula, or you will end up with a coarse sauce. Note that once "cooked," the normally peppery and pungent arugula flavor softens. Ideally, the dish should be prepared with top-quality fresh homemade pasta. Although the sauce is prepared in the food processor, the garlic should be minced by hand.

4 plump fresh garlic cloves, degermed and minced

Sea salt

2 cups (1 bunch, or about 1½ ounces; 45 g) loosely packed arugula leaves, washed and dried

1 cup (250 ml) heavy cream

1 pound (500 g) fresh tagliatelle or fettuccine (page 126)

½ cup (2 ounces; 60 g) freshly grated Italian Parmigiano-Reggiano cheese

1. In a food processor, combine the minced garlic, a pinch of salt, and the arugula, and process to a rough purée. Add the cream, and process to a rough sauce. Taste for seasoning. Set aside.

2. In a large pot, bring 6 quarts (6 l) of water to a rolling boil. As the water is heating, place a large serving bowl over the pot to warm the bowl. When the water is boiling, add 3 tablespoons salt and the pasta, stirring to prevent the pasta from sticking. Cook until tender but firm. Drain thoroughly.

3. Just before serving, transfer the arugula sauce to the warmed bowl and stir in the cheese, blending thoroughly. Add the drained pasta, and toss to blend thoroughly. Transfer to warmed shallow soup bowls and serve immediately.

■ Yield: 4 to 6 servings ■

WINE SUGGESTION: With this cream-based sauce, I relish an Italian white, such as a Chardonnay, an Orvieto, or a Frascati.

Tagliatelle with Porcini Mushroom Sauce

Tagliatelle alla Boscaiola

This elegant, rich, woodsy-flavored pasta dish reminds me of a cold winter's night, a roaring fire, and an intimate group of very hungry friends or family. It's an ideal first-course pasta, followed by a simple roasted chicken and a crisp green salad.

1¼ cups (1½ ounces; 45 g) dried porcini mushroom slices

2 cups (500 ml) boiling water

½ cup (2 ounces; 60 g) pancetta or ham, cut into thin strips

1 shallot, peeled and minced

2 tablespoons extra-virgin olive oil

Sea salt

Freshly ground black pepper

1 cup (250 ml) heavy cream

Freshly grated nutmeg to taste

8 ounces (250 g) fresh tagliatelle or fettuccine (page 126)

Freshly grated Italian Parmigiano-Reggiano cheese, for the table (optional)

1. In a small bowl, combine the mushrooms and boiling water. Soak the mushrooms for at least 30 minutes, preferably for 2 hours. Using your hands, lift the mushrooms from the water, squeezing out as much water at possible. Unless they are perfectly clean, rinse the mushrooms under cold running water. If there are still pieces of soil embedded in the mushrooms, use a small knife to scrape off the soil. Pat them dry with paper towels. If the mushroom slices are unusually large, chop them coarsely. Transfer the mushrooms to a small bowl and set aside. Strain the soaking liquid—rich with porcini flavor—through several thicknesses of moistened cheesecloth. Set aside.

2. In a large skillet, combine the pancetta, shallot, oil, and salt and pepper to taste over moderate heat and cook until the shallot is golden and translucent, 3 to 4 minutes. Add the mushrooms and cook until the mushrooms become fragrant, 3 to 4 minutes more. Add the cream and nutmeg, cook for 2 minutes, then add the mushroom liquid. Cook over very low (diffused) heat until the sauce has reduced to the consistency of heavy cream, from 20 to 25 minutes, stirring regularly. Do not overcook, or the sauce will be too thick to coat the pasta.

3. Meanwhile, in a large pot, bring 6 quarts (6 l) of water to a rolling boil. Add 3 tablespoons salt and the pasta, stirring to prevent the pasta from sticking. Cook until tender. Drain thoroughly.

4. Add the drained pasta to the sauce in the skillet, and toss to blend. Remove from the heat, cover, and let rest for 1 to 2 minutes to allow the pasta to absorb the sauce. Transfer to warmed shallow soup bowls and serve immediately, passing freshly grated Parmesan cheese if desired.

■ Yield: 4 servings ■

WINE SUGGESTION: If you're in the mood for a white wine, drink a nicely oaky Chardonnay. If a red, go for a Chianti Riserva.

Lasagne with Tomato-Cream Sauce and Mozzarella

Pasta al Forno Trattoria Diva

One spring afternoon, after a long morning of interviews, we arrived at Trattoria Diva in the Tuscan village of Montepulciano in a state of nearly terminal hunger. It was almost 2 P.M., and after a fifteen-minute wait for a table to clear, we proceeded to order nearly every pasta on the menu. The slightly grumpy but always accommodating padróne scurried about, quickly mollifying us with a basket of thickly sliced Tuscan bread and a bottle of 1986 Montepulciano from the Fattoria del Cerro, a wine that's dark, deep, and serious. The daily special was, quite simply, pasta al forno, the most succulent and fresh lasagne imaginable: thin, thin sheets of fresh homemade egg pasta, interlayered with the lightest of tomato and meat sauces, and a simple topping of mozzarella. The dish made me want to rush home and dust off my pasta machine! We thought of opting for seconds, but then we saw the giant bowls of freshly picked cherries and knew it was time to surrender our forks.

After I did get home and dust off my pasta machine, I decided I best loved the version made with a fragrant tomato and cream sauce between the layers of fresh pasta. But any favorite tomato-based sauce can be used.

A few words of advice: This dish is somewhat long in the making, but not difficult and definitely worth the effort. In fifteen years of marriage, this was one dish that made me say to my husband, "You'd better appreciate all the work that went into this." He did, of course, and now there are regular requests for "fresh lasagne." The fact is, this dish just doesn't sing if it's made with thick, dried lasagne. Its charm is the delicate lightness of the pasta, layered with an equally delicate sauce. Note that the pasta will be easier to handle if it is cut into smaller rectangles.

Butter for preparing the baking dish

Grated zest (yellow peel) of 1 lemon

1 tablespoon extra-virgin olive oil

Sea salt

1 recipe Fresh Egg Pasta (page 126), rolled as thin as possible and cut into 3½- by 4½-inch (9- by 12-cm) rectangles

1 recipe Tomato and Cream Sauce (page 259)

8 ounces (250 g) fresh whole-milk mozzarella, thinly sliced

1. Preheat the oven to 350°F (175°C; gas mark 4/5).

2. Butter a 9- by 14-inch (23- by 36-cm) baking dish and sprinkle with the grated zest.

3. Precook the lasagne: In a large bowl, combine 2 quarts (2 l) of cold water and the oil. In a large pot, bring 6 quarts (6 l) of water to a rolling boil. Add 3 tablespoons salt and slide 4 to 5 rectangles of pasta into the boiling water. Cover and cook for 1 minute. Using a slotted spoon, retrieve the pasta rectangles and transfer to the bowl of cold water for about 30 seconds, just to stop the cooking. (Do not use ice water, or the ice will make holes in the pasta.) Immediately transfer the pasta squares to a clean, damp cloth. Continue until all the pasta is cooked.

4. Spoon about ½ cup (125 ml) of the sauce over the bottom of the baking dish. Cover with 4 slices of precooked pasta. Continue layering the lasagne and sauce in this manner until all of the sauce and the pasta have been used, ending with a layer of pasta. Cover the top with the mozzarella slices.

5. Place the baking dish in the center of the oven and bake until the cheese is melted and the dish is fragrant and bubbling, about 20 minutes. Remove from the oven and let sit for 10 minutes before cutting. Since this dish tends to release a fair amount of liquid, it is best to serve with a slotted spoon.

▪ Y i e l d : 6 t o 8 s e r v i n g s ▪

WINE SUGGESTION: With the tomato and cream sauce, a pleasing white, such as a Soave Classico, is nice.

Fettuccine with Butter and Parmesan

Fettuccine al Burro e Parmigiano

"Everything should be made as simple as possible, but not simpler."

ALBERT EINSTEIN

When I'm exhausted and hungry as a wolf, yet the pantry and refrigerator are nearly bare, this is the pasta dish that comes to mind. Soothing, filling, quick, and simple, this traditional trattoria dish is a winning late-night supper for one or for a crowd. It's best prepared with fresh pasta, but in a pinch, dried fettuccine or tagliatelle make respectable substitutes.

Sea salt

1 pound (500 g) fresh tagliatelle or fettuccine (page 126) (or substitute dried Italian tagliatelle or fettuccine)

8 tablespoons (4 ounces; 125 g) unsalted butter, at room temperature

2 cups (8 ounces; 250 g) freshly grated Italian Parmigiano-Reggiano cheese

Freshly ground black pepper to taste

1. In a large pot, bring 6 quarts (6 l) of water to a rolling boil. As the water is heating, place a large serving bowl over the pot to warm the bowl. When the water is boiling, add 3 tablespoons salt and the pasta, stirring to prevent the pasta from sticking. Cook fresh pasta until tender, dried pasta until tender but firm to the bite. Drain thoroughly.

2. Transfer the drained pasta to the warmed bowl, add the butter, and toss thoroughly. Add about half the cheese and toss again. Transfer the pasta to warmed shallow soup bowls and sprinkle with the rest of the cheese. Serve immediately, passing the pepper mill.

▪ Yield: 4 to 6 servings ▪

WINE SUGGESTION: I enjoy this dish with a full-bodied white, such as a Vernaccia di San Gimignano from Teruzzi & Puthod, or with a quality Sauvignon Blanc.

Citrus-Infused Baked Tagliatelle

Pasta al Forno

Nine times out of ten, baked pasta is a big disappointment. Why? It's inevitably too dry, with burnt or overcooked sauce and dried-out pasta on the top, and a layer of undersauced and often undercooked pasta below. To ensure a deliciously moist baked pasta—a dish that can't be beat when it's fragrant and warm—do what you should always do in the kitchen: Pay attention to details! There's no secret to success here. Just be sure to take care to make a perfect white sauce, be certain that your sauce is not too thick, and don't overdrain the pasta once it's cooked. The best baked pasta—pasta al forno—I've ever sampled came from the kitchen of Pierro Giannacci at Le Quattro Stagioni ("the four seasons") in Florence.

I spent a morning with him and his staff in his tiny, super-busy kitchen, where varied pastas were boiling away, a chef was frying a mix of zucchini and artichokes, and coming from the oven was the warming fragrance of giant veal chops, roasting away. One of the day's specials was oven-baked pasta. Chef Giannacci's twist is to use fresh tagliolini (thin tagliatelle), layered in a large rectangular pan with a flavorful white sauce and a lovely meat sauce, all topped with freshly grated cheese. The finer egg pasta—either fresh or dried—takes better to baking and dries out less than traditional lasagne or penne.

The end result: a lovely, layered "sandwich" of pasta, moist with an herb-rich white sauce, rich with meat sauce, all accented by the wholesomeness of the Parmesan cheese.

Butter for preparing the baking dish

Grated zest (orange peel) of 1 orange

Sea salt

1 pound (500 g) fresh tagliatelle or fettuccine (page 126) (or substitute dried Italian tagliatelle or fettuccine)

3 cups (750 ml) warm Meat Sauce (page 268)

2 cups (500 ml) warm White Sauce (page 270)

2 cups (8 ounces; 250 g) freshly grated Italian Parmigiano-Reggiano cheese

1. Preheat the oven to 350°F (175°C; gas mark 4/5).

2. Butter a 9- by 14-inch (23- by 36-cm) baking dish and sprinkle with the grated zest. Set aside.

3. In a large pot bring 6 quarts (6 l) of water to a rolling boil. Add 3 tablespoons salt and the pasta, stirring to prevent the pasta from sticking. Cook until tender. Drain, leaving a few drops of water clinging to the pasta.

4. Add about one-third of the drained pasta to the baking dish in an even layer. Cover evenly with half of the meat sauce. Add another one-third of the pasta, and cover with all of the white sauce and about half of the cheese. Add the rest of the pasta and cover with the rest of the meat sauce. Sprinkle with the rest of the cheese.

5. Place the baking dish in the center of the oven and bake until the cheese is melted and the dish is fragrant and bubbling, 15 to 20 minutes. Remove from the oven. Serve immediately, spooning out the pasta with a very large serving spoon.

■ Yield: 6 to 8 servings ■

WINE SUGGESTION: A crisp white, such as Pinot Grigio.

For Citrus-Infused Baked Pasta or Rice

I crave the deep, lingering flavor of citrus zest, whether it be lemon, lime, or orange. One day while paging through an Italian cooking magazine, I found this tip on the letters-to-the-editor page: For more flavorful baked pasta dishes, rub your baking pan with butter, then sprinkle with grated lemon or orange zest. The final dish will be infused with the welcoming citrus flavor, giving a special twist to new dishes, as well as old favorites. Now, whenever I bake pasta or rice, lemon or orange comes first!

HOMAGE TO ITALIAN RICE

.

Creamy Italian Arborio rice deserves a menu of its own—a soup, a risotto, a dessert. Try a mature Gattinara or a Chianti Riserva with this menu.

MILANESE VEGETABLE SOUP
OSSO BUCO (BRAISED VEAL SHANKS WITH LEMON AND PARSLEY GARNISH)
SAFFRON RISOTTO
RISOTTO ICE CREAM

RICE AND POLENTA

Lemon Risotto

Risotto al Limone

Creamy, pale yellow risotto is appealing cool weather fare, and the addition of flecks of fresh green herbs and lemon make for a dish that's very spring-like and refreshing. This version was inspired by Roberto Fontana at Milan's Casa Fontana, where twenty-three different risottos can be found on the menu at any given time. Remember three important rules for making perfect risotto: Do not add too much liquid at once; do not add more liquid until the previous addition has been absorbed; and stir, stir, stir.

About 5 cups (1.25 l) chicken stock, preferably homemade (page 272)

Sprig of fresh mint

Sprig of fresh rosemary

Sprig of fresh sage

Grated zest (yellow peel) of 1 lemon

4 tablespoons (2 ounces; 60 g) unsalted butter

1 tablespoon extra-virgin olive oil

2 shallots, minced

Sea salt to taste

1½ cups (270 g) Italian Arborio rice

3 tablespoons freshly squeezed lemon juice

½ cup (2 ounces; 60 g) freshly grated Italian Parmigiano-Reggiano cheese, plus additional for the table

1. In a large saucepan, heat the stock and keep it simmering, at barely a whisper, while you prepare the risotto.

2. Stem the fresh herbs. Combine the leaves with the lemon zest and, with a large chef's knife, chop finely. Set aside.

3. In a large heavy-bottomed saucepan, combine 2 tablespoons of the butter, the oil, shallots, and salt over moderate heat. Cook, stirring, until the shallots are soft and translucent, about 3 minutes. (Do not let the shallots brown.) Add the rice, and stir until the rice is well coated with the fats, glistening and semitranslucent, 1 to 2 minutes. (This step is important for good risotto: The heat and fat will help separate the grains of rice, ensuring a creamy consistency in the end.)

4. When the rice becomes shiny and partly translucent, add a ladleful of the stock. Cook, stirring constantly, until the rice has absorbed most of the stock, 1 to 2 minutes. Add another ladleful of the simmering stock, and stir regularly until all of the broth is absorbed. Adjust the heat as necessary to maintain a gentle simmer. The rice should cook slowly and should always be covered with a veil of stock. Continue adding ladlefuls of stock, stirring frequently and tasting regularly, until the rice is almost tender but firm to the bite, about 17 minutes total. The risotto should have a creamy, porridge-like consistency.

5. Remove the saucepan from the heat and stir in the remaining 2 tablespoons butter, the lemon zest and herbs, lemon juice, and the Parmesan. Cover and let stand off the heat for 2 minutes, to allow the flavors to blend. Taste for seasoning. Transfer to warmed shallow soup bowls, and serve immediately, passing additional cheese. Risotto waits for no one.

▪ Y i e l d : 4 t o 6 s e r v i n g s ▪

WINE SUGGESTION: Serve this with a pale, golden Vernaccia from the hilltop village of San Gimignano in Tuscany.

Risotto with Tomatoes and Parmesan

Risotto alla Cardinale

Festive, bright, and restorative, this is a perfect main-dish rice meal, named after the red color of the cardinal's robes. The recipe was shared with me by Chef Walter Tripodi of the restored monastery-inn La Frateria di Padre Eligio, in Cetona, in Tuscany.

About 4 cups (1 l) vegetable broth or chicken stock, preferably homemade (pages 274 and 272)

2 cups (500 ml) Tomato Sauce (page 256)

4 tablespoons (2 ounces; 60 g) unsalted butter

2 tablespoons extra-virgin olive oil

1 shallot, minced

4 bay leaves, preferably fresh

Fine sea salt to taste

2 cups (360 g) Italian Arborio rice

½ cup (2 ounces; 60 g) freshly grated Parmigiano-Reggiano cheese, plus additional for the table

1. In a large saucepan, combine the broth or stock and tomato sauce and keep the liquid simmering, at barely a whisper, while you prepare the risotto.

2. In a large heavy-bottomed saucepan, combine 2 tablespoons of the butter, the oil, shallot, bay leaves, and salt over moderate heat. Cook, stirring, until the shallot is soft and translucent, about 3 minutes. (Do not let the shallot brown.) Add the rice, and stir until the rice is well coated with the fats, glistening and semitranslucent, 1 to 2 minutes. (This step is important for good risotto: The heat and fat will help separate the grains of rice, ensuring a creamy consistency in the end.)

3. When the rice becomes shiny and somewhat translucent, add a ladleful of the simmering liquid. (The pan will sizzle as you add the liquid.) Stirring constantly, cook until the rice has absorbed most of the liquid, 1 to 2 minutes. Add another ladleful of the liquid, and stir constantly until all of the liquid is absorbed. Adjust the heat as necessary to maintain a gentle simmer. The rice should cook slowly and should always be covered with a veil of liquid. Continue adding ladlefuls of warm stock, stirring constantly and tasting regularly, until the rice is almost tender but firm to the bite, about 17 minutes total. The risotto should have a creamy, porridge-like consistency.

4. Remove the saucepan from the heat and stir in the remaining 2 tablespoons butter and the cheese. Cover and let stand off the heat for 2 minutes, to allow the flavors to blend and the rice to finish cooking. Taste for seasoning. Remove and discard the bay leaves. Transfer to warmed shallow soup bowls, and serve immediately, passing additional cheese.

▪ Yield: 6 to 8 servings ▪

WINE SUGGESTION: I like a mature red with this dish, such as a Gattinara, Barbaresco, or Chianti Riserva.

Eating Risotto

If you thought there was ritual involved in making risotto, just wait until you eat it! Italian etiquette even covers proper risotto consumption. The cooked risotto should always be mounded, steaming hot, in the center of warmed individual shallow bowls. As you eat the risotto, use a fork to push the grains of cooked rice out slightly toward the edge of the bowl, eating from the thinned-out ring of rice. Continue spreading from the center and eating around the edges in a circle. The mound in the middle will keep the risotto hot as you savor the just-right risotto around the rim. Some Milanese risotto-eating purists even advocate spoons instead of forks.

Saffron Risotto

Risotto alla Milanese

T he traditional accompaniment to the delicately flavored osso buco, braised veal shanks, saffron risotto is one of the world's most elegantly beautiful dishes, seductive with its creamy texture, joyful with its hopeful golden hues. For simple meals, I often serve it as a main dish, the better to savor it all on its own. Here, go easy on the saffron: The color should be golden, the flavor faint with saffron. And be sure to use the best-quality saffron you can find.

About 5 cups (1.25 l) chicken stock, preferably homemade (page 272)

A pinch (about ¼ teaspoon) saffron threads

4 tablespoons (2 ounces; 60 g) unsalted butter

1 tablespoon extra-virgin olive oil

2 shallots, minced

Sea salt to taste

1 ½ cups (270 g) Italian Arborio rice

½ cup (2 ounces; 60 g) freshly grated Italian Parmigiano-Reggiano cheese, plus additional for the table

1. In a large saucepan, heat the stock and keep it simmering, at barely a whisper, while you prepare the risotto.

2. In a measuring cup with a spout, combine ½ cup (125 ml) of the hot stock and the saffron. Stir to infuse and set aside.

3. In a large heavy-bottomed saucepan, combine 2 tablespoons of the butter, the oil, shallots, and salt over moderate heat. Cook, stirring, until the shallots are soft and translucent, about 3 minutes. (Do not let the shallots brown.) Add the rice, and stir until the rice is well coated with the fats, glistening and semitranslucent, 1 to 2 minutes. (This step is important for good risotto: The heat and fat will help separate the grains of rice, ensuring a creamy consistency in the end.)

4. When the rice becomes shiny and partly translucent, add a ladleful of the stock. Cook, stirring constantly, until the rice has absorbed most of the stock, 1 to 2 minutes. Add another ladleful of the simmering stock, and stir regularly until all of the broth is absorbed. Adjust the heat as necessary to maintain a gentle simmer. The rice should cook slowly and should always be covered with a veil of stock. Repeat the procedure, stirring frequently and tasting regularly, until the rice is almost tender but firm to the bite, about 17 minutes total. Add the saffron stock at the end. The risotto should have a creamy, porridge-like consistency.

5. Remove the saucepan from the heat and stir in the remaining 2 tablespoons butter and the cheese. Cover and let stand off the heat for 2 minutes, to allow the flavors to blend. Taste for seasoning. Transfer to warmed shallow soup bowls, and serve immediately, passing additional cheese.

▪ Yield: 4 to 6 servings ▪

WINE SUGGESTION: I serve this, and the osso buco, with a fine Gattinara.

On Leftover Risotto

Should you find yourself with leftover risotto, prepare risotto al salto, a dish historically served in country inns. When travelers in a hurry were unwilling to wait the twenty-five to thirty minutes it took to create a risotto from scratch, innkeepers tossed already prepared risotto into a buttered or oiled frying pan and made it "jump," or *salto,* offering diners a quick, hot meal.

Orange, Sage, and Mushroom Risotto

Risotto Sforzesco

A combination of fresh and woodsy flavors—the orange zest and the wild mushrooms—makes for a dish that's not just zesty and perky from the first bite, but has a depth of flavor that wears well on the palate. Be sure to use fresh sage, and taste it first to make sure it does not have a strong, medicinal flavor. If you can't secure fresh sage, substitute parsley. This recipe was shared with me by Roberto Fontana, owner and chef of Milan's temple to risotto, Casa Fontana.

1 cup (1½ ounces; 45 g) dried porcini mushroom slices

5 cups (1.25 ml) boiling water

Grated zest (orange peel) of 1 orange

¼ cup (60 ml) fresh sage leaves (or substitute fresh flat-leaf parsley leaves)

4 tablespoons (2 ounces; 60 g) unsalted butter

1 tablespoon extra-virgin olive oil

2 shallots, minced

Sea salt to taste

1½ cups (270 g) Italian Arborio rice

3 tablespoons freshly squeezed orange juice

½ cup (2 ounces; 60 g) freshly grated Italian Parmigiano-Reggiano cheese, plus additional for the table

1. In a large bowl, combine the mushrooms and the boiling water. Soak the mushrooms for at least 30 minutes, preferably for 2 hours. Using your hands, lift the mushrooms from the water, squeezing out as much water at possible. Unless they are perfectly clean, rinse the mushrooms under cold running water. If there are still pieces of soil embedded in the mushrooms, use a small knife to scrape off the soil. Pat them

dry with paper towels. If the mushroom slices are unusually large, chop them coarsely. Transfer the mushrooms to a small bowl and set aside. Strain the soaking liquid—rich with porcini flavor—through several thicknesses of moistened cheesecloth.

2. In a large saucepan, heat the mushroom liquid and keep it simmering, at barely a whisper, while you prepare the risotto.

3. Combine the orange zest and sage leaves and, with a large chef's knife, chop fine. Set aside.

4. In a large heavy-bottomed saucepan, combine 2 tablespoons of the butter, the oil, shallots, and salt over moderate heat. Cook, stirring, until the shallots are soft and translucent, about 3 minutes. (Do not let the shallots brown.) Add the rice, and stir until the rice is well coated with the fats, glistening and semitranslucent, 1 to 2 minutes. (This step is important for good risotto: The heat and fat will help separate the grains of rice, ensuring a creamy consistency in the end.)

5. When the rice becomes shiny and partly translucent, add a ladleful of the mushroom liquid. Cook, stirring constantly, until the rice has absorbed most of the liquid, 1 to 2 minutes. Add another ladleful of the simmering liquid, and stir regularly until all of the liquid is absorbed. Adjust the heat as necessary to maintain a gentle simmer. The rice should cook slowly and should always be covered with a veil of liquid. Repeat this procedure, stirring frequently and tasting regularly, until the rice is almost tender but firm to the bite, about 17 minutes total. About 2 minutes before the rice is cooked, add the mushrooms along with a ladleful of the liquid. The risotto should have a creamy, porridge-like consistency.

6. Remove the saucepan from the heat and stir in the remaining 2 tablespoons butter, the orange zest and sage, the orange juice, and cheese. Cover and let stand off the heat for 2 minutes, to allow the flavors to blend. Taste for seasoning. Transfer to warmed shallow soup bowls, and serve immediately, passing additional cheese.

■ Yield: 4 to 6 servings ■

WINE SUGGESTION: Serve this with a rich, dark Barolo, from the Piedmont.

Baked Risotto with Asparagus, Spinach, and Parmesan

Risotto Verde

I love preparing (as well as eating) this spring-inspired specialty of Ugo Salis, cook and owner of the no-frills Trattoria da Graziella, situated in the hills above Florence, in Maiano. We sampled this verdant, fragrant, and soul-satisfying baked rice as a side dish, part of a huge, multi-course Sunday lunch one warm August afternoon. At home, I usually make it a quick main dish, served with a mixed salad alongside. Note that here both the spinach and the asparagus are added raw. They cook up nicely along with the rice.

1 tablespoon extra-virgin olive oil

1 small onion, minced

Sea salt to taste

1 cup (180 g) Italian Arborio rice

2 cups (500 ml) vegetable broth or chicken stock, preferably homemade (pages 274 and 272)

4 cups (about 3 ounces; 90 g) loosely packed fresh spinach leaves, rinsed, dried, and finely chopped

10 thin spears fresh asparagus, rinsed, trimmed, and cut into thin diagonal slices

¼ teaspoon freshly grated nutmeg

½ cup (2 ounces; 60 g) freshly grated Italian Parmigiano-Reggiano cheese

1. Preheat the oven to 400°F (200°C; gas mark 6/7).

2. In a 1½-quart (1½-liter) saucepan, combine the oil, onion, and salt over moderate heat. Stir to coat with the oil and cook until the onion is soft and translucent, 3 to 4 minutes. Add the rice, stirring to coat with the oil. Add the chicken stock, spinach, asparagus, nutmeg, and salt, and bring just to a simmer over moderate heat. Stir in half the cheese. Transfer to a 1-quart (1-l) soufflé dish, and smooth out the top with the back of a spoon. Sprinkle with the remaining Parmesan. Cover the soufflé dish.

3. Place the soufflé dish in the center of the oven. Bake until the rice is cooked through and has absorbed most of the liquid, 35 to 40 minutes. The baked rice should be moist but not soupy. Serve immediately, as a vegetable side dish or a main dish.

■ Yield: 4 to 6 servings ■

WINE SUGGESTION: With this, sample a crisp Italian white: a Pinot Grigio, Galestro, or Orvieto Secco.

The Secrets of Arborio

The highly glutinous Arborio rice—the quintessential risotto grain—is grown in Italy's Po River Valley, which runs through the Piedmont, Lombardy, Emilia, and Veneto regions. This medium-grain rice produces a risotto that is creamy on the outside and al dente—firm to the bite—on the inside.

During cooking, proper risotto rice releases a starch that helps the grains to cling together without becoming sticky or glutinous, resulting in the characteristic creamy texture. Italian Arborio also has the ability to absorb great quantities of liquid without losing its shape and turning to mush. Even the best American short-grained rice cannot make the same claim.

Baked Risotto with
Tomato Sauce and Pecorino

Risotto Rosso

O ne evening I found myself with some leftover tomato sauce and a small chunk of top-quality Italian sheep's milk cheese (pecorino). I immediately remembered Ugo Salis—owner of Trattoria da Graziella in the hills of Florence—talking about his baked rice with tomatoes and pecorino, a dish that pays homage to his native Sicily. So, here it is, in all its simple glory, a homey baked rice that goes well with roast chicken, or stands on its own with a mixed salad of greens and vegetables.

1 tablespoon extra-virgin olive oil

1 small onion, minced

Sea salt to taste

1 cup (180 g) Italian Arborio rice

1½ cups (375 ml) vegetable broth or chicken stock, preferably homemade (pages 274 and 272)

½ cup (125 ml) Tomato Sauce (page 256)

¾ cup (3 ounces; 90 g) freshly grated Italian sheep's milk cheese (pecorino), such as Romano

1. Preheat the oven to 400°F (200°C; gas mark 6/7).

2. In a 1-quart (1-l) heatproof baking dish, combine the oil, onion, and salt over moderate heat. Stir to coat the onion with the oil and cook until the onion is soft and translucent, 3 to 4 minutes. Add the rice, stirring to coat with the oil, and cook for 1 minute. Add the chicken stock and tomato sauce, and bring just to a simmer over moderate heat. Stir in half the cheese and smooth out the top with the back of a spoon. Sprinkle with the remaining cheese. Cover the baking dish.

3. Place the baking dish in the center of the oven. Bake until the rice is cooked through and has absorbed most of the liquid, 30 to 35 minutes. The baked rice should be moist but not soupy. Serve immediately, as a vegetable side dish or a main dish.

▪ Yield: 4 to 6 servings ▪

WINE SUGGESTION: With this, try a crisp Italian white, such as a Pinot Grigio, Galestro, or Orvieto Secco.

Tips for Better Risotto

▪ Rice is highly porous, and quickly absorbs odors, many of them unwanted. Store rice with a few fresh and fragrant bay leaves in an airtight container in a cool location. For best results, use rice within one year.

▪ Never rinse rice for risotto: You'll be washing away the highly desirable starch.

▪ Allow the raw rice to cook in fat (butter or oil) for at least one minute, so the grains are well coated and nicely flavored, and do not stick.

▪ Acid—wine or tomatoes, for example—inhibits rice's ability to take in water. You may need more liquid, and more time, when cooking with acidic ingredients.

Risotto with Bay Leaves and Parmesan

Risotto all'Alloro

I am a bona fide bay leaf lover. I have a huge pot of it growing at all times and use the fragrant, elegant leaf whenever and wherever it seems to make sense. I sampled this risotto at La Frateria di Padre Eligio in Cetona, in Tuscany, where Chef Walter Tripodi oversees the small restaurant on the grounds of the monastery. Rice easily absorbs any odors with which it comes in contact, so I always stuff a few fresh bay leaves into a new bag of rice and seal it carefully, then add another batch of fresh leaves when I prepare this ultimately simple and subtle dish. If you cannot secure fresh bay leaves, use the freshest of dried leaves you can find in your market. This risotto makes a nice first course when served with roast chicken or Parmesan-coated lamb chops (page 238).

About 5 cups (1.25 l) vegetable broth or chicken stock, preferably homemade (pages 274 and 272)

4 tablespoons (2 ounces; 60 g) unsalted butter

2 tablespoons extra-virgin olive oil

1 shallot, minced

4 bay leaves, preferably fresh

Sea salt to taste

2 cups (360 g) Italian Arborio rice

½ cup (125 ml) flowery white wine, such as a Vernaccia di San Gimignano

½ cup (2 ounces; 60 g) freshly grated Italian Parmigiano-Reggiano cheese, plus additional for the table

1. In a large saucepan, heat the broth or stock and keep it simmering, at barely a whisper, while you prepare the risotto.

2. In a large heavy-bottomed saucepan, combine 2 tablespoons of the butter, the oil, shallot, bay leaves, and salt over moderate heat. Cook, stirring, until the shallot is soft and translucent, about 3 minutes. (Do not let the shallot brown.) Add the rice, and stir until the rice is well coated with the fats, glistening and semitranslucent, 1 to 2 minutes. (This step is important for good risotto: The heat and fat will help separate the grains of rice, ensuring a creamy consistency in the end.)

3. When the rice becomes shiny and partly translucent, add the wine. Cook, stirring constantly, until the rice has absorbed most of the wine, 1 to 2 minutes. Add a ladleful of the simmering stock, and stir regularly until all of the liquid is absorbed. Adjust the heat as necessary to maintain a gentle simmer. The rice should cook slowly and should always be covered with a veil of stock. Continue adding ladlefuls of warm stock, stirring frequently and tasting regularly, until the rice is almost tender but firm to the bite, about 17 minutes total. The risotto should have a creamy, porridge-like consistency.

4. Remove the saucepan from the heat and stir in the remaining 2 tablespoons butter and the Parmesan. Cover and let stand off the heat for 2 minutes, to allow the flavors to blend. Taste for seasoning. Remove and discard the bay leaves. Transfer to warmed shallow soup bowls, and serve immediately, passing additional cheese.

■ Yield: 6 to 8 servings ■

WINE SUGGESTION: A pale golden Vernaccia from the hilltop village of San Gimignano in Tuscany.

Polenta

Along with pasta and rice, fluffy, golden cooked cornmeal, or polenta, is Italy's favorite starch. I love to serve it with roast meats and poultry, and never mind having some leftover polenta waiting in the refrigerator, ready to prepare grilled snacks. Here I offer two methods: The first, the traditional method, does require a bit of patience and attention but results in a superior product. The second method, cooking in a double boiler, results in a polenta that is quite delicious and a reasonable compromise.

1 quart (1 l) water

1½ teaspoons fine sea salt

1 cup (150 g) coarse-grained Italian yellow cornmeal

TRADITIONAL METHOD:

1. In a large heavy saucepan, bring the 1 quart (1 l) water to a boil over high heat. Add the salt and stir vigorously to create a vortex in the center of the water. Very slowly add the cornmeal in a thin, steady stream, stirring constantly with a whisk as you add the cornmeal, making sure that the water continues to boil. (Should any lumps form, press them against the side of the pot and they will disappear.)

2. Once all the cornmeal has been added, reduce the heat to low and stir constantly with a long-handled wooden spoon, bringing the mixture up from the bottom and scraping the sides of the pot. Continue to stir until the cornmeal forms a mass that pulls cleanly away from the sides of the pot, 40 to 45 minutes. Remove from the heat.

Double-Boiler Method:

1. Fill the bottom half of a large double boiler with water, adding as much water as will be needed to touch the bottom of the top half of the boiler. Bring to a simmer over moderate heat.

2. Pour the 1 quart (1 l) of water into the top half of the double boiler. On another burner, bring this water to a boil over high heat. Add the salt and stir vigorously to create a vortex in the center of the water. Very slowly add the cornmeal in a thin, steady stream, stirring constantly with a whisk, making sure that the water continues to boil.

3. Once all the cornmeal has been added, reduce the heat to low and stir constantly with a long-handled wooden spoon, bringing the mixture up from the bottom and scraping the sides of the pot, for 2 minutes.

4. Cover the top half of the double boiler and set it over the bottom half of the double boiler. Every 10 minutes, uncover and stir for 1 full minute, cooking until the cornmeal forms a mass that pulls cleanly away from the sides of the pan, about 1 to 1½ hours. (Between stirring times, keep the pot covered.)

Serving Methods:

■ To serve the polenta warm, in an attractive mound, rinse the inside of a 1½-quart (1½-l) stainless steel bowl with water. Pour the hot polenta into the bowl and let set for 10 minutes. Turn the bowl out onto a large round platter, and serve.

■ To prepare polenta ahead of time, for cutting into squares and broiling or grilling later on, line a 9- by 5-inch (23- by 13-cm) baking pan with plastic wrap. Pour the hot polenta into the pan and smooth out the top with a spatula. Let cool. Cover and refrigerate for up to 3 days.

To serve, remove the chilled polenta from the refrigerator about an hour before serving. Preheat a broiler or grill. Invert the polenta onto a cutting board and remove the plastic wrap. Cut the polenta into 3- to 4-inch (8- to 10-cm) rectangles or diamonds. To broil, place on a baking sheet and broil until lightly browned on each side, about 3 minutes per side. Alternatively, grill on a gas, electric, or charcoal grill for about 2 minutes per side.

■ Yield: 6 to 8 servings ■

Pilaf with Tomato Sauce, Porcini Mushrooms, and Parmesan

Riso con Funghi Porcini

I think it would actually be a toss-up as to which I could consume every day, rice or pasta. As much as I love risotto, I sometimes revel in the leisure of letting a dish cook away without constant attention, and it's on days like that that I turn to this special rice dish taught to me by Chef Pierro Giannacci at Le Quattro Stagioni in Florence. Here Italian Arborio rice is cooked like a pilaf, in that it is browned lightly in fats, then cooked, covered, atop the stove, with a blend of flavorful liquids. I love the depth of flavors in this dish, and the way the intensity of the dried mushrooms manages to penetrate the rice.

⅓ cup (½ ounce; 15 g) dried porcini mushroom slices

2 cups (500 ml) boiling water

1 tablespoon extra-virgin olive oil

3 tablespoons (1½ ounces; 45 g) unsalted butter

1 shallot, minced

Sea salt to taste

1½ cups (270 g) Italian Arborio rice

3 bay leaves, preferably fresh

1 cup (250 ml) chicken stock, preferably homemade (page 272)

2 cups (500 ml) Tomato Sauce (page 256)

¼ cup (60 ml) fresh flat-leaf parsley leaves, snipped with a scissors

¼ cup (60 ml) fresh basil leaves, snipped with a scissors

1. In a large bowl, combine the mushrooms and the boiling water. Soak the mushrooms for at least 30 minutes, preferably for 2 hours. Using your hands, lift the mushrooms from the water, squeezing out as much water at possible. Unless they are perfectly clean, rinse the mushrooms under cold running water. If there are still pieces of soil embedded in the mushrooms, use a small knife to scrape off the soil. Pat them dry with paper towels. If the mushroom slices are unusually large, chop them coarsely. Transfer the mushrooms to a small bowl and set aside. Strain the soaking liquid—rich with porcini flavor—through several thicknesses of moistened cheesecloth. Set aside.

2. In a large heavy-bottomed saucepan, combine the oil, 1 tablespoon of the butter, the shallot, and salt over moderate heat. Cook, stirring, until the shallot is soft and translucent, about 3 minutes. (Do not let the shallot brown.) Add the rice, and stir until the rice is well coated with the fats, glistening and semitranslucent, 1 to 2 minutes.

3. Add the bay leaves, the mushrooms and reserved mushroom liquid, the chicken stock, and tomato sauce, and bring just to a boil over high heat. Reduce the heat to a gentle simmer, cover, and cook until the rice is almost tender but firm to the bite, about 15 minutes total. The pilaf should have a creamy, porridge-like consistency, and the liquid should not be completely absorbed. Taste for seasoning. Remove the bay leaves.

4. Remove the pan from the heat, and stir in the remaining 2 tablespoons butter, the parsley, and basil. Transfer to warmed shallow soup bowls and serve immediately. (Cheese is not traditionally served with this dish.)

■ Yield: 4 to 6 servings ■

WINE SUGGESTION: With the heartiness of the mushrooms and the tomatoes, I enjoy a big wine, such as a Gattinara, from the Piedmont.

Bay Leaves: Grow Your Own!

I often feel as though I am waging a one-woman campaign to make fresh bay leaves fashionable. Why would anyone ever think that grayish-brown, dead, dried bay leaves could perk up a sauce? Particularly when fresh bay leaves grow so easily indoors, in window boxes, in pots on the balcony, or outdoors, in the earth.

With their distinctively fragrant, almost citrus-like aroma, fresh bay leaves add a singularly herbal, intense flavor to sauces and stews. The essence of the bay lies in its fragrant oils: Just try rubbing a fresh bay leaf between your fingers, then inhale deeply, and you'll become an instant convert.

Since it is the oil that actually infuses the sauce, the bay leaf instantly flavors quick sauces that cook in a matter of minutes. (Once the bay has done its job, the leaf should be removed: You don't want to feed yourself, or your guests, a mouthful of leaves.)

Since antiquity the evergreen shrub sweet bay—known as *laurus nobilis*—has been used to crown the heads of heroes. English myth suggests that bay leaves ward off disease, witchcraft, and lightning. In the Mediterranean region, bay leaf trees—often towering, bushy shrubs—grow to forty feet high. In Italy, bay trees often form a fragrant and protective hedge around country properties.

When grown in pots, the bay leaf plant requires minimal care: The plants are naturally pruned and trimmed by regular harvesting for the kitchen, and demand only a rich, well-drained soil and moderate sunlight. Bay trees will freeze in cold weather, but can be brought indoors during the colder months. While most recipes are miserly with bay, suggesting a single leaf, I often prepare an aromatic herb bundle consisting of a dozen or more bay leaves and thick bunches of thyme.

BREADS AND

PIZZAS

Basic Bread and Pizza Dough

Pasta per Pane e Pizza

This is my basic bread and pizza dough recipe, one I've developed over time, finding that it makes an incomparably moist, flavorful, and dependable dough. The fact that there is a very small amount of yeast, combined with the long, cool rise, makes for pizzas and breads that are not overly yeasty. I almost always have some in my refrigerator, for making last-minute pizza, focaccia, rolls, or bread. For best results, be sure to check the expiration date on your package of yeast and make sure it's still viable. The dough may be prepared with unbleached all-purpose flour, but bread flour will ensure a finer flavor and texture. (If you are in a hurry, the dough can be allowed to rise at room temperature, but the result will be less moist and flavorful.)

1 teaspoon active dry yeast

1 teaspoon sugar

1⅓ cups (330 ml) lukewarm water

2 tablespoons extra-virgin olive oil

1 teaspoon fine sea salt

About 3¾ cups (1 pound; 500 g) bread flour

1. In a large bowl, combine the yeast, sugar, and water, and stir to blend. Let stand until foamy, about 5 minutes. Stir in the oil and salt.

2. Add the flour, a little at a time, stirring until most of the flour has been absorbed and the dough forms a ball. Transfer the dough to a floured work surface and knead until soft and satiny but still firm, 4 to 5 minutes, adding additional flour as needed to keep the dough from sticking.

3. Transfer the dough to a bowl, cover tightly with plastic wrap, and place in the refrigerator. Let the dough rise in the refrigerator until doubled or tripled in bulk, 8 to 12 hours. (The dough can be kept for 2 to 3 days in the refrigerator. Simply punch down the dough as it doubles or triples.)

Proceed with the individual recipes for rolls and pizza.

■ Y i e l d :   E n o u g h   d o u g h   f o r   1 5 ■
r o l l s   o r   4   s m a l l   p i z z a s

All-Purpose Versus Bread Flour

While all-purpose flour is suitable for all types of baking, bread flour—with a higher level of protein—is preferred for breads. When the protein in flour is mixed with liquid, it forms gluten. And it is the gluten in flour that gives dough its resiliency and elasticity. Thus the higher the protein, the greater the gluten, and the better the bread.

Tips for Better Bread, a Crustier Loaf

■ Use a minimal amount of yeast: Too much yeast makes for a puffy, bloated loaf.

■ Let the bread rise slowly, two to three times. Each rise will give the bread additional character.

■ Begin baking in a very hot—500°F (260°C; gas mark 9)—oven to create a thick, dense, chewy crust.

■ Use a baking stone: It helps to create a dense and deeply colored loaf.

■ Spray the bottom and sides of the oven with water three to four times during the first six minutes of baking. The steam created by the spray will help give the dough a boost during rising, will improve the crust, and will give the loaf a lovely, finished sheen.

■ Cool the bread on a rack. Circulating air will cool it down more quickly and help maintain a crisp bottom crust.

■ Do not slice the bread until it has cooled for at least one hour. Slicing warm bread makes for a doughy loaf.

Rustic Whole Wheat Bread

Pane Integrale

"The vintage has begun here at San Gimignano. . . . On the roads the heavy farm carts drawn by two oxen are already lumbering from the vineyards to the fattoria, each fully loaded with tubs. It is a festal time. . . . The vineyards, till now so silent and empty, are echoing with laughter and sprinkled with happy people, men and women, boys and girls, and children too, from neighboring poderi, have come to help the gathering, not for pay or wages but in expectations of similar assistance in their turn. The contadino provides these helpers with bread, fruit, and thin wine through the long day."

EDWARD HUTTON,
SIENA AND SOUTHERN TUSCANY, *1955*

This recipe makes a lovely, moist, and springy whole wheat bread, much like those I sampled all over Italy. Yes, the slow rise overnight in the refrigerator may seem long, but in the end, you'll be rewarded with a healthful, wheaty loaf. The use of a very small amount of yeast makes for a bread with the character of wheat, not yeast.

(continued)

1 teaspoon active dry yeast

1 teaspoon sugar

1⅓ cups (330 ml) lukewarm water

1 teaspoon fine sea salt

1½ cups (225 g) whole wheat flour

About 1½ cups (225 g) bread flour

1. In a large bowl, combine the yeast, sugar, and water, and stir to blend. Let stand until foamy, about 5 minutes. Stir in the salt.

2. Add the whole wheat flour a little at a time, stirring to blend after each addition. Add 1½ cups bread flour a little at a time, stirring until most of the flour has been absorbed and the dough begins to form a ball. Transfer the dough to a lightly floured work surface and knead until soft and satiny, about 5 to 10 minutes, adding additional bread flour as needed to keep the dough from sticking.

3. Return the dough to the bowl, cover tightly with plastic wrap, and refrigerate. Let the bread rise in the refrigerator until doubled in bulk, about 8 hours. (You can let the dough rise and then punch it down several times, for a more flavorful and finely textured loaf.)

4. The next day, remove the dough from the refrigerator, punch it down, and cover again with plastic wrap. Let rise at room temperature until about doubled in size, 2 to 3 hours.

5. Shape the loaf: Punch down the dough and knead it for about 30 seconds. Shape the dough into a tight rectangular loaf by rolling the ball of dough and folding it over itself. Place a large floured cloth in a large loaf pan or rectangular basket and place the dough, smooth side down, in the pan or basket. Loosely fold the cloth over the dough. Let rise at room temperature until doubled in size, about 1 hour and 15 minutes.

6. At least 40 minutes before placing the dough in the oven, preheat the oven to 500°F (260°C; gas mark 9). If using a baking stone, place it in the oven to preheat.

7. Bake the bread: Lightly flour a baking paddle or rimless baking sheet, and turn the dough over onto the paddle or sheet. Slash the top of the dough several times with a razor blade, so it can expand evenly during baking. With a quick jerk of the wrists, slide the dough onto the baking stone. Using a garden mister, generously spray the bottom and sides of the oven with water. Then spray 3 more times during the first 6 minutes of baking. (The steam created will help give the loaf a good crust and will give the dough a boost during rising.) Once the bread is lightly browned—after about 10 minutes—reduce the heat to 400°F (200°C; gas mark 6/7), and rotate the loaf so that it browns evenly. Bake until the crust is a dark golden brown and the loaf sounds hollow when tapped on the bottom, about 10 minutes more, for a total baking time of about 20 minutes. Transfer to a baking rack to cool. Do not slice the bread for at least 1 hour, for it will continue to bake as it cools.

■ Y i e l d : 1 l o a f ■

Sardinian Parchment Bread

Carta da Musica

ITALIAN PROVERB:

"Give me eggs of an hour, bread of a day, wine of a year, a friend of thirty years."

A s with all unleavened bread, this Sardinian flat bread is thin, crunchy, and irresistible. I've sampled this sort of dimpled, bumpy bread at trattorias all over Italy, and always go back for more. I particularly love it as a "salad" antipasto, served on a platter and showered with a mixed salad of cucumbers and tomatoes tossed with olive oil and vinegar. The dough should be rolled as thin as a *carta da musica*, or sheet of music paper. This cracker-like bread (which is also called pane carasau), is an obvious relative of other unleavened breads, such as matzo, Indian flat breads, and Armenian cracker bread. Because it will keep for a long time, it is the traditional bread of the shepherds, who take it with them while tending their flocks of sheep.

I know of few recipes in the world that make a cook feel so accomplished: Guests are always in awe of the fact that the bread is homemade. Actually, it is child's play. I prepare little "bread kits" in advance, mixing the flours and salt and storing them in zipper-lock bags. When I want to make a quick bread, I turn on the oven, prepare the dough, and within an hour's time, we're snacking on these golden, no-fat, crusty cracker breads.

I have baked carta da musica on a baking stone and on baking sheets with equally successful results. Parchment bread baked on a baking stone will puff up a bit more, but will not be measurably better than the bread baked on a baking sheet. A good heavy-duty rolling pin is helpful, though not essential.

2 cups (265 g) unbleached all-purpose flour

1 cup (180 g) fine semolina flour

1½ teaspoons fine sea salt

About 1¼ cups (300 ml) lukewarm water

1. At least 40 minutes before placing the dough in the oven, preheat the oven to 450°F (230°C; gas mark 9). If using a baking stone, place it in the oven to preheat.

2. In a large shallow bowl, combine the flours and salt. With your fingers, mix the ingredients thoroughly. Slowly add the water, stirring with a wooden spoon, until the mixture forms a soft dough. (You may not need to add all the water.) With your hands, work the dough into a ball. Transfer to a clean, floured work surface, and knead gently for about 1 minute. The dough should be firm and pliable, and not sticky.

3. Divide the dough evenly into 12 balls. Place the balls on a lightly floured surface and cover with a clean, damp cloth. Flatten each ball into a thick 4-inch (10-cm) pancake. Generously flour the work surface, and, with a heavy-duty rolling pin, roll each portion of the dough as thin as possible into an 8- to 9-inch (20- to 23-cm) round. These breads are meant to be rather roughly shaped. Thinness is more important than shape here: The dough should be thin enough to see your hand through it.

4. Place several rounds of dough on an ungreased baking sheet and place in the oven, or place on the baking stone with a baker's paddle. Bake just until the top of the bread is firm and lightly browned, 3 to 4 minutes. (Baking time will vary from oven to oven and will depend upon the number of breads placed in the oven. I find variations between 2 and 5 minutes, depending upon the oven. Watch the bread carefully, and don't leave the room to answer the telephone! It helps if you have an oven with a glass window.) With tongs or with your fingers, turn the bread over and bake until the other side is lightly browned, 3 to 4 minutes more. (The bread should be rather bumpy, puffy, and irregular, with occasional huge pockets full of air.) Transfer the bread to a wire rack to cool. Repeat the procedure with the remaining rounds of dough.

5. The bread cools almost instantly and can be served immediately. To serve, stack the breads in the center of the table or in a large basket. (The bread may be stored for 3 to 4 days, in a metal container or zipper-lock bag at room temperature.)

■ Yield: 12 breads ■

· · ·

Sourdough Bread

Pane Campagnolo

For any cook who also loves bread and baking, there is no joy like perfecting the art of baking natural yeast, or sourdough, bread. It takes time, trial and error, patience and stick-to-itiveness, but the simple, slightly sour aroma of that loaf rising in your hot, steamy oven is enough to make you forgive it any anguish or frustration it may have caused.

This is a recipe I have worked on over the years, and it works for me and for others with whom I have shared the process. But each cook will find that perfecting and keeping the starter, and finding the right flour, the exact oven temperature, even the spot in the oven that produces the best bread, becomes a very personal affair.

My own habits change from month to month. Sometimes I bake this "pure," with just a single type of flour, and the same proportions of salt and water. Other times, I add olives or walnuts for a bread to pair with cheese. Or I add toasted pumpkin seeds, sunflower seeds, sesame seeds, or a bit of whole wheat or rye flour, alone or in tandem with other ingredients.

The point is to make it *your* loaf, *your* bread, the one with which you identify, and the one that fits into your schedule and life-style. I often like to prepare the bread at night, then bake it first thing in the morning, so the oven is heating as I sip my first cup of espresso.

Some helpful tips I've found over the years: Bread flour offers more consistent results than all-purpose flour. The starter remains more active, while with all-purpose flour it can turn lifeless. Also, don't be impatient with the rising time. The bread will pay you back with flavor.

And when I slice the bread, toast it, and spread it with tuna mousse or tomato and artichoke sauce, or simply drizzle it with golden freshly pressed olive oil, I know no greater gastronomic high.

SOURDOUGH STARTER

1 cup (125 ml) water, at room
  temperature

2 cups (270 g) bread flour, at
  room temperature

FOR THE FINAL LOAF

3 cups (750 ml) water, at room
  temperature

1 tablespoon fine sea salt

7 to 8 cups (945 g to 1 kg 80 g)
  bread flour, at room
  temperature

1. DAY 1 THROUGH 4: Prepare the starter: In a small bowl, combine ¼ cup (60 ml) of the water and ½ cup (70 g) of the flour and stir until the water absorbs all of the flour and forms a soft dough. Transfer the dough to a lightly floured work surface and knead into a smooth ball. It should be fairly soft and sticky. Return the starter to the bowl, cover with plastic wrap, and set aside at room temperature for 24 hours. The starter should rise slightly and take on a faintly acidic aroma. Repeat this for 3 days, each day adding an additional ¼ cup (60 ml) water and ½ cup (70 g) flour to the starter. Each day the starter should rise slightly, and it should become more acidic with time. (If it does not progress as described, and turns unpleasantly sour or greyish, toss it out and begin again, until you have a pleasantly fragrant starter.)

2. DAY 5: You are ready to make bread. Transfer the starter to a large shallow bowl. Add the 3 cups (750 ml) water and the salt, and, with a wire whisk, stir for about 1 minute to thoroughly dissolve the starter. The mixture should be very thin and very bubbly. Add the flour a bit at a time, stirring with a wooden spoon after each addition. Once you have added about 6 cups (810 g) flour, the dough should be firm enough to knead. Lightly flour a large clean work surface, and turn the dough onto the floured surface. (If your bowl is large and shallow enough, you can knead the bread right in the bowl, reducing cleanup later.) Begin kneading, at first actually folding the dough over itself to incorporate air (it may be too soft to knead), adding additional flour as needed. Knead for a full 10 minutes, until the dough is nicely elastic and soft, but still firm enough to hold itself in a ball. (Set a timer, to be sure there's no cheating!)

(continued)

3. Form the loaf and reserve the starter: Pinch off a handful of dough, about 8 ounces (250 g), to set aside for the next loaf. Transfer this starter to a medium-size covered container (see Note). Shape the remaining dough into a tight ball by folding it over itself. Place a large floured cloth in a shallow bowl or basket—one about 10 inches (25 cm) in diameter works well—and place the dough, smooth side down, in the cloth-lined bowl or basket. Loosely fold the cloth over the dough. Let rise at room temperature for 6 to 12 hours. (You have a lot of flexibility here. A 6-hour rise is the minimum, but I often prepare bread in the evening and bake the next morning, allowing the dough to rise for up to 12 hours. I have even forgotten the bread, finally baking it 24 hours later, and it was deliciously light and airy.) The dough will rise very slowly, but a good loaf should just about double in size.

4. At least 40 minutes before placing the dough in the oven, preheat the oven to 500°F (260°C; gas mark 9). If using a baking stone, place it in the oven to preheat.

5. Lightly flour a baking peel or paddle (or a rimless baking sheet). Turn the loaf out onto the peel, and lightly slash the top of the bread several times with a razor blade held at an angle, so it can expand evenly during baking. With a quick jerk of the wrists, shuffle the bread onto the baking stone. Using a garden mister, generously spray the bottom and sides of the oven with water. Then spray 3 more times during the first 6 minutes of baking. (The steam created will help give the loaf a good crust and will give the dough a boost during rising.) The bread will rise very slowly, but will rise to its full height during the first 15 minutes of baking. Once the bread has begun to brown nicely—after about 15 minutes—reduce the heat to 425°F (220°C; gas mark 8) and continue baking until the crust is a deep golden brown and the loaf sounds hollow when tapped on the bottom, 20 to 25 minutes more (for a total baking time of 35 to 40 minutes). Transfer to a baking rack to cool. Do not slice the bread for at least 1 hour, for it will continue to bake as it rests. Store the bread in a paper, cloth, or plastic bag. The bread should remain deliciously fresh for 3 to 4 days.

■ Y i e l d : 1 l o a f ■

NOTE: After you have made your first loaf and have saved the starter, begin at Step 2 for subsequent loaves. Proceed normally through the rest of the recipe, always remembering to save about 8 ounces (250 g) of the starter. The starter may be stored at room temperature (in a covered plastic container) for 1 or 2 days, or refrigerated for up to 1 week. If not using the starter regularly, reactivate it once a week by thoroughly mixing in ¼ cup (60 ml) of water and ½ cup (70 g) of flour. Do not use more than 8 ounces (250 g) of starter per loaf. (If you find you can't bake bread every week and you end up with excess starter, offer the excess to a friend, add it to a yeast dough, or—as a last resort—discard it.)

If the starter has been refrigerated, remove it from the refrigerator at least 2 hours before preparing the dough. Although starter can be frozen, I find it takes so long to reactivate that one might just as well begin with a new starter.

•••

Italian Slipper Bread

Ciabatta

Ciabatta—the flat, airy, puffy, light loaf known as Italian slipper bread—is ideal for those who want great flavor in a hurry. I often make it in the morning and have it ready for lunchtime. Once you've made ciabatta, you'll make it again and again, for it fills a lot of our requirements for a good homemade bread: It's got a golden but light crust, with a soft and moist interior. It's multipurpose in that it's great as is or toasted. The bread itself may not fit the portrait of the picture-perfect loaf, for the dough is so soft and batter-like, it does not resemble other yeast doughs of our acquaintance. And the final loaf may not rise to more than 2 inches (5 cm).

Ciabatta really does need to be made in a heavy-duty mixer, unless you are willing to stand and beat with a wooden spoon for about twenty minutes to get all that gluten working and turn the mixture into a highly elastic dough. Note that the dough is too sticky and soft to let rise in a traditional basket. I prefer to give the dough a single rise on a piece of cornmeal-dusted waxed paper, then turn it onto a baking stone for baking. My favorite memories of this bread come from two weeks we once spent at Lake Garda: Each morning we would walk to town for the makings of lunch, then take the long trek back up the hill home, picking bay leaves from high hedges along the road and clutching our ciabatta under our arms.

Cornmeal for dusting

1 teaspoon active dry yeast

1 teaspoon sugar

2 cups (500 ml) lukewarm water

1½ teaspoons fine sea salt

About 4 cups (530 g) bread flour

1. Place a piece of waxed paper on a baking sheet. Dust it with cornmeal and set aside.

2. In the bowl of a heavy-duty electric mixer fitted with a paddle, combine the yeast, sugar, and water, and stir to blend. Let stand until foamy, 5 to 10 minutes. Stir in salt.

3. Slowly add the flour to the mixing bowl in general additions, stirring well after each addition. Mix until all of the flour has been absorbed, 1 to 2 minutes. (The dough will not form a ball.) Knead at the lowest speed for about 7 minutes, or just until the dough cleans the bowl and forms an elastic mass around the paddle.

4. Using a pastry scraper, transfer the dough (it will be loose, and stringy with gluten) to the waxed paper. Let it spread out in a rectangular mass. Cover with a clean cloth and let rise at room temperature until doubled and puffy, about 2 hours.

5. At least 40 minutes before baking, preheat the oven to 500°F (260°C; gas mark 9). If using a baking stone, place it in the oven to preheat. If you do not have a baking stone, preheat a baking sheet.

6. Using 2 heavy oven mitts, remove the baking stone, or the baking sheet, from the oven. Invert the dough onto the baking stone or baking sheet. Inevitably, some of the moist dough will stick to the waxed paper, despite the cornmeal dusting. Use a pastry scraper to peel away the dough. At this point the dough will look like a rather unpromising mass. Don't despair. Return the baking stone or sheet and dough to the oven. Using a garden mister, generously spray the bottom and sides of the oven with water. Then spray 3 more times during the first 6 minutes of baking. (The steam created will help give the loaf a good crust and will give the dough a boost during rising.) Once the bread has risen slightly and is a nice golden brown—about 10 minutes—reduce the heat to 425°F (220°C; gas mark 8). Continue baking until the crust is a deep, golden brown and the loaf sounds hollow when tapped on the bottom, 25 to 30 minutes total. Transfer to a baking rack to cool. Do not slice the bread for at least 1 hour, for it will continue to bake as it rests. Store the bread in a paper, cloth, or plastic bag for up to 3 days. It is delicious toasted.

■ Yield: 1 loaf ■

Individual Olive Rolls

Pane alle Olive

One sunny August afternoon, we lunched on the terrace of Il Vescovino, a white, pristine little restaurant set off a narrow, meandering street in the village of Panzano in Chianti, and luxuriated in the spectacular, slumber-inducing view of the Tuscan hillsides. After an all-morning auto trip, the restaurant offered a welcoming sense of lushness and country calm as we sat on the hazelnut tree–shaded terrace facing a landscape filled with jolts of blue, yellow, green, and red. The inevitable Tuscan haze lent a sense of softness and roundness that was instantly comforting. From the brief handwritten menu we selected a simple homemade pasta dressed with melted butter flavored with lemon, served with delicious homemade rolls, fresh from the oven. To our surprise and delight, they were studded with rich chunks of pitted black olives. I can't offer you that afternoon or the view, but hope that you'll create your own sense of the country when you prepare these at home, enjoying the rich olive flavor that permeates the small golden breads.

1 recipe Basic Bread and Pizza Dough (page 178),
prepared through Step 3

About 30 salt-cured black olives, such as Italian Gaeta or French Nyons
olives (see page 280), pitted and halved

Cornmeal, for dusting

1. Three hours before you plan to bake the rolls, remove the dough from the refrigerator. Punch down the dough and let it rise until about doubled in size, 2 to 3 hours.

2. Preheat the oven to 450°F (230°C; gas mark 9).

3. Divide the dough evenly into 15 portions, each weighing about 2 ounces (60 g). Press several olive halves into each portion of dough and shape each portion into a neat round, pulling the dough around itself to form a tight ball. Place the balls of dough on a baking sheet dusted with cornmeal. Cover with a clean towel and let rise for 30 minutes.

4. Place the rolls on the baking sheet in the center of the oven. Using a garden mister, generously spray the bottom and sides of the oven with water. Then spray 3 more times during the first 6 minutes of baking. (The steam created will give the dough a boost during rising and will help give the roll a good crust.) Bake until the rolls are a deep golden brown, 20 to 25 minutes, turning the baking sheet from time to time for even browning.

5. Remove the rolls from the oven and transfer to a rack to cool. The rolls should be consumed the day they are prepared.

■ Yield: 15 rolls ■

VARIATION: In place of olives, you may wish to fill these light little rolls with freshly chopped rosemary or leaves of fresh thyme.

Pizza with Red Onions, Rosemary, and Hot Pepper

Pizza Fiamma

One Sunday afternoon in January, we reserved a table at the modern, lively, family-style pizzeria Il Mozzo, in Milan. The nondescript pizzeria was bustling with large multigeneration families, all feasting on different fare. The children drank Coke and the parents sipped beer or wine, while salads, pizzas, and steaming bowls of pasta were rushed by. We devoured a variety of luscious pizzas, including this thin-crusted version topped with slices of red onions marinated in oil, rosemary, and hot red pepper flakes. The name—*fiamma*, or flame—comes from the fact that the cooked onions resemble bright red flames. I find that the unleavened dough for Sardinian Parchment Bread works great as a thin-crusted pizza.

4 small red onions (about 10 ounces/300 g total), sliced into
very thin rings

½ cup (125 ml) extra-virgin olive oil

¼ teaspoon fine sea salt

½ teaspoon crushed red peppers (hot red pepper flakes), or to taste

¼ cup (60 ml) fresh rosemary leaves, finely minced

1 recipe Sardinian Parchment Bread (page 184)

1. In a large bowl, combine the onions, oil, salt, crushed red peppers, and rosemary. Stir to coat the onions with oil and to distribute the herbs and spices evenly. Set aside to marinate for at least 1 hour, or up to 4 hours. The marinating will soften the flavor of the onions and will give the topping a more finished flavor.

2. At least 40 minutes before placing the pizza in the oven, preheat the oven to 450°F (230°C; gas mark 9). If using a baking stone, place it in the oven to preheat.

3. Divide the dough evenly into 4 balls, and flatten each ball into a thick 4-inch (10-cm) pancake. On a generously floured work surface, roll each portion of the dough into an even 8-inch (20-cm) round.

4. Place the rounds of dough on a baking sheet or a baking stone. Spoon the onion topping onto the dough, spreading it out evenly with the back of the spoon. Bake until the dough is crusty and browned and the onions are sizzling, about 10 minutes. Serve immediately.

▪ Yield: 4 servings ▪

WINE SUGGESTION: Any drinkable red would do here, such as a good-quality Chianti.

Pizza Five Ways

In traveling throughout Italy, the finest pizzas I sampled were the most classic: thin, moist, flavorful dough topped frugally and simply with top-quality ingredients. The following are suggestions for classic Italian pizza.

Tomato and Mozzarella Pizza/Pizza Margherita

Brush the dough with oil, then top, in concentric circles, with a single, slightly overlapping layer of thin, half-moon slices of tomato and mozzarella. Drizzle with oil and sprinkle with crumbled leaf oregano. After baking, drizzle with oil. A few torn leaves of fresh basil can also be added after baking.

White Pizza/Pizza Bianca

Brush the dough with oil, sprinkle with freshly grated Italian Parmigiano-Reggiano cheese, and top with wedges of drained marinated artichokes. Drizzle with oil before baking.

Vegetarian Pizza/Pizza Vegetariana

Brush the dough with oil, then top, in concentric circles, with a single, slightly overlapping layer of thin slices of tomato, eggplant, and mozzarella. Sprinkle with dried leaf oregano.

Four Seasons' Pizza/Pizza alle Quattro Stagioni

Brush the dough with a thin layer of tomato sauce, and sprinkle with strips of prosciutto and slices of fresh mushrooms tossed in olive oil, slices of marinated artichokes, and halved and pitted black olives.

Tomato, Caper, Olive, and Anchovy Pizza/Pizza Marinara

Brush the dough with a thin layer of tomato sauce, and sprinkle with rinsed and drained capers, halved and pitted black olives, minced anchovies, fine sea salt, and freshly ground black pepper.

1 recipe Basic Bread and Pizza Dough (page 178)
Coarse cornmeal for dusting

1. At least 40 minutes before placing the assembled pizzas in the oven, preheat the oven to 500°F (260°C; gas mark 9). If using a baking stone, preheat it.

2. Punch down the prepared dough and divide it evenly into 4 pieces. Shape each piece into a ball. On a lightly floured surface, roll each ball of dough into an 8-inch (20-cm) round.

3. Sprinkle a wooden pizza peel (or a baking sheet) with coarse cornmeal and place the rounds of dough on the peel or sheet. Working quickly to keep the dough from sticking, assemble the pizzas.

4. Slide the pizzas off the peel and onto the baking stone (or place the baking sheet on a rack in the oven). Bake until the dough is crisp and golden, 10 to 15 minutes.

5. Remove from the oven, transfer to a cutting board, and cut into wedges. Serve immediately.

■ Yield: Four 8-inch pizzas ■

Tips for Better Pizza

■ To ensure a fine flavor and better texture, use bread flour when preparing pizza dough. Remember that in pizza, the crust should be the star, the topping simply an accent.

■ For a moist and flavorful dough, let it rise slowly, several times, preferably overnight in the refrigerator. Soft, moist dough makes for a light and crispy crust.

■ Keep the topping simple and light in weight: Try to limit yourself to three to five top-quality ingredients.

■ For an even flavor, coat the dough with a thin layer of either oil or sauce before assembling.

■ Toss any ingredients that tend to dry out in baking—such as sliced fresh mushrooms or strips of prosciutto—with oil a few minutes before assembling the pizza.

■ Be prepared: Arrange all topping ingredients in a series of small bowls, with the ingredients already chopped, sliced, or drained, so that pizza assembly is leisurely and fun, not frantic.

■ Distribute toppings evenly, so that each bite of pizza contains a bit of each flavor. Leave a margin around the edges to form a crust and to keep toppings from seeping out onto the baking stone.

■ Preheat the oven to its highest temperature at least 40 minutes before baking.

■ Bake on a baking stone. Additionally, you may want to use a wooden peel to slide the pizza into the oven, a metal peel for retrieving it from the oven.

■ If you want crisp crust, eat pizza hot. The crust will soften as it sits after baking.

FISH AND

SHELLFISH

Spicy Squid Salad

Insalata di Calamari

Variations of this spicy, refreshing salad show up often on antipasto tables all over Italy. Personally, I can make a meal out of this, as long as there is plenty of crusty bread for absorbing the sauce, and a chilled white wine, such as a Pinot Grigio or Orvieto. Just make sure your squid is perfectly fresh (it should have almost no odor—only a "fresh" one). And it should be springy and just pleasantly chewy once cooked. Make the salad as hot as you like, but don't let the red pepper overwhelm it. What I love most about this salad is the texture: the pleasant chewiness of the squid, the crunch of the celery, the softness of the green olives. You might even consider this salad a contender for the Italian flag food award: the white of the squid, the red of the peppers, the green of the olives.

⅔ *cup (150 ml) extra-virgin olive oil*

⅓ *cup (80 ml) freshly squeezed lemon juice, or to taste*

4 *plump fresh garlic cloves, degermed and minced*

¾ *teaspoon crushed red peppers (hot red pepper flakes), or to taste*

4 *ribs celery hearts with leaves, minced*

20 *drained pimento-stuffed green olives, quartered crosswise*

2 *pounds (1 kg) very fresh squid*

Sea salt

1. In a medium-size bowl, stir together the oil, lemon juice, garlic, and crushed red pepper. Stir in the celery and olives. Taste for seasoning. Set aside.

2. Clean the squid: Rinse and drain thoroughly. Slice off the tentacles just above the eyes. Squeeze out and discard the hard little beak just inside the tentacles at the point where they join the head. Pull out the innards and the cuttlebone, or quill, from the body and discard. Do not worry about removing the skin, which is edible. Slice the squid bodies crosswise into ¼-inch rings. Cut the tentacles in half lengthwise. Rinse the squid again and drain thoroughly.

3. In a large saucepan, bring 3 quarts (3 l) of water to a rolling boil. Add 2 tablespoons salt. Add the squid rings and tentacles and cook just until they turn opaque—not more than 1 minute, or they will toughen. (I usually begin tasting the squid about 30 seconds after it has hit the water.) Drain but do not rinse: Hot squid will absorb the sauce better. Transfer the hot squid to the olive dressing and toss to coat. Taste for seasoning. Cover and refrigerate for at least 3 hours, or overnight.

4. At serving time, taste for seasoning. Serve as part of an antipasto platter, or on small salad plates as a first course, accompanied by plenty of crusty bread.

■ Yield: 6 servings as a ■
first course

WINE SUGGESTION: With this, try a crisp Italian white, such as Pinot Grigio or Orvieto.

• • •

Fried Calamari

Calamari Fritti

A quick and delicious first course, these tiny fried calamari can be found in trattorias all over Italy, and are a particular favorite along the coastline. I like using a light semolina flour here, for it makes a very rustic, crispy, golden coating. If you do not have semolina on hand, use superfine flour for an even more refined coating. Be sure to soak the squid in ice water for at least ten minutes before cooking so the shock of the cold against the heat of the oil stops the squid from absorbing the fat and they remain pure in flavor.

1 pound (500 g) very fresh squid, or calamari

¾ cup (100 g) semolina flour or superfine flour, such as Wondra

Fine sea salt

1 to 1½ quarts (1 to 1.5 l) peanut or sunflower oil, for deep-frying

Lemon wedges, for garnish

1. Clean the squid: Rinse and drain thoroughly. Slice off the tentacles just above the eyes. Squeeze out and discard the hard little beak just inside the tentacles at the point where they join the head. Pull out the innards and the cuttlebone, or quill, from the body and discard. The mottled red skin is edible, though it may be removed if desired. Slice the squid bodies crosswise into ¼-inch (½-cm) rings. Cut the tentacles in half lengthwise. Rinse the squid again and drain thoroughly. Place in a large bowl of ice water for at least 10 minutes before cooking.

2. In a clean paper or plastic bag, combine the flour and 1 teaspoon salt and shake to blend.

3. Preheat the oven to 200°F (100°C; gas mark 1).

4. Pour the oil into a wide 6-quart (6-1) saucepan, or use a deep-fat fryer. (The oil should be at least 2 inches [5 cm] deep.) Place a deep-fry thermometer in the oil and heat the oil to 375°F (190°C).

5. Drain the squid and dry thoroughly with a clean towel. Dip a handful of squid into the bag of flour and shake to coat with flour. Transfer the coated squid to a fine-mesh sieve and shake off excess flour. Carefully drop the squid by small handfuls into the hot oil. Cook until lightly browned, 1 to 2 minutes. With a wire skimmer, lift them from the oil, drain, and transfer to paper towels. Immediately season with sea salt, then place in the oven—with the door slightly ajar—to keep warm. Continue frying until all of the squid are cooked, allowing the oil to return to 375°F (190°F) each time before adding another batch. Serve immediately, with lemon wedges.

■ Yield: 4 servings as a ■
first course

WINE SUGGESTION: This dish makes me think of Venice, and so why not a light white Sauvignon Blanc, such as one from Italy's northeast, Friuli-Venezia Giulia?

Shrimp with Garlic, Oil, and Hot Peppers

Scampi alla Veneziana

Venice is a haven for fish and shellfish, and this simple and beautiful dish is typical of the specialties found in waterside trattorias throughout that magic city. Pink, fresh, and glistening, this is a glorious dish, made for sharing at the table, along with crusty bread for soaking up the garlic-rich sauce, and a nicely chilled white wine, such as a Pinot Grigio. I like to serve it as a first course, followed by a simple grilled or roasted chicken with potatoes and Garlic Mayonnaise (page 275).

16 to 20 (about 1 pound; 500 g) large shrimp, in their shells

½ cup (125 ml) extra-virgin olive oil

4 plump fresh garlic cloves, minced

2 teaspoons fresh thyme leaves

¼ teaspoon crushed red peppers (hot red pepper flakes), or to taste

Coarse sea salt

¼ cup (60 ml) fresh flat-leaf parsley leaves, snipped with a scissors

1. Rinse the shrimp, pat them dry, and set aside. In a skillet large enough to hold all the shrimp in a single layer, heat the oil over moderately high heat. When the oil is hot but not smoking, add the garlic, thyme, crushed red peppers, and shrimp. Toss to coat with oil, and cook, stirring occasionally, just until the shrimp are pink, 4 to 5 minutes.

2. Remove the pan from the heat and, with a slotted spoon, transfer the shrimp to a warmed serving platter or to warmed individual plates. Pour the sauce over the shrimp, sprinkle lightly with coarse salt, and shower with the parsley. Serve immediately, offering a finger bowl and an extra napkin for each guest.

■ Yield: 4 servings as a
first course ■

WINE SUGGESTION: Try a nicely chilled Pinot Grigio or a Chardonnay.

MEMORIES OF VENICE

■

The elegance and the subtlety of Venice are reflected in the city's food, which displays a great sense of grace and prosperity. Make the wine a crisp white Sauvignon from the Veneto.

MIXED FRIED FISH
TAGLIATELLE WITH FRESH CRABMEAT
TIRAMSÙ (COFFEE AND MASCARPONE LADYFINGER CREAM)

Baked Swordfish with Tomatoes and Green Olives

Pesce Spada alla Marinara

Swordfish—known as *pesce spada* in Italian—is a popular Mediterranean fish, and takes well to baking with tomatoes and salty ingredients such as olives. With its off-white, sometimes pinkish, flesh and fine grain, swordfish is a firm, medium-fat fish most often cut into steaks for grilling or for baking. This uncomplicated, distinctly Italian dish can also be prepared with tuna steaks.

3 tablespoons extra-virgin olive oil

1 fresh swordfish or tuna steak, cut ¾ inch (2 cm) thick (about 1 pound; 500 g)

Sea salt and freshly ground black pepper to taste

1 small onion, minced

1 rib celery, cut into thin slices

One 28-ounce (765-g) can peeled Italian plum tomatoes in juice or one 28-ounce (765-g) can crushed tomatoes in purée

¼ teaspoon crushed red peppers (hot red pepper flakes), or to taste

⅓ cup (80 ml) drained pitted green olives

1. Preheat the oven to 450°F (230°C; gas mark 9).

2. In a large skillet, heat the oil over moderately high heat until hot but not smoking. Add the swordfish and brown lightly, 2 to 3 minutes per side, seasoning each side with salt and pepper after browning. Using a large flat spatula, transfer to a baking dish just large enough to hold the fish. Set aside.

3. In the same skillet, cook the onion and celery over moderate heat until translucent, 4 to 5 minutes. If using whole canned tomatoes, place a food mill over the skillet and purée the tomatoes directly into it. Crushed tomatoes can be added directly from the can. Add the crushed red peppers. Stir to blend, cover, and simmer until the sauce begins to thicken, about 15 minutes. Stir in the olives. Taste for seasoning.

4. Spoon the sauce over the fish. Cover the dish with foil, place in the center of the oven, and bake for 30 minutes. To serve, quarter the fish, remove the skin, and transfer to warmed individual dinner plates. Use a slotted spoon to transfer the sauce to the plates, since the sauce has a tendency to thin out while baking.

Yield: 4 servings as a
main course

WINE SUGGESTION: Swordfish cries out for a dry white, such as a Sauvignon Blanc or Chardonnay.

Sea Bass with Potatoes and Tomatoes in Parchment

Branzino in Cartoccio

O n one trip to Florence, I sampled this lovely baked fish dish—with very minor variations—on two consecutive evenings. This version comes from La Capannina di Sante, where the whole baked fish is brought to the table in its bundle of parchment, allowing diners to enjoy the fragrance of the fresh fish as the package is slit open. While Chef Sante Collesano used branzino, the firm-fleshed sea bass, here you might also use a super-fresh whole snapper. Be sure to peel and slice the potatoes at the last minute or they will discolor.

1 whole sea bass, Alaska snapper, or red snapper (about 2 pounds; 1 kg), scaled and cleaned, but with head and tail on, rinsed, and patted dry

Sea salt and freshly ground black pepper to taste

Several sprigs fresh thyme

4 bay leaves, preferably fresh

2 large potatoes, peeled and sliced paper-thin

1 medium onion, sliced paper-thin

10 cherry tomatoes, halved lengthwise

3 tablespoons extra-virgin olive oil

Lemon wedges, for garnish

Extra-virgin olive oil, for the table

1. Preheat the oven to 450°F (230°C; gas mark 9).

2. Place the fish on one half of a sheet of baking parchment large enough to comfortably wrap the fish. Season the fish inside and out with salt and pepper. Tuck the thyme and bay leaves inside. Spread the potatoes, onion, and tomatoes over and around the fish. Drizzle with the oil.

3. Carefully fold the other half of the paper over the fish, closing it like a book. To seal the package, double-fold the edges and secure each side with several staples.

4. Place the package on a large baking sheet and place in the center of the oven. Bake for 25 minutes.

5. Remove the package from the oven and cut it open with a scissors. Let sit for about 3 minutes to allow the fish to firm up enough to fillet. To serve, fillet the fish and divide it among 4 warmed dinner plates. Arrange several spoonfuls of the potato-tomato mixture alongside. Pass the lemon wedges and a cruet of olive oil.

Yield: 4 servings as a
main course

WINE SUGGESTION: A white, for sure, such as a Pinot Grigio or Vernaccia di San Gimignano.

...

Baked Sea Bass with Artichokes

Branzino coi Carciofi

I adore whole baked fish and have an equal passion for artichokes, so whenever I see this dish on a trattoria menu, I simply have to have it. It's amazing what an affinity sea bass and artichokes have for one another. The only secret here is to slice the artichokes very thin, so once they are precooked, they take about the same cooking time as the sea bass.

4 globe artichokes

1 lemon

½ cup (125 ml) plus 3 tablespoons extra-virgin olive oil

Sea salt and freshly ground black pepper to taste

1 whole sea bass, Alaska snapper, or red snapper (about 2 pounds; 1 kg), scaled and cleaned, but with head and tail on, rinsed, and patted dry

4 sprigs fresh rosemary

3 tablespoons freshly squeezed lemon juice

Lemon wedges, for garnish

Extra-virgin olive oil, for the table

1. Preheat the oven to 450°F (230°C; gas mark 9).

2. Prepare the artichokes: Halve the lemon, squeeze the juice, and place the halved lemon and the juice in a bowl filled with cold water. Rinse the artichokes under cold running water. Using a stainless steel knife to minimize discoloration, trim the stem of an artichoke to about 1 inch (2.5 cm) from the base. Carefully trim off and discard the stem's fibrous exterior. Bend back the tough outer green leaves one at a time, and snap them off at the base. Continue snapping off the leaves until only the central cone of

 TOMATO AND MOZZARELLA PIZZA *(PAGE 196)*, TOP; WHITE PIZZA *(PAGE 196)*

 CHICKEN WITH A CONFIT OF RED BELL PEPPERS AND ONIONS *(PAGE 221)*; **BROILED POLENTA** *(PAGE 172)*

**BRAISED VEAL SHANKS WITH LEMON AND PARSLEY
GARNISH** *(PAGE 244)*; **SOURDOUGH BREAD** *(PAGE 186)*

 HAZELNUT AND ORANGE BISCOTTI *(PAGE 314)*, LEFT
ALMOND AND VANILLA BISCOTTI *(PAGE 315)*

yellow leaves with pale green tips remains. Trim the top of the cone of leaves to just below the green tips. Trim any dark green areas from the base. Halve the artichoke lengthwise. With a small spoon, scrape out, and discard, the hairy choke. Cut each artichoke half lengthwise into paper-thin slices, and place immediately in the acidulated water. Repeat with the remaining 3 artichokes.

3. Drain the artichokes. In a large skillet, combine the artichoke slices, ½ cup (125 ml) of the oil, and the salt, and toss to coat the artichokes with the oil. Over moderate heat, sauté the artichoke slices just until soft, 3 to 4 minutes. Taste for seasoning. Set aside.

4. Season the fish generously inside and out with salt and pepper. Place it in a shallow baking dish that will hold it snugly. Stuff the cavity with the sprigs of rosemary and as many artichoke slices as will fit. Scatter the remaining artichokes, as well as the juices, around the fish. Drizzle with the lemon juice and the remaining 3 tablespoons olive oil.

5. Place the baking dish in the center of the oven and bake, uncovered, until the fish is opaque through but not dry, 30 to 40 minutes, basting every 10 minutes. (The baking time will vary according to the size of the fish.)

6. Remove the fish from the oven and let sit for about 3 minutes to allow the fish to firm up enough to fillet.

7. To serve, fillet the fish and divide it among 4 warmed dinner plates. Arrange several spoonfuls of artichokes alongside. Pass the lemon wedges and a cruet of olive oil.

Yield: 4 servings as a
main course

WINE SUGGESTION: With this dish, I've thoroughly enjoyed a chilled white Vernaccia di San Gimignano from the Tuscan hill town of San Gimignano. One of the finest is the oak-aged Terre di Tufo, from Teruzzi & Puthod.

Mixed Fried Fish

Frittura di Pesce

S ome of the best fried fish I ever ate was prepared by Cesare Benelli, chef and owner of Venice's small forty-seat restaurant Al Covo. Cesare grew up in a Venetian restaurant family (his father had a trattoria on the Lido) and remembers when, as a child, he could go snorkeling in the area and within a few minutes spear with a kitchen fork more baby sole than the family could possibly consume. Today—due to pollution and overfishing—sole are a sacred delicacy in Venice. But with a combination of rigorous standards and careful shopping Cesare manages to obtain a variety of sparkling fresh fish for Al Covo diners. One of his specialties is delicate first-course servings of mixed fried fish, light as pillows and deliciously crispy and fresh. He confided his personal frying secret: Once the fish is rinsed, he lets it rest in ice water for a few minutes. With the temperature difference between the cold fish and the hot oil, the fish has less of a tendency to absorb the oil.

1 pound (500 g) mixed small fish or shellfish, such as anchovies, sardines, smelt, whitebait, tiny sole fillets, rings or tentacles of small squid, shrimp, or cubed fish fillets

1 cup (135 g) superfine flour, such as Wondra

Fine sea salt and freshly ground black pepper to taste

⅛ teaspoon cayenne pepper

1 to 1½ quarts (1 to 1.5 l) oil, for deep-frying

Lemon wedges, for garnish

1. Rinse the fish or shellfish and soak in ice water for at least 10 minutes before cooking.

2. In a paper or plastic bag, combine the flour, ¼ teaspoon salt, pepper, and cayenne and shake to blend.

3. Preheat the oven 200°F (100°C; gas mark 1).

4. Pour the oil into a wide 6-quart (6-l) saucepan, or use a deep-fat fryer. (The oil should be at least 2 inches [5 cm] deep.) Place a deep-fry thermometer in the oil and heat the oil to 375°F (190°C).

5. Drain the fish or shellfish and dry well. Dip a handful of fish or shellfish into the bag of flour and shake to coat with flour. Transfer the coated fish to a fine-mesh sieve and shake off excess flour. Carefully drop the fish or shellfish by small handfuls into the hot oil. Cook until lightly browned, 1 to 2 minutes. With a wire skimmer, lift from the oil, drain, and transfer to paper towels. Immediately season with sea salt, then place in the oven—with the door slightly ajar—to keep warm. Continue frying until all of the fish or shellfish are cooked, allowing the oil to return to 375°F (190°C) each time before adding another batch. Serve immediately, with lemon wedges.

Yield: 4 to 6 servings as a first course

WINE SUGGESTION: Light fried foods suggest a light white Sauvignon Blanc, such as one from Italy's northeast, the Friuli-Venezia-Giulia region.

Small-Fry Fish Tips

■ Use any kind of small whole fish, fish fillets, or steaks, but try to keep them small and uniform: no more than 1½ inches (4 cm) thick and no more than 5 to 6 inches (13 to 15 cm) long. (Small fish with bones—such as whitebait, smelt, anchovies, or sardines—can be fried whole, for the deep-frying will soften the bones and make them edible.)

■ Use a very light coating of seasoned flour only, to help the fish retain its moisture on the inside and form a nice crispy crust on the outside.

■ Preheat the oil to 375°F (190°C). Hotter oil will burn the fish and coating; cooler oil will make for soggy fish that have absorbed the oil.

■ Soak the rinsed fish in ice water, and dry thoroughly before coating.

POULTRY AND

MEATS

Sautéed Chicken Breasts with Fresh Sage

Petti di Pollo alla Salvia

ITALIAN DICTUM:

"Spring is for looking, autumn for tasting."

Ever since I sampled this ultimately simple chicken preparation one weekday afternoon at Antico Fattore in Florence, it's been a family favorite. I love to "stretch" a whole chicken by reserving the breasts for this recipe and using the remaining bird to prepare chicken stock. The quick marinade of lemon juice, oil, and fresh sage helps infuse the chicken with flavor, and tenderizes it at the same time. Be sure to sample the sage leaves before using: Even fresh sage can taste bitter. Do not use dried sage. Worthy substitutes include fresh rosemary or fresh tarragon. Serve this with Lemon Risotto (page 158) or simple buttered pasta.

4 *skinless, boneless chicken breast halves (each 6 ounces; 180 g)*

3 *tablespoons freshly squeezed lemon juice*

5 *tablespoons extra-virgin olive oil*

28 *fresh sage leaves*

3 *tablespoons (1½ ounces; 45 g) unsalted butter*

Sea salt and freshly ground black pepper to taste

2 *lemons, halved, for garnish*

1. Place the chicken breasts in a glass baking dish. Add the lemon juice, 3 tablespoons of the oil, and the sage leaves. Turn the chicken to coat evenly, cover, and set aside at room temperature for 30 minutes.

2. Remove the chicken from the marinade and pat dry. Strain the marinade into a small bowl; reserve the sage leaves separately.

3. In a large skillet, melt the butter in the remaining 2 tablespoons of oil over moderately high heat until hot and bubbly. Add the chicken breasts, smooth side down, and cook until evenly browned, about 5 minutes. Turn the breasts, and season the cooked side generously with salt and pepper. Tuck the reserved sage leaves around the chicken and cook until the chicken is browned on the bottom and just white throughout but still juicy, 5 to 10 minutes more. Do not scorch the sage.

4. Remove the skillet from the heat. Transfer the chicken to a cutting board and season the bottom side with salt and pepper. Slice the chicken breasts on the diagonal into thick slices, and arrange on a warmed serving platter. Place the sage leaves over the chicken. Cover loosely with foil.

5. Discard the fat from the skillet. Heat the skillet over moderately high heat until hot. Add the reserved marinade and stir with a wooden spoon, scraping up the brown bits from the bottom of the pan. The sauce will boil almost immediately. As soon as it reduces to a brown glaze (less than 1 minute), pour the sauce over the chicken. Garnish with the lemon halves. Serve immediately.

■ Yield: 4 servings ■

WINE SUGGESTION: At the Antico Fattore, we sampled the trattoria's house wine, Ruffino Chianti Riserva Ducale, a worthy local choice from Tuscany.

Chicken Cooked Under Bricks

Pollo al Mattone

This is the dish that inspired this book. On a trip to Italy some years ago, I sampled this delicious, crispy-skinned chicken at a popular Tuscan trattoria named Da Giulio, in Lucca. I loved it so, I had to have the recipe, and, almost like a sleepwalker, I found myself in the kitchen, pen and notebook in hand. The move to the kitchen was so instinctive and spontaneous, I wasn't even quite sure how I had arrived there. But as I was writing down the details of the preparation of this traditional Italian dish, a light bulb went off in my head, and I knew that trattoria cooking would be the subject of my next book. I now jokingly call the dish "smashed" chicken, for it is really fried chicken that is split and cooked whole, held tight to the skillet and the heat by a heavy weight, usually bricks, or *mattone*. In fact, kitchenware shops in Italy sell special heavy-duty glazed pottery two-piece cookers, with flat, heavy tops, made for cooking the chicken. The weights are not just a gadget or gimmick, but actually help the chicken cook more evenly and make for a very pleasantly crispy but not fatty crust, for all the fat is pressed out of the chicken. The weighting technique can also be used in grilling chicken over medium-hot coals. Cooked whole and on the bone, the chicken has more flavor than chicken parts or boned chicken. I like to serve this with steamed new potatoes and garlic mayonnaise as an accompaniment. The only trick here is to properly regulate the heat as you cook, for you can't actually keep an eye on the chicken skin, but want to make sure it does not burn. I work by aroma and keep the heat beneath the skillet between moderate and moderately high, making sure the chicken cooks evenly but not too quickly.

1 free-range roasting chicken (4 pounds; 2 kg)

½ cup (125 ml) extra-virgin olive oil

Sea salt and freshly ground black pepper to taste

1. Place the chicken breast side down on a flat surface. With a pair of poultry shears, split the bird lengthwise along the backbone. Open it flat, and press down with the heel of your hand to flatten completely. Turn the chicken skin side up, and, with a sharp knife, make slits in the skin near the tail and tuck the wing tips in to secure. The bird should be as flat as possible to ensure even cooking.

2. In a 12-inch skillet, or one large enough to hold the chicken flat, heat the oil over moderately high heat. When the oil is hot but not smoking, place the chicken skin side down in the skillet. Put a lid or another skillet over the chicken, then weight it down with a 10-pound (5-kg) weight. (Use bricks or heavy rocks.) Cook over medium-high heat until the skin is golden brown, about 12 minutes. Remove the lid and weights. Using tongs so that you do not pierce the meat, turn the chicken over, and season generously with salt and pepper. Replace the lid and weights, and cook for 12 minutes more. To test for doneness, pierce the thigh with a skewer. The chicken is done when the juices run clear.

3. Transfer the chicken to a platter and place it at an angle against the edge of an overturned plate, with its head down and tail in the air. (This heightens the flavor by allowing the juices to flow down through the breast meat.) Cover the chicken loosely with foil. Let it rest for at least 10 minutes and up to 30 minutes. (If desired, keep the chicken warm by placing it in a warm oven.)

4. To serve, carve the chicken and arrange it on a warmed platter. (The chicken can also be served at room temperature.)

■ Yield: 4 to 6 servings ■

WINE SUGGESTION: Any good, sturdy wine goes well with this. If you prefer white, try a Vernaccia di San Gimignano; for a red, try a Tuscan Cabernet from Antinori.

Tomatoes: Canned versus Fresh, Whole versus Crushed

For preparing tomato sauce, canned tomatoes are often better than fresh. Fresh tomatoes vary dramatically in quality and price, and are quite labor-intensive, while quality canned tomatoes remain a year-round, dependable pantry staple. Brands of canned tomatoes vary widely, and most markets offer a choice. Do a test at home, preparing the same sauce with several brands of both domestic and imported tomatoes to decide which flavor you prefer. A good canned tomato should produce a rich sauce with a fresh tomato flavor, with no tomato paste aftertaste.

The recipes here offer a choice of peeled plum tomatoes in their juice and crushed tomatoes in purée, since these are the most readily available choices. I prefer whole canned plum tomatoes in juice: They produce a more refined sauce, with a pure tomato flavor. Pass the whole canned tomatoes through a food mill to remove bits of tough pulp as well as most of the seeds, which can turn the sauce bitter. Crushed tomatoes tend to be thicker than whole tomatoes, so adjust cooking times accordingly.

Chicken with a Confit of Red Bell Peppers and Onions

Pollo alla Peperonata

Some dishes become such challenges that a cook can't rest until the result truly sings! One August evening in Florence, at the very simple Osteria del Cinghiale Bianco, I ordered an uncomplicated dish of sautéed chicken with red peppers and onions, a garnish traditionally called a peperonata. The thin strips of red peppers, shining with their own natural oils, were all intertwined with the strands of onions, which had melted away to almost nothing. The entire dish was a symphony of aromatic flavors of the Mediterranean. It was hard to believe these vegetables weren't meat, because they were so serious and stood out on their own. At home, I began to play around with the combination, and, time after time, hit a false chord. Too bland. No life. Nothing special. Finally, I carefully thought it all through, and came up with a dish that, I hope, sings on key as well as the one I sampled that evening in Florence. The trick is to slice the onions very, very thin, so they cook up quickly, taking on a soft, sweet essence. The peppers are cooked separately, just until they wilt and give up their natural oils. The balsamic vinegar at the end gives the final dish a unity and a special boost. I like to serve this with wedges of grilled polenta(page 172), preceded by a salad of tomatoes and mozzarella or a light pasta, finishing off with a bowl of fresh fruit.

(continued)

4 red bell peppers (about 1½ pounds; 750 g), cut into thin lengthwise strips

Sea salt to taste

7 tablespoons extra-virgin olive oil

2 medium onions (about 10 ounces; 300 g), peeled

1 tablespoon finely minced fresh rosemary leaves

1 tablespoon sugar

1 chicken (3 to 4 pounds; 1.5 to 2 kg), at room temperature, cut into 8 serving pieces

Freshly ground black pepper to taste

1 tablespoon unsalted butter

One 16-ounce (480-g) can peeled Italian plum tomatoes in juice or one 16-ounce (480-g) can crushed tomatoes in purée

Several sprigs of fresh parsley, bay leaves, sprigs of fresh rosemary, and celery leaves, tied in a bundle with cotton twine

½ cup (125 ml) chicken stock, preferably homemade (page 272)

1 tablespoon balsamic vinegar

1. In large skillet, combine the peppers, a pinch of salt, and 2 tablespoons of the oil. Toss to coat the peppers with oil and cook, covered, over very low heat, stirring from time to time, until soft and glazed, about 15 minutes. Remove from the heat and set aside.

2. Meanwhile, slice an onion in half lengthwise. Place each half, cut side down, on a cutting board and cut crosswise into very thin slices. Slice the remaining onion in this manner, for about 2 cups (500 ml) sliced onions.

3. In another large skillet, combine the onions, a pinch of salt, the rosemary, sugar, and 2 tablespoons of the oil. Toss to coat the onions with oil and cook, uncovered, over very low heat, stirring from time to time, until the onions are very soft and glazed, about 10 minutes. Remove from the heat and set aside.

4. Season the chicken liberally with salt and pepper. In a third skillet, combine the remaining 3 tablespoons oil and the butter over high heat. When hot, add several pieces of chicken and cook on the skin side until it turns an even golden brown, about 5 minutes. Turn the pieces and brown them on the other side, about 5 minutes more. Do not crowd the pan; brown the chicken in several batches. Carefully regulate the heat to avoid scorching the skin. When all the pieces are browned, return all the chicken to the pan. If using whole canned tomatoes, place a food mill over the skillet and purée the tomatoes directly into it. Crushed tomatoes can be added directly from the can. Add the herb bundle, stir to blend, and simmer for about 5 minutes. Add the chicken stock, onions, and peppers, and simmer, partially covered, until the chicken is cooked through, 25 to 30 minutes more. Remove and discard the herb bundle. Stir in the vinegar and cook for 1 minute more.

5. Transfer the chicken to warmed dinner plates, along with the sauce. Serve immediately.

▪ Yield: 4 to 6 servings ▪

WINE SUGGESTION: This assertive, acidic dish is not tailor-made for wines, so go with something uncomplicated, such as a young Chianti or Valpolicella.

. . .

Grilled Chicken with Lemon, Oil, and Black Pepper

Pollo alla Diavola

"Amarone . . . a wine of incredible depth, bouquet, and breed. Forget about that, however, and listen to the name, preferably pronounced by Luciano Pavarotti—Am-mahr-roh-nay; a siren song, a seduction."

LEONARD BERNSTEIN, THE OFFICIAL GUIDE TO WINE SNOBBERY

"Hot as the devil," or *alla diavola,* is perhaps the most classic of all Italian trattoria dishes. The fire comes from coarsely ground black peppercorns, which radiate their fire as the chicken marinates, along with extra-virgin olive oil and freshly squeezed lemon juice. Roasting only heightens their flavor, making a superbly intense dish that's great during hot or cold weather. I like to serve this with Seared and Roasted Tomatoes (page 13) and perhaps Panfried Potatoes with Black Olives (page 52).

1 free-range roasting chicken (about 3 pounds; 1.5 kg)

5 tablespoons freshly squeezed lemon juice

3 tablespoons extra-virgin olive oil

About 1 tablespoon whole black peppercorns, coarsely crushed

Fine sea salt to taste

1. Prepare the chicken: Place the chicken breast side down on a flat surface. With a pair of poultry shears, split the bird lengthwise along the backbone. Open it flat, and press down with the heel of your hand to flatten completely. Turn the chicken skin side up, and, with a sharp knife, make slits in the skin near the tail and tuck the wing tips in to secure. The bird should be as flat as possible to ensure even cooking.

2. Place the chicken in a deep dish and add the lemon juice, oil, and 1 tablespoon crushed black peppercorns. Cover and marinate at room temperature for 30 minutes, turning the chicken with tongs from time to time.

3. Preheat the oven broiler. Or prepare a wood or charcoal fire. The fire is ready when the coals glow red and are covered with ash.

4. Season the chicken generously with salt. With the skin side toward the heat, place beneath the broiler or on the grill about 5 inches (13 cm) from the heat so that the poultry cooks evenly without burning. Cook until the skin is evenly browned, basting occasionally with the marinade, about 15 minutes. Using tongs so you do not pierce the meat, turn and cook the other side, basting occasionally, about 15 minutes more. To test for doneness, pierce the thigh with a skewer. The chicken is done when the juices run clear.

5. Remove the chicken from the heat and season once more with salt and with additional cracked pepper if desired. To serve, quarter the chicken and slice the breast meat, arranging it on a warmed serving platter.

▪ Y i e l d : 4 t o 6 s e r v i n g s ▪

WINE SUGGESTION: A full-bodied white, such as a Chardonnay, or a soft red, such as Recioto della Valpolicella Amarone.

Poached Chicken with Fresh Herb Sauce

Pollo Bollito in Salsa Verde

T ender, moist poached chicken and the vibrant herb sauce known as salsa verde seem to be made for one another. The delicate flavor of the chicken takes well to the strength and freshness of the tangy herb sauce. Poaching a chicken is the age-old way of preparing chicken broth and dinner at the same time.

1 chicken (3 to 4 pounds; 1.5 to 2 kg), at room temperature, trussed

2 large onions, halved and stuck with 2 whole cloves

3 plump fresh garlic cloves

Several parsley stems, celery leaves, and sprigs of thyme, wrapped in several ribs of celery and tied in a bundle with cotton twine

4 large carrots, trimmed, peeled, and tied in a bundle

4 ribs celery, trimmed and tied in a bundle

6 whole black peppercorns

1 recipe Fresh Herb Sauce (page 266)

1. Place all the ingredients except the herb sauce in an 8-quart (8-l) stockpot. Add cold water to cover and bring just to a simmer over high heat. Skim off the impurities that rise to the surface. Reduce the heat and simmer very gently for 3 hours, skimming as necessary.

2. To serve: Remove the chicken and set aside to drain. With a slotted spoon, remove the vegetables. Discard the herb bundle and the garlic. Untie the chicken and vegetables. Carve the chicken and place on a large warmed platter. Surround the chicken with the vegetables. Serve immediately, with the herb sauce.

3. To prepare chicken broth from the remaining liquid: Line a fine-mesh sieve with dampened cheesecloth and set over a large bowl. Ladle—do not pour—the liquid into the bowl. The broth can be stored, covered, in the refrigerator for up to 3 days or frozen for up to 1 month.

Yield: 6 servings of chicken, and about 3 ½ quarts (3.5 l) chicken broth

WINE SUGGESTION: I like red with this dish, a good Chianti Classico or Barbera d'Alba.

For Quick Crushed Peppercorns

Although one can purchase crushed or cracked peppercorns, home-crushed pepper will have a far superior, fresher, more pungent flavor. Here's how: On a work surface, carefully crush the peppercorns with a heavy mallet or with the bottom of a heavy skillet. Alternatively, crush the peppercorns in a mortar with a pestle.

Chicken Cacciatora

Pollo alla Cacciatora

Is there a chicken dish more universal and more universally loved than chicken cacciatora? When I was a child, this was one of my favorite dishes, and I can still smell my mother's chicken simmering in her ever-sturdy skillet on the stove. Today, all I have to say is "chicken cacciatora," and everyone's in a cheery mood. Like many dishes in the book, this version is a personal composite of many I sampled along the trattoria trail, and shows off to good advantage one of the Italian national vegetables, celery. Be sure to buy the best chicken you can find: It will make all the difference between an everyday dish and one that's really special. I like to serve this on its own as a main course, but you may want to accompany it with steamed or boiled potatoes.

1 chicken (3 to 4 pounds; 1.5 to 2 kg), at room temperature, cut into 8 serving pieces

Sea salt and freshly ground black pepper to taste

3 tablespoons extra-virgin olive oil

1 tablespoon unsalted butter

1 small onion, minced

2 ribs celery, thinly sliced

¼ teaspoon crushed red peppers (hot red pepper flakes), or to taste

One 28-ounce (765-g) can peeled Italian plum tomatoes in juice or one 28-ounce (765-g) can crushed tomatoes in purée

Several sprigs of fresh parsley, bay leaves, sprigs of fresh rosemary, and celery leaves, tied in a bundle with cotton twine

1. Season the chicken liberally with salt and pepper. In a large skillet, combine the oil and butter over high heat. When hot, add several pieces of chicken and cook on the skin side until it turns an even golden brown, about 5 minutes. Turn the pieces and brown them on the other side, about 5 minutes more. Do not crowd the pan; brown the chicken in several batches. Carefully regulate the heat to avoid scorching the skin. When all the pieces are browned, transfer them to a platter.

2. Add the onion, celery, crushed red peppers, and salt to the fat in the pan and cook over moderate heat until the onion and celery are soft and translucent, 4 to 5 minutes. If using whole canned tomatoes, place a food mill over the skillet and purée the tomatoes directly into it. Crushed tomatoes can be added directly from the can. Add the herb bundle, stir to blend, and simmer for about 5 minutes. Bury the chicken in the sauce, and simmer, partially covered, until the chicken is cooked through, 25 to 30 minutes more. Remove and discard the herb bundle.

3. Transfer the chicken to warmed dinner plates, along with the sauce. Serve immediately.

■ Yield: 4 to 6 servings ■

WINE SUGGESTION: For a favorite dish, drink your favorite red, such as a simple Chianti Classico.

"There was one more formative experience, and that was Italy. It was still a country where wine was a part of life—we picked the grapes from the roadside vineyard to quench our thirst as the Eighth Army clanked and rumbled its way northwards in a cloud of dust—and where men grew wine as a matter of course, and put grimy carafes of it on the table at every meal, equally a matter of course."

CYRIL RAY, RAY ON WINE

Peperoncini: Trip to Hell!

The Italians call them *peperoncini;* we know the dried, fiery red spice as chile peppers. On one visit to the large indoor fruit and vegetable market in Florence, I spied a merchant who sold baskets of whole, cayenne-like peperoncini with a sign in English warning: "Trip to Hell!" Hot peppers are used sparingly in most Italian cooking, generally in stews and sauces, or to flavor olives or oil. Chile peppers not only make a dish fiery, but also serve as subtle flavor enhancers, adding a stimulating layer of flavor. Since the intensity of hot peppers varies, the recipes here suggest quantities for crushed red peppers, readily found packed in glass jars and sold in supermarket spice departments. Internationally, the intensity of peppers is rated on a Scoville heat indicator scale, and these crushed red peppers range from 25,000 to 50,000 heat units. (As a point of comparison, cayenne pepper is rated at 40,000 heat units, jalapeño peppers at 55,000 heat units. On the scale, a pepper rated at 20,000 heat units is half as hot as a pepper rated at 40,000.) Bottled crushed red peppers may come from China, India, Pakistan, or all three. Generally the entire pepper is dried and then crushed, with the stem and cap removed. Dried chile peppers—capsicum—should be refrigerated, to maintain their shiny, fresh color. Crushed red peppers have a shelf life of about two years. Discard any dried peppers that have become cloudy or dark red, for they can turn a dish harshly bitter. When traveling in Italy, purchase whole dried peperoncini, and use to taste in any recipe calling for crushed red peppers. Just be sure to remove the whole peppers at serving time!

Rabbit with Red Peppers and Polenta

Coniglio alla Cacciatora

Even though I first sampled this dish one golden, sunny evening in May in the Piedmont, it always reminds me of Christmastime, especially when the polenta is prepared with a pale ivory-white cornmeal. The reddish hues of the cacciatora—a brilliant red sauce of red peppers punctuated with a touch of tomatoes and an avalanche of herbs—are gorgeous alongside a little "fence" of steaming polenta. The recipe was shared with me by the hard-working, enthusiastic Claudia Verro. Along with her husband, Tonino, she runs one of Italy's more charming and romantic restaurants, La Contea, in the hamlet of Neive. When preparing this dish, be sure that you carefully stem the thyme and finely chop the rosemary beforehand, or you're likely to feed your guests, and yourself, sticks of herbs.

(continued)

1 whole fresh rabbit (about 3 pounds; 1.5 kg), cut into serving pieces

9 tablespoons extra-virgin olive oil

5 tablespoons freshly squeezed lemon juice

2 tablespoons fresh thyme leaves

2 tablespoons fresh rosemary leaves, finely chopped

3 bay leaves, preferably fresh

4 red bell peppers, thinly sliced

One 16-ounce (480-g) can peeled Italian plum tomatoes in juice or one 16-ounce (480-g) can crushed tomatoes in purée

Sea salt and freshly ground black pepper to taste

1 recipe freshly made Polenta (page 172), served warm in a mound

1. In a large shallow dish, marinate the rabbit pieces in 3 tablespoons of the oil, the lemon juice, and the herbs at room temperature for at least 1 hour, and up to 3 hours, turning from time to time.

2. In a skillet large enough to hold the rabbit later on, heat 3 tablespoons of the oil until hot but not smoking. Add the sliced peppers, toss to coat with oil, and reduce the heat to low. Cover and cook until soft, stirring from time to time, about 10 minutes. Do not let the peppers burn. If using whole canned tomatoes, place a food mill over the skillet and purée the tomatoes directly into it. Crushed tomatoes can be added directly from the can. Season with salt and pepper, stir to blend, cover, and continue cooking for 30 minutes more, stirring from time to time.

3. Meanwhile, cook the rabbit: Pat the rabbit pieces dry and season with salt and pepper. Reserve the marinade. In a very large skillet, heat the remaining 3 tablespoons oil over moderately high heat. When hot but not smoking, add the rabbit pieces. Reduce the heat immediately to low (to keep the rabbit meat from drying out), cover, and cook, shaking the pan from time to time, until the rabbit is lightly browned but still moist, about 5 minutes per side. (Cooking time will vary according to the size of the pieces.)

4. Add the rabbit pieces to the simmering pepper mixture, burying the rabbit in the sauce. Deglaze the pan in which the rabbit was cooked with the reserved marinade, scraping up any bits that cling to the bottom. Add this liquid to the pepper and rabbit mixture. Stir, cover, and cook for 30 minutes more. Remove and discard the bay leaves.

5. To serve, place the rabbit pieces on warmed dinner plates, cover with the pepper-tomato sauce, and place a dollop of polenta alongside. Serve immediately.

■ Yield: 4 to 6 servings ■

WINE SUGGESTION: With this meal, we sampled a spectacular old Barolo, and you should do so if your pocketbook can afford it. Otherwise, any robust red would go well with this brilliant dish.

Cutting up a Rabbit

To cut up a whole rabbit for cooking: Place the rabbit belly side down on a clean work surface. Trim off the flaps of skin, the tops of the forelegs, and any excess bone. With a cleaver or a heavy knife, divide the carcass crosswise into three sections: hind legs, saddle, and forelegs, including the rib cage. Cut between the hind legs to separate them into two pieces. Split the front carcass into two pieces to separate the forelegs. Split the saddle crosswise into three even pieces.

Cubed Pork with Garlic, Spinach, and Spicy Chick Peas

Ceciata di Suino della Casa

Bright and colorful, with the spring green of spinach, the mahogany hues of the pork, and the wheaty tones of garbanzo beans, this unusual pork dish is a specialty of Trattoria Cammillo, a popular Florentine trattoria that's been a gathering spot since 1946. Then, Cammillo Masiero ran a little one-room trattoria where people came to drink, play cards, and eat whatever was being served that day. Today, Cammillo's grandson, Francesco Masiero, runs a bustling, multiroom spot filled with an equal mix of tourists and locals. On my last visit, I spent part of the evening in the tiny kitchen that's open to view, and the cooks kindly walked me through this dish, which has been a family favorite ever since. This dish is equally delicious prepared with cubed lamb.

1½ cups (8 ounces; 250 g) dried chick peas (garbanzo beans)

6 tablespoons extra-virgin olive oil

Several parsley stems, several sprigs of thyme, a bay leaf, and several celery leaves, tied in a bundle with cotton twine

Sea salt to taste

1 pound (500 g) fresh spinach, washed and stemmed

1 pound (500 g) pork loin, cut into 1-inch (2.9-cm) cubes

8 to 12 plump fresh garlic cloves, cut into slivers, or to taste

½ teaspoon crushed red peppers (hot red pepper flakes), or to taste

Freshly ground black pepper to taste

1. Rinse and drain the chick peas, picking them over to remove any pebbles. Place the beans in a large bowl, add boiling water to cover, and set aside for 1 hour. Drain and rinse the beans, discarding the water.

2. Transfer the chick peas to a medium-size saucepan. Add 2 tablespoons of the oil and the herb bundle, cover with fresh cold water, and bring to a simmer over moderate heat. Reduce the heat to low, cover, and simmer for 1 hour. Add the salt and continue simmering until tender, about 1 hour more. (Check the water level in the pan every half hour, adding water as needed.) Once cooked, taste for seasoning, then drain. Remove and discard the herb bundle. Set aside.

3. In a large deep skillet, combine the spinach and salt, cover, and steam the leaves over moderate heat, tossing them from time to time, until they wilt, about 5 minutes. Strain and squeeze the spinach, removing every drop of water you can. Using a stainless steel knife to prevent discoloration, chop very fine. Set aside.

4. In another large deep skillet, combine the pork, the remaining 4 tablespoons olive oil, the garlic, and the crushed red peppers, tossing to coat with oil. Sauté the pork over moderate heat, adjusting the heat as necessary so the garlic does not scorch. The small cubes of pork should cook in 5 to 7 minutes. Season to taste with salt and pepper. Add the chick peas, then the spinach, and cook just until warmed through. Taste for seasoning. Transfer to warmed dinner plates and serve immediately.

■ Yield: 4 to 6 servings ■

WINE SUGGESTION: With this dish, sample an equally energetic wine, such as a Vino Nobile di Montepulciano.

Rosemary Roast Loin of Pork

Arista

"The roast, flavored with herbs and larded with olive oil, mounts to the nostrils and invites to the feast."

ANGELO PELLEGRINI, THE UNPREJUDICED PALATE

The secret to this succulent roast pork is to begin roasting at high heat—to sear and brown the surface and help seal in the juices—then continue to roast gently at a lower heat, to cook it slowly without drying out. Don't cheat and cook the roast without the bones: They add essential flavor to the meat. But do not forget to ask your butcher to crack the bones, to make it easy for you to slice the roast into thick chops at serving time. I always serve this Florentine specialty with oven-roasted potatoes. An incredible version of this pork roast can be found at Florence's Coco Lezzone, an elbow-to-elbow trattoria where one sits at long, shared tables. The waiter there confided that their secret was to cook the meat on the bone, using the high-heat/low-heat method for a crusty exterior and moist interior. As an extra touch, the roast is studded with whole sprigs of fragrant, fresh rosemary. Any leftover pork is great the next day, as picnic fare or cold cuts.

1 loin-end pork roast, bones split (about 5 pounds; 2.5 kg)

2 sprigs of fresh rosemary

Sea salt and freshly ground black pepper to taste

About 1½ cups (375 ml) dry white wine, such as a Pinot Grigio

About 1½ cups (375 ml) water

1. Preheat the oven to 400°F (200°C; gas mark 6/7).

2. Using a thick needle or a sharp knife, pierce the meatiest center portion of the pork at either end. Carefully insert a sprig of rosemary in each slit. Season the roast generously with salt and pepper. Place the pork, fat side up, bone portion down, on a roasting rack in a roasting pan. Place in the center of the oven and roast until the skin is crackling and brown and the meat begins to exude fat and juices, about 30 minutes.

3. Reduce the heat to 325°F (160°C; gas mark 3/4) and baste with the juices from the pan, adding about ½ cup (125 ml) each of the white wine and water to the pan juices. Continue to add wine and water as needed, to maintain a thin layer of liquid in the pan at all times, and baste at 20 minute intervals. Roast the pork for about 25 minutes per pound, until an instant-read meat thermometer registers 155°F (85°C). (Alternatively, insert a standard meat thermometer into the roast, making sure it does not touch the bone.) Remove from the oven, season immediately with salt and pepper, and cover loosely with foil. Let stand for 15 minutes, or until the thermometer registers 160°F (90°C). (Since the meat will continue to cook—and eventually dry out—as it rests, do not allow the pork to rest for more than 15 minutes.)

4. Meanwhile, skim the fat from the pan juices, and place the roasting pan over moderate heat, scraping up any bits that cling to the bottom. If necessary, add several tablespoons of cold water to deglaze (hot water would cloud the sauce), and bring to a boil. Cook, scraping and stirring, until the liquid is almost caramelized, 2 to 3 minutes. Do not let it burn. Spoon off and discard any excess fat. Reduce the heat to low and simmer just until thickened, 2 to 3 minutes more. Strain the sauce through a fine-mesh sieve and pour into a warmed sauceboat.

5. Carve the roast into thick chops and serve immediately on warmed dinner plates, passing the sauce.

■ Yield: 6 servings ■

WINE SUGGESTION: This succulent roast calls for a deep ruby red, such as a Vino Nobile di Montepulciano.

Parmesan-Breaded
Baby Lamb Chops

Costolettine d'Abbacchio Fritte

"When I am asked, as I sometimes am, what is the bottle of wine I have most enjoyed, I have to answer that it was probably some anonymous Italian 'fiasco' that I drank one starlit Tyrrhenian night under a vine-covered arbor, while a Neapolitan fiddler played 'Come Back to Sorrento' over the veal cutlet of the young woman I had designs on. . . ."

CYRIL RAY, RAY ON WINE

The Italians are true masters at gently coating meats, poultry, and vegetables with beautifully seasoned combinations of ingredients, and the results could convert anyone who thinks that breaded food has to be heavy. There is nothing tricky or complicated about properly breading and panfrying these lamb chops: Simply pay attention, and follow directions to the letter. I love these on a cool day, served with oven-roasted potatoes (page 54) or fried zucchini blossoms or artichokes (page 56). As a first course, serve sautéed red peppers (page 4). These lamb chops actually benefit from being breaded one to two hours in advance so that the coating dries a bit and better adheres to the meat. I sampled these one December evening in Rome.

8 single-rib lamb chops, bone ends trimmed of fat and meat (about ½ inch/1 cm thick; 6 to 9 ounces/180 to 270 g per chop)

¾ cup (3 ounces; 90 g) freshly grated Italian Parmigiano-Reggiano cheese, in a shallow dish

2 large eggs, lightly beaten, in a shallow dish

1 cup (250 ml) very fine fresh bread crumbs, in a shallow dish

About 2 cups (500 ml) peanut oil, for frying

Fine sea salt and freshly ground black pepper to taste

2 lemons, quartered, for garnish

1. With a meat pounder or the side of a heavy cleaver, gently pound the meat of the lamb chops, to form an even chop, not to flatten to a pancake. Hold a lamb chop by the bone and turn each side of the chop in the cheese, shaking off the excess. Immediately turn each side of the chop in the beaten eggs, shaking off the excess. Turn each side of the chop in the bread crumbs, shaking off the excess. Transfer to a platter. Repeat with the remaining lamb chops. (The lamb may be coated up to 1 hour in advance and held at room temperature, or up to 4 hours in advance and refrigerated; remove from the refrigerator 1 hour before preparation.)

2. In a large skillet, heat the oil over moderately high heat until hot but not smoking. Place as many lamb chops as possible in the skillet without crowding. Cook until golden brown, about 4 minutes. Using tongs, carefully turn each chop, trying not to upset the coating, and season the browned side with salt and pepper. Continue cooking until the second side is golden brown, about 4 minutes more. Season the second side and drain briefly on paper towels. Transfer to warmed dinner plates and serve immediately, with the wedges of lemon for squeezing over the lamb chops.

■ Yield: 4 servings ■

WINE SUGGESTION: This sturdy dish deserves a good, tasty red, such as a Montepulciano d'Abruzzo.

Grilled Marinated Lamb Chops with Lemon and Oil

Costolette d'Agnello a Scottadito

The secret of these light, thin chops is to dry-cook them over high heat, producing a crisply golden and crusty exterior, and a tender, meaty interior. For this style of dry-cooking, you need tender meat, such as top-quality lamb chops or slices of leg of lamb. To keep the interior of the meat moist during cooking (and to keep it from sticking to the unoiled pan or grill), the meat is marinated first, in a classic mixture of top-quality olive oil and freshly squeezed lemon juice. Do not salt the meat, or it will draw out the flavorful juices. These thin and dainty lamb chops—meant to be eaten with your hands—are justly called *scottadito*, or "burning fingers." I've sampled them at several trattorias in Rome. My favorite accompaniments are Roasted Rosemary Potatoes (page 54) and Sautéed Spinach with Garlic, Lemon, and Oil (page 62).

8 single-rib lamb chops, bone ends trimmed of fat and meat (about ½ inch/1 cm thick; 6 to 9 ounces/180 to 270 g per chop)

¼ cup (60 ml) extra-virgin olive oil

3 tablespoons freshly squeezed lemon juice

Fine sea salt and freshly ground black pepper to taste

Lemon wedges, for garnish

1. In a large shallow dish, marinate the lamb chops in the oil and lemon juice at room temperature for at least 1 hour, and up to 3 hours, turning from time to time. Pat the meat dry and set aside.

2. Preheat a heavy-duty cast-iron skillet or cast-iron griddle pan over high heat for 5 minutes. Or prepare a wood or charcoal fire. The fire is ready when the coals glow red and are covered with ash.

3. If cooking in a skillet or a griddle, reduce the heat to moderate, add the lamb chops, and cook until nicely browned, about 2 minutes per side for rare. Season each side with salt and pepper after cooking. Transfer to warmed dinner plates and serve immediately, with the wedges of lemon for squeezing over the lamb chops.

■ Yield: 4 servings ■

WINE SUGGESTION: Set aside a fine, tasty red, such as a Montepulciano d'Abruzzo, for these deliciously simple lamb chops.

Testing for Doneness

To test for doneness when cooking lamb chops or other meat, press the meat with the tip of your finger. If the meat is very soft, the lamb chops are rare. If the meat is medium-soft, the lamb chops are medium-rare. If the meat is very firm, the lamb chops are well-done.

Lamb Braised in White Wine, Garlic, and Hot Peppers

Abbàcchio alla Cacciatora

> *"When in Rome, live as the Romans do; when elsewhere, live as they live elsewhere."*
>
> ADVICE TO SAINT AUGUSTINE FROM SAINT AMBROSE, BISHOP OF MILAN (A.D. 339–397)

Fragrant and bold, with a surprising complexity of flavors, this wintry dish from Trattoria Checchino dal 1887 in Rome is the sort of dish I love to make when I'm at home all day, puttering around the house. The aromatic combination of vinegar, white wine, anchovies, hot peppers, and garlic is an unusual one, but the flavor results are supremely satisfying, as each ingredient manages to have its say. While white wine is traditionally used in this classic dish, I've prepared it with both red and white with equal success. Likewise, the cut of lamb used here should be dictated by your pocketbook. I've prepared the dish with leg of lamb and neck of lamb, and one could also use a good lamb shoulder.

3 tablespoons extra-virgin olive oil

10 flat anchovy fillets in olive oil, drained and minced

¾ teaspoon crushed red peppers (hot red pepper flakes), or to taste

2½ pounds (1.25 kg) lamb (leg of lamb, shoulder, or neck), cut into 3-inch (8-cm) cubes

Sea salt and freshly ground black pepper to taste

1 cup (250 ml) dry white wine, preferably a young Chardonnay

¼ cup (60 ml) best-quality red wine vinegar

3 plump fresh garlic cloves, minced

½ teaspoon dried leaf oregano

1 teaspoon superfine flour, such as Wondra

1. In a 6-quart (6-l) flameproof casserole with a lid, combine the oil, 5 of the anchovy fillets, and ½ teaspoon of the crushed red peppers over moderate heat. Cook just until the peppers begin to color the oil, 2 to 3 minutes. Add the cubed lamb and brown carefully on all sides, about 3 to 4 minutes per side. Do not crowd the pan: The pieces of meat should not touch as they brown. The lamb may need to be browned in batches. Season the lamb generously with salt and pepper. Add the wine, vinegar, and garlic. Cover, reduce the heat to low, and cook until tender, about 1 hour.

2. To serve, remove the pieces of meat to a warmed serving platter. Cover the platter with foil to keep the lamb warm. Leave the sauce in the casserole, keeping it warm over very low heat.

3. Finish the sauce: In a small bowl, combine the oregano, flour, the remaining ¼ teaspoon crushed red peppers, the remaining 5 anchovy fillets, and about 3 tablespoons of the sauce. Stir to blend. Whisk this mixture into the sauce in the casserole, stirring over low heat until the flour is cooked and the sauce begins to thicken, 1 to 2 minutes. Taste the sauce, and adjust the seasoning as necessary. Serve the lamb on warmed dinner plates, spooning the sauce on top of the meat.

▪ Yield: 4 to 6 servings ▪

WINE SUGGESTION: This dish deserves an 8- or 9-year-old Chianti Riserva, such as Chianti Ruffino Riserva.

· · ·

Braised Veal Shanks with Lemon and Parsley Garnish

Osso Buco

A great osso buco can bring you to your knees. With veal that's chewy, meaty, yet pale and tender—braised to perfection—topped by a lively and colorful blend of lemon zest, parsley, and garlic, it is one of the world's great dishes. As an added advantage, the core of the bones of the veal shank is filled with delicious marrow, which can be scooped out with a tiny spoon. But the greatness of an osso buco (literally, "hollow bone") lies in securing top-quality ingredients, then following through with careful and attentive braising. The lemon and parsley garnish—known as *gremolata*—helps transform what could be an ordinary dish into an unforgettable one. Ask your butcher to cut a meaty veal shank crosswise into very thick (3-inch; 8-cm) slices. The hindshank will produce more marrow and meatier pieces of veal. Traditionally, osso buco is served with Saffron Risotto (page 162).

4 tablespoons (2 ounces; 60 g) unsalted butter

3 tablespoons extra-virgin olive oil

Six 3-inch (8-cm) meaty veal shanks (about 6 pounds; 3 kg)

Sea salt and freshly ground black pepper to taste

1 medium onion, minced

1 rib celery, minced

1 carrot, minced

3 to 4 cups (750 ml to 1 l) chicken stock, preferably homemade (page 272)

4 medium tomatoes, peeled, cored, seeded, and chopped

LEMON AND PARSLEY GARNISH

2 plump fresh garlic cloves, degermed and minced

½ cup (125 ml) minced fresh flat-leaf parsley leaves

Grated zest (yellow peel) of 2 lemons

1. In a very large deep skillet, heat the butter and oil until hot. Working in batches, add the veal to the skillet and brown over moderate heat until browned on all sides, about 10 minutes. As the veal pieces are browned, transfer to a platter and season with salt and pepper. Set aside.

2. Add the onion, celery, and carrot to the skillet, and cook until softened but not browned, about 5 minutes. Deglaze with about ½ cup (125 ml) of the stock, using a metal spatula to scrape up any bits that cling to the bottom of the pan. Cook until most of the liquid has evaporated. (This will help begin building the flavor base of the sauce.)

3. Return the veal to the skillet. Add the tomatoes, along with enough stock to almost immerse the veal. Cover the skillet and simmer very gently for 1½ hours, or until the meat is almost falling off the bone. Watch the pan carefully, keeping the heat as low as possible, and adding liquid as necessary to keep the pan moist. The resulting sauce should be thick and gelatinous.

4. As the veal cooks, finely chop together the garlic, parsley, and grated lemon zest to make the gremolata. Five minutes before serving, sprinkle some of the gremolata over the meat, allowing its aromas to penetrate the tender veal.

5. Serve the veal shanks on 4 warmed dinner plates, spooning the sauce over the veal, and garnishing with the remaining gremolata if desired.

■ Yield: 4 servings ■

WINE SUGGESTION: A cold-weather red wine, such as a Gattinara or a Chianti Riserva, is my choice for this wintry dish.

The Art of Braising

Braising is an art unto itself. Proper braising turns tough and fibrous cuts of meat into soft and pleasantly chewy meat. After preliminary browning, the meat is cooked slowly in a bit of liquid, while its natural gelatin helps to form a thick and luxurious sauce. A braise is not a stew: The meat cooks in just enough liquid to keep the meat moist. Since the pan is covered during braising, the steam that rises helps keep the exposed portion of meat moist. Either water, wine, or stock can be used for braising, but remember that it is the braising liquid that will add body to the final sauce.

• • •

Beef Braised in Barolo Wine

Brasato al Barolo

ITALIAN PROVERB:

*"One barrel of wine can work more miracles
than a church full of saints."*

This moist, fragrant, and flavorful braised beef is perfect for entertaining. Everything can be prepared a day in advance and reheated at serving time. The whole piece of beef is marinated, then slowly braised, making for meat that is full of a complex combination of flavors, including rosemary, wild mushrooms, and rich red wine. This recipe comes from Angelo Maionchi, chef at Turin's historic Del Cambio. Although Barolo is the traditional wine for this dish, the less-expensive Barbera can be substituted without loss of quality. Traditionally, beef in Barolo is served with slices of polenta alongside (page 172). Other accompaniments might include baked and seasoned rice or pasta tossed with butter and fresh rosemary (page 134).

(continued)

2 pounds (1 kg) boneless braising beef in one piece (such as chuck roast)

1 bottle tannic red wine, preferably a Barolo or Barbera

¼ cup (60 ml) brandy

¼ cup (60 ml) plus 3 tablespoons extra-virgin olive oil

2 carrots, cut into thin rounds

2 bay leaves, preferably fresh

3 ribs celery, cut into thin slices

1 sprig fresh rosemary

1 teaspoon whole black peppercorns, crushed

½ cinnamon stick

4 whole cloves

2 medium onions, peeled and halved

Sea salt and freshly ground black pepper to taste

1. In a 3-quart (3-l) heavy-duty flameproof casserole with a lid, combine the beef, wine, brandy, ¼ cup (60 ml) of the olive oil, the carrots, bay leaves, celery, rosemary, crushed peppercorns, and cinnamon stick. Stick a clove into each of the onion halves and add to the casserole. Cover and refrigerate for at least 12 hours or up to 24 hours. Turn the meat from time to time.

2. At least 2 hours before cooking the meat, remove the casserole from the refrigerator to bring the meat to room temperature.

3. Remove the beef from the marinade and pat dry. Pour the marinade into another container and reserve. Rinse and dry the casserole. Add the remaining 3 tablespoons of olive oil to the casserole and heat the oil over moderately high heat. When the oil is hot but not smoking, add the beef and brown evenly on all sides, 3 to 4 minutes per side. Season the beef generously with salt and pepper. Add the marinade to the casserole. Bring just to a simmer, then reduce heat to very low and cover. Simmer gently until the beef is fork-tender, about 3 hours. Turn the meat from time to time, and do not let much of the liquid evaporate. You should have about 1 cup (250 ml) of liquid remaining for the sauce.

4. To serve: Transfer the beef to a carving board. Cut the beef against the grain into thick slices, and transfer the slices—slightly overlapping them—to a warmed serving platter. Cover with foil to keep warm. Strain the cooking liquid through a fine-mesh sieve and discard the solids. Return the liquid to the casserole, taste for seasoning, and, if necessary, boil until reduced to 1 cup (250 ml). Spoon the sauce over the beef and serve immediately. (Alternatively, the beef may be sliced and refrigerated for 1 more day, along with the sauce. Simply reheat at serving time.)

■ Y i e l d : 4 t o 6 s e r v i n g s ■

WINE SUGGESTION: A Barolo or Barbera, both from the Piedmont, are not only traditional but also ideal red wines for this dish. Their acidity, combined with a flavor that's full-bodied and robust, makes for a lovely food and wine marriage.

New Life for a Rich Sauce

Should you have even a few tablespoons of the delicious Barolo sauce left over, use it to add to a flavorful vinaigrette to toss over cold beef or salad greens. If you can't use the sauce right away, freeze in a small container and save for a rainy day.

On Warmed Plates

Nothing is more frustrating than to be served a succulent portion of meat on a cold plate. Since meat tends to cool as it rests, it is rarely piping hot at serving time. To best enjoy a warm meal, be sure to make a practice of warming plates ahead of time. You don't need to install an "official" plate-warming drawer (though if you're planning a kitchen, don't leave it out!). Here are some tips on do-it-yourself plate warmers:

▪ Warm plates in a warm oven.

▪ Or warm them in a microwave, heated at a low setting.

▪ If you have a free sink, you can fill it or a large wash basin with hot tap water, and keep the plates warm in the hot water.

▪ If at serving time you find you've forgotten to do any of the above, run very hot tap water over the plates to warm them, and dry before serving.

Braised Oxtail with Tomatoes, Onions, and Celery

Coda alla Vaccinara

Dark, rich, and meaty, this warming, wintry dish is typical of the hearty cold-weather fare served at Checchino dal 1887, near the old Roman stockyards. The restaurant all but put this dish on the map, although it has been changed and lightened over the years. Rather than pork rind and pork fat, only pancetta and olive oil are now used to cook the oxtail. Believe me, there is no loss of flavor, and the dish is a good deal more healthful. To reduce the fat even further, prepare the dish a day or two in advance, chill it, and then remove any solidified fat that has risen to the surface. The final dish should not be swimming in fat, but there should be enough tomato sauce left to use your homemade bread for sopping up! *Coda*, by the way, is Italian for oxtail. *Vaccinara* is the old name for a butcher in the Roman dialect: They worked in the stockyards, often lunching on oxtail stew at a nearby trattoria. This recipe dates from 1887, when Checchino was founded. The inclusion of golden raisins, pine nuts, and a touch of chocolate dates from the original recipe, but is optional. This is the type of main dish that demands no vegetable accompaniment. I like to precede it with Celery Salad with Anchovy Dressing (page 34).

(continued)

3 tablespoons extra-virgin olive oil

½ cup (2 ounces; 60 g) minced pancetta (see Note)

5 pounds (2.5 kg) oxtail, cut into 4-inch (10-cm) pieces (about 15 pieces)

Sea salt and freshly ground black pepper to taste

2 whole cloves

3 small onions, peeled and halved

3 plump fresh garlic cloves, minced

2 cups (500 ml) dry white wine, preferably a Chardonnay

One 28-ounce (785-g) can peeled Italian plum tomatoes in juice or one 28-ounce (785-g) can crushed tomatoes in purée

8 ribs celery, trimmed to 6-inch (15-cm) lengths

1 ounce (30 g) unsweetened chocolate, grated (optional)

2 tablespoons pine nuts (optional)

2 tablespoons golden raisins (optional)

1. In a 6-quart (6-l) flameproof casserole with a lid, combine the oil and pancetta over moderate heat. Cook the pancetta just until browned and crisp, 3 to 4 minutes. Remove the pancetta with a slotted spoon and set aside. Add the oxtail pieces and brown thoroughly on all sides, about 15 minutes. This may have to be done in batches. Do not crowd the meat in the pan, and do not allow the pieces of oxtail to touch. Once the meat is browned, season it generously with salt and pepper. Stick the cloves into two of the onion halves and add to the casserole. Add the remaining onions, the browned pancetta, and the garlic, and cook for 2 to 3 minutes. Add the wine and stir to incorporate. If using whole canned tomatoes, place a food mill over the casserole and purée the tomatoes directly into it. Crushed tomatoes can be added directly from the can. Cover, and bring just to a simmer over moderate heat. Reduce the heat to very low and simmer gently until the oxtail is fork-tender and the meat is falling off the bones, about 4 hours. Turn the meat two or three times during the cooking period. (The stew may be prepared up to this point 1 day in advance. Remove the casserole from the heat and allow to cool for several hours. Cover and refrigerate. At serving time, remove the casserole from the refrigerator and, with a small spoon, remove and discard any fat that has solidified on the surface. Bring to a simmer before proceeding with the recipe.)

2. Add the celery, slipping it under the pieces of oxtail so it cooks in the sauce. Simmer until the celery is tender, about 30 minutes. About 10 minutes before the celery is cooked, stir in the chocolate and add the pine nuts and raisins, if using. Taste the sauce, seasoning it as necessary. To serve, transfer the pieces of oxtail to individual warmed dinner plates. Spoon several tablespoons of the sauce around the meat, and arrange the pieces of celery alongside.

■ Yield: 4 to 6 servings ■

WINE SUGGESTION: With this dish, we drank a 4-year-old Colle Picchioni, considered one of the best wines of the Castelli Romani area of Rome. Its sturdiness stands up well to the robust flavors of the oxtail.

NOTE: There is no American equivalent for pancetta, the unsmoked Italian bacon that is cured with salt and mild spices and rolled. Pancetta is prized for its subtle, delicate flavor, and can be found in most Italian specialty shops. If you cannot find it, substitute a very lean, top-quality bacon. Blanch it for 1 minute in boiling water, then drain thoroughly. Blanching will remove the smoked flavor from the bacon without cooking it.

HERE'S TO THE

RICCI HERITAGE

■

Thank you Mom, Grandpa Felix, Aunt Ella, and Aunt Flora, for the great taste memories of childhood. The flavors of these trattoria dishes take me back to the days when we crowded around the table—aunts, uncles, and cousins—to share in a family feast. We didn't drink wine back then, but today, I'd opt for a gutsy, ruby Montepulciano d'Abruzzo Rosso, in honor of our Abruzzi ancestry.

AUNT FLORA'S OLIVE SALAD

CHICKEN CACCIATORA

**BAKED RISOTTO WITH ASPARAGUS,
SPINACH, AND PARMESAN**

SOURDOUGH BREAD

FRAGRANT ORANGE AND LEMON CAKE

SAUCES, BROTHS, SPREADS, AND CONDIMENTS

Tomato Sauce

Salsa di Pomodoro

Tomato sauce should be rich, elegant, smooth, and redolent of herbs, even if it's to figure as the centerpiece of a basically peasant dish. Here is my everyday favorite. I sometimes make up a double batch, so there's always some in the freezer for those days I don't have time to cook.

¼ cup (60 ml) extra-virgin olive oil

1 small onion, minced

3 plump fresh garlic cloves, minced

Sea salt to taste

One 28-ounce (765 g) can peeled Italian plum tomatoes in juice or one 28-ounce (765 g) can crushed tomatoes in purée

Several sprigs of fresh parsley, bay leaves, and celery leaves, tied in a bundle with cotton twine

In a large unheated saucepan, combine the oil, onion, garlic, and salt, and stir to coat with oil. Cook over moderate heat just until the garlic turns golden but does not brown, 2 to 3 minutes. If using whole canned tomatoes, place a food mill over the skillet and purée the tomatoes directly into it. Crushed tomatoes can be added directly from the can. Add the herb bundle, stir to blend, and simmer, uncovered, until the sauce begins to thicken, about 15 minutes. For a thicker sauce, for pizzas and toppings, cook for 5 minutes more. Taste for seasoning. Remove and discard the herb bundle. The sauce may be used immediately, stored in the refrigerator for up to 2 days, or frozen for up to 2 months. If small quantities of sauce will be needed for pizzas or other toppings, freeze in ice cube trays.

■ Yield: About 3 cups (750 ml) sauce ■

Tomato-Mushroom Sauce

Sugo di Pomodoro e Funghi

This simple tomato and mushroom sauce is ideal for serving alongside Tartra (page 30), the savory, herb-flecked custard from the Piedmont. The sauce can also serve as an all-purpose sauce for pasta or as a vegetarian sauce for a baked pasta dish.

- ¼ cup (60 ml) extra-virgin olive oil
- 1 small onion, minced
- 3 plump fresh garlic cloves, minced
- 1 small carrot, minced
- 1 small rib celery, minced

Sea salt to taste

- 8 ounces (250 g) fresh brown mushrooms, such as cremini, rinsed, dried, and thinly sliced
- One 28-ounce (765-g) can peeled Italian plum tomatoes in juice or one 28-ounce (765-g) can crushed tomatoes in purée
- Several sprigs of fresh parsley, bay leaves, and celery leaves, tied in a bundle with cotton twine

In a large unheated saucepan, combine the oil, onion, garlic, carrot, celery, and salt, and stir to coat with oil. Cook over low heat just until the vegetables are soft but not brown, 5 to 6 minutes. Add the mushrooms and cook until soft, 3 to 4 minutes more. If using whole canned tomatoes, place a food mill over the skillet and purée the tomatoes directly into it. Crushed tomatoes can be added directly from the can. Add the herb bundle, stir to blend, and increase the heat to moderate. Simmer, uncovered, until the sauce begins to thicken, about 15 minutes. Taste for seasoning. Remove and discard the herb bundle. The sauce may used immediately, refrigerated for 1 to 2 days, or frozen for up to 1 month.

■ Yield: About 3 cups (750 ml) sauce ■

Fantastic Tomato-Artichoke Sauce

Salsa Fantastica

O n one of my trips to Italy, I picked up a little jar of this orange-red sauce, partially because I loved the name. Of course, the fact that it included two of my favorite foods—tomatoes and artichokes—didn't hurt a bit. The combination may at first seem unusual, but once you taste the union, with the tomato's light acidity and the creamy richness of the puréed artichokes, I think you'll agree that they're a marriage made in heaven. This sauce can be spread on toasted homemade bread, used as a dip for vegetables, or tossed with hot pasta, such as farfalle.

½ cup (125 ml) Tomato Sauce (page 256)

4 small Marinated Baby Artichoke Hearts Preserved in Oil (page 18), drained and quartered

2 tablespoons extra-virgin olive oil

2 teaspoons fresh thyme leaves

Sea salt to taste

Combine all the ingredients in the bowl of a food processor and purée. Taste for seasoning. Transfer to a small bowl. The sauce can be refrigerated for up to 3 days. Bring to room temperature and stir before serving.

▪ Yield: ¾ cup (185 ml) sauce ▪

Tomato and Cream Sauce

Sugo di Pomodoro e Panna

omato sauce all on its own is wonderful. Add a touch of cream and you've moved into another, more elegant world altogether. The garlic and peppers can be adjusted to personal tastes, though I enjoy a sauce with a good hit of both.

¼ cup (60 ml) extra-virgin olive oil

4 plump fresh garlic cloves, minced

Sea salt to taste

½ teaspoon crushed red peppers (hot red pepper flakes), or to taste

One 28-ounce (765-g) can peeled Italian plum tomatoes in juice or one 28-ounce (765-g) can crushed tomatoes in purée

1 cup (250 ml) heavy cream

In a large unheated skillet, combine the oil, garlic, salt, and crushed red peppers, stirring to coat with oil. Cook over moderate heat just until the garlic turns golden but does not brown, 2 to 3 minutes. If using whole canned tomatoes, place a food mill over the skillet and purée the tomatoes directly into it. Crushed tomatoes can be added directly from the can. Stir to blend and simmer, uncovered, until the sauce begins to thicken, about 15 minutes. Add the cream, stir, and heat for 1 minute. Taste for seasoning. Use over pasta, or with fresh lasagne (page 126).

■ Yield: 1 quart (1 l) sauce ■

Pungent Parsley Sauce

Salsa Prezzemolo

T oss this with hot spaghetti, spread it on bread, use it as a dip for raw celery or strips of fresh fennel. Pungent and verdant parsley "pesto" can be found in specialty stores in Italy, but the sauce can be prepared at home in a minute.

3 to 5 plump fresh garlic cloves (to taste), degermed and minced

½ teaspoon fine sea salt, or to taste

6 flat anchovy fillets in olive oil, drained and minced

3 cups (750 ml) loosely packed fresh flat-leaf parsley leaves

3 tablespoons freshly squeezed lemon juice

½ cup (125 ml) extra-virgin olive oil

Place the minced garlic, salt, and anchovies in the bowl of a food processor. Add the parsley and pulse 2 or 3 times, or until the sauce is homogeneous. Scrape down the sides of the bowl. With the machine running, add the lemon juice and then the olive oil in a slow, steady stream. Taste for seasoning. Transfer to a small bowl. Serve immediately. Do not prepare more than 1 hour in advance. As the sauce sits, the anchovy flavor quickly overpowers the fresh parsley taste, so I prefer to prepare it at the last moment.

Yield: 1 cup (250 ml) sauce, or enough
to sauce 1 pound (500 g) of pasta

• • •

Basil and Garlic Sauce

Pesto

Basil and garlic sauce, known as pesto, is so satisfying that one can never have too many versions. Here is the most traditional, including nothing more than basil, garlic, extra-virgin olive oil, and salt, enriched with Parmigiano-Reggiano cheese. Although most traditional recipes call for pine nuts, I choose not to use them here: They are expensive, and the few tablespoons used in most recipes don't play much of a role in enhancing flavor. I've included instructions for preparing the pesto by hand, using a mortar and pestle, or in the food processor.

4 plump fresh garlic cloves, degermed and minced

Fine sea salt to taste

2 cups (500 ml) loosely packed fresh basil leaves and flowers

½ cup (125 ml) extra-virgin olive oil

½ cup (2 ounces; 60 g) freshly grated Italian Parmigiano-Reggiano cheese

BY HAND: Place the garlic and salt in a mortar and mash with a pestle to form a paste. Add the basil, little by little, pounding and turning the pestle with a grinding motion to form a paste. Continue until all the basil has been used and the paste is homogeneous. Add the oil, little by little, working it in until you have a fairly fluid paste. Stir in the cheese. Taste for seasoning. Transfer to a bowl. Stir again before serving.

IN A FOOD PROCESSOR: Place the garlic, salt, and basil in the bowl of a food processor and process to a paste. Add the oil, and process again. Transfer to a bowl and stir in the cheese. Taste for seasoning. Stir again before serving.

Serve immediately, or cover and refrigerate. (The sauce can be stored in the refrigerator for up to 1 day. Bring to room temperature and stir before serving.)

▪ Y i e l d : ¾ c u p (1 8 5 m l) s a u c e ▪

Basil, Garlic, and Tomato Sauce

Pesto alla Santa Margherita

Few flavors offer more pleasure than a fresh, vibrant tomato pesto sauce, a pungent combination of garlic, basil, oil, a touch of tomato, and a good hit of Parmigiano-Reggiano cheese. Lighter than the better-known sauce prepared without tomatoes, this version offers a welcome change of pace. I sampled it at a small restaurant off the harbor of Santa Margherita, in Liguria, the home of the traditional basil-rich sauce. I prefer to prepare the sauce by hand, using a mortar and pestle—it seems more genuine and earthy. But I'm aware not everyone wants to go that extra step, so I also offer instructions for preparing it in a food processor.

4 plump fresh garlic cloves, degermed and minced

Fine sea salt to taste

2 cups (500 ml) loosely packed fresh basil leaves and flowers

1 firm medium-size ripe tomato, peeled, cored, seeded, and chopped

½ cup (125 ml) extra-virgin olive oil

½ cup (2 ounces; 60 g) freshly grated Italian Parmigiano-Reggiano cheese

BY HAND: Place the garlic and salt in a mortar and mash with a pestle to form a paste. Add the basil, little by little, pounding and turning the pestle with a grinding motion to form a paste. Continue until all the basil has been used and the paste is homogeneous. Alternately add the tomato and oil in several additions, working until you have a fairly fluid paste. Stir in the cheese. Taste for seasoning. Transfer to a small bowl. Stir again before serving.

IN A FOOD PROCESSOR: Place the garlic, salt, and basil in the bowl of a food processor and process to a paste. Add the tomatoes and oil, and process again. Transfer to a small bowl and stir in the cheese. Taste for seasoning. Stir again before serving.

Serve immediately, or cover and refrigerate. (The sauce can be refrigerated for up to 1 day. Bring to room temperature and stir before serving.)

■ Yield: ¾ cup (185 ml) sauce ■

Red Pesto Sauce

Pesto Rosso

I first sampled this sauce one summer in the Lake Garda village of Torri del Benaco, where the local wine and specialty shop offered an abundant assortment of sauces, unusual jams, dried mushrooms, oils, vinegars, and, of course, great wines. Each day we walked to town to put together the makings of a grand lunch, and each day I would purchase a different item for sampling. This red pesto sauce is one of my favorites, for it symbolizes much of what is great about Italian cuisine: The combination of sun-dried tomatoes, black olives, a generous hit of fresh herbs, and a touch of garlic and hot pepper makes for a sauce that is complex in its simplicity. When you toss a tablespoon or two with hot spaghetti, or spread it very lightly on freshly toasted bread, you arrive at a symphony of flavors that is distinctly Italian, distinctly pleasurable. I prepare this with tomatoes I have dried at home. Of course the sauce can be made with purchased dried tomatoes or those in oil. The taste won't always be the same, so I will leave it up to you to taste carefully as you prepare the sauce. Whatever you use, the sauce should be pungent, surprising, and complex, filled with the flavors of fresh herbs, excellent oil, and olives. The tomatoes, in effect, serve as only a backdrop, but a vital one. I prefer this sauce with a good bite of crushed red peppers, but if you prefer a tamer sauce, leave them out.

10 Sun-Dried Tomatoes (page 284)

1 plump fresh garlic clove, degermed and minced

½ teaspoon crushed red peppers (hot red pepper flakes), or to taste

6 tablespoons extra-virgin olive oil

About 20 salt-cured black olives, such as Italian Gaeta or French Nyons olives (see page 280), pitted

2 teaspoons minced fresh thyme leaves

1 tablespoon minced fresh rosemary leaves

In the bowl of a food processor, combine all the ingredients and process until the sauce is lightly emulsified but still quite coarse and almost chunky. (You do not want a smooth sauce.) The sauce can be stored in a jar in the refrigerator for up to 1 month. If you do so, first cover the pesto with a film of olive oil.

▪ Y i e l d :   ½   c u p   (1 2 5   m l)   s a u c e ▪

...

Fresh Herb Sauce for Meats and Poultry

Salsa Verde

Vivid green and just slightly piquant, this is a fabulous all-purpose sauce for serving as one would mayonnaise, with boiled meats, poultry, or fish. I first sampled this sauce in Florence, at Sostanza, the quintessential old-fashioned trattoria. There it was served with boiled beef, but the sauce can also be served with poached chicken or fish. There are countless versions of salsa verde, which means you can customize the recipe to your tastes, mood, and what you have at hand. Since my garden is full of arugula, parsley, and sorrel, this is what I prefer. If you have only parsley, that's fine, too. (Just be sure to stem it thoroughly, using just the leaves for the sauce.) Red wine vinegar may be substituted for the lemon juice, capers may be added, and garlic may or may not be included in the repertoire, according to your whim. The only constants are top-quality extra-virgin olive oil and fresh greens. The sauce can be made ahead of time and refrigerated for up to one day. I like the old-fashioned quality of the mortar and pestle and the rather rough texture that results. Those interested in saving time can prepare the sauce in the food processor.

2 *plump fresh garlic cloves, degermed*

½ *teaspoon fine sea salt, or to taste*

4 *flat anchovy fillets in olive oil, drained and minced*

2 *cups (500 ml) loosely packed fresh flat-leaf parsley leaves*

1 *cup (250 ml) loosely packed sorrel leaves, coarsely chopped*

1 *cup (250 ml) loosely packed arugula leaves, coarsely chopped*

2 *tablespoons freshly squeezed lemon juice*

½ *cup (125 ml) extra-virgin olive oil*

BY HAND: Place the garlic and salt in a mortar and mash with a pestle to form a paste. Add the anchovies and pound to a paste. Add the greens, little by little, pounding into a thick paste. Continue until all the greens have been used and the paste is homogeneous. Slowly add the lemon juice, and then the oil, stirring until the sauce is well blended. Taste for seasoning.

IN A FOOD PROCESSOR: With the food processor running, add the garlic cloves and mince them. Add the salt and anchovies and pulse 2 or 3 times. Add the greens and pulse 2 or 3 times, or until the sauce is homogeneous. With the processor running, add the lemon juice and then the olive oil in a slow, steady stream. Taste for seasoning.

Transfer to a small bowl. Serve immediately. (The sauce can be stored, covered and refrigerated, for 1 day. Before serving, bring to room temperature. Stir, and taste for seasoning.)

■ Yield: 1 cup (250 ml) sauce ■

Meat Sauce

Ragù

Ll too often one thinks of ragù, or meat sauce, as a heavy, meaty sauce with just a tinge of red sauce. In most trattorias in Italy, what you find is quite the opposite: a very flavorful, light tomato sauce flecked with just a suggestion of meat. A sauce like that allows the pasta to star, not the meat or the sauce. Sometimes you'll find it flecked with bits of pungent hot peppers, sometimes the meat is so close to nonexistent it could almost fool a vegetarian, and other times the meat is noticeably chunky though never dominant. This all-purpose sauce is simply a suggestion: One can use a mix of meats, including nicely seasoned Italian sausage, or a mix of freshly ground pork, beef, and veal. Although I'm aware that many people try to stay away from any visible fat in their food, the best sauces do have a nice, overt veil of fat, so don't use meat that is too, too lean. For a vegetarian version, substitute half an ounce (15 g) of dried porcini mushrooms soaked in hot water and drained (see page 148). I generally make a large batch so I can have some for the freezer. Depending upon how I intend to use the sauce, I add more or less hot pepper.

3 tablespoons extra-virgin olive oil

1 small onion, minced

1 rib celery, minced

1 carrot, minced

¼ cup (60 ml) fresh flat-leaf parsley leaves, snipped with a scissors

Sea salt to taste

About 8 ounces (250 g) bulk sausage meat or a mix of chopped beef, pork, and veal

Two 28-ounce (765-g) cans peeled Italian plum tomatoes in juice, or two 28-ounce (765-g) cans crushed tomatoes in purée

1 teaspoon crushed red peppers (hot red pepper flakes), or to taste (optional)

In a large skillet, combine the oil, onion, celery, carrot, parsley, and salt, and toss to coat with oil. Cook over moderate heat until the vegetables are soft and fragrant, 3 to 4 minutes. Add the meat, making sure to break it up into small bits with a spatula, and toss to blend. Reduce the heat to low, and cook until the meat changes color, about 5 minutes more. If using whole canned tomatoes, place a food mill over the skillet and purée the tomatoes directly into it. Crushed tomatoes can be added directly from the can. Stir to blend, and, if using, add the crushed red peppers. Cover, and cook until the sauce begins to thicken, about 20 minutes. Taste for seasoning. The sauce can be refrigerated, covered, for up to 3 days or frozen for up to 1 month.

■ Yield: About 1 ½ quarts (1.5 l) ■
sauce

···

White Sauce

Salsa Balsamella

A classic white sauce of milk, butter, and flour is used often in Italian cooking. It's a quick, easy, flavorful sauce to prepare at the last minute. The worst versions can taste like paste. The best serve as delicate binders that add fresh and refreshing herbal flavors and a pleasantly creamy texture to a dish. This recipe makes a medium-thick white sauce, ideal for baked pasta (page 154). Since milk so readily absorbs the flavors of fresh herbs, I like to infuse my white sauce with fresh bay leaves and rosemary.

2 cups (500 ml) whole milk

2 bay leaves, preferably fresh

1 tablespoon minced fresh rosemary leaves

4 tablespoons (2 ounces; 60 g) unsalted butter

3 tablespoons all-purpose flour

¼ teaspoon fine sea salt

1. In a medium-size saucepan, scald the milk over high heat, bringing it just to the boiling point. Add the bay leaves and rosemary, cover, and set aside to infuse for 10 minutes. Strain the milk through a fine-mesh sieve into a measuring cup with a pouring spout. Discard the herbs.

2. In a heavy medium-size saucepan, melt the butter over moderate heat. Whisk in the flour and cook, stirring constantly, for 1 minute. Do not let the flour brown. Remove the saucepan from the heat and whisk in the hot strained milk a few tablespoons at a time, stirring constantly until all the milk has been incorporated into the flour and butter.

3. Return the saucepan to the heat, add the salt, and whisk constantly until the sauce thickens, 2 to 3 minutes. Taste for seasoning. (White sauce should not be held for more than a few hours. If you prepare it an hour or two in advance, rub the surface with a lump of butter so that it melts and forms a thin coating that will prevent the sauce from drying out. At serving time, simply warm the sauce on the top of a double boiler, stirring the butter into the sauce.)

■ Yield: About 2 cups (500 ml) sauce ■

Tips For a Better White Sauce

■ Do not allow the flour to brown when cooking it with the butter, or the sauce will taste burnt.

■ Add the hot milk off the heat, and very gradually, to obtain a very smooth sauce without lumps.

■ Whisk, whisk, and whisk again, without distraction.

Chicken Stock

Brodo di Pollo

This is a light, freshly flavored chicken stock, similar to those found in Italian kitchens. I always have stock on hand in the freezer, a way to cut down on cooking time and to enrich my larder, enlarging my repertoire on days I don't have time to cook. And I adore the way my house smells when the stock is simmering away.

4 pounds (2 kg) chicken parts (necks and wings), rinsed

1 tablespoon sea salt, or to taste

1 large onion, halved, each half stuck with a whole clove

3 plump fresh garlic cloves

Several parsley stems, celery leaves, bay leaves, and sprigs of thyme, wrapped in the green part of a leek and tied in a bundle with cotton twine

4 large carrots, peeled

4 leeks, white and tender green part, trimmed and well rinsed

4 whole black peppercorns

1. In a large stockpot, combine the chicken parts, salt, and cold water to cover. Bring to a boil over high heat, skimming off any impurities that rise to the surface. With a slotted spoon, transfer the chicken parts to a large sieve. Rinse, drain, and set aside. Discard the blanching liquid.

2. Rinse out the stockpot, and add the blanched chicken parts and all the remaining ingredients. Add cold water to cover and bring just to a simmer over moderately high heat. Skim off the impurities that rise to the surface. Reduce the heat and simmer—as gently as possible—for 3 hours, skimming as necessary.

3. Line a fine-mesh sieve with dampened cheesecloth and set over a large bowl. Ladle—do not pour—the liquid into the bowl. Measure the liquid. If it exceeds 2 quarts (2 l), return to a saucepan and reduce over moderate heat. Cool to room temperature. Skim off and discard any fat that rises to the surface. The stock may be refrigerated for up to 2 days or frozen for up to 2 months.

▪ Y i e l d :   A b o u t   2   q u a r t s   (2   l)   l i g h t   s t o c k ▪

Vegetable Broth

Brodo di Verdura

I once spent a very profitable and informative day in the kitchens with Chef Walter Tripodi at La Frateria di Padre Eligio in Tuscany. This is the quick vegetable stock he uses when preparing risotto. It can cook while you're doing other kitchen tasks, and should be ready just about the time you're ready to prepare your risotto. Because the broth is thin, delicate, and without the fat that would give it holding power, it should be used the same day it is prepared.

2 quarts (2 l) cold water

3 onions, halved

3 carrots, peeled

2 ribs celery, rinsed

Several parsley stems, celery leaves, bay leaves, and sprigs of fresh thyme, wrapped in the green part of a leek and tied in a bundle with cotton twine

½ teaspoon sea salt

In a large stockpot, combine all the ingredients and bring to a simmer over moderate heat. Simmer, uncovered, for about 1 hour. With a slotted spoon, remove and discard the vegetables and herb bundle. The broth can be used immediately.

▪ Yield: About 1 ½ quarts (1.5 l) broth ▪

Garlic Mayonnaise

Aïoli

One evening in the waterside town of Santa Margherita—not too far from the French border—I was served a side dish of sliced steamed potatoes and offered this garlic-pungent, French-inspired mayonnaise to toss with the potatoes. The combination has been a family favorite ever since. This is one sauce that simply cannot be made properly in the food processor: It turns to glue and is not worthy of its name. Serve garlic mayonnaise with any steamed or boiled vegetables, grilled fish, cold meats, and poultry.

6 plump fresh garlic cloves, degermed

½ teaspoon fine sea salt

2 large egg yolks, at room temperature

1 cup (250 ml) extra-virgin olive oil

1. Pour boiling water into a large mortar to warm it. Discard the water and dry the mortar. Place the garlic and salt in the mortar and mash together with a pestle to form as smooth a paste as possible. (The fresher the garlic, the easier it will be to crush.)

2. Add 1 of the egg yolks. Stir, pressing slowly and evenly with the pestle, always in the same direction, to thoroughly blend the garlic and yolk. Add the second yolk and repeat until well blended.

3. Very slowly work in the oil, drop by drop, until the mixture thickens. Then gradually whisk in the remaining oil in a slow, thin stream until the sauce is thickened to a mayonnaise consistency. (Garlic mayonnaise can be refrigerated, covered and well sealed, for up to 1 day. Bring to room temperature to serve.)

▪ Yield: About 1 cup (250 ml) mayonnaise ▪

• • •

Artichoke Cream

Crema di Carciofi

To anyone who loves artichokes as I do, this soothing, golden cream is like a taste of heaven on earth. Spread it on freshly grilled homemade bread, scoop it up with a wand of crisp celery, toss it with a bit of spaghetti for a very quick, nourishing meal. I first sampled this one summer along Lake Garda, where each day brought a new, specific flavor to my life, along with bold swaths of sunshine, friendship, and fun. Don't use just any marinated artichokes for this recipe: Make your own!

1 cup (250 ml) drained Marinated Baby Artichokes Preserved in Oil (page 18), marinade reserved

About ¼ cup (60 ml) reserved marinade liquid

Fine sea salt to taste (optional)

In the bowl of a food processor, combine the artichokes and about 2 tablespoons of the marinade liquid. Purée, adding additional liquid as necessary to create a light, fluffy cream. Taste for seasoning. This sauce will keep for several days, well covered, in the refrigerator. Bring to room temperature before serving.

▪ Yield: About 1 cup (250 ml) cream ▪

···

Black Olive Spread

Olivada

This is one of my favorite Mediterranean flavors, a sunny blend of top-quality black olives, fresh herbs, oil, and a touch of capers and anchovies. Keep it on hand for those days you don't have time to cook and need a zesty pick-me-up.

2 plump fresh garlic cloves, degermed and minced

1 teaspoon fresh thyme leaves

2 tablespoons drained capers, rinsed

4 anchovy fillets, drained, rinsed, and coarsely chopped

2 tablespoons extra-virgin olive oil

1 tablespoon rum

2 cups (8 ounces; 250 g) salt-cured black olives, such as Italian Gaeta or French Nyons olives (see page 280), pitted

Combine all the ingredients except the olives in the bowl of a food processor and process just until blended. Add the olives and pulse about 10 times. The mixture should be fairly coarse. The spread can be used to dress pasta, or spread on toast. It can be stored in a jar in the refrigerator for up to 1 month. If you plan to do so, cover the olivada with a film of olive oil.

■ Y i e l d : 2 c u p s (5 0 0 m l) s p r e a d ■

Green Olive Spread

Olivada Verde

While black olive spreads are well known in Italy and in France, the green olive version is, unfortunately, less common. I love this bright, brilliantly flavored spread, and think it's best when freshly made, when you have the slightly tangy mix of green olives, capers, anchovies, and a good hit of freshly ground black pepper, all flavors at their peak. I like to serve it as a condiment to spread on toast with a first course of Seasoned Raw Beef (page 12), or as a dip for wands of fresh celery.

2 tablespoons drained capers, rinsed

4 anchovy fillets, drained, rinsed, and coarsely chopped

¼ cup (60 ml) extra-virgin olive oil

Freshly ground black pepper to taste

2 cups (10 ounces; 300 g) drained pitted green olives

Combine all the ingredients except the olives in the bowl of a food processor and process just until blended. Add the olives and pulse about 10 times. The mixture should be just slightly chunky, but very spreadable, and should have a good flavor of freshly ground pepper. The spread can be used to dress pasta, or spread on toast. It can be stored in a jar in the refrigerator for up to 1 week, covered with a film of olive oil. (If stored longer than a week, the spread loses its bright, fresh flavor.)

▪ Yield: 2 cups (500 ml) spread ▪

. . .
Salt-Cured Anchovies

Acciughe sotto Sale

Pungent with the rich, salty flavors of the sea, salt-cured anchovies are essential to the Italian larder. Anchovies turn up in pasta sauces, are an essential ingredient in Lamb Braised in White Wine, Garlic, and Hot Peppers (page 242), and are delicious all on their own, on top of a slice of freshly grilled bread. When you find anchovies in your fish market, eat some fresh, marinated in lemon juice, herbs, and oil (see page 20). With the rest, prepare these simple cured anchovies for all-year eating.

2 pounds (1 kg) very fresh anchovies

2 pounds (1 kg) coarse sea salt

About 20 bay leaves, preferably fresh

About 20 sprigs fresh thyme

1. Lightly rinse (don't wash or soak) the anchovies. Individually head and gut the fish: Hold each anchovy firmly just beneath the head, and gently pull off the head and attached entrails. Discard the head and entrails. Place a thick layer of coarse salt on the bottom of a large sterilized canning jar. Place a layer of anchovies side by side on the salt, and sprinkle a thin layer of salt on top of the anchovies. Continue layering until all the anchovies and salt have been used, ending with a layer of salt. Every layer or so, add several bay leaves and sprigs of fresh thyme. Cover securely and store in a cool, dark place, or refrigerate. A brine will form as the anchovies absorb the salt and are cured. This should not be discarded. The anchovies are ready to eat in about 2 weeks and they can be stored for up to 1 year, in a cool, dark spot.

2. Properly cured anchovies are a deep mahogany color, much like ham. To eat, simply fillet the anchovies and use in any dish calling for cured anchovies.

■ Yield: 2 quarts (2 l) anchovies ■

Salt-Cured Black Olives

Olive Nere

OLIVE LORE:

*Venetians say an olive branch on the chimney piece wards off
lightning, while throughout Italy, a branch over the door is said to
keep out witches and wizards.*

Pure, salt-cured black olives are one of life's simple pleasures. Served as an appetizer with a glass of wine, tossed with pasta or with salads, stuffed into breads or rolls, or sprinkled on pizza, they are an integral part of Italian cooking.

As varieties of olives vary according to region, so do methods of curing them, and one finds Umbrian specialties flavored with orange zest, garlic, and bay leaves, as well as Sicilian salads of black olives crushed with oil, garlic, vinegar, and oregano, or with licorice-like fennel and orange and lemon zest. Note that different varieties are destined for different uses. However, as smooth green olives are those that were picked before they were ripe, smooth purple olives just as they ripened, and wrinkled black olives when they were overripe, in principle, all three could come from the same tree.

Although we don't all have olive trees in our backyards, fresh, uncured olives can be secured in parts of California, and on wintertime trips to Italy or France. Come harvesttime in December, markets in the south of Italy and France offer the bitter, wrinkled ripe olives for curing at home. Here is the simplest and most classic method: The olives are pricked all over with a small fork, then tossed in sea salt, using a traditional ratio of 2 pounds (1 kg) olives to 3½ ounces (100 g) coarse sea salt. Bay leaves, thyme, rosemary, and black peppercorns can also be added at this point. After being tossed daily, the olives are ready to be consumed in ten to fifteen days, and will stay fresh-tasting for about six months. If you're new to uncured olives, don't make the mistake so many people do, and pop one in your mouth. Uncured olives are unforgivably bitter and unpleasant.

2 pounds (1 kg) ripe uncured
black olives
3½ ounces (100 g) coarse sea salt

OPTIONAL FLAVORINGS

Sprigs of fresh thyme
Sprigs of fresh rosemary
Fresh bay leaves

Extra-virgin olive oil
Whole black peppercorns
Minced fresh garlic
Red wine vinegar
Grated lemon zest (yellow peel)
Grated orange zest (orange peel)
Dried red chile peppers

1. Do not wash the olives. If there are any leaves or stems still attached to the olives, remove and discard them. With a small seafood fork, prick each olive three or four times, all the way through to the center. (The pricking allows the olives to quickly absorb the salt.) Place the olives in a large, shallow bowl and add the salt. Toss with your hands to coat the olives with salt. Add, according to taste, sprigs of fresh thyme or rosemary, bay leaves, and/or a teaspoon of whole black peppercorns.

2. Leave the olives uncovered, at room temperature, tossing them once or twice a day. After 10 days, sample the olives. If they are still too bitter, let them cure for several more days. By this time much of the salt should be absorbed into the olives, but there may still be salt and some liquid at the bottom of the bowl. Do not discard this brine, for it will eventually be absorbed into the olives.

3. Once the olives are edible, pack them in pint (half-liter) glass jars, with any salt still attached, layering them with herbs as desired. Sprinkle with just enough olive oil to moisten. (Do not add citrus zest or garlic at this time, or the olives will lose their fresh, vibrant flavor.) Cover the jars and store, at room temperature, for up to 6 months.

4. At serving time, sample the olives. Should they be overly salty (though they shouldn't be), they can be rinsed in cold water. To serve, add, to taste, freshly minced garlic, a drop or two of red wine vinegar, citrus zest, black peppercorns, or hot red chile peppers.

▪ Yield: 2 pounds (1 kg) olives ▪

Brine-Cured Black Olives

Olive Nere

"If I could paint and had the necessary time, I should devote myself for a few years to making pictures only of olive trees."

ALDOUS HUXLEY

Brine-cured black olives are the "backbone" of cured olives, for they are basically indestructible. The ripe olives are simply placed in a 10 percent brine solution until they are edible, a process that takes several months. Unlike salt-cured olives, which are pricked with a fork to allow the salt to penetrate the meat of the olive, brine-cured olives are simply cured whole in the brine. The process can take three to four months, depending upon the size of the olives and the age of the brine. Once cured, they can be kept indefinitely, though I've never cured enough to last me more than a year.

2 pounds (1 kg) ripe uncured black olives

3½ ounces (100 g) fine sea salt

1 quart (1 l) water

OPTIONAL FLAVORINGS

Sprigs of fresh thyme

Sprigs of fresh rosemary

Fresh bay leaves

Extra-virgin olive oil

Whole black peppercorns

Red wine vinegar

Minced fresh garlic

Grated lemon zest (yellow peel)

Grated orange zest (orange peel)

Dried red chile peppers

1. Do not wash the olives. If there are any leaves or stems still attached to the olives, remove and discard them. In a large crock, combine the salt and water and stir to dissolve. Add the olives, cover, and set aside in a cool spot for several months, stirring from time to time. The olives should be covered with a small plate, to keep them all immersed in the brine. A scum will form on top, but it is harmless and should not be discarded, for it is the sign of a healthy, active brine. Do not add any seasonings other than salt and water to the brine. For specific flavorings, add those at serving time. When starting with a fresh brine, olives will take 3 to 4 months of curing before they are edible. Once cured, the olives can be kept indefinitely. Never discard a salt brine for olives, which will become black and inky. It can be used indefinitely, year after year.

2. To serve, remove the olives from the brine with a slotted spoon or wooden olive scoop with holes. Taste the olives. If they are excessively salty, they can be rinsed or soaked in cold water to remove some of the saltiness. Serve as they are, or season with any of the optional seasonings.

■ Yield: 2 pounds (1 kg) olives ■

● ● ●

Sun-Dried Tomatoes

Pomodori Secchi

These tomatoes are the essence of summer, serving to warm your soul in the depths of winter. Although sun-dried tomatoes are certainly more popular in America than in their native Italy (where they are sold at markets, but seldom show up in restaurant kitchens), they do have their place in the Italian kitchen. And the price of commercially sun-dried tomatoes makes it practical and economical to prepare them. I like to snip the dried tomatoes and toss them into a green salad for a cold-weather pick-me-up. They can also be tossed into a last-minute pasta dish, or served as is, with cheese.

It goes without saying that it's best to dry these on a dry day. I've had success with drying tomatoes in gas, electric, and convection ovens. The drying time will vary according to the size and moistness of the tomatoes, your oven, and the outdoor temperature. Since they are dried at the lowest possible oven temperature, there is little danger of burning the tomatoes. Be sure not to underdry: The tomatoes should be perfectly dry, with no interior moisture, and darkened to deep vermilion shades. Generally, there is no need to place a broiler pan beneath the tomatoes on the racks: They will dry better if there is good oven circulation. Look for small tomatoes of equal size so they will dry more quickly and in the same time. To ensure that the tomatoes do not fall through the racks as they dry and shrink, place them on cake racks set on the oven racks.

5 pounds (2.5 kg) Roma (oval) tomatoes

Fine sea salt

1. Preheat the oven to 200°F (100°C; gas mark 1), or the lowest setting possible. Remove the oven racks.

2. Trim and discard the stem ends of the tomatoes. Halve each tomato lengthwise. Arrange the tomatoes, cut side up, side by side and crosswise on cake racks set on the oven racks. Do not allow the tomatoes to touch one another. Sprinkle lightly with salt.

3. Place in the oven and bake until the tomatoes are shriveled and feel dry, anywhere from 6 to 12 hours. Check the tomatoes from time to time: They should remain rather flexible, not at all brittle. Once dried, remove the tomatoes from the oven and allow them to thoroughly cool on cake racks. (Smaller tomatoes will dry more quickly than larger ones. Remove each tomato from the oven as it is dried.)

4. Transfer the tomatoes to zipper-lock bags. The tomatoes will last indefinitely.

■ Yield: About 2 cups (500 ml) ■
dried tomatoes

THANK YOU, PIEMONTE

•

The Piedmont is certainly one of Italy's most appealing regions, with great wine, great food, and a solid, robust personality all of its own. This menu is a "thank-you note" to the cooks and restaurateurs who so kindly opened their doors to me on the trattoria tour. With the meal, I'd drink a young red Dolcetto d'Alba, while with dessert, I'd search out a good bottle of sweet, muscat-fragrant Moscato d'Asti.

SEASONED RAW BEEF

TAJARIN WITH ROSEMARY-INFUSED BUTTER

RABBIT WITH RED PEPPERS AND POLENTA

PANNA COTTA (ALMOND-VANILLA CREAMS)

.

DESSERTS,

GRANITA,

SORBET, AND

ICE CREAMS

.

Summer Peaches and Raspberries

Insalata di Pesche e Lampone

O ne sunny August afternoon we sat on the shaded terrace of La Stalla, a tranquil, country brick farmhouse-trattoria in the village of Gardone Riviera, on Lake Garda. We feasted on giant mixed salads of greens and vegetables, slices of grilled fontina cheese served with wild mushrooms, and this delightfully refreshing and simple dessert. The dish consists of a lightly sweetened peach purée that's spread out to fill the bottom of a flat serving plate, then edged with slices of fragrant ripe peaches. The center may be filled with raspberries or other small seasonal berries. Serve this with a bubbly white, such as the charming Italian Prosecco, and pass a basket of biscotti (page 314). Sort of makes you want to kick up your heels and dance!

5 ripe peaches (about 2 pounds; 1 kg)

¼ cup (50 g) Vanilla Sugar (page 324)

2 cups (8 ounces; 250 g) fresh raspberries

1. Peel the peaches: Bring a large pot of water to a rolling boil. Drop the peaches in, one by one, and scald just until the skins are softened, 1 to 2 minutes. Using a slotted spoon, remove the peaches and plunge directly into cold water, for easier handling. Once cool enough to handle, peel the peaches using the tip of a small sharp knife. Discard the skins.

2. Coarsely chop 3 of the peeled peaches and place in the bowl of a food processor. Add 2 tablespoons of the sugar and purée. Spoon the purée into a 10½-inch (27-cm) round porcelain baking dish and, using a spatula, evenly spread the purée over the bottom of the dish.

3. Cut the remaining 2 peeled peaches into 16 even slices. Place in a bowl and toss with 1 tablespoon of the sugar. Evenly arrange the peach slices, slightly overlapping, on top of the peach purée, forming a ring of peaches around the edge of the dish. Cover and refrigerate for up to 4 hours.

4. At serving time, toss the raspberries with the remaining 1 tablespoon sugar. Carefully spoon the berries on top of the peach purée, filling in the center of the dish. Serve immediately. Carefully spoon a few peach slices onto each dessert plate, and place a spoonful of purée, then a spoonful of berries, alongside.

■ Yield: 4 to 6 servings ■

WINE SUGGESTION: I love this with a fruity, fizzy Prosecco from the Veneto.

Almond Macaroons

Amaretti

I adore these fetching macaroons, rich with almond flavor. The cookies are prepared with simple pantry ingredients one is always likely to have on hand. They've become part of my repertoire, and make a great dessert, particularly when paired with biscotti. With this recipe, I find better results if one measures the egg whites (since volume varies), and prefer baking on parchment paper, for easier removal.

> ¾ *cup (3½ ounces; 105 g) blanched almonds, ground to a fine powder*
>
> ¾ *cup (150 g) sugar*
>
> ⅓ *cup (80 ml) egg whites (about 2 large whites), at room temperature*
>
> ½ *teaspoon pure almond extract*

1. Preheat the oven to 350°F (175°C; gas mark 4/5). Line 3 baking sheets with baking parchment. Set aside.

2. In a large bowl, combine the almonds and sugar, and stir to blend.

3. In another bowl, whisk together the egg whites and almond extract. Add the egg whites to the almond mixture, and stir to form a soft, sticky batter.

4. With a teaspoon, drop about ½ teaspoon of batter per cookie onto the baking sheets, spacing them slightly apart, for about 12 cookies per sheet.

5. Place the baking sheets in the center of the oven and bake until the macaroons are lightly browned around the edges and slightly firm to the touch, about 15 minutes. Remove the baking sheets from the oven and transfer the sheets of parchment to racks to cool until the cookies begin to firm up, just 3 to 4 minutes. With a sharp knife, gently lift the cookies from the parchment and transfer to racks to cool thoroughly. (The cookies can be stored in a tin box in a cool, dry spot for up to about 10 days.)

■ Yield: About 36 cookies ■

Tip

When buttering a pan, place about 1 tablespoon of softened butter in the pan, place your hand inside a small plastic sandwich bag, and carefully butter the pan. Discard the bag and come out with spotless hands. Likewise, keep a small shaker full of all-purpose flour at hand, using that to sprinkle the pan with flour.

Baked Peaches with Almond Macaroons

Pesche Ripiene alla Piemontese

ITALIAN DICTUM:

*"Peel a fig for a friend,
and a peach for an enemy."*

While this remarkably simple and delicious peach dessert is generally associated with the Piedmont, I've sampled these baked peaches stuffed with ground amaretti cookies all over Italy. Although a topping of whipped cream is not traditional here, remember, in your kitchen, you and your palate are boss.

Unsalted butter for preparing the baking dish

6 ripe peaches

*10 full-size (5 pairs) amaretti cookies (Italian macaroons; see Mail Order
Sources, page 325)*

¼ cup (50 g) Vanilla Sugar (page 324)

1 large egg yolk

2 tablespoons (1 ounce; 30 g) unsalted butter

1. Preheat the oven to 350°F (175°C; gas mark 4/5).

2. Lightly butter a baking dish large enough to hold the peaches in a single layer. Rinse, halve, and stone the peaches, cutting along the natural line of the fruit. With a small spoon, scoop out about 1 teaspoon of the peach pulp from each half to enlarge the cavity. Reserve the scooped-out pulp. Place the peach halves, pitted side up, side by side in the buttered baking dish.

3. In the bowl of a food processor, process the amaretti to fine crumbs. (Do not process to a paste.) Transfer the mixture to a small mixing bowl. Stir in the reserved peach pulp, the sugar, and egg yolk, mixing thoroughly.

4. Spoon the filling into the cavity of each peach half, distributing it as evenly as possible. Dot each half with a bit of butter.

5. Place the dish in the center of the oven and bake until the peaches are soft and the filling firms up and begins to form a crust, about 40 minutes. Serve warm or at room temperature, transferring the peaches to an attractive serving platter or individual dessert bowls or plates.

■ Yield: 6 servings ■

NOTE: If using miniature amaretti, use 20 cookies, about 2 ounces (30 g).

Coffee and Mascarpone Ladyfinger Cream

Tiramisù

Tiramisù—that hyper-rich layered modern Italian dessert—has become an international favorite, appealing particularly to sweet tooths that love soft, creamy, and cool. The name means "pick me up," and that's what happens when you offer your body a jolt of this brandy- and chocolate-flavored cream. It's easy as pie to make, especially when you begin with imported Italian ladyfingers, or sponge biscuits known as savoiardi. Although some versions call for sprinkling the top with cocoa powder, I always find this disagreeable: Inevitably a coughing fit follows the first bite. I prefer grated bitter chocolate. Perhaps the best version of this dish I ever sampled was in Venice, at the tiny, folksy trattoria Antica Besseta.

3 tablespoons very strong espresso coffee

1 tablespoon brandy or grappa

3 large eggs, at room temperature, separated

½ cup (100 g) Vanilla Sugar (page 324)

8 ounces (250 g) mascarpone, at room temperature

About 24 savoiardi or Italian lady fingers (see Mail Order Sources, page 325)

1 ounce (30 g) bittersweet chocolate, preferably Lindt Excellence

1. In a small bowl, combine the coffee and brandy. Set aside.

2. In the bowl of an electric mixer fitted with a whisk, whisk the egg whites until stiff and glossy but not dry. Set aside.

3. In a clean bowl of an electric mixer fitted with a whisk, combine the egg yolks and the vanilla sugar, and whisk until thick and lemon colored, about 2 minutes. Add the mascarpone and whisk to blend. With a large spatula, carefully fold the egg whites into the mascarpone mixture.

4. Place a single layer of 12 savoiardi (two rows of six each, end to end) in a 10-inch (25-cm) square baking dish or on a flat platter. Dip a pastry brush into the coffee mixture and soak the biscuits with the liquid. Spread about one-half of the mascarpone cream over the biscuits. Sprinkle with about one-half of the grated chocolate. Repeat with a second layer of biscuits, brush the biscuits with the remaining liquid, and cover with the remaining mascarpone cream. Reserve the final sprinkling of chocolate for serving time. Cover and refrigerate for at least 3 hours, allowing the cream to firm up slightly and the biscuits to absorb some of the liquid. (The tiramisù can also be prepared 1 day in advance). To serve, divide the tiramisù into rectangular slices and transfer to chilled dessert plates. Sprinkle with the remaining grated chocolate and serve immediately.

■ Yield: 4 to 8 servings ■

WINE SUGGESTION: Although tiramisù is almost too sweet for any wine, a velvety, amber Vin Santo would be right at home here.

A FROZEN TREAT: Haste in the kitchen usually spells disaster. But one day I prepared a last-minute tiramisù and slipped the prepared dessert in the freezer for about 30 minutes to help it stiffen up a bit before serving. The result was a pleasantly firm, chilled dessert, which I actually prefer to traditional tiramisù. The chilling somehow cuts through the richness of the mascarpone, making for a dessert that—to the palate at least—appears lighter than it really is.

Almond-Vanilla Creams

Panna Cotta

There's no question in my mind that panna cotta—an almond-and-vanilla-flavored cream—is up there among the top ten Italian desserts. It's so rich and creamy, this specialty of the Piedmont makes almost any other cream dessert taste like diet food! Panna cotta is a cinch to make, and can be even prepared a day or so in advance, a great boon to those looking for do-ahead desserts. And although the dessert is called "cooked cream," the cream is really only brought to a boil, to help dissolve the sugar and bring out the almond and vanilla flavors. For this dish, search out the freshest, most delicious cream you can find. One spring morning Chef Angelo Maionchi of Turin's Del Cambio kindly allowed me to work with him and his staff in their kitchen, and shared this recipe with me. Panna cotta can be prepared in a long loaf pan, set on a caramel base, or all on its own, in individual ramekins. I prefer the "special-ness" of having your very own dessert, and love the potential play of colors with the ivory-white panna cotta and brilliant and shiny fresh fruits, such as strawberries, raspberries, cherries, or raspberries.

Unsalted butter for preparing the ramekins

4 *teaspoons (about 2 packages) unflavored gelatin*

2 *cups (500 ml) whole milk*

1 *cup (120 g) confectioner's sugar*

2 *cups (500 ml) heavy cream*

1 *teaspoon pure vanilla extract*

½ *teaspoon pure almond extract*

Assorted soft fresh fruits, for garnish

1. Butter eight ½-cup (125-ml) ramekins. Place on a tray.

2. In a small bowl, sprinkle the gelatin over ¼ cup (60 ml) of the milk and stir to blend. Set aside until the gelatin completely absorbs the milk, 2 to 3 minutes.

3. In a large saucepan, combine the remaining 1¾ cups (440 ml) milk, the confectioner's sugar, and the cream, and bring just to a boil over moderate heat, whisking to dissolve the sugar. Remove from the heat, add the softened gelatin and milk and the vanilla and almond extracts, and whisk to completely dissolve the gelatin. Strain the mixture through a fine-mesh sieve into a large measuring cup or bowl with a pouring spout. Pour the mixture into the ramekins. Cover with plastic wrap and refrigerate until set, about 4 hours. (The panna cotta can be prepared up to 1 day in advance. Refrigerate until serving time.)

4. To serve, run a sharp knife along the inside of each ramekin, to help loosen the cream. Dip the bottom of each ramekin into a bowl of hot water, shaking to completely loosen the cream. Invert onto chilled dessert plates and serve with fresh fruit alongside, such as sliced strawberries, raspberries, or cherries.

■ Yield: 8 servings ■

WINE SUGGESTION: This dreamy dessert deserves a fine, sweet wine, such as a Moscato d'Asti from the Piedmont.

VARIATION: While almond is the traditional flavoring for panna cotta, one also finds in Italy pure vanilla versions. For a more intense vanilla flavor, prepare the panna cotta with 2 teaspoons of pure vanilla extract, and omit the almond extract.

Individual Chocolate Flans

Budini di Cioccolato

ich, creamy, and not overly sweet, these easy chocolate flans come from Cibrèo, a favorite Tuscan trattoria near the outdoor food market in Florence. I like to serve them with Hazelnut and Orange Biscotti (page 314), for I love the happy marriage of chocolate, coffee, and orange.

Unsalted butter for preparing the
   ramekins

8 ounces (250 g) bittersweet
   chocolate, preferably Lindt
   Excellence, finely chopped

⅔ cup (160 ml) whole milk

2 cups (500 ml) heavy cream

¼ cup (60 ml) espresso or very
   strong coffee (do not use
   instant)

1 tablespoon Vanilla Sugar (page
   324)

2 large eggs, at room
   temperature, lightly beaten

2 large egg yolks, at room
   temperature, lightly beaten

1. Preheat the oven to 325°F (165°C; gas mark 4).

2. Cut 3 slits in a piece of waxed paper and use it to line a roasting pan. Butter six 1-cup (250-ml) ramekins and place them in the pan, on top of the paper. Set aside. (The paper will prevent the water added to the pan from boiling and splashing up on the custards.)

3. Prepare a large kettle of boiling water for the water bath; set aside.

4. In a large saucepan, combine the chocolate and milk. Melt the chocolate over low heat, stirring from time to time. Remove from the heat and set aside to cool for 5 minutes. Add the remaining ingredients, whisking to blend.

5. Divide the chocolate mixture evenly among the ramekins. Add enough boiling water to the roasting pan to reach about halfway up the sides of the ramekins. Place in the center of the oven and bake until the flans are just set at the edges but still trembling in the center, 45 to 50 minutes.

6. Remove from the oven and carefully remove the ramekins from the water. Serve in the ramekins warm or at room temperature, but not chilled.

■ Yield: 6 servings ■

Ricotta Cheesecake with Pine Nuts and Raisins

Torta di Ricotta

This is a delicate, tenderly sweet, and crustless ricotta cheesecake, studded lightly with pine nuts and raisins, and harboring a faint hint of lemon, orange, and spice. I frankly prefer it to the heavier, richer American-style cheesecake, and strongly recommend that cheesecake lovers add it to their repertoire. I sampled this dessert one sunny Saturday in December, at the excellent family-run trattoria Checchino dal 1887, in Rome.

Unsalted butter and all-purpose flour for preparing the cake pan

1 cup (200 g) Vanilla Sugar (page 324)

⅓ cup (45 g) all-purpose flour, sifted

½ cup (2 ounces; 60 g) pine nuts

½ cup (70 g) golden raisins

¼ teaspoon fine sea salt

2 pounds (1 kg) whole-milk ricotta (or two 15-ounce containers)

6 large eggs, at room temperature, lightly beaten

1 teaspoon ground cinnamon

1 teaspoon freshly ground nutmeg

2 teaspoons pure vanilla extract

Grated zest (yellow peel) of 1 lemon

Grated zest (orange peel) of 1 orange

Confectioner's sugar, for garnish

1. Preheat the oven to 300°F (150°C; gas mark 3/4).

2. Generously butter and flour a 9-inch (23-cm) springform pan, tapping out any excess flour. Set aside.

3. In a small bowl, stir together the vanilla sugar, flour, pine nuts, raisins, and salt. Set aside.

4. In the bowl of an electric mixer fitted with a paddle, gently beat the ricotta at low speed until smooth. Add the beaten eggs little by little, then add the vanilla sugar mixture and gently mix to blend. Add the spices, vanilla, and zests. Mix to blend thoroughly.

5. Pour the batter into the prepared cake pan. Place the pan in the center of the oven and bake until the cheesecake is a deep golden brown and fairly firm in the center, and a toothpick inserted in the center comes out clean, about 1 hour and 30 minutes. Transfer to a baking rack to cool. Once cooled, cover the cheesecake with plastic wrap and refrigerate until serving time. (The cake can be made up to 1 day in advance.)

6. To serve, release the sides of the springform pan, leaving the cheesecake on the pan base. Sprinkle the top generously with confectioner's sugar and serve, cutting into very thin wedges.

■ Y i e l d : 1 6 t o 2 0 s e r v i n g s ■

Toasted Hazelnut Cake

Torta di Nocciole

The buttery, hazelnut-rich aroma that wafts from the oven as this cake bakes is enough of a reason to prepare it. The proof, as ever, is in the eating, and this is a moist, full-flavored, and satisfying cake. I first sampled torta di nocciole at the Panettiere Cravero, a bakery in the wine-making village of Barolo, where fresh rounds of the cake are stacked at the counter. Their version has the unusual addition of cocoa powder, which adds a lovely, rich touch to a cake that is already quite luxurious. Do take the time to toast the hazelnuts so they will better release their full, rich flavor and aroma. And don't be tempted to embellish this simple cake with frosting or even a dusting of confectioner's sugar. You'll see, it's perfectly fine all on its own.

Unsalted butter and all-purpose
  flour for preparing the cake
  pan

10 tablespoons (5 ounces; 150 g)
  unsalted butter, at room
  temperature

1¼ cups (250 g) Vanilla Sugar
  (page 324)

3 large eggs, separated, at room
  temperature

1 teaspoon pure vanilla extract

1 cup (4 ounces; 125 g)
  hazelnuts (filberts), toasted
  and finely chopped

2 cups (265 g) all-purpose flour

1 tablespoon baking powder

1 tablespoon cocoa powder,
  preferably Dutch process

¼ teaspoon fine sea salt

1. Preheat the oven to 350°F (175°C; gas mark 4/5).

2. Generously butter and flour a round 9-inch (23-cm) cake pan, preferably nonstick, tapping out any excess flour. Set aside.

3. In the bowl of an electric mixer fitted with a paddle, cream the butter and 1 cup (200 g) of the sugar until light and fluffy, 3 to 4 minutes. Add the egg yolks one by one, beating thoroughly after each addition. Beat in the vanilla and hazelnuts.

4. Sift the flour, baking powder, cocoa, and salt into a large bowl. Then sift the mixture over the hazelnut batter, folding gently until thoroughly blended. The batter will be very stiff, almost like a cookie dough.

5. In a clean bowl of an electric mixer fitted with a whisk, whisk the egg whites until fluffy. Add the remaining ¼ cup (50 g) of the sugar, a bit at a time, and whisk until the mixture is stiff and glossy but not dry.

6. With a large spatula, carefully fold the egg whites into the batter until there are no white patches in the batter.

7. Spoon the batter into the prepared cake pan, smoothing out the top with a spatula. Place the pan in the center of the oven. Bake until the cake is an even golden brown and a toothpick inserted in the center comes out clean, 40 to 50 minutes. Transfer to a wire rack to cool and firm up in the pan for 10 minutes. Then turn out of the pan and reinvert onto a serving plate. To serve, slice in wedges, for breakfast, as a snack, or for dessert.

▪ Yield: 12 servings ▪

WINE SUGGESTION: This cake is delicious with a slightly sweet muscatel, such as a Moscato d'Asti from the Piedmont.

Toasting for Flavor

Toasting nuts, such as hazelnuts, helps bring them to life, enhancing their flavor and reducing their moisture content. To toast: Spread the nuts out on a baking sheet. Toast in a preheated 350°F (175°C; gas mark 4/5) oven until lightly browned, about 10 minutes. Check every few minutes to avoid burning the nuts. Certain sophisticated recipes demand that the skins of hazelnuts be removed, by rubbing the nuts in a towel to loosen the skin. I don't find it necessary in this rustic cake. The nuts can easily be chopped in the food processor: Just be careful not to overprocess, or they turn to a paste. Hazelnuts are especially perishable and should be kept refrigerated or frozen, then thawed at room temperature before using.

Fragrant Orange and Lemon Cake

Torta di Arancio e Limone

One of the most romantic spots in Tuscany is the Locanda dell' Amorosa in Sinalunga. More than just an inn, this is a tiny U-shaped hamlet, set on a hill at the end of a long alley of cypresses. We stayed there one evening in late May, awakening to the sounds of thousands of tiny birds singing their little hearts out. Breakfast included cups of thick black coffee, unsalted Tuscan bread with homemade jams and honey, an assortment of sausages and cheeses, cereals, fruits, yogurt, and this magnificently golden cake studded with the flavorful zest of both lemon and orange. This version is based on the recipe that Chef Walter Ridaelli kindly shared with me that very day. This simple Italian delight, without the complication of a frosting, is always enjoyed with a beverage, making it the biscotti of the cake family. For special occasions, you may want to drizzle the cake with a simple icing, or serve it with sliced fruit macerated in rum or sweet wine.

(continued)

Unsalted butter and all-purpose flour for preparing the cake pan

3 cups (405 g) all-purpose flour

1½ teaspoons baking powder

½ teaspoon baking soda

¼ teaspoon fine sea salt

Grated zest and juice of 1 orange

Grated zest and juice of 1 lemon

¾ cup (185 ml) whole milk

16 tablespoons (8 ounces; 225 g) unsalted butter, softened

1½ cups (300 g) Vanilla Sugar (page 324)

5 large eggs

Confectioner's sugar for dusting (optional)

1. Preheat the oven to 350°F (175°C; gas mark 4/5).

2. Evenly coat the interior of a 10-inch (12-cup; 2.5 l) Bundt pan with butter. Dust lightly with flour, shaking out the excess flour. Set aside.

3. Sift the flour, baking powder, baking soda, and salt into a large bowl. Stir in the orange and lemon zests.

4. Combine the orange and lemon juices and the milk, and set aside to "sour" the milk.

5. In a large bowl, using a hand-held electric mixer at high speed, beat the butter and vanilla sugar until light and fluffy, about 2 minutes. One at a time, beat in the eggs, mixing well after each addition. The mixture will looked curdled—don't worry. Alternating in thirds, add the flour and milk mixtures, beating well after each addition, and scraping down the sides of the bowl with a rubber spatula as needed.

 COFFEE AND MASCARPONE LADYFINGER CREAM *(PAGE 294)*

 ALMOND-VANILLA CREAMS *(PAGE 296)*

 LAKE GARDA APPLE CAKE *(PAGE 308)*, TOP;
RICOTTA CHEESECAKE WITH PINE NUTS AND RAISINS *(PAGE 300)*

FRAGRANT ORANGE AND LEMON CAKE *(PAGE 305)*

6. Pour the batter into the prepared pan and place in the center of the oven. Bake until the cake is an even golden brown and a toothpick inserted in the center comes out clean, 45 to 55 minutes. (Don't worry if the cracks in the top of the cake don't look dry—use the toothpick test to test for doneness.) Transfer to a wire rack to cool in the pan for 10 minutes. Then turn out onto a serving plate. If desired, sift confectioner's sugar over the top of the cake. Slice into wedges and serve for breakfast, as a snack, or for dessert.

■ Yield: 10 to 14 servings ■

BEVERAGE RECOMMENDATION: I love this cake for breakfast, with a steaming cup of lemon-verbena herb tea. It could also be served for dessert, with a tiny glass of Tuscany's Vin Santo.

For Juicier Lemons

Place a room-temperature lemon on a flat work surface and, pressing down firmly with the palm of your hands, roll the lemon back and forth. This helps lemons give up the maximum amount of juice.

Lake Garda Apple Cake

Torta di Mele

Concentric circles of sweet, soft, golden-brown apples baked atop a vanilla-rich pound cake is one of my favorite Italian desserts. Late one summer on Lake Garda, a different version of torta di mele seemed to turn up every day, in trattorias, at open-air markets, at the local pastry shops along quiet side streets. This recipe is a montage, actually, a homage to all those apple cakes eaten in reality, or only with my eyes. I like its simple, old-fashioned quality, and enjoy it as much for breakfast or tea as for dessert after a meal.

Unsalted butter and all-purpose flour for preparing the baking pan

8 tablespoons (4 ounces; 125 g) unsalted butter, at room temperature

1 cup (200 g) Vanilla Sugar (page 324)

1 teaspoon pure vanilla extract

3 tablespoons whole milk

Grated zest (yellow peel) of 1 lemon

3 large eggs, at room temperature

1½ cups (200 g) all-purpose flour

¾ teaspoon baking powder

¼ teaspoon fine sea salt

½ teaspoon ground cinnamon

3 Golden Delicious apples (about 1½ pounds; 750 g)

1. Preheat the oven to 350°F (175°C; gas mark 4/5).

2. Generously butter and flour a 9-inch (23-cm) springform pan, tapping out any excess flour. Set aside.

3. In the bowl of an electric mixer fitted with a paddle, cream the butter, ¾ cup (150 g) of the vanilla sugar, the vanilla, milk, and lemon zest until light and fluffy, 1 to 2 minutes. Add the eggs one by one, beating thoroughly after each addition.

4. Sift the flour, baking powder, and salt into a large bowl. Spoon the mixture into the batter and mix until thoroughly blended. Scrape down the sides of the bowl and mix once more. Set aside for 10 minutes to allow the flour to absorb the liquids.

5. In a large bowl, toss together 2 tablespoons of the vanilla sugar and ¼ teaspoon of the cinnamon. Peel and core the apples, and cut each lengthwise into 16 even wedges. Transfer the apples to the bowl, tossing with the cinnamon sugar. Set aside.

6. Spoon the batter into the prepared cake pan, smoothing out the top with a spatula. Starting just inside the edge of the pan, neatly overlap the wedges of apples—thicker curved outer side against the side of the pan—in 2 or 3 concentric circles, working toward the center. Fill in the center with the remaining apples. Toss together the remaining 2 tablespoons sugar and ¼ teaspoon cinnamon, and sprinkle over the apples.

7. Place the pan in the center of the oven. Bake until the apples are a deep golden brown and the cake feels quite firm when pressed with a fingertip, about 1 hour. Remove to a baking rack to cool. After 10 minutes, run a knife along the sides of the pan. Release and remove the side of the springform pan, leaving the cake on the pan base. Serve at room temperature, cutting into thin wedges.

■ Yield: 8 to 12 servings ■

● ● ●

Golden Lemon-Rice Cake

Torta di Riso

Golden and welcoming, with a happy, healthy glow, this lovely rice cake is one of my favorite desserts. I first sampled this spring-yellow cake in a Florentine café. I make it often, for it is inexpensive and usually can be prepared with ingredients I have on hand, and, what's more, it's a dessert that can easily be made a day in advance. This popular dessert gives proof to the fact that the Italians do much more with their marvelous Arborio rice than make risotto, for their repertoire of desserts is filled with a great variety of rice puddings and cakes. It is equally delicious prepared with orange zest and orange juice.

1 cup (180 g) Italian Arborio rice (see Mail Order Sources, page 325)

1 quart (1 l) whole milk

Pinch of fine sea salt

¾ cup (150 g) Vanilla Sugar (page 324)

Unsalted butter for preparing the pan

1 tablespoon semolina flour

3 large eggs, at room temperature

Grated zest (yellow peel) of 1 lemon

3 tablespoons freshly squeezed lemon juice

Confectioner's sugar, for garnish

1. In a 6-quart (6-l) saucepan, combine the rice, milk, salt, and ½ cup (100 g) of the sugar. Stir to blend, and bring to a simmer over moderate heat, stirring regularly to keep the rice from sticking to the bottom of the pan. (Watch the pan carefully so the milk does not boil over.) Reduce the heat to low and simmer until the rice is tender, most of the milk is absorbed, and the mixture is thick and porridge-like, 15 to 20 minutes. Continue stirring regularly to keep the rice from sticking to the bottom of the pan. Transfer the mixture to a bowl to cool for at least 1 hour.

2. Preheat the oven to 325°F (165°C; gas mark 4).

3. Thoroughly butter the bottom and sides of a 10-inch (25-cm) springform pan. Dust lightly with the semolina, tapping the sides to distribute it evenly. Shake out the excess, and set aside.

4. In the bowl of an electric mixer fitted with a whisk, combine the eggs and the remaining ¼ cup (50 g) sugar and whisk until thick and lemon-colored, about 2 minutes. Add the zest and lemon juice, and mix thoroughly. Stir in the rice mixture and blend thoroughly. Pour the mixture into the prepared cake pan, smoothing out the top with the back of a spoon.

5. Place the pan in the center of the oven and bake until the rice cake is a deep golden color and firm in the center, 25 to 30 minutes. Remove from the oven and transfer to a baking rack to cool. Once cooled, cover the rice cake with plastic wrap until ready to serve. (The cake can be made 1 day in advance.)

6. To serve, run a knife along the sides of the pan, and release and remove the side of the springform pan, leaving the rice cake on the pan base. Sprinkle the top generously with confectioner's sugar and serve, cutting into very thin wedges.

■ Yield: 8 to 12 servings ■

Lemon-Rice Tea Cakes

Budini di Riso

The first time I sampled these sweet and golden little rice cakes—purchased at a bakery in Florence—they were a true revelation. Moist, golden, and not overly sweet, these soothing, gentle, flourless cakes (like elegant rice pudding in tea cake form) have become a favorite, perfect as a late morning or afternoon snack with a thick cup of espresso. They are most easily made in small muffin or cupcake tins. To ensure that the cakes do not stick to the pan, use paper liners (just be sure to remove the liners before serving). These tea cakes are best served at room temperature, sprinkled with a bit of confectioner's sugar. They look lovely arranged on a paper doily set on a footed cake stand.

3 cups (750 ml) whole milk

½ cup (90 g) Italian Arborio rice (see Mail Order Sources, page 325)

Pinch of fine sea salt

½ cup (100 g) Vanilla Sugar (page 324)

4 tablespoons (2 ounces; 60 g) unsalted butter

Grated zest (yellow peel) of 2 lemons

2 large eggs, separated, at room temperature

1 tablespoon freshly squeezed lemon juice

2 tablespoons dark rum

Confectioner's sugar, for garnish

1. Line fifteen ¼-cup (60-ml) muffin molds with cupcake liner papers. Set aside.

2. Rinse a 3-quart (3-l) saucepan with water, leaving a few drops of water clinging to the bottom of the pan. (This will help to keep the milk from sticking to the pan.) Add the milk and scald over high heat, bringing it just to the boiling point. Reduce the heat to medium, stir in the rice and salt, and bring to a simmer. Simmer for 10 minutes, stirring regularly. Add the sugar and butter, and simmer until the rice is cooked, about 7 minutes more, stirring regularly. At this point, the liquid should have the consistency of heavy cream. Transfer the mixture to a bowl to cool for at least 1 hour. During this time, the rice will absorb most of the liquid.

3. Preheat the oven to 375°F (190°C; gas mark 5).

4. When the rice mixture is cool, stir in the grated lemon zest, then the egg yolks, lemon juice, and rum, stirring to blend thoroughly.

5. In the bowl of an electric mixer fitted with a whisk, beat the egg whites until stiff but not dry. Gently fold the whites into the rice mixture. Spoon the mixture into the muffin molds, filling them to the top.

6. Place the muffin tins in the center of the oven and bake until the rice cakes are a deep golden brown and firm in the center, and a toothpick inserted in the center comes out clean, 25 to 30 minutes. Transfer the cakes to a wire rack to cool in the pan for 10 minutes. (Do not let sit longer, or the cakes are likely to stick to the paper.) Remove the cakes from the papers and place upright on a wire rack to cool completely. The cakes will sink slightly as they cool. At serving time, sprinkle generously with confectioner's sugar.

■ Yield: 15 tea cakes ■

Hazelnut and Orange Biscotti

Biscotti means "twice cooked," and that's just what you do to these fragrant, crunchy cookies studded with toasted whole nuts and scented with lemon zest, pure vanilla essence, and a warming touch of almond. Biscotti are made with no added fat—save for that in the eggs and nuts—and are intentionally dry, giving them an extra-long shelf life. During the day, they are ideal for dipping in a cup of espresso (softening them, and adding a zesty coffee flavor), while at dinnertime, they seem destined to be paired with a tiny glass of sherry, port, or the traditional Vin Santo.

About 2 cups (265 g) unbleached all-purpose flour

¼ teaspoon baking powder

¼ teaspoon baking soda

¼ teaspoon fine sea salt

3 large eggs

⅔ cup (135 g) Vanilla Sugar (page 324)

1 teaspoon pure vanilla extract

¾ teaspoon pure almond extract

Grated zest (yellow peel) of 1 lemon

Grated zest (orange peel) of 1 orange

1 cup (4 ounces; 125 g) hazelnuts (filberts), toasted and cooled

1 egg mixed with ¼ teaspoon salt, for egg wash

1. Preheat the oven to 350°F (175°C; gas mark 4/5). Line a baking sheet with baking parchment. Set aside.

2. In a sifter, combine 2 cups (265 g) flour, the baking powder, baking soda, and salt, and sift onto a clean work surface.

3. In a small bowl, combine the eggs, sugar, vanilla and almond extracts, and lemon and orange zest. Make a well in the center of the flour mixture, and slowly add the liquids to the well, drawing the flour mixture into the liquid and mixing gently with your hands. If necessary, add additional flour to form a firm and workable dough. Add the hazelnuts and work them evenly into the dough.

4. Divide the dough into 2 equal pieces. Flour your hands to keep the dough from sticking, and, with your palms, carefully roll each piece into an oval cylinder about 2 inches (5 cm) wide and 12 inches (31 cm) long. Carefully transfer each cylinder to the parchment-lined baking sheet. Evenly brush the dough with the egg wash.

5. Place the baking sheet in the center of the oven and bake until the dough is slightly risen and an even light golden brown, 25 to 30 minutes. Remove the baking sheet from the oven and transfer the cylinders to a cooling rack for 10 minutes.

6. Transfer each cylinder to a cutting board and slice the biscotti on a sharp diagonal (45-degree angle) at ½-inch (1-cm) intervals. Stand the biscotti upright on the baking sheet, about ½ inch (1 cm) apart. Return the baking sheet to the center of the oven and bake until the biscotti are a deep golden brown, about 15 minutes more. Remove from the oven, and transfer the biscotti to a rack to cool thoroughly. The cookies should be dry and crispy. (Once cooled, the biscotti can be stored in an airtight container for up to 1 month.)

■ Yield: About 50 biscotti ■

VARIATION: For Almond and Vanilla Biscotti, substitute 1 cup toasted unblanched whole almonds for the hazelnuts, and omit the orange zest.

WINE SUGGESTION: Vin Santo is the classic partner. If not available, try either port or sherry.

...

Light Orange Ice Cream

Gelato d'Arancia

Light fruit ice cream—which is what good fruity gelato should resemble—is a real favorite. It offers the richness of ice cream without the heaviness, and the fruitiness of a true sorbet with just a little boost from the cream. This orange gelato, in fact, reminds me of the orange Creamsicles that were a favorite of my childhood. Now I eat it out of a pretty little bowl, rather than lick it from a stick! Serve these with Almond and Vanilla Biscotti (page 315) and maybe a few sips of Vin Santo.

2 cups (500 ml) water

Grated zest (orange peel) of 2 oranges

1 cup (200 g) Vanilla Sugar (page 324)

1 cup (250 ml) freshly squeezed orange juice

1 cup (250 ml) heavy cream

1. In a small saucepan, combine the water, orange zest, sugar, and orange juice, and bring to a boil over moderate heat. Boil vigorously for 2 minutes. Place a sieve over a bowl and strain the syrup through the sieve. Discard the solids. Let the syrup cool to room temperature. (To speed cooling, place the bowl inside a larger bowl filled with ice cubes and water. Stir occasionally until cold to the touch.)

2. When the syrup is thoroughly cooled, stir in the heavy cream. Transfer to an ice cream maker and freeze according to the manufacturer's instructions.

▪ Yield: About 1 quart (1 l) ice cream ▪

. . .

Lemon Ice Cream

Gelato di Limone

*"What one feels always in Italy is an extraordinary and direct min-
gling of freshness and repose, as though all life were sunset and sun-
rise, winter and spring."*

CHARLES MORGAN, REFLECTIONS IN A MIRROR, *1944*

Lemon is one of my favorite flavors, and when I find myself at a gelato stand in
Italy, it never takes me long to decide which flavor I'll select. The clerk behind
the counter is always sad when I order a single flavor. I like my tastes pure! This is a
simple, light, surefire recipe bound to thrill any lemon lover.

2¼ cups (560 ml) water

Grated zest (yellow peel) of 6 lemons

1½ cups Vanilla Sugar (page 324)

1 cup (250 ml) freshly squeezed lemon juice (about 8 to 10 lemons)

1 cup (250 ml) heavy cream

1. In a small saucepan, combine the water, lemon zest, sugar, and lemon juice, and
bring to a boil over moderate heat. Boil vigorously for 2 minutes. Place a sieve over
a bowl and strain the syrup through the sieve. Discard the solids. Let the syrup cool
to room temperature. (To speed cooling, place the bowl inside a larger bowl filled with
ice cubes and water. Stir occasionally until cool.)

2. When the syrup is thoroughly cooled, stir in the heavy cream. Transfer to an ice
cream maker and freeze according to the manufacturer's instructions.

▪ Y i e l d : A b o u t 1 q u a r t (1 l) i c e c r e a m ▪

• • •

Lemon Granita

Granita di Limone

I adore lemons. In the heat of summer, nothing is more refreshing than freshly squeezed lemon juice, a touch of sugar to chase away the pucker, and crushed ice. The Italians are masters at granita, a sort of half-ice, half-liquid drink made up of very fine-grained frozen crystals of fruit syrup. Whenever I see this "fruit slush" on the menu in Italian cafés, I order it. The best trattorias serve the granita from tall glasses with a straw and a long thin spoon, for sipping or for eating by the spoonful. At home, it's nice to serve out of shallow champagne coupes as an afternoon pick-me-up or a light dessert.

¼ cup (50 g) sugar

½ cup (125 ml) freshly squeezed lemon juice

1½ cups (375 ml) water

1. In a saucepan, combine the sugar, lemon juice, and water. Bring to a boil, stirring until the sugar is dissolved. Remove from the heat and allow the syrup to cool thoroughly.

2. Transfer the mixture to a shallow metal pan (or 2 metal ice cube trays with the grids removed). Carefully place in the freezer. After 15 minutes, remove from the freezer (be careful not to splash the sticky liquid!) and, with a fork, stir the contents to crush any lumps and break up the ice crystals, being sure to reach the crystals formed along the side and bottom of the pan. Return to the freezer and freeze for 15 minutes more. Stir the granita. Continue this every 15 minutes or so, for up to 3 hours. Spoon into chilled glasses and serve immediately: The granita will melt very quickly!

• Yield: 4 to 6 servings •

Pineapple Sorbet

Sorbetto di Ananas

Pineapple is among the most refreshing flavors I know, and this light, delicious, satisfying sorbet is as pure as you can get. Punctuated with just a hint of vanilla to accentuate the pineapple essence, it's a dessert I could easily enjoy on a regular basis. I sampled a superb version one sunny Sunday afternoon in the historic town of Sirmione, at the tip of Lake Garda. If you want to consume this as is, as a *frulatto*, or whip, simply eat it chilled as a snack or dessert, without freezing.

4 cups (1 l) cubed fresh pineapple (about 1 medium pineapple)

⅓ cup (65 g) Vanilla Sugar (page 324)

1 teaspoon pure vanilla extract

1. In the bowl of a food processor, combine the pineapple, sugar, and vanilla extract, and purée until light and fluffy.

2. Transfer to an ice cream maker and freeze according to the manufacturer's instructions.

▪ Yield: About 1 quart (1 l) sorbet ▪

Ricotta Gelato

T he tartness and tang of ricotta cheese blended with the richness of a traditional custard creates a sublime and elegant gelato. I love the purity of flavors here, unmarred by excess. Although the gelato appears to be extravagantly rich, the richness comes from the ricotta, not a heavy dose of cream.

CUSTARD BASE

1 plump, moist vanilla bean

2 cups (500 ml) whole milk

6 large egg yolks, at room temperature

¾ cup (150 g) Vanilla Sugar (page 324)

2 cups (500 ml) whole-milk ricotta

1 tablespoon dark rum

¼ cup (60 ml) heavy cream

1. Prepare the custard base: Flatten the vanilla bean and cut in half lengthwise. With a small spoon, scrape out the seeds and place them in the bowl of an electric mixer. Set aside. In a large saucepan, combine the milk and vanilla pods over moderate heat. Scald the milk (cook until tiny bubbles form around the edges), and remove from the heat. Cover and set aside to infuse for 15 minutes. Remove the vanilla pods.

2. In the bowl of an electric mixer, whisk the vanilla seeds, egg yolks, and sugar until thick, fluffy, and a pale-lemon color, 2 to 3 minutes. When you lift the whisk, the mixture should form a trail, or ribbon, on the surface. Set aside.

3. Return the simmered and infused milk to high heat and bring just to a boil. Remove from the heat and pour one-third of the boiling milk into the egg yolk mixture in the bowl, whisking constantly. Return the milk and egg yolk mixture to the saucepan, reduce the heat to low, and cool, stirring constantly with a wooden spoon, until the mixture thickens to a creamy consistency. To test, run your finger down the back of the wooden spoon: If the mixture is sufficiently cooked, the mark will hold. The entire process should take less than 5 minutes. (Alternatively, use a candy thermometer to test the custard. Cook until the thermometer registers 185°F; 85°C.)

4. Remove from the heat and immediately pass through a fine-mesh sieve into a bowl. Set the bowl in a larger bowl filled with ice cubes and water, and stir occasionally until the mixture is cold to the touch. The cooling process should take about 30 minutes.

5. In the bowl of an electric mixer fitted with a paddle, combine the ricotta, rum, and heavy cream, and mix until smooth. Fold the ricotta mixture into the cooled custard. Transfer to an ice cream maker and freeze according to the manufacturer's instructions.

■ Y i e l d : A b o u t 1 q u a r t (1 l) i c e c r e a m ■

Risotto Gelato

Gelato di Riso

Imagine cold, soothing rice pudding and you have it in this rich and memorable risotto ice cream. And if you come to love this delicious gelato as I do, almost anything short of a trip to Florence is worth the effort to sample it just one more time. Serve this with tiny almond macaroon cookies (page 290). The amount of rice seems infinitesimal, but the cooked rice swells up greatly, making for a rich vanilla ice cream flecked with sweet rice.

RICE

¼ cup (45 g) Italian Arborio rice (see Mail Order Sources, page 325)

1½ cups (375 ml) whole milk

½ cup (100 g) Vanilla Sugar (page 324)

¼ teaspoon fine sea salt

CUSTARD BASE

2 plump, moist vanilla beans

2 cups (500 ml) whole milk

6 large egg yolks, at room temperature

¾ cup (150 g) Vanilla Sugar (page 324)

1 cup (250 ml) heavy cream

1. Prepare the rice: In a large saucepan, combine the rice, milk, sugar, and salt. Bring to a simmer over moderate heat, stirring often to keep the rice from sticking to the bottom of the pan. Reduce the heat and simmer until the rice is cooked, about 20 minutes, stirring often to keep the rice from sticking to the bottom of the pan. Transfer the mixture to a bowl to cool to room temperature. (To speed cooling, place the bowl inside a larger bowl filled with ice cubes and water. Stir occasionally until cool.) Place a sieve over a bowl and strain the cooled rice and liquid through the sieve. Discard the cooking liquid. Set the rice aside.

2. Prepare the custard base: Flatten the vanilla beans and cut in half lengthwise. With a small spoon, scrape out the seeds and place them in the bowl of an electric mixer. Set aside. In a large saucepan, combine the milk and vanilla pods over moderate heat. Scald the milk, and remove from the heat. Cover and set aside to infuse for 15 minutes. Remove the vanilla pods.

3. In the bowl of an electric mixer, whisk the vanilla seeds, egg yolks, and sugar until thick, fluffy, and a pale-lemon color, 2 to 3 minutes. When you lift the whisk, the mixture should form a trail, or ribbon, on the surface. Set aside.

4. Return the simmered and infused milk to high heat and bring just to a boil. Remove from the heat and pour one-third of the boiling milk into the egg yolk mixture in the bowl, whisking constantly. Return the milk and egg yolk mixture to the saucepan, reduce the heat to low, and cook, stirring constantly with a wooden spoon, until the mixture thickens to a creamy consistency. To test, run your finger down the back of the wooden spoon: If the mixture is sufficiently cooked, the mark will hold. The entire process should take less than 5 minutes. (Alternatively, use a candy thermometer to test the custard. Cook until the thermometer registers 185°F; 85°C.)

5. Remove from the heat and immediately stir in the heavy cream to stop the cooking. Pass through a fine-mesh sieve into a bowl, and let cool completely. (To speed cooling, place the bowl inside a larger bowl filled with ice cubes and water. Stir occasionally until cool.)

6. When thoroughly cooled, stir in the cooled rice. Transfer to an ice cream maker and freeze according to the manufacturer's instructions.

▪ Y i e l d : A b o u t 1 q u a r t (1 l) i c e c r e a m ▪

Vanilla Sugar

Zucchero di Vaniglia

anilla plays a lovely supporting role in many Italian desserts. Little packages of vanilla sugar can be purchased in the supermarket for scenting traditional desserts, including the famed sponge cake *pan di spagna*, as well as milk-and-egg-based desserts such as *budini*. Whenever I bake, I use vanilla sugar, rather than the "unscented" variety. You'll be well rewarded for the few seconds it takes to prepare a batch at home.

4 plump moist vanilla beans

4 cups (800 g) sugar

Flatten the vanilla beans and cut them in half lengthwise. With a small spoon, scrape out the seeds and place them in a small bowl; reserve the seeds for another use. Combine the pods and sugar in a jar. Cover securely and set aside at room temperature for several days to scent and flavor the sugar. Use in place of regular sugar when preparing desserts. Vanilla sugar can be stored indefinitely. As the vanilla sugar is used, replace with new sugar. (When baking, I return rinsed and thoroughly dried pods to the sugar mixture as an added boost.)

■ Yield: 4 cups (800 g) vanilla sugar ■

Mail Order Sources

The following companies offer mail-order catalogs and/or newsletters, for purchasing specialty and hard-to-find ingredients and equipment. There is a charge for most catalogs.

APPLE PIE FARM, INCORPORATED (THE HERB PATCH), Union Hill Road #5, Malvern, PA 19355. Tel: (215) 933–4215.

THE CHEF'S CATALOG, 3215 Commercial Avenue, Northbrook, IL 60062–1900. Tel: (800) 338–3232; Fax: (708) 480–8929.

A COOK'S WARES, 211 37th Street, Beaver Falls, PA 15010–2103. Tel: (412) 846–9490.

CORTI BROTHERS, 5810 Folsom Boulevard, P.O. Box 191358, Sacramento, CA 95819. Tel: (916) 736–3800; Fax: (916) 736–3807.

LA CUISINE, 323 Cameron Street, Alexandria, VA 22314. Tel: (800) 521–1176.

DEAN & DELUCA, 560 Broadway, New York, NY 10012. Tel: (800) 221–7714 or (212) 431–1691.

FERRARA, 195 Grand Street, New York, NY 10013. Tel: (212) 226–6150.

G. B. RATTO & COMPANY, 821 Washington Street, Oakland, CA 94607. Tel: (800) 325–3483; Fax: (510) 836–2250.

GIANT ARTICHOKE, 11241 Merritt Street, Castroville, CA 95012. Tel: (408) 633–2778.

THE HERBFARM, 32804 Issaquah–Fall City Road, Fall City, WA 98024. Tel: (800) 866–HERB.

HERB GATHERING INC., 5742 Kenwood Avenue, Kansas City, MO 64110. Tel: (816) 523–2653.

IDEAL CHEESE, 1205 Second Avenue, New York, NY 10021. Tel: (212) 688–7579.

ISLAND FARMCRAFTERS, Waldron Island, WA 98297. Tel: (206) 739–2286.

KERMIT LYNCH WINE MERCHANT, 1605 San Pablo Avenue, Berkeley, CA 94702–1317. Tel: (510) 524–1524; Fax: (510) 528–7026.

KING ARTHUR FLOUR BAKER'S CATALOGUE, P.O. Box 876, Norwich, VT 05055. Tel: (800) 827–6836.

MOZZARELLA COMPANY, 2944 Elm Street, Dallas, TX 75226. Tel: (800) 798–2954 or (214) 741–4072; Fax: (214) 741–4076.

NICHOLS GARDEN NURSERY, 1190 North Pacific Highway, Albany, OR 97321. Tel: (503) 928–9280.

NORTHWEST SELECT, 14724 184th Street NE, Arlington, WA 98223. Tel: (800) 852–7132 or (206) 435–8577.

PENZEY'S SPICE HOUSE, 1921 S. West Avenue, Waukesha, WI 53186. Tel: (414) 574-0277; Fax: (414) 574-0278.

SHEPHERD'S GARDEN SEEDS, 6116 Highway 9, Felton, CA 95018. Tel: (408) 335–6910.

TODARO BROTHERS, 555 Second Avenue, New York, NY 10016. Tel: (212) 679–7766.

WALNUT ACRES ORGANIC FARMS, Walnut Acres, Penns Creek, PA 17862. Tel: (800) 433–3998.

WELL-SWEEP HERB FARM, 317 Mount Bethal Road, Port Murray, NJ 07865. Tel: (908) 852–5390.

WILLIAMS-SONOMA, Mail Order Department, P.O. Box 7456, San Francisco, CA 94120–7456. Tel: (415) 421–4242; Fax: (415) 421–5153.

HARD-TO-FIND PANTRY ITEMS

SALTED ANCHOVIES

Todaro Brothers

AMARETTI COOKIES

Ferrara

G. B. Ratto & Company

BEANS, DRIED ASSORTED

(borlotti, cannellini, cranberry)

Corti Brothers

Dean & Deluca

G. B. Ratto & Company

SALTED CAPERS, ANCHOVIES IN OIL

Dean & Deluca

Todaro Brothers

OLIVES, OILS, VINEGARS

Corti Brothers

Dean & Deluca

G. B. Ratto & Company

Kermit Lynch Wine Merchant

IMPORTED PASTA

G. B. Ratto & Company

POLENTA, ARBORIO RICE

Dean & Deluca

G. B. Ratto & Company

Williams-Sonoma

DRIED PORCINI MUSHROOMS,

DRIED TOMATOES

Dean & Deluca

G. B. Ratto & Company

RED PEPPER FLAKES, PURE

SPANISH SAFFRON

Penzey's Spice House

SAVOIARDI BISCUITS

Ferrara

SEMOLINA FLOUR

G. B. Ratto & Company

King Arthur Flour Baker's Catalogue

IMPORTED TUNA IN OIL

G. B. Ratto & Company

PURE VANILLA EXTRACT, VANILLA BEANS

Penzey's Spice House

Williams-Sonoma

WHEAT BERRIES

King Arthur Flour Baker's Catalogue

Walnut Acres Organic Farms

■

CHEESE, CURED MEATS, AND PRODUCE

FRESH BABY ARTICHOKES

Giant Artichoke

Northwest Select

PLUMP ORGANIC GARLIC BRAIDS

Island Farmcrafters

DOMESTIC GOAT'S CHEESE, MASCARPONE, MOZZARELLA, PECORINO, RICOTTA

Mozzarella Company

FRESH HERB AND BAY LEAF PLANTS

The Herbfarm

Nichols Garden Nursery

Well-Sweep Herb Farm

FRESH-CUT HERBS

Apple Pie Farm, Inc.

Herb Gathering Inc.

IMPORTED ITALIAN CHEESES

Ideal Cheese

PANCETTA

Dean & Deluca

PARMIGIANO-REGGIANO

Dean & Deluca

G. B. Ratto & Company

Ideal Cheese

■

GARDEN SEEDS FOR GROWING IT YOURSELF

FLAT-LEAF PARSLEY, SAGE, AND OTHER HERBS

Herb Gathering Inc.

Shepherd's Garden Seeds

Well-Sweep Herb Farm

ARUGULA, CHILES, RADICCHIO, BORLOTTI BEANS

Shepherd's Garden Seeds

■

KITCHEN AND BAKING EQUIPMENT

The Chef's Catalog

Dean & Deluca

King Arthur Flour Baker's Catalogue

La Cuisine

Williams-Sonoma

Index

Basil *(continued)*
   and garlic sauce, 261
   tagliatelle with tricolor peppers
     and, 136–137
Bass, sea:
   baked, with artichokes,
     210–211
   with potatoes and tomatoes in
     parchment, 208–209
Bay leaves, 176
   herb-infused savory custards,
     30–31
   rice stored with, 169, 170
   risotto with Parmesan and,
     170–171
   white sauce, 270–271
Bean(s), 39, 41
   acidic ingredients and, 74
   cubed pork with garlic,
     spinach, and spicy chick
     peas, 234–235
   five, Tuscan soup, 70–71
   fresh fava, and pecorino salad,
     38–39
   freshness of, 74
   Milanese vegetable soup,
     80–81
   and pasta soup, 75–77
   raw vegetables dipped in olive
     oil, 25
   red, and onion salad, 40–41
   salt added to, 74
   water for cooking, 74
   and wheat berry soup, Tuscan,
     78–79
   white, green olive, tuna, celery,
     and red pepper salad with,
     37
   white, salad with fresh sage
     and thyme, 42–43
Beef:
   braised in Barolo wine,
     247–249
   seasoned raw, 12
   see also Meat sauce
Biscotti:
   almond and vanilla, 315
   hazelnut and orange, 314–315
Biscotti, 314–315
Blenders, hand, 77
Borlotti bean(s):
   and onion salad, 40–41
   and pasta soup, 75–77
   Tuscan five-bean soup,
     70–71
Braising, 246
Branzino coi carciofi, 210–211
Branzino in cartoccio, 208–209
Brasato al barolo, 247–249
Bread and tomato salad, 44–45

Breads:
   basic dough for, 178–179
   individual olive rolls, 192–193
   Italian slipper, 190–191
   rustic whole wheat, 181–183
   Sardinian parchment, 184–185
   sourdough, 186–189
   tips for, 180
   see also Pizzas
Broad (fava) bean(s):
   fresh, and pecorino salad,
     38–39
   raw vegetables dipped in olive
     oil, 25
Brodo di pollo, 272–273
Brodo di verdura, 274
Broiled vegetables, tips for, 17
Broth:
   chicken stock, 272–273
   poached chicken with fresh
     herb sauce, 226–227
   vegetable, 274
Bucatini alla gricia, 104–105
Bucatini with pancetta, pecorino,
   and black pepper,
   104–105
Budini di riso, 312–313
Budini di cioccolato, 298–299
Butterflies, saffron, 108–109
Buttering pans, 291

Cakes:
   fragrant orange and lemon,
     305–307
   golden lemon-rice, 310–311
   Lake Garda apple, 308–309
   lemon-rice tea, 312–313
   toasted hazelnut, 302–303
Calamari, fried, 202–203
Calamari fritti, 202–203
Cannellini bean(s):
   Milanese vegetable soup,
     80–81
   and pasta soup, 75–77
   salad with fresh sage and
     thyme, 42–43
   Tuscan five-bean soup, 70–71
   and wheat berry soup, Tuscan,
     78–79
Capellini d'angelo al
   prezzemolo, 124
Capellini with pungent parsley
   sauce, 124
Caper(s):
   spaghetti with olives, tomatoes,
     hot peppers and, 94–95
   tomato, olive, and anchovy
     pizza, 196
Carciofi:
   alla Romana, 66–67

   saltati, 64–65
   sott'olio, 18–19
Carta da musica, 184–185
Ceciata di suino della casa,
   234–235
Celery:
   braised oxtail with tomatoes,
     onions, and, 251–253
   green olive, tuna, and red
     pepper salad, 37
   and meat sauce, rigatoni with,
     112–113
   raw vegetables dipped in olive
     oil, 25
   salad with anchovy dressing,
     34–35
   spicy squid salad, 200–201
Cheese:
   coffee and mascarpone
     ladyfinger cream, 294–295
   gemelli with eggplant,
     tomatoes, and mozzarella,
     100–101
   goat, and garlic spread, 2–3
   goat, and thyme spread, 3
   individual eggplant Parmesans,
     14–15
   individual Gorgonzola soufflés,
     28–29
   lasagne with tomato-cream
     sauce and mozzarella,
     150–151
   ricotta cheesecake with pine
     nuts and raisins, 300–301
   ricotta gelato, 320–321
   speedy lasagne, 122–123
   tomato and mozzarella pizza,
     196
   vegetarian pizza, 196
   see also Parmigiano-Reggiano
     cheese; Pecorino cheese
Cheesecake, ricotta, with pine
   nuts and raisins, 300–301
Chicken:
   breasts with fresh sage,
     sautéed, 216–217
   cacciatora, 228–229
   with a confit of red bell
     peppers and onions,
     221–223
   cooked under bricks, 218–219
   grilled, with lemon, oil, and
     black pepper, 224–225
   poached, with fresh herb sauce,
     226–227
   stock, 272–273
Chick pea(s):
   and pasta soup, 82–83
   spicy, cubed pork with garlic,
     spinach, and, 234–235

chicken with a confit of red
  bell peppers and onions,
  221–223
cubed pork with garlic,
  spinach, and spicy chick
  peas, 234–235
fresh artichoke omelet, 22–23
grilled chicken with lemon, oil,
  and black pepper,
  224–225
grilled marinated lamb chops
  with lemon and oil,
  240–241
individual eggplant Parmesans,
  14–15
lamb braised in white wine,
  garlic, and hot peppers,
  242–243
Parmesan-breaded baby lamb
  chops, 238–239
poached chicken with fresh
  herb sauce, 226–227
rabbit with red peppers and
  polenta, 231–233
rosemary roast loin of pork,
  236–237
saffron risotto, 162–163
sautéed chicken breasts with
  fresh sage, 216–217
sea bass with potatoes and
  tomatoes in parchment,
  208–209
tomato and bread salad, 44–45
see also Pasta, dried; Pasta,
  fresh; Soups
Marjoram, 15
Mascarpone and coffee ladyfinger
  cream, 294–295
Mayonnaise, garlic, 275
Meats:
  beef braised in Barolo wine,
    247–249
  braised oxtail with tomatoes,
    onions, and celery,
    251–253
  braised veal shanks with lemon
    and parsley garnish,
    244–245
  braising of, 246
  cubed pork with garlic,
    spinach, and spicy chick
    peas, 234–235
  grilled marinated lamb chops
    with lemon and oil,
    240–241
  lamb braised in white wine,
    garlic, and hot peppers,
    242–243
  Parmesan-breaded baby lamb
    chops, 238–239

rabbit with red peppers and
  polenta, 231–233
rosemary roast loin of pork,
  236–237
seasoned raw beef, 12
to test for doneness, 241
see also Pancetta; Prosciutto
Meat sauce, 268–269
  celery and, rigatoni with,
    112–113
  citrus-infused baked tagliatelle,
    154–155
  spicy, spaghetti with, 114–115
Minestrone alla Milanese,
  80–81
Minestrone alla Toscana, 70–71
Mousse di tonno, 8
Mozzarella cheese:
  eggplant Parmesan, 59–61
  gemelli with eggplant,
    tomatoes, and, 100–101
  individual eggplant Parmesans,
    14–15
  lasagne with tomato-cream
    sauce and, 150–151
  and tomato pizza, 196
  vegetarian pizza, 196
Mushroom(s):
  four seasons' pizza, 196
  orange, and sage risotto,
    164–165
  porcini, pilaf with tomato
    sauce, Parmesan and,
    174–175
  porcini, sauce, 268–269
  porcini, sauce, tagliatelle with,
    148–149
  tomato sauce, 257
Mussel(s):
  spaghetti with shrimp, clams,
    and, in tomato sauce,
    117–119
  tomato soup, spicy, 72–73

Navy bean(s):
  Milanese vegetable soup,
    80–81
  salad with fresh sage and
    thyme, 42–43
  Tuscan five-bean soup, 70–71
  and wheat berry soup, Tuscan,
    78–79
Nut(s):
  almond and vanilla biscotti,
    315
  almond macaroons, 290–291
  fusilli with walnut and garlic
    sauce, 110–111
  hazelnut and orange biscotti,
    314–315

pine, arugula, and Parmesan
  salad, 46–47
pine, ricotta cheesecake with
  raisins and, 300–301
toasted hazelnut cake, 302–303
toasting of, 304
walnut and pecorino salad, 48

Olivada, 277
Olivada verde, 278
Olive nere, 280–281, 282–283
Olive(s):
  black, panfried potatoes with,
    52
  black, spread, 277
  brine-cured black, 282–283
  four seasons' pizza, 196
  green, baked swordfish with
    tomatoes and, 206–207
  green, spread, 278
  green, tuna, celery, and red
    pepper salad, 37
  red pesto sauce, 264–265
  rolls, individual, 192–193
  salad, Aunt Flora's, 36
  salt-cured black, 280–281
  spaghetti with capers,
    tomatoes, hot peppers
    and, 94–95
  spicy squid salad, 200–201
  tomato, caper, and anchovy
    pizza, 196
Omelets:
  fresh artichoke, 22–23
  spinach-Parmesan frittata,
    26–27
  tips for, 24
Onion(s):
  braised oxtail with tomatoes,
    celery and, 251–253
  chicken with a confit of red
    bell peppers and, 221–223
  raw vegetables dipped in olive
    oil, 25
  red, pizza with rosemary, hot
    pepper and, 194–195
  and red bean salad, 40–41
Orange:
  and hazelnut biscotti, 314–315
  ice cream, light, 316
  -infused baked pasta or rice,
    technique for, 156
  -infused baked tagliatelle,
    154–155
  and lemon cake, fragrant,
    305–307
  rice cake, golden, 310–311
  sage, and mushroom risotto,
    164–165
  zest, 132, 156

Vegetable(s):
asparagus with butter and
Parmesan, 50–51
braised artichokes with garlic
and parsley, 66–67
broiled ("grilled"), tips for, 17
broth, 274
caponata, 9
deep-fried zucchini and
zucchini blossoms, 56–57
eggplant Parmesan, 59–61
grilled zucchini with fresh
thyme, 16–17
marinated baby artichokes
preserved in oil, 18–19
panfried potatoes with black
olives, 52
raw, dipped in olive oil, 25
roasted rosemary potatoes,
54–55
sautéed baby artichokes, 64–65
sautéed spinach with garlic,
lemon, and oil, 62–63
seared and roasted tomatoes,
13
silky sautéed red peppers, 4–5
soup, Milanese, 80–81

Vinegar:
in bean dishes, 74
lamb braised in white wine,
garlic, and hot peppers,
242–243
Vodka, penne with spicy
tomato-cream sauce and,
92–93

Walnut:
and garlic sauce, fusilli with,
110–111
and pecorino salad, 48
Warmed plates, 250
Watercress, tonnarelli with
tomatoes, shaved
Parmesan and, 128–129
Wheat berry and bean soup,
Tuscan, 78–79
White bean(s):
Milanese vegetable soup,
80–81
and pasta soup, 75–77
salad with fresh sage and
thyme, 42–43
Tuscan five-bean soup,
70–71

and wheat berry soup, Tuscan,
78–79
Wine:
Barolo, beef braised in,
247–249
in rice dishes, 169
white, lamb braised in garlic,
hot peppers and, 242–243

Zest, *see* Citrus zest
Ziti:
with eggplant, tomatoes, and
mozzarella, 100–101
tossing of, 107
Zucchero di vaniglia, 324
Zucchini:
deep-fried zucchini blossoms
and, 56–57
grilled, with fresh thyme,
16–17
penne with spicy pizza sauce
and, 98–99
tagliatelle with fresh parsley
and, 142–143
Zucchini alla griglia,
16–17
Zuppa di cozze, 72–73